FUNDAMENTALS
OF CONTEMPORARY
MASS SPECTROMETRY

THE WILEY BICENTENNIAL–KNOWLEDGE FOR GENERATIONS

Each generation has its unique needs and aspirations. When Charles Wiley first opened his small printing shop in lower Manhattan in 1807, it was a generation of boundless potential searching for an identity. And we were there, helping to define a new American literary tradition. Over half a century later, in the midst of the Second Industrial Revolution, it was a generation focused on building the future. Once again, we were there, supplying the critical scientific, technical, and engineering knowledge that helped frame the world. Throughout the 20th Century, and into the new millennium, nations began to reach out beyond their own borders and a new international community was born. Wiley was there, expanding its operations around the world to enable a global exchange of ideas, opinions, and know-how.

For 200 years, Wiley has been an integral part of each generation's journey, enabling the flow of information and understanding necessary to meet their needs and fulfill their aspirations. Today, bold new technologies are changing the way we live and learn. Wiley will be there, providing you the must-have knowledge you need to imagine new worlds, new possibilities, and new opportunities.

Generations come and go, but you can always count on Wiley to provide you the knowledge you need, when and where you need it!

WILLIAM J. PESCE
PRESIDENT AND CHIEF EXECUTIVE OFFICER

PETER BOOTH WILEY
CHAIRMAN OF THE BOARD

FUNDAMENTALS OF CONTEMPORARY MASS SPECTROMETRY

CHHABIL DASS
University of Memphis

WILEY-INTERSCIENCE

A JOHN WILEY & SONS, INC., PUBLICATION

Copyright © 2007 by John Wiley & Sons, Inc. All rights reserved.

Published by John Wiley & Sons, Inc., Hoboken, New Jersey.
Published simultaneously in Canada.

No part of this publication may be reproduced, stored in a retrieval system, or transmitted in any form or by any means, electronic, mechanical, photocopying, recording, scanning, or otherwise, except as permitted under Section 107 or 108 of the 1976 United States Copyright Act, without either the prior written permission of the Publisher, or authorization through payment of the appropriate per-copy fee to the Copyright Clearance Center, Inc., 222 Rosewood Drive, Danvers, MA 01923, (978) 750-8400, fax (978) 750-4470, or on the web at www.copyright.com. Requests to the Publisher for permission should be addressed to the Permissions Department, John Wiley & Sons, Inc., 111 River Street, Hoboken, NJ 07030, (201) 748-6011, fax (201) 748-6008, or online at http://www.wiley.com/go/permission.

Limit of Liability/Disclaimer of Warranty: While the publisher and author have used their best efforts in preparing this book, they make no representations or warranties with respect to the accuracy or completeness of the contents of this book and specifically disclaim any implied warranties of merchantability or fitness for a particular purpose. No warranty may be created or extended by sales representatives or written sales materials. The advice and strategies contained herein may not be suitable for your situation. You should consult with a professional where appropriate. Neither the publisher nor author shall be liable for any loss of profit or any other commercial damages, including but not limited to special, incidental, consequential, or other damages.

For general information on our other products and services or for technical support, please contact our Customer Care Department within the United States at (800) 762-2974, outside the United States at (317) 572-3993 or fax (317) 572-4002.

Wiley also publishes its books in a variety of electronic formats. Some content that appears in print may not be available in electronic formats. For more information about Wiley products, visit our web site at www.wiley.com.

Wiley Bicentennial Logo: Richard J. Pacifico

Library of Congress Cataloging-in-Publication Data:

Dass, Chhabil.
   Fundamentals of contemporary mass spectrometry / Chhabil Dass.
      p. cm.
   Includes bibliographical references.
   ISBN 978-0-471-68229-5
   1. Mass spectrometry. 2. Spectrum analysis—Instruments. 3. Molecular spectroscopy. 4. Biomolecules—Analysis. I. Title.
   QC454.M3D37 2007
   543′.65—dc22

                                                                                                      2006027635

Printed in the United States of America

10 9 8 7 6 5 4 3 2

Dedicated to All My Teachers

CONTENTS

PREFACE xix

I INSTRUMENTATION 1

1 BASICS OF MASS SPECTROMETRY 3

    1.1 Brief History of Mass Spectrometry 3
    1.2 Desirable Features of Mass Spectrometry 5
    1.3 Basic Principles of Mass Spectrometry 5
    1.4 Anatomy of a Mass Spectrum 7
    1.5 Atomic and Molecular Masses 9
            1.5.1   Mass-to-Charge Ratio 10
    1.6 General Applications 11
          Overview 12
          Exercises 12
          References 13

2 MODES OF IONIZATION 15

    2.1 Why Ionization Is Required 15
    2.2 General Construction of an Ion Source 16
          Gas-Phase Ionization Techniques 17
    2.3 Electron Ionization 17

2.4		Chemical Ionization	20
	2.4.1	Charge-Exchange Chemical Ionization	24
	2.4.2	Negative-Ion Chemical Ionization	25
2.5		Photoionization	26
2.6		Field Ionization	28
2.7		Metastable Atom Bombardment Ionization	28

Condensed-Phase Ionization Techniques: Ionization of Solid-State Samples   29

2.8		Field Desorption	29
2.9		Plasma Desorption Ionization	30
2.10		Secondary-Ion Mass Spectrometry	31
2.11		Fast Atom Bombardment	32
2.12		Laser Desorption/Ionization	35
2.13		Matrix-Assisted Laser Desorption/Ionization	35
	2.13.1	MALDI Analysis of Low-Molecular-Mass Compounds	42
	2.13.2	Atmospheric-Pressure MALDI	43
	2.13.3	Surface-Enhanced Laser Desorption/Ionization	44
	2.13.4	Material-Enhanced Laser Desorption/Ionization	45

Condensed-Phase Ionization Techniques: Ionization of Liquid-State Samples   45

2.14		Thermospray Ionization	45
2.15		Atmospheric-Pressure Chemical Ionization	46
2.16		Atmospheric-Pressure Photoionization	47
2.17		Electrospray Ionization	48
	2.17.1	Mechanism of Electrospray Ionization	52
	2.17.2	Sample Consideration	54
	2.17.3	Nanoelectrospray Ionization	54
2.18		Desorption Electrospray Ionization	55
	2.18.1	DART Ion Source	56
		Overview	57
		Exercises	58
		Additional Reading	60
		References	60

3 MASS ANALYSIS AND ION DETECTION   67

3.1	Mass Resolving Power	68
3.2	Kinetic Energy of Ions	70
	Mass Analyzers	70

3.3	Magnetic-Sector Mass Spectrometers		70
	3.3.1	Working Principle of a Magnetic Analyzer	70
	3.3.2	Working Principle of an Electrostatic Analyzer	73
	3.3.3	Working Principle of Double-Focusing Magnetic-Sector Mass Spectrometers	73
	3.3.4	Performance Characteristics	74
3.4	Quadrupole Mass Spectrometers		75
	3.4.1	Working Principle	76
	3.4.2	Performance Characteristics	79
	3.4.3	RF-Only Quadrupole	80
3.5	Time-of-Flight Mass Spectrometers		80
	3.5.1	Working Principle	81
	3.5.2	Delayed Extraction of Ions	83
	3.5.3	Reflectron TOF Instrument	84
	3.5.4	Orthogonal Acceleration TOF Mass Spectrometer	85
	3.5.5	Performance Characteristics	86
3.6	Quadrupole Ion-Trap Mass Spectrometers		86
	3.6.1	Working Principle	87
	3.6.2	Operational Modes	89
	3.6.3	Performance Characteristics	90
3.7	Linear Ion-Trap Mass Spectrometers		92
	3.7.1	Rectilinear Ion Trap	94
3.8	Fourier-Transform Ion Cyclotron Resonance Mass Spectrometers		94
	3.8.1	Working Principle	95
	3.8.2	Performance Characteristics	98
3.9	Orbitrap Mass Analyzers		99
3.10	Ion Mobility Mass Spectrometers		101
3.11	Detectors		103
	3.11.1	Faraday Cup Detector	103
	3.11.2	Electron Multipliers	104
	3.11.3	Photomultiplier Detectors	105
	3.11.4	Postacceleration Detectors	105
	3.11.5	Low-Temperature Calorimetric Detectors for High-Mass Ions	106
	3.11.6	Focal-Plane Detectors	107
	Overview		108
	Exercises		110

		Additional Reading	112
		References	112
4	**TANDEM MASS SPECTROMETRY**		**119**
	4.1	Basic Principles of Tandem Mass Spectrometry	119
	4.2	Types of Scan Functions	121
	4.3	Ion Activation and Dissociation	123
		4.3.1 Collision-Induced Dissociation	124
		4.3.2 Surface-Induced Dissociation	125
		4.3.3 Absorption of Electromagnetic Radiations	126
		4.3.4 Electron-Capture Dissociation	127
	4.4	Reactions in Tandem Mass Spectrometry	128
	4.5	Tandem Mass Spectrometry Instrumentation	129
		4.5.1 Magnetic-Sector Tandem Mass Spectrometers	129
		4.5.2 Tandem Mass Spectrometry with Multiple-Quadrupole Devices	132
		4.5.3 Tandem Mass Spectrometry with Time-of-Flight Instruments	133
		4.5.4 Tandem Mass Spectrometry with a Quadrupole Ion-Trap Mass Spectrometer	136
		4.5.5 Tandem Mass Spectrometry with an FT–ICR Mass Spectrometer	138
		4.5.6 Tandem Mass Spectrometry with Hybrid Instruments	138
		Overview	143
		Exercises	145
		Additional Reading	146
		References	146
5	**HYPHENATED SEPARATION TECHNIQUES**		**151**
	5.1	Benefits of Coupling Separation Devices with Mass Spectrometry	152
	5.2	General Considerations	153
		5.2.1 Characteristics of an Interface	153
		5.2.2 Mass Spectral Data Acquisition	153
		5.2.3 Characteristics of Mass Spectrometers	155
	5.3	Chromatographic Properties	155
	5.4	Gas Chromatography/Mass Spectrometry	158
		5.4.1 Basic Principles of Gas Chromatography	158

		5.4.2	Interfaces for Coupling Gas Chromatography with Mass Spectrometry	159
	5.5	Liquid Chromatography/Mass Spectrometry		161
		5.5.1	Basic Principles of HPLC Separation	161
		5.5.2	Fast-Flow Liquid Chromatography	162
	5.6	Interfaces for Coupling Liquid Chromatography with Mass Spectrometry		163
		5.6.1	Moving-Belt Interface	164
		5.6.2	Direct-Liquid Introduction Probe	165
		5.6.3	Continuous-Flow Fast Atom Bombardment Interface	165
		5.6.4	Thermospray Interface	166
		5.6.5	Particle–Beam Interface	167
		5.6.6	Electrospray Ionization Interface	168
		5.6.7	Atmospheric-Pressure Chemical Ionization Interface	171
		5.6.8	Atmospheric-Pressure Photoionization Interface	171
		5.6.9	Coupling LC with TOF–MS	171
		5.6.10	Coupling LC with MALDI–MS	172
	5.7	Multidimensional LC/MS		173
	5.8	Capillary Electrophoresis/Mass Spectrometry		174
		5.8.1	Basic Principles of Capillary Electrophoresis	175
		5.8.2	Interfaces for Coupling Capillary Electrophoresis with Mass Spectrometry	177
	5.9	Affinity Chromatography/Mass Spectrometry		181
	5.10	Supercritical-Fluid Chromatography/Mass Spectrometry		183
	5.11	Coupling Planar Chromatography with Mass Spectrometry		183
		Overview		185
		Exercises		186
		Additional Reading		187
		References		187

II   ORGANIC AND INORGANIC MASS SPECTROMETRY 195

6   ORGANIC MASS SPECTROMETRY 197

	6.1	Determination of Molecular Mass		198
		6.1.1	Molecular Mass Measurements at Low-Mass Resolving Power	198
		6.1.2	Molecular Mass Measurements at High-Mass Resolving Power	198
		6.1.3	Molecular Mass Measurements by ESI and MALDI	200

		6.1.4 Mass Calibration Standards	201
	6.2	Molecular Formula from Accurate Mass Values	201
	6.3	Molecular Formula from Isotopic Peaks	203
	6.4	General Guidelines for Interpretation of a Mass Spectrum	210
		6.4.1 Odd- and Even-Electron Ions	210
		6.4.2 Recognizing the Molecular Ion	211
		6.4.3 Nitrogen Rule	211
		6.4.4 Value of the Rings Plus Double Bonds	214
		6.4.5 Systematic Steps in Interpreting a Mass Spectrum	215
		6.4.6 Mass Spectral Compilations	216
	6.5	Fragmentation Processes	216
		6.5.1 Simple Bond-Cleavage Reactions	219
		6.5.2 Rearrangement Reactions	223
		6.5.3 Fragmentation of Cyclic Structures	227
		6.5.4 Differentiation of Isomeric Structures	232
		6.5.5 Structurally Diagnostic Fragment Ions	235
	6.6	Fragmentation Reactions of Specific Classes of Compounds	238
		6.6.1 Hydrocarbons	238
		6.6.2 Alcohols	240
		6.6.3 Ethers	241
		6.6.4 Aldehydes and Ketones	242
		6.6.5 Carboxylic Acids	242
		6.6.6 Esters	243
		6.6.7 Nitrogen-Containing Compounds	244
		6.6.8 Sulfur-Containing Compounds	246
		6.6.9 Halogen-Containing Compounds	246
	6.7	Theory of Ion Dissociation	247
	6.8	Structure Determination of Gas-Phase Organic Ions	250
		Overview	254
		Exercises	255
		Additional Reading	259
		References	259
7	**INORGANIC MASS SPECTROMETRY**		**263**
	7.1	Ionization of Inorganic Compounds	263
	7.2	Thermal Ionization Mass Spectrometry	264
	7.3	Spark-Source Mass Spectrometry	265
	7.4	Glow Discharge Ionization Mass Spectrometry	267

	7.5	Inductively Coupled Plasma Mass Spectrometry	268
		7.5.1 Inductively Coupled Plasma Ion Source	268
		7.5.2 Coupling an ICP Source with Mass Spectrometry	269
		7.5.3 Sample Introduction Systems for an ICP Source	270
		7.5.4 Spectral Interferences	271
		7.5.5 Laser Ablation–ICP–MS	273
	7.6	Resonance Ionization Mass Spectrometry	273
	7.7	Isotope Ratio Mass Spectrometry	275
		7.7.1 Isotope Ratio MS Systems	277
		7.7.2 Applications of Isotope Ratio MS	277
	7.8	Accelerator Mass Spectrometry	278
	7.9	Isotope Dilution Mass Spectrometry	280
		Overview	281
		Exercises	282
		Additional Reading	283
		References	283

III BIOLOGICAL MASS SPECTROMETRY   287

8 PROTEINS AND PEPTIDES: STRUCTURE DETERMINATION   289

8.1	Structure of Proteins	290
8.2	Determination of the Sequence of a Protein	292
8.3	General Protocol for Amino Acid Sequence Determination of Proteins	294
	8.3.1 Homogenization and Subcellular Fractionation	295
	8.3.2 Enrichment and Purification of Proteins	295
8.4	Molecular Mass Measurement of Proteins	297
8.5	Peptide Mass Mapping	298
	8.5.1 Reduction and Carboxymethylation	299
	8.5.2 Cleavage of Proteins	299
	8.5.3 Mass Spectrometric Analysis of Peptide Maps	302
8.6	Proteomics	303
	8.6.1 Strategies for Proteomics	304
8.7	Quantitative Proteomics	310
8.8	Biomarker Discovery	314
8.9	De Novo Protein Sequencing	316
8.10	Determination of the Amino Acid Sequence of Peptides	316

		8.10.1	Peptide Fragmentation Rules	317
		8.10.2	Mass Spectrometry Techniques for Sequence Determination of Peptides	322
		8.10.3	Guidelines for Obtaining the Amino Acid Sequence from a Mass Spectrum	327
	Overview			332
	Exercises			333
	Additional Reading			336
	References			336

9 PROTEINS AND PEPTIDES: POSTTRANSLATIONAL MODIFICATIONS 343

	Disulfide Bonds in Proteins		345
9.1	Traditional Approaches to Identify Disulfide Bonds		346
9.2	Mass Spectrometry–Based Methods to Identify Disulfide Bonds		346
	9.2.1	Determination of the Number of Disulfide Bonds	347
	9.2.2	Generation of Disulfide-Containing Peptides	347
	9.2.3	Identification of Disulfide-Containing Peptides by FAB–MS	348
	9.2.4	Identification of Disulfide-Containing Peptides by MALDI–MS	349
	9.2.5	Identification of Disulfide-Containing Peptides by Electron-Capture Dissociation	350
	9.2.6	Identification of Disulfide-Containing Peptides by Tandem MS	350
	Analysis of Phosphoproteins and Phosphoproteomics		352
9.3	^{32}[P] Labeling for the Analysis of Phosphoproteins		353
9.4	Mass Spectrometry Protocol for the Analysis of Phosphoproteins		354
	9.4.1	Cleavage of Purified Phosphoproteins	355
	9.4.2	Fractionation of Peptide Fragments in the Digest	355
	9.4.3	Determination of the Average Number of Phosphate Groups	358
	9.4.4	Identification of Phosphopeptides	358
	9.4.5	Identification of Phosphorylation Sites	361
	Analysis of Glycoproteins		364
9.5	Structural Diversity of Glycoproteins		365
9.6	Analysis of Glycoproteins		366

		9.6.1 Molecular Mass Determination of Glycoproteins	366

        9.6.1   Molecular Mass Determination of Glycoproteins   366
        9.6.2   Identification of Glycosylation   368
        9.6.3   Site of Glycosylation   369
      Overview   370
      Exercises   370
      References   371

10 PROTEINS AND PEPTIDES: HIGHER-ORDER STRUCTURES   379

   10.1 Charge-State Distribution   380
   10.2 Hydrogen–Deuterium Exchange to Study Conformational
         States of Proteins   383
         10.2.1 Folding and Unfolding Dynamics of Proteins   385
         10.2.2 Experimental Measurements of Amide Hydrogen
                 Isotopic Exchange   386
   10.3 Chemical Cross-Linking as a Probe for the
         Three-Dimensional Structure of Proteins   391
   10.4 Ion Mobility Measurements to Study Protein
         Conformational Changes   391
         Overview   392
         Exercises   392
         Additional Reading   393
         References   393

11 CHARACTERIZATION OF OLIGOSACCHARIDES   397

   11.1 Structural Diversity in Oligosaccharides   398
   11.2 Classes of Glycans   400
   11.3 Mass Spectrometric Methods for Complete Structure
         Elucidation of Oligosaccharides   401
         11.3.1 Release of Glycans   402
         11.3.2 Derivatization of Carbohydrate Chains   402
         11.3.3 Composition Analysis by GC/MS   403
         11.3.4 Linkage Analysis by GC/MS   403
         11.3.5 Rapid Identification by a Precursor-Ion Scan   403
         11.3.6 Composition Analysis by Direct
                 Mass Measurement   403
         11.3.7 Structure Determination of Oligosaccharides by
                 Sequential Digestion   406
         11.3.8 Tandem Mass Spectrometry for Structural
                 Analysis of Carbohydrates   408

		Overview	416
		Exercises	417
		References	418

12 CHARACTERIZATION OF LIPIDS — 423

- 12.1 Classification and Structures of Lipids — 423
- 12.2 Mass Spectrometry of Fatty Acids and Acylglycerols — 428
  - 12.2.1 Analysis of Fatty Acids — 428
  - 12.2.2 Analysis of Acylglycerols — 430
- 12.3 Mass Spectrometry of Phospholipids — 433
- 12.4 Mass Spectrometry of Glycolipids — 436
- 12.5 Analysis of Bile Acids and Steroids — 440
- 12.6 Analysis of Eicosanoids — 441
- 12.7 Lipidomics — 442
- Overview — 446
- Exercises — 447
- References — 447

13 STRUCTURE DETERMINATION OF OLIGONUCLEOTIDES — 453

- 13.1 Structures of Nucleotides and Oligonucleotides — 453
- 13.2 Mass Spectrometry Analysis of Nucleosides and Nucleotides — 457
- 13.3 Cleavage of Oligonucleotides — 458
- 13.4 Molecular Mass Determination of Oligonucleotides — 459
  - 13.4.1 Electrospray Ionization for Molecular Mass Determination — 459
  - 13.4.2 Matrix-Assisted Laser Desorption/Ionization for Molecular Mass Determination — 461
  - 13.4.3 Base Composition from an Accurate Mass Measurement — 463
- 13.5 Mass Spectrometry Sequencing of Oligonucleotides — 464
  - 13.5.1 Gas-Phase Fragmentation for Oligonucleotide Sequencing — 465
  - 13.5.2 Solution-Phase Techniques for Oligonucleotide Sequencing — 471
- Overview — 476
- Exercises — 477
- References — 477

14 QUANTITATIVE ANALYSIS — 485

- 14.1 Advantages of Mass Spectrometry — 486
- 14.2 Data Acquisition — 486
  - 14.2.1 Selected-Ion Monitoring — 487
  - 14.2.2 Selected-Reaction Monitoring — 487
- 14.3 Calibration — 488
  - 14.3.1 External Standard Method — 488
  - 14.3.2 Standard Addition Method — 489
  - 14.3.3 Internal Standard Method — 489
- 14.4 Validation of a Quantitative Method — 491
- 14.5 Selected Examples — 492
  - 14.5.1 Applications of Gas Chromatography/Mass Spectrometry — 493
  - 14.5.2 Applications of Liquid Chromatography/Mass Spectrometry — 493
  - 14.5.3 Applications of MALDI–MS — 494
  - Overview — 495
  - Exercises — 497
  - Additional Reading — 498
  - References — 498

15 MISCELLANEOUS TOPICS — 501

- 15.1 Enzyme Kinetics — 501
  - 15.1.1 Theory — 501
  - 15.1.2 Reaction Monitoring — 504
- 15.2 Imaging Mass Spectrometry — 507
  - 15.2.1 Imaging with SIMS — 508
  - 15.2.2 Imaging with MALDI–MS — 509
- 15.3 Analysis of Microorganisms — 511
  - 15.3.1 Bacterial Identification — 511
  - 15.3.2 Analysis of Viruses — 513
- 15.4 Clinical Mass Spectrometry — 513
  - 15.4.1 Low-Molecular-Mass Compounds as Biomarkers of Disease — 514
  - 15.4.2 Analysis of DNA to Diagnose Genetic Disorders — 514
  - 15.4.3 Proteins as Biomarkers of Disease — 515
- 15.5 Metabolomics — 517
- 15.6 Forensic Mass Spectrometry — 517
  - 15.6.1 Analysis of Banned Substances of Abuse — 518

		15.6.2	Analysis of Explosives	519
		15.6.3	Analysis of Glass and Paints	519
		15.6.4	Authenticity of Questioned Documents	519
		15.6.5	Mass Spectrometry in Bioterror Defense	520
	15.7	Screening Combinatorial Libraries		520
		15.7.1	Combinatorial Synthetic Procedures	521
		15.7.2	Screening Methods	522
		Additional Reading		526
		References		526

Appendix A:	ABBREVIATIONS	533
Appendix B:	PHYSICAL CONSTANTS, UNITS, AND CONVERSION FACTORS	541
Appendix C:	ISOTOPES OF NATURALLY OCCURRING ELEMENTS AND THEIR ABUNDANCES	543
Appendix D:	REFERENCE IONS AND THEIR EXACT MASSES	551
Appendix E:	INTERNET RESOURCES	555
Appendix F:	SOLUTIONS TO EXERCISES	557
INDEX		**577**

PREFACE

For over 100 years, mass spectrometry has played a pivotal role in a variety of scientific disciplines. With a small beginning in the late nineteenth century as a tool to detect cathode rays, mass spectrometry currently has assumed a major role in identification of proteins in biological specimens, with the aim of unraveling their functional role and detecting biomarkers of a specific disease. Mass spectrometry has become an integral part of proteomics and the drug development process. Several diverse fields, such as physics, chemistry, medicinal chemistry, pharmaceutical science, geology, cosmochemistry, nuclear science, material science, archeology, petroleum industry, forensic science, and environmental science, have benefited from this highly sensitive and specific instrumental technique.

With the expansion of activity in mass spectrometry, an impressing need is felt to teach and train diversified and ever-increasing numbers of users of this somewhat esoteric analytical technique. This book is intended to fulfill this need by providing a well-balanced and in-depth discussion of the basic concepts and latest developments over a range of important topics in modern mass spectrometry. The material in the book has evolved from my experience of more than 20 years in teaching mass spectrometry courses at the undergraduate and graduate levels. Writing an earlier book, *Principles and Practice of Biological Mass Spectrometry* (Wiley-Interscience, 2001), was also of immense help in preparing the present volume. The previous book was well accepted by the mass spectrometry community, which encouraged me to undertake this project.

For convenience, the book is organized into three parts and 15 chapters. Part I has five chapters that provide a detailed description of the instrumentation aspects of mass spectrometry. Topics in this section include modes of ionization (Chapter 2), mass analysis and ion detection (Chapter 3), tandem mass spectrometry (Chapter 4), and hyphenated separation techniques (Chapter 5). Mass spectrometry has long made a valuable contribution to the identification of small (<1000 Da) organic compounds (Chapter 6) and the characterization of inorganic materials (Chapter 7). These two important topics are discussed in Part II. The protocol for interpretation of the electron ionization mass spectrum of organic compounds and the rules of their fragmentation are described in Chapter 6. Currently, the role of mass spectrometry has expanded to the biological sciences. Keeping this aspect in mind, a large portion of this book (Part III) is devoted to the field of biological mass spectrometry. This section contains eight chapters. The analysis of proteins and peptides, which is a major focus of biological mass spectrometry, is dealt with at length; three chapters (Chapters 8 to 10) are devoted to this topic. Also discussed are oligosaccharides (Chapter 11), lipids (Chapter 12), and oligonucleotides (Chapter 13). The field of quantitative analysis is reviewed separately in Chapter 14. Chapter 15 covers a range of miscellaneous topics, including enzyme kinetics, imaging mass spectrometry, analysis of microorganisms, clinical mass spectrometry, metabolomics, forensic analysis, and combinatorial chemistry. Several appendixes provide additional helpful material. A comprehensive up-to-date list of references is included at the end of each chapter.

As an aid to better understanding of the concepts and to improve problem-solving skills, several worked-out examples are included in most chapters. Another novel feature of the book is an overview of each chapter, which provides a concise survey of the concepts discussed in the chapter. Also, the practice exercises included at the end of the chapter will help readers grasp the material. Solutions to the exercises are given in Appendix F.

It is hoped that the book will be a good teaching tool of the principles of mass spectrometry to undergraduates and graduates as well as to those with no background in mass spectrometry. The practitioner of mass spectrometry at all levels should also enjoy reading the book.

I would like to express my gratitude to Drs. Dominic M. Desiderio, Chris G. Enke, Michael L. Gross, Nico M. M. Nibbering, and Kenneth B. Tomer for their valuable expert opinion. They have all read the text completely or in part and have provided valuable insight and suggestions. I also acknowledge the assistant of Hari Kosanam and Tarun Gheyi in preparing the manuscript. The editorial staff at Wiley-Interscience also deserves my appreciation for the excellent appearance of the book. Finally, I lack the words to express my full appreciation to my wife, Asha, for her love, encouragement, and sacrifice during the writing of the book.

Most of the EI mass spectra of organic compounds in Chapter 6 are reproduced from the *NIST Chemistry WebBook*. I am highly indebted to the NIST for the use of these spectra. Several figures in this book are reproduced from my earlier book, *Principles and Practice of Biological Mass Spectrometry*, for which I am grateful to Wiley-Interscience.

CHHABIL DASS

Memphis, Tennessee

PART I

INSTRUMENTATION

CHAPTER 1

BASICS OF MASS SPECTROMETRY

Mass spectrometry (MS) is an analytical technique that measures the molecular masses of individual compounds and atoms precisely by converting them into charged ions. Quite often, the structure of a molecule can also be deduced. Mass spectrometry is also uniquely qualified to provide quantitative information of an analyte at levels of structure specificity and sensitivity that are beyond imagination (e.g., in the zeptomole range). In addition, mass spectrometry allows one to study reaction dynamics and chemistry of ions, to provide data on physical properties such as ionization energy, appearance energy, enthalpy of a reaction, proton and ion affinities, and so on, and to verify molecular orbital calculations-based theoretical predictions. Thus, mass spectrometry probably is the most versatile and comprehensive analytical technique currently at the disposal of chemists and biochemists. Several areas of physics, chemistry, medicinal chemistry, pharmaceutical science, geology, cosmochemistry, nuclear science, material science, archeology, petroleum industry, forensic science, and environmental science have benefited from this highly precise and sensitive instrumental technique.

1.1. BRIEF HISTORY OF MASS SPECTROMETRY

As early as 1898, Wien demonstrated that canal rays could be deflected by passing them through superimposed parallel electric and magnetic fields. Sir Joseph J. Thomson (1856–1940) is credited with the birth of mass spectrometry through his work on the analyses of negatively charged cathode ray particles [1] and of positive rays with a parabola mass spectrograph [2]. His prophesy was that this

---

Fundamentals of Contemporary Mass Spectrometry, by Chhabil Dass
Copyright © 2007 John Wiley & Sons, Inc.

new technique would play a profound role in the field of chemical analysis. In the next two decades, however, the developments of mass spectrometry continued in the hands of renowned physicists like Aston, Dempster, Bainbridge, and Nier [3]. During this time, mass spectrometry played a pivotal role in the discovery of new isotopes and in determining their relative abundances and accurate masses. In the 1940s, mass spectrometry played a major role in the Manhattan Project, a wartime program to separate on a preparative scale the fissionable ^{235}U isotope and as a leak detector in a UF_6 gaseous diffusion plant.

In the 1940s, chemists ultimately recognized the potential of mass spectrometry as an analytical tool and applied it to monitor a petroleum refinery stream. The first commercial mass spectrometer became available in 1943 through the Consolidated Engineering Corporation. The principles of time-of-flight (TOF) and ion cyclotron resonance (ICR) mass spectrometry were introduced in 1946 and 1948, respectively [4,5]. Applications to organic chemistry began in the 1950s and exploded in the 1960s and 1970s. Double-focusing high-resolution mass spectrometers, which became available in the early 1950s, paved the way for accurate mass measurements of a variety of compounds. The concept of quadrupole mass analyzer and ion traps as mass detectors was described by Wolfgang Paul et al. in 1953 [6,7]. The development of gas chromatography (GC)/MS in the 1960s marked the beginning of the analysis of seemingly complex mixtures by mass spectrometry [8,9]. The 1960s also witnessed the development of tandem mass spectrometry (MS/MS) [10]; the emergence of this technique is a high point in the field of structure analysis and unambiguous quantification by mass spectrometry. Chemical ionization, a "soft" mode of ionization, was also introduced during this period [11].

By the 1960s, mass spectrometry had become a standard analytical tool in the analysis of organic compounds. Its applications to biological fields were, however, miniscule, owing primarily to the lack of suitable ionization techniques for fragile and nonvolatile compounds of biological origin. Over the last two decades, that situation has changed. Several unique developments in gentler modes of ionization have allowed the production of ions from compounds of large molecular mass and compounds of biological relevance. These methods include fast atom bombardment (FAB) in 1981 [12], electrospray ionization (ESI) (in 1984–1988) [13], and matrix-assisted laser desorption/ionization (MALDI) in 1988 [14,15]. The last two methods have extended the upper mass range beyond 100 kilodaltons (kDa) and had an enormous impact on the use of mass spectrometry in biology and the life sciences. Concurrent with these developments, several innovations in mass analyzer technology, such as the introduction of high-field and superfast magnets and improvements in the TOF and Fourier transform (FT) ion cyclotron resonance–mass spectrometry (ICR–MS) analysis concepts, have also improved the sensitivity and upper mass range amenable to mass spectrometry. The current decade has seen the introduction of two new types of ion traps, the quadrupole linear ion trap (LIT) and the orbitrap, for mass spectrometric analysis [16,17]. A variety of hybrid tandem mass spectrometry systems are available for enhanced performance in tandem mass spectrometry (see Chapter 4). The coupling of high-performance liquid chromatography (HPLC) with mass spectrometry, first demonstrated in the 1970s [18,19] and later optimized with an ESI interface [20], is another high point that has provided chemists and biochemists with one of their most useful instruments. Improvements in detection devices and the introduction of fast data systems have also paralleled these developments. Currently, mass spectrometry has found a niche in the biomedical field and life sciences and is at the forefront of proteomics techniques.

1.2. DESIRABLE FEATURES OF MASS SPECTROMETRY

The wide popularity of mass spectrometry is the result of its unique capabilities:

- It provides unsurpassed molecular specificity because of its unique ability to measure accurate molecular mass and to provide information on structurally diagnostic fragment ions of an analyte.
- It provides ultrahigh detection sensitivity. In theory, mass spectrometry has the ability to detect a single molecule; the detection of molecules in attomole and zeptomole amounts has been demonstrated.
- It has unparalleled versatility to determine the structures of most classes of compounds.
- It is applicable to all elements.
- It is applicable to all types of samples: volatile or nonvolatile; polar or nonpolar; and solid, liquid, or gaseous materials.
- In combination with high-resolution separation devices, it is uniquely qualified to analyze "real-world" complex samples.

1.3. BASIC PRINCIPLES OF MASS SPECTROMETRY

Mass spectrometry measurements deal with ions because unlike neutral species, it is easy to manipulate the motion and direction of ions experimentally and detect them. Three basic steps are involved in mass spectrometry analysis (Figure 1.1):

1. The first step is ionization that converts analyte molecules or atoms into gas-phase ionic species. This step requires the removal or addition of an

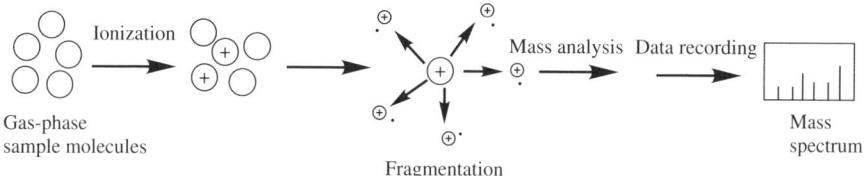

Figure 1.1. Basic concept of mass spectrometry analysis. (Reproduced from C. Dass, *Principles and Practice of Biological Mass Spectrometry*, Wiley-Interscience, 2001.)

6 BASICS OF MASS SPECTROMETRY

electron or proton(s). The excess energy transferred during an ionization event may break the molecule into characteristic fragments.
2. The next step is the separation and mass analysis of the molecular ions and their charged fragments on the basis of their m/z (mass-to-charge) ratios.
3. Finally, the ion current due to these mass-separated ions is measured, amplified, and displayed in the form of a mass spectrum.

The first two steps are carried out under high vacuum, which allows ions to move freely in space without colliding or interacting with other species. Collisions may lead to fragmentation of the molecular ions and may also produce a different species through ion–molecule reactions. These processes will reduce sensitivity, increase ambiguity in the measurement, and decrease resolution. In addition, the atmospheric background will introduce interference.

A simplistic view of the essential components of a mass spectrometer is given in Figure 1.2. These components are:

- *An inlet system:* transfers a sample into the ion source. An essential requirement is to maintain the integrity of the sample molecules during their transfer from atmospheric pressure to the ion-source vacuum.
- *An ion source:* converts the neutral sample molecules into gas-phase ions. Several ionization techniques have been developed for this purpose (see Chapter 2).
- *A mass analyzer:* separates and mass-analyzes the ionic species. Magnetic and/or electric fields are used in mass analyzers to control the motion of ions. A magnetic sector, quadrupole, time-of-flight, quadrupole ion trap, quadrupole linear ion trap, orbitrap, and Fourier transform ion cyclotron resonance instrument are the most common forms of mass analyzers currently in use (discussed in Chapter 3).

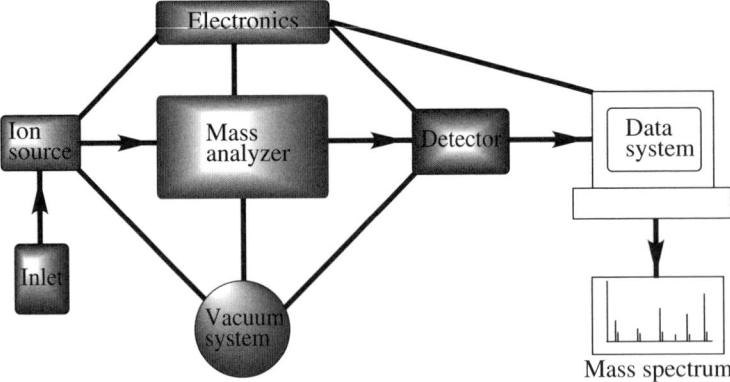

Figure 1.2. Basic components of a mass spectrometer. (Reproduced from C. Dass, *Principles and Practice of Biological Mass Spectrometry*, Wiley-Interscience, 2001.)

- *A detector:* measures and amplifies the ion current of mass-resolved ions.
- *A data system:* records, processes, stores, and displays data in a form that a human eye can easily recognize (computer screen or printer output).
- *A vacuum system:* maintains a very low pressure in the mass spectrometer. The ion source region is usually maintained at a pressure of 10^{-4} to 10^{-8} torr; somewhat lower pressure is required in the mass analyzer region (around 10^{-8} torr). Most instruments use a differential pumping system to maintain an optimal vacuum.
- *Electronics:* controls the operation of various units.

1.4. ANATOMY OF A MASS SPECTRUM

The simplest and most common means of ion formation in mass spectrometry is to bombard the gas-phase sample molecules with a beam of electrons. During this process, an electron is removed from the highest-occupied molecular orbital (HOMO) of the sample molecule to form a positively charged molecular ion:

$$M + e^- \rightarrow M^{+\bullet} + 2e^- \tag{1.1}$$

Fragment ions may also form. The data output is in the form of a mass spectrum.

It is essential to become familiar with a mass spectrum. A common form, the computer-generated bar-graph plot, is shown in Figure 1.3. It is a plot of m/z values (on the x-axis) of all ions (i.e., the molecular ion and its fragment ions, plus background ions, if any) that reach the detector versus their abundances (on the y-axis). The spectrum in Figure 1.3 is a positive ion electron ionization–generated mass spectrum of acetophenone. A mass spectrum is usually characterized by a molecular ion region (i.e, the molecular ion signal plus

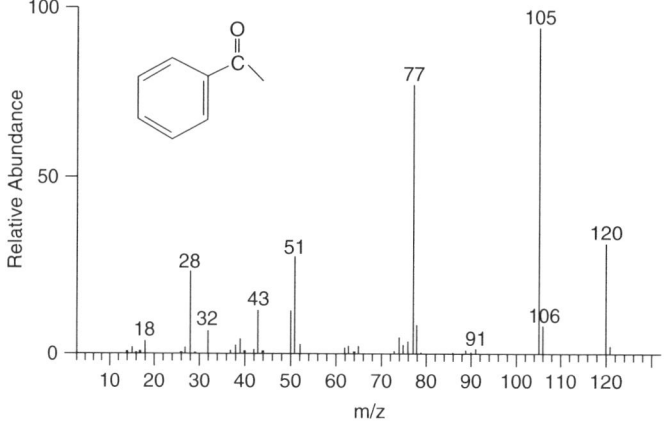

Figure 1.3. The 70-eV EI mass spectrum of acetophenone (MW = 120 u).

8 BASICS OF MASS SPECTROMETRY

associated heavy isotope satellite ions) and a fragment ion region (the sample molecule-related fragments). It is general practice to designate the most abundant ion in the spectrum as the base peak (here, m/z 105), which is arbitrarily assigned a relative height of 100. The abundances of all other ions in the spectrum are reported as percentage abundances relative to this base peak. Before the advent of computers, a photographic chart paper recorder was used to provide a mass spectrum. An advantage of this type of recording was that information derived from metastable ion fragmentation could also be retrieved. The ions from a metastable fragmentation appear as diffuse peaks at nonintegral masses. The computer-generated spectrum can also be presented in list format, a tabulation of all ion m/z values versus their abundances.

A mass spectrum is a useful source of structure-specific information, the most important datum being the molecular mass of the analyte. The molecular mass of an analyte can readily be inferred from the molecular ion because this ion represents the intact molecule minus an electron [reaction. (1.1)]. The molecular ion usually is the largest peak among the high-mass cluster of peaks in the spectrum (e.g., the m/z 120 in Figure 1.3). From the m/z values of the fragment ions, the structure of the analyte can be deduced. The structure determination of organic compounds by mass spectrometry is discussed in more detail in Chapter 6. In some ionization techniques (see Chapter 2), the molecular ion is obtained as a protonated or deprotonated molecule (i.e., $[M + H]^+$ or $[M - H]^-$). The molecular mass from those spectra is obtained by subtracting the mass of a proton from the m/z value of the $[M + H]^+$ ion. For example, Figure 1.4 is the ESI

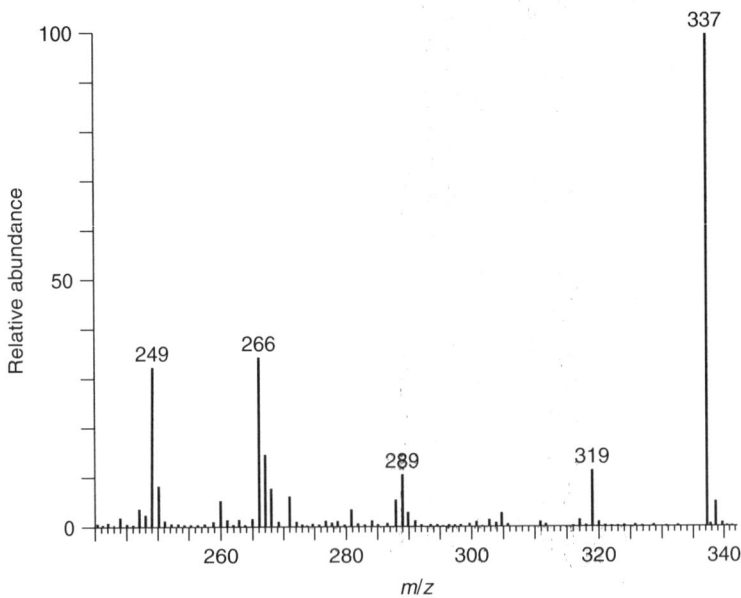

Figure 1.4. ESI mass spectrum of acetebutol (MW = 336 u).

mass spectrum of a β-blocker, acetebutol (molecular mass = 336.205 u). From the m/z observed for the $[M + H]^+$ ion (i.e., 337), the molecular mass of acetebutol can readily be determined. The molecular mass can also be obtained from a negative-ion spectrum. The structural information is, however, very sparse in negative-ion spectra because negative ions are more stable. The structural information is also sparse in a mass spectrum that is acquired by ionizing the molecule with a gentler mode of ionization such as FAB, ESI, or MALDI. The structural information from FAB-, ESI-, and MALDI-generated ions is obtained by subjecting the molecular ions to collision-induced dissociation (CID) and acquiring the MS/MS spectrum (see Chapter 4 for more details).

1.5. ATOMIC AND MOLECULAR MASSES

The SI base unit of mass is the kilogram (kg). The mass of microscopic species, such as atoms and molecules, is very small, and to express this small quantity in kilograms is cumbersome. For example, to denote the mass of a single carbon atom as 1.99266×10^{-26} kg is very inconvenient. Therefore, the mass of atoms and molecules is expressed in terms of the unified atomic mass unit (u). By international agreement, the mass of one atom of the ^{12}C isotope is assigned the exact value 12. One unified atomic mass unit is defined as equal to $\frac{1}{12}$ the mass of a single atom of the ^{12}C isotope. Alternatively, the dalton (Da) is also used in place of u, especially when expressing the mass of large biomolecules. The mass of other atoms is expressed relative to the mass of the ^{12}C isotope. Thus,

$$1 \text{ u} = 1 \text{ Da} = 1.6605402 \times 10^{-27} \text{ kg}$$

Several different molecular mass terms are in use:

- *Nominal ion mass:* the mass of the ion for a given empirical formula, calculated by adding the integer mass of the most abundant isotope of each element (e.g., $^1H = 1$ and $^{12}C = 12$).
- *Monoisotopic ion mass:* the mass of the ion for a given empirical formula, calculated from the exact mass of the most abundant isotope of each element (e.g., $^1H = 1.007825$ and $^{12}C = 12.000000$). The exact mass of the elements and their isotopes are provided in Appendix C.
- *Most abundant ion mass:* the mass that corresponds to the most abundant peak in the isotopic cluster of the ion of a given empirical formula.
- *Average mass:* the mass of an ion for a given empirical formula calculated with the atomic weight of each element (e.g., C = 12.01115 and H = 1.00797): that is, the average of the isotopic masses of each element, weighted for isotopic abundance. The average mass represents the centroid of the distribution of the isotopic peaks of the molecular ion and is used by chemists in stoichiometric calculations.

▶ **Example 1.1** Calculate the nominal and exact mass of acetophenone.

Solution The nominal mass of acetophenone (C_8H_8O) = $(8 \times 12) + (8 \times 1) + (1 \times 16) = 120$ u.
The exact mass of acetophenone = $(8 \times 12.000000) + (8 \times 1.007825) + (1 \times 15.994915) = 120.0575$ u.

The monoisotopic mass is meaningful for low-mass compounds because the elemental composition can be determined from a well-defined isotopic pattern of the molecular ion. The nominal and monoisotopic masses can both be correlated with the most abundant peak in the isotopic cluster. As the mass of a compound, however, increases, the isotopic pattern becomes more symmetrical and extends over many masses [21]. Also, the monoisotopic peak becomes difficult to identify. For high-mass compounds (e.g., proteins and oligonucleotides), the molecular ion profile measured coalesces and becomes a single asymmetric peak. For such compounds, the average mass value is accepted as the molecular mass.

1.5.1. Mass-to-Charge Ratio

As mentioned above, the mass spectrometry data are presented as the *mass-to-charge ratio*, which by definition is the mass of the ion (m) divided by the number of charges (z) the ion carries. The total charge on the ion is represented by $q = ze$, where e is the charge on an electron [$e = 1.602 \times 10^{-19}$ coulomb (C)]. The unit of the mass-to-charge ratio is the thomson (Th) [22]. However, use of the unitless term m/q is common practice in the literature, and that practice is followed in this book. In the past, m/e had been used in place of m/z. The term m/e assumes that all ions in the spectrum are singly charged, whereas z can be a multiple integer.

Multiply charged cations are formed by attachment of several protons. This process usually occurs for biomolecules in the ESI mode of ionization. The corresponding ions will appear at $[M + nH]^{n+}/n$, where M is the molecular mass of the biomolecule, n the number of protons it can accept, and H the mass of a proton. Thus, M and m/q can have two distinct values. These values are identical only for singly charged ions. This distinction is clearly explicable in Example 1.2.

▶ **Example 1.2** The nominal mass of acetophenone is 120 u (see Example 1.1). In the mass spectrum, $C_8H_8O^+$ ion will appear at $m/q = 120$ and $C_8H_8O^{2+}$ at $120/2 = 60$.
Similarly, a protein that has a mass of 50,000 Da and can accept 25 protons to produce an $[M + 25H]^{25+}$ ion displays an m/q at $(50{,}000 + 25 \times 1)/25 = 2001$.

1.6. GENERAL APPLICATIONS

Mass spectrometry plays a central role in almost every field of science. This distinction is the result of the high level of molecular specificity, detection sensitivity, and availability of ionization techniques for all classes of compounds. Some of the major areas of applications are:

Physics:
- Determination of the accurate masses of elements and abundances of isotopes

Chemistry:
- Accurate mass measurement of atoms and molecules
- Structure analysis of organic compounds
- Quantitative analysis of inorganic and organic compounds
- Fundamentals of gas-phase ion chemistry
- Measurement of physical properties of ions
- Elemental analysis
- Precise isotope ratio measurements

Environmental science:
- Analysis of environmental pollutants in air, water, and soil
- Study of Earth's atmosphere and water resources (lakes, rivers, oceans)

Medicine and life sciences:
- Molecular mass measurement of large biological compounds
- Simultaneous separation and detection of complex mixtures of biological compounds
- Quantitative analysis of a variety of compound types in biological tissues and fluids
- Amino acid sequence determination of proteins and peptides
- Higher-order structures of proteins and peptides
- Covalent complexes of biomolecules
- Identification of specific diseases
- Structural characterization of lipids and oligosaccharides
- Sequence determination of oligonucleotides
- Profiling of bacteria and viruses
- Study of functional aspects of biomolecules
- Clinical studies
- Measurement of isotope ratios for biological tracer studies

Pharmaceutical sciences:
- Analysis of isolated and synthesized drugs
- Pharmacodynamic and pharmacokinetic evaluation of new and old drugs

Geology:
  Determination of age and composition of rocks and other geological species
Industry:
  Monitoring of process streams such as a refinery stream in the petroleum industry
Forensic science:
  Analysis of explosives and banned substances
Material science:
  Analysis of metals, alloys, semiconductors, and polymers

This list is by no means complete. Mass spectrometry will continue to become an integral part of many diverse fields. With continued developments in the future, we will witness an expanded role for mass spectrometry in many unchartered territories, especially in offering new perspectives on solving real-world problems. With sensitive and faster analysis methods at hand, the role of mass spectrometry will expand to studies related to human health and safety.

OVERVIEW

In this introductory chapter, some basic concepts of mass spectrometry were discussed. A brief history of mass spectrometry was presented. Mass spectrometry has its roots in early work with the cathode-ray tube but now it is a more mature discipline and an indispensable analytical tool. It is used primarily to determine the mass of atomic and molecular species and to structurally characterize and quantify a very broad range of compounds. The major assets of this technique are specificity, sensitivity, and ability to analyze real-world samples.

Essential steps of mass spectrometric analysis are ionization, separation of ions on the basis of m/z ratio, and detection of the ion current of separated ions. To perform these functions, a mass spectrometer is made of an ion source, a mass analyzer, a detector, a data system, a vacuum system, and electronic control units. The data are presented in the form of a mass spectrum, which is a plot of m/z values on the x-axis versus their abundances on the y-axis. From this spectrum, the mass of the target species and its structure can be determined.

Depending on the size of the molecule, mass spectrometry provides information on the nominal, monoisotopic, and average masses. The nominal mass and monoisotopic mass information is obtained for low-mass compounds, whereas, for high-mass compounds, the average mass value is measured.

EXERCISES

1.1. List the basic steps involved in mass spectrometric analysis.

1.2. Why is mass spectrometric analysis performed under high vacuum?

1.3. What are the essential components of a mass spectrometer, and what is the function of each?

1.4. Calculate the nominal mass, monoisotopic mass, and average mass of the tranquilizer diazepam, $C_{16}H_{13}N_2OCl$.

1.5. The molecular mass of a peptide is 2051 Da. Calculate the m/z value of the triply protonated peptide.

1.6. Calculate the mass of diazepam in kg.

REFERENCES

1. J. J. Thomson, *Philos. Mag. V* **44**, 293 (1897).
2. J. J. Thomson, *Rays of Positive Electricity and the Application to Chemical Analyses*, Longmans, Green & Co., London, 1913.
3. M. A. Grayson, ed., *Measuring Mass from Positive Rays to Proteins*, Chemical Heritage Press, Philadelphia, PA, 2002, pp. 1–149.
4. W. E. Stephens, A pulsed mass spectrometer with time dispersion, *Phys. Rev.* **69**, 691 (1946).
5. H. Sommer, H. A. Thomas, and J. A. Hipple, Measurement of e/m by cyclotron resonance, *Phys. Rev.* **82**, 697–702 (1951).
6. W. Paul and H. Steinwedel, A new mass spectrometer without magnetic field, *Z. Naturforsch.* **8a**, 448–450 (1953).
7. W. Paul, P. Reinhard, and O. Zahn, The electric mass filter as mass spectrometer and isotope separator, *Z. Phys.* **152**, 143–182 (1958).
8. R. Ryhage, MS as a detector for GC, *Anal. Chem.* **36**, 759–764 (1964).
9. J. T. Watson and K. Biemann, High-resolution MS of GC effluents, *Anal. Chem.* **37**, 844–851 (1965).
10. K. R. Jennings, Collision-induced decompositions of aromatic molecular ions, *Int. J. Mass Spectrom. Ion Phys.* **1**, 227–235 (1968).
11. M. S. B. Munson and F. H. Field, Chemical ionization mass spectrometry, I: General introduction, *J. Am. Chem. Soc.* **88**, 2621–2630 (1966).
12. M. Barber, R. S. Bordoli, R. D. Sedgwick, and A. N. Tyler, Fast atom bombardment of solids (F.A.B.): a new ion source for mass spectrometry, *J. Chem. Soc. Chem. Commun.*, 325–327 (1981).
13. J. B. Fenn, M. Mann, C. K. Meng, S. F. Wong, and C. M. Whitehouse, Electrospray ionization for mass spectrometry of large biomolecules, *Science* **246**, 64–71 (1989).
14. M. Karas and F. Hillenkamp, Laser desorption ionization of proteins with molecular masses exceeding 10,000 Daltons, *Anal. Chem.* **60**, 2299–2301 (1988).
15. K. Tanaka, H. Waki, H. Ido, S. Akita, and T. Yoshida, Protein and polymer analyses up to m/z 100,000 by laser ionization time-of-flight mass spectrometry, *Rapid Commun. Mass Spectrom.* **2**, 151–153 (1988).
16. J. W. Hager, A new linear ion trap mass spectrometer, *Rapid Commun. Mass Spectrom.* **16**, 512–526 (2002).
17. A. Makarov, Electrostatic axially harmonic orbital trapping: high-performance technique of mass analysis, *Anal. Chem.* **72,** 1156–1162 (2000).

18. M. A. Baldwin and F. W. McLafferty, Liquid chromatography–mass spectrometry interface, I: Direct introduction of liquid solutions into a chemical ionization mass spectrometer, *Org. Mass Spectrom.* **7**, 1111–1112 (1973).
19. W. H. McFadden, H. L. Schwartz, and D. C. Bradford, Direct analysis of liquid chromatographic effluents, *J. Chromatogr.* **122**, 389–396 (1976).
20. T. R. Covey, E. C. Huang, and J. D. Henion, Structural characterization of protein tryptic peptides via liquid chromatography/mass spectrometry and collision-induced dissociation of their doubly charged molecular ions, *Anal. Chem.* **63**, 1193–1200 (1991).
21. J. Yergy, D. Heller, G. Hansen, R. J. Cotter, and C. Fenselau, Isotopic distributions in mass spectra of large molecules, *Anal. Chem.* **55**, 353–356 (1983).
22. R. G. Cooks and A. L. Rockwood, The "Thomson": suggested unit for mass spectroscopists, *Rapid Commun. Mass Spectrom.* **5**, 93 (1991).

CHAPTER 2

MODES OF IONIZATION

Ionization of the analyte is the first crucial and challenging step in the analysis of any class of compounds by mass spectrometry. The key to a successful mass spectrometric experiment lies to a large extent in the approach to converting a neutral compound to a gas-phase ionic species. A wide variety of ionization techniques have become available over the years, but none has universal appeal. In some techniques, ionization is performed by ejection or capture of an electron by an analyte to produce a radical cation $[M^{+\bullet}]$ or anion $[M^{-\bullet}]$, respectively. In others, a proton is added or subtracted to yield $[M + H]^+$ or $[M - H]^-$ ions, respectively. The adduction with alkali metal cations (e.g., Na^+ and K^+) and anions (e.g., Cl^-) is also observed in some methods. The choice of a particular method is dictated largely by the nature of the sample under investigation and the type of information desired. Table 2.1 lists some of the methods currently in vogue. Some methods are applicable to the atomic species, whereas others are suitable for molecular species. Also, some methods require sample molecules to be present in the ion source as gas-phase species, whereas others can accommodate condensed-phase samples. The methods that are applicable to molecular species are the subject of the present chapter; those applicable to atomic species are described in Chapter 7.

2.1. WHY IONIZATION IS REQUIRED

Mass spectrometry measurements are performed with charged particles because it is easy to manipulate experimentally the motion and direction of ions. By applying electric and magnetic forces, the energy and velocity of ionic species

Fundamentals of Contemporary Mass Spectrometry, by Chhabil Dass
Copyright © 2007 John Wiley & Sons, Inc.

TABLE 2.1. Modes of Ionization

Atomic Ionization	Molecular Ionization		
	Sample Phase	Mode	Pressure[a]
Thermal ionization	Gas phase	Electron ionization	HV
Spark source		Chemical ionization (CI)	IV
Glow discharge		Photoionization (PI)	HV
Inductively coupled plasma		Field ionization	HV
Resonance ionization		Metastable atom bombardment	HV
	Solution phase	Thermospray	LV
		Atmospheric-pressure CI	AP
		Atmospheric-pressure PI	AP
		Electrospray	AP
	Solid phase	Plasma desorption	HV
		Field desorption	HV
		Secondary-ion MS	HV
		Fast atom bombardment	HV
		Matrix-assisted laser desorption	HV

[a] HV, high vacuum; IV, intermediate vacuum; LV, low vacuum; AP, atmospheric pressure.

can be controlled, and both help in their separation and detection. In contrast, neutral gas-phase species move randomly and aimlessly. Their separation by gravitational force is highly impractical, and if attempted at all, then it might require an extremely long flight path, perhaps miles.

2.2. GENERAL CONSTRUCTION OF AN ION SOURCE

The function of an ion source is to convert sample molecules or atoms into gas-phase ionic species. Several different types of ion-source designs are in use, some operating at very low pressures and some at atmospheric pressure, and not all are alike in construction. Some common elements of these sources are (1) a source block, (2) a source of energy (e.g., an electron, particle, or ion beam), (3) a source heater, (4) a short ion-extraction region that accelerates the ions to a specified fixed kinetic energy, and (5) an exit slit assembly. The accelerating potential is set to several kilovolts in magnetic-sector and time-of-flight (TOF) instruments, but to only a few volts in quadrupole-based mass spectrometers.

The ion source should have the following desirable characteristics: (1) high ionization efficiency (a requirement for high detection sensitivity), (2) a stable ion beam, (3) a low-energy spread in the secondary-ion beam, (4) minimum background ion current, and (5) minimum cross-contamination between successive samples.

GAS-PHASE IONIZATION TECHNIQUES

2.3. ELECTRON IONIZATION

Electron ionization (EI) is one of the oldest modes of ionization, first used by Dempster in 1918 [1]. It is the most popular means of ionization for organic compounds with molecular mass less than 600 Da. Several other classes of compounds can also be analyzed conveniently by EI–MS. It is, however, restricted to thermally stable and relatively volatile compounds. Many solids and liquids are quite volatile at the prevailing vacuum of the instrument, whereas others must first be vaporized at elevated temperatures.

In the *EI process,* the vaporized sample molecules are bombarded with a beam of energetic electrons at low pressure (ca. 10^{-5} to 10^{-6} torr). An electron from the target molecule (M) is expelled during this collision process to convert the molecule to a positive ion with an odd number of electrons. This positive ion, called a *molecular ion or radical cation,* is represented by the symbol $M^{+\bullet}$:

$$M + e^- \longrightarrow M^{+\bullet} + 2e^- \quad (2.1)$$

For ionization to occur, it is essential that the kinetic energy of the bombarding electrons exceed the ionization energy (IE) of the sample molecule. Conventional wisdom is to employ a beam of 70-eV electrons. The energy gained by the ionized molecule in excess of the IE promptly causes it to dissociate into structurally diagnostic smaller-mass-fragment ions, some of which may still have sufficient energy to fragment further to second-generation product ions. The fragmentation pattern thus obtained is diagnostic of the structure of the sample molecule (see Figure 1.3). Fragmentation of molecular ions occurs primarily within the ion-source region. The efficiency of ionization and of subsequent fragmentation increases with increased electron energy and reaches a plateau at 50 to 100 eV; at these electron energies, the EI spectrum becomes a "fingerprint" of the compound being analyzed (several EI spectra of organic compound are shown in Chapter 6). Because the mass of the "lost" electron is negligible, the mass-to-charge (m/z) value of the molecular ion is a direct measure of its molecular mass. Negative ions can be formed via the capture of an electron by a neutral molecule.

A schematic of a prototypical EI source is shown in Figure 2.1. The main body of the source is a metal block with holes drilled in it. Electrons are produced by heating to an incandescent temperature a thin filament (cathode) of rhenium wire, and are allowed to enter the ionization chamber through a slit. The applied potential difference (usually 70 V) between the filament and the ion-source block accelerates the electrons to the kinetic energy required. An electron trap (anode) placed just outside the ionization chamber opposite the cathode is held at a slightly positive potential with respect to the ion-source block. After traveling

18   MODES OF IONIZATION

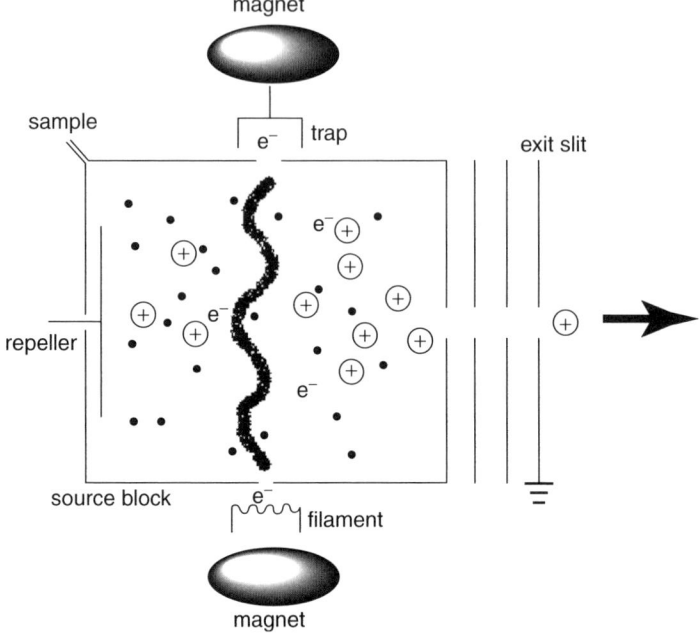

Figure 2.1. Schematic diagram of an electron ionization source.

the entire length of the ionization chamber, the unused electrons, are collected by the trap. Sample molecules are introduced into the ionization chamber through another hole as a gas-phase stream. To increase the probability of ionization, electrons are made to travel in a narrow helical trajectory of increased path length by applying a weak magnetic field parallel to the direction of the electron beam. The positive ions thus formed are pushed through a small slit into the accelerating region by applying a positive potential to a repeller electrode. Before entering the mass analyzer, all ions must be accelerated to a certain fixed kinetic energy by applying a potential to the ion-source block and holding the exit slit at the ground potential. The accelerating potential is set to several kilovolts in magnetic-sector and time-of-flight instruments, but to only a few volts in quadrupole-based mass spectrometers. The source block is usually heated up to 300°C to prevent condensation of the sample.

In an attempt to achieve higher ionization efficiency, filament current, emission current, and ionizing current must be optimized. The *filament current* is the current supplied to the filament to heat it to incandescent. The *emission current*, a measure of the rate of electron emission from the filament, is the current measured between the filament and the electron entry slit. The *ionizing current*, the rate of electron arrival at the trap, is a direct measure of the number of electrons in the chamber that are available for ionization.

▶ **Example 2.1** If a sample is ionized with a 70-eV electron beam and the sample ions attain 8000 eV of kinetic energy after exiting the ion source, at what potentials will various components of this ion source be maintained?

Solution Ion-source block = 8000 V; electron filament (cathode) = 7930 V; electron trap = about 8100 V; ion repeller = about 8020 V; and the exit slit at ground potential (i.e., at 0 V).

The sample ion current I^+, which is a measure of the ionization rate, can be enhanced by manipulation of the ion extraction efficiency β, the total ionizing cross section Q_i, the effective ionizing path length L, the concentration of the sample molecules $[N]$, and the ionizing current I_e:

$$I^+ = \beta Q_i L [N] I_e \tag{2.2}$$

whereas the ionizing cross section depends on the chemical nature of the sample molecules and the energy of the ionizing electrons, the ion extraction efficiency can be enhanced by increasing the repeller voltage and acceleration voltage. As mentioned above, the effective ionizing path length will increase by applying a weak magnetic field. The most effective way to increase the sample ion current is, however, to operate the ion source at higher values of the ionizing current. A typical value of the sample ion current at 100 μA ionizing current is about 10^{-7} A.

Sample Introduction Three devices used frequently to introduce sample molecules into the EI ion source are:

1. *A vacuum manifold:* used for highly volatile samples. The gaseous samples and volatile liquids are first expanded into a heated sample reservoir of the vacuum manifold and leaked into the ionization chamber through a pinhole.
2. *A direct insertion probe:* used for solids and relatively less volatile liquids. The probe, a long metal shaft that terminates in a small well for holding a sample capillary, can be inserted into the ion-source block through a vacuum lock. The sample is placed in a glass capillary and loaded into the well. Sudden vaporization of the sample is achieved by rapid heating of the probe to a desired temperature. With this probe, the sample can be positioned very close to the ionization region.
3. *Gas chromatography* (GC): used for complex mixtures. EI and GC are a perfect match because both methods deal with relatively volatile samples.

▶ **Example 2.2** In a typical EI–MS analysis, a 40-eV electron beam is used for ionization. Is the energy of this electron beam large enough to cause fragmentation in an organic compound whose IE is 10.0 eV?

Solution First calculate the energy per mole of electrons and compare it to typical bond energies of organic compounds.

Because 10.0 eV is used up in ionization of the analyte, the remaining 30.0 eV excess energy is available for fragmentation.

$$\text{kinetic energy (of a single electron)} = 30 \text{ eV} = (30)(1.602 \times 10^{-19}) \text{ C} \cdot \text{V}$$

$$\text{energy per mole electron} = (30)(1.602 \times 10^{-19}) \text{ C} \cdot \text{V}$$
$$\times (6.02 \times 10^{23}) \text{ e}^-\text{mol}^{-1}(1 \text{ J}/1 \text{ C} \cdot \text{V})$$
$$= 2.89 \times 10^6 \text{ J (kJ/1000 J)}$$
$$= 2.89 \times 10^3 \text{ kJ}$$

Bond energies of organic compounds usually vary in the range 200 to 600 kJ mol^{-1}. Therefore, the 40-eV electron beam is sufficient to cause fragmentation in this organic compound.

Advantages and Limitations of Electron Ionization EI is simple to use and provides library-searchable fingerprint spectra for most organic compounds, from which the molecular mass and structure of target compounds can be deduced. Several spectral libraries with over 100,000 spectral entries are available commercially or can be accessed over the Internet (see Section 6.4.6). A major limitation of EI is, however, the requirement that analytes be available as gas-phase molecules prior to ionization. This restriction makes many thermally labile and nonvolatile compounds inaccessible to EI. The volatility of some compounds can be augmented by chemical derivatization. This additional sample-handling step is, however, not a practical solution for trace amounts of samples. Also, derivatization may create uncertainty in determining the molecular mass of the original compound when derivatization is incomplete and the number of derivatized sites is not known. Another serious limitation of EI is that many compounds are not stable under EI conditions. Because of a highly energetic process, certain types of molecules fragment extensively upon EI, with the consequence that the molecular ion peak is either absent or of very little significance in their spectra. Under those conditions, the determination of molecular mass is a challenge. Extensive fragmentation may also complicate interpretation of a mass spectrum. A useful mass range for compounds that are amenable to EI is rather low (ca. <1000 Da).

2.4. CHEMICAL IONIZATION

Chemical ionization (CI) is a relatively less energetic mode of ionization. Therefore, it is a feasible option for compounds that fail to yield molecular ion signal in the EI mode. Credit for its discovery goes to Munson and Field [2,3]. Basically, CI shares a common ion-source assembly with EI, but after making certain modifications. First, to maintain the higher pressure (ca. 1 torr) needed for effective ion–molecule reactions, the ionization chamber is made as gastight as possible by reducing the apertures of the various slits. Second, electrons are produced from a sturdier metal filament or ribbon, and the electron energy is increased to 500 eV so that the electron beam can penetrate a reasonable length into the ion chamber. Third, the permanent magnet and the electron trap are eliminated because the electron beam cannot travel all the way to the other end of the ionization chamber. Finally, a more efficient pumping system is employed to remove the CI gas and to maintain a source pressure to below 10^{-4} torr, and the analyzer region is pumped differentially.

Conceptually, the *CI process* is quite distinct from EI. CI is accomplished through gas-phase acid–base reactions between the sample molecules and the reagent gas ions. Three steps are involved in the CI process. In the first step, the reagent gas, the partial pressure of which is 10 to 100 times greater than the sample pressure, is ionized at a pressure of 0.1 to 1 torr by bombardment with a beam of 200- to 500-eV electrons. In the second step, one or more stable reagent ions are produced by ion–molecule reactions. Finally, the sample molecules are ionized by gas-phase ion–molecule reactions with the stable reagent ions.

Consider the example of methane as the CI reagent gas; ionization of the sample molecule M proceeds through the following reactions:

Step 1: ionization of the reagent gas

$$CH_4 + e^- \longrightarrow CH_4^{+\bullet} + 2e^- (+ CH_3^+ + \text{other ions}) \quad (2.3)$$

Step 2: formation of stable reagent ions

$$CH_4 + CH_4^{+\bullet} \longrightarrow CH_5^+ + CH_3^{\bullet} \quad (2.4)$$

$$CH_4 + CH_3^+ \longrightarrow C_2H_5^+ + H_2 \quad (2.5)$$

Step 3: ionization of the sample molecules

Proton transfer: $\quad M + CH_5^+ \longrightarrow [M+H]^+ + CH_4 \quad (2.6)$

Adduct formation: $\quad M + CH_5^+ \longrightarrow [M+CH_5]^+ \quad (2.7)$

Adduct formation: $\quad M + C_2H_5^+ \longrightarrow [M+C_2H_5]^+ \quad (2.8)$

Hydride ion abstraction: $\quad M + C_2H_5^+ \longrightarrow [M-H]^+ + C_2H_6 \quad (2.9)$

Reagent ions formed in reactions (2.4) and (2.5) act as Brønsted acids. Ionization of the sample molecules occurs through a variety of ion–molecule reactions, the prominent reaction being an acid–base proton transfer reaction to produce $[M+H]^+$ ions [reaction (2.6)]. That reaction occurs when the proton affinity of

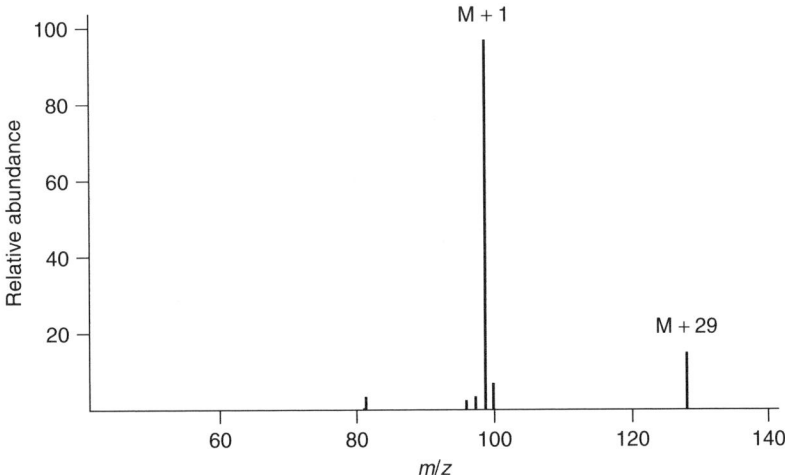

Figure 2.2. CH_4–CI mass spectrum of cyclohexanone. Compare it with the EI mass spectrum (Figure 6.9). (Redrawn from ref. 3.)

the analyte molecule exceeds that of methane. Adduct ions of type $[M + CH_5]^+$ or $[M + C_2H_5]^+$ are also formed [see reactions (2.7) and (2.8), and Figure 2.2], but to a lesser extent. Finally, the hydride ion abstraction to form $[M - H]^+$ [reaction (2.9)] also occurs for certain compounds. Unlike the EI-formed ions, the ions formed by CI are even-electron ions (meaning that, no electron is removed from the neutral molecule). The molecular mass of analyte can be deduced from the m/z values of those adduct ions after accounting for the mass of attached or abstracted species.

Isobutane and ammonia are the other two common CI reagent gases. The corresponding ionization reactions are:

Isobutane CI: $\quad C_4H_{10} + e^- \longrightarrow C_4H_{10}^{+\bullet} + 2e^- (+ C_3H_7^+ \text{ and other ions})$

$\quad\quad\quad\quad\quad\quad C_4H_{10} + C_3H_7^+ \longrightarrow C_4H_9^+ + C_3H_8$

$\quad\quad\quad\quad\quad\quad M + C_4H_9^+ \longrightarrow [M + H]^+ + C_4H_8$

Ammonia CI: $\quad NH_3 + e^- \longrightarrow NH_3^{+\bullet} + 2e^-$

$\quad\quad\quad\quad\quad\quad NH_3 + NH_3^{+\bullet} \longrightarrow NH_4^+ + NH_2$

$\quad\quad\quad\quad\quad\quad M + NH_4^+ \longrightarrow [M + H]^+ + NH_3$

Energetically, CI-formed ions are quite different from EI-formed ions. First, ionization occurs due to thermal energy collisions between the sample molecules and the reagent ions. Second, the energy of CI-formed ions is quickly quenched through collisions with the neutral reagent gas molecules. Third, even-electron ions are generally more stable than odd-electron ions. As a consequence, fragmentation of the CI-formed ions is suppressed drastically to yield a simple spectrum that contains primarily the molecular ion signal and perhaps a few fragment ions. This aspect is demonstrated in Figure 2.2, which is the CH_4–CI mass spectrum of cyclohexanone. Compared to the EI mass spectrum shown in Figure 6.9, there is almost no fragmentation. Thus, the use of CI is restricted primarily to determining or confirming the molecular mass of volatile compounds.

Also, the extent of fragmentation can be controlled by a proper choice of a reagent gas. The amount of energy transferred to the ionized sample molecule under CI conditions is a function of the exothermicity, $\Delta H°$, of the acid–base proton transfer reaction. Consider the reaction

$$M + [B + H]^+ \longrightarrow [M + H]^+ + B \qquad (2.10)$$

The energy transferred during this reaction is given by the difference in proton affinity (PA) of the sample molecule and that of the neutral reagent gas, B:

$$\Delta H° = -[PA(B) - PA(M)] \qquad (2.11)$$

Thus, the reaction is more efficient when PA(M) > PA(B). The PA is measured as the heat liberated during the protonation of a molecule. The CH_4–CI transfers more energy than does the i-C_4H_{10}–CI; therefore, the former produces more fragment ions.

Another useful feature of CI is selectivity in ionization. A number of reagent gases with a wide range of PA values are available; some of them are listed in Table 2.2 along with their proton affinities and the corresponding reagent ions. By proper choice of a reagent gas, only those compounds with PA values greater than that of the reagent gas are ionized. As an example, a reagent gas with a high PA value, such as ammonia or methylamine, is an ineffective CI reagent for many organic compounds.

TABLE 2.2. Reagents for Positive-Ion Chemical Ionization

Reagent Gas	Reagent Ion	Proton Affinity[a] (kcal mol^{-1})
H_2	H_3^+	101
CH_4	CH_5^+	132
H_2O	H_3O^+	167
CH_3OH	$CH_3OH_2^+$	182
C_2H_5OH	$C_2H_5OH_2^+$	186
i-C_4H_{10}	i-$C_4H_9^+$	196
$(CH_3)_2CO$	$(CH_3)_2COH^+$	197
NH_3	NH_4^+	204
CH_3NH_2	$CH_3NH_3^+$	211

[a] From ref. 4.

▶ **Example 2.3** Suppose that a compound with PA = 205 kcal mol^{-1} is ionized by CI with either H$_2$ or isobutane as the reagent gas. Which gas will provide more energetic ionization?

Solution By using Eq. (2.11), the values of $\Delta H°$ in these two cases are given by $-(101 - 205) = 104$ kcal mol^{-1} and $-(196 - 205) = 9$ kcal mol^{-1}, respectively. Thus, the [M + H]$^+$ ions produced via CI with H$_3^+$ ions would be more energetic than those produced by CI with C$_4$H$_9^+$ ions.

▶ **Example 2.4** Suppose that a sample contains three compounds: trifluoroacetic acid (PA = 176.0 kcal mol^{-1}), tetrahydrofuran (PA = 199.6 kcal mol^{-1}), and aniline (PA = 211.5 kcal mol^{-1}). Which of the two reagent gases, methane or ammonia, will ionize aniline selectively?

Solution Because PA (= 132.0 kcal mol^{-1}) of methane is less than all three components of the mixture, all of them will be ionized by methane. In contrast, ammonia (PA = 204.0 kcal mol^{-1}) will ionize aniline only because its PA is lower than aniline and higher than those of trifluoroacetic acid and tetrahydrofuran.

2.4.1. Charge-Exchange Chemical Ionization

Conceptually, *charge-exchange (CE) ionization* is similar to the normal mode of CI, but with the difference that the odd-electron molecular ions are the product of CE ionization [5]. As with CI, the reagent gas is first ionized via EI to produce reagent gas ions, G$^{+\bullet}$, which in turn abstract an electron from a neutral analyte molecule to convert it into a radical cation:

$$M + G^{+\bullet} \longrightarrow M^{+\bullet} + G \quad (2.12)$$

The internal energy, E, of the molecular ion is given by the difference between the recombination energy (RE) of G$^{+\bullet}$ and the IE of the reagent gas molecules:

$$E(M^{+\bullet}) = RE(G^{+\bullet}) - IE(M) \quad (2.13)$$

The RE of G$^{+\bullet}$ is usually approximated to the negative value of the vertical IE of the neutral reagent gas molecules with the stipulation that G$^{+\bullet}$ is formed in the ground electronic state.

Similar to the normal mode of CI, ions of a well-defined internal energy can be produced by a careful selection of a reagent gas. Also, by a judicial choice of a reagent gas, the degree of fragmentation can be controlled. The greater the difference in RE($G^{+\bullet}$) and IE(M), the more extensive is fragmentation. Thus, an EI-like spectrum can be generated with a reagent gas of higher RE value. Some common reagent gases with increasing RE are toluene, benzene, NO, CS_2, COS, Xe, CO_2, CO, N_2, Ar, and He.

Often, it is beneficial to acquire a spectrum that has the features of EI and CI (i.e., it contains molecular ion as well as fragment ions). Such a mixed spectrum can provide molecular mass and structural information in a single analysis. A spectrum with those types of features can be acquired by using a mixture of reagent gases that promotes conventional CI and charge-exchange CI (e.g., a mixture of Ar/H_2O or Ar/i-C_4H_{10}).

2.4.2. Negative-Ion Chemical Ionization

Two different CI reaction schemes have been proposed to generate negative ions of volatile compounds. In the first scheme, an electron is captured directly by the sample molecule to produce a molecular anion:

$$\text{Resonance electron capture: } AB + e^- (\approx 0 \text{ eV}) \longrightarrow AB^{-\bullet} \quad (2.14)$$

This process, termed *resonance electron capture* (REC), occurs with high efficiency when electrons of near-thermal energies are available. These electrons are conveniently generated in the EI source in the presence of a moderating gas such as H_2, CH_4, i-C_4H_{10}, NH_3, N_2, or Ar. With electrons of higher energies, the dissociative electron capture [reaction (2.15)] and ion-pair formation [reaction (2.16)] processes are also operative.

$$\text{Dissociative electron capture: } AB + e^- (0 \text{ to } 15 \text{ eV}) \longrightarrow A^- + B^\bullet \quad (2.15)$$

$$\text{Ion-pair formation: } AB + e^- (> 10 \text{ eV}) \longrightarrow A^- + B^+ + e^- \quad (2.16)$$

The negative-ion REC–CI is more practical for high-electron-affinity compounds such as those that contain a nitro group, a halogen atom, or a conjugated π-electron system. An increase of 100- to 1000-fold in detection sensitivity is realized when these compounds are analyzed in the negative-ion mode. One practical way to impart electrophilic character to nonelectrophilic compounds is to derivatize them with a perfluoroacyl, perfluorobenzoyl, pentafluorobenzyl, or nitrobenzoyl group.

The second scheme is similar to the positive-ion CI in that negative ions are generated via ion–molecule reactions with specific reagent anions as exemplified in the reaction

$$B^- + M \longrightarrow [M - H]^- + BH$$

This acid–base reaction occurs because the B^- ion is a stronger Brønsted base than the sample molecules. Some of the reagent ions that produce abundant negative molecular ions, with decreasing PA, include NH_2^-, OH^-, $O^{-\bullet}$, CH_3O^-, F^-, $O_2^{-\bullet}$, and Cl^- ions. Among these ions, the OH^-, $O^{-\bullet}$, and CH_3O^- ions have been employed more frequently; schemes for their generation are depicted in reactions (2.17), (2.18), and (2.19), respectively:

N_2O/CH_4 or $N_2O/i\text{-}C_4H_{10}$ mixture:
$$N_2O + e^- \longrightarrow O^{-\bullet} + N_2 \quad (2.17)$$
$$O^{-\bullet} + CH_4 \longrightarrow OH^- + CH_3^{\bullet}$$

N_2O/N_2 mixture:
$$N_2O + e^- \longrightarrow O^{-\bullet} + N_2 \quad (2.18)$$
$$CH_3NO_2 + e^- \longrightarrow CH_3O^- + NO \quad (2.19)$$

As an example, the OH^- ions are produced by EI of N_2O in the presence of CH_4 or $i\text{-}C_4H_{10}$.

A major asset of CI is to provide molecular mass information. As with EI, nonvolatile compounds, however, cannot be analyzed with CI, and thus many compounds with polar functional groups are excluded from the CI analysis protocol. The upper mass limit of the compound that can be accessed with CI is about 1000 Da. Another obvious limitation is the lack of structural information in the CI spectrum.

2.5. PHOTOIONIZATION

Photoionization (PI), another efficient process to ionize small molecules [6], is accomplished when a gaseous molecule is irradiated with a beam of photons of known energy. The design of a PI source is similar to that used in EI, except that a beam of photons is used in place of an electron beam. Most organic compounds require 8 to 10 eV energy for the removal of an electron from the highest occupied molecular orbital (HOMO). Therefore, the photon beam is usually obtained from a vacuum ultraviolet (VUV) radiation source such as a deuterium lamp, Kr-excimer lamp, or 118-nm Nd:YAG laser. An excimer lamp has the advantage that by exchanging the rare-gas mixture in the lamp, radiation of a wavelength desired can be generated. As an example, Ne, Ar, Kr, or Xe as the fill gas will produce radiation at 83, 126, 147, or 172 nm, respectively. Because photons of a desired wavelength, the energy of which is known precisely, can be selected in PI, the degree of fragmentation and to some extent selectivity can be controlled as desired. Because of the high PI cross section of organic molecules in the VUV range, PI is a highly sensitivity mode of ionization.

$$M + h\nu \longrightarrow M^{+\bullet} + e^- \quad (2.20)$$

Alternatively, tunable lasers that produce longer-wavelength photons (e.g., UV, visible, or infrared radiation) can be employed in a two-step process known as *multiphoton ionization* (MPI) [7]. In the first step, one or more photons excite a ground-state electron of the molecule to one of the higher electronic states, and in the second step, an additional photon of the same or different energy ultimately expels the excited electron from the molecule to leave behind a radical cation. MPI is practiced in several different ionization schemes, the most common are shown in Figure 2.3. In the *resonant two-photon ionization* (R2PI) scheme, one photon excites a molecule precisely to a known excited electronic state, and the second photon completes ionization of the molecule. In this scheme, the sum of the energy of these two photons must exceed the IE of the molecule (i.e., $2 h\nu >$ IE). R2PI can be operated as either a single- or a two-color scheme. In a single-color scheme, the two photons are of the same frequency (i.e., of the same $h\nu$; see Figure 2.3a), whereas in a two-color scheme, the two photons are of different frequencies (Figure 2.3b). Both of these processes are known collectively as *resonance-enhanced MPI* (REMPI). Because of the resonance photon absorption process, the cross section for ionization and the spectroscopic selectivity in REMPI both increase dramatically. MPI is also operated in the *nonresonant mode*, in which a single-frequency photon, whose energy does not match any of the real excited electronic states of the molecule, is used (Figure 2.3c). Such a single-color multiphoton process is nonselective and has a much lower probability of ionization than REMPI.

A precondition for PI is that the analyte molecule must be present in the gas phase. Solid samples must be desorbed or vaporized prior to photoionization. Thermal vaporization [8] or laser desorption (LD) [9] are convenient means to convert the solid samples into a gas-phase plume. A combination of LD and laser ionization has been used successfully to analyze organic contaminants in water and soils [9]. A proper choice of the wavelength will exclude ionization of the bulk components of a real-world sample (e.g., N_2, O_2, CO_2, or water).

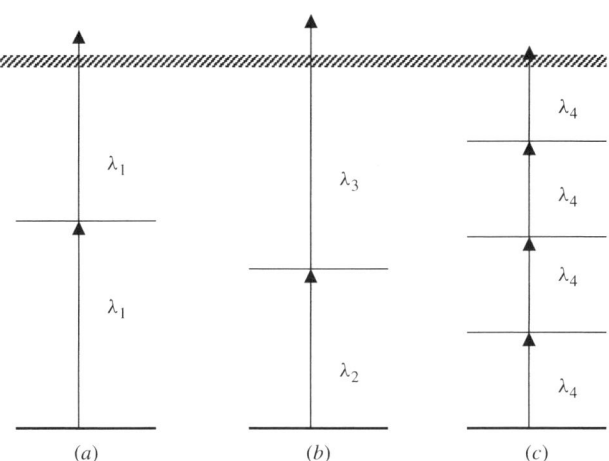

Figure 2.3. Various MPI schemes. (*a*) Single-color R2PI, (*b*) two-color R2PI, (*c*) single-color nonresonant MPI.

▶ **Example 2.5** Calculate the energy (in eV) of a 100-nm photon.

Solution

$$E = h\nu = \frac{hc}{\lambda} = \frac{(6.626 \times 10^{-34} \text{ J} \cdot \text{s})(3.00 \times 10^8 \text{ m/s})}{100 \text{ nm}(10^{-9} \text{ m/nm})} \times \frac{1 \text{ eV}}{1.602 \times 10^{-19} \text{ J}}$$
$$= 12.41 \text{ eV}$$

2.6. FIELD IONIZATION

Field ionization occurs when gas-phase sample molecules are interjected in a strong electrical field that is on the order of 10^8 Vcm^{-1}. The field distorts the electron cloud around the sample molecule and lowers the barrier for the removal of an electron. The quantum mechanical tunneling of this electron from the molecule to the conduction bands of the emitter produces M$^{+\bullet}$ ions [10]. The heart of the FI ion source is an emitter electrode made from a sharp metal object such as a razor blade or thin wire. The emitter electrode is placed approximately 1 mm away from the cathode. The field is produced by applying a high potential (10 to 20 kV) to the tip of the emitter electrode. FI is a very soft ionization technique that produces primarily a molecular ion signal. It is applicable to volatile samples only.

2.7. METASTABLE ATOM BOMBARDMENT IONIZATION

Metastable atom bombardment (MAB) is another suitable ionization option for gas-phase organic compounds [11,12]. In this approach, ionization occurs through Penning processes of the type shown in reactions (2.21) and (2.22) by bombarding the sample with a beam of metastable species. The process depicted in reaction (2.21) involves an electrophilic reaction of the metastable atom (A*) to produce an odd-electron molecular ion of the analyte. The excitation energy of A* must exceed the IE of the analyte molecule. With much higher excitation energy, dissociative ionization process also takes place [reaction (2.22)].

$$M + A^* \longrightarrow M^{+\bullet} + e^- \qquad (2.21)$$
$$M + A^* \longrightarrow [M - n]^{+\bullet} + n + e^- \quad \text{or} \quad M + A^* \longrightarrow [M - n]^+ + n^{\bullet} + e^- \qquad (2.22)$$

The corona discharge of rare-gas atoms (e.g., He, Ne, Ar, Kr, or Xe) in a MAB gun external to the ionization chamber provides a beam of metastable atoms. The excited-state nitrogen molecules have also been used. Because each metastable species has a different operable energy regime, a wide range of tunable excitation energy is accessible. As an example, with rare-gas atoms, the available energy range is 8.3 to 20.6 eV, whereas the metastable N_2 atoms have the energy range 8.5 to 11.9 eV. Thus, with MAB an opportunity exists for selective ionization and tunable fragmentation of analyte molecules.

CONDENSED-PHASE IONIZATION TECHNIQUES: IONIZATION OF SOLID-STATE SAMPLES

The ionization methods discussed so far are restricted to volatile compounds. A large number of nonvolatile and thermally labile compounds (e.g., of biological origin) cannot be ionized by these methods. A variety of desorption and ionization techniques has emerged for such condensed-phase samples. These techniques are radically different from gas-phase methods in that they enable simultaneous volatilization and ionization of the sample molecules. A common feature of these ionization techniques, except for field desorption, is that the sample is mixed with a suitable matrix and bombarded with a high-energy beam of ions, atoms, or photons to produce a high concentration of gas-phase neutral and ionic species. The underlying rationale in those sudden-energy methods is that by depositing a large amount of energy into the analyte molecule on a time scale that is fast compared with the vibrational time period, vaporization of the analyte will occur before its thermal decomposition begins. These developments have opened a flood gate of applications of mass spectrometry to a wide variety of biomolecules and synthetic polymers with molecular mass beyond imagination ($>100,000$ Da). The methods discussed in this group are applicable to samples that are converted to a solid state or dissolved in nonvolatile matrices irrespective of the original state of the sample.

2.8. FIELD DESORPTION

Field desorption (FD) was developed in 1969 by Beckey [13]. It was the first serious attempt to ionize nonvolatile and thermally labile substances. Unlike other desorption ionization methods, no primary beam is used in FD to bombard a sample. Instead, the nonvolatile sample is applied to sharp microneedlelike structures or "whiskers" that are grown on a thin metallic wire filament, and desorbed as intact gas-phase ions by applying a high electric field gradient (10^8 Vcm^{-1}). Heating the emitter to a certain temperature, known as the *best emitter temperature* (BET), might be necessary to bring some samples to the molten state and to facilitate their ionization.

As in the case of the FI process (Section 2.6), ionization of the sample molecules in FD also involves the loss of an electron via quantum-mechanical tunneling to produce $M^{+\bullet}$ ions, which after conversion to $[M + H]^+$ and $[M + Na]^+$ types of ions are desorbed from the emitter surface under the influence of a strong electric field. Negative-ion formation requires the capture of an electron by the sample molecule from the negatively charged emitter. Because almost no vibrational excitation occurs in the ionized molecule, fragmentation is barely

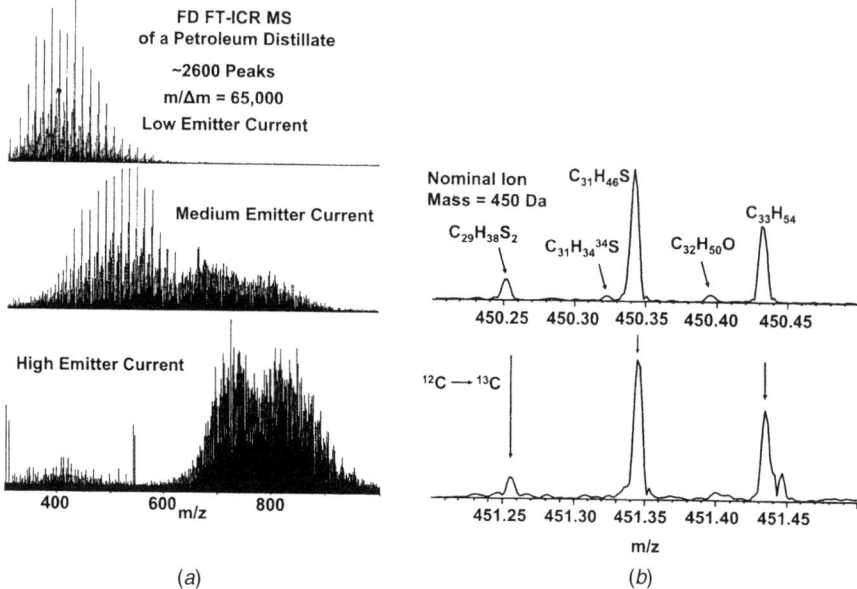

Figure 2.4. (*a*) Positive-ion time-resolved FD FT-ICR mass spectra of mid-boiling crude oil distillate that was acquired at lowmedium, and high emitter current settings; (*b*) Mass scale-expanded spectrum shows compositionally resolved ions. (Reproduced from ref. 14 by permission of the American Chemical Society, Washington, DC, copyright © 2003.)

detectable. Thus, a major application of FD is to determine the molecular mass of nonvolatile compounds such as carbohydrates, peptides, organometallics, sugar moieties, and industrial polymers. A recent noteworthy application of FD–MS is high-resolution mass analysis of nonpolar molecules such as C60 and a mid-boiling crude oil distillate [14]. Figure 2.4 shows positive-ion time-resolved FD FT–ICR mass spectra of mid-boiling crude oil distillate that were acquired by ramping the emitter heating current; the spectra shown were acquired at low- mid-, and high-emitter current settings. More than 2600 ions were identified compositionally at a resolution of about 65,000. An example of compositionally resolved ions is shown in the right panel of Figure 2.4. Many species present in this crude oil distillate are not accessible to other soft ionization techniques.

2.9. PLASMA DESORPTION IONIZATION

Californium-252 (^{252}Cf) *plasma desorption* (PD) *ionization*, introduced in 1974 by Torgerson and co-workers [15], soon found applications for the analysis of large nonvolatile, polar, and thermally labile molecules [16,17]. Because of the pulsed nature of the plasma beam, PD is well adapted to TOF mass spectrometry. The basic concept of PD–TOFMS is illustrated in Figure 2.5. In this technique, the sample is deposited as a solid film on a thin aluminum foil or other suitable

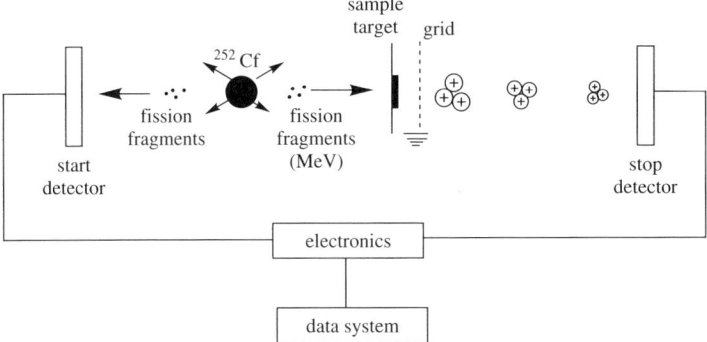

Figure 2.5. Schematic diagram of ^{252}Cf plasma desorption ionization and time-of-flight mass spectrometry. (Reproduced from C. Dass, *Principles and Practice of Biological Mass Spectrometry*, Wiley-Interscience, 2001.)

material, and is bombarded from the back side by a beam of plasma of MeV-range ions to cause emission of secondary ions and neutrals. The plasma is produced from the spontaneous fission of a ^{252}Cf radionuclide. Each fission event simultaneously releases several pairs of fission particles with a range of masses. For example, one such pair is ^{106}Tc^{22+} and ^{142}Ba^{18+}, with kinetic energies of 100 and 80 MeV, respectively. One fission fragment bombards the sample target and the other travels in the opposite direction to activate the start detector. A potential of 10 to 20 kV is applied between the sample target and a grounded grid to accelerate the secondary ions emitted into the flight tube of the TOF instrument.

Sample preparation is a critical step in PD–MS analysis. Alkali metal salts and other impurities must be removed because they might reduce or quench the analyte signal. Deposition of the sample on a suitable matrix, such as nafian, Mylar, or nitrocellulose, followed by washing with ultrapure water, removes the alkali metal ion impurities and produces a much sharper analyte signal. Plasma desorption has been used primarily for determination of the molecular mass of biomolecules. The peaks observed in a PD spectrum of proteins are the protonated molecular species with different charged states. PD has also been a great asset in determining posttranslational modifications in proteins and peptides and in obtaining information about tryptic maps. One major accomplishment of PD–MS is that it showed conclusively that large molecules such as proteins could be introduced successfully into the gas phase.

2.10. SECONDARY-ION MASS SPECTROMETRY

Secondary-ion mass spectrometry (SIMS) is used largely as a surface characterization technique [18]. It is practiced in two formats, dynamic SIMS and static SIMS. In both formats, a beam of keV-energy ions, usually of Ar$^+$ ions, bombards the surface of the sample directly. The momentum transfer from the beam to the

32 MODES OF IONIZATION

point of impact on the sample surface results in sputtering of secondary ions to reflect the chemical composition of the surface. The sputtered secondary ions are mass-analyzed (e.g., by a magnetic sector or quadrupole mass analyzer). In *dynamic SIMS*, the current densities (ca. 1 $\mu A\,cm^{-2}$) of the primary beam are high. As a consequence, there is a possibility of rapid damage to the sample surface and that the secondary-ion signal will be short-lived. In contrast, *static SIMS* uses a low ion-current density (<1 nA cm^{-2}) primary beam [18], which prolongs the sample ion current but at the cost of detection sensitivity. Organic compounds can be analyzed by depositing them on etched metal surfaces, the bombardment of which results in the emission of protonated or cationized sample molecules. An application of current interest is biomedical field imaging of the tissue samples using a Bi^{3+} ion beam [19].

2.11. FAST ATOM BOMBARDMENT

The discovery of *fast atom bombardment* (FAB) in 1980 by Barber and colleagues was a major breakthrough in the analysis of condensed-phase samples [20]. FAB is an improved version of SIMS. In this approach, the sample is first dissolved in a polar, relatively less volatile, and inert liquid matrix (e.g., glycerol), and the sample–matrix mixture is bombarded with a beam of high-energy (in keV) atoms or ions (see Figure 2.6). This innovation overcomes the sample-damaging effect of the primary beam observed in SIMS. In the seminal work on FAB, a beam of high-energy argon atoms was used as the primary beam, and hence the name *FAB* was coined. Later, it was discovered that the production of secondary ions is equally efficient when the sample–matrix mixture is bombarded with a beam of ions [21]. A distinct name, *liquid secondary-ion mass spectrometry (liquid SIMS)*, was given to this technique, even though conceptually and in terms of outcome, it is identical to FAB. The only difference between the two techniques is the nature of the primary beam (i.e., fast atoms versus fast ions). In liquid SIMS, the primary-ion beam consists of Cs^+ ions, which are generated by heating to an incandescent temperature a pellet of cesium aluminum silicate. The acceleration of Cs^+ ions through the applied electrical field between the emitter electrode and the grounded exit electrode provides a beam of focused high-energy (25 to 40 keV) Cs^+ ions. The use of Cs^+ ions as a primary-ion beam leads to better sensitivity and high-mass capability than that obtained with the FAB.

A scheme for the generation of a fast atom beam is

$$\underset{\text{slow atoms}}{Xe} \xrightarrow{\text{ionization}} \underset{\text{slow ions}}{Xe^+} \xrightarrow{\text{acceleration}} \underset{\text{fast ions}}{Xe^+} \xrightarrow{\text{neutralization}} \underset{\text{fast atoms}}{Xe^\circ} \quad (2.23)$$

Here the source of primary beam is xenon atoms, which because of higher mass and momentum provides better sensitivity than that of the beam of argon atoms used in the original discovery. The xenon atoms are first ionized in the FAB gun by collisions with electrons that are moving in a saddle-field configuration.

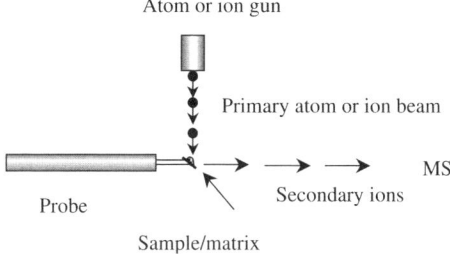

Figure 2.6. Conceptual depiction of fast atom bombardment ionization process. (Reproduced from C. Dass, *Principles and Practice of Biological Mass Spectrometry*, Wiley-Interscience, 2001.)

Ionized xenon atoms are accelerated to the required potential (2 to 10 kV). Next, the fast-moving Xe^+ ions are neutralized in a dense cloud of excess neutral gas atoms to generate a continuous stream of high-translational-energy xenon atoms. Any residual undischarged ions are deflected with a positive potential onto a deflector plate.

The exact mechanism of sample ionization remains a mystery. The current hypothesis is, however, that the ions observed in a FAB mass spectrum are the sum of three distinct processes: (1) the preformed ions present in the matrix, (2) the ions formed in the selvedge region (a cavity formed in the liquid matrix at the point of impact of the beam), and (3) the ions formed in the matrix solution due to the acid–base and electrochemical reactions. The molecular ions of the type $[M + H]^+$ or $[M - H]^-$ are the predominant species in the FAB spectra. In addition, in the positive-ion mode, metal ion clusters with the analyte molecule can be observed, owing to the presence of metallic salts as impurities. In special cases, radical cations have also been observed [22]. During FAB ionization–desorption, the energy gained by many samples is large enough to induce their fragmentation. Thus, in favorable cases, a FAB mass spectrum may include fragments of structural significance [23]. This phenomenon is shown clearly in Figure 2.7, which is the spectrum of an octapeptide Tyr–Gly–Gly–Phe–Met–Arg–Gly–Leu.

A liquid matrix plays several significant roles in the FAB analysis. It provides sustained sample ion current, reduces the damage to the sample by absorbing the impact of the primary beam, and keeps aggregation of the sample molecules to a minimum. More important, it provides a medium where ionization of the sample can be promoted. Any liquid substance that is reasonably viscous, chemically inert, nonvolatile, mass spectrally transparent, and exhibits good solvent and electrolytic properties, can act as a matrix. A variety of matrices is available [24]. Glycerol is by far the most widely accepted matrix. Other commonly used matrices are α-thioglycerol, a mixture of dithiothreitol (DTT) and dithioerythritol (DTE; 5:1 v:v), thiodiglycol, 3-nitrobenzyl alcohol (NBA), tetraglyme, sulfolane, diethanolamine, and triethanolamine. The $[M + H]^+$ ion signal can be enhanced in acidic matrices (e.g., glycerol, α-thioglycerol, and DTT/DTE). Similarly, basic matrices (e.g., diethanolamine and triethanolamine) facilitate

34 MODES OF IONIZATION

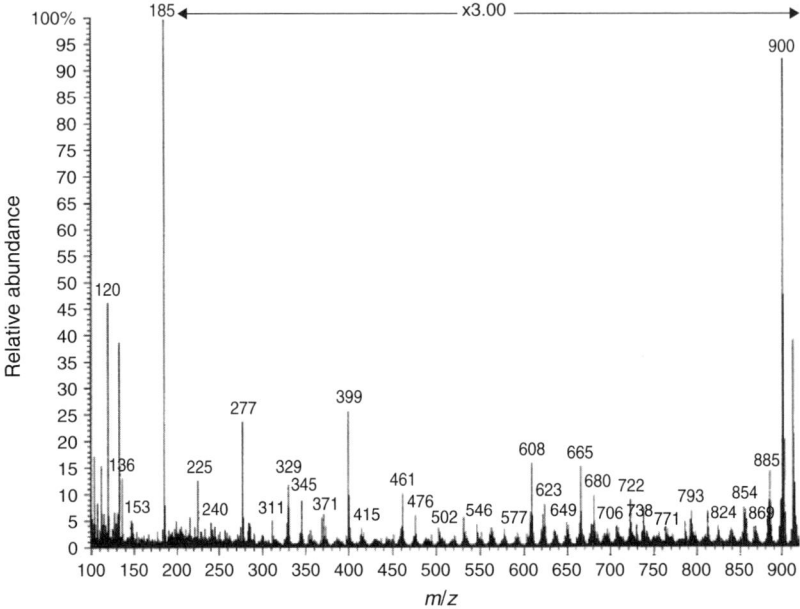

Figure 2.7. Positive-ion FAB mass spectrum of Tyr–Gly–Gly–Phe–Met–Arg–Gly–Leu acquired using glycerol as the matrix.

$[M - H]^-$ ion production. The matrix can also influence the extent of fragmentation of the analyte ions [25]. Glycerol is known to promote fragmentation, whereas α-thioglycerol or a DTT/DTE mixture stabilizes the molecular ions.

The detection sensitivity in FAB–MS analysis is a function of the chemical composition of the sample–matrix mixture and of the presence of other unwanted impurities. The surfactancy of a solute and the matrix also influences the analyte signal. Because hydrophobic compounds tend to occupy the upper layer of the hydrophilic matrix, they are ionized preferentially relative to the hydrophilic compounds. In contrast, hydrophilic compounds exhibit poor response, because they remain buried within the lower layers of the matrix. Also, alkali salts are known to suppress ionization. Therefore, to obtain a sufficiently high ion current of the target compound, the matrix surface composition must be optimized by adjusting its pH or by the addition of surfactants.

A normal method of *sample introduction* into the FAB ion source is via a solid insertion probe. The matrix and sample are loaded onto a specially designed stainless steel sample target of the probe, which is inserted into the ion source through a vacuum lock. Although this procedure is very simple, it is fraught with poor detection sensitivity owing to excessive matrix background, which is the result of a high matrix-to-sample ratio used in the analysis. An alternative sample introduction device called a *continuous-flow* (CF)–FAB *probe* was developed to minimize that problem [26]. It typically consists of a hollow stainless steel shaft that contains a coaxial narrow fused-silica capillary. A stream of very dilute aqueous solution of the matrix (usually 1 to 10% by volume of glycerol) flows continuously through the capillary over a FAB target. The matrix forms a very thin layer at the probe tip. The sample is injected into this stream, and the stream is bombarded by a high-energy beam as usual. The matrix background is reduced significantly because of the reduced matrix-to-sample ratio; consequently, the detection sensitivity increases severalfold. The CF–FAB probe is also a useful adjunct for the online direct monitoring of biochemical reactions and as an interface for online coupling of HPLC, capillary electrophoresis (CE), and a microdialysis probe with mass spectrometry (see Chapter 5).

The molecular mass measurement and structural characterization of a wide range of inorganic, organic, and biochemical compounds (e.g., peptides, antibiotics, fatty acids, lipids, carbohydrates, nucleotides and nucleic acids, organometallics, and surfactants) are strong suits of FAB–MS. Another major area of application of FAB–MS is characterization of peptide fragments of proteolytic digests of proteins, prefiguring modern proteomics. The molecular mass measurement is restricted to compounds below 5000 Da.

2.12. LASER DESORPTION/IONIZATION

In early mass spectrometry applications of lasers, the sample was irradiated directly by a laser beam to desorb intact sample-related ions [27]. In this direct mode, termed *laser desorption/ionization* (LDI), the extent of energy transfer is, however, difficult to control and often leads to excessive thermal degradation. Also, not all compounds absorb radiation at the laser wavelength and thus are not amenable to LDI. Only those compounds that have mass below 1000 Da can be analyzed by LDI. Analytical sensitivity is also poor. A key contribution of LDI experiments is the observation that the desorption efficiency of amino acids and peptides that absorb the laser light beam is greater than those without the chromophore [28]. IR lasers (e.g., an Nd:YAG laser at 1.06 μm and a pulsed CO_2 laser at 10.6 μm) and UV lasers (frequency-quadrupled Nd:YAG laser at 266 nm) have all been used. The detection of malaria parasites in blood by LDI with an N_2 laser has been demonstrated [29].

2.13. MATRIX-ASSISTED LASER DESORPTION/IONIZATION

Matrix-assisted laser desorption/ionization (MALDI) was developed nearly simultaneously by two research groups, Karas and Hillenkamp [30] in Germany and Tanaka and co-workers in Japan [31]. Tanaka is a recipient of the 2002 Nobel Prize in Chemistry. MALDI has significantly revolutionized approaches to the study of large biopolymers. That landmark development has provided a unique opportunity to apply mass spectrometry to the analysis of proteins and other biomolecules with masses in excess of 200 kDa and with an improved sensitivity of several orders of magnitude.

36 MODES OF IONIZATION

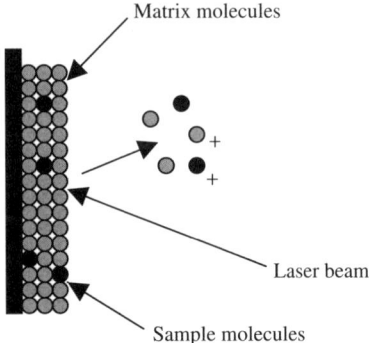

Figure 2.8. Schematic representation of the MALDI process.

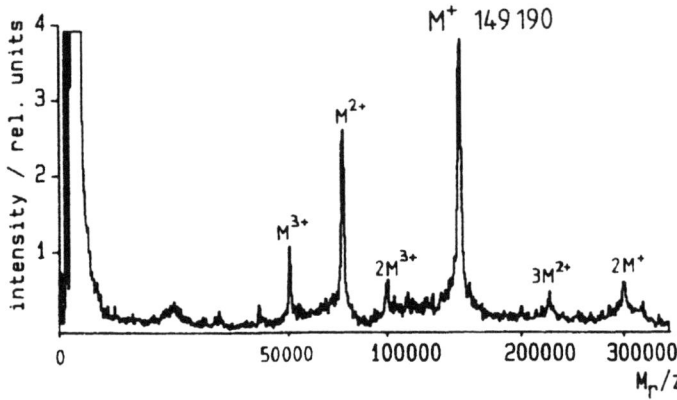

Figure 2.9. MALDI mass spectrum of a monoclonal antibody. (Reproduced from ref. 32 by permission of Elsevier, copyright © 1990.)

MALDI is initiated by mixing the sample solution with a large molar excess (ca. 10,000:1) of the host matrix material and depositing the mixture on a specially designed MALDI sample target. After evaporation of the solvent, the sample–matrix crystals are irradiated with a laser beam of high irradiance power (10^6 Wcm^{-2}) and short pulse widths (few nanoseconds) to simultaneously desorb and ionize the sample and matrix molecules into the gas phase (see Figure 2.8). A key ingredient to the success of MALDI is a matrix that is able to absorb a large amount of energy at the wavelength of the laser radiation, and subsequently relays it to the sample molecules in a controlled manner to permit desorption of even massive molecules as intact gas-phase ions. The high-mass capability of MALDI is demonstrated in the spectrum of the monoclonal antibody shown in Figure 2.9. As shown, the MALDI-generated ions are mainly singly protonated molecules. Oligomeric ions and doubly and triply charged protonated ions are also formed. In addition, the adducts of the sample molecule with Na$^+$ and K$^+$

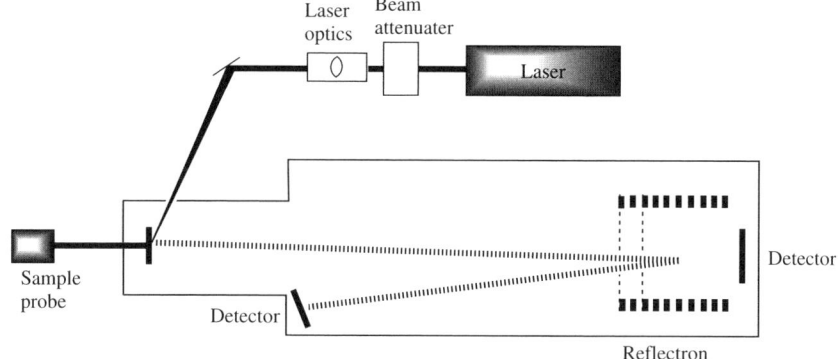

Figure 2.10. Schematic diagram of MALDI–TOF–MS instrumentation. (Reproduced from C. Dass, *Principles and Practice of Biological Mass Spectrometry*, Wiley-Interscience, 2001.)

are also a common feature of the MALDI spectra of samples of biological origin. Negative-ion analysis can also be performed with MALDI.

Because the irradiating laser beam is pulsed, MALDI is optimally combined with a TOF mass analyzer. The unlimited mass range of TOF and its ability to acquire the entire spectrum from a single laser pulse event are other factors in favor of the MALDI/TOF-MS combination. MALDI–TOF has become a well-known acronym for many researchers. Quadrupole, ion trap, and Fourier transform ion cyclotron resonance (FT–ICR) instruments have also been modified to accommodate MALDI. A schematic diagram of MALDI/TOF-MS is presented in Figure 2.10. A variety of laser systems has found applications in MALDI analysis, and the most common ones use UV lasers such as the N_2 laser (337 nm), the frequency-tripled (355 nm) and frequency-quadrupled (266 nm) Nd:YAG laser, and the ArF excimer laser (193 nm). IR lasers have also been used to produce the MALDI effect. The transversely excited atmospheric (TEA) CO_2 laser (10.6 μm), the Q-switched Er:YAG laser (2.94 μm), and the Cr:LiSAF or Nd:YAG pumped optical parametric oscillator (OPO) laser (3.28 μm) are the common IR lasers. UV and IR lasers yield similar spectra for proteins, although better resolution has been obtained for some proteins with an IR laser.

How Analyte Ions Are Formed Several diverse views have been expressed on this topic [33–38]. No single model provides a complete picture of MALDI ionization. The widely accepted current view is that the analyte ionization is a two-step process: a primary ionization event, followed by in-plume secondary ion–molecule reactions [34–36]. In the first step, reactive matrix ionic species are generated. The analyte ions are produced in the expanding gas plume (an area just above the matrix surface) via extensive secondary ion–molecule charge-transfer reactions between the primary matrix ions and neutral analyte molecules:

$$M + S^+ \longrightarrow M^+ + S \tag{2.24}$$

where M is the analyte species and S^+ is the matrix-related active species, which could be a protonated, sodiated, or deprotonated matrix molecule or a radical cation. These species might participate in proton transfer, cation transfer, electron transfer, or electron-capture reactions. In this mechanism, the in-plume reactions play a dominant role in analyte ionization. Primary matrix ions are generated by laser excitation of individual matrix molecules or an exciton-pooling process, in which two nearby excited matrix molecules redistribute their energy and concentrate it on one molecule. In this model, matrix–analyte clusters and free electrons play a very small role in analyte ionization [36]. Photoelectrons are also generated from the sample substrate and might be important in influencing the relative yields of multiply charged ions.

An alternative mechanism for analyte ionization is via cluster formation between matrix and analyte molecules, followed by charging of the clusters [37,38]. Two pathways have been proposed to explain charging of the matrix–analyte clusters. One pathway assumes that the analyte exists as multiply charged ions and that the clusters are formed by trapping those ions along with their counterions in a large chunk of matrix. The clusters are released in the desorption plume by laser irradiation. The charge separation in the clusters that have a deficit or excess of ions produces charged clusters. The analyte ions are then formed by evaporation from these matrix-rich clusters of the excess matrix molecules or of other neutrals formed from trapped ionic species by proton transfer reactions:

$$[\underset{\text{cluster}}{(M + nH)^{n+} + (n-1)A^- + xS}] \longrightarrow MH^+ + (n-1)A + xS \quad (2.25)$$

where A is a counter-ion and S is a matrix. The second pathway of charging the matrix-analyte cluster is photoionization. It has been shown that the ionization energy of the matrix-analyte cluster is reduced compared to the free matrix molecules to facilitate two-photon ionization by the normally used laser systems (e.g., 337 nm nitrogen laser or 355 nm frequency-tripled Nd : YAG laser) [35,36]. The role of free electrons on the formation and nature of ions has also been suggested [37,38].

Matrices for MALDI The purpose of a matrix is twofold. First, it absorbs photon energy from the laser beam and transfers it into the excitation energy of the solid system. Second, it serves as a solvent for the analyte so that the intermolecular forces are reduced and aggregation of analyte molecules is held to a minimum. The desirable characteristics of a matrix are (1) strong absorption of radiation at the laser wavelength; (2) good mixing and solvent compatibility with the analyte so that well-defined microcrystals of a proper size are formed; (3) a low sublimation temperature, to allow the formation of an instantaneous high-pressure plume of the matrix–sample material during the laser pulse duration; and (4) participation in some kind of photochemical reactions so that the sample molecules can be protonated or deprotonated with high efficiency.

TABLE 2.3. Matrices Used in MALDI–MS

Matrix	Mass (Da)	Solvents[a]	Laser λ (nm)	Applications
3-Amino-4-hydroxybenzoic acid	153	ACN, water		Oligosaccharides
2,5-Dihydroxybenzoic acid (DHB)	154	ACN, water, methanol, acetone, chloroform	337	Oligosaccharides, peptides, nucleotides, oligonucleotides
5-Hydroxy-2-methoxybenzoic acid	168	ACN, water		Lipids
2[4-Hydroxyphenylazo]benzoic acid (HABA)	242	ACN, water, methanol	266, 337	Proteins, lipids
Cinnamic acid	148	ACN, water		General
α-Cyano-4-hydroxycinnamic acid	189	ACN, water, ethanol, acetone	337, 355	Peptides, lipids, nucleotides
4-Methoxycinnamic acid	178	ACN, water	337	Proteins
Sinapinic acid (3,5-dimethoxy-4-hydroxycinnamic acid)	224	ACN, water, acetone, chloroform	266, 337	Lipids, peptides, proteins
Ferlulic acid (4-hydroxy-3-methoxycinnamic acid)	194	ACN, water, propanol	337	Proteins
6,7-Dihydroxycoumarin (esculetin)	178	Water		Lipids, peptides
3-Hydroxypicolinic acid (HPA)	139	Ethanol	337, 355	Oligonucleotides
Picolinic acid (PA)	123	Ethanol	266	Oligonucleotides
3-Aminopicolinic acid	138	Ethanol	266, 337, 355	Oligonucleotides
6-Aza-2-thiothymine (ATT)	143	ACN, water, methanol	266, 337, 355	Oligonucleotides, lipids
2,6-Dihydroxyacetophenone	152	ACN, water	337, 355	Proteins, oligonucleotides
2,4,6-Trihydroxyacetophenone	168	ACN, water	337, 355	Oligonucleotides
Nicotinic acid	123	Water	266, 337, 355	Proteins, oligonucleotides
3-Aminoquinoline	145		337	Oligosaccharides
1,5-Diaminonaphthalene	158	ACN, water, methanol	337	Lipids
Dithranol(1,8,9-trihydroxy anthracene)	226	DCM	337	

[a] ACN, acetonitrile; DCM, dichloromethane; the laser light of 337 nm is produced with a nitrogen laser, and 266 and 355 nm with frequency-quadrupled and frequency-tripled Nd:YAG lasers, respectively.

The four different categories of matrices that have found applications in MALDI analysis of various types of molecules include:

1. *Solid organic matrices.* Solid organic matrices are the most common type and consist of light-absorbing aromatic ring–containing compounds [39]. Some commonly used matrices, along with the wavelengths at which they are used, the solvents in which they can be dissolved, and fields of their applications, are listed in Table 2.3. Their structures are shown in Figure 2.11. Of these, α-cyano-4-hydroxycinnamic acid (CHCA) and

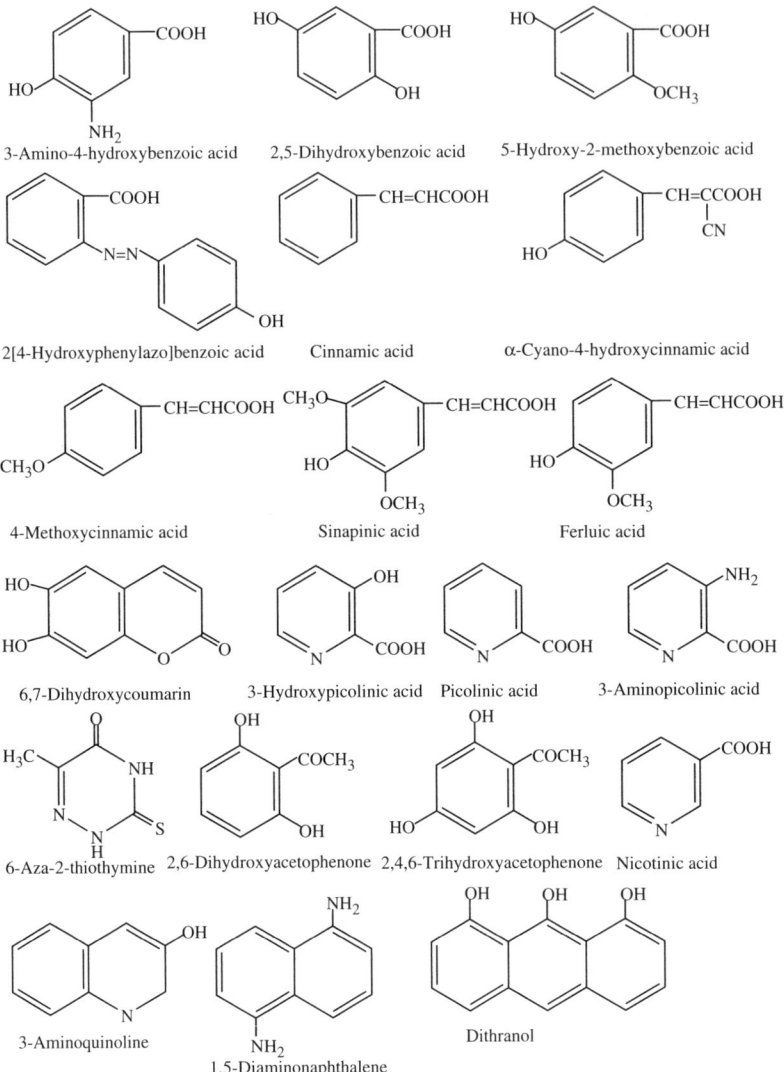

Figure 2.11. Structures of common organic MALDI matrices.

2,5-dihydroxybenzoic acid (DHB) are the most versatile; a wide range of compounds can be analyzed with these two matrices. A major problem with solid organic matrices is an inhomogeneous distribution of analyte and matrix molecules in the solid sample preparation obtained after evaporation of the solvent from the matrix–analyte solution. This inhomogeneity creates sample-to-sample and spot-to-spot variations in the analyte signal.

2. *Liquid organic matrices.* To overcome the inhomogeneous sample preparation problem, organic liquids and ionic liquids have been used in

place of traditional solid organic matrices. 2-Nitrophenyloctyl ether and 3-nitrobenzyl alcohol are the common liquid matrices [40].
3. *Ionic liquids.* Ionic liquids are formed by an equimolar mixture of a traditional solid matrix, such as CHCA and DHB, with an organic base (e.g., butylamine) [41–44]. These cation–anion pairs allow homogeneous sample preparation, which improves reproducibility and increases signal intensity.
4. *Inorganic materials.* Similar improvement has been realized with the use of inorganic materials such as metals (Cu, Co, Al, Mn, W, etc.), metal oxides, and graphite [45,46]. These materials are dispersed in a suitable nonvolatile liquid (e.g., glycerol and liquid paraffin). Graphite, active carbon, carbon nanotubes, fullerenes, and textured silica can also serve as a MALDI matrix [47–49]. They behave as a suitable trap for analyte molecules and as a receptacle of laser energy.

Sample Preparation A critical problem during sample preparation for MALDI–MS analysis is the possibility of segregation of the analyte and matrix molecules during cocrystallization process, resulting in inhomogeneous sample preparations. This inhomogeneity compels one to search for the *sweet spot*, a portion of the sample preparation that yields a strong analyte signal. Several sample preparation techniques are available, all with the aim of achieving fine-grained homogeneous crystal formation. No universal sample preparation for a broad type of analyte molecules, however, exists. These techniques include:

1. *Dried-droplet technique.* The *dried-droplet technique* (also referred to as the one-layer spot) is the most widely practiced mode of sample preparation [30,50]. In this technique, a few microliters of the sample solution (e.g., in 0.1% aqueous TFA) is mixed with an equal volume of the saturated matrix solution (prepared in the same medium) in a molar ratio of 1 : 1000 to 10,000. A drop of that mixture is applied onto the MALDI target and is dried slowly in the ambient air or by a gentle stream of cold air. Drying the sample spot under vacuum or in a refrigerator can improve the homogeneity of the sample preparation.
2. *Fast-evaporation technique.* Another way to improve homogeneity is to use the *fast-evaporation technique*, in which a small drop of the matrix solution prepared in a highly volatile solvent, such as acetone, is placed on the probe. Rapid evaporation of the solvent occurs and a thin homogeneous film of small crystals of matrix is left behind [51]. The analyte solution prepared in water is deposited on this film and dried slowly in the ambient air. This technique has been alternatively called the *two-layered technique* [52].
3. *Sandwich matrix technique.* In the *sandwich matrix technique*, a thin layer of matrix is created first, followed by addition and drying of the small volumes of 0.1% aqueous TFA, the sample solution, and an additional matrix solution [53].

4. *Spin-dry technique.* In the *spin-dry technique*, a solution that contains equal volumes of the nitrocellulose (NC) membrane and the matrix is prepared first. This solution is applied to a rotating target. This spin-drying process produces a uniform NC–matrix surface onto which is added a small drop of the sample solution [54].
5. *Seed-layer technique.* In the *seed-layer technique*, a seed layer of matrix is prepared by depositing a droplet (0.5 µL) of the matrix solution on a sample target and allowing it to dry in the ambient air [55]. A small volume of the 1 : 1 (v : v) analyte–matrix mixture is deposited on top of the seed layer and allowed to dry as usual in the ambient air.
6. *Electrospray deposition.* In *electrospray deposition*, the analyte–matrix solution is passed through a capillary held at high potential and is sprayed over the MALDI plate to provide a much-improved homogeneous sample preparation [52].
7. *Solvent-free sample preparation.* Solvent-free sample preparation is a relatively new approach termed *solvent-free or solvent-less MALDI* [56]. In this approach, the matrix, sample, and a salt are mixed together in an appropriate proportion and ground with either a mortar and pestle or a ball mill. The dry mixture is applied to the MALDI target with a double-sided sticky tape. The solvent-free procedure is more universal because it may also be used with insoluble analytes.

2.13.1. MALDI Analysis of Low-Molecular-Mass Compounds

Traditionally, MALDI has been applied to high-molecular-mass biopolymers. Analysis of small molecules (<500 Da) is a challenge for MALDI, primarily because an excessive matrix background seriously interferes in this mass range. The following innovative approaches are available for low-mass compounds.

1. *Suppression of the matrix signal.* A practical approach to the analysis of low-mass compounds is to suppress the matrix background. One way to achieve this suppression is to add a suitable surfactant to the regular matrix material. For example, by mixing α-CHCA with cetrimonium bromide in a 1000 : 1 concentration ratio, a high-quality spectrum with suppressed matrix background can be acquired [57]. An example of this approach is shown in Figure 2.12, which is the MALDI mass spectrum of a low-mass compound, 2-chloro-10-(6-chlorohexyl)-10H-phenoxazine. Reduction in the matrix background can also be achieved through the *matrix suppression effect* (MSE) phenomenon. To realize this effect, a higher proportion of the analyte is mixed with the matrix and the mixture is bombarded with lower-than-normal laser fluence [58].
2. *Use of high-mass matrices.* Another simple approach for the analysis of low-mass compounds is to shift the matrix background to a higher mass range with high-molecular-mass matrices [59]. A typical example is the

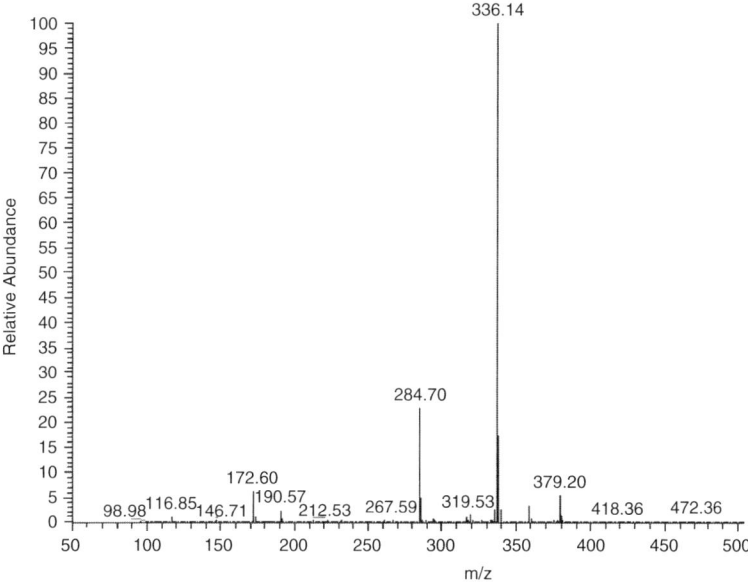

Figure 2.12. MALDI mass spectrum of 2-chloro-10-(6-chlorohexyl)-10H-phenoxazine (MW = 335 Da). The ion of *m/z* 284.70 is due to cetrimonium bromide.

analysis of alkylphenol ethoxylates with *meso*-tetrakis (pentafluorophenyl) porphyrin as a matrix [59].

3. *Surface-assisted laser desorption/ionization (SALDI)*. SALDI is a matrix-free approach for the analysis of low-mass molecules. In this innovative approach, the sample solution is placed directly onto a solid surface prepared by depositing an active material, such as powdered graphite, active carbon, carbon nanotubes, or silica sol-gel, onto a suitable substrate (e.g., Al foil or Cu tape) and bombarded with a laser beam [47,48,60].

4. *Desorption/ionization on silicon*. Another matrix-free approach for the analysis of small molecules is to use modified porous silicon surfaces [61]. A separate name, *desorption/ionization on silicon* (DIOS), has been coined for this technique [62]. In this approach, the sample is dispensed directly onto a modified porous silicon surface (e.g., silylated porous silicon) and laser-desorbed from the surface. The silicon surface is prepared from flat crystalline silicon by galvanostatic etching.

2.13.2. Atmospheric-Pressure MALDI

Atmospheric-pressure MALDI (AP–MALDI) is conceptually similar to normal MALDI except that it is performed outside the source vacuum at atmospheric pressure [63,64]. Similar to other AP ion sources (e.g., ESI), orthogonal acceleration TOF, quadrupole ion trap, linear ion trap (LIT), and FT–ICR instruments are ideal MS systems for coupling with AP–MALDI. Ion guides made of multipoles (e.g., rf-only quadrupoles, hexapoles, or octopoles) are employed to steer AP–MALDI-formed ions into the mass analyzer. This combination limits AP–MALDI to compounds that fall within the mass range transmittable through these ion guides. Because of increased collisional cooling, the molecular ion species are of low internal energy and relatively less prone to fragmentation. The clustering of cooler molecular species with the matrix ions is, however, a problem. Another advantage of AP–MALDI is that the mass resolution of the instrument is not affected by the initial conditions of the ionization process.

2.13.3. Surface-Enhanced Laser Desorption/Ionization

Surface-enhanced laser desorption/ionization (SELDI) is a novel alternative to conventional MALDI for the analysis of target compounds in a mixture [65,66]. It is a unique sample preparation platform that combines the concepts of liquid chromatography separation and MALDI analysis in one. In this approach, an analyte or a group of analytes are captured on a solid-phase chip via adsorption, partition, electrostatic interaction or affinity chromatography. A matrix is added to the solid surface next, and analyzed as usual by MALDI–TOF approach.

Consider the analysis of proteins in a mixture: The active surface for the capture of target proteins can take the form of various chromatographic surfaces, such as those modified chemically with cationic, anionic, hydrophobic, or hydrophilic groups or with metal ions (Figure 2.13). Active spots can also be prepared by biochemical modifications with antibodies, receptors, DNA molecules, or enzymes. A small volume of the crude sample (e.g., serum, urine, or plasma) is applied directly to the spot and washed with an appropriate solvent to remove unretained proteins and other constituents of the biological matrix. The MALDI matrix is applied to the dried surface, and bound species are analyzed as usual. Protein chips with multiple active spots can be created for high-throughput analysis.

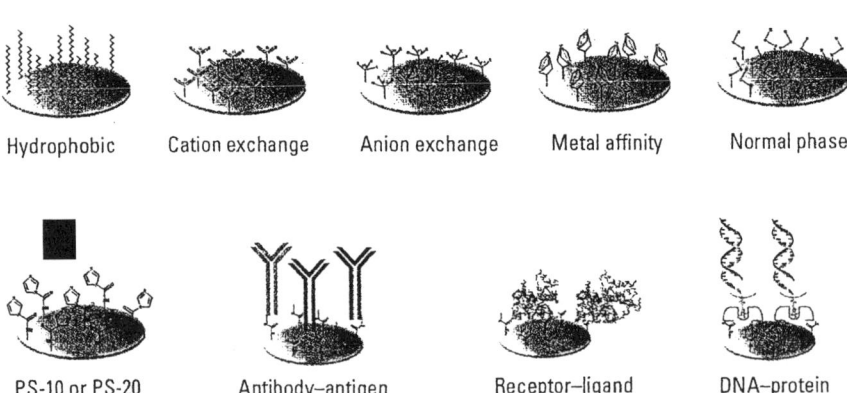

Figure 2.13. Various types of microarrays employed in SELDI analysis. (Reproduced from ref. 66 by permission of the American Chemical Society, Washington, DC, copyright © 2003.)

2.13.4. Material-Enhanced Laser Desorption/Ionization

A new technique, known as material-enhanced laser desorption/ionization (MELDI), has become available for profiling proteins in complex mixtures [67]. It uses derivatized carrier materials, such as cellulose, silica, poly(glycidyl methylacrylate/divinylbenzene), or diamond powder, to bind proteins from the sample. The carrier particles are derivatized with iminodiacetic acid (IDA) and loaded with Cu^{2+} ions to form [carrier-IDA-Cu^{2+}] complex, which is then added to the protein sample. The protein-bound carrier materials are spotted on the MALDI target, mixed with a MALDI matrix and analyzed by MALDI-TOF-MS.

CONDENSED-PHASE IONIZATION TECHNIQUES: IONIZATION OF LIQUID-STATE SAMPLES

The direct sampling of solutions is often necessary in a variety of situations such as biological fluids and eluants from liquid chromatography and capillary electrophoresis separation devices. Liquid solutions are difficult to handle by the mass spectrometry vacuum system and require some novel introduction and ionization systems. The last two decades have witnessed the development of some unique ionization methods that are suitable for direct analysis of sample solutions; the important ones are discussed below.

2.14. THERMOSPRAY IONIZATION

Thermospray, a predecessor of currently popular and more versatile ESI, has proven to be a convenient mode of ionization for liquid-phase samples, especially for HPLC effluents [68,69]. The liquid stream is passed through a heated capillary at a flow rate of 0.5 to 2 mL min^{-1} to produce a spray of fine liquid droplets. Ionization occurs through three distinct processes:

1. The first process is direct desorption of the preformed sample ions into the gas phase from the liquid droplets after solvent evaporation.
2. The second process involves ionization by acid–base proton transfer reactions with the ionic components of the buffer (e.g., NH_4^+ and CH_3COO^- ions) or solvent ions. The RP–HPLC solvent (water mixed with ammonium acetate buffer and an appropriate composition of methanol, acetonitrile, or isopropanol) is a convenient source of these ions.
3. In the third process, sample molecules are ionized via CI with a plasma of the solvent reagent ions. The plasma is generated by EI of the ambient solvent molecules. This mode of ionization is also called *filament-on operation* or *plasmaspray ionization*.

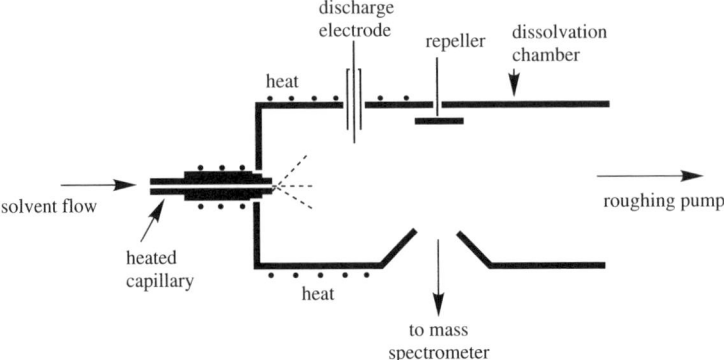

Figure 2.14. Block diagram of thermospray ionization source. (Reproduced from C. Dass, *Principles and Practice of Biological Mass Spectrometry*, Wiley-Interscience, 2001.)

Figure 2.14 is a schematic representation of a typical thermospray source. It is a modified form of a conventional CI source and contains a heated capillary, a low-current discharge electrode (for creating plasma of the solvent ions), a repeller, and a high-throughput mechanical pump placed directly opposite the capillary. An EI gun is also placed inside the ion source to create the plasma of the solvent reagent ions.

Thermospray is a gentle mode of ionization. The spectrum contains mostly the intact molecular ion and its adduct with buffer components. Because fragmentation of the molecular ion species is relatively sparse, structural information is usually absent. Several classes of compounds, such as peptides, dinucleotides, prostaglandins, diquaternary ammonium salts, pesticides, drugs, dyes, and environmental pollutants, are accessible to thermospray analysis. Also, it is an effective interface for LC/MS applications. Conventional columns with flow rates up to 2 mL min^{-1} can be coupled to this interface.

2.15. ATMOSPHERIC-PRESSURE CHEMICAL IONIZATION

The technique of *atmospheric-pressure chemical ionization* (APCI) also serves to analyze LC effluents by mass spectrometry. It is applicable to relatively less polar and thermally stable compounds with an upper mass range of 1500 Da. The principle of ionization in APCI is identical to that described for conventional CI, with the difference that APCI is performed at atmospheric pressure, at which many more ion–molecule collisions can occur between the sample molecules and reagent ions. Therefore, the ionization efficiency and detection sensitive are improved significantly.

A schematic diagram of a typical APCI ion source is shown in Figure 2.15. It consists of three main parts: a removable heated nebulizer probe (350 to 500°C), an ionization region, and an intermediate-pressure ion-transfer region. The LC effluent flows through a fused-silica capillary tube, and the nebulizer gas and

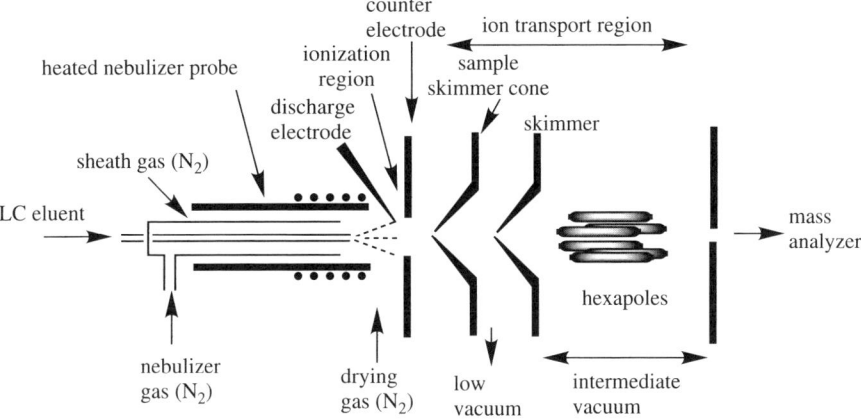

Figure 2.15. Block diagram of atmospheric pressure chemical ionization source. (Reproduced from C. Dass, *Principles and Practice of Biological Mass Spectrometry*, Wiley-Interscience, 2001.)

makeup gas flow coaxially. A mist of fine droplets emerges from the central tube. The combination of heat and gas flow converts the droplets and the analytes into a gas stream. The ionization region contains a discharge electrode (corona pin), which is held at 2 to 3 kV potential with respect to the exit aperture of the counter-electrode. The ions formed in this region are sampled through a small-orifice skimmer, called the *nozzle* or sample *skimmer cone*, and are transported through additional skimmer and ion-focusing lenses into a mass analyzer for subsequent mass analysis.

It is believed that the primary ions of the type N^{2+}, N^{4+}, and H_2O^+ are first formed by interaction of the corona-created electrons with the buffer gas. Charge-transfer reactions of these primary ions with H_2O, methanol, and acetonitrile, present in the aerosol, produce a set of solvated charged species $H_3O^+(H_2O)_n$, $CH_3OH_2^+(CH_3OH)_n$, and $CH_3CN^+(CH_3CN)_n$, respectively. Finally, the transfer of a proton from these solvated ions ionizes the sample molecules to produce $[M + H]^+$ ions. In the negative-ion mode, $[M - H]^-$ ions are produced via the ion–molecule reactions of the solvated oxygen anions. Typical applications of the APCI technology include the analysis of steroids, pesticides, and pharmaceutical drugs.

2.16. ATMOSPHERIC-PRESSURE PHOTOIONIZATION

Atmospheric-pressure photoionization (APPI) is a relatively new technique suited to analyze LC effluents that contain nonpolar and polar low-mass compounds [70–74]. In this methodology, the liquid effluent is vaporized in a heated nebulizer similar to the one described above for the APCI source to generate a dense cloud of gas-phase analytes. Ionization is initiated by a beam of photons emitted from a UV lamp (usually, a krypton discharge lamp emitting photons at 10 eV). APPI is practiced in two formats: (1) direct APPI and (2) dopant-assisted APPI (DA–APPI). In *direct APPI* an analyte molecule absorbs a photon directly; an electron is ejected from the molecule to form a radical cation [reaction (2.20)]. Direct APPI occurs when the IE of the analyte molecule is lower than the photon energy. In the presence of protic solvents (e.g., water and methanol), the dominant ion observed is MH^+, which is formed via a hydrogen atom abstraction by $M^{+\bullet}$:

$$M^{+\bullet} + S \longrightarrow MH^+ + (S - H) \qquad (2.26)$$

In the *DA–APPI mode,* a large molar concentration of a dopant is infused into the ionization region. The dopant radical cations are first formed by absorption of photons:

$$D + h\nu \longrightarrow D^{+\bullet} + e^- \qquad (2.27)$$

which in turn ionize the sample molecules via one of the two mechanistic pathways. For nonpolar molecules, ionization occurs by direct-charge exchange with the dopant radical cations, provided that the IE of the analyte is lower than that of the dopant:

$$D^{+\bullet} + M \longrightarrow M^{+\bullet} + D \qquad (2.28)$$

The second mechanistic pathway is the solvent-mediated ionization, in which the dopant radical cation first ionizes the solvent molecule by a proton transfer reaction to produce an SH^+ ion [reaction (2.29)], which in turn ionizes the analyte molecule by a second proton transfer reaction [reaction (2.30)]:

$$D^{+\bullet} + S \longrightarrow SH^+ + (D - H) \qquad (2.29)$$

$$SH^+ + M \longrightarrow MH^+ + S \qquad (2.30)$$

This alternative pathway will operate only when the PA of the solvent is greater than that of the dopant and the PA of the analyte is greater than that of the solvent. Toluene, being less toxic, is the dopant of choice for many applications. Its IE value is 8.83 eV. Acetone and anisole have also been used as dopants. Direct photoionization of common LC solvents is avoided when the 10-eV UV photon lamp is used as a photon source; thus, the solvent background, common to many other modes of ionization, is absent, which helps improve detection limits.

APPI can also be performed in the negative-ion mode [74]. In that case, phenol is used as a dopant. In this mode, compounds with high gas-phase acidity are ionized as deprotonated ions by proton transfer reactions with O_2^- or $[S - H]^-$ ions, whereas compounds of positive electron affinity form negative molecular ions either by electron capture or by charge exchange with the phenoxide ion.

2.17. ELECTROSPRAY IONIZATION

Electrospray ionization (ESI) is also an API technique that is applicable to a wide range of liquid-phase samples. In particular, it has made an enormous impact in the characterization of large biomolecules. It has also become the most successful interface for LC/MS and CE/MS applications [75,76]. Although the concept of electrospray was put forward by Malcom Dole in 1968 [77], the development of ESI–MS is credited to John Fenn [78,79], who was awarded the 2002 Nobel Prize in Chemistry for that contribution.

As the name implies, electrospray ionization is a process that produces a fine spray of highly charged droplets under the influence of an intense electric field. Evaporation of the solvent converts those charged droplets into gas-phase ions. A simplistic view of the ESI process is depicted schematically in Figure 2.16. The sample solution in a suitable solvent mixture flows continuously through a stainless steel capillary tube whose tip is held at a high potential (3 to 4 kV) with respect to the walls of the surrounding atmospheric-pressure region, called a *counter-electrode*. The solvent consists of a 1 : 1 (v : v) mixture of water and acetonitrile and typically contains <1% acetic acid, formic acid, or trifluoroacetic acid. The potential difference between the tip of the capillary and the counter-electrode produces an electrostatic field that is sufficiently strong to disperse the emerging solution into a fine mist of charged droplets. Evaporation of the charged droplets is assisted by a flow of hot nitrogen. During the process of droplet evaporation, some of the dissolved ions are released into the atmosphere. The resulting ions are transported from the atmospheric-pressure region to the high vacuum of the mass analyzer via a series of pressure-reduction stages. Two designs of the transport region are commonplace in the commercial version of the ESI source; one consists of a heated metal or glass capillary several centimeters long (shown in Figure 2.17) and the other of small-orifice skimmer lenses (Figure 2.18) similar to those outfitted in an APCI source (see Figure 2.14). To improve the ion-transmission efficiency, radio-frequency (rf) multipoles (quadrupoles, hexapoles, or octopoles) or "ion funnels" are placed between the ESI source and the mass analyzer.

Optimum operation of a normal ESI source is achieved at flow rates of 2 to 10 µL min. For stable operation at higher flow rates (0.2 to 1.0 mL min^{-1}; e.g., effluents from narrow- and wide-bore analytical HPLC columns), some form of an additional source of energy, such as heat or a high-velocity annular flow of gas, is supplemented to disperse the liquid into fine droplets.

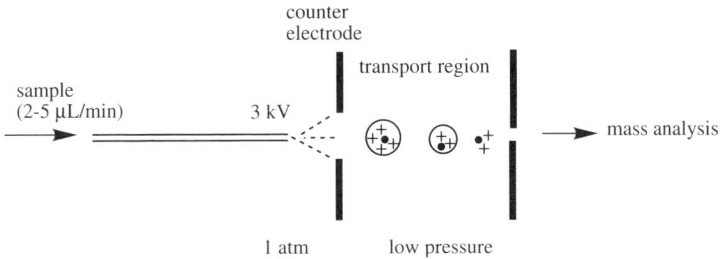

Figure 2.16. Basic components of electrospray ionization. (Reproduced from C. Dass, *Principles and Practice of Biological Mass Spectrometry*, Wiley-Interscience, 2001.)

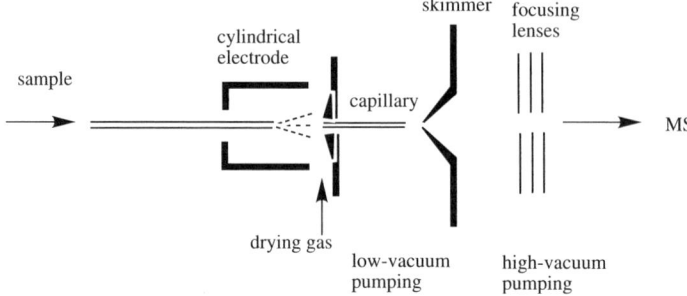

Figure 2.17. Heated capillary design of an electrospray ionization source. (Reproduced from C. Dass, *Principles and Practice of Biological Mass Spectrometry*, Wiley-Interscience, 2001.)

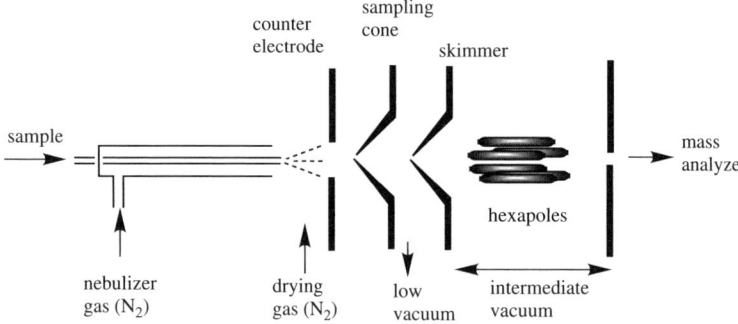

Figure 2.18. Block diagram of an electrospray ionization source that uses skimmer lenses. (Reproduced from C. Dass, *Principles and Practice of Biological Mass Spectrometry*, Wiley-Interscience, 2001.)

Electrospray analysis can be performed in positive and negative ionization modes. The polarity of the ions to be analyzed is selected by the capillary voltage bias. A novel feature of the ESI mass spectrum is the formation of intact molecular ions of the analyte. Fragmentation, if desired, can be induced in the ion-transport region of the ESI source by increasing the sampling cone voltage. This process is known as *in-source collision-induced dissociation* (CID) or *nozzle–skimmer* (NS) *dissociation*.

ESI of polymeric species yields a bell-shaped spectrum that contains a series of multiply charged ions of the general nature $[M + nH]^{n+}$ or $[M - nH]^{n-}$. The spectrum of horse heart myoglobin shown in Figure 2.19 is an illustrative example of a typical ESI mass spectrum of a protein. Each peak in that spectrum differs from its neighbor by one charge. Because a mass spectrometer analyzes ions based on their m/z ratios rather than on their actual mass, multiple charging has made possible analysis of high-mass compounds even with ordinary mass spectrometers of a limited mass range.

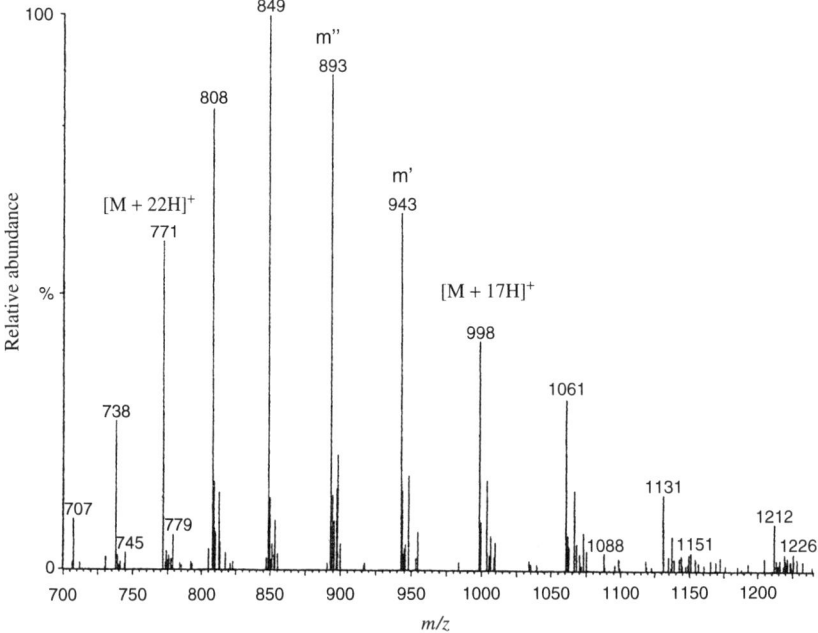

Figure 2.19. ESI mass spectrum of horse heart myoglobin. (Reproduced from C. Dass, *Principles and Practice of Biological Mass Spectrometry*, Wiley-Interscience, 2001.)

Consider two ions $[M + nH]^{n+}$ and $[M + (n+1)H]^{n+1}$, with their measured masses m' and m'' ($m' > m''$) given by Eqs. (2.29) and (2.30) (H = 1.007825 is the mass of the proton), respectively (see Figure 2.19):

$$m' = \frac{[M + nH]}{n} \tag{2.29}$$

$$m'' = \frac{[M + (n+1)H)]}{n+1} \tag{2.30}$$

By solving those two simultaneous equations, one can derive Eqs. (2.31) and (2.32), which can be used to calculate the charge state of a particular peak and the molecular mass of a biopolymer, respectively. In favorable cases, the molecular mass of macromolecules can be calculated with a precision of within ±0.005%.

$$n = \frac{m'' - H}{m' - m''} \tag{2.31}$$

$$M = n(m' - H) \tag{2.32}$$

The coupling of an ESI source with a quadrupole mass analyzer is the most successful ESI–MS combination. Since the development of the orthogonal ion extraction in TOF mass analyzers, coupling of ESI with TOF–MS has also become feasible. ESI–QIT, ESI–LIT, ESI-Orbitrap and ESI/FT–ICRMS are other successful combinations. An unprecedented range of compounds is being analyzed using these ESI–MS systems. This list includes proteins, peptides, combinatorial libraries, drug and drug metabolites, clinical samples, nucleotides, oligonucleotides, DNA adducts, oligosaccharides, synthetic polymers, organometallics, environmental pollutants, and many more.

2.17.1. Mechanism of Electrospray Ionization

The mechanism of ESI is a highly debated topic [79–85]. It is generally believed that ionization in electrospray involves three different processes: droplet formation, droplet shrinkage, and desorption of gaseous ions. At the onset of the electrospray process, the electrostatic force on the liquid leads to a partial separation of charges. In the positive-ion mode, cations concentrate at the tip of the metal capillary and tend to migrate toward the counter-electrode, whereas anions migrate inside the capillary away from the tip (Figure 2.20a). The migration of the accumulated positive ions toward the counter-electrode is counterbalanced by the surface tension of the liquid, giving rise to a Taylor cone at the tip of the capillary (Figure 2.20b). If the applied electric force is large enough, a thin cylinder of the liquid extends from this cone and breaks into a mist of fine droplets. The continuous production of charged species is assisted by electrochemical redox processes. In the positive-ion mode, electrochemical oxidation occurs in solution at the metal contact of the sprayed solution and reduction at the counter-electrode. In the negative-ion mode, the migration of anions and direction of electrochemical processes are reversed. A number of factors, such as applied potential, flow rate of the solvent, capillary diameter, and solvent characteristics influence the size of the droplets formed initially. Evaporation of the solvent from these droplets leads to droplet shrinkage. A cascade of droplet-fission processes follows. As the droplets shrink in size, the charge density on their surface increases until it reaches the Rayleigh instability limit. At this point, the repulsive coulombic forces exceed the droplet surface tension, causing fission of the droplets into smaller and highly charged offspring droplets. Further evaporation of the solvent results in Coulomb fission of the offspring droplets into second-generation droplets.

Two mechanisms have been proposed to explain ion desorption from the droplets: the charge-residue model (CRM) and the ion-desorption model (IDM) [81]. According to the CRM proposal (see Figure 2.21a), the sequence of solvent evaporation and droplet fission is repeated several times, until the drop size becomes so small that it contains only one solute molecule. As the last of the solvent is evaporated, that molecule is dispersed into the ambient gas, retaining the charge of the droplets. As shown in Figure 2.21b, the IDM proposal also relies on the sequence of solvent evaporation and droplet fission. Expulsion of the solvated ions into the gas phase, however, takes place at some intermediate droplet size (10 to 20 nm in radius) when the electric field due to the surface charge density is sufficiently high but less than the Rayleigh instability limit.

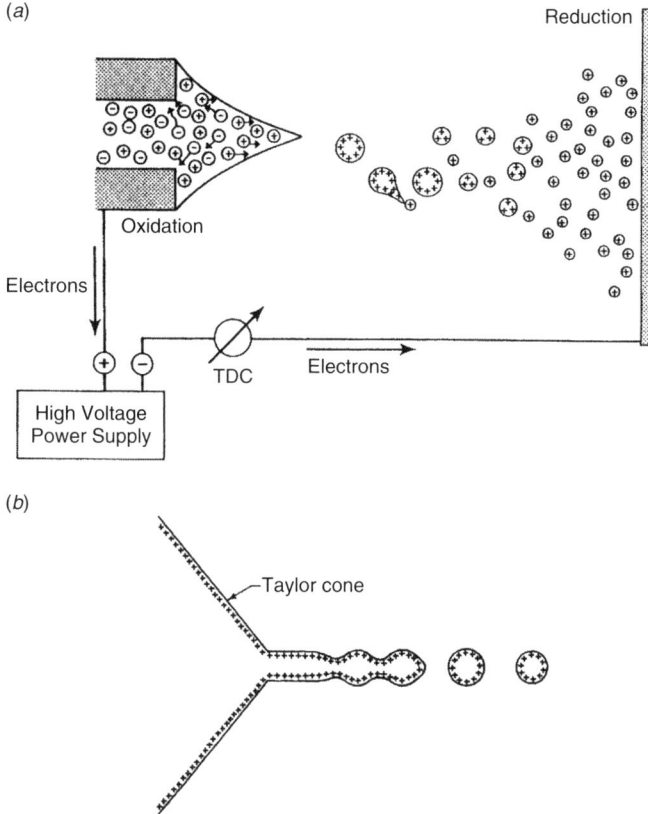

Figure 2.20. Depiction of electrochemical processes and Taylor cone formation in the ESI. (Adapted from ref. 84 by permission of Wiley-Interscience, copyright © 2000.)

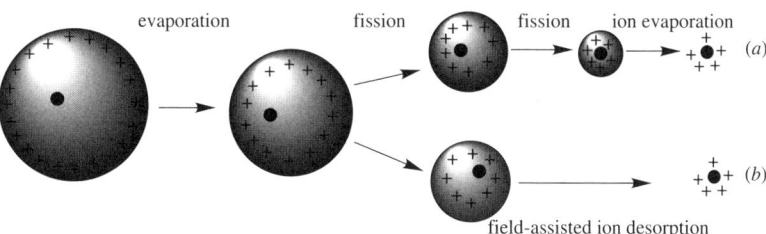

Figure 2.21. Desorption of ions from charged droplets into the gas phase: (*a*) charge-residue model; (*b*) ion-desorption model (From ref. 81.)

Currently, it is generally believed that the IDM model applies to ions with significant surface activity (i.e., hydrophobic molecules) and the CRM model applies largely to hydrophilic species. Thus, proteins and metal cations may follow CRM, whereas peptide and fatty acids follow IDM. As a consequence, the ESI response depends on the nature of the analyte. Enke has put forward an equilibrium partitioning model to explain differences in ESI response [86]. This model predicts that surface-active analytes compete favorably with excess droplet surface charge and thus exhibit a higher ESI response. Because of this relationship between the nonpolar character of the analytes and their ESI response [87], a correlation exists between ESI response and the HPLC retention time [88].

2.17.2. Sample Consideration

Ultrahigh detection sensitivity (attomole to femtomole) is the hallmark of ESI–MS. Several factors must, however, be controlled to achieve such high levels of sensitivity. One determinant of the sensitivity of an analysis is the ease with which charge-transfer processes are operative. The analyte concentration, its pK_a (or pK_b) and PA values, its surface activity, the presence of electrolytes or other species, the pH, and the shape and size of the sample molecule are some of the factors that generally influence the analyte ion current. The analyte detection sensitivity first increases up to about 10^{-5}, then decreases as the total electrolyte concentration increases in the electrospray solvent. The presence of dilute acids (acetic, formic) and bases (NH_4OH) increases the sample ion current substantially in the positive- and negative-ion modes, respectively. The high concentrations of salts, ion-pairing reagents, plasticizers, and detergents that, however, are often encountered in protein chemistry are detrimental to the ESI process.

2.17.3. Nanoelectrospray Ionization

Nanoelectrospray (nano-ES) is a miniaturized version of standard ESI and is designed to operate at submicroliter flow rates [89]. There are several practical differences in the operation of those two modes of ESI. First, nano-ESI uses a gold-coated fine-drawn capillary with a spraying tip aperture of only 1 to 3 μm rather than the 100-μm-i.d. capillary used in a standard ESI source. Second, the stable spray in nano-ESI is achieved at much lower capillary voltages (500 to 800 V). Another difference is that no external pumping is required to maintain a continuous solvent flow because the solution emerges from the tip itself at the rate of 10 to 20 nL min^{-1}. The emerging droplets are small, typically less than 200 nm in diameter. Finally, the nano-ES capillary is usually mounted within 1 to 2 mm of the nozzle orifice of an ES source. This arrangement improves ionization, desolvation, and transfer efficiencies. Sample utilization is also very efficient in the nano-ES design. The sample volume typically is as small as 1 μL, which can last over 30 minutes. The extended lifetime of the analyte signal allows prolonged accumulation of the ion current and an opportunity to conduct different types of experiments on the same sample. Because the rate of desorption of ions from small droplets is much higher than that from large droplets, the mass or molar sensitivity of ESI is inversely proportional to flow rate. Therefore, nano-ESI provides several orders of increased sensitivity over standard ESI. It should, however, be remembered that because ESI is a concentration-sensitive detector, the sensitivity remains unaffected with changes in the solvent flow rate.

2.18. DESORPTION ELECTROSPRAY IONIZATION

Recently, a new method of desorption ionization has emerged that combines the features of ESI with the concept of desorption ionization [90,91]. The technique is termed *desorption electrospray ionization* (DESI). A useful feature of this technique is that it allows ambient mass spectrometry; unlike most other ionization processes, ionization in DESI is accomplished by placing the sample in the ambient air outside the mass spectrometry vacuum. In this process, the sample solution is deposited on an insulating surface such as poly(tetrafluoroethylene) (PTFE) and subjected to a beam of electrosprayed charged microdroplets, ionic clusters, and gas-phase ions of solvent. The sprayed solution usually consists of a 1:1 (v:v) methanol–water mixture that contains 0.1% acetic acid and flows through a stainless steel capillary at the rate of 3 to 15 $\mu L\,min^{-1}$ under the influence of high voltage (e.g., 4 kV). The spray is assisted by a nebulizing gas. The desorbed ions are transferred to a mass spectrometer through an atmospheric-pressure flexible ion-transfer line made of either metal or an insulator. Similar to the conventional ESI process, the ions produced are singly or multiply charged ions of the analyte.

Schematics of a DESI experiment and a prototype source are shown in Figures 2.22 and 2.23, respectively. The ion source consists of a sprayer assembly and a surface assembly. The former is a pneumatically assisted microelectrospray source that is a part of a vertical rotating stage and a three-dimensional linear moving stage. The surface assembly is mounted on a separate three-dimensional moving stage. For high-throughput applications, the surface assembly can be replaced by a moving belt system.

The DESI process can be observed with conducting and insulating surfaces. DESI is applicable to polar and nonpolar molecules as well as to condensed-phase

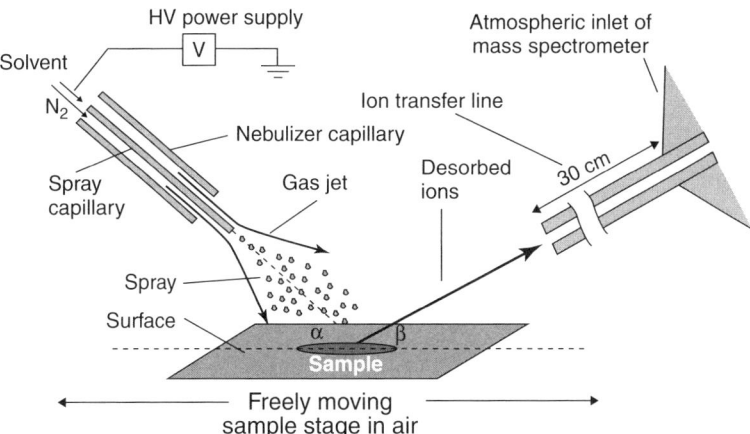

Figure 2.22. Schematic of a DESI experiment. (Reproduced from ref. 90 by permission of the American Association for the Advancement of Science, Danvers, MA, copyright © 2004.)

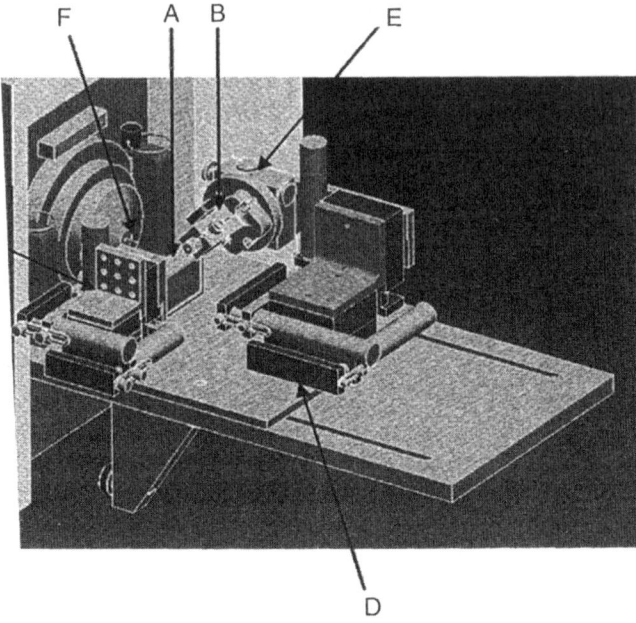

Figure 2.23. Schematic of a prototype DESI ion source: A, surface holder block; B, sprayer; C, three-dimensional moving stage for surface assembly; D, three-dimensional moving stage for sprayer alignment; E, rotating stage for sprayer; F, mass spectrometer inlet. (Reproduced from ref. 91 by permission of Wiley-Interscience, copyright © 2005.)

samples, frozen solutions, and adsorbed gases. Thus, a broad range of molecules as small as amino acids, drugs, and alkaloids, and as large as peptides and proteins, can be analyzed with this technique. The spectrum of equine cytochrome c that was desorbed from a surface deposition is shown in Figure 2.24. Other unique aspect of DESI is that it does not require any sample preparation; thus, the technique appears to be highly promising for field applications. For future applications, it will be feasible to sniff luggage at an airport for traces of explosives and drugs, test fruits in an orchard for pesticide residues, or test dried blood at a crime scene. National security may benefit highly from this development. Other novel applications of DESI are monitoring of living tissues such as plant or animal tissues or human skin.

2.18.1. DART Ion Source

Direct analysis in real time (DART) is another ambient mass spectrometry technique that has emerged recently [92]. Electronically excited helium or vibronic excited-state species of nitrogen (i.e., metastable helium atoms or nitrogen molecules) are, however, the primary beam particles in this approach. These species are produced by electrical discharge of the respective gases. The beam is directed toward the mass spectrometer orifice or reflected off a sample surface into the

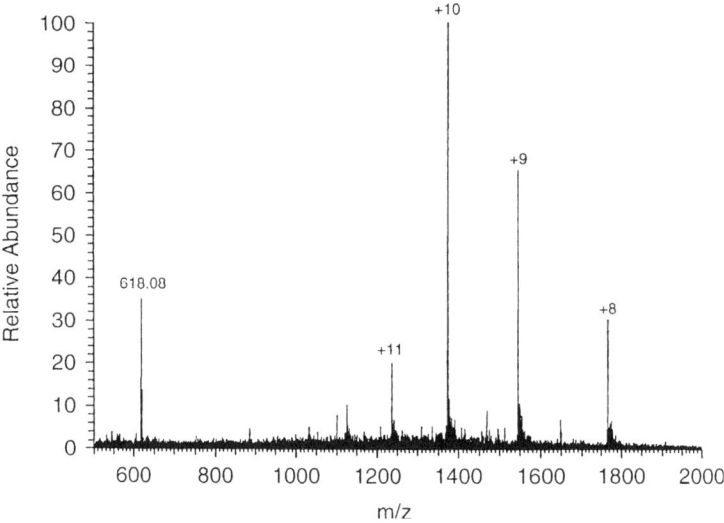

Figure 2.24. DESI mass spectrum of equine cytochrome c. (Reproduced from ref. 91 by permission of Wiley-Interscience, copyright © 2005.)

mass spectrometer. The ionized species are sampled by the atmospheric-pressure interface of the mass spectrometer. Similar to DESI, this technique is applicable to solids, liquids, and gases and to species that are deposited on surfaces without the need for prior sample treatment. Proton transfer is the dominant mechanism of positive ionization. The reagent ion species are the protonated water clusters that are formed by interaction of the metastable helium with atmospheric water. $[M - H]^-$ anions are the dominant species in the negative-ion mode. These ions are formed by the ion–molecule reactions of negative-ion clusters of oxygen and water. For some compounds (e.g., nitrobenzene), $M^{-\bullet}$ ions are also formed.

The technique has been used successfully in the analysis of organic compounds, biomolecules, pharmaceuticals, organometallics, drugs of abuse, explosives, and toxic chemicals on a variety of unusual surfaces, such as concrete, asphalt, human skin, currency, boarding passes, business cards, fruits, vegetables, body fluids, cocktail glasses, and clothing.

OVERVIEW

Ionization is the first essential step in mass spectrometry. The most popular mode of ionization is electron ionization, in which a gas-phase sample is bombarded by a beam of energetic electrons. It is applicable to relatively volatile and thermally stable compounds. Thermally labile compounds are preferentially analyzed with chemical ionization, which makes use of acid–base ion–molecule reactions between reactive ions and sample molecules. Other gas-phase ionization methods are photoionization, metastable atom bombardment, and field ionization, in which ionization occurs by irradiation with energetic photons, bombardment with a beam of metastable species, and interjection of gas-phase species into a high electric field, respectively.

Ionization of condensed-phase analytes occurs by mixing a sample in a suitable matrix and bombarding the matrix–analyte mixture with an energetic beam made of either laser photons as in MALDI, high-energy fission particles as in ^{252}Cf plasma desorption, or high-energy fast atoms or ions (FAB or liquid SIMS). When an analyte is present in a solution, such as an effluent from a separation device, it can be ionized via thermospray ionization, atmospheric-pressure chemical ionization, atmospheric-pressure photoionization, or electrospray ionization. Desorption electrospray ionization and direct analysis in real time are new modes of ionization that are accomplished in ambient air.

EXERCISES

2.1. Draw a schematic picture of an EI source and describe the function of each component.

2.2. A sample was ionized with a 50-eV electron beam and the ions were accelerated out of the ion source at a potential of 6000 V. What voltages will be applied to various components of this EI source?

2.3. What different ionization methods can be used to produce negative ions?

2.4. What modifications must be made to an EI source assembly for it to be used as a CI source?

2.5. Explain:
  (a) Why in electron ionization, the energy of the electron beam is set to 70 eV.
  (b) Why chemical ionization is preferred over electron ionization.
  (c) How a mass spectrometer, the upper mass limit of which is 4000 mass units, is able to analyze a protein of mass 50,000 Da using an ESI technique.
  (d) Why a matrix such as 2,5-dihydroxybenzoic acid is used in MALDI–MS analysis.
  (e) Why a large biopolymer can be ionized intact with MALDI–MS.
  (f) Why flow FAB provides better sensitivity than static FAB.
  (g) Why a TOF is the most appropriate mass analyzer for PD and MALDI modes of ionization.

2.6. An organic compound needs to be ionized by one of the following CI reagent gases: (a) methane (CH_4) positive-ion CI, (b) methane (CH_4)

negative-ion CI, (**c**) N_2 positive-ion CI, (**d**) Ar/isobutane (C_4H_{10}) positive-ion CI, and (**e**) N_2O/CH_4 negative-ion CI. Name the reagent ions that are involved in CI reactions.

2.7. A long-chain fatty alcohol of molecular mass 242 Da is ionized via positive-ion chemical ionization with methane, isobutane, hydrogen, water, or methanol as the reagent gas. The two major ions in the spectrum are m/z 243 and 225 (the loss of water from the protonated molecule). Predict how the ratio of ion abundances of m/z 243 versus m/z 225 will change with those different reagent gases. Rationalize your answers with a reasonable explanation. The proton affinity of this alcohol is 200 kcal mol^{-1}.

2.8. The ionization energy of cyclopentene is 9.02 eV. What wavelength photon will be required to photoionize it?

2.9. How many photons of 650-nm wavelength are required to ionize cyclopentene in a multiphoton ionization process?

2.10. Cyclopentene is ionized by a 10.0-eV photon. Would you observe any fragmentation?

2.11. If cyclopentene is ionized by a beam of 25.0-eV electrons, would any fragmentation be observed?

2.12. Two adjacent ions appear in the bell-shaped ESI mass spectrum of a protein at m/z 1618.1 and 1658.5. Calculate the charge states of these two ions and the molecular mass of the protein.

2.13. LDI is limited to compounds of less than 1000 Da, whereas MALDI can be used for compounds of much higher mass. Why?

2.14. What unique features are required in a MALDI matrix?

2.15. MALDI is a common practice to analyze large-molecular-mass compounds. Recently, MALDI has also been used to analyze small molecules. What strategies are used in the analysis of small molecules?

2.16. Sample preparation is a critical step in MALDI–MS analysis. What are different modes of sample preparation?

2.17. Name ionization techniques that you would prefer for the analysis of (**a**) anthracene, (**b**) a peptide, (**c**) a large protein, and (**d**) a large oligonucleotide.

2.18. How does DESI differ from the normal mode of ESI?

2.19. Toluene (IE = 8.83 eV) is ionized by two color R2PI multiphoton ionization scheme. The first photon is derived from a 337-nm nitrogen laser. What should be the wavelength of the second photon?

ADDITIONAL READING

1. C. Dass, *Principles and Practice of Biological Mass Spectrometry*, Wiley-Interscience, New York, 2001.
2. A. E. Ashcroft, *Ionization Methods in Organic Mass Spectrometry*, Royal Society of Chemistry, Cambridge, England, 1997.
3. M. L. Vestal, Methods for ion generation, *Chem. Rev.* **101**, 361–375 (2001).
4. A. G. Harrison, *Chemical Ionization Mass Spectrometry*, CRC Press, Boca Raton, FL, 1992.
5. H. D. Beckey, *Principles of Field Ionization and Field Desorption Mass Spectrometry*, Pergamon Press, Oxford, England, 1977.
6. L. Prokai, *Field Desorption Mass Spectrometry*, Marcel Dekker, New York, 1990.
7. A. Benninghoven, F. G. Rudenauer, and H. W. Werner, *Secondary Ion Mass Spectrometry:Basic Concepts, Instrumental Aspects, Applications and Trends*, Wiley, New York, 1987.
8. R. M. Caprioli, *Continuous-Flow Fast Atom Bombardment Mass Spectrometry*, Wiley, New York, 1990.
9. A. P. Snyder, ed., *Biochemical and Biotechnological Applications of Electrospray Ionization Mass Spectrometry*, American Chemical Society, Washington, DC, 1995.
10. R. B. Cole, ed., *Electrospray Ionization Mass Spectrometry: Fundamentals, Instrumentation, and Applications*, Wiley-Interscience, New York, 1997.

REFERENCES

1. A. J. Dempster, *Phys. Rev.* **18**, 398 (1921).
2. M. S. B. Munson and F. H. Field, Chemical ionization mass spectrometry I: General introduction, *J. Am. Chem. Soc.* **88**, 2621–2630 (1966).
3. B. Munson, Chemical ionization mass spectrometry, *Anal. Chem.* 43 (13), 28A–43A (1971).
4. G. G. Lias, J. E. Bartmess, J. F. Liebman, J. L. Holmes, R. D. Levin, and W. G. Mallard, Gas-phase ion and neutral thermochemistry, *J. Phys. Chem. Ref. Data* **17**, Suppl. 1 (1988).
5. N. Einolf and B. Munson, High-pressure charge exchange mass spectrometry, *Int. J. Mass Spectrom. Ion Phys.* **9**, 141–160 (1972).
6. H. Herzeler, M. G. Inghram, and J. D. Morrison, Photon impact studies of molecules with a mass spectrometer, *J. Chem. Phys.* **28**, 76–82 (1958).
7. J. Grotemeyer and E. W. Schlag, Multiphoton ionization mass spectrometry, *Angew. Chem.* **27**, 461–474 (1988).
8. D. C. Sykes, E. Woods III, et al, Thermal vaporization–vacuum ultraviolet visible time-of-flight mass spectrometry of single aerosol particles, *Anal. Chem.* **74**, 2048–2052 (2002).
9. C. Weickhartdt, K. Tönnies, and D. Globig, Detection and quantification of aromatic contaminants in water and soil samples by means of laser desorption laser mass spectrometry, *Anal. Chem.* **74**, 4861–4867 (2002).

10. H. D. Beckey, Determination of the structures of organic molecules and quantitative analyses with the field ionization mass spectrometer, *Angew. Chem.* **8**, 623–639 (1969).
11. D. Faubert, G. J. C. Paul, J. Giroux, and M. J. Bertrand, Selective fragmentation and ionization of organic compounds using an energy-tunable rare-gas metastable beam source, *Int. J. Mass Spectrom. Ion Phys.* **124**, 69–77 (1993).
12. M.-E. Dumas, L. Debrauwer, et al, Analyzing the physiological signature of anabolic steroids in cattle urine using pyrolysis/MAB MS and pattern recognition, *Anal. Chem.* **74**, 5393–5404 (2002).
13. H. D. Beckey, Field desorption mass spectrometry: a technique for the study of thermally unstable substances of low volatility, *Int. J. Mass Spectrom. Ion Phys.* **2**, 500–503 (1969).
14. T. M. Schaub, C. L. Hendrickson, K. Qian, J. P. Quinn, and A. G. Marshall, High-resolution field desorption/ionization FT ICR mass analysis of nonpolar molecules, *Anal. Chem.* **75**, 2172–2176 (2003).
15. D. F. Torgerson, R. P. Skowronski, and R. D. Macfarlane, New approach to the mass spectroscopy of nonvolatile compounds, *Biochem. Biophys. Res. Commun.* **60**, 616–621 (1974).
16. B. Sundqvist and R. D. Macfarlane, Californium-252-plasma desorption mass spectrometry, *Mass Spectrom. Rev.* **2**, 421–460 (1985).
17. R. J. Cotter, Plasma desorption mass spectrometry: coming of age, *Anal. Chem.* **60**, 781A–793A (1988).
18. A. Benninghoven and W. K. Sichtermann, Detection, identification, and structural investigation of biologically important compounds by secondary ion mass spectrometry, *Anal. Chem.* **50**, 1180–1184 (1978).
19. D. Touboul, F. Kollmer, et al, Improvement of biological time-of-flight-secondary ion mass spectrometry imaging with a bismuth cluster ion source, *J. Am. Soc. Mass Spectrom.* **16**, 1608–1618 (2005).
20. M. Barber, R. S. Bordoli, R. D. Sedgwick, and A. N. Tyler, Fast atom bombardment of solids (F.A.B.): a new ion source for mass spectrometry, *J. Chem. Soc. Chem. Commun.*, 325–327 (1981).
21. W. Aberth, K. Straub, and A. L. Burlingame, Secondary ion mass spectrometry with cesium ion primary beam and liquid target matrix for analysis of bioorganic compounds, *Anal. Chem.* **54**, 2029–2034 (1982).
22. C. Dass, Fast atom bombardment combined with mass spectrometry for characterization of polycyclic aromatic hydrocarbons, *J. Am. Soc. Mass Spectrom.* **1**, 405–412 (1990).
23. C. Dass and D. M. Desiderio, Fast atom bombardment mass spectrometry analysis of opioid peptides, *Anal. Biochem.* **163**, 52–66 (1987).
24. E. DePauw, A. Agnello, and F. Derwa, Liquid matrices for liquid secondary ion mass spectrometry–fast atom bombardment [LSIMS-FAB]: an update, *Mass Spectrom. Rev.* **10**, 283–301 (1991).
25. C. Dass, The role of a liquid matrix in controlling FAB-induced fragmentation, *J. Mass Spectrom.* **31**, 77–82 (1996).
26. R. M. Caprioli, Continuous-flow fast atom bombardment mass spectrometry, *Anal. Chem.* **62**, 477A–485A (1990).

27. R. J. Cotter, Lasers and mass spectrometry, *Anal. Chem.* **56**, 485A–504A (1984).
28. M. Karas, D. Bachmann, and F. Hillenkamp, Influence of the wavelength in high-irradiance ultraviolet laser desorption mass spectrometry of organic molecules, *Anal. Chem.* **57**, 2935–2939 (1985).
29. P. A. Demirev, A. B. Feldman, D. Kongkasuriyachai, P. Scholl, D. Sullivan, Jr., and N. Kumar, Detection of malaria parasites in blood by laser desorption mass spectrometry, *Anal. Chem.* **60**, 3262–3266 (2002).
30. M. Karas and F. Hillenkamp, Laser desorption ionization of proteins with molecular masses exceeding 10,000 daltons, *Anal. Chem.* **60**, 2299–2301 (1988).
31. K. Tanaka, H. Waki, et al, Protein and polymer analysis up to m/z 100,000 by laser ionization time-of-flight mass spectrometry, *Rapid Commun. Mass Spectrom.* **2**, 151–153 (1988).
32. F. Hillenkamp and M. Karas, Mass spectrometry of peptides and proteins by matrix-assisted ultraviolet laser desorption/ionization, in J. A. McCloskey, ed., *Methods in Enzymology,* Vol. 193, Academic Press, San Diego, CA, 1990, pp. 280–295.
33. K. Dreisewerd, M. Schuerenberg, M. Karas, and F. Hillenkamp, Influence of the laser intensity and spot size on the desorption of molecules and ions in matrix-assisted laser desorption/ionization with a uniform beam profile, *Int. J. Mass Spectrom. Ion Proc.* **141**, 127–148 (1995).
34. R. Knochenmuss, A quantitative model of ultraviolet matrix-assisted laser desorption/ionization including analyte ion generation, *Anal. Chem.* **75**, 2199–2207 (2003).
35. R. Knochenmuss and R. Zenobi, MALDI ionization: the role of in-plume processes, *Chem. Rev.* **103**, 441–452 (2003).
36. R. Knochenmuss, Photoionization pathways and free electrons in UV-MALDI, *Anal. Chem.* **76**, 3179–3184 (2004).
37. M. Karas, M. Glückmann, and J. Schäfer, Ionization in matrix-assisted laser desorption/ionization: singly charged molecular ions are the lucky survivors, *J. Mass Spectrom.* **35**, 1–12 (2000).
38. M. Karas and R. Krüger, Ion formation in MALDI: the cluster ionization mechanism, *Chem. Rev.* **103**, 427–439 (2003).
39. F. Hillenkamp, M. Karas, R. C. Beavis, and B. T. Chait, Matrix-assisted laser desorption/ionization mass spectrometry of biopolymers, *Anal. Chem.* **63**, 1193A–1203A (1991).
40. A. Overberg, M. Karas, et al., MALDI of large biomolecules with a TEA-carbon dioxide-laser, *Rapid Commun, Mass Spectrom*, **5**, 128–131 (1991).
41. D. W. Armstrong, L.-K, Zhang, L. He, and M. L. Gross, Ionic liquids as matrices for matrix-assisted laser desorption/ionization mass spectrometry, *Anal. Chem.* **73**, 3679–3686 (2001).
42. S. Carda-Broch, A. Berthod, and D. W. Armstrong, Ionic matrices for matrix-assisted laser desorption/ionization time-of-flight detection of DNA oligomers, *Rapid Commun. Mass Spectrom.* **17**, 553–560 (2003).
43. M. Mank, B. Stahl, and G. Boehm, 2,5-Dihydroxybenzoic acid butylamine and other liquid matrices for enhanced MALDI analysis of biomolecules, *Anal. Chem.* **76**, 2938–2950 (2004).
44. J. L. Anderson, D. W. Armstrong, and G.-T. Wei, Ionic liquids in analytical chemistry, *Anal. Chem.* **78**, 2892–2902 (2006).

45. T. Kinumi, T. Saisu, M. Takayama, and H. Niwa, Matrix-assisted laser desorption/ionization time-of-flight mass spectrometry for small molecules, *J. Mass Spectrom.* **35**, 417–422 (2000).
46. M. J. Dale, R. Knochenmuss, and R. Zenobi, Graphite/liquid mixed matrices for laser desorption/ionization mass spectrometry, *Anal. Chem.* **68**, 3321–3329 (1996).
47. S. Xu, Y. Li, H. Zou, J. Qiu, Z. Guo, and B. Guo, Carbon nanotubes as assisted matrix for laser desorption/ionization time-of-flight mass spectrometry, *Anal. Chem.* **75**, 6191–6195 (2003).
48. T. T. Hoang, Y. Chen, S. W. May, and R. F. Browner, Analysis of organoselenium compounds using active carbon and chemically modified silica sol-gel by surface-assisted laser desorption/ionization time-of-flight mass spectrometry, *Anal. Chem.* **76**, 2062–2070 (2004).
49. M. V. Ugarov, T. Egan, et al., MALDI matrices for biomolecular analysis based on functionalized carbon nano-materials, *Anal. Chem.* **76**, 6734–6742 (2004).
50. S. J. Doktycz, P. J. Savickas, and D. A. Krueger, Matrix/sample interactions in ultraviolet laser-desorption of proteins, *Rapid Commun. Mass Spectrom.* **5**, 145–148 (1991).
51. O. Vorm, P. Roepstorff, and M. Mann, Improved resolution and very high sensitivity in MALDI TOF of matrix surfaces made by fast evaporation, *Anal. Chem.* **66**, 3281–3287 (1994).
52. J. Axelsson, A.-M. Hoberg, C. Waterson, P. Myatt, G. L. Shield, J. Varney, D. M. Haddelton, and P. J. Derrick, Improved reproducibility and increased signal intensity in matrix-assisted laser desorption/ionization as a result of electrospray sample preparation, *Rapid Commun. Mass Spectrom.* **11**, 209–213 (1997).
53. M. Kussmann, E. Nordhoff, H. Rahbek-Nielsen, S. Haebel, M. Rossel-Larsen, L. Jakobsen, J. Gobom, E. Mirgorodskaya, A. Kroll-Kristensen, L. Palm, and P. Roepstorff, Matrix-assisted laser desorption/ionization mass spectrometry sample preparation techniques designed for various peptide and protein analytes, *J. Mass Spectrom.* **32**, 593–601 (1997).
54. I. K. Perera, J. Perkins, and S. Kantartzoglou, Spin-coated samples for high-resolution matrix-assisted laser desorption/ionization time-of-flight mass spectrometry of large proteins, *Rapid Commun. Mass Spectrom.* **9**, 180–187 (1995).
55. A. Westman, C. L. Nilsson, and R. Ekman, Matrix-assisted laser desorption/ionization time-of-flight mass spectrometry analysis of proteins in cerebrospinal fluid, *Rapid Commun. Mass Spectrom.* **12**, 1092–1098 (1998).
56. S. Trimpin and M. L. Deinzer, Solvent-free MALDI–MS for the analysis of biological samples via a mini-ball mill approach, *J. Am. Soc. Mass Spectrom.* **16**, 542–547 (2005).
57. Z. Guo, Q. Zhang, H. Zou, B. Guo, and J. Ni, A method for the analysis of low-mass molecules by MALDI-TOF mass spectrometry, *Anal. Chem.* **74**, 1637–1641 (2002).
58. G. McCombie and R. Knochenmuss, Small molecule MALDI using the matrix suppression effect to reduce or eliminate background interferences, *Anal. Chem.* **76**, 4990–4997 (2004).
59. F. O. Ayordinde, P. Hambright, T. N. Poerter, and Q. L. Keith, Jr., Use of meso-tetrakis(pentafluorophenyl)porphyrin as a matrix for low molecular weight alkylphenol ethoxylates in laser desorption/ionization time-of-flight mass spectrometry, *Rapid Commun. Mass Spectrom.* **13**, 2474–2479 (1999).

60. S.-F. Ren, L. Zhang, Z.-H Cheng, and Y.-L Guo, Immobilized carbon nanotubes for MALDI–TOF-MS analysis: applications to neutral small carbohydrates, *J. Am. Soc. Mass Spectrom.* **16**, 333–339 (2005).
61. S. A. Trauger, E. P. Go, et al., Sensitivity and analyte capture with desorption/ionization on silylated porous silicon, *Anal. Chem.* **76**, 4484–4489 (2004).
62. G. Siuzdak, J. Wei, and J. M. Buriak, Desorption–ionization mass spectrometry on porous silicon, *Nature* **399**, 243–246 (1999).
63. V. V. Laiko, M. A. Baldwin, and A. L Burlingame, Atmospheric pressure matrix-assisted laser desorption/ionization time-of-flight mass spectrometry, *Anal. Chem.* **72**, 652–657 (2000).
64. S. C. Moyer and R. J. Cotter, Atmospheric MALDI, *Anal. Chem.* **74**, 469A–479A (2002).
65. T. W. Hutchens and T. T. Yip, New desorption strategies for mass spectrometry analysis of macromolecules, *Rapid Commun. Mass Spectrom.* **7**, 576–580 (1993).
66. H. J. Issaq, T. P. Conrads, D. A. Prieto, R. Tirumalai, and T. D. Veenstra, SELDI-TOF for diagnostic proteomics, *Anal. Chem.* **75**, 149A–155A (2003).
67. I. Feuerstein, et al., Material-enhanced laser desorption/ionization (MELDI)—A new protein profiling tool utilizing specific carrier materials for time-of-flight mass spectrometric analysis, *J. Am. Soc. Mass Spectrom.* **17**, 1203–1208 (2006).
68. C. R. Blakley and M. L. Vestal, Thermospray interface for liquid chromatography/mass spectrometry, *Anal. Chem.* **55**, 750–754 (1983).
69. P. Arpino, Combined liquid chromatography mass spectrometry, II: Techniques and mechanisms of thermospray, *Mass Spectrom. Rev.* **9**, 631–669 (1990).
70. D. B. Robb, T. R. Covey, and A. P. Bruins, Atmospheric pressure photoionization: an ionization method for liquid chromatography–mass spectrometry, *Anal. Chem.* **72**, 3653–3659 (2000).
71. A. Raffaelli and A. Saba, Atmospheric pressure photoionization, *Mass Spectrom. Rev.* **22**, 318–331 (2003).
72. J. A. Syage, Mechanism of $[M + H]^+$ formation in photoionization mass spectrometry, *J. Am. Soc. Mass Spectrom.* **15**, 1521–1533 (2004).
73. J. M. Purcell, C. L. Hendrickson, R. P. Rogers, and A. G. Marshall, Atmospheric pressure photoionization Fourier-transform ion cyclotron resonance mass spectrometry for complex mixture analysis, *Anal. Chem.* **78**, 5906–5912 (2006).
74. T. Kauppila, T. Kotiaho, R. Kostiainen, and A. P. Bruins, Negative ion-atmospheric pressure photoionization mass spectrometry, *J. Am. Soc. Mass Spectrom.* **15**, 203–211 (2004).
75. R. D. Smith, J. A. Loo, R. R. Ogorzalek Loo, M. Busman, and H. R. Udseth, Principles and practice of electrospray ionization for mass spectrometry of large polypeptides and proteins, *Mass Spectrom. Rev.* **10**, 359–451 (1991).
76. C. Dass, Recent developments and applications of high-performance liquid chromatography–electrospray ionization mass spectrometry, *Curr. Org. Chem.* **3**, 193–209 (1999).
77. M. Dole, L. L. Mack, R. L. Hines, R. C. Mobley, L. D. Ferguson, and M. B. Alice, Molecular beams of macroions, *J. Chem. Phys.* **49**, 2240–2249 (1968).
78. J. B. Fenn, M. Mann, C. K. Meng, S. F. Wong, and C. M. Whitehouse, Electrospray ionization for mass spectrometry of large biomolecules, *Science* **246**, 64–71 (1989).

79. J. B. Fenn, M. Mann, C. K. Meng, S. F. Wong, and C. M. Whitehouse, Electrospray ionization:principles and practice, *Mass Spectrom. Rev.* **9**, 37–70 (1990).
80. M. S. Wilm and M. Mann, Electrospray and Taylor–Cone theory: Dole's beam of macromolecules at last? *Int. J. Mass Spectrom. Ion Proc.* **136**, 167–180 (1994).
81. J. B. Fenn, J. Rosell, T. Nohmi, S. Shen, and F. J. Banks, Jr., Electrospray ion formation: desorption versus desertation, in A. P. Snyder, ed., *Biochemical and Biotechnological Applications of Electrospray Ionization Mass Spectrometry*, American Chemical Society, Washington, DC, 1995, pp. 60–80.
82. M. H. Amad, N. B. Cech, G. S. Jackson, and C. G. Enke, Importance of gas-phase proton affinities in determining the electrospray ionization response, *J. Mass Spectrom.* **35**, 784–789 (2000).
83. M. Gamero-Castano and J. F. de la Mora, Kinetics of small ion evaporation from the charge and mass distribution of multiply charged clusters in electrosprays, *J. Mass Spectrom.* **35**, 790–803 (2000).
84. P. Kebarle, A brief overview of the present status of the mechanisms involved in electrospray mass spectrometry, *J. Mass Spectrom.* **35**, 804–817 (2000).
85. J. F. de la Mora, G. J. Van Berkel, C. G. Enke, M. Martinez-Sanchez, and J. B. Fenn, Electromechanical processes in electrospray ionization mass spectrometry, *J. Mass Spectrom.* **35**, 939–952 (2000).
86. C. G. Enke, A predictive model for matrix and analyte effects in electrospray ionization of singly-charged analytes, *Anal. Chem.* **69**, 4885–4893 (1997).
87. N. B. Cech and C. G. Enke, Relating electrospray ionization response to nonpolar character of small peptides, *Anal. Chem.* **72**, 2717–2723 (2000).
88. N. B. Cech, J. R. Krone, and C. G. Enke, Predicting electrospray response from chromatographic retention time, *Anal. Chem.* **73**, 208–213 (2001).
89. M. Wilm and M. Mann, Analytical properties of the nanoelectrospray ion source, *Anal. Chem.* **68**, 1–8 (1996).
90. Z. Takáts, J. M. Wiseman, B. Gologan, and R. G. Cooks, Mass spectrometry sampling under ambient conditions with desorption electrospray ionization, *Science* **306**, 471–473 (2004).
91. Z. Takáts, J. M. Wiseman, and R. G. Cooks, Ambient mass spectrometry using desorption electrospray ionization (DESI): instrumentation, mechanisms, and applications in forensics, chemistry, and biology, *J. Mass Spectrom.* **40**, 1261–1275 (2005).
92. R. B. Cody, J. A. Larame, and H. D. Durst, Versatile new ion source for the analysis of materials in open air under ambient conditions, *Anal. Chem.* **73**, 2297–2302 (2005).

CHAPTER 3

MASS ANALYSIS AND ION DETECTION

In this chapter a detailed account is given of the basic concepts and latest developments of several different types of mass analyzers and detectors. The mass analyzer is the heart of a mass spectrometer. The performance of a mass spectrometer depends largely on the design of the mass analyzer and associated ion optics. A mass analyzer performs two vital functions: (1) to disperse all ions in terms of their mass-to-charge (m/q) ratio, and (2) to focus all mass-resolved ions at a single focal point. These two functions are parallel to the dispersive action of a monochromator and the focusing action of a lens assembly of conventional optical instruments. In addition, a mass analyzer maximizes the transmission of all ions that enter it from the ion source. A moving charged particle can be distinguished from another ion on the basis of differences in their momentum, kinetic energy, and velocity. A mass analyzer makes use of one or more of these properties to mass-resolve and to focus various ions. Several different types of mass analyzers are available that use these basic properties. Some popular designs are magnetic sector, quadrupole, quadrupole ion trap (QIT), quadrupole linear ion trap (LIT), orbitrap, time-of-flight (TOF), and ion cyclotron resonance (ICR).

The performance of a mass analyzer is evaluated on the basis of the following desirable features:

- *Mass range:* the maximum allowable m/q ratio amenable to analysis. A higher value is an asset for the analysis of high-mass compounds.
- *Resolution:* the ability to separate two neighboring mass ions (see Section 3.1).

Fundamentals of Contemporary Mass Spectrometry, by Chhabil Dass
Copyright © 2007 John Wiley & Sons, Inc.

- *Efficiency:* the transmission multiplied by the duty cycle (defined as the fraction of ions of interest formed in a single ionization event).
- *Mass accuracy:* the measured error in m/q divided by the accurate m/q.
- *Linear dynamic range:* the range over which an ion signal is linear with the analyte signal.
- *Speed:* the number of spectra acquired per unit time. A fast scan speed is required for rapidly changing events: for example, to monitor eluants from a chromatography step or to record a pulsed event. A slow speed is desired in accurate mass measurement experiments.
- *Sensitivity:* expressed as abundance or detection sensitivity. *Abundance sensitivity* is the inverse of a quantity obtained by dividing the abundance of a large peak by the abundance of a background peak one m/q lower or higher. *Detection sensitivity* is the smallest amount of an analyte that can be detected at a certain defined confidence level.
- *Adaptability:* the possibility of outfitting with certain ionization techniques and other ancillary devices, such as multichannel array detectors and chromatographic devices. Coupling a high-resolution separation device to a mass spectrometer helps in the analysis of real-world complex mixtures of samples.

Tandem MS (MS/MS) capability, small size, and lower cost are other desirable characteristics of a mass analyzer. Miniaturization of mass spectrometers is a growing area of interest for field applications. Tandem MS is helpful in the analysis of complex mixtures.

Before enumerating the basic principles of various types of mass analyzers, it is pertinent to outline the concept of resolution and kinetic energy as applicable to mass spectrometry.

3.1. MASS RESOLVING POWER

High resolution is a desirable figure of merit of a mass spectrometer because it helps to (1) perform accurate mass measurements, (2) resolve isotopically labeled species when the percent incorporation of the label is to be determined, (3) resolve an isotopic cluster when the charge state of high-mass compounds is to be determined, (4) enhance the accuracy of quantification, and (5) unambiguously mass-select precursor ions in MS/MS experiments.

By definition, the *mass resolution* of a mass spectrometer is its ability to distinguish between two neighboring ions that differ only slightly in their mass (Δm). Mathematically, it is the inverse of resolving power (RP), given as

$$\mathrm{RP} = \frac{m}{\Delta m} \qquad (3.1)$$

where m is the average of, and Δm is the difference in, the accurate masses of two neighboring ions. A larger RP means that a smaller mass differences can be resolved. According to the 10% valley definition (depicted in Figure 3.1),

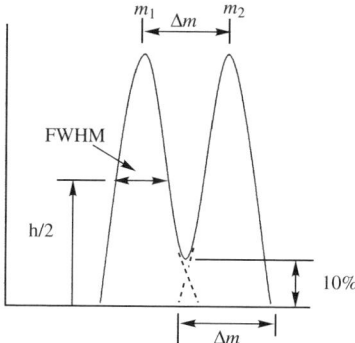

Figure 3.1. Resolving power: 10% valley definition (depicted by separation of two ions of mass m_1 and m_2) and FWHM definition ($h/2$ is half the height of the peak).

the two equal-height peaks are considered resolved when the valley between the two ions is 10% of the height of either one (i.e., each contributes 5% to the valley). The value of RP can also be expressed in terms of the width of a single symmetrical well-resolved peak. Here m is the m/q value at the apex of the peak and Δm is the width of the peak at a specified height. The peak-width definition is depicted pictorially on the right-hand side of Figure 3.1. Use of the 10% valley definition is a common practice with magnetic-sector and quadrupole instruments. In TOF, QIT, and FT-ICR mass analyzers, the peaks separated by 50% valley are considered as resolved. In such cases, resolution is reported in terms of full width at half maximum (FWHM; see Figure 3.1). The RP value in FWMM definition is larger by a factor of 2 than the 10% valley definition.

▶ **Example 3.1** Calculate the resolving power required to resolve two singly charged ions of elemental compositions C_3H_7COOH and $C_3H_7CONH_3$.

Solution First calculate the accurate mass of the ions, using the accurate mass of the most abundant isotope of each element:

$$m(C_3H_7COOH) = (12.00000 \times 4) + (1.007825 \times 8) + (15.994915 \times 2)$$
$$= 88.05243 \text{ u}$$

$$m(C_3H_7CONH_3) = (12.00000 \times 4) + (1.007825 \times 10) + (15.994915 \times 1)$$
$$+ (14.003074 \times 1) = 88.076239 \text{ u}$$

$$RP = \frac{m}{\Delta m} = \left(\frac{88.05243 + 88.076239}{2}\right)\left(\frac{1}{(88.076239 - 88.05243)}\right)$$

$$= \frac{176.128669}{(2)(0.023809)} = 3698.8$$

3.2. KINETIC ENERGY OF IONS

When a charged particle traverses in the direction of an electric field, it experiences acceleration, as is the case when ions travel in the accelerating region of an ion source. The kinetic energy of the accelerated ion is given as

$$\text{KE} = qV = \frac{mv^2}{2} \tag{3.2}$$

where m is the mass of the ion in kilograms (1 u = 1 Da = 1.6605×10^{-27} kg), v is the velocity of the ion in m/s, q is the charge on the ion ($q = ze$; z is the integral number of units of charge and e is the fundamental unit of charge = 1.602×10^{-19} C), and V, in volts, is the ion acceleration voltage.

▶ **Example 3.2** Calculate the kinetic energy of a singly charged ion of 78-u mass after it is accelerated through a potential difference of 6000 V.

Solution

$$\text{KE} = qV = zeV = 6000 \text{ eV} = (1 \times 1.602 \times 10^{-19} \text{ C} \times 6000\text{V})(1\text{J}/1\text{C} \cdot \text{V})$$
$$= 9.6 \times 10^{-16} \text{ J}$$

MASS ANALYZERS

3.3. MAGNETIC-SECTOR MASS SPECTROMETERS

Magnetic-sector instruments are the oldest type of mass spectrometers (see Table 3.1). The instrument used by J. J. Thomson in the 1910s for the analysis of positive rays was a sector instrument that employed magnetic and electric fields [1]. Current magnetic-sector (or simply sector) instruments are two types, single and double focusing. Single-focusing instruments are constructed with only a magnet field, whereas in double-focusing sector instruments, an electrostatic analyzer is incorporated along with a magnetic analyzer [2, 3].

3.3.1. Working Principle of a Magnetic Analyzer

A *magnetic sector* separates ions of different m/q values via momentum dispersion and directional focusing. When ions of different m/q but identical kinetic energy enter the magnetic field, they travel in different circular paths; ions of higher m/q value move in a larger trajectory, and ions of lower m/q move in a lower trajectory and thus are separated from each other (Figure 3.2).

MAGNETIC-SECTOR MASS SPECTROMETERS

TABLE 3.1. Comparison of Various Types of Mass Analyzers

Characteristic	Magnetic	Quadrupole	QIT	LIT	TOF	FT–ICR
Mass range (Da)	15,000	4000	4000	4000	Unlimited	$>10^4$
Resolving power	10^2-10^5	4000	10^3-10^4	10^3-10^4	15,000	$>10^6$
Mass accuracy (PPM)	1–5	100	50–100	50–100	5–50	1–5
Abundance sensitivity	10^6-10^9	10^4-10^6	10^3	10^3-10^5	up to 10^6	10^2-10^5
Speed (Hz)	0.1–20	1–20	1–30	1–300	10^1-10^6	$10^{-2}-10^1$
Efficiency%	<1	<1–95	<1–50	<1–99	1–100	<1–95
Dynamic range	10^9	10^7	10^2-10^5	10^2-10^5	10^2-10^6	10^2-10^5
MS/MS	Excellent	Great	Great	Excellent	Great	Great
LC (CE)/MS	Poor	Excellent	Excellent	Excellent	Good	Good
Cost	$$$$	$	$	$$	$$-$$$	$$$$

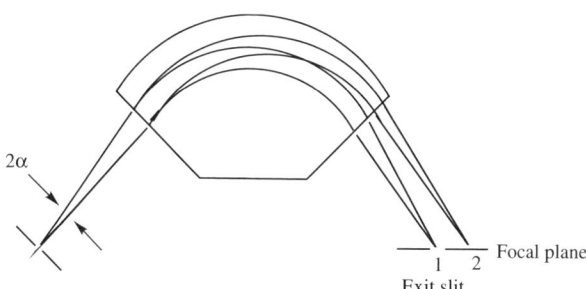

Figure 3.2. Separation of two ions of different m/q by a magnetic sector. The two ions that emanate from a point source with an angular divergence of 2α are brought to focus on the focal plane at points 1 and 2. Changing the magnetic field allows ions to exit the slit sequentially for their detection.

In a magnetic field of strength B, an ion of charge q and velocity v experiences a force given by

$$F_B = Bqv$$

The force is perpendicular to the direction of the magnetic field as well as to the direction of the motion of the ion. As a result, the ion travels in a curved path that has a radius r in a plane perpendicular to the direction of the magnetic field. The force F_B is counterbalanced by the centrifugal force such that

$$Bqv = \frac{mv^2}{r} \quad \text{or} \quad \frac{mv}{q} = Br \tag{3.3}$$

where the unit of B is the tesla (T), mass is in kilograms, velocity is in meters/second, and radius is in meters. From this equation it is obvious that the magnetic sector acts as a momentum analyzer, and disperses each ion according to its momentum-to-charge ratio. When the value of v from Eq. (3.2) is substituted into (3.3), it becomes clear that a magnetic sector also separates ions in terms of their m/q (or m/z) ratios:

$$\frac{m}{q} = \frac{B^2 r^2}{2V} \tag{3.4}$$

Equation (3.4) implies that all ions of the same m/q value will follow the same trajectory provided that they have the same kinetic energy and will be focused at a single focal point. Another function of the magnetic field is *directional focusing*; that is, a divergent beam of the same m/q ions (e.g., 2α in Figure 3.2) can come to focus at a single focal point. The focal length of the magnetic sector is determined by the geometry of its poles.

To acquire a mass spectrum, a common practice is to scan the magnetic field while keeping V constant. Alternatively, all ions can be collected simultaneously in the same plane by keeping B fixed, as is the usual practice in Mattauch–Herzog geometry double-focusing mass spectrometers. The third, not so common option is to keep B constant and vary the value of V.

Dividing both sides of Eq. (3.4) by q and rearranging it in terms of radius r gives

$$r = \frac{\sqrt{2mqV}}{Bq}$$

which implies that all ions of the same m/q but of different kinetic energy (qV) cannot be brought to focus at a single point. Thus, a magnetic sector has limitation of resolution. A requirement for high resolution is that all ions that enter the magnetic sector be homogeneous in energy. This requirement is fulfilled to a certain extent by accelerating the ions to high potentials (usually, 8 to 10 kV). A better approach is, however, to incorporate an energy-focusing device such as an electrostatic analyzer (ESA) into the design of a magnetic-sector instrument.

▶ **Example 3.3** The flight tube of a magnetic-sector instrument has a radius of 0.40 m. After acceleration to 8000 V a singly charged ion was detected at a magnetic field strength of 0.65 T. Calculate the mass of the ion.

Solution

$$m(\text{kg}) = \frac{zeB^2r^2}{2V} = m(\text{u}) = \frac{(1)(1.602 \times 10^{-19} \text{ C})(0.65 \text{ T})^2(0.4 \text{ m})^2}{(1.66 \times 10^{-27} \text{ kg})(2)(8000 \text{ V})} = 407.7 \text{ u}$$

All units will cancel out and the mass of the ion will be calculated in the unified mass unit (u) because $T = \text{kg/s} \cdot \text{C}$ and $\text{C} \cdot \text{V} = \text{J} = \text{kg} \cdot \text{m}^2/\text{s}^2$.

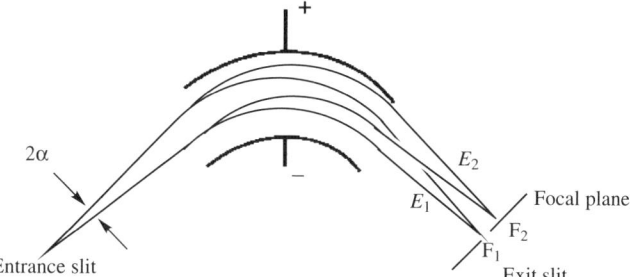

Figure 3.3. Separation of ions by an ESA on the basis of differences in their kinetic energies. The same kinetic energy ions that enter the ESA with an angular divergence of 2α are brought to focus on the focal plane at a single point. For example, F_1 is the focal point of ions with kinetic energy E_1, and F_2 that of energy E_2.

3.3.2. Working Principle of an Electrostatic Analyzer

An *electrostatic analyzer* (ESA) consists of two coaxial condenser plates (Figure 3.3), one held at a positive potential and the other at a negative potential of equal magnitude. Its purpose is to produce ions homogeneous in kinetic energy (i.e., ions that have the same energy after traversing through this device are focused at a single point). When charged particles enter this radial electric field, they are forced to follow a circular trajectory, where the centrifugal field is exactly balanced by the electrical field:

$$\frac{mv^2}{r} = qE \quad \text{and} \quad r = \frac{2V}{E} \tag{3.5}$$

and ions are dispersed according to their kinetic energy. Thus, an ESA can be used as an energy filter to produce an ion beam of nearly homogeneous energy. In Eq. (3.5), r is the radius of curvature in meters, E the electric field strength in volts/meter, and m, v, z, and V are as defined above. An ESA also acts as a direction-focusing device. All ions of identical energy content that emanate from the same point with a small angular divergence are brought to focus at a single point (e.g., at points F_1 and F_2 in Figure 3.3).

3.3.3. Working Principle of Double-Focusing Magnetic-Sector Mass Spectrometers

Double-focusing mass spectrometers use a proper combination of an ESA and a magnetic analyzer to provide very high resolution. Ionization of the sample molecules produces a population of ions with different energy and spatial distribution. As a consequence, all ions of the same m/q value will not be focused to a single point by a single-focusing magnetic-sector instrument; the result is poor resolution. By the combined energy-focusing action of an ESA and directional focusing of a magnetic analyzer, however, the directional and

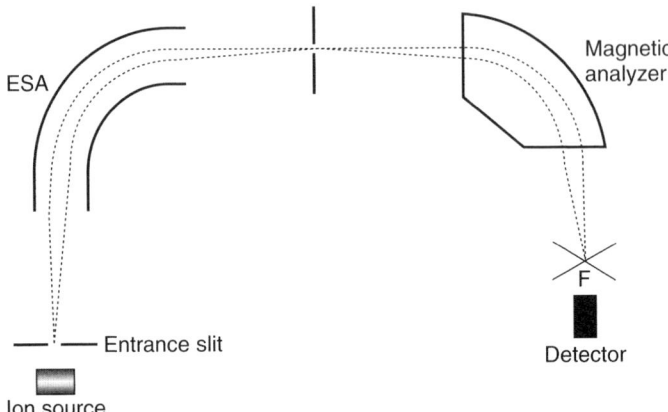

Figure 3.4. Double-focusing forward-geometry Nier–Johnson geometry instrument. Ions traverse in the C-shaped trajectory and are collected at a single focal point.

velocity inhomogeneities can both be corrected. The ESA is designed in such a way that it counterbalances the kinetic energy spread caused by the magnetic sector.

The forward-geometry Nier–Johnson instrument (EB) is the most common type of double-focusing sector instrument (Figure 3.4). In this design, the ESA (E) is placed before the magnetic analyzer (B), and the two fields deflect the ions in the same direction (i.e., have a C-shaped trajectory) [3]. The condition of double focusing is achieved at a single focal point where the velocity- and direction-focusing planes intersect (e.g., point F in Figure 3.4). The Nier–Johnson instrument is a scanning type of mass spectrometer, meaning that ions are detected sequentially one at a time. In a reverse-geometry instrument, a magnetic analyzer precedes an ESA (i.e., the BE geometry). In Mattauch–Herzog geometry (Figure 3.5), the two fields deflect the ion beam in opposite directions (i.e., have an S-shaped geometry), and the double focusing of all ions occurs in the same plane [2]. In this dispersion type of instrument, all mass-resolved ions are detected simultaneously by placing a focal-plane detector (e.g., a photographic plate or a multichannel array detector) at the exit boundary of the magnet. Because of the property of simultaneous detection of all ions, these instruments provide increased sensitivity and accuracy over those of Nier–Johnson instruments.

3.3.4. Performance Characteristics

Double-focusing sector instruments have long enjoyed status as state-of-the-art high-performance machines. They have a reputation for providing high resolution, high precision, and accuracy in mass determination; reasonable high-mass range; acceptable scan speed; and high dynamic range. At 2.0-T field strength and maximum acceleration potential (e.g., of 10 kV), a mass range above 15,000 u

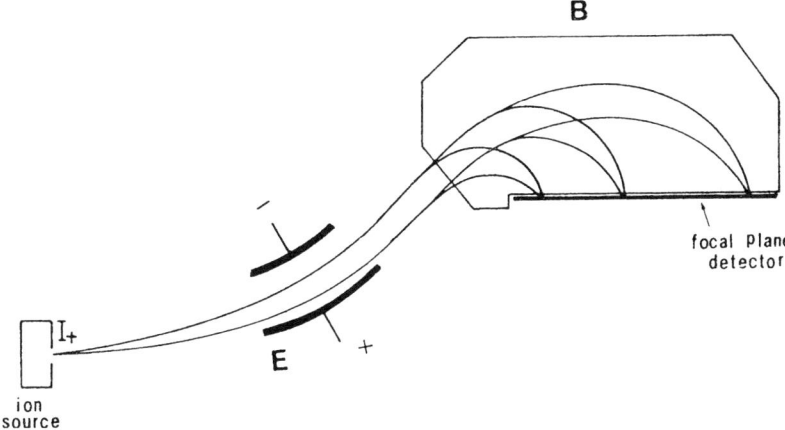

Figure 3.5. Mattauch–Herzog geometry instrument. (Reproduced from C. Dass, in D. M. Desiderio, ed., *Mass Spectrometry: Clinical and Biomedical Applications*, Vol. 2, by permission of Plenum Press, copyright © 1994.)

can be achieved. These instruments are capable of providing a mass resolution of over 100,000 (10% valley definition) and a mass accuracy of better than 1 ppm (with the peak-matching technique). The transmission and sensitivity both increase with the acceleration potential. Another strong point of these instruments is their ability to perform a range of reactions in field-free regions (FFRs). For example, in addition to high-energy collision-induced dissociation (CID) reactions, charge-permutation reactions, reactions of metastable ions, and kinetic energy release (KER) experiments can be investigated. The exorbitant cost of these instruments is, however, a major concern. An other disadvantage is their lower transmission efficiency. In addition, the high-voltage environment in the ion source is a hindrance to their coupling with atmospheric-pressure ionization (API) techniques and liquid chromatography. Also, these instruments are suited only for continuous-beam ion sources.

3.4. QUADRUPOLE MASS SPECTROMETERS

Quadrupole instruments are probably the most widely used type of mass spectrometer. A *quadrupole* consists of four precisely matched parallel metal rods (Figure 3.6). The mass separation is accomplished by the stable vibratory motion of ions in a high-frequency oscillating electric field that is created by applying direct-current (dc) and radio-frequency (rf) potentials to these electrodes [4–6]. Under a set of defined dc and rf potentials, ions of a specific m/q value pass through the geometry of quadrupole rods. A mass spectrum is obtained by changing both the dc and rf potentials while keeping their ratio constant.

76 MASS ANALYSIS AND ION DETECTION

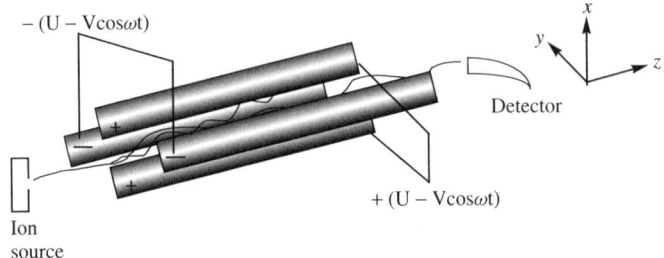

Figure 3.6. Quadrupole mass analyzer. At a certain values of dc potential U and rf potential V, ions of a specific range of m/q value have stable trajectories in the x- and y-directions and travel in the z-direction within the quadrupole field to reach the detector; all other ions are lost because they follow unstable trajectories.

3.4.1. Working Principle

Four metallic rods (electrodes) are arranged symmetrically in a square array (Figure 3.6). Ideally, the rods should be of hyperbolic geometry, but for convenience, precisely machined circular rods are acceptable. The field within the square array is created by connecting opposite pairs of electrodes electrically. One pair receives a superimposed positive dc potential U and a time-dependent rf potential $V \cos \omega t$, where ω is the angular frequency (in rad s^{-1}) of the applied rf voltage, V its amplitude, and t the time. The angular frequency is related to the radio-frequency f (in hertz) by $\omega = 2\pi f$. The other adjacent pair of rods receives a dc potential $-U$ and an rf potential of the same magnitude, $V \cos \omega t$, but out of phase by 180° [i.e., $-(U - V \cos \omega t)$]. Application of these voltages creates an oscillating field within the rods that is given by

$$\Phi_{(x,y)} = \Phi_0 \frac{x^2 - y^2}{r_0^2} = (U - V \cos \omega t) \frac{x^2 - y^2}{r_0^2} \qquad (3.6)$$

where Φ_0 is the applied potential (i.e., $U - V \cos \omega t$), r_0 the inscribed radius (i.e., one-half the distance between the opposite electrodes), and x and y the distances from the center of the field. Ions are injected at one end of the quadrupole structure in the direction of the quadrupole rods (the z-direction). Separation of ions of different m/q value is accomplished through the criterion of path stability within the quadrupole field. At a given set of operating parameters, ions of a narrow but adjustable m/q range have stable trajectories (i.e., their motion is confined within the field-defining electrodes), whereas the remainder of the ions will have unstable trajectories (i.e., the amplitude of their motion exceeds the boundaries of the electrodes). Thus, the mass separation action of a quadrupole is similar to that of a narrow bandpass filter rather than a conventional mass spectrometer (i.e., ions of a narrow mass window can survive within the quadrupole geometry). To obtain a mass spectrum, the quadrupole field is varied to force other mass window ions to sweep through the quadrupole.

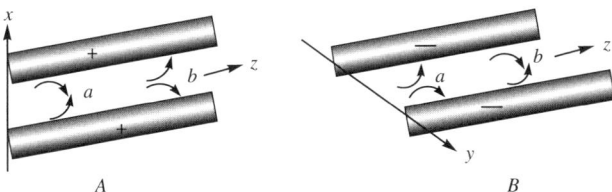

Figure 3.7. Operation of a quadrupole in (*A*) the xz-plane and (*B*) the yz-plane. In the xz-plane, the positive rods act as a high-pass filter for positive ions. The ions converge toward the z-axis (at point a) due to the action of the positive dc potential and are defocused due to the action of the negative half-cycle of the rf potential. Similar actions make the negative pair of rods low-pass filter in the yz-plane.

The mass-filtering action of a quadrupole can be explained as follows [6]: Consider first the action of a positive pair of electrodes (i.e., those acting in the xz-plane). As shown in Figure 3.7A, the positive dc potential applied to these rods will accelerate positive ions toward the central axis (point a). The simultaneous action of the rapidly changing rf potential during its negative half-cycle will accelerate these ions toward the rods (point b); the low-m/q ions will be accelerated to the highest velocities on each half cycle and will ultimately be eliminated from the field-defining space. In contrast, the higher-m/q ions will respond sluggishly to this rf potential and will remain confined within the boundaries of the rods. Thus, a positive pair of rods acts as a high-pass filter for positive ions. Now consider the action of the negative pair of electrodes the ($y - z$-plane). Because of the continuous action of the negative dc potential, all positive ions will be attracted toward the rods (point a in Figure 3.7B). During the short period of the positive half-cycle of the rf potential, only the motion of lower-m/q ions will be reversed toward the center of the rods (point b). Because of their inertia, the higher-m/q ions will be lost to the rapidly changing positive rf potential. Thus, the negative pair of electrodes acts as a low-pass filter. The combination of these actions creates a stability window for ions of a narrow m/q range to travel through the rods (in the z-direction).

The motion of an ion in the x- and y-directions is described by using a quadratic equation that is commonly known to mathematicians as the *Mathieu equation*:

$$\frac{d^2 u}{d^2 \xi} + (a_u - 2q_u \cos 2\xi)u = 0 \tag{3.7}$$

where u is the transverse displacement in the x- and y-directions from the center of the field and ξ is equal to $\omega t/2$. Solution of the Mathieu equation provides two important dimensionless parameters a and q, given by

$$a_u = a_x = -a_y = \frac{8zeU}{m\omega^2 r_0^2} \tag{3.8}$$

$$q_u = q_x = -q_y = \frac{4zeV}{m\omega^2 r_0^2} \tag{3.9}$$

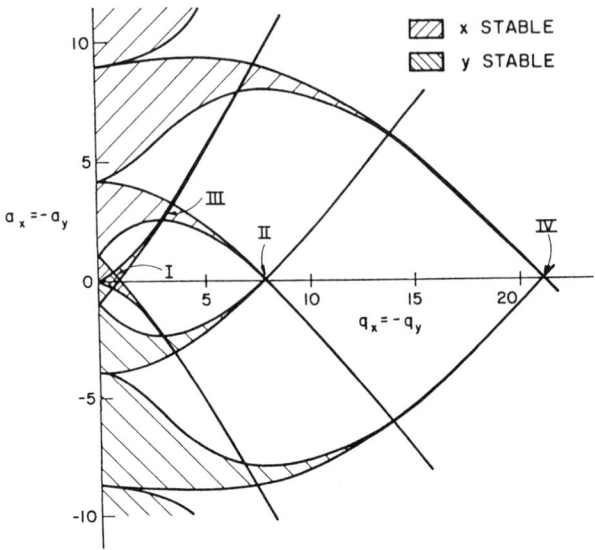

Figure 3.8. The a/q stability diagram for a quadrupole mass filter. The stability regions for the x- and y-directions are shown as shaded portions, and the regions of simultaneous stability are at I, II, III, and IV. Of these, region I is the most used. (Reproduced from ref. 4 by permission of Wiley-Interscience, copyright © 1986.)

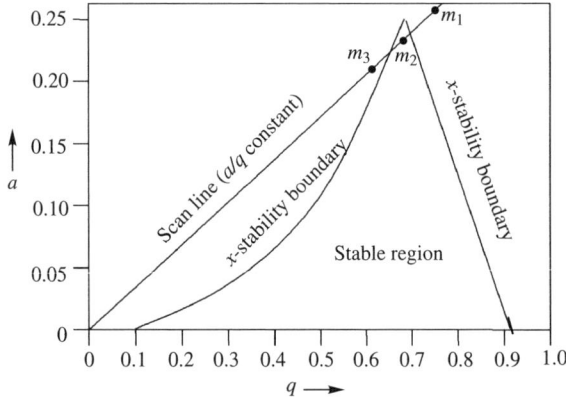

Figure 3.9. First stability region for a quadrupole mass filter.

A plot of the Mathieu parameters a_u and q_u provides several regions of simultaneous stable oscillations in the x- and y-directions (see Figure 3.8). The region closest to the origin (shown in detail in Figure 3.9) is, however, the most useful. In this region, the maximum value of q_u is 0.908. A mass spectrum can be acquired by sweeping U and V, simultaneously while keeping their ratio and f constant. A line with the slope $a/q = 2U/V$ that passes through the origin of the a–q diagram is the operating line or mass-scan line. All ions that fit with points on this line will travel the entire length of the quadrupole assembly without interruption. The mass resolution is an inverse function of the width of the stable region. The optimum mass separation is obtained when the mass-scan line is made to intersect the tip of the stability diagram. At that point, the U/V ratio is 0.167, and ions of only a single m/q value pass through the quadrupole assembly.

▶ **Example 3.4** Suppose that a quadrupole mass filter is operated at the tip of the first stability region, where the Mathieu parameters a_u and q_u are equal to 0.237 and 0.706, respectively. If the maximum rf potential is 6000 V, the rf frequency is 0.5 MHz, and r_0 is 1.0 cm, what maximum m/z value can be analyzed with this instrument?

Solution Because the value of V is known, Eq. (3.9) can be used to calculate the m/z value. Here q_u is 0.706. After rearranging Eq. (3.9), we get

$$\frac{m}{z} = \frac{4\,eV}{q_u \omega^2 r_0^2} = \frac{4\,eV}{q_u (2\pi f)^2 r_0^2}$$

$$= \left[\frac{(4)(1.602 \times 10^{-19}\text{ C})(6000\text{ V})}{(0.706)(2 \times 3.14 \times 0.5 \times 10^6 \text{s}^{-1})^2 (0.01\text{ m})^2}\right]\left(\frac{1\text{ u}}{1.66 \times 10^{-27}\text{ kg}}\right) = 3327\text{ u}$$

Remember: $C \cdot V = J = kg \cdot m^2/s^2$.

3.4.2. Performance Characteristics

Quadrupoles are compact benchtop instruments, and the mass range, mass resolution, and mass accuracy are significantly lower than the corresponding parameters of sector instruments. A practical limit of upper m/z is 4000. This upper limit can, however, be extended by increasing the amplitude of the rf signal, decreasing its frequency, and using narrower-diameter rods. An m/q value of 45,000 has been achieved by reducing the operating frequency to 262 kHz, but at the cost of resolution [7]. The primary use of quadrupole mass filters is for the analysis of low-mass compounds. With the advent of ESI and its inherent ability to generate multiply charged ions, however, high-mass bipolymers can also be analyzed conveniently with quadrupoles.

In a quadrupole field, resolution is proportional to the square of the number (n) of rf cycles to which an ion is subjected (i.e., $\propto n^2$) and is given by the expression

$$R = \frac{mf^2 L^2}{qV} \qquad (3.10)$$

where V is the accelerating potential and qV is the kinetic energy of the ions. Thus, the resolution of a quadrupole device can be improved by increasing the frequency of the rf signal, operating it at lower acceleration potential (i.e., by decreasing the initial velocity of ions), and increasing the length of the rods. Because of the field imperfections of the rods, a practical limit of resolution is unit mass. With the use of precisely machined hyperbolic rods, however, higher-resolution (ca. 15,000) can be achieved. The operation of quadrupoles in higher-stability regions (e.g., second and third regions) can also improve resolution [8,9]. For example, the isobaric $^{56}Fe^+$ and $^{40}Ar^{16}O^+$ ions have been separated at a resolving power of 9000 (FWHM) [9]. Resolution of a quadrupole mass filter also depends on the number of oscillations that ions make within the quadrupole field (i.e., $R \propto n^2$). Therefore, the ions that enter the quadrupole structure must be accelerated to very low potentials (usually, 10 to 20 V) so that their velocity can be kept low. Reflecting the ions back and forth within the quadrupoles can also increase the number of oscillations and subsequently, the resolution [10].

Low cost, mechanical simplicity, high scan speeds, high transmission, increased sensitivity, independence of the initial energy distribution of ions, and linear mass range are other useful attributes of a quadrupole mass filter. In addition, the ion source can be operated at relatively higher pressures, making practically feasible the coupling of quadrupoles with inductively coupled plasma (ICP), gas chromatography (GC), liquid chromatography (LC), and capillary electrophoresis (CE). In addition, because of the compactness and lightweight features, quadrupoles are ideally suited for field applications, study of the upper atmosphere, and onboard space exploration research.

3.4.3. RF-Only quadrupole

For many applications, quadrupoles are operated in the rf-only mode. When the dc component of Φ_0 is made zero (i.e., $U = 0$), then under all operating conditions the Mathieu coordinate a_u also becomes zero. At that point the mass-scan line lies along the q_u-axis in the stability diagram (see Figure 3.9), and the quadrupole behaves as a wideband mass filter, meaning that ions of a wide m/q range, for which q_u is <0.908, can be contained within the rf-only quadrupole field. The rf-only quadrupole has been used as an intermediate reaction region in modern triple-quadrupole tandem mass spectrometers; as an ion-containment region and as an ion guide in TOF, QIT, LIT, and FT–ICR–MS instruments; and as pre- and postfilters in high-performance quadrupole mass analyzers.

3.5. TIME-OF-FLIGHT MASS SPECTROMETERS

A *time-of-flight* (TOF) *mass spectrometer*, one of the simplest mass-analyzing devices, is currently in high demand. In combination with matrix-assisted laser desorption/ionization (MALDI), it has emerged as a mainstream technique for the analysis of biomolecules. For further reading, several reviews [11–16] and the book cited at the end of the chapter may be consulted.

3.5.1. Working Principle

The basic principle of ion separation by TOF mass spectrometry was conceptualized by Stephens in 1946 [17]. This mass analyzer consists of a long (ca. 100 cm in length) field-free flight tube in which ions are separated on the basis of their velocity differences. A short pulse of ions of defined kinetic energy [given by Eq. (3.2)] is dispersed in time when it travels a long flight tube (of length L). The velocities, v, of ions are an inverse function of the square root of their (m/q or m/z) values:

$$v = \sqrt{\frac{2qV}{m}} \tag{3.11}$$

Therefore, the lower-m/q ions travel faster and reach the detector earlier than the higher m/q ions. Thus, a short pulse of ions is dispersed into packets of isomass ions (Figure 3.10). The time of arrival of an ion is given by

$$t = \frac{L}{v} = L\sqrt{\frac{m}{2qV}} \tag{3.12}$$

The measured arrival times of all ions provides a time spectrum that is converted into a mass spectrum by calibrating the instrument. A generally accepted calibration equation is

$$\frac{m}{q} = at^2 + b \tag{3.13}$$

which is valid with currently popular delayed-extraction TOF instruments (see Section 3.5.2). In these instruments, the linear relation between m/q and t^2 no longer exists. In this equation, a is the constant of proportionality between the arrival time of an ion and its m/q, and b is a time offset that arises from the difference in time between the ion extraction and the data acquisition start pulse. These constants are determined with two different ions of known m/q values.

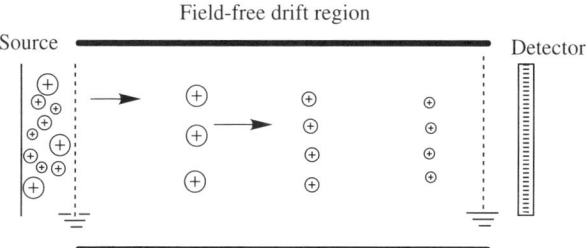

Figure 3.10. Principle of mass separation by a time-of-flight mass analyzer. Ions are separated on the basis of their m/q; high m/q ions (big circles) travel slower than the lower m/q ions (smaller circles). (Reproduced from C. Dass, *Principles and Practice of Biological Mass Spectrometry*, Wiley-Interscience, New York, 2001.)

Because a defined start–stop signal is required for the measurement of time, an essential prerequisite in the TOF operation is that all ions enter the flight tube at exactly the same time. This arrangement would also avoid any artifact left behind from the previous ionization event. For this reason, TOF instruments are optimally combined with pulsed-mode ion sources such as ^{252}Cf plasma desorption or MALDI. Alternatively, pulsing the accelerating potential can provide a pulsed ion beam from continuous ion beam sources (e.g., electron ionization, and electrospray ionization).

▶ **Example 3.5** Suppose that a singly charged toluene ion is subjected to an acceleration potential of 10 kV. How much time will it take to travel through the 100-cm flight tube of a TOF instrument?

Solution

$$v = \sqrt{\frac{2zeV}{m}} = \sqrt{\frac{(2)(1)(1.602 \times 10^{-19} \text{ C})(10000 \text{ V})}{(92)(1.66 \times 10^{-27} \text{ kg})}} = 1.45 \times 10^5 \text{ m/s}$$

Remember: $C \cdot V = J = kg \cdot m^2/s^2$.

$$t = \frac{L}{v} = \frac{1.0 \text{ m}}{1.45 \times 10^5 \text{ m/s}} = 6.9 \times 10^{-6} \text{ s} = 6.9 \text{ μs}$$

In the past, conventional TOF–MS, popularly known as *linear TOF–MS*, found limited application owing primarily to its poor resolution (<500) and incompatibility with continuous-beam ion sources. The limitation of poor resolution has been circumvented to some extent by the current developments described next.

Mass Resolution In TOF–MS, mass resolution is related to the temporal width of the isomass ions packet when that packet arrives at the detector (i.e., $R = t/\Delta t$). In the ion source, ions are accelerated out of the source region with inherent dispersion in time (instant of ion formation), space (location of ion at the time of acceleration), and velocity (owing to differences in the initial kinetic energy of ions). These are the three primary factors that limit the resolution in a TOF instrument [11, 13, 16]. The initial kinetic energy (KE) of ions (i.e., KE before acceleration) is given by $KE = \frac{1}{2}mv_0^2$, where v_0 is the initial velocity, which will be in a random direction; after acceleration, $KE = qV + \frac{1}{2}mv_0^2$. The temporal dispersion creates uncertainty in the ions' arrival time at the detector. The contribution of this factor can be minimized by the use of a very short ionization pulse and/or a fast-rise ion-extraction pulse, and also by increasing the flight path. Multiturn and multipass research TOF mass spectrometers are available to increase the ion flight path without increasing the size of the instrument significantly [18,19]. The spatial distribution gives rise to differences in kinetic energies and source exit times of ions after their acceleration. Consider two same-mass ions formed at different locations in the ion source. Because the ion formed to the left of the central axis is subjected to a higher potential, it will be accelerated to a higher velocity than the ion formed to the right of the central line, but it will exit the source later. Convergence of spatially dispersed ions is achieved at the *space focus plane*, where the faster later ions catch up with the slower earlier ones. The spatial distribution of MALDI-formed ions is inherently low because the plane of ion formation is well defined.

The initial kinetic energy spread $\frac{1}{2}mv_0^2$ is the dominating factor that restricts resolution in TOF instruments. The spatial distribution and initial kinetic-energy spread of ions are minimized by incorporating delayed-extraction and reflectron devices. With gas-phase ionization techniques, the initial kinetic energy variations in both magnitude and direction gives rise to a resolution factor called the *turnaround time*, the extra time that an ion traveling initially away from the exit slit must take to exit the ion source. Because this ion must reverse its direction before it begins its journey toward the detector, it lags behind other ions of identical initial velocity that were moving toward the exit slit. Longer flight tubes and longer flight times can reduce the effect of the turnaround time. The difference in the arrival times of ions that differ in mass by 1 u (say, 2000 and 2001 u) is very short (in nanosections). Therefore, the mass resolution of TOF instruments is also limited by the time-resolving power of the ion detection system.

3.5.2. Delayed Extraction of Ions

Delayed extraction (DE), the principle of which was first enumerated by Wiley and McLaren in 1956 in the form of *time-lag focusing* [20], is one way to improve the resolution of linear TOF mass spectrometers [21–23]. This procedure uses a dual-stage ion-extraction optics with two distinct extraction and acceleration regions (Figure 3.11). During the ionization pulse, no potential is applied to the extraction region. Therefore, ions drift in this region in a field-free environment with their initial velocities v_0. After a short delay of a few hundred nanoseconds, the acceleration potential is applied to extract the ions from the source. During this delay period, the slow-moving ions lag behind the fast-moving ions. As mentioned above, ions near the repeller electrode (i.e., farther from the extraction grid) will be subjected to a greater electrical potential than will ions closer to the extraction grid. Because of this difference in the accelerating field, the ions that were lagged behind are accelerated to a higher velocity. With a proper setting of the delay time and amplitude of the extraction pulse, all ions of a particular mass but of different initial kinetic energy can be made to reach the detector at the same time.

A DE device has also been useful to gain knowledge of the structure-specific fragment ions [23,24]. That information is missing in a conventional linear TOF mass spectrum because the molecular ions are usually promptly extracted from the ion source before they have a chance to fragment. During the delayed-extraction

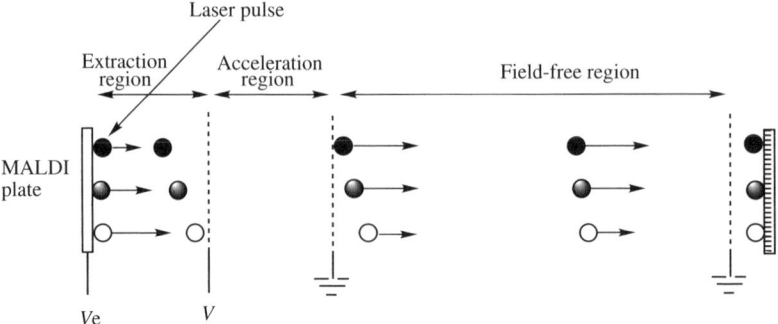

Figure 3.11. Principle of time-lag focusing. Ions with initial velocity dispersion drift to a different extent in the field-free environment of the extraction region; the slower-moving ions lag behind the faster-moving ions. After a short delay, when an acceleration pulse is applied, the slower ions are subjected to a higher field than the faster ions and thus are accelerated to a higher velocity than the faster ions. In the FFR region, all isomass ions will get closer to each other, and ultimately will arrive at the detector simultaneously.

period, the energetic ions are given the opportunity to fragment and subsequently are mass-analyzed.

3.5.3. Reflectron TOF Instrument

A *reflectron* is an energy-correcting device that can minimize the effects of initial spatial and energy spreads [25]. This electrostatic mirror consists of grids and a series of ring electrodes, each with a progressively increasing repelling potential (Figure 3.12). The mirror is placed at the end of the flight tube [i.e., the first field-free region (FFR) of length L_1] and works on the principle that after entering this device, ions are slowed down by the repelling electric field until they come to rest; subsequently, their direction of motion is reversed, and they are

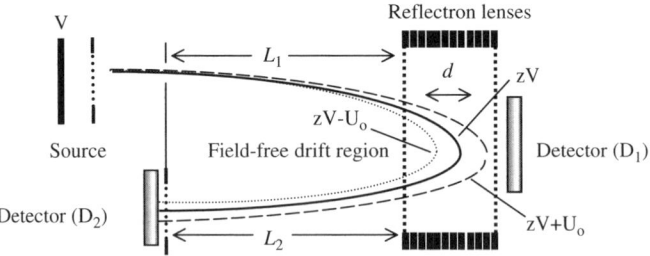

Figure 3.12. Reflectron time-of-flight mass analyzer. All ions of the same mass, but that differ in kinetic energy, are made to arrive at the same time at a detector (D_2) that is located at the end of the second field-free region (L_2). (Reproduced from C. Dass, *Principles and Practice of Biological Mass Spectrometry*, Wiley-Interscience, 2001.)

reaccelerated into a second FFR of length L_2. In principle, an ion with excess energy $qV + U_0$ (where $U_0 = \frac{1}{2}(mv_0^2)$) will arrive earlier at the reflectron but will spend more time in the reflecting field as it penetrates to a greater depth (d) than does an ion with average energy qV. Thus, with a proper setting of ring voltages, the shorter flight time of the faster ion in the drift regions is compensated by this extra time in the mirror. As a consequence, all ions of the same m/q value arrive simultaneously at the detector placed at the end of the second FFR; the result is an improvement in mass resolution. The additional flight path due to the second FFR also contributes to improved resolution. The mirror also recreates the space focus plane at a useful distance from the source. It allows use of a high extraction field, which reduces the turn around time.

Linear-field reflectrons with a single or dual stage and nonlinear-field reflectrons have been described in the literature [25–27]. A single-stage reflectron consists of an entrance grid electrode and a series of ring electrodes and uses a single retarding or reflecting field to provide first-order correction of the kinetic-energy distribution. The dual-stage device contains two linear retarding-voltage regions that are separated by an additional grid.

3.5.4. Orthogonal Acceleration TOF Mass Spectrometer

The *orthogonal acceleration* (oa) feature of a TOF mass analyzer enables it to be used with continuous-beam ion sources [28–31]. The ion beam from the external source enters an ion acceleration region from a direction perpendicular to the main axis of the TOF instrument (see Figure 3.13). A short pulse of an orthogonal accelerating field is applied to eject the ions efficiently in a section of

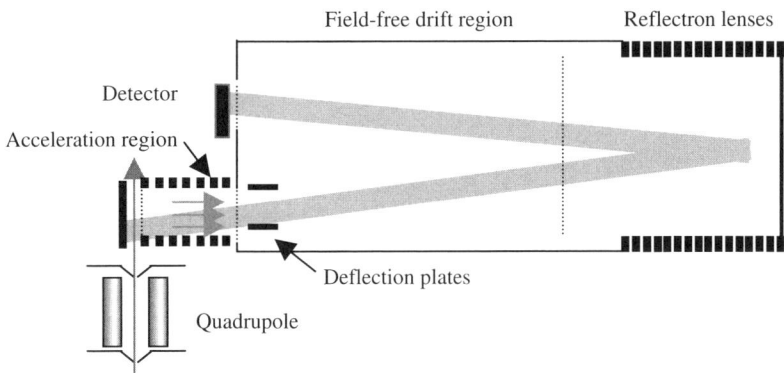

Figure 3.13. Principle of an orthogonal time-of-flight mass analyzer. A beam of ions from a continuous ionization source, such as ESI, are transmitted between the orthogonal ion extraction plate and grid. A segment of the beam is then pushed into the field-free region by a pulse in the orthogonal direction. (Reproduced from C. Dass, *Principles and Practice of Biological Mass Spectrometry*, Wiley-Interscience, 2001.)

the beam out of the ion-sampling region. This approach provides high efficiency (a high duty cycle) to gate ions from an external source. It also reduces the spatial and energy spreads of source-formed ions because of minimization of the turnaround effect [32]. Coupling an external continuous-beam ion source with the ion-sampling region of an oa–TOF mass spectrometer is achieved via tube lenses, rf-only multipoles, a QIT, or a linear ion trap. In each case, the intent is to minimize an ion beam's divergence and to increase ion-transport efficiency.

3.5.5. Performance Characteristics

Time-of-flight mass spectrometers have the number of attractive features. First, they have a potential of an unlimited mass range. Restriction to the upper mass limit is imposed, however, owing to the difficulty in the detection of high-m/q ions, broadness of peaks caused by isotopic distribution, and use of a reflectron. High ion-transmission efficiency, extremely high spectrum-acquisition rate, multiplex-detection capability, simplicity in instrument design and operation, and low cost are other strong features of these instruments. High transmission efficiency and multiplex detection both translate into much higher detection sensitivity than that of scanning-type instruments. Mass resolution is also reasonably high. A major asset of TOF mass spectrometers is their ability to record a complete mass spectrum in intervals as short as 25 μs. This feature ensures that mass spectra of even transient species are free of spectral skew. Because of these useful features, TOF–MS has become a workhorse for the analysis of macromolecules such as proteins, carbohydrates, and oligonucleotides. These instruments are used routinely to determine the molecular mass of proteins of >300-kDa mass with an accuracy >0.1% and with detection sensitivity in the attomole to femtomole range; also for peptide mapping, assessing glycosyl, phosphoryl, and other posttranslational heterogeneity; and for the determination of cleavage sites in protein processing. TOF instruments can also be used for accurate mass measurement with <5 ppm accuracy of organic compounds that have molecular mass <500 u.

3.6. QUADRUPOLE ION-TRAP MASS SPECTROMETERS

The *Paul trap*, popularly known as a *quadrupole ion trap* (QIT), was introduced in 1958 by Paul and colleagues [33]. This contribution was recognized by the award of the 1989 Nobel Prize for Physics to Wolfgang Paul. Because it is a three-dimensional analog of a quadrupole mass filter, it is also called a three-dimensional ion trap to distinguish it from the two-dimensional ion trap described in Section 3.7. The QIT became popular as a mass spectrometer after development of the *mass-selective instability* mode of mass analysis by Stafford and co-workers [34]. For further reading, several review articles [35–41] and books are cited at the end of the chapter.

3.6.1. Working Principle

Unlike beam-type instruments, mass separation in a QIT is achieved by storing the ions in the trapping space and by manipulating their motion in time rather than in space. This task is accomplished with an oscillating electric field that is created within the boundaries of a three-electrode structure. The mass spectrum is acquired by changing the applied rf field to eject ions sequentially from the trapping field.

The three-electrode structure of a QIT consists of a doughnut-shaped central ring electrode and two identical end-cap electrodes, each with a hyperbolic geometry (Figure 3.14). One of the end-cap electrodes has a small aperture through which an electron beam (for in situ ionization of the sample molecules) or an externally formed ion beam can be gated periodically into the trap. The other end-cap electrode has several perforations for the ejection of ions toward an externally located detector. The three-dimensional quadrupole field is created by applying a potential $\Phi_0 = U - V \cos \omega t$ to the ring electrode and maintaining the end-cap electrodes at ground potential. Here, U and V are the amplitudes of the dc and rf potentials, respectively, and ω is the angular frequency. For a pure quadrupole field, the dimension of each electrode is chosen so that the internal radius, r_0, of the central ring electrode is related to the closest distance, z_0, from the center to one of the end-cap electrodes by the expression $r_0^2 = 2z_0^2$. Assuming a cylindrical symmetry, the potential at any point within the space bounded by the three electrodes in terms of the radial (r) and axial (z) coordinates is given by

$$\Phi_{r,z} = \frac{U - V \cos \omega t}{2} \frac{r^2 - 2z^2}{r_0^2} + \frac{U - V \cos \omega t}{2} \quad (3.14)$$

The radial coordinate r is related to the x- and y-coordinates by $r^2 = x^2 + y^2$. The quadrupole field forces ions that have a broad m/q range to be trapped within the boundaries of the electrodes. Helium, at a pressure of 10^{-3} torr, is introduced into the trap to cool the ions collisionally and to confine them in the center of the trap. The trapped ions precess in the trapping field with a frequency that is

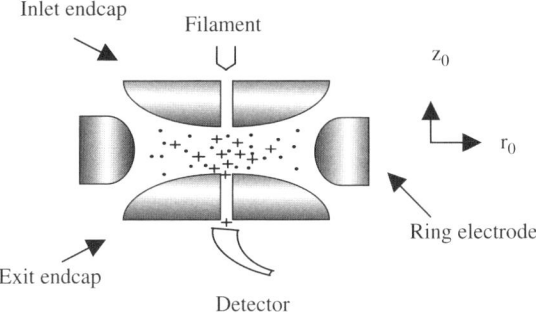

Figure 3.14. Quadrupole ion trap mass analyzer. (Reproduced from C. Dass, *Principles and Practice of Biological Mass Spectrometry*, Wiley-Interscience, 2001.)

dependent upon their m/q ratio. The mass spectrum of the trapped ions is obtained by using the *mass-selective instability* (also known as *mass-selective axial ejection*) mode of mass analysis. Increasing the magnitudes of the dc and rf voltages and the frequency of the rf signal, either singly or in combination, forces ions of higher m/q values to become sequentially unstable in the axial direction and to be ejected out of the trap for external detection.

The Mathieu equation (3.7) of the form discussed in Section 3.4.1 can also be written to describe the motion of ions in a three-dimensional quadrupole field. The corresponding Mathieu coordinates a_z and q_z are given by

$$a_z = -2a_r = \frac{-8zeU}{m\omega^2 r_0^2} \quad (3.15)$$

$$q_z = 2q_r = \frac{4zeV}{m\omega^2 r_0^2} \quad (3.16)$$

The secular frequency of ion motion in the xy-plane is given by $\beta_{zr}\omega/2$, where β_{zr} is a proportionality constant that can be calculated from a_z and q_z. The term β_z is equal to $q_z/\sqrt{2}$ for the mass-selective instability mode of operation.

Commercial ion traps are not of ideal geometry, but are stretched, for which the expression $r_0^2 = 2z_0^2$ no longer holds. The values of the Mathieu parameters for such traps are obtained by replacing $2r_0^2$ with $r_0^2 + 2z_0^2$ in Eqs. (3.15) and (3.16). The plot of a and q gives several Mathieu stability regions (shown in Figure 3.15).

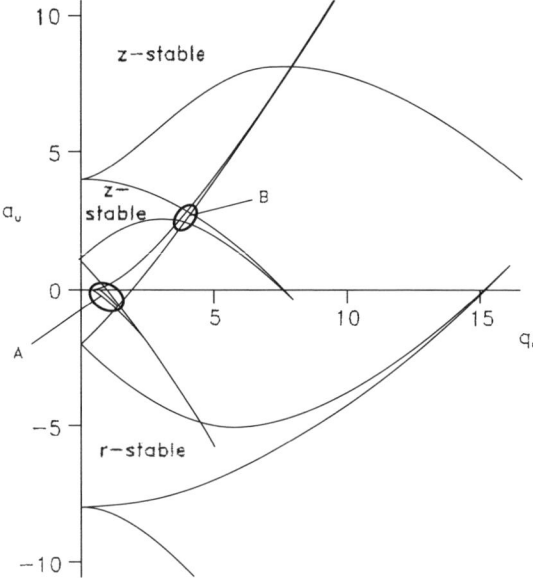

Figure 3.15. Stability diagram in (a_z, q_z) space for the quadrupole ion trap in the r- and z-directions. A and B are regions of simultaneous stability. (Reproduced from ref. 38 by permission of Wiley-Interscience, copyright © 1997.)

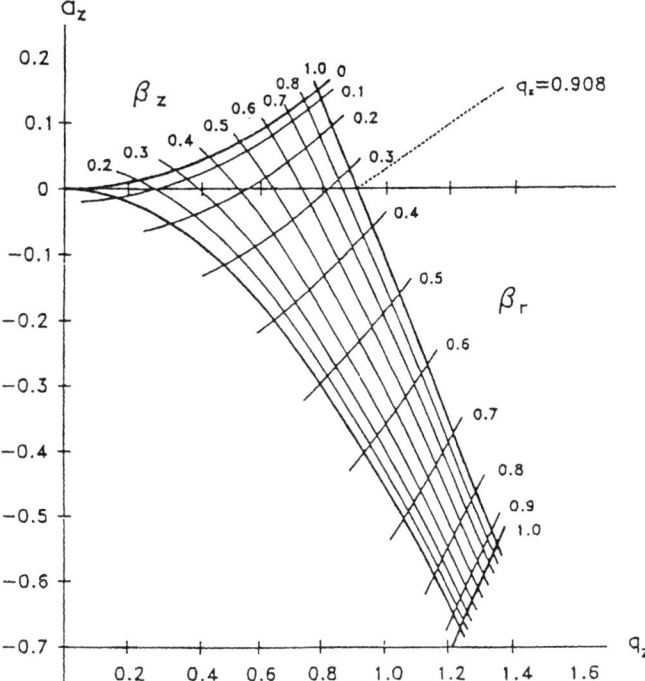

Figure 3.16. Region of simultaneous stability closest to the origin for the quadrupole ion trap in the r- and z-directions. (Reproduced from ref. 38 by permission of Wiley-Interscience, copyright © 1997.)

Ions of a certain m/q range can be confined within the volume of the trap only when their radial and axial trajectories are both stable simultaneously (e.g., the circled regions A and B in Figure 3.15). The region A closest to the origin, shown in expanded form in Figure 3.16, is of practical interest.

3.6.2. Operational Modes

Mass-Selective Instability Mode In a simple and most popular mode of mass analysis, known as *mass-selective axial instability*, only an rf voltage is applied to the ring electrode and the end caps are held at ground potential (i.e., U and a_z are both zero), so that the containing field is purely oscillatory. Most commercial instruments employ this mass analysis approach. In this *rf-only mode* of operation, the locus of all trapped ions is on the q_z-axis. Consequently, the trap behaves as a total ion-storage device. At a given value of V, the storage voltage, all ions above a certain m/q value are trapped within the quadrupole field. The maximum value of q_z is 0.908 and corresponds to the low-mass cutoff value, and the m/z is given by

$$\frac{m}{z} = \frac{4eV}{q_{\max}\omega^2 r_0^2} \tag{3.17}$$

By ramping the rf voltage V, ions of successively increasing m/z are forced to adopt unstable trajectories and are ejected out of the trap in the axial direction for external detection.

Resonant Excitation All ions within the boundaries of the trap have a natural tendency to undergo oscillation with a characteristic secular frequency in the radial and axial directions. This natural motion can be excited when it is brought in to resonance with a bipolar supplementary rf potential (also known as *tickle voltage*) applied to the end-cap electrodes. The energy absorbed by an ion from the resonant signal promotes it to higher orbits. One of the uses of resonant excitation is the ejection of ions from the trap along the axial direction when the axial amplitude of the oscillatory motion exceeds the geometry of the ion trap. This operational mode, known as *resonance ion ejection*, is used for the selective *ion isolation* of species of interest [36]. To isolate an ionic species, the amplitude of the main rf signal is scanned first in the reverse direction and then in the forward direction while the supplementary resonance signal is being applied. A reverse scan ejects ions with m/q values greater than those of the ion of interest, whereas a forward scan removes smaller m/q ions.

Mass-Selective Stability Mode This mode is analogous to the operation of a quadrupole mass filter [41]. The quadrupole field is generated by applying a potential Φ_0 to the ring electrode and $-\Phi_0$ to the end-cap electrodes. The applied potential is chosen such that the q_z and a_z values are close to the apex of the stability diagram, the consequence of which is that only one species is trapped at a time. This step can be used to isolate a selected ion for multistage mass spectrometry (MS^n) or to obtain a mass spectrum by sweeping the dc and rf voltages while holding their ratio constant.

Ion Detection A conventional approach to ion detection is to place a detector just outside the end-cap electrode with more openings. Another means of ion detection is to monitor the image current. Using the principle of resonant excitation, the ions are promoted to higher oscillations with the use of a supplementary fast dc pulse to bring them closer to the detection plates. This nondestructive procedure of detection of ion current is conceptually similar to that used in FT–ICR–MS (Section 3.8.1) [42].

3.6.3. Performance Characteristics

Like quadrupole mass filters, QIT mass spectrometers are compact, simple to operate, relatively inexpensive, and have high scan speed. Because of their small size and low pumping requirements, these instruments have gained wide acceptance for field applications. High efficiency and sensitivity are other hallmarks of ion-trap instruments. An ion trap is also a useful device to conduct multistage MS experiments (MS^n with $n = 12$ has been envisioned) and gas-phase ion–molecule reactions. It can serve as an effective detector for high-resolution separation techniques (such as GC, HPLC, and CE). Its major drawback is its poor mass measurement accuracy. Another serious shortcoming is its limited ion storage capacity, which results in a lower dynamic range. The ion-trapping efficiency for externally injected ions is also poor. Also, it is not an ideal system for quantitative analysis.

Extension of Mass Range and Resolution Although under normal operating conditions the upper m/q limit (ca. 650) and mass resolution are low, both can be extended by the manipulation of certain operating parameters. As can be envisioned from Eq. (3.17), the upper mass limit can be extended by operating the trap at higher rf voltages, reducing the rf frequency, reducing the dimensions (i.e., r_0), and forcing instability at lower q_z values than the usual stability limit of 0.908. On practical grounds, however, only the last approach yields promising results. With the resonance ion ejection, the ions are forced to adopt unstable trajectories at q_z values that are lower than 0.908, resulting in extension of the mass range of analysis to much higher values [43].

Although the mass resolution of commercial instruments is poor (unit mass resolution), substantial improvements can be realized on research instruments by reducing the scan speed and the frequency and amplitude of the resonance ejection signal [44,45]. For example, a resolution of 1.13×10^6 was achieved for a cluster of CsI ions at 3510 u with a 2000-fold decrease in the scan speed [44]. The reduction of the scan rate to 0.1 m/q unit/s has led to a further improvement in resolution to 1.2×10^7 for m/q 614 [45]. A *zoom scan*, in which a narrow window of masses is scanned, is another alternative to enhance resolution.

External Ion Injection The analysis of volatile samples can be accomplished by in situ ionization with a gated electron beam. This mode of ionization is, however, not applicable to the analysis of biological macromolecules. For such compounds, the two modern ionization methods, ESI and MALDI, must be coupled with QITs [46–49]. As with oa–TOF instruments, this coupling is achieved with the help of ion guides, which consist of either tube lenses or rf-only multipoles. One such arrangement for the injection of ions is shown in Figure 3.17.

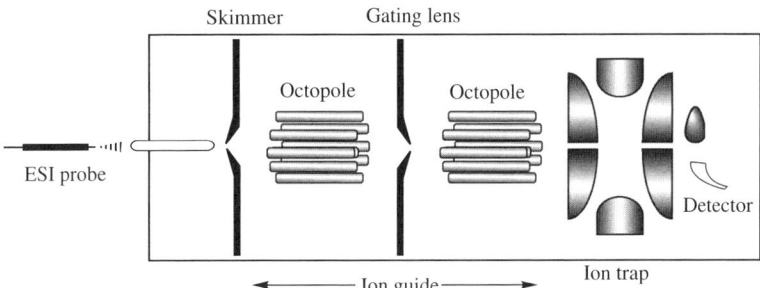

Figure 3.17. Schematic representation of external ion injection of ESI-produced ions into a quadrupole ion trap.

The ESI-produced ions are collected and transferred through a skimmer and two rf-only octopole assemblies. The space-charging effect (i.e., a buildup of excessive charge) is a serious problem in ion traps. Therefore, there is a need to restrict the number of ions entering the trap. In Figure 3.17, a gating lens is added to control the transfer of ions to an acceptable value.

3.7. LINEAR ION-TRAP MASS SPECTROMETERS

A quadrupole linear ion trap (LIT or LTQ), also known as a two-dimensional quadrupole ion trap, is a newer addition to the family of mass spectrometers. In essence, it is identical in construction to the quadrupole mass analyzer (i.e., it is also made of four parallel rods and uses rf and dc potentials for mass separation and analysis). Its operational modes are, however, similar to those of the three-dimensional ion trap discussed in Section 3.6. A typical three-section design is shown in Figure 3.18. It consists of two pairs of hyperbolic rods, each segmented into three axial sections. The central section is about three times larger than the two end sections. A small slit is cut along the length of one of the rods of the central section to eject ions for their external detection. In each section, the opposite rods are connected electrically. Ions are trapped in the axial direction (yz-plane) by applying three separate dc voltages to the three sections of the trap; for radial (in the xy-plane) trapping of ions, the rf trapping field is generated by applying rf potentials (known as the *main rf voltage*) between the x- and y-electrode pairs. In addition, two phases of supplemental ac voltage are applied to the x-electrodes (exit rods) for ion activation, isolation, and ejection. The Mathieu coordinates a_z and q_z and the fundamental frequency of ion motion are given by Eq. (3.15) to (3.17), respectively. As with the three-dimensional trap, mass

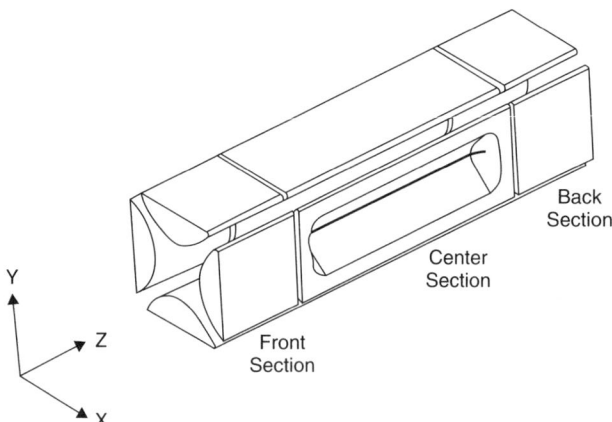

Figure 3.18. Basic design of a two-dimensional linear ion trap. (Reproduced from ref. 50 by permission of Elsevier Science, copyright © 2002 American Society for Mass Spectrometry.)

analysis of the trapped ions is performed by using the mass-selective instability mode with a radial ejection of the trapped ions. To perform this operation, the main rf voltage is increased at a constant rate so that ions of increasing mass become successively unstable in the radial direction and are ejected from the slots in the exit rods. The supplemental ac signal is also used for resonance excitation and ejection of ions during the main rf voltage scan [50]. Resonance excitation occurs when the frequency of the supplemental ac field coincides with the fundamental frequency of the trapped ions, leading to an increased radial amplitude of ion motion and subsequent ion ejection. In a commercial design, two parallel slots are cut in the central section, where two detectors are placed to double the detection sensitivity of the radially ejected ions.

A one-section design of an LIT has also been described [51]. This device also contains two pairs of x- and y-electrodes and two entrance and exit aperture lenses (z-electrodes). Ions are trapped by the rf voltage that operates in the radial direction through x- and y-electrodes and by the dc voltage that operates in the axial direction through the aperture lenses. For mass analysis, ions are axially ejected mass selectively (rather than radially as in the three-section design) through the exit aperture by scanning the rf potential and auxiliary ac voltages that are applied to the x- and y-electrodes. The rf-only central quadrupole and the rf/dc third quadrupole of a triple-sector quadrupole can both be made to perform as a linear ion trap [52].

Axial ejection of ions can also be achieved by the axial resonant excitation (AREX) procedure [53]. For this operation, a set of vane lenses are inserted between each set of quadrupole rods to create a dc potential that is harmonic along the central axis of the linear trap. AREX LIT is shown in Figure 3.19. It consists of four quadrupole rods (x- and y-electrodes), an in-cap lens, an end-cap lens, and eight vane lenses. The latter are divided into two sets, front and rear, each consisting of four lenses, and each set is maintained at the same dc potential. As usual, ions enter the LIT through a pretrap and get trapped in it axially by applying a potential between in-cap and end-cap electrodes, and radially between quadrupole rods. Within the harmonic potential, ions of a specific m/q range can

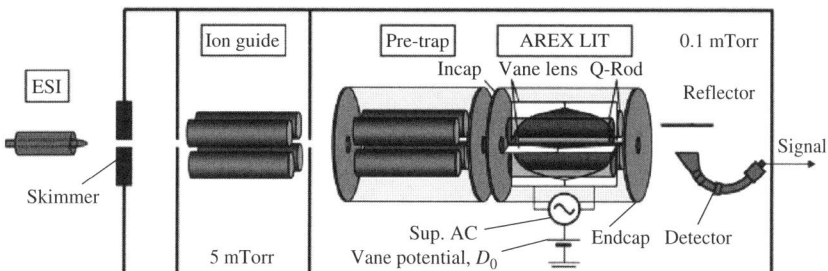

Figure 3.19. Schematic of an AREX linear ion trap. In this design, ions are axially ejected by inserting two sets of eight vane lenses into quadrupole electrodes. (Reproduced from ref. 53 by permission of Elsevier Science, copyright © 2006 American Society for Mass Spectrometry.)

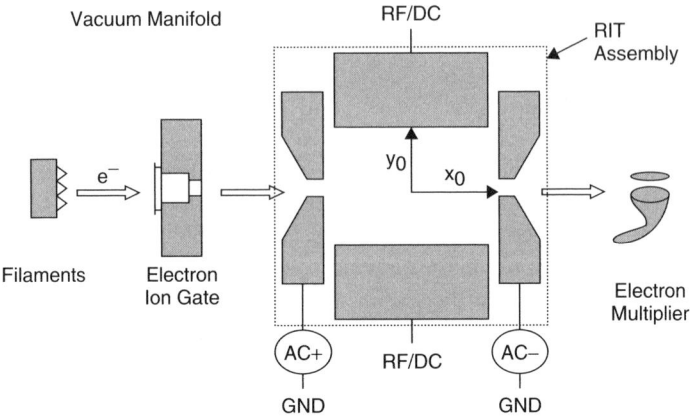

Figure 3.20. Basic design of a rectilinear ion trap. (Reproduced from ref. 54 by permission of the American Chemical Society, Washington, DC. copyright © 2004.)

be resonantly excited in the axial direction with a superimposing supplemental ac field applied to the vane lenses to facilitate their ejection in the axial direction. This procedure provides much higher ejection efficiency at a higher scan rate than those of the two methods discussed above.

An LIT has the advantage over a three-dimensional ion trap of much-improved trapping efficiency, increased ion-storage capacity, improved ion-ejection efficiency, faster scan speeds, and improved detection sensitivity.

3.7.1. Rectilinear Ion Trap

The *rectilinear ion trap* (RIT) is a simplified version of an LIT [54]. The x- and y-electrodes have a rectangular geometry (Figure 3.20) instead of the hyperbolic geometry of the LIT. The RIT also contains two apertureless z-electrodes, which are supplied with a dc potential for the axial trapping of ions. For the radial trapping of ions, the trapping field is created by applying an rf signal with a floated dc voltage to the y-pair of electrodes and grounding the x-electrodes. As with the three-section LIT, the x-electrodes of the RIT also contain slits for the entrance and exit of ions. Mass analysis involves the ejection of the trapped ions with the sweep of the rf signal and ac potential that are applied to the x-electrodes.

3.8. FOURIER-TRANSFORM ION CYCLOTRON RESONANCE MASS SPECTROMETERS

Fourier transform ion-cyclotron resonance mass spectrometry (FT–ICR–MS) is also an ion-trap technique. The concept of ICR spectroscopy was first described in 1930 by Lawrence and Edlefsen [55] and was later developed as a mass spectrometry in 1949 by Sommer and colleagues [56]. The development of FT–ICR–MS took place in 1974 through the efforts of Comisarow and Marshall [57]. Since then, FT–ICR–MS has matured into a state-of-the-art high-resolution mass spectrometry instrument for the analysis of organic and biomolecules. Several reviews on this subject are worth reading [58–63].

The heart of this instrument is a cell (also known as a *Penning trap*), which is placed in a strong magnetic field. Ions are confined laterally in this trap by a static magnetic field and axially by a static electric field. Ions are excited by a broadband rf pulse to a coherent orbital motion, and from the frequency of this motion, the m/q of the trapped ions can be obtained. Over a dozen designs of cell geometry exists [64]. The cubic cell is shown in Figure 3.21.

3.8.1. Working Principle

The mass analysis in ICR relies on two basic concepts: First, an ion traveling in a magnetic field will precess at a frequency that is characteristic of its m/q value. Second, energy can be absorbed by the precessing ion from an external rf source when its cyclotron frequency coincides with the rf signal being used. The forces acting on the ion can be described by Eq. (3.3). The time to complete a single revolution ($2\pi r$) is given by

$$t = \frac{2\pi r}{v} = 2\pi \frac{m}{qB} \quad (3.18)$$

(r, v, q, m, and B have been defined in Section 3.3.1), from which the cyclotron frequency ω_c, defined as the number of revolutions per second, is calculated according to

$$\omega_c = \frac{qB}{2\pi m} \quad (3.19)$$

A typical ICR experiment consists of four time-spread events: quenching, ion formation/injection, excitation, and detection. All of these functions are performed in the ICR cell, which consists of three pairs of opposing plates; each pair performs a distinct function: trapping, excitation, or detection of ions [58]. First, a quench pulse is applied before the start of mass analysis to empty the cell of any ions that might have been left from the previous experiment. This task is performed by applying antisymmetric voltages (e.g., +10 V and −10 V) to the trapping electrodes (front and rear plates in Figure 3.21). The ions are either formed in situ in the pulsed mode or are pulsed into the cell from the external source and are trapped by applying a few volts (1 to 10 V) of electric potential to the trapping electrodes (Figure 3.21a). Concurrent with the trapping voltage, a short burst of a bath gas is introduced into the cell. The gas cools the ions and improves the trapping efficiency of the cell. Next, an excitation pulse is applied through the excitation electrodes (side plates in Figure 3.21b) and ions, whose precessional frequency matches that of the excitation pulse, absorbing energy from the external excitation pulse and getting promoted to larger orbits. Finally, the excited

96   MASS ANALYSIS AND ION DETECTION

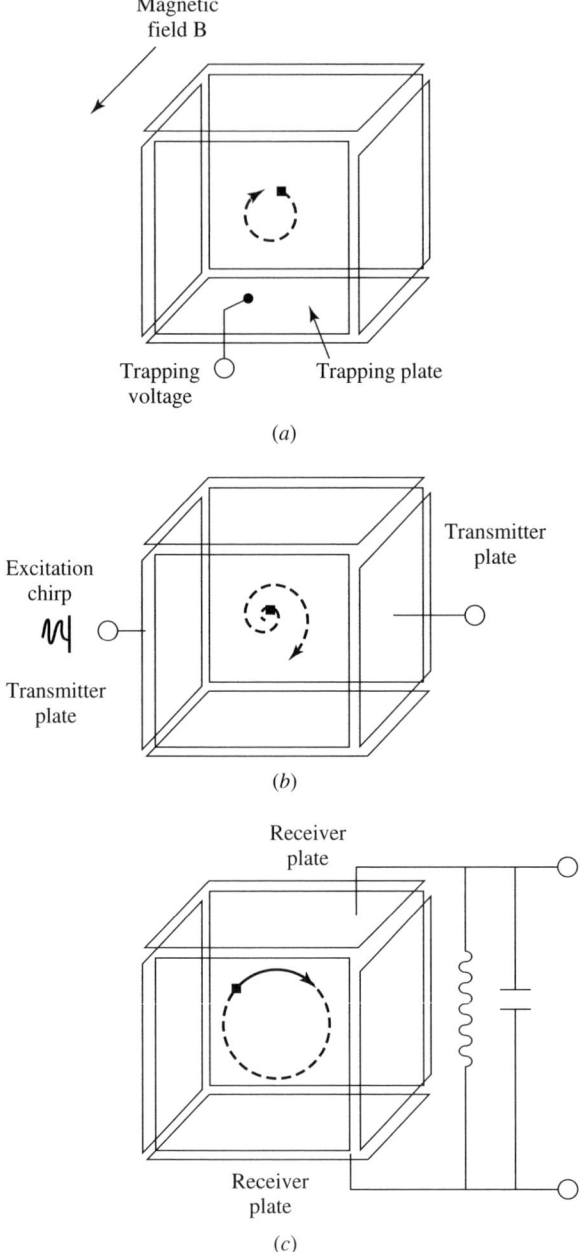

Figure 3.21. Schematic depiction of the operation of FT-ICR/MS. The mass analysis involves three steps: (*a*) ion formation and storage; (*b*) excitation of the trapped ions by an external broad-frequency range pulse; (*c*) detection of the ions by measuring their image current [58].

ions are detected by measuring the amplitude and frequency of the image current in the receiving plates (top and bottom plates in Figure 3.21c) as they orbit the ICR cell.

In the FT–MS version of ICR–MS, the excitation pulse is in the form of a *chirp*, which is a fast sweep of frequencies over a broadband (e.g., in the range 10 kHz to 1 MHz) that can encompass the cyclotron frequencies of the m/q range desired. Ions of a given m/q range, whose cyclotron frequencies fall within the frequencies applied, simultaneously absorb energy from the chirp signal and are forced to move in phase-coherent packets of larger orbital radii; each different m/q ion packet orbits at its own characteristic frequency. When in the proximity of the receiving plates, these coherently moving ion packets transmit a complex rf signal (i.e., the image current) that contains frequency components characteristic of each ion. After passing through an impedance and amplifying circuit, the image current, is converted to a time-domain free-ion decay signal (Figure 3.22a). The application of a Fourier transform converts this complex time-domain signal into a frequency-domain signal. A mass calibration file finally retrieves a normal mass spectrum (Figure 3.22b) from the frequency-domain signal. Thus, the FT–MS version allows simultaneous detection of all ions across a wide mass range. Another common excitation method is a *stored waveform inverse Fourier transform* (SWIFT). In this mode, the m/q range of analysis is first defined and converted to the frequency band required. Finally, use

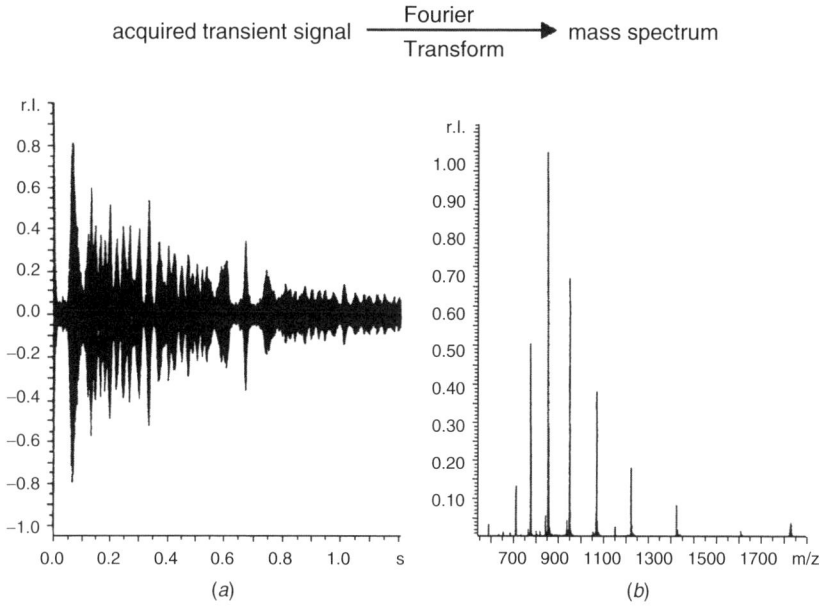

Figure 3.22. (*a*) Time-domain free-ion decay signal and (*b*) Fourier-transformed normal mass spectrum. (Reproduced from ref. 61 by permission of Wiley-Interscience, copyright © 1996.)

of an inverse Fourier transform converts this frequency band to a time-domain excitation signal.

▶ **Example 3.6** Suppose that singly charged toluene ions are trapped in an ICR cell placed in a 2.4-T magnet. Calculate the ICR frequency at which the toluene ion will precesse in the cell.

Solution

$$\omega_c = \frac{Bze}{2\pi m} = \frac{(2.4\ T)(1.602\times 10^{-19}\ C)(1)}{(2)(3.14)(92)(1.66\times 10^{-27}\ kg)} = 4.0\times 10^5\ Hz = 0.4\ MHz$$

(*Note*: $T = kg/s \cdot C$ and $Hz = s^{-1}$.)

External Ion Injection Similar to QIT, the analysis of volatile samples in a Penning trap is accomplished by in situ ionization with a gated electron beam. For the analysis of biomolecules, ESI and MALDI are coupled by using ion guides of the type shown in Figure 3.17. This way externally formed ions are transported into the cell for mass analysis and detection. MALDI can also be performed within the cell by aiming the laser beam at a sample target.

3.8.2. Performance Characteristics

FT–ICR–MS, particularly, is highly promising for high-mass analysis [59,62]. The m/q range is, however, a function of the magnetic field strength. With a commercial 9.4-T instrument, the upper m/q range is 10,000. Another useful attribute of FT–ICR–MS is its ability to provide remarkably high resolving power because of the fact that the physical quantity measured is frequency, which can be determined with high precision. Mass resolution degrades rapidly, however, as the mass of the analyte increases. With an external accumulation of ions from an ESI source, the isotopic fine structure of proteins with molecular mass as high as 15.8 kDa has been observed at 8,000,000 resolving power with a 9.4-T instrument [65]. The mass accuracy of FT–ICR–MS is far superior to that of other mass analyzers. A mass accuracy of 1 ppm is routine for large molecules and much better for ions with $m/q < 300$. The detection limit (in attomoles) of these instruments is also commendable. Other advantages of FT–ICR–MS include multiplex detection (the Fellgett advantage), high scan speed, improved signal-to-noise ratio (S/N), multistage tandem mass spectrometry, and the ability to trap ions for extended periods of time. In addition, FT–ICR–MS is a powerful tool for conducting ion–molecule reactions and for structure elucidation studies. Because of these useful features, FT–ICR–MS in conjunction with ESI has emerged as the most powerful form of mass spectrometry for the analysis of biomolecules. Disadvantages include exorbitant cost, limited dynamic range, and slow spectrum generation when high resolution and mass accuracy are desired.

3.9. ORBITRAP MASS ANALYZERS

A new type of mass analyzer, known as the *orbitrap*, has recently been introduced [66–68]. It works on the principle of an orbital trapping of ions around an axial central electrode with a purely electrostatic field. Orbital trapping was first proposed by Kingdon in 1923 [69]. An orbitrap consists of an axial spindlelike central electrode and a coaxial barrel-like outer electrode (see Figure 3.23). The trapped ions undergo rotation around the central electrode and harmonic oscillations along its length. The m/q values of the trapped ions are related to the frequencies of their harmonic oscillations. Similar to an FT–ICR mass analyzer, ion frequencies are obtained from the image current that is induced between the axial halves of the outer electrode and are converted to mass spectra using fast FT.

The electrostatic field between two electrodes creates a quadro-logarithmic potential distribution inside the trap that is given by

$$U(r,z) = \frac{k}{2}\left(z^2 - \frac{r^2}{2}\right) + \frac{k}{2}(R_m)^2 \ln\left\{\frac{r}{R_m}\right\} + C \qquad (3.20)$$

where r and z are cylindrical coordinates; $z = 0$ is the plane of symmetry of the field, C is a constant, k is the field curvature, and R_m is the characteristic radius. The trajectory of the trapped ions is an intricate spiral that is a combination of rotation around the central electrode and oscillations along this central axis (Figure 3.23). Thus, three sets of frequencies can be identified along the three polar coordinates r, φ, and z: that is, the frequency of radial oscillations ω_r, the frequency of rotation ω_φ, and the frequency of axial oscillations ω. Only the axial frequency, given by

$$\omega = \sqrt{\frac{q}{mk}} \qquad (3.21)$$

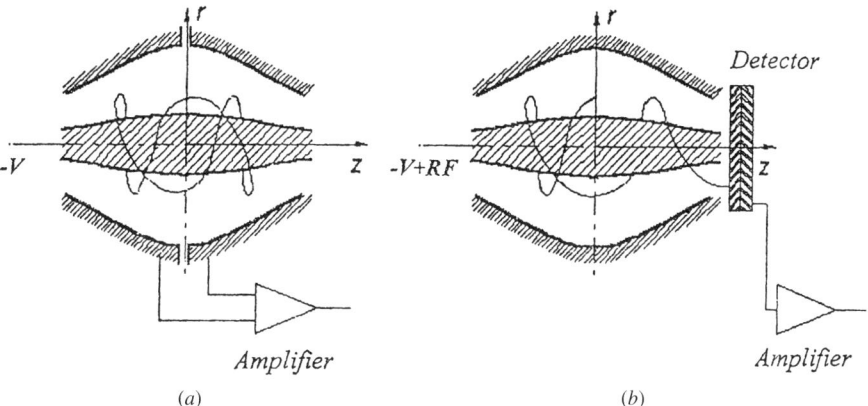

Figure 3.23. Shape of an orbitrap, and ion motion and modes of mass analysis: (*a*) image-current detection mode and (*b*) mass-selective instability detection. (Reproduced from ref. 66 by permission of the American Chemical Society, Washington, DC, copyright © 2000.)

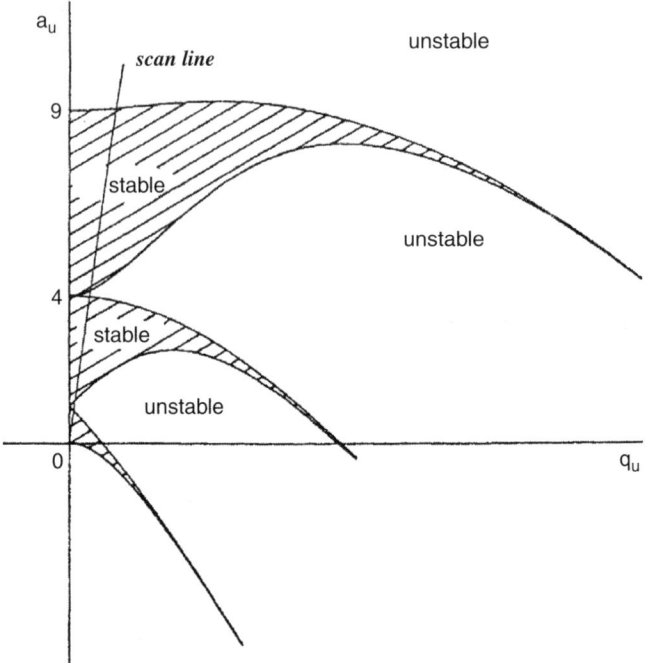

Figure 3.24. Stability diagram and scan line of an orbitrap in the mass-selective instability mode. (Reproduced from ref. 66 by permission of the American Chemical Society, Washington, DC, copyright © 2000.)

is used for mass analysis because it is completely independent of the energy and spatial distributions of the ions. The stability diagram in the axial direction takes the shape shown in Figure 3.24. MALDI and electrospray modes of ionization have been combined for the analysis of a variety of compounds with an orbitrap [66,67]. An orbitrap has also been combined with an LIT to provide a hybrid tandem mass spectrometry system (Section 4.5.6) [70,71].

In practice, ions are injected into the trap through a narrow ion-injection channel with a velocity perpendicular to the long axis (i.e., z-axis) and at a point far displaced from its equator (i.e., $z = 0$). The ions injected are squeezed closer to the center of the trap by increasing the electric field strength. Once in the trap, each m/q ion packet begins to move in coherent axial oscillations without the need for additional excitation. All ions have the same amplitude but different frequencies of oscillations. The ESI-formed ions are injected as a fast pulse via an LIT storage device [68]. For image-current detection, the outer electrode is split in half at $z = 0$, and the ion current from each split electrode is amplified differentially (Figure 3.23a) and processed as in FT–ICR to obtain time-domain image current transients. A mass spectrum of the trapped ions is obtained by Fourier-transforming the transient. Although not a common procedure, the mass-selective instability mode, as used in QITs, can be implemented for ion detection.

In this mode, the rf voltage is applied to the central electrode. As a consequence, the axial coordinate of the ions increases until ions are ejected from the trap for detection by an external detector along the axis of the orbitrap (Figure 3.23*b*).

The orbitrap has several desirable features, including high-mass resolution (ca. 150,000 FWHM), high mass accuracy (2 to 5 ppm), large space-charge capacity, high dynamic range (>10) relative to other trap instruments, good sensitivity, and an m/q range up to 6000.

3.10. ION MOBILITY MASS SPECTROMETERS

Ion mobility (IM) *mass spectrometers* are hybrid instruments that combine an IM separation system with conventional MS systems. An *ion mobility spectrometer* (IMS) can also serve as a stand-alone ion detection system [72]. An IMS uses gas-phase mobility rather than the m/q ratio as a criterion to separate ions [73,74]. The mobility of ions is measured under the influence of an electrical field gradient and cross-flow of a buffer gas, and depends on ion's collision cross section and net charge.

A typical IMS is shown in Figure 3.25. It consists of a reaction region and a much longer drift region; both regions contain a series of uniformly spaced electrodes that are connected via a series of high resistors to provide a uniform electric field strength. The two regions are separated by an electrical shutter. Buffer gas is also circulated in the drift tube. The ions are generated in the reaction chamber and are allowed to enter the drift region by opening the electrical shutter for a brief period. Under the influence of an electrical field, ions drift into the drift tube, where they are separated according to their size-to-charge ratio. The mobility of ions is a combined effect of ion acceleration by the electric field and retardation by collisions with the buffer gas. At the end of the drift region is placed an ion detector (e.g., a Faraday cup) for the detection of the separated ions. This type of instrument has been highly successful for national security in the detection of explosives in commercial aviation, mass transportation, and urban centers [72].

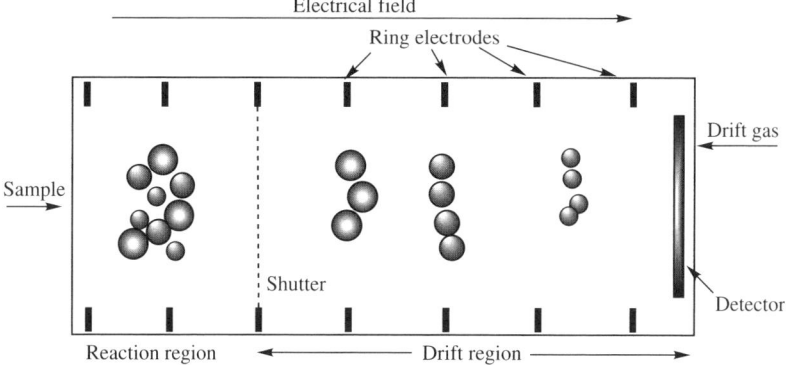

Figure 3.25. Depiction of the principle of ion mobility spectrometry.

For more accurate mass analysis, an IMS is coupled to a quadrupole or TOF mass analyzer [75–79]. Similar to LC–MS systems, the IMS serves as a separation device and the quadrupole or TOF mass analyzer as a detection device, but has the added advantage that separation times are in milliseconds. ESI and MALDI ion sources have both been coupled to IM–MS instruments [75–78]. Such systems can be employed for the analysis of mixtures of proteins and tryptic peptides [75,77,78]. An instrument that depicts the coupling of IM with TOF mass spectrometry is shown in Figure 3.26. It consists of a MALDI source, an ion mobility cell, a CID cell, and an oa–TOF mass analyzer. Ions exit the drift tube when the axial field strength of the ion mobility cell is ramped up, and enter the source region of an oa–TOF instrument, where the ions are detected intact in the usual manner (see Section 3.5.4). Alternatively, ions can be fragmented in the CID prior to their detection by the TOF–MS.

The resolution of an IMS, usually very low (10 to 12) can be increased to 200 to 400 by increasing the pressure of the buffer gas, connecting the ion source directly with the drift tube, increasing the length of the drift tube, and increasing the electric field gradient of the drift tube. Unlike LC/MS separation, an ion chromatogram can be obtained within 1 s.

Other applications of IM–MS include the detection of drugs, chemical warfare agents, and environmental pollutants; size distribution of aerosol particles; structure information of gas-phase clusters; and conformational studies of proteins and oligonucleotides [79].

A variation of IMS is field-ion spectrometry (FIS), which functions as an ion filter to allow one type of ion to be transmitted continuously [80]. In FIS the

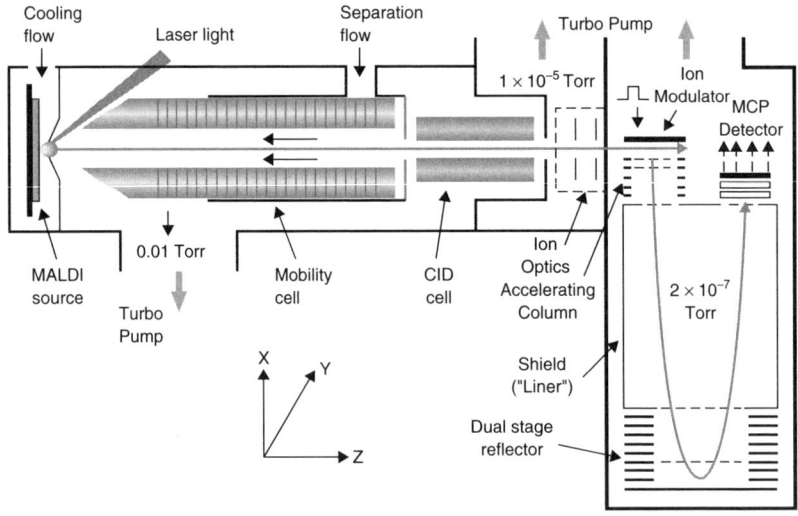

Figure 3.26. Schematic of the ion mobility/oa–TOF instrument. (Reproduced from ref. 75 by permission of Elsevier Science, copyright © 2006 American Society for Mass Spectrometry.)

electric field is applied as a high-frequency asymmetric waveform rather than as a dc voltage. Ions travel in the axial direction of the drift tube in a flowing stream of a buffer gas, and the electric field is applied perpendicular to the direction of the gas flow.

3.11. DETECTORS

The function of a detector is to improve our chemical vision. A detector provides information on the ion fluxes or abundances of the ions after their exit from the mass analyzer. A detector converts the beam of ions into an electrical signal that can be amplified, stored, and displayed by the data system into a form that is readily perceived by human eyes or a series of ion arrival pulses that can be timed and/or counted. Sensitivity, accuracy, resolution, response time, stability, wide dynamic range, and low noise are the most important characteristics that are sought in any ion detector. Two major categories of detectors exist: focal-point detectors and focal-plane detectors [81,82]. The former detect ions one m/q at a time and are used with scanning mass analyzers. The ion current due to ions that are not being measured is wasted. In contrast, focal-plane detectors monitor all ions all the time upon their arrival along a plane and are used with m/q spatial dispersion mass analyzers; this arrangement results in improved detection sensitivity and mass measurement accuracy. Ion-arrival-time detectors, on the other hand, count the ions that arrive at each specific time interval after the ion acceleration event. They are also very efficient, although limited to low ion fluxes.

3.11.1. Faraday Cup Detector

A *Faraday cup detector* is a very simple and robust device that detects ions by direct charge measurement with a conducting electrode, which usually is made of a conically shaped metal cup or an inclined collector electrode surrounded by a metal cage (known as a Faraday cage; Figure 3.27). The collector electrode is connected to a high-impedance amplifier via a large feedback resistance. An incoming ion beam transfers its charge upon impact with the collector electrode; a voltage drop develops across a large feedback resistance (10^{11} ohms), and the

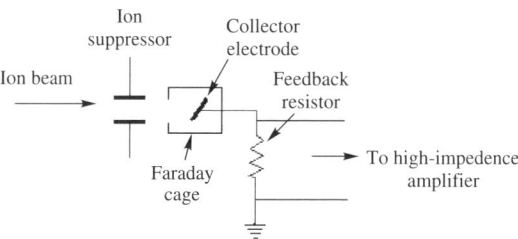

Figure 3.27. Faraday cup detector.

resulting current is amplified by the high-impedance amplifier. Electrodes with a high depth-to-width ratio are preferred because of their increased ion capture efficiency and minimal scattering losses. A Faraday detector exhibits slow but stable response because of the very large feedback resistor used with low ion fluxes. Therefore, this detector is unsuitable for scanning mass spectrometers but is highly compatible with isotope-ratio mass spectrometers, where high signal stability is critical (see Section 7.7). The response of the detector, however, is independent of the energy of the incoming ions.

3.11.2. Electron Multipliers

The *electron multiplier* or *secondary-electron multiplier* (SEM) is the most common ion detector in mass spectrometry. It works on the phenomenon of secondary electron emission. A variety of designs are in common use. A discrete-dynode SEM consists of a series of dynodes (generally, 16 or 20), usually made of copper–beryllium, that are connected together via a chain of resistors of equal value (Figure 3.28). A high voltage (up to -3000 V) is applied between the first dynode (the conversion dynode) and the last dynode (the anode); each dynode is maintained at a higher positive potential than the preceding one. When a beam of fast-moving ions strikes a specially coated conversion dynode, several secondary electrons are emitted. The electrons emitted strike the second dynode, where again, several more secondary electrons are emitted for each electron that has struck the dynode. This process is repeated at subsequent dynodes to cause an amplification of secondary electrons at each successive stage. Finally, a cascade of electron current that arrives at the anode is amplified to provide a gain in excess of 10^7 electrons for each incident ion. The overall amplification of the ion current is a factor of the number of dynodes, their composition (i.e., work function), and the bias voltage. The detector can be used in an analog or pulse counting mode. Fast response time, high sensitivity, and high gain are the characteristics of these detectors. The yield of secondary electrons falls off exponentially as the velocity of the striking ions decreases [83]. They are, however, prone to damage from overload and, in any case, have a limited lifetime.

Another common version of the SEM is a horn-shaped assembly, popularly known as a *channel electron multiplier* (CEM) or *continuous-dynode EM*

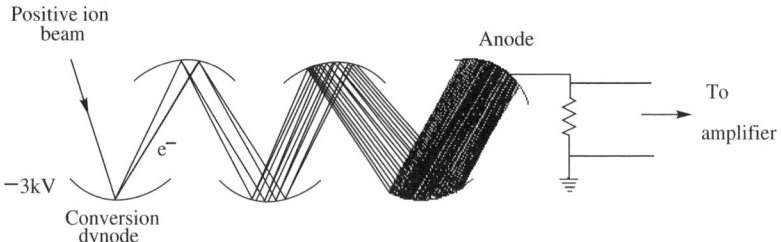

Figure 3.28. Principle of ion detection by an electron multiplier.

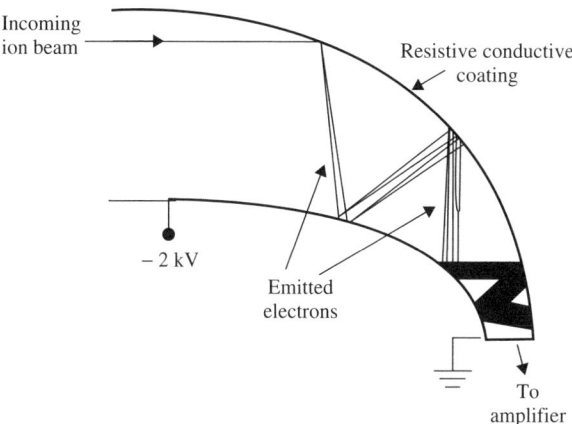

Figure 3.29. Simplified sketch of a continuous dynode electron multiplier.

(CDEM), in which a conducting surface acts as an array of continuous electrodes (Figure 3.29). It is constructed from a special type of glass that has either been heavily doped with lead or whose inner surface is coated with beryllium. A high voltage (1.8 to 2 kV) is impressed across the two ends of the tube to create a uniform field throughout the length of the tube. The beam of ions separated is made to strike the surface near the entrance of the detector, and the electrons ejected are reflected toward the opposite surface until they reach the other end of the detector. With each impact on the inner surface of the tube, the current is amplified, due to the emission of several more electrons for each striking electron. A CEM is more compact and less expensive than a discrete EM and provides gains as high as 10^8.

3.11.3. Photomultiplier Detectors

In *photomultiplier-based detectors*, the incoming ion beam is first converted to a photon beam when ions strike a scintillation material. The emitted photons are amplified and detected by a conventional photomultiplier. The construction of a photomultiplier is similar to that of an EM, except that the conversion dynode, called the *photocathode*, is coated with a photoemissive material that emits electrons when struck by photons. Photomultipliers are usually employed in combination with postacceleration devices, which are discussed next.

3.11.4. Postacceleration Detectors

The principle of secondary electron emission eventually breaks down as the mass of the ion increases. As shown in Eq. (3.2), the velocity of an ion is an inverse function of its mass. Therefore, the detection efficiency decreases with an increase in the mass of the incoming beam. In particular, the detection of high-mass biomolecules with EMs poses a problem. A solution is to augment

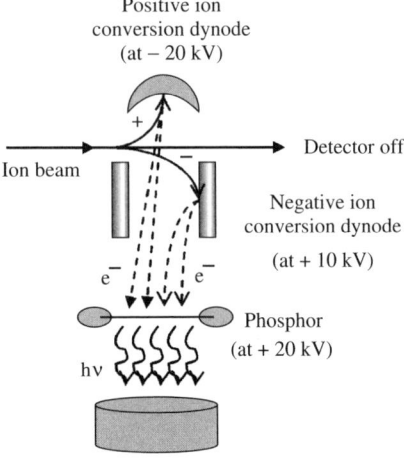

Figure 3.30. Conceptual sketch of a combined positive- and negative-ion postacceleration detector.

the velocity of the high-mass incident ions by a *postacceleration device* before their detection by EMs. In this device the positive-ion beam is first deflected toward a cathode held at a potential of -5 to -30 kV. The secondary electrons emitted from this cathode are amplified and are detected in the usual manner by an EM. A postacceleration device is especially useful for the detection of negative ions because they normally slow down considerably as they approach the first dynode of a conventional EM.

A postacceleration-based detector that can be used for the detection of positive and negative ions is shown in Figure 3.30. It contains two conversion dynodes, one for positive ions and one for negative ions, a scintillation material (phosphor), and a photomultiplier. For positive-ion detection, the conversion dynode is maintained at -10 to -20 kV. For the detection of negative ions, the incoming beam is first deflected toward a cylindrical conversion dynode that is held at half the phosphor voltage. The secondary electrons emitted in both cases are accelerated toward the phosphor. The photons ($h\nu$) released from the phosphor are transmitted to the photomultiplier for further detection.

3.11.5. Low-Temperature Calorimetric Detectors for High-Mass Ions

Low-temperature calorimetric detectors, also known as *cryogenic detectors* [84], are based on the principle that the kinetic energy E of the impinging ions is transformed into heat; the rise in temperature is given by

$$\Delta T = \frac{E}{C_T} \tag{3.22}$$

where C_T is the heat capacity, which at low cryogenic temperatures has a $1/T^3$ dependence for dielectrics and superconductors [84]. The response of calorimetry detectors is 100% efficient and mass-independent, has no upper mass limit, and is related to the kinetic energy of ions.

A typical design uses a superconductor–insulator–superconductor (SIS) junction, which consists of two superconducting films interspersed by a thin insulator made of the oxide of the base metal (e.g., $Nb-Al_2O_3-Nb$ and $Sn-SnO_x-Sn$) [85]. The detector is operated at cryogenic temperatures (ca. 2 K). When a pulse of energetic ions impinges on a superconducting film, nonthermal phonons are created in the film. Photons in turn break apart the *Cooper pairs*, an assemblage of weakly bound electrons that are formed at cryogenic temperatures. These free electrons are excited to Fermi levels, where they quantum-mechanically tunnel through the insulator to produce a large tunnel current. A kinetic energy of only a few meV is required to break a Cooper pair, which means that an incident ion of 25-keV energy will produce millions of excited electrons. The detector works with 100% efficiency because every impact transfers energy to the junction and generates a current. In combination with MALDI–TOFMS, cryogenic detectors can be used to detect megadalton ions [86].

3.11.6. Focal-Plane Detectors

In *focal-plane detectors*, the simultaneous detection of all spatially dispersed (*multiplex detection*) ions is achieved, resulting in several orders of improvement in detection efficiency over that of focal-point detectors [87]. A photographic plate is the earliest example of focal-plane detection which in the past was used in conjunction with Mattauch–Herzog double-focusing mass spectrometers.

Multichannel Plate Detector A multichannel plate (MCP) assembly is a multichannel version of CEM that provides a two-dimensional image of the impinging signal [88]. This assembly consists of a large number of individual microchannels, each made of very small diameter ($\approx$10 µm) tubes of metal-doped glass (Figure 3.31). Small-diameter channels provide a much higher gain

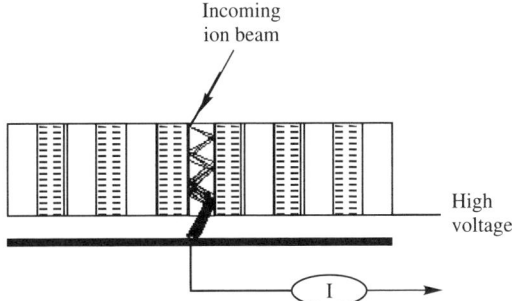

Figure 3.31. Conceptual diagram of a multichannel plate detector.

108 MASS ANALYSIS AND ION DETECTION

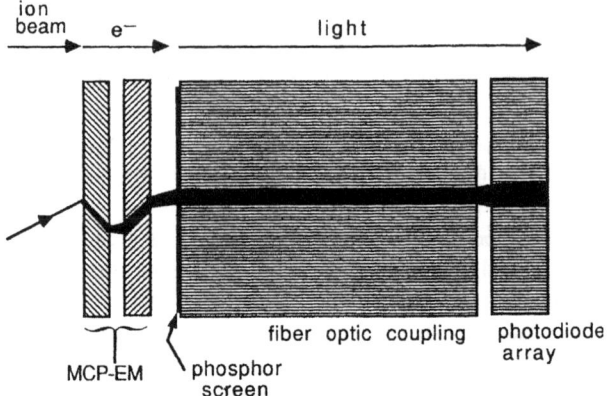

Figure 3.32. Conceptual diagram of a multichannel array detector. (Reproduced from C. Dass, in D. M. Desiderio, ed., *Mass Spectrometry: Clinical and Biomedical Applications*, Vol. 2, by permission of Plenum Press, copyright © 1994.)

than that of conventional CEM. A burst of electrons is produced when an incoming ion strikes the emissive surface near the entrance of the microchannel. Two or more MCPs can be arranged in a chevron configuration to provide single-ion detection when pulse counting is used. An MCP detector is frequently used with TOF mass spectrometers.

Multichannel Array Detector A typical multichannel array-based FPD consists of an MCP assembly and a one-dimensional array of detector electrodes. Each electrode acts as an independent detector and contains its own charge sensor, 8-bit counter, control logic, and bus interface. The ion beam that emerges from a mass analyzer first strikes the entrance face of the MCP assembly to emit secondary electrons. The MCP output falls onto the electrodes and is detected independently. A schematic of an integrating electrooptical multichannel array detector is shown in Figure 3.32. It employs a CEM made of a pair of MCPs arranged in a chevron configuration, a phosphor screen, a fiber-optic coupling, and a 1024-channel photodiode array. The photons emitted from the phosphor screen travel through the fiber-optic cables into a charge- (CCD) or plasma-coupled device (PCD), where they are converted to electric charge. This 25-mm array detector simultaneously covers 4% of the mass range of a magnetic-sector mass spectrometer.

OVERVIEW

In this chapter the principles of various mass analyzers were enumerated. The function of a mass analyzer is to disperse all ions (in space, time, or frequency) in terms of their m/q ratio. This work is accomplished by manipulating their momentum, kinetic energy, and velocity. Magnetic-sector instruments are one of the most successful mass spectrometers. They separate ions via momentum dispersion and directional focusing. Ions travel in the magnetic field in distinct circular paths that depend on an ion's m/q ratio and thus are separated from each other. To improve the mass separation, an ESA is added to a magnetic sector to provide double-focusing action (i.e., energy and direction focusing).

Quadrupoles are the most widely used type of mass spectrometer. A quadrupole is constructed from four parallel metal electrodes. The opposite electrodes are connected by dc and rf potentials. Mass separation is accomplished by the stable vibratory motion of ions in the high-frequency oscillating electric field thus created. Under the set of defined dc and rf potentials, ions of a specific m/q pass through the geometry of quadrupole rods. A mass spectrum is obtained by changing dc and rf potentials while keeping their ratio constant.

A TOF mass analyzer separates ions on the basis of their velocity differences. A short pulse of ions of a defined kinetic energy is allowed to drift in a long field-free flight tube in which the lighter ions travel faster and reach the detector earlier than do the heavier ones. Some innovative developments are incorporated into the linear TOF design to improve resolution (e.g., the addition of a reflectron and the use of delayed extraction of ions) and make it suitable for a continuous mode of ion beams.

The quadrupole ion trap has become another popular mass analyzer. It consists of a central ring electrode and two end-cap electrodes. Mass separation in a QIT is achieved by storing the ions in the trapping space and by manipulating their motion in time rather than in space. The field is created by applying an rf field to the ring electrode. The mass spectrum is acquired in the instability mode of operation, in which the applied rf field is changed sequentially to destabilize the trapped ions out of the trapping field. Its cousin the LIT, also works on the principle of ion trapping. It is also made of four parallel rods and uses rf and dc potentials for mass separation and analysis. As with QIT, mass analysis of the trapped ions is performed in the mass-selective instability mode by ramping the main rf voltage.

FT–ICR–MS is also an ion-trap technique. The heart of this instrument is a cell that consists of three pairs of opposing plates (trapping, excitation, and detection). The cell is placed in a strong magnetic field. Ions are confined laterally in this trap by a static magnetic field and axially by a static electric field. Ions are excited by a broadband rf pulse to coherent orbital motion and are detected by measuring the image current. The application of a Fourier transform converts this complex time-domain signal into a normal mass spectrum.

The orbitrap is a newer member of the ion-trap family of mass analyzers. It consists of an axial spindlelike central electrode and a coaxial barrel-like outer electrode. The trapped ions undergo rotation and harmonic oscillations along the central electrode. The m/q values of the trapped ions are related to the frequencies of their harmonic oscillations along the central axis. Mass analysis is performed by measuring the image current that is induced in the outer electrode and converting the time-domain signal into mass spectra using fast FT.

Ion mobility spectrometry instruments work on the principle of ion migration under the influence of an electrical field gradient against a flow of buffer gas. The mobility of ions depends on an ion's collision cross section and net charge. The instrument consists of two sections (reaction region and drift region) that are made of a series of uniformly spaced electrodes. Ions drift under the influence of an electrical field in the drift tube, where they are separated according to their size-to-charge ratio and are detected by placing a detector at the end of the drift region. Alternatively, a conventional mass analyzer is coupled to IMS for more accurate mass analysis of the separated ions.

Ion detection is provided by Faraday cups, electron multipliers, channel electron multipliers, cryogenic detectors, multichannel plate detectors, and electrooptical detectors.

EXERCISES

3.1. Answer briefly why:
  (a) an electrostatic analyzer is incorporated into a double-focusing sector instrument.
  (b) the Mattauch–Herzog geometry instrument provides better detection sensitivity than the Nier–Johnson instrument.
  (c) in a quadrupole mass spectrometer, the ions need to be accelerated to very low velocities.
  (d) it is essential to use a pulsed ion beam for TOF mass analysis.
  (e) a reflectron is incorporated into the TOF mass analyzer.
  (f) focal-plane detectors provide better detection sensitivity than focal-plane detectors.

3.2. Calculate the resolving power required to separate the molecular ions of (a) C_3H_6S and $C_5H_{12}D$, (b) $C_{12}H_{24}$ and $C_6H_4N_2O_4$, and (c) $C_{18}H_{18}O$ and $C_{18}H_{20}N$.

3.3. What strategies can be applied in a magnetic-sector instrument to enhance its mass range?

3.4. The upper mass limit of a double-focusing magnetic-sector instrument is 2500 Da. What would be the upper mass limit when the strength of the magnetic field and the radius of the flight tube are both doubled?

3.5. Calculate the potential that would be required to accelerate a singly charged ion of a polypeptide of mass 5000 Da into a sector instrument that has a 1.5-T magnet and a 50-cm radius flight tube.

3.6. Calculate the difference in the time taken by two singly charged ions of mass 2500 and 2000 Da to travel through a 1-m-long flight tube. The acceleration potential was set to 6000 V.

3.7. The spectrum of benzene (C_6H_6) was obtained with a double-focusing magnetic-sector instrument that is equipped with a 12.5-KG magnet. Ions exited the source at 6000 V. What would be the radius of trajectory for a $C_6H_6^{2+}$ ion in the magnetic field?

3.8. What would be the velocity of $C_6H_6^{2+}$ ions when they are accelerated to a potential of 8000 V?

3.9. Suppose that a quadrupole mass filter is operated at the tip of the first stability region, where the Mathieu parameters a_u and q_u are equal to 0.256 and 0.750, respectively. If the maximum dc potential is 500 V, the rf frequency is 0.5 MHz, and r_0 is 1.0 cm, then what is the maximum m/z value that can be analyzed with this instrument?

3.10. A sample contains a mixture of three compounds, the molecular ions of which can be separated by a mass spectrometer at a resolution of 9500. Your laboratory is equipped with the following instruments: **(a)** a double-focusing magnetic-sector spectrometer, **(b)** a single-quadrupole spectrometer, **(c)** a quadrupole ion-trap spectrometer, **(d)** a linear time-of-flight (TOF) spectrometer, **(e)** a reflectron–TOF spectrometer, and **(f)** a Fourier transform ion cyclotron mass spectrometer. Suggest all possible choices among these instruments that you can employ for the analysis of this mixture.

3.11. What are the causes of poor resolution in a linear TOF instrument?

3.12. What strategies can be applied to improve the resolution of a TOF instrument?

3.13. Enumerate the principle of time-lag focusing in TOF mass analyzers.

3.14. How are ions trapped and detected in an FT–ICR mass spectrometer?

3.15. The cubic cell of an FT–ICR–MS instrument is placed in a magnetic field of 1.0 T. What would be the cyclotron frequency (ω_c) of the molecular ion of benzoic acid (C_6H_5COOH)?

3.16. If the doubly charged ion of Exercise 3.15 was trapped in an FT–ICR cell placed in a 12.5-KG magnet, what would be the cyclotron frequency of the ion motion?

3.17. How are externally formed ions injected into a quadrupole ion trap?

3.18. Your laboratory is equipped with a QIT. The i.d. of its central electrode is 1.0 cm, and an rf signal of 0.5 MHz is applied to this electrode. The mass analysis is performed by varying the rf potential from 100 to 3000 V. What would by the lower and upper mass limits of the trap?

3.19. Why is the detection efficiency of an electron multiplier low for high-mass ions?

3.20. How can the detection of high-mass ions and negative ions by an electron multiplier be improved?

ADDITIONAL READING

1. C. Dass, in D. M. Desiderio, ed., *Mass Spectrometry: Clinical and Biomedical Applications*, Vol. 2, Plenum Press, New York, 1994, pp. 1–52.
2. C. Dass, *Principles and Practice of Biological Mass Spectrometry*, Wiley-Interscience, New York, 2001.
3. J. Roboz, *Introduction to Mass Spectrometry*, Interscience, New York, 1968.
4. C. A. McDowell, *Mass Spectrometry*, McGraw-Hill, New York, 1963.
5. P. H. Dawson, *Quadrupole Mass Spectrometry and Its Applications*, American Institute of Physics, AIP-AVS Classics Series, AIP Press, Woodbury, NY, 1994.
6. P. K. Ghosh, *Ion Traps*, Clarendon Press, Oxford, England, 1995.
7. R. E. March and J. F. J. Todd, *Practical Aspects of Ion Trap Mass Spectrometry: Chemical, Environmental, and Biomedical Applications*, CRC Press, Boca Raton, FL, 1996, p. 544.
8. R. E. March and J. F. J. Todd, *Quadrupole Ion Trap Mass Spectrometry*, Wiley, Hoboken, NJ, 2005.
9. R. J. Cotter, *Time-of-Flight Mass Spectrometry: Instrumentation and Applications in Biological Research*, American Chemical Society, Washington, DC, 1997.
10. M. V. Buchanan, ed., *Fourier Transform Mass Spectrometry: Evolution, Innovations, and Applications*, American Chemical Society, Washington, DC, 1987.
11. K. Birkinshaw, ed. Special Issue: Detectors and the measurement of mass spectra, *Int. J. Mass Spectrom.* **215** (2002).

REFERENCES

1. J. J. Thomson, *Rays of Positive Electricity and the Application to Chemical Analyses*, Longmans, Green & Co., London, 1913.
2. J. Mattauch and R. Herzog, Double-focusing mass spectrograph and the masses of N^{15} and O^{18}, *Phys. Rev.* **44**, 617 (1936).
3. E. G. Johnson and A. O. Nier, Angular aberrations in sector shaped electromagnetic lenses for focusing beams of charged particles, *Phys. Rev.* **84**, 10 (1953).
4. P. H. Dawson, Quadrupole mass analyzers: performance, design, and some recent applications, *Mass Spectrom. Rev.* **5**, 1–37 (1986).
5. J. E. Campana, Elementary theory of quadrupole mass spectrometry, *Int. J. Mass Spectrom. Ion Proc.* **33**, 101–117 (1980).
6. P. E. Miller and M. Bonner Denton, The quadrupole mass filter: basic operating concepts, *J. Chem. Educ.* **63**, 617–622 (1986).
7. B. E. Winger, K. J. Light-Wahl, R. R. O. Loo, H. R. Udseth, and R. D. Smith, Observations and implications of high-mass-to-charge ratio ions from electrospray ionization mass spectrometry, *J. Am. Soc. Mass Spectrom.* **4**, 536–545 (1993).
8. Z. Duo, T. N. Olney, and D. J. Douglas, Inductively coupled plasma mass spectrometry with a quadrupole mass filter operated in a third stability region, *J. Am. Soc. Mass Spectrom.* **8**, 1230–1236 (1997).
9. Z. Duo and D. J. Douglas, High resolution inductively coupled plasma mass spectra with a quadrupole mass filter, *Rapid Commun. Mass Spectrom.* **10**, 649–652 (1996).

10. M. H. Amad and R. S. Houk, High-resolution mass spectrometry with a multiple pass quadrupole mass analyzer, *Anal. Chem.* **70**, 4885–4889 (1998).
11. R. J. Cotter, Time-of-flight mass spectrometry for the structural analysis of biological molecules, *Anal. Chem.* **64**, 1027A–1039A (1992).
12. H. Wollnik, Time-of-flight mass analyzers, *Mass Spectrom. Rev.* **12**, 89–114 (1993).
13. M. Guilhaus, Principles and instrumentation in time-of-flight mass spectrometry: Physical and instrumental concepts, *J. Mass Spectrom.* **30**, 1519–1532 (1995).
14. M. Guilhaus, V. Mlynski, and D. Selby, Perfect timing: time-of-flight mass spectrometry, *Rapid Commun. Mass Spectrom.* **11**, 951–962 (1997).
15. C. Weickhardt, F. Moritz, and J. Grotemeyer, Time-of-flight mass spectrometry: state-of-the-art in chemical analysis and molecular sciences, *Mass Spectrom. Rev.* **15**, 139–162 (1996).
16. R. J. Cotter, The new time-of-flight mass spectrometry, *Anal. Chem.* **71**, 445A–451A (1999).
17. W. E. Stephens, A pulsed mass spectrometer with time dispersion, *Phys. Rev.* **69**, 691 (1946).
18. M. Toyoda, M. Ishihara, S. Yamaguchi, H. Ito, T. Matsuo, R. Roll, and H. Rosenbauer, Construction of a new multi-turn time-of-flight mass spectrometer, *J. Mass Spectrom.* **35**, 163–167 (2000).
19. C. K. Piyadasa, P. Hakansson, and T. R. Ariyarathe, A high resolving power multiple reflection matrix-assisted laser desorption/ionization time-of-flight mass spectrometer, *Rapid Commun. Mass Spectrom.* **13**, 620–624 (1999).
20. W. C. Wiley and I. H. McLaren, Time-of-flight mass spectrometer with improved resolution, *Rev. Sci. Instrum.* **26**, 1150–1157 (1955).
21. R. S. Brown and J. J. Lenon, Mass resolution improvement by incorporation of pulsed ion extraction in a matrix-assisted laser desorption/ionization linear time-of-flight mass spectrometer, *Anal. Chem.* **67**, 1998–2003 (1995).
22. R. M. Whittal, L. M. Russon, S. R. Weinberger, and L. Li, Functional wave time-lag focusing matrix-assisted laser desorption/ionization in a linear time-of-flight mass spectrometer: improved mass accuracy, *Anal. Chem.* **69**, 2147–2153 (1997).
23. M. L. Vestal, P. Juhasz, and S. A. Martin, Delayed extraction matrix-assisted laser desorption time-of-flight mass spectrometry, *Rapid Commun. Mass Spectrom.* **9**, 1044–1050 (1995).
24. R. S. Brown and J. J. Lenon, Sequence-specific fragmentation of matrix-assisted laser-desorbed protein/peptide ions, *Anal. Chem.* **67**, 3990–3999 (1995).
25. B. A. Mamyrin, Laser assisted reflectron time-of-flight mass spectrometry, *Int. J. Mass Spectrom. Ion Proc.* **131**, 1–19 (1994).
26. T. J. Cornish and R. J. Cotter, High-order kinetic energy focusing in an end cap reflectron time-of-flight mass spectrometer, *Anal. Chem.* **69**, 4615–4618 (1997).
27. J. Zhang and C. G. Enke, Simple cylindrical ion mirror with three elements, *J. Am. Soc. Mass Spectrom.* **11**, 759–764 (2000).
28. J. H. J. Dawson and M. Guilhaus, Orthogonal-acceleration time-of-flight mass spectrometer, *Rapid Commun. Mass Spectrom.* **3**, 155–159 (1989).
29. A. N. Verentchikov, W. Ens, and K. G. Standing, Reflecting time-of-flight mass spectrometer with an electrospray ion source and orthogonal extraction, *Anal. Chem.* **66**, 126–133 (1994).

30. I. V. Chernushevich, W. Ens, and K. G. Standing, Orthogonal injection TOFMS for analyzing biomolecules, *Anal. Chem.* **71**, 452A–461A (1999).
31. M. Guilhaus, D. Selby, and V. Mlynski, Orthogonal-acceleration time-of-flight mass spectrometry, *Mass Spectrom. Rev.* **19**, 65–107 (2000).
32. A. N. Krutchinsky, I. V. Chernushevich, V. L. Spicer, W. Ens, and K. G. Standing, Collision damping interface for an electrospray ionization time-of-flight mass spectrometer, *J. Am. Soc. Mass Spectrom.* **9**, 569–579 (1998).
33. W. Paul, H. P. Reinhard, and O. Zahn, The electric mass filter as mass spectrometer and isotope separator, *Z. Phys.* **152**, 143–182 (1958).
34. G. C. Stafford, P. E. Kelley, J. E. P. Syka, W. E. Reynolds, and J. F. J. Todd, Recent improvements in and analytical applications of advanced ion trap technology, *Int. J. Mass Spectrom. Ion Proc.* **60**, 85–98 (1984).
35. R. G. Cooks, G. L. Glish, S. A. McLuckey, and R. E. Kaiser, Ion trap mass spectrometry, *Chem. Eng. News*, 26–41 (1991).
36. R. G. Cooks and R. E. Kaiser, Quadrupole ion trap mass spectrometry, *Acc. Chem. Res.* **23**, 213–219 (1990).
37. J. F. J. Todd, Ion trap mass spectrometer: past, present, and future(?), *Mass Spectrom. Rev.* **10**, 3–52 (1991).
38. R. E. March, Introduction to quadrupole ion trap mass spectrometry, *J. Mass Spectrom.* **32**, 351–369 (1997).
39. R. E. March, Ion trap mass spectrometry, *Int. J. Mass Spectrom. Ion Proc.* **118–119**, 71–135 (1992).
40. K. R. Jonscher and J. R. Yates III, The quadrupole ion trap mass spectrometer: A small solution to a big challenge, *Anal. Biochem.* **244**, 1–15 (1997).
41. R. E. March and R. J. Hughes, *Quadrupole Storage Mass Spectrometry*, Wiley Interscience, New York, 1989.
42. R. K. Julian, M. Nappi, C. Weil, and R. G. Cooks, Multiparticle simulation of ion motion in the ion trap mass spectrometer: resonant and direct current pulse excitation, *J. Am. Soc. Mass Spectrom.* **6**, 57–70 (1995).
43. U. P. Schlunegger, Jr., M. Stoeckli, and R. M. Caprioli, Frequency scan for the analysis of high mass ions generated by matrix-assisted laser desorption/ionization in a Paul trap, *Rapid Commun. Mass Spectrom.* **13**, 1792–1796 (1999).
44. J. D. Williams, K. A. Cox, R. G. Cooks, and J. C. Schwartz, High mass-resolution using a quadrupole ion-trap mass spectrometer, *Rapid Commun. Mass Spectrom.* **5**, 327–329 (1991).
45. F. A. Londry, G. J. Wells, and R. E. March, Enhanced mass resolution in a quadrupole ion trap, *Rapid Commun. Mass Spectrom.* **7**, 43–45 (1993).
46. G. J. Van Berkel, S. A. McLuckey, and G. L. Glish, Electrospray ionization combined with ion trap mass spectrometry, *Anal. Chem.* **62**, 1284–1295 (1990).
47. J. C. Schwartz and I. Jardine, High-resolution parent-ion selection/isolation using a quadrupole ion-trap mass spectrometer, *Rapid Commun. Mass Spectrom.* **6**, 313–317 (1992).
48. K. Jonscher, G. Currie, A. L. McCormack, and J. R. Yates III, Matrix-assisted laser desorption of peptides and proteins on a quadrupole ion trap mass spectrometer, *Rapid Commun. Mass Spectrom.* **7**, 20–26 (1993).

49. J. C. Schwartz and M. E. Bier, Matrix-assisted laser desorption of peptides and proteins using a quadrupole ion trap mass spectrometer, *Rapid Commun. Mass Spectrom.* **7**, 27–32 (1993).
50. J. C. Schwartz, M. W. Senko, and J. E. P. Syka, A two dimensional quadrupole ion trap mass spectrometer, *J. Am. Soc. Mass Spectrom.* **13**, 659–669 (2002).
51. J. W. Hager, A new liner ion trap mass spectrometer, *Rapid Commun. Mass Spectrom.* **16**, 512–526 (2002).
52. G. Hofgartner, E. Varesio, V. Tschäppät, C. Grivet, E. Bourgogne, and L. A. Leuthold, Triple quadrupole linear ion trap mass spectrometer for the analysis of small molecules and macromolecules, *J. Mass Spectrom.* **39**, 845–855 (2004).
53. Y. Hashimoto, H. Hasegawa, T. Baba, and I. Waki, Mass selective ejection by axial resonant excitation from a linear ion trap, *J. Am. Soc. Mass Spectrom.* **17**, 685–690 (2006).
54. Z. Ouyang, G. Wu, Y. Song, H. Li, W. R. Plass, and R. G. Cooks, Rectilinear ion trap: concepts, calculations, and analytical performance of a new mass analyzer, *Anal. Chem.* **76**, 4595–4605 (2004).
55. E. O. Lawrence and N. E. Edlefsen, *Science* **72**, 376 (1930).
56. H. Sommer, H. A. Thomas, and J. A. Hipple, Measurement of e/m by cyclotron resonance, *Phys. Rev.* **82**, 697–702 (1951).
57. M. B. Comisarow and A. G. Marshall, Fourier transform ion cyclotron resonance mass spectroscopy, *Chem. Phys. Lett.* **25**, 282 (1974).
58. J. H. Holland, C. G. Enke, J. Allison, J. T. Stults, J. D. Pinkston, B. Newcombe, and J. T. Watson, Mass spectrometry on the chromatographic time scale: realistic expectations, *Anal. Chem.* **55**, 997A–1012A (1983).
59. M. L. Gross and D. L. Rempel, Fourier transform mass spectrometry, *Science* **226**, 261–268 (1984).
60. A. G. Marshall and P. B. Grosshans, Fourier transform ion cyclotron resonance mass spectrometry: the teenage years, *Anal. Chem.* **63**, 215A–229A (1991).
61. I. J. Amster, Fourier transform mass spectrometry, *J. Mass Spectrom.* **31**, 1325–1337 (1996).
62. C. L. Holliman, D. L. Rempel, and M. L. Gross, Detection of high mass-to-charge ions by Fourier transform mass spectrometry, *Mass Spectrom. Rev.* **13**, 105–132 (1994).
63. T. Dienes, S. J. Pastor, S. Schürch, J. R. Scott, J. Yao, S. Cui, and C. A. Wilkins, Fourier transform mass spectrometry-advancing years (1992–mid. 1996), *Mass Spectrom. Rev.* **15**, 163–211 (1996).
64. S. Guan and A. G. Marshall, Ion traps for Fourier transform ion cyclotron resonance mass spectrometry: principles and design of geometric and electric configurations, *Int. J. Mass Spectrom. Ion Proc.* **146–147**, 261–296 (1995).
65. S. D. Shi, C. L. Hendrickson, and A. G. Marshall, Counting individual sulfur atoms in a protein by ultrahigh-resolution Fourier-transform ion cyclotron resonance mass spectrometry: experimental resolution of isotropic fine structure in proteins, *Proc. Natl. Acad. Sci. USA* **95**, 11532–37 (1998).
66. A. Makarov, Electrostatic axially harmonic orbital trapping: high-performance technique of mass analysis, *Anal. Chem.* **72**, 1156–1162 (2000).

67. M. Hardman and A. Makarov, Interfacing the orbitrap mass analyzer to an electrostatic ion source, *Anal. Chem.* **75,** 1699–1701 (2003).
68. Q. Hu, R. J. Noll, H. Li, A. Makarov, M. Hardman, and R. G. Cooks, The orbitrap: a new mass spectrometer, *J. Mass Spectrom.* **40,** 430–443 (2005).
69. K. H. Kingdon, A method for the neutralization of electron space charge by positive ionization at very low gas pressure, *Phys. Rev.* **21,** 408–418 (1923).
70. A. Makarov, E. Denisov, et al., Performance evaluation of a hybrid linear ion trap/orbitrap mass spectrometer, *Anal. Chem.* **78,** 2113–2120 (2006).
71. J. R. Yates, D. Cociorva, L. Liao, and V. Zabrouskov, Performance of a linear ion trap–orbitrap hybrid for peptide analysis, *Anal. Chem.* **78,** 493–500 (2006).
72. G. A. Eiceman and J. A. Stone, Ion mobility spectrometers in national defence, *Anal. Chem.* **76,** 391A–396A (2004).
73. C. Wu, W. F. Siems, G. R. Asbury, and H. H. Hill, Jr., Electrospray ionization high-resolution ion mobility spectrometry–mass spectrometry, *Anal. Chem.* **70,** 4929–4938 (1998).
74. D. E. Clemmer and M. F. Jarrold, Ion mobility measurements and their applications to clusters and biomolecules, *J. Mass Spectrom.* **32,** 577–592 (1997).
75. A. Laboda, Novel ion mobility setup combined with collision cell and time-of-flight mass spectrometry, *J. Am. Soc. Mass Spectrom.* **17,** 691–699 (2006).
76. C. S. Hoaglund-Hyzer and D. E. Clemmer, An ion trap/ion mobility/quadrupole/time-of-flight mass spectrometry for peptide mixture analysis, *Anal. Chem.* **73,** 177–184 (2001).
77. B. T. Ruotolo, K. J. Gillig, E. G. Stone, D. H. Russell, K. Fuhrer, M. Gonin, and J. A. Schultz, Analysis of protein mixtures by matrix-assisted laser desorption ionization–ion mobility–orthogonal time-of-flight mass spectrometry, *Int. J. Mass Spectrom.* **219,** 253–267 (2002).
78. B. T. Ruotolo, J. A. McLean, K. J. Gillig, E. G. Stone, and D. H. Russell, The influence and utility of varying field strength for the separation of tryptic peptides by ion mobility–mass spectrometry, *J. Am. Soc. Mass Spectrom.* **16,** 158–165 (2005).
79. C. A. Srebalus Barnes and D. E. Clemmer, Assessment of purity and screening of peptide libraries by nested ion mobility–TOFMS: identification of RNase S-protein binders, *Anal. Chem.* **73,** 424–433 (2001).
80. R. Guevremont and R. W. Purves, High field asymmetric waveform ion mobility–mass spectrometry: an investigation of leucine enkephalin ions by electrospray ionization, *J. Am. Soc. Mass Spectrom.* **10,** 492–501 (1999).
81. S. Evans, Detectors, in J. A. McCloskey, ed., *Methods in Enzymology*, Vol. 193, Academic Press, 1990, San Diego, CA, pp. 61–68.
82. D. W. Kopenaal, C. J. Baringa, et al., MS detectors, *Anal. Chem.* **77,** 418A–427A (2005).
83. R. J. Beuhler and L. Friedman, Low noise, high voltage secondary emission ion detector for polyatomic ions, *Int. J. Mass Spectrom. Ion Phys.* **23,** 81–97 (1977).
84. N. E. Booth, Calorimetric detectors for high mass ions, *Rapid Commun. Mass Spectrom.* **11,** 943–947 (1997).
85. M. Frank, S. E. Labov, G. Westmacott, and W. H. Bener, Energy-sensitive cryogenic detectors for high-mass biomolecule mass spectrometry, *Mass Spectrom. Rev.* **18,** 155–186 (1999).

86. R. J. Wenzel, U. Matter, L. Schultheis, and R. Zenobi, Analysis of megadalton ions using cryodetection MALDI time-of-flight mass spectrometry, *Anal. Chem.* **77,** 4329–4337 (2005).
87. K. Birkinshaw, Fundamentals of focal plane detectors, *J. Mass Spectrom.* **32**, 795–806 (1997).
88. W. Aberth, An imaging detector system for mass spectrometry, *Int. J. Mass Spectrom. Ion Phys.* **37**, 379–382 (1981).
89. J. S. Cottrell and S. Evans, Characteristics of microchannel electrooptical detection system and its application to the analysis of large molecules by fast atom bombardment mass spectrometry, *Anal. Chem.* **59**, 1990–1995 (1987).

CHAPTER 4

TANDEM MASS SPECTROMETRY

Tandem mass spectrometry (MS/MS) has attained an enviable status as an analytical tool to identify and quantify compounds in complex mixtures. MS/MS refers to the coupling of two stages of mass analysis, either in time or space. Of all the ionization techniques, only electron ionization (EI) provides abundant structural information. To obtain additional structure-specific information by other ionization techniques, it has become essential to perform MS/MS experiments [1,2]. MS/MS was first used in the late 1960s [3]. Since that time, its applications and popularity have continued to grow. Its major contributions are in the fields of structure elucidation of unknown compounds, identification of compounds in complex mixtures, elucidation of fragmentation pathways, and quantification of compounds in real-world samples. In recent times, several new generations of instruments have become available for tandem mass spectrometry applications. Basic concepts of tandem mass spectrometry and an account of these new developments are presented in this chapter. Additional reading material is listed at the end of the chapter.

4.1. BASIC PRINCIPLES OF TANDEM MASS SPECTROMETRY

The concept of *in-space tandem mass spectrometry* is illustrated in Figure 4.1. It involves two mass spectrometry systems; the first system (MS-1) performs the mass selection of a desired target ion from a stream of ions produced in the ion source. This mass-selected ion undergoes either unimolecular fragmentation [reaction (4.1)] or a chemical reaction in the intermediate region. The second MS

Fundamentals of Contemporary Mass Spectrometry, by Chhabil Dass
Copyright © 2007 John Wiley & Sons, Inc.

TANDEM MASS SPECTROMETRY

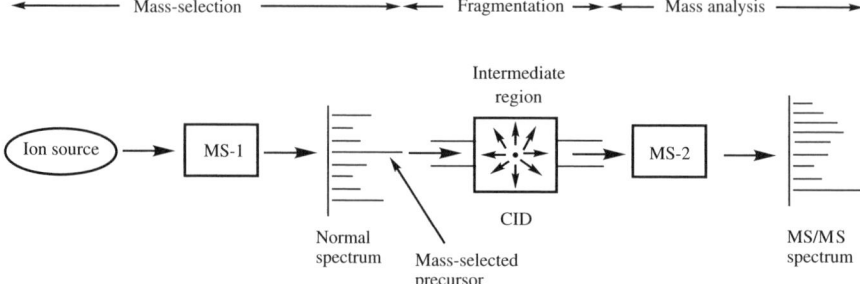

Figure 4.1. Basic principle of tandem mass spectrometry. (Reproduced from C. Dass, *Principles and Practice of Biological Mass Spectrometry*, Wiley-Interscience, 2001.)

system (MS-2) performs the mass analysis of the product ions that are formed in the intermediate step.

$$m_p^+ \longrightarrow m_1^+ + m_n'$$
$$\longrightarrow m_2^+ + m_n''$$
$$\longrightarrow m_3^+ + m_n''' \tag{4.1}$$

The MS/MS operation can be likened to the sorting of cards in a playing card deck. First, cards can be sorted according to their rank (e.g., aces from rest of the cards). After all four aces have been sorted, it is important to know their identity: that is, the suit. The MS-1 operation is similar to the sorting of cards according to their rank (e.g., aces here), whereas the MS-2 function is identification of the suit of each ace. The operation of tandem MS is also similar to the function of the gas chromatography (GC)-MS system (see Section 5.4). The first stage of MS/MS separates a mixture of ions into individual components as GC resolves a mixture of compounds, and the second stage acts as an identification system for the mass-resolved ions.

By convention, the mass-selected ion is called the *precursor ion* (m_p^+) and its fragments are called *product ions* [e.g., m_1^+, m_2^+, m_3^+, etc. in reaction (4.1)]. The m_n species in reaction (4.1) are neutral losses. Because of the incontrovertible link between the precursor ion and all of its product ions, the molecular specificity of MS/MS approaches unambiguity. This unique attribute of tandem mass spectrometry is a highly useful feature that plays a role in the unequivocal identification of a target compound in real-world samples.

Tandem mass spectrometry is not restricted to two stages of mass analysis (i.e., MS/MS or MS^2); it is also possible to perform *multistage MS* (i.e., higher-order MS) experiments, abbreviated as MS^n. These experiments can determine the genealogical relation between a precursor and its ionic products. For example, MS^3 indicates three stages of tandem mass spectrometry, which involves mass selection of one of the products, say either m_1^+, m_2^+, or m_3^+, formed from the precursor m_p^+ of the first-stage MS [reaction (4.1)], and determination of the second-generation products of that mass-selected ion. Multistage MS experiments are performed mostly with ion-trapping instruments. A maximum of 12 MS/MS experiments has been envisioned with a quadrupole ion trap. Beam-type instruments can also be used for MS^n experiments but require as many discrete mass analyzers as there are number of stages in the experiment, making it difficult to perform more than four stages of MS/MS experiments. As an example, a three-sector magnetic field instrument can perform up to MS^3 experiments.

4.2. TYPES OF SCAN FUNCTIONS

Practical applications of tandem mass spectrometry require data to be acquired in the following four scan modes. A pictorial representation of these scans and their symbolism is given in Figure 4.2.

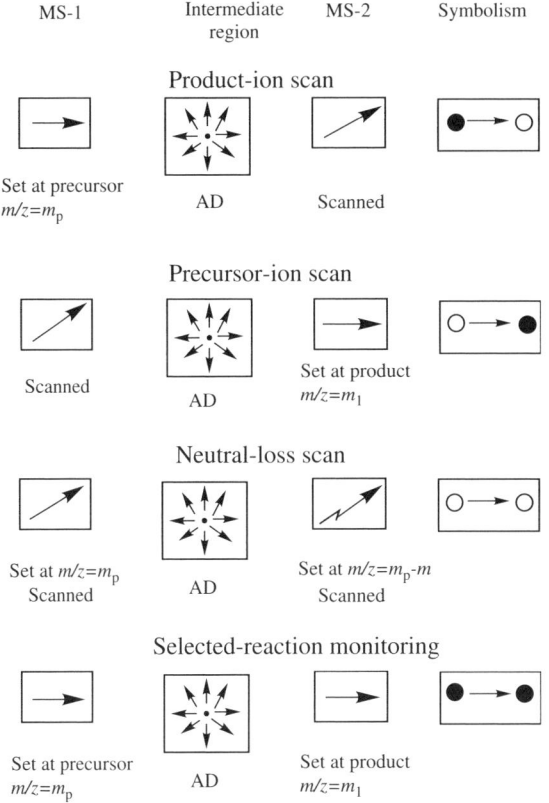

Figure 4.2. Pictorial representation of four scan modes of tandem mass spectrometry. AD refers to ion activation and dissociation, and the filled and open circles stand for fixed and scanning mass analyzers, respectively.

1. The *product-ion scan* (the old, now-unaccepted term, still used by some, is *daughter-ion scan*) is the most common mode of MS/MS operation. That spectrum is useful in the structure elucidation of a specified analyte. Information obtained in this scan is similar to that derived from a normal mass spectrum, except that the spectrum contains only those product ions that are formed exclusively from a mass-selected precursor ion. To acquire this spectrum, the first mass analyzer is set to transmit only the precursor ion chosen, and the second mass spectrometer is scanned over a required m/z range. As an example, the ketene radical cation reacts with neutral ketene to form a cycloadduct, the structure of which was delineated by mass-selecting the m/z of the adduct (at 84) and acquiring the product-ion scan (see Figure 4.3).

2. Another popular MS/MS scan is the *precursor-ion scan* (the past, now-unaccepted term is *parent-ion scan*). It provides a spectrum of all precursor ions that might fragment to a common, diagnostic product ion. The spectrum is obtained by adjusting the second mass spectrometer to transmit a chosen product ion (e.g., m_1) and scanning the first mass analyzer over a certain m/z range to transmit only those precursor ions that fragment to yield the chosen product ion. This scan is useful for the identification of a closely related class of compounds in a mixture. A typical example is the detection of phosphopeptides from biological samples. A major product of fragmentation of phosphopeptides is the PO_3^- ion at m/z 79. The precursor-ion scan of this m/z value will detect the presence of phosphopeptides selectively in a mixture of peptides from biological samples (see Figure 4.4).

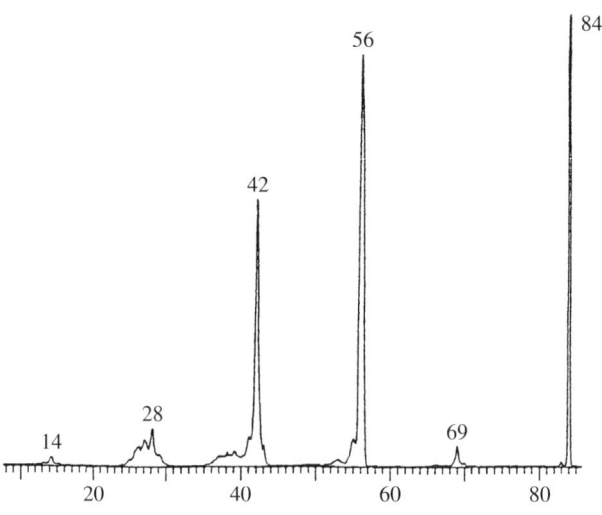

Figure 4.3. Product-ion mass spectrum of the [ketene$^{+\cdot}$ + ketene] cycloadduct. (Reproduced from ref. 4 by permission of Wiley-Interscience, copyright © 1993.)

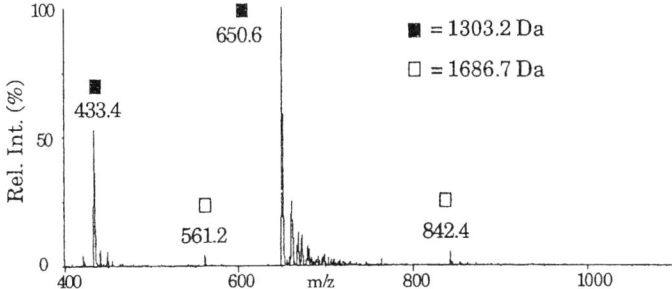

Figure 4.4. Precursor-ion scan of *m/z* 79 for selective detection of phosphopeptides. Two tyrosine phosphorylated peptides (1303.2 and 1686.7 Da) are detected in an LC fraction. (Reproduced from ref. 5 by permission of the American Chemical Society, Washington, DC, copyright © 1993.)

3. In a *constant-neutral-loss scan*, all precursors that undergo the loss of a specified common neutral are monitored. To obtain this information, both mass analyzers are scanned simultaneously, but with a mass offset that correlates with the mass of the specified neutral. Similar to the precursor-ion scan, this technique is also useful in the selective identification of closely related class of compounds in a mixture. For example, the loss of 44 Da is a common reaction of carboxylic acids. Through the constant-neutral loss scan, the identity of all carboxylic acids present in a complex mixture can be revealed. Similarly, by monitoring the 98-Da neutral loss, the presence of phosphopeptides can be detected in a complex mixture [6].

4. The fourth scan, *selected-reaction monitoring* (SRM), is useful in quantitative measurements of analytes present in complex mixtures (see Chapter 14). Conceptually, this scan mode is similar to the product-ion scan. However, instead of scanning the second mass spectrometer in a broad mass range, the two mass analyzers are adjusted to monitor one or more chosen precursor–product pairs of the analyte. This operation is identical to the selected-ion monitoring mode (SIM; see Chapter 14) of data acquisition. Monitoring more than one reaction is termed *multiple-reaction monitoring* (MRM).

All four scan modes can be implemented with magnetic sector- and quadrupole-based true or hybrid tandem instruments. Time-of-flight (TOF) and tandem-in-time devices are also suitable for product scan experiments, but they are unable to perform the other three scans.

4.3. ION ACTIVATION AND DISSOCIATION

To gather information about structure-specific fragment ions in tandem mass spectrometry, ion activation and dissociation of the mass-selected precursor is essential. There are several reasons for this activation step: (1) the abundance of the ion source–activated ions that fragment spontaneously in the field-free regions (FFRs) (i.e., the metastable ions) is usually low; (2) only a few selected reaction pathways are detectable from those spontaneously fragmenting ions; and (3) ions that are generated by currently popular softer modes of ionization, matrix-assisted laser desorption/ionization (MALDI) and electrospray ionization (ESI), are either stable or exhibit a weaker fragmentation pattern for them to yield any meaningful structural information. Ion activation not only increases the population of precursor ions with energies beyond the threshold of dissociation, but the number of dissociation channels increase as well, to provide useful structural information. A variety of ion-activation techniques are available; some of them are accessible with a specific instrumental combination only [7]. A concise description of some common modes of ion activation is presented here.

4.3.1. Collision-Induced Dissociation

By far the most common means of ion activation and dissociation of ions of organic and bioorganic molecules is *collision-induced dissociation* (CID), also known as *collisionally activated dissociation* (CAD) [8,9]. This technique, introduced in 1968 [3,10], is a two-step process: collision activation and unimolecular dissociation. In the collision activation step, the fast-moving mass-selected precursor ions are excited to higher-energy states by collisions with atoms of an inert gas such as helium or argon [reaction (4.2), where m_p^+ is the mass-selected precursor ion and N and N' are pre- and postcollision forms of the neutral target]. During this excitation process, a part of the initial translational energy of the fast-moving precursor ion is converted into the excitation energy of the ion to cause it to fragment [reaction (4.3); m_1^+ and m_n are fragments]. The maximum amount of kinetic energy that can be converted into internal energy during collision depends on the masses of the collision partners and is given by Eq. (4.4), where E_{lab} is the ion's kinetic energy in the laboratory frame (e.g., it is 10 keV when the accelerating potential is 10 kV), E_{com} is the center-of-mass kinetic energy of the ion, m_p is the mass of the precursor chosen, and N is the mass of the neutral target. A major drawback of CID is a gradual fall-off in the ion-dissociation efficiency with an increase in the mass of the precursor ion.

First step of CID (collision activation): $\qquad m_p^+ + N \longrightarrow [m_p^+]^* + N'$ (4.2)

Second step of CID (unimolecular dissociation): $\qquad [m_p^+]^* \longrightarrow m_1^+ + m_n$ (4.3)

$$E_{\text{com}} = \frac{N}{m_p + N} E_{\text{lab}} \qquad (4.4)$$

The subsequent step, unimolecular dissociation, has been explained fully in terms of RRKM and quasiequilibrium theory (QET) theories (see Section 6.7).

Depending on the instrument, CID can be performed in either a high- or a low-energy regime. In high-energy collisions, implemented primarily in magnetic-sector and TOF tandem instruments, ions have translational energies in the keV range (3 to 10 keV). Those collisions raise an ion to higher-energy excited electronic states via a single collision event and fastest activation time ($<10^{-10}$ s). The excitation energy is rapidly converted into vibrational energy, which is distributed statistically among all covalent bonds. Fragmentation of the excited ion occurs when the excitation energy exceeds the energy required to cleave a particular bond. During the unimolecular dissociation step, a part of the excess energy and the bond recoil energy is released as translational energy, which causes a broadening of peaks in the CID spectrum. Poor resolution in the second stage of mass analysis is the consequence in some types of tandem instruments. In high-energy CID, the products that result from consecutive fragmentations dominate the spectra. Also, the higher-energy fragmentation pathways, such as remote-site fragmentation and side-chain fragmentations of peptides, can be accessed only in this energy regime.

In low-energy collisions, first employed by Enke and colleagues, ions have energies above 200 eV [11]. Low-energy CID is popularly practiced in quadrupole- and ion trap (QIT, LIT, and FT–ICR–MS)-based tandem instruments. The low-energy CID process differs in several ways from high-energy CID. First, the ion activation time is somewhat longer (milliseconds to seconds), and excitation occurs primarily to higher vibrational states. Second, the collision region is maintained at a relatively higher pressure to allow more ions to participate in the collision process. Third, multiple collisions occur that lead to stepwise activation of the precursor. Fourth, because low-energy collisions excite ions only to vibrational states, the CID spectra are dominated by the products formed via the low-activation-energy fragmentation pathways. Fifth, at very low kinetic energies the products of ion–molecule reactions might also appear in those spectra. Finally, the mass of the target gas has a profound effect on the nature of the MS/MS spectrum. Ar and Xe are the preferred CID gases in this energy regime because they permit transfer of more energy to the dissociating ion.

4.3.2. Surface-Induced Dissociation

Ion activation is also accomplished by collisions of the fast-moving precursor ions with a solid surface [12,13]. This ion-surface collision technique, known as *surface-induced dissociation* (SID), can be implemented on a variety of tandem mass spectrometry systems, such as magnetic sector, TOF, quadrupoles, ion traps, and FT–ICR–MS, by placing a solid surface in the path of the ions [13]. The surface can be a bare metal (e.g., a stainless-steel plate) or a metal covered with self-assembled monolayers [14]. Ion-surface collisions are more efficient in terms of internal energy conversion because of the greater mass of the colliding surface [see Eq. (4.4)]. Consequently, compared to the CID process, in which a serious decrease in the dissociation efficiency is observed, ions of much higher mass can be fragmented by SID. Also, SID yields a narrow distribution of internal energies of the excited ions. A contemporary noteworthy application of SID is described in peptide sequencing and proteomics [15].

4.3.3. Absorption of Electromagnetic Radiations

The energy required for activation of ions and their subsequent fragmentation can also be made available by absorption of ultraviolet (UV) or infrared (IR) photons or by absorption of heat. These processes are more commonly practiced in ion-trapping instruments.

Ultraviolet Photodissociation (PD)   Absorption of a UV photon by an ion is an attractive alternative to CID. Absorption of a single photon deposits enough energy (e.g., 6.42 eV from a photon at 193 nm) to induce fragmentation in a variety of molecular ions [16–18]. For example, the carbonyl and aryl groups in peptides strongly absorb UV radiation, and thus oligopeptides are efficiently fragmented with 193-nm radiation from an ArF excimer laser. The deposition of a large amount of energy at that wavelength, however, leads to indiscriminate fragmentation.

Infrared Multiphoton Dissociation (IRMPD)   This mode of dissociation is a slow heating process that usually is performed in the ICR cell; the ions trapped are bombarded with IR photons typically from a continuous-wave (CW) CO_2 laser with irradiation times in the millisecond range [19,20]. The mechanism of ion activation involves slow, sequential multiphoton absorption through IR-active modes to enable a stepwise vibrational excitation to the continuum beyond the dissociation threshold. During this process, the lowest-energy decomposition pathways are sampled to produce structurally more useful fragmentation. However, for efficient ion activation and dissociation, the rate of the competitive collisional cooling of and photoemission from the excited ions must be lower than the rate of the energy gain by photon absorption. Because the dissociation probabilities of the isolated ions are photon energy dependent, IRMPD provides a means of selective ion dissociation. As an example, the phosphate group is a powerful chromophore for 10.6-μm photons (e.g., from a CO_2 laser). Therefore, ion activation with these photons will allow the selective detection of phosphopeptides in a mixture [21]. Other advantages of IRMPD over CID are greater control over energy deposition, on-axis fragmentation, lower mass discrimination, and higher efficiency in fragmentation. As a result, IRMPD permits increased sequence coverage for peptides over CID. In recent years, this ion dissociation approach has gained increased popularity because of its ease of implementation in ion-trapping instruments.

Blackbody-Induced Radiative Dissociation (BIRD)   In this technique also, ion dissociation occurs by absorption of IR photons but without the use of lasers [22]. Instead, the trapped ions are activated by absorption of blackbody radiations that are emitted from the walls of a heated vacuum chamber (at $<10^{-8}$ torr) of an FT–ICR mass spectrometer [21]. The energy deposition and fragmentation can be controlled by varying the temperature. Therefore, BIRD is a highly selective process with nearly 100% dissociation efficiency. As with other slow heating processes, low-energy fragmentations tend to dominate the BIRD spectra. Applications of BIRD, first demonstrated for small weakly bound cluster ions, have been extended to molecules as large as bovine ubiquitin (8.6 kDa) [23]. Consistent with other low-energy fragmentation processes, the BIRD spectra of peptides are dominated by structurally useful *y*- and *b*-type sequence ions as well as by ions formed by small-molecule losses (e.g., water and ammonia).

4.3.4. Electron-Capture Dissociation

Electron-capture dissociation (ECD), developed by McLafferty, Zubarev, and colleagues, is a recent addition to the repertoire of ion-activation methods [24,25]. At present, it is solely applicable to the ESI-produced multiply protonated biopolymeres, $[M + nH]^{n+}$ [24–27]. The mechanism of ECD involves an excitation of the mass-selected protonated ion by the capture of a low-energy (<0.2 eV) electron and subsequent fragmentation of the resulting odd-electron ion $[M + nH]^{(n-1)+\bullet}$. The capture of an electron adds 5 to 7 eV of recombination energy to the odd-electron ion. Because the process is nonergodic (i.e., a process that does not involve any intramolecular vibrational-energy distribution), fragmentation of a large protein ion can occur even before the deposited energy has a chance to randomize over a large number of degrees of freedom. Hydrogen rearrangement in these hydrogen-abundant species plays a significant role in their fragmentation. For example, in peptides, the migration of a hydrogen atom to the backbone carbonyl group and subsequent homolytic cleavage of the $NC-C_\alpha$ bond results in the formation of amino acid sequence-specific (e.g., *c*- and *z*-type) ions in large abundance. ECD produces far more backbone cleavages than CAD and can be used to sequence larger peptides and intact proteins.

ECD has been implemented primarily with FT–ICR–MS instruments. A low-energy electron beam is generated by a conventional filament source, or better, with a dispenser cathode. The latter has the advantages of a higher flux of electrons at lower surface temperatures and a larger emitting area. A recent study also reports the coupling of ECD with a three-dimensional quadrupole ion trap [28]. A unique feature of this coupling is the use of magnetized electrons, which can be trapped in the QIT for an extended time.

Although the electron-capture process is highly efficient with electrons of nearly thermal energies (e.g., 0.2 eV), higher-energy electrons can also serve the purpose. For example, in a technique termed *hot ECD* (HECD), electrons of about 7.0 eV energy have been used, albeit with slightly lower efficiency than that of thermal electrons [26]. Uses of ECD include polypeptide sequencing, DNA sequencing, study of gas-phase conformations of polypeptides, and identification of sites of phosphorylation, glycosylation, sulfation, γ-carboxylation, fatty acid attachment, and S–S bond sites.

Electron-Transfer Dissociation A new technique, electron-transfer dissociation (ETD), has been introduced for fragmentation of multiply charged peptide ions [29–31]. The process is similar to ECD but uses an ion–ion reaction to transfer an electron to the peptide ion. Anthracene anions generated in a CI source are employed as electron donors. ETD methodology has been adapted in a quadrupole linear ion-trap instrument. Analogous to the ECD process, the transfer of an electron induces fragmentation in the peptide backbone to produce sequence-specific c- and $z^{\bullet}$-type ions.

4.4. REACTIONS IN TANDEM MASS SPECTROMETRY

Apart from the endoergic dissociation reactions mentioned above, a variety of other reactions can also be accessed in tandem mass spectrometry. In many such reactions, collectively termed *charge-permutation reactions*, there is a change in the charge of the colliding species during the collision-activation step. Some of the reactions are discussed here. For further reading, an excellent review on the subject can be consulted [32].

Charge-exchange ionization occurs by abstraction of an electron during collision between a fast-moving ion and a neutral molecule [reaction (4.5)]. In this reaction the charged precursor is transformed to a neutral species and the neutral target to an ionized species. Another endoergic ionization event is *charge stripping*, whereby a singly charged ion is converted to a doubly charged ion by the removal of an electron [reaction (4.6)]. The *charge inversion reaction* refers to the removal of two electrons from a negatively charged ion during a collision event to convert it to a positively charged ion [reaction (4.7)]. The charge inversion of positive ions can also occur during their collisions with a neutral target (reaction (4.8)]. The *collision ionization* refers to ionization of a fast-neutral species during a high-energy collision event [reaction (4.9)]. *Neutralization-reionization mass spectrometry* (NRMS) is an extension of this process in which a fast-moving high-energy ionic species is mass-selected and neutralized in a collision cell by charge-exchange collisions with a target gas (e.g., Xe) or with metal vapors (e.g., from Na, K, Mg) [33]. Residual ions are deflected away, and the beam of fast atoms is further subjected to CID reaction in the second collision cell. The CID products are mass-analyzed by MS-2. NRMS is an ideal approach to probe several elusive transient as well as stable species formed by neutralization. The *electron-transfer reaction* occurs when a high-translational-energy positive ion collides with a neutral species to form a cation with one charge state higher than its precursor and a radical anion (reaction (4.10)]. These reactions have played a significant role in structure elucidation of organic ions. Most of the examples cited in the literature have used sector instruments to perform these reactions.

Charge exchange: $\quad m^{+\bullet} + N \longrightarrow m + N^{+}$ (4.5)

Charge stripping: $\quad m^{+\bullet} + N \longrightarrow m^{++} + N + e^{-}$ (4.6)

Charge inversion: $\quad m^{-\bullet} + N \longrightarrow m^{+\bullet} + N + 2e^{-}$ (4.7)

$\qquad\qquad\qquad\quad m^{+\bullet} + N \longrightarrow m^{-\bullet} + N^{2+}$ (4.8)

Collision ionization: $\quad m + N \longrightarrow m^{+\bullet} + N + e^-$ (4.9)

Electron transfer: $\quad m^{n+} + N \longrightarrow m^{(n+1)+\bullet} + N^{-\bullet}$ (4.10)

Under some conditions, exoergic ion–molecule reactions have provided useful structural information as well as fundamental aspects of gaseous ions [34]. An essential requirement in these reactions is the formation of a relatively long-lived activated complex through collisions between low-translational-energy target ions with a neutral reactant.

4.5. TANDEM MASS SPECTROMETRY INSTRUMENTATION

Tandem mass spectrometry instruments are classified into two broad categories: tandem in-space and tandem in-time. In *tandem in-space instruments*, the three steps of MS/MS (i.e., mass selection, fragmentation, and mass analysis) are carried out in three discrete regions. In this category, two mass spectrometers are arranged sequentially. Multisector magnetic, triple-quadrupole, TOF-based instruments, and hybrid instruments are examples of tandem in-space instruments. *Tandem in-time instruments*, on the other hand, perform these steps in the same region but by using a temporal sequence. QIT, LIT, and FT–ICR mass spectrometers fall into the latter category of tandem instruments.

4.5.1. Magnetic-Sector Tandem Mass Spectrometers

Since the 1960s, magnetic-sector instruments have enjoyed the status of mainstream tandem mass spectrometers. Their contribution to the chemistry of gas-phase ions and to the field of peptides and other biomolecules has been remarkable. These instruments are highly suited to study of high-energy charge-permutation reactions. Ions typically in the range 3 to 10 keV are sampled. Various different arrangements of electrostatic (E) and magnetic analyzers (B) have produced several diverse types of tandem instruments [35]. Initially, only double-focusing instruments arranged either in forward (EB type) or reverse (BE type) geometry were available for MS/MS studies. Their major shortcoming, however, was poor resolution either in precursor-ion selection or in product-ion mass analysis. Later, to ameliorate this situation, three-, four-, and even five-sector instruments of various combinations were developed; the most common being the EBE, BEB, EBEB, BEEB, and BEBE geometries. Three-sector instruments (EBE and BEB) have the ability to mass-select precursor ions at high resolving power (over 30,000), but product ion analysis is still limited to low-to-modest resolution. In contrast, four-sector instruments perform precursor-ion selection and product-ion analysis at higher resolutions.

A major advantage of these sector instruments is that all four types of scan functions (Section 4.2) can be implemented. Multisector instruments also have the facility to perform MS^n experiments. The number of stages, n, of MS analysis is, however, restricted to the number of sectors in that tandem instrument. In addition, the charge-permutation reactions of fast-moving species, mentioned in

Section 4.4, can be studied conveniently in these sector machines. Sector instruments also play an important role in *translational-energy spectroscopy*. This field involves measurements of angles and energies of the reaction products to provide information on the energetics and dynamics of a reaction [36]. The high cost and poor MS/MS efficiency are serious limitations of these instruments. Also, their coupling with high-pressure ionization sources and high-resolution separation devices is a problem.

Tandem MS Scans for Magnetic-Sector Instruments   In magnetic-sector instruments, fragmentation reactions that occur in the FFRs in front of an electric or a magnetic analyzer are monitored by using specifically designed scan laws [37].

1. *The B-Scan.* If the precursor ion fragments in front of a magnetic analyzer, the resulting fragment ion will not be detected at its true mass values but at some noninteger mass given by

$$m^* = \frac{m_1^2}{m_p} \qquad (4.11)$$

   Scanning the magnet field strength provides the product-ion mass spectrum. The product-ion peaks are broad and of low intensity.

2. *E-scan.* If the precursor ion fragments in front of an electric sector, the kinetic energy of the resulting product ion is lower than its precursor, owing to the fact that the velocity of the product ion is conserved but its mass is reduced. The newly formed ion m_1 can be transmitted through the electric sector only after the electric field E_p is reduced commensurately to a new value E_1, according to

$$E_1 = \frac{m_1}{m_p} E_p \qquad (4.12)$$

   Thus, by scanning the electric sector (*E-scan*), the product-ion spectrum, also known as the *ion kinetic energy* (IKE) spectrum, is obtained, but without any information about mass values.

3. *V-scan.* The scan of the accelerating voltage (*V*), known as a *V-scan*, has also served to provide a precursor-ion spectrum. When an ion fragments in the first FFR of an EB instrument, its charged fragment will not be transmitted through the electric sector. Scanning *V* upward will, however, provide information on all precursors that fragment to yield a mass-selected (by adjusting *B*) fragment. The accelerating voltage V_p at which the precursor of a chosen fragment is detected is given by

$$V_p = \frac{m_1}{m_p} V_1 \qquad (4.13)$$

   where V_1 is the normal accelerating voltage.

4. *Mass-analyzed ion kinetic energy spectroscopy (MIKES)*. This scan is implemented with reverse-geometry (BE-type) instruments, in which the front-end magnetic sector allows exclusive mass selection of a precursor ion [37]. Fragmentation occurs in a region between the two analyzers (second FFR). A scan of the electric sector [Eq. (4.12)] yields a product-ion spectrum. MIKES also allows direct measurement of kinetic-energy release values.

5. *Linked-field scan at constant B/E*. This scan mode is used to obtain a product-ion spectrum on forward- and reverse-geometry instruments [37]. In this scan, the value of V is fixed and B and E are both scanned while the ratio B/E is held constant. Fragmentation occurs in the first FFR. The mass resolution of the product-ion analysis is much higher; selection of the precursor ion is at a lower resolution.

6. *Linked-field scan at constant B^2/E*. A precursor-ion spectrum of a chosen fragment formed in the first FFR of an EB or BE instrument can be acquired with a linked-field scan, in which the B- and E-fields are scanned simultaneously (keeping V fixed) according to the relation $B^2/E =$ constant.

7. *Constant-neutral-loss scan*. A neutral loss that occurs in the first FFR of both EB and BE instrument can be monitored when the B- and E-fields are scanned simultaneously according to the expression $B^2(1 - E')/E'^2 =$ constant (where B is the magnetic field required to transmit m_2 and $E' = E_2/E_1$).

8. *Linked-field scan at constant B^2E*. To obtain a precursor-ion spectrum on a BE instrument, B is scanned upward and E downward, holding the product B^2E constant; precursor-ion selection and fragmentation occur in the second FFR (i.e., in front of the electric sector).

▶ **Example 4.1** With the help of fundamental equations, show that to acquire a product-ion spectrum, it is necessary to keep the B/E ratio constant during scanning of the magnetic and electrostatic fields.

Solution First, remember that if a precursor ion (m_p) fragments in the first FFR of an EB or BE instrument, the velocity of the product ion (m_1) is the same as that of its precursor. Now, let us assume that E_p and E_1 are the electric-field strengths required to transmit these ions through the ESA and that B_p and B_1 are the corresponding values of the magnetic fields, respectively. By applying Eq. (3.3), we get $m_p v = B_p rq$ and $m_1 v = B_1 rq$, and finally, $m_p/m_1 = B_p/B_1$. Similarly, from Eq. (3.5), we get $m_p v^2 = E_p rq$ and $m_1 v^2 = E_1 rq$, and finally, $m_p/m_1 = E_p/E_1$. By equating these two terms, we get $m_p/m_1 = B_p/B_1 = E_p/E_1 = B/E$. Thus, to transmit precursor and its product ions through the electric and magnetic sectors, it is necessary to scan B- and E-fields simultaneously while keep their ratio constant.

4.5.2. Tandem Mass Spectrometry with Multiple-Quadrupole Devices

The wide popularity of tandem mass spectrometry can be attributed largely to the emergence in 1978 of a triple-quadrupole instrument [38]. In this simple device [39], three quadrupoles are arranged sequentially (Figure 4.5). The first (Q_1) and last (Q_3) quadrupoles both operate as normal mass filters (i.e., for mass analysis) and the middle one as a radio-frequency (rf)-only quadrupole (Q_2), the working principles of which were described in Section 3.4. The direct-current (dc) and rf potentials both control the operation of Q_1 and Q_3, whereas Q_2 is operated with only the rf potential. The rf-only mode of operation of Q_2 allows all ions to pass through it. The Q_2 also serves as a total ion containment region and a collision cell. To improve the ion-transport efficiency, some contemporary instruments are outfitted with an octopole or hexapole collision cell in place of a conventional quadrupole cell. Because ions in the range 0 to 100 eV can be transmitted through quadrupoles, the MS/MS spectra obtained with triple-quadrupole instruments contains ions formed via low-energy fragmentation processes. The collision energy can be controlled by offsetting the voltage between the ion source and Q_2. A pentaquadrupole instrument, comprising three normal quadrupoles and two rf-only quadrupoles, has also been designed with the aim to perform MS^3-type experiments [40].

All four types of scan laws discussed in Section 4.2 can be implemented with a triple-quadrupole instrument. For example, to acquire a product-ion spectrum, Q_1 is set to transmit ions of a specified m/z value into Q_2, where they undergo a CID process. Q_3 is scanned to mass-analyze the products formed in Q_2. A precursor-ion spectrum is acquired by reversing this procedure; that is, Q_3 is set to transmit just the m/z value of a desired product ion, and Q_1 is scanned to transmit all precursors of this chosen product ion. As compared to the magnetic sector–based tandem instruments, a simple scan law is used in the triple-quadrupole instruments to monitor the loss of a neutral. The fields of Q_1 and Q_3 are both scanned in tandem, but with an offset value related to the mass of the neutral.

Although triple-quadrupole instruments are inferior to the four-sector magnetic field instruments with respect to mass resolution and mass range, they have the advantages of low cost, operational simplicity, straightforward scan laws, and a linear mass scale. In addition, the MS/MS efficiency of modern triple-quadrupole instruments is very high. They have unmatched detection sensitivity in the SRM mode and quantification. Also, these instruments permit the study of low-energy ion–molecule reactions. *Energy-resolved mass spectrometry* experiments can also

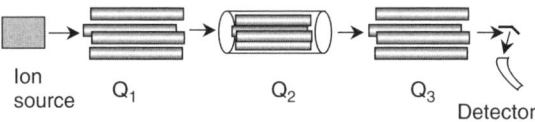

Figure 4.5. Schematic diagram of a triple-quadrupole tandem mass spectrometer. (Reproduced from C. Dass, *Principles and Practice of Biological Mass Spectrometry*, Wiley-Interscience, 2001.)

be performed with triple-quadrupole instruments. In these experiments, MS/MS spectra can be acquired as a function of the collision energy [41]. Other limitations of these instruments are a gradual fall-off in performance above m/z 1000, and their inability to perform certain types of charge-permutation reactions is discussed in Section 4.4.

4.5.3. Tandem Mass Spectrometry with Time-of-Flight Instruments

A linear time-of-flight (TOF) mass spectrometer is not suited to MS/MS experiments. Considering the wide popularity of TOF instruments, it was natural that a tandem MS instrument based on this geometry is developed. Some of the developments in this field are described below.

Post-source Decay One of the attempts in this direction is the development of a technique known as *post-source decay* (PSD) [42]. A substantial portion of ion source–formed ions undergoes metastable dissociation in the flight tube FFR. Normally, these fragments escape detection in a linear TOF mass analyzer because the charged fragment and the corresponding neutral formed in a metastable decay both retain the velocity of their precursor and thus arrive at the detector along with their precursor. The kinetic energy of the product ions, however, is reduced in proportion to the change in their mass. As discussed in Section 3.5, the energy-resolving characteristic of a reflectron TOF (RTOF) instrument has the ability to distinguish ions on the basis of differences in their kinetic energy. Thus, a reflectron can serve as a tandem MS to detect the flight-tube fragmentation reactions.

To enhance the FFR metastable fragmentation of MALDI-produced ions, a PSD spectrum is usually acquired using "hot" MALDI matrices, such as α-cyano-4-hydroxycinnamic acid. Figure 4.6 shows a conceptual diagram of the PSD method. A fast computer-controlled ion gating provides mass selection of the precursor ion. The ion gate is a timed-deflection device that remains active all the time except for a very brief period to allow passage of ions of chosen m/z value. Scanning of the spectrum is initiated when the neutral products strike the detector placed behind the reflectron. The charged products of the metastable decay of the precursor ion are mass-resolved by the reflectron and are detected by a detector located at the end of the second drift region. Because the energy range of PSD products far exceeds the bandwidth of a common reflectron [see Eq. (4.12)], at a specific reflectron voltage setting (potentials U_1 and U_2 in Figure 4.6), only a small segment of the product-ion spectrum can be brought to focus at the detector. A complete spectrum, however, can be obtained by piecing together several segments, each acquired by stepping up the reflectron voltage. The quality of the mass spectral data is modest: first, because the mass resolution of the ion gate–selection is poor (ca. 40 to 70 at FWHM), and second, because not all fragmentation channels are accessible, owing to the inherent inefficiency of the metastable fragmentation process. Bleeding air or a collision gas into the flight tube might enhance the fragment ion yield.

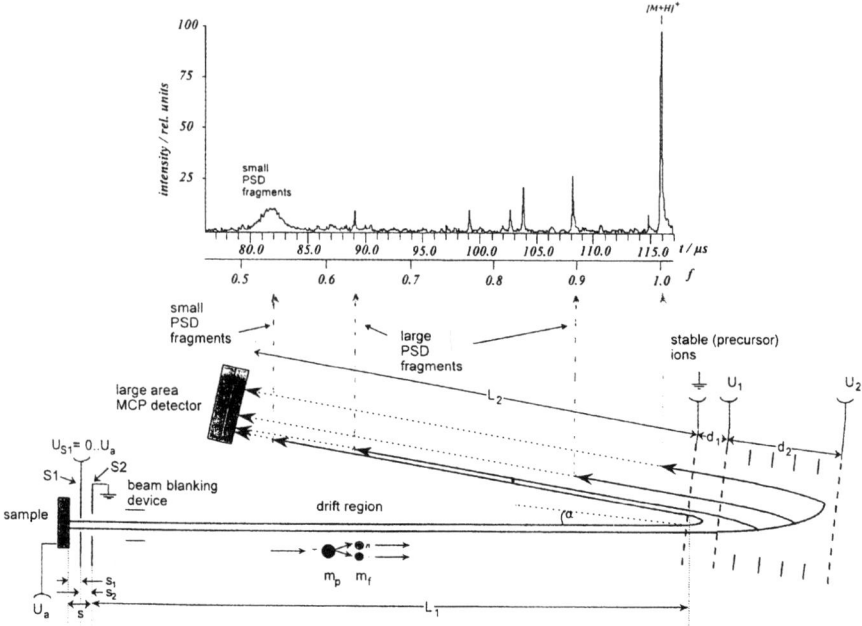

Figure 4.6. Post-source decay measurements with a reflectron TOF mass spectrometer. (Reproduced from ref. 42 by permission of Wiley-Interscience, copyright © 1997.)

Two alternative approaches have been suggested as a possible solution to the mismatch between the wide energy range of the product ions and the reflectron bandwidth. First, a wideband reflectron can be employed to acquire a complete mass spectrum with a single set of reflectron potentials. A curved-field reflectron, which has a wide energy-acceptance window, can be employed for this purpose [43–45]. Second, the precursor ions are slowed down to lower kinetic energies by using a lower accelerating potential. An alternative approach to slow precursor ions is to decelerate them to acceptable energies prior to dissociation and to reaccelerate all ions for the mass analysis before their entry into the reflectron [46]. The latter approach is useful specifically for CID-formed products.

Tandem TOF Instruments A significant step to overcome limitations of the PSD approach is to combine two discrete TOF mass analyzers into a true tandem instrument [45–47]. An ion activation–fragmentation region is added between the two analyzers. All permutations of TOF and RTOF combinations have been tested (e.g., TOF/TOF, TOF/RTOF, and RTOF/RTOF). The design of a commercial version is shown in Figure 4.7. Primarily, it consists of three sections: a linear TOF for the MS-1 (TOF1) function, a collision region that serves as an ion source for the MS-2 section, and a reflectron that acts as an MS-2. The precursor ions are generated in the TOF1 and focused in the center of a timed-ion selector (TIS) device by setting the appropriate delay between the ion-extraction

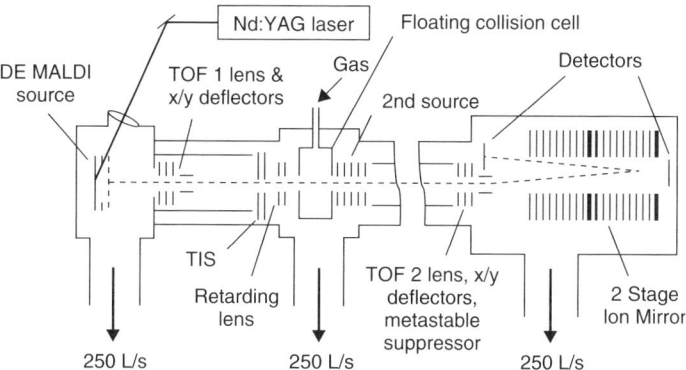

Figure 4.7. Schematic of MALDI–TOF/TOF instrument. (Reproduced from ref. 47 by permission of Elsevier Science, copyright © 2002 American Society for Mass Spectrometry.)

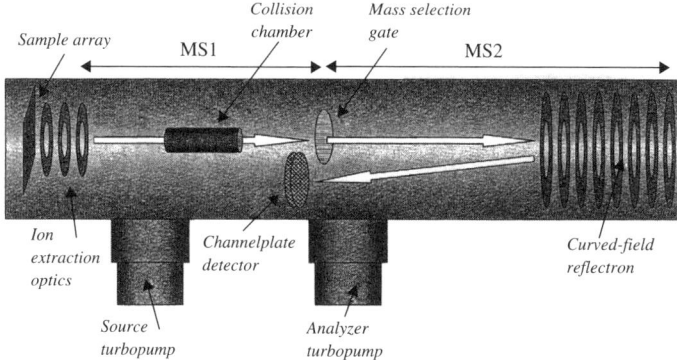

Figure 4.8. Schematic of a TOF/TOF tandem instrument with a curved field reflectron as the MS-2. The mass-selection ion gate is located at the focal plane of MS-1. The MS-1 region also contains a collision cell. (Reproduced from ref. 45 by permission of the American Chemical Society, Washington, DC, copyright © 2004.)

pulse and the laser-triggering event. The TIS is a double-sided deflection gate, the first gate of which acts as a high-pass filter, and the second gate, as a low-pass filter. With a proper setting of voltages at predetermined times, the double-sided gate prevents all m/z values that are lower and higher than the precursor ion m/z to enter the collision cell. The gate-selected ions are decelerated to 1- to 2-keV energies prior to their entry into the collision cell. After the CID event, ions exit the collision cell and travel through a short FFR region at reduced velocities before they are reaccelerated into the TOF2 region, where the mass dispersion and energy focusing of product ions are achieved. With this arrangement, it is feasible to obtain a complete MS/MS spectrum with a single extraction pulse.

This commercial design has recently been modified to incorporate a curved field reflectron as a TOF2 (Figure 4.8). With this modification, there is no need to decelerate the ions prior to CID and to reaccelerate the CID products because the curved field reflectron can accept broad-energy-range ions. As a consequence, CID can be performed at the full 20-keV collision energy to increase the extent of fragmentation.

Another tandem TOF/TOF design uses a LIFT cell for post-acceleration to higher energy [48]. This instrument also consists of a linear TOF as an MS-1 and a gridless space-angle reflectron as an MS-2. In addition, it contains a TIS for gated-ion selection, a LIFT cell to raise the potential energy, a postlift metastable-suppression device to remove unfragmented precursors. A collision cell is also added between the first ion source and the TIS gate to acquire a true CID spectrum. Raising the potential with a lift device allows the mass analysis of fragment ions in a single spectrum.

▶ **Example 4.2** The loss of $^{\bullet}C_3H_7$ is a common fragmentation reaction of the 2-pentanone$^{+\bullet}$ radical cation $(C_5H_{10}O^{+\bullet} \longrightarrow CH_3CO^+ + {^{\bullet}C_3H_7})$. Calculate the velocity of CH_3CO^+ if it is formed (a) in the ion source and (b) in the flight tube of a TOF instrument. The ions were extracted out of the ion source with an accelerating voltage pulse of 10 kV.

Solution

(a) From Eq. (3.11), the velocity of an ion is given by $v = \sqrt{2qV/m}$. Therefore, with $q = 1.602 \times 10^{-19}$ C, $V = 10,000$ V, and the mass of the ion $= (43)(1.66 \times 10^{-27}$ kg), and remembering that $C \cdot V = J = kg \cdot m^2/s^2$, the velocity of the ion source–formed CH_3CO^+ is given by

$$v = \sqrt{\frac{(2)(1.602 \times 10^{-19} \text{ C})(10,000 \text{ V})}{(43)(1.66 \times 10^{-27} \text{ kg})}} = 2.12 \times 10^5 \text{ m/s}$$

(b) Remember that a flight tube fragment will retain the velocity of its precursor. Therefore, the velocity of the flight tube–formed CH_3CO^+ is given by

$$v = \sqrt{\frac{(2)(1.602 \times 10^{-19} \text{ C})(10,000 \text{ V})}{(86)(1.66 \times 10^{-27} \text{ kg})}} = 1.50 \times 10^5 \text{ m/s}$$

4.5.4. Tandem Mass Spectrometry with a Quadrupole Ion-Trap Mass Spectrometer

The quadrupole ion trap (QIT) has developed into a highly successful MS/MS device [49]. It is a tandem-in-time mass spectrometer, meaning that all steps of MS/MS are performed in the same space but with a temporal sequence. A typical

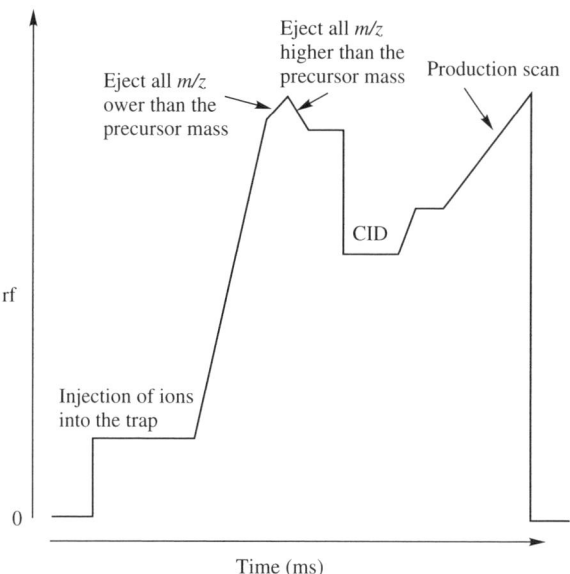

Figure 4.9. Pictorial depiction of various steps of MS/MS analysis in a QIT. (Reproduced from C. Dass, *Principles and Practice of Biological Mass Spectrometry*, Wiley-Interscience, 2001.)

MS/MS experiment in this device involves three time-separated steps: ion isolation, ion excitation/dissociation, and mass analysis of the product ions. Typical scan functions for MS/MS operation are depicted in Figure 4.9. After the initial steps of ion formation and ion confinement in the trapping field, all ions except the precursor are ejected from the trap using the resonance-ejection procedure [50]. Next, an excitation pulse of such a low amplitude that it does not eject any precursor ions out of the trap is applied to facilitate their dissociation via collisions with an inert bath gas. Finally, the mass analysis of the product ions is achieved by the conventional rf amplitude scan of the ring electrode. One possible method to isolate a precursor ion is to scan the fundamental rf voltage in the reverse direction to eject all ions with m/z values greater than that of the precursor ion [51]. After this step, the fundamental rf voltage is scanned in the forward direction to eject all low-mass ions. During these reverse and forward scans, an auxiliary resonance signal is also ramped. Alternative ion-isolation methods include (1) use of a broadband waveform [52] and (2) use of combined rf and dc potentials to bring the q_z and a_z values of the ion to the apex of the stability diagram (see Section 3.6.2).

One major drawback of QIT is its inability to perform precursor-ion and neutral-loss scans. The feasibility of MS^n experiments, however, is a remarkable feature of this instrument. After dissociation of the precursor ions, a specific product ion is mass-selected by ejecting all other ions present in the trap, and the ion-excitation process is repeated for the new precursor. The sequential product spectra can be acquired up to any MS^n stage until the ion population permits ion detection with useful ion statistics. A practical limit of tandem mass spectrometry with current commercial instruments is MS^{10}.

A significant enhancement in the ion signal and dynamic range is realized by use of the sustained off-resonance irradiation (SORI) mode of ion activation [53]. In this modified CID approach, the isolated precursor ion is excited slightly off-resonance. As a result, the ions undergo many cycles of acceleration to larger orbits, and each time, deceleration to smaller orbits during the excitation period. With this approach, ion activation occurs for much longer times to facilitate a greater number of collisions and subsequent enhancement of the product-ions-population.

4.5.5. Tandem Mass Spectrometry with an FT–ICR Mass Spectrometer

High-mass, high-resolution, high-efficiency, and MS^n capability have made FT–ICR instruments highly desirable tandem mass spectrometers. Similar to QIT, FT–ICR–MS is, however, limited to acquiring the product-ion spectrum. Conceptually, the FT–ICR scan functions for the MS/MS operation are similar to those used in a QIT [53–55]. The sequence of an MS/MS experiment is shown in Figure 4.10. First, a quench pulse is applied to eject from the cell any residual ion from the preceding experiment. Next, an ionization pulse is applied, and the precursor ion is isolated with a resonant excitation signal that contains all frequencies except the one that is related to the precursor ion; as a result, all other ions are ejected out of the FT–ICR cell. After mass selection, the precursor is translationally excited to a bigger orbit, and helium is pulsed into the cell concurrent with this excitation pulse to allow CID. The CID products are mass-analyzed as usual: that is, by excitation of the products to orbits of larger radius and detection of their image current. The FT–ICR has the unique distinction that in addition to SORI–CID, a variety of other fast and slow ion-activation methods discussed in Section 4.3 can be implemented easily with this device.

4.5.6. Tandem Mass Spectrometry with Hybrid Instruments

Hybrid tandem mass spectrometers are constructed by coupling two different types of mass analyzers with the objective to accrue the benefits of the best performance features of each analyzer type. This arrangement enhances the capability

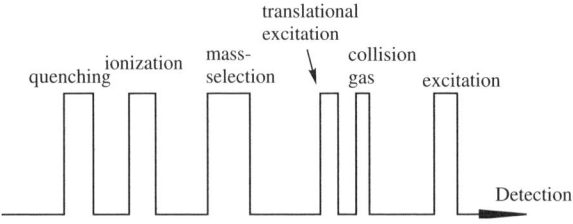

Figure 4.10. Sequence of events in a typical MS/MS experiment in an FT–ICR.

of a single type of mass analyzer. In the past, efforts were directed to construct hybrid instruments made of a magnetic-sector mass spectrometer as MS-1 and a quadruple as MS-2 [56]. Some common designs of this type of instruments are B-qQ, EB-qQ, BE-qQ, and EBE-qQ. With these instruments, especially the last three, mass selection of the precursor ion is achieved at high resolution and product-ion analysis at unit mass resolution. Their major strength is the accessibility of low- and high-energy CID reactions on a single instrument. An EB–QIT combination has also been developed for MS^n analysis [57]. Today, these instruments are not much in demand. Hybrid instruments that are currently popular are discussed next.

Quadrupole–Orthogonal Acceleration TOF Instrument As of today, quadrupole (Q)–orthogonal acceleration (oa) TOF (oa–TOF) instrument is the most popular hybrid instrument [58,59]; a simplistic pictorial representation is shown in Figure 4.11. The quadrupole section consists of a normal mass-resolving quadrupole and an rf-only quadrupole. The latter serves as a collision cell and as an ion-accumulation device. For precursor-ion scan, the ions desired are mass-selected by the main quadrupole, accumulated in the collision cell, and a packet of the CID product ions is pushed into the TOF analyzer. The precursor-ion selection by the quadrupole is at a medium resolution, but the product-ion analysis by the TOF section is at a reasonably high resolution.

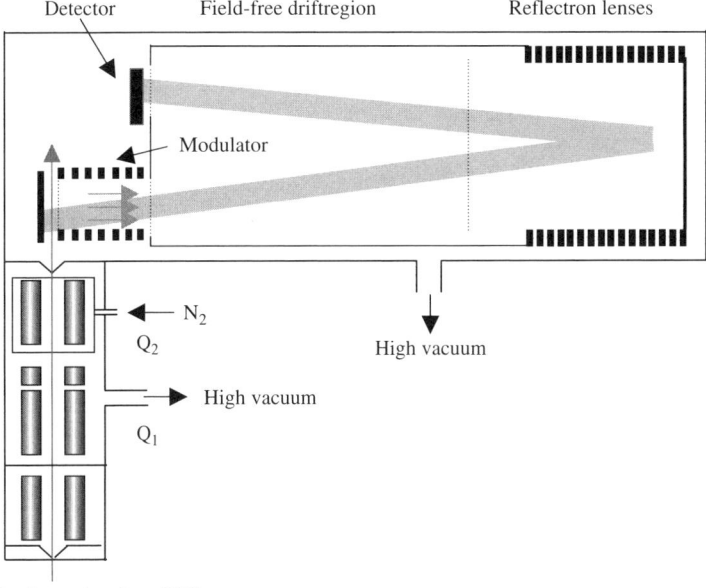

Figure 4.11. Schematic diagram of a quadrupole-orthogonal acceleration time-of-flight hybrid tandem mass spectrometer. (Reproduced from C. Dass, *Principles and Practice of Biological Mass Spectrometry*, Wiley-Interscience, 2001.)

The duty cycle of a Q–oa–TOF can reach up to 100% over a narrow mass range by appropriate setting of the TOF injection pulse with respect to the ion gating pulse. High transmission, ultrahigh sensitivity (low femtomoles for peptide sequencing), good mass accuracy, and very high spectral generation rate are the other useful attributes of the Q–oa–TOF design. In addition, the TOF section can be operated as a stand-alone mass spectrometer when the main quadrupole is switched to the rf-only mode. An improved version of Q–TOF capable of transmission and analysis of large macromolecular assemblies became available recently [60]. By reducing the rf frequency applied to the quadrupole rods and optimizing the pressure in various regions of the instrument, individual ionic species up to m/z 22,000 can be isolated with the quadrupole.

In a normal operation, Q–oa–TOF is suitable only for acquiring a product-ion scan. The instrument can, however, be programmed to emulate precursor-ion and neutral-loss scans [61]. These scans are made possible because of the innate capability of the Q–TOF to record in quick succession a normal mass spectrum and a product-ion spectrum. In this novel procedure, alternate mass spectra are acquired at high (30 to 70 eV; to induce fragmentation) and low (<10 eV) collision energies with the quadrupole operating in the wideband mode. The lower-energy spectrum provides a list of all the precursors present in the sample injected. Upon detection of a specified product ion, the Q–TOF is programmed to switch to the MS/MS mode to record just the product-ion spectra of those precursor ions. By setting a specified neutral loss, a precursor–product combination can be identified. For the neutral-loss experiments, the data system first calculates the masses of all putative precursors in the low-energy spectrum, and then examines the high-energy data for neutral losses of specified m/z differences. Once the neutral loss is identified, MS/MS spectrum of the specified precursor is aquired to confirm its identity.

Linear Ion Trap–oa–TOF Instrument A useful alternative design combines a quadrupole linear ion trap (LIT) with an oa–TOF mass analyzer [62]. This instrument consists of a short rf quadrupole ion guide, a quadrupole mass filter that operates in the rf-only mode, and an oa-TOF mass analyzer. The quadrupole mass filter is operated as an LIT to serve as an ion-accumulation and precursor-ion selection device. The mass-selected precursor is activated via a resonant excitation mode in the LIT, while the TOF instrument provides high-speed mass analysis of the resulting product ions. MS^3 experiments have also been demonstrated with this hybrid system. A combination of QIT with an oa–TOF mass analyzer has also proven to be an effective tandem MS system [63,64]. The QIT performs tasks similar to those performed by an LIT in the LIT–oa–TOF design (i.e., its functions are ion accumulation, mass separation, and ion activation.

Magnetic Sector–Orthogonal Acceleration (oa) TOF instrument A magnetic sector–based MS-1 has also been combined with an oa–TOF analyzer to serve as a tandem instrument [65]. The double-focusing action of the magnetic-sector MS-1 provides high-resolution mass selection of the precursor

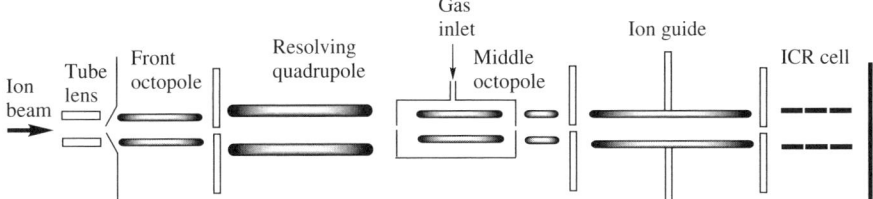

Figure 4.12. Schematic diagram of a hybrid quadrupole–FTICR tandem mass spectrometer. (From ref. 66.)

ion; its fragmentation takes place via high-energy CID to provide MS/MS spectra that are similar in quality to those obtained with four-sector tandem instruments but without incurring the exorbitant cost of those four-sector instruments.

Quadrupole–Fourier Transform Ion Cyclotron Resonance Instrument
Another purposeful hybrid tandem instrument has been constructed with a quadrupole at the front end and an FT–ICR as a second-stage mass analysis device [66,67] (Figure 4.12). Only mass-selected target ions are trapped in the ICR cell, thus reducing the space-charging effect and improving the dynamic range and sensitivity of FT–ICR–MS analysis. The MS-1 section of this instrument is constructed from several different multipole devices, such as an rf-only front multipole (octopole or hexapole), a normal mass-resolving quadrupole mass filter, an rf-only middle multipole for external accumulation and activation, and an rf-only transfer multipole to serve as an ion guide (Figure 4.12). The front multipole provides accumulation of ESI- or MALDI-generated ions. Mass selection of the desired precursor is accomplished by the mass-resolving quadrupole. The middle multipole serves as a collision cell for CID of the mass-selected precursor. The CID products are gated into the ICR cell via an rf-only transfer multipole for further mass analysis. If desired, the mass-selected precursors can be transported directly into the ICR cell through the middle and transfer multipoles for ion activation and dissociation and subsequent product ion detection. The latter arrangement has the advantage that ion activation is not limited to CID only but can be performed by a suite of techniques that were discussed in Section 4.3.

Linear Ion Trap–Fourier Transform Ion Cyclotron Resonance Instrument
To reap the benefits of the high ion-trapping capacity, MS^n capability, and automatic gain control of a linear quadrupole ion trap and the high-mass accuracy and high-resolving power of FT–ICR–MS, a hybrid combination of these two mass spectrometry instruments has also emerged [68]. A commercial version of this instrument (abbreviated LTQ/FT–MS) is available through Thermo Electron Corporation. Figure 4.13 is the schematic of a prototype LTQ/FT–MS. The main components of this instrument are an API ion source, an LTQ, an ion guide, and an ICR cell. The LTQ consists of an array of four hyperbolic rods, each segmented into four axial sections. The ICR cell is an open, cylindrical Penning trap

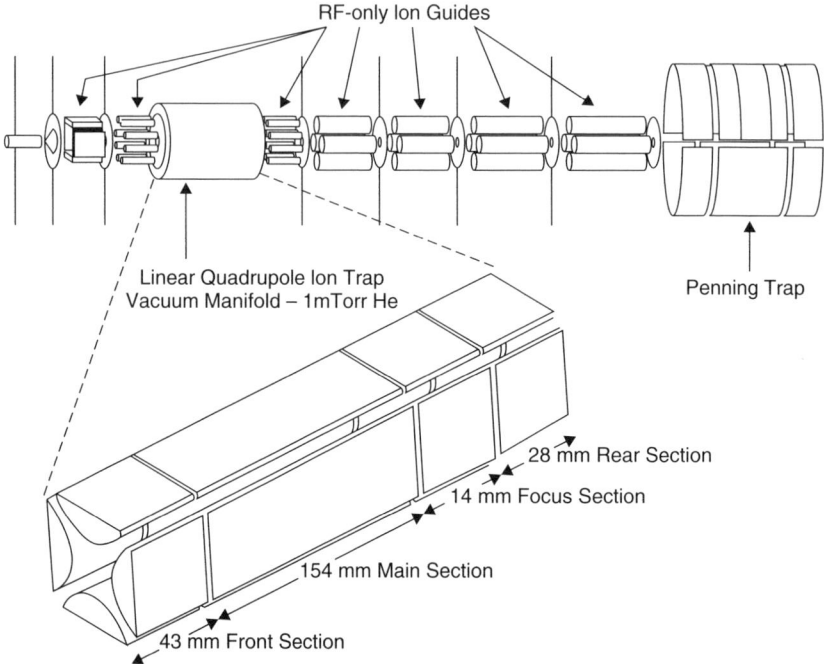

Figure 4.13. Schematic of a hybrid LTQ/FT–MS tandem instrument. (Reproduced from ref. 68 by permission of the American Chemical Society, Washington, DC, copyright © 2004.)

that sits within the bore of a superconducting solenoid magnet. Ions produced in an ESI source are transferred to the LTQ through a flat rod quadrupole and an octopole assembly. The LTQ is used for ion accumulation, precursor isolation, and subsequent CID of precursor ions. The mass-selected precursor ion and its CID products are transported through a series of five rf-only multipole ion guides into the Penning trap for m/z detection. MS/MS spectra can be recorded with either LTQ or FTMS acting as a mass analyzer. This hybrid instrument can acquire spectra in data-dependent mode at high resolution (ca. 100,000) with a mass measurement accuracy of 1 to 2 ppm. Amino acid sequence analysis of peptides at zeptomole sensitivity has been demonstrated with this instrument [68].

Linear Ion Trap–Orbitrap Mass Spectrometer A recent addition to the family of hybrid tandem mass spectrometers is a linear ion trap (LTQ)–orbitrap mass spectrometer [69,70]. Conceptually, this instrument is similar to the LTQ/FT–ICR tandem mass spectrometer discussed above but uses an orbitrap in place of a Penning trap, thus avoiding the complexity and cost of a superconducting magnet. The working principle of an orbitrap was described in Section 3.9 [71,72]. A schematic diagram of the LTQ/orbitrap mass spectrometer is shown in Figure 4.14. This hybrid tandem mass spectrometer consists of an

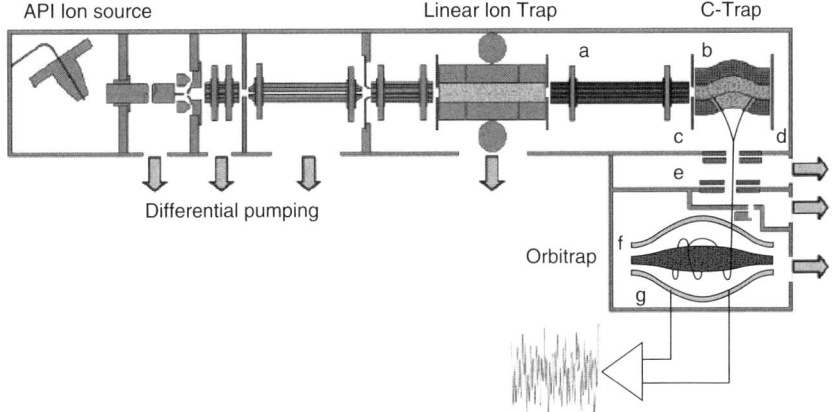

Figure 4.14. Schematic diagram of the LTQ/orbitrap hybrid tandem mass spectrometer: a, transfer octopole; b, C-trap; c, front gate electrode; d, rear trap electrode; e, ion optics; f, inner orbitrap electrode; g, outer orbitrap electrode. (Reproduced from ref. 69 by permission of the American Chemical Society, Washington, DC, copyright © 2006.)

ion source, an LTQ as an MS-1, an orbitrap as an MS-2, and the usual multipole ion guides. A unique feature of this instrument is a C-trap that collects ions from the LTQ and injects them orthogonally into the orbitrap. The C-trap is a curved rf-only quadrupole that is enclosed by two flat lenses; the front-plate lens has an aperture and is used as a gate electrode for entry of ions from the LTQ via a transfer octopole, and the rear plate is used as a trapping electrode. Upon entry into the C-trap, ions are cooled collisionally by the presence of the nitrogen buffer gas at low pressure. Stored ions are pulsed out of the C-trap orthogonally and injected into the orbitrap, wherein they oscillate around an inner electrode. The detection of these ions is achieved through image current that develops in the outer electrodes and Fourier transformation of the time-domain transient signal. The instrument also contains several multipole ion guides for transfer of ions. As with the LTQ/FT–MS, all modes of LTQ operation are accessible and are complemented by the high-resolution mass detection with an orbitrap. The utility of this instrument has been demonstrated for proteomic applications [70].

As a quick guide, a comparison of several important characteristics of some of the instruments that are employed for MS/MS applications is given in Table 4.1.

OVERVIEW

Tandem mass spectrometry is a multistage mass spectrometry technique that can precisely ascertain the progeny of ions. MS/MS involves mass selection of a precursor ion, activation and fragmentation of the mass-selected precursor, and mass analysis of the charged products. This technique can be applied to

TABLE 4.1. Comparative Assessment of Various Tandem Instrument

Instrument	Precursor-Ion Resolution	Product-Ion Resolution	Scan Type	Cost
EB (linked scans)	Poor	High	All four types	$$$
BE (linked scans) (MIKES)	Poor Medium	High Poor	All four types Product ion	$$$
EBE	Very high	Poor	All four types	$$$$
EBEB	Very high	Very high	All four types	$$$$$
QqQ	Medium	Medium	All four types	$$$
R–TOF (PSD)	Poor	High	Product ion	$$
R–TOF/R–TOF	High	High	Product ion	$$$$
QIT	Medium	Medium	Product ion	$$
LIT	Medium	Medium	Product ion	$$
FT–ICR	Very high	Very high	Product ion	$$$$
Hybrid instruments				
EB–qQ	Very high	Medium	All four types	$$$$
EBE–oa–TOF	Very high	High	Product ion	$$$$
Qq–oa–TOF	Medium	High	Product ion precursor ion	$$$
QIT–oa–TOF	Medium	High	Product ion	$$$
Q–FT–ICR	Medium	Very high	Product ion	$$$$$
LTQ–FT–ICR	Medium	Very high	Product ion	$$$$$
LTQ–Orbitrap	Medium	Very high	Product ion	$$$$$

obtain a product-ion scan (i.e., information about the products ions of a mass-selected precursor), a precursor-ion scan (i.e., information about all precursor ions that dissociate to form a preselected product ion), and a neutral-loss scan (i.e., information about all precursor ions that exhibit loss of a preselected neutral moiety). Tandem MS is also useful in quantification of target analytes in real-world samples using the multiple-reaction monitoring approach. Several ion-activation techniques are available, including collision-induced dissociation, surface-induced dissociation, photodissociation with UV-visible and IR photons, blackbody-induced radiative dissociation, and electron-capture dissociation.

Tandem mass spectrometry instruments are of two types: tandem in-space, in which the three steps of MS/MS are carried out in three discrete regions, and tandem in-time, which perform the MS/MS steps in the same region by employing a temporal sequence. Tandem in-space instruments are constructed from several sectors of the same types of mass spectrometers (e.g., multisector magnetic, triplequadrupoles, and TOF/TOF) or are of hybrid design, in which two different types of mass spectrometry concepts are coupled (e.g., magnetic sector–quadrupole, magnetic sector–TOF, quadrupole–TOF, quadrupole–LIT, and quadrupole–FT–ICR). The tandem in-time principle is applicable to ion-trapping devices such as QIT, LIT, and FT–ICR mass spectrometers.

EXERCISES

4.1. Explain the basic concepts of tandem mass spectrometry.

4.2. How does tandem-in-time MS/MS differ from tandem-in-space MS/MS?

4.3. What makes tandem mass spectrometry a truly structure-specific technique?

4.4. What different scan functions are available for the tandem MS/MS operation, and how are they implemented?

4.5. Calculate the COM kinetic energy available to m/z 1000 in the rf-only collision cell of a triple-sector quadrupole tandem instrument. Ions were accelerated out of the ion source at 200 V, and argon was used as the collision gas.

4.6. What ion-activation techniques are available to fragment mass-selected precursor ions in tandem MS/MS analysis?

4.7. What types of charge-permutation reactions can be implemented on magnetic-sector tandem MS/MS instruments?

4.8. Similar to Example 4.2, prove that to acquire a precursor-ion spectrum on a forward-geometry sector instrument, the term B^2/E must be held constant during scans of the magnetic and electric fields.

4.9. How are (**a**) precursor-ion and (**b**) neutral-loss spectra acquired on a triple-quadrupole tandem instrument?

4.10. How is a PSD spectrum acquired on a TOF instrument?

4.11. Calculate the velocity of the b_4 ion (m/z 425) formed via PSD of the $[M + H]^+$ ion (m/z 556) leucine enkephalin (TyrGlyGlyPheLeu) in the flight tube of a TOF instrument.

4.12. How is a precursor ion isolated in a quadrupole-ion-trap mass analyzer?

4.13. A Q–TOF has become a highly useful MS/MS device. Explain how the product and precursor-ion spectra are acquired on this instrument.

4.14. Name all types of tandem MS instruments that can be employed to acquire a precursor-ion spectrum.

4.15. Your laboratory is equipped with a BE reverse-geometry magnetic-sector instrument. How will you acquire (**a**) a product-ion spectrum and (**b**) a precursor-ion spectrum with this instrument?

4.16. Upon electron ionization, 2-methylphenol (MW = 108) undergoes the loss of water to produce m/z 90. If this fragmentation occurs in an FFR in front of the magnetic sector, at what m/z value will the product ion be seen?

4.17. You are asked to quantify a drug sample in a plasma extract. The specificity of the analysis is critical. Therefore, you decide to use a selected-reaction monitoring procedure. How will you perform this experiment on a triple-quadrupole tandem instrument?

4.18. How are ions generally transferred between the two MS devices of ion trap–based hybrid tandem MS/MS instruments?

4.19. Why is an ion activation step required in the MS/MS operation?

4.20. The precursor ion of Exercise 4.16 was accelerated out of the ion source of an EB double-focusing instrument at 6000 V. The electric field strength E of the ESA was set at 400 V/m. If the precursor ion fragments in from of the ESA, then what will be value of E required to transmit the production of m/z 90?

4.21. Name all types of tandem instruments that can be used for product ion analysis.

ADDITIONAL READING

1. F. W. McLafferty, *Tandem Mass Spectrometry*, Wiley-Interscience, New York, 1983.
2. K. L. Busch, G. L. Glish, and S. A. McLuckey, *Mass Spectrometry/Mass Spectrometry: Techniques and Applications of Tandem Mass Spectrometry*, VCH Publishers, New York, 1988.
3. R. G. Cooks, ed., *Collision Spectroscopy*, Plenum Press, New York, 1978.
4. R. G. Cooks, J. H. Beynon, R. M. Caprioli, and G. R. Lester, *Metastable Ions*, Elsevier, Amsterdam, The Netherlands, 1973.
5. R. G. Cooks and G. L. Glish, Mass spectrometry/mass spectrometry, *Chem. Eng. News* **59**, 40–52 (1981).
6. C. Dass, *Principles and Practice of Biological Mass Spectrometry*, Wiley-Interscience, New York, 2001.

REFERENCES

1. R. A. Yost and D. D. Fetterolf, Tandem mass spectrometry (MS/MS) instrumentation, *Mass Spectrom. Rev.* **2**, 1–45 (1983).
2. E. de Hoffmann, Tandem mass spectrometry: a primer, *J. Mass Spectrom.* **31**, 129–137 (1996).
3. K. R. Jennings, Collision-induced decompositions of aromatic molecular ions, *Int. J. Mass Spectrom. Ion Phys.* **1**, 227–235 (1968).
4. C. Dass, Dimerization of ketene radical cation in the gas phase, *Rapid Commun. Mass Spectrom.* **7**, 95–98 (1993).
5. G. Neubauer and M. Mann, Mapping of phosphorylation sites of gel-isolated proteins by nanoelectrospray tandem mass spectrometry: potentials and limitations, *Anal. Chem.* **71**, 235–242 (1999).

6. A. Schlosser, R. Ripkorn, D. Bossemeyer, and W. D. Lehmann, Analysis of protein phosphorylation by a combination of elastase digestion and neutral loss tandem mass spectrometry, *Anal. Chem.* **73**, 170–176 (2001).
7. L. Sleno and D. A. Volmer, Ion activation methods for tandem mass spectrometry, *J. Mass Spectrom.* **39**, 1091–1112 (2004).
8. L. Levsen and H. Schwarz, Gas-phase chemistry of collisionally activated ions, *Mass Spectrom. Rev.* **2**, 77–148 (1983).
9. A. K. Shukla and J. H. Futrell, Collisional activation and dissociation of polyatomic ions, *Mass Spectrom. Rev.* **12**, 211–255 (1993).
10. W. F. Haddon and F. W. McLafferty, Metastable ion characteristics, VII: Collision-induced metastables, *J. Am. Chem. Soc.* **90**, 4745–4746 (1968).
11. R. A. Yost, C. G. Enke, D. C. McGilvery, D. Smith, and J. D. Morrison. High efficiency collision-induced dissociation in an RF-only quadrupole, *Int. J. Mass Spectrom. Ion Proc.* **30**, 127–136 (1979).
12. M. A. Mabud, M. J. Dekrey, and R. G. Cooks, Surface-induced dissociation of molecular ions, *Int. J. Mass Spectrom. Ion Proc.* **67**, 285–294 (1985).
13. V. Grill, J. Shen, C. Evans, and R. G. Cooks, Collisions of ions with surfaces at chemically relevant energies: instrumentation and phenomenon, *Rev. Sci. Instrum.* **72**, 3149–3179 (2001).
14. Z. Herman, Collisions of slow polyatomic ions with surfaces: the scattering methods and results, *J. Am. Soc. Mass Spectrom.* **14**, 1360–1372 (2003).
15. F. M. Fernandez, L. L. Smith, K. Kuppannan, X. Yang, and V. H. Wysocki, Peptide sequencing using a patchwork approach and surface-induced dissociation in sector-TOF and dual quadrupole mass spectrometers, *J. Am. Soc. Mass Spectrom.* **14**, 1387–1041 (2003).
16. R. E. Tecklenburg, Jr., and D. H. Russell, An evaluation of the analytical utility of the photodissociation of fast ion beams, *Mass Spectrom. Rev.* **9**, 405–451 (1990).
17. J.W. Morgan and D.H. Russell, Comparative studies of 193–nm photodissociation and TOF-TOFMS analysis of bradykinin analogues: The effect of charge site(s) and fragmentation timescales, *J. Am. Soc. Mass Spectrom.* **17**, 721–729 (2006).
18. L. M. Nuwaysir and C. L. Wilkins, Photodissociation of laser-desorbed ions as a structure determination tool, *Anal. Chem.* **61**, 689–694 (1989).
19. K. M. Choi, S. H. Yoon, et al., Characteristics of photodissociation at 193 nm of singly protonated peptides generated by MALDI, *J. Am. Soc. Mass Spectrom.* **17**, 1643–1653 (2006).
20. D. P. Little, J. P. Spier, M. W. Senko, P. B. O'Conner, and F. W. McLafferty, Infrared multiphoton dissociation of large multiply charged ions for biomolecule sequencing, *Anal. Chem.* **66**, 2809–2815 (1994).
21. M. C. Crowe and J. S. Brodbelt, Infrared multi-photon dissociation (IRMPD) and collisionally activated dissociation of peptides in a quadrupole ion trap (QIT) with selective IRMPD of phosphopeptides, *J. Am. Soc. Mass Spectrom.* **15**, 1581–1592, (2004).
22. R. C. Dunbar and T. B. McMahon, Activation of unimolecular reactions by ambient blackbody radiation, *Science* **279**, 194–197 (1998).

23. W. D. Price, P. D. Schnier, and E. R. Williams, Tandem mass spectrometry of large biomolecule ions by blackbody infrared radiative dissociation, *Anal. Chem.* **68**, 859–866 (1996).
24. R. A. Zubarev, N. K. Kelleher, and F. W. McLafferty, Electron capture dissociation of multiply charged protein cations: a nonergodic process, *J. Am. Chem. Soc.* **120**, 3265–3266 (1998).
25. R. A. Zubarev, N. A. Kruger, et al., Electron capture dissociation of gaseous multiply-charged proteins is favored at disulfide bonds and other sites of high hydrogen atom affinity, *J. Am. Chem. Soc.* **121**, 2857–2862 (1999).
26. F. Kjeldsen, K. F. Haselmann, B. A. Budnik, F. Jensen, and R. A. Zubarev, Dissociation capture of hot (3–13 eV) electrons by polypeptide polycations: an efficient process accompanied by secondary fragmentation, *Chem. Phys. Lett.* **356**, 201–206 (2002).
27. E. A. Syrstad and F. Tureček, Toward a general mechanism of electron capture dissociation, *J. Am. Soc. Mass Spectrom.* **16**, 208–224 (2005).
28. O. A. Silivra, F. Kjeldsen, I. A. Ivonin, and R. A. Zubarev, Electron capture dissociation of polypeptides in a three-dimensional quadrupole ion trap: instrumentation and first results, *J. Am. Soc. Mass Spectrom.* **16**, 22–27 (2005).
29. J. E. P. Syka, J. J. Coon, M. J. Schroeder, J. Shabanowitz, and D. F. Hunt, Peptide and protein sequence analysis by electron transfer dissociation mass spectrometry, *Proc. Natl. Acad. Sci. USA* **101**, 9528–9533 (2004).
30. J. J. Coon, J. Shabanowitz, D. F. Hunt, and J. E. P. Syka, Electron transfer dissociation of peptide anions, *J. Am. Soc. Mass Spectrom.* **16**, 880–882 (2005).
31. P. H. Gunawardena, F. Joshua, and S. A. McLuckey, Phosphopeptide anion characterization via sequential charge inversion and electron-transfer dissociation, *Anal. Chem.* **78**, 3788–3793 (2006).
32. M. He and S. A. McLuckey, Charge permutation reactions in tandem mass spectrometry, *J. Mass Spectrom.* **39**, 1231–1259 (2004).
33. P. O. Danis, C. Wesdemiotis, and F. W. McLafferty, Neutralization–reionization mass spectrometry (NRMS), *J. Am. Chem. Soc.* **105**, 7454–7456 (1983).
34. S. Gronert, Mass spectrometric studies of organic ion/molecule reactions, *Chem. Rev.* **101**, 329–360 (2001).
35. M. L. Gross, Tandem mass spectrometry: multisector magnetic instruments, in J. A. McCloskey, ed., *Methods in Enzymology*, Vol. 193, Academic Press, San Diego, CA, 1990, pp. 131–153.
36. A. G. Brenton and C. M. Lock, The design and performance of a high resolution translational energy loss spectrometer, *Rapid Commun. Mass Spectrom.* **11**, 1155–1170 (1997).
37. R. K. Boyd, Linked-scan techniques for MS/MS using tandem-in-space instruments, *Mass Spectrom. Rev.* **13**, 359–410 (1994).
38. R. A. Yost and C. G. Enke, Selected ion fragmentation with a tandem quadrupole mass spectrometer, *J. Am. Chem. Soc.* **100**, 2274–2275 (1978).
39. R. A. Yost and C. G. Enke, Triple quadrupole mass spectrometry for direct mixture analysis and structure elucidation, *Anal. Chem.* **51**, 1251A–1264A (1979).
40. M. N. Eberlin, Triple-stage pentaquadrupole (QqQqQ) mass spectrometry and ion/molecule reactions, *Mass Spectrom. Rev.* **16**, 113–144 (1997).

41. A. G. Harrison, Energy-resolved mass spectrometry: a comparison of quadrupole cell and cone-voltage collision-induced dissociation, *Rapid. Commun. Mass Spectrom.* **13**, 1663–1670 (1999).
42. B. Spengler, Post-source decay analysis in matrix-assisted laser desorption/ionization mass spectrometry of biomolecules, *J. Mass Spectrom.* **32**, 1019–1036 (1997).
43. T. J. Cornish and R. J. Cotter, High-order kinetic energy focusing in an end cap reflectron time-of-flight mass spectrometer, *Anal. Chem.* **69**, 4615–4618 (1997).
44. T. J. Cornish and R. J. Cotter, A curved field reflectron time-of-flight mass spectrometer for the simultaneous focusing of metastable product ions, *Rapid Commun. Mass Spectrom.* **8**, 781–785 (1994).
45. R. J. Cotter, B. D. Gardner, S. Ilchenko, and R. D. English, Tandem time-of-flight mass spectrometry with a curved field reflectron, *Anal. Chem.* **76**, 1976–1981 (2004).
46. D. J. Beussman, P. R. Vlasak, R. D. McLane, M. A. Seeterlin, and C. G. Enke, Tandem reflectron time-of-flight mass spectrometer utilizing photodissociation, *Anal. Chem.* **67**, 3952–3957 (1995).
47. A. L. Yergey, J. R. Coorssen, P. S. Backlund, Jr., P. S. Blank, G. A. Humphrey, J. Zimmerberg, J. M. Campbell, and M. L. Vestal, De novo sequencing of peptides using MALDI/TOF-TOF, *J. Am. Soc. Mass Spectrom.* **13**, 784–791 (2002).
48. D. Suckau, A. Rosemann, M. Schuerenberg, P. Hufnegel, J. Franzen, and A Holle, A novel LIFT-TOF/TOF mass spectrometer for proteomics, *Anal. Bioanal. Chem.* **376**, 952–965 (2003).
49. J. N. Louris, R. G. Cooks, J. E. P. Syka, P. E. Kelley, G. C. Stafford, Jr., and J. F. J. Todd, Instrumentation, applications, and energy deposition in quadrupole ion-trap tandem mass spectrometry, *Anal. Chem.* **59**, 1677–1685 (1987).
50. R. G. Cooks and R. E. Kaiser, Quadrupole ion-trap mass spectrometry, *Acc. Chem. Res.* **23**, 213–218 (1992).
51. K. R. Jonscher and J. R. Yates III, The quadrupole ion trap mass spectrometer: a small solution to a big challenge, *Anal. Biochem.* **244**, 1–15 (1997).
52. R. E. March, An introduction to quadrupole ion-trap mass spectrometry, *J. Mass Spectrom.* **32**, 351–369 (1997).
53. J. W. Gauthier, T. R. Trautman, and D. B. Jacobson, Sustained off-resonance irradiation for collision-activated dissociation involving Fourier transform mass spectrometry: collision-activated dissociation technique that emulates infrared multiphoton dissociation, *Anal. Chim. Acta* **246**, 211–225 (1991).
54. H. A. Hofstadler, J. H. Wahl, et al., Capillary electrophoresis Fourier transform ion cyclotron mass spectrometry with sustained off-resonance irradiation for the characterization of proteins and peptides, *J. Am. Soc. Mass Spectrom.* **5**, 894–899 (1994).
55. S. C. Beu, M. W. Senko, et al., Fourier transform electrospray instrumentation for tandem high-resolution mass spectrometry of large molecules, *J. Am. Soc. Mass Spectrom.* **4**, 557–565 (1993).
56. S. A. McLuckey, G. L. Glish, and R. G. Cooks, Kinetic energy effects in mass spectrometry/mass spectrometry using a sector/quadrupole tandem instrument, *Int. J. Mass Spectrom. Ion Phys.* **39**, 219–230 (1981).
57. J. A. Loo and H. Muenster, Magnetic sector–ion trap mass spectrometry with electrospray ionization for high sensitivity peptide sequencing, *Rapid Commun. Mass Spectrom.* **13**, 54–60 (1999).

58. H. R. Morris, T. Paxton, A. Dell, J. Langhorne, M. Berg, R. S. Bordoli, J. Hoyes, and R. H. Bateman, High sensitivity collisionally-activated decomposition tandem mass spectrometry on a novel quadrupole/orthogonal-acceleration time-of-flight mass spectrometer, *Rapid Commun. Mass Spectrom.* **10**, 889–896 (1996).
59. A. V. Loboda, A. N. Krutchinsky, W. Ens, and K. G. Standing, A tandem quadrupole/time-of-flight mass spectrometer with a matrix-assisted laser desorption/ionization source: design and performance, *Rapid Commun. Mass Spectrom.* **14**, 1047–1057 (2000).
60. F. Sobott, H. Hernandez, M. G. McCammon, M. A. Tito, and C. V. Robinson, A tandem mass spectrometer for improved transmission and analysis of large macromolecular assemblies, *Anal. Chem.* **74**, 1402–1507 (2002).
61. R. H. Bateman, R. Carruthers, J. B. Hoyes, C. Jones, J. I. Langridge, A. Miller, and J. P. C. Vissers, A novel precursor ion discovery method on a hybrid quadrupole orthogonal acceleration time-of-flight (Q-TOF) mass spectrometer for studying protein phosphorylation, *J. Am. Soc. Mass Spectrom.* **13**, 792–803 (2002).
62. B. A. Collings, J. M. Campbell, D. Mao, D. J. Douglas, A combined linear ion trap time-of-flight system with improved performance and MS^n capabilities, *Rapid Commun. Mass Spectrom.* **15**, 1777–1795 (2001).
63. M. G. Quin and D. M. Lubman, A marriage made in MS, *Anal. Chem.* **65**, 234A–242A (1995).
64. S. M. Michael, B. M. Chien, and D. M. Lubman, Detection of electrospray ionization using a quadrupole ion trap storage/reflectron time-of-flight mass spectrometer, *Anal. Chem.* **65**, 2614–2620 (1993).
65. R. H. Bateman, M. R. Green, G. Scott, and E. Clayton, A combined magnetic sector-time-of-flight mass spectrometer for structural determination studies by tandem mass spectrometry, *Rapid Commun. Mass Spectrom.* **9**, 1227–1233 (1995).
66. K. Hakansson, M. J. Chambers, J. P. Quinn, M. A. McFarland, C. L. Hendrickson, and A. G. Marshall, Combined electron capture and infrared multiphoton dissociation for multistage MS/MS in a Fourier-transform ion cyclotron resonance mass spectrometer, *Anal. Chem.* **75**, 3256–3262 (2003).
67. A. Brock, D. M. Horn, E. C. Peters, C. M. Shaw, C. Ericson, Q. T. Fung, and A. R. Salomon, An automated matrix-assisted laser desorption/ionization quadrupole-Fourier-transform ion cyclotron resonance mass spectrometer, *Anal. Chem.* **75**, 3419–3428 (2003).
68. J. E. P. Syka, J. A. Marto, et al., Novel linear quadrupole ion trap/FT mass spectrometer: performance characterization and use in the comparative analysis of histone H3 post-translational modifications, *J. Proteome Res.* **3**, 621–626 (2004).
69. A. Makarov, E. Denisov, et al., Performance evaluation of a hybrid linear ion trap/orbitrap mass spectrometer, *Anal. Chem.* **78**, 2113–2120 (2006).
70. J. R. Yates, D. Cociorva, L. Liao, and V. Zabrouskov, Performance of a linear ion trap–orbitrap hybrid for peptide analysis, *Anal. Chem.* **78**, 493–500 (2006).
71. A. Makarov, Electrostatic axially harmonic orbital trapping: a high-performance technique of mass analysis, *Anal. Chem.* **72**, 1156–1162 (2000).
72. Q. Hu, R. J. Noll, H. Li, A. Makarov, M. Hardman and R. G. Cooks, The orbitrap: a new mass spectrometer, *J. Mass Spectrom.* **40**, 430–443 (2005).

CHAPTER 5

HYPHENATED SEPARATION TECHNIQUES

The coupling of chromatography with mass spectrometry is an area that has served a critical need in the analysis of real-world samples, such as environmental samples and biological tissues and fluids. These samples are complex by nature, as they are made up of hundreds and thousands of components. Unequivocal characterization of such complex samples is a monumental task for any stand-alone analytical technique. One needs either infinite separation capability of a separation system or infinite selectivity of an identification method. In practice, both situations are intangible. One solution is to subject the sample to an exhaustive purification protocol to isolate the analyte of interest from a complex real-world mixture. Such off-line purification procedures are, however, laborious and time consuming and entail losses of precious samples. A more sensible approach is to combine a separation technique with a compound-specific detector that can identify the separated components. Mass spectrometry offers an opportunity for such an ideal detection system. A range of separation techniques, including gas chromatography (GC), high-performance liquid chromatography (HPLC), capillary electrophoresis (CE), supercritical-fluid chromatography (SFC), size-exclusion chromatography, and thin-layer chromatography (TLC), has been combined with mass spectrometry. Some of these important developments are highlighted in this chapter. The coupling of various separation techniques with mass spectrometry has been reviewed [1].

Fundamentals of Contemporary Mass Spectrometry, by Chhabil Dass
Copyright © 2007 John Wiley & Sons, Inc.

5.1. BENEFITS OF COUPLING SEPARATION DEVICES WITH MASS SPECTROMETRY

The role of GC, HPLC, and CE in high-resolution separation of complex mixtures is unquestionable. Similarly, mass spectrometry has attained an indisputable position in analytical chemistry as a highly structure-specific technique that can provide structural identity of a wide range of compounds. Chromatography and mass spectrometry both, however, have their limitations in stand-alone operation. First, the separation power of any chromatography system is finite. It will be nearly impossible to achieve complete separation of all components of a complex mixture. Second, identification of a compound in chromatography is less than reliable because of marginal information content. The basis of identification of the target analyte in chromatography is comparison of its retention time with that of a reference material. Therefore, the common separation techniques of GC, LC, and CE cannot provide unequivocal identity of the analyte when used with conventional detection systems. Uncertainty may arise because another component of the mixture may elute at the desired retention time. A compound-specific detector is thus an essential adjunct to characterize unambiguously the components that elute from any separation system. In this respect, mass spectrometry offers the unique advantages of high molecular specificity, detection sensitivity, and dynamic range. Only mass spectrometry has the ability to provide confirmatory evidence of an analyte because of its ability to distinguish closely related compounds on the basis of the molecular mass and structure-specific fragment ion information. The confidence in identification of a target compound, however, diminishes when it is present in a mixture. Because of the universal nature of mass spectrometry detection, the data obtained might also contain signal due to other components of the mixture. The coupling of a separation device with mass spectrometry thus benefits mutually. The result is a powerful two-dimensional analysis approach, where the high-resolution separation and the highly sensitive and structure-specific detection are both realized simultaneously. Following are some of the benefits that accrue when a separation technique and mass spectrometry are coupled.

- The capabilities of the techniques are enhanced synergistically. As a consequence, both instruments may be operated at subpar performance levels without compromising the data outcome.
- The high selectivity of mass spectrometry detection allows one to identify coeluting components.
- The certainty of identification is enhanced further because, in addition to the structure-specific mass spectral data, the chromatographic retention time is also known.
- Multicomponent samples can be analyzed directly without prior laborious off-line separation steps, resulting in a minimal sample loss and saving of time.
- The sensitivity of analysis is improved because the sample enters the mass spectrometer in the form of a narrow focused band.

- Less sample is required than the amount required for off-line analysis by the two techniques separately.
- Because of the removal of interferences, the quality of mass spectral data is improved and any mutual signal suppression is minimized.
- The confidence in quantitative analysis is increased because mass spectrometry permits the use of a stable isotope analog of the analyte as an internal standard.

5.2. GENERAL CONSIDERATIONS

5.2.1. Characteristics of an Interface

A chromatography–mass spectrometry system consists of three major components: a separation device, an interface, and a mass spectrometer (Figure 5.1). The purpose of an interface is to transport the separated components into the ion source of a mass spectrometer for their identification. An ideal interface should not affect the performance capabilities of either the chromatography or the mass spectrometry system. Other important characteristics of an interface are that it should have a high sample transport efficiency, should not alter the chemical identity of the analyte, should be mechanically simple and chemically inert and have a low chemical background, should have the ability to remove most of the mobile phase while transferring the maximum amount of eluted component, and should not degrade the ionizing efficiency and vacuum of the mass spectrometry system.

5.2.2. Mass Spectral Data Acquisition

The determination of molecular mass, identification of compounds, and quantification are three broad areas of application of a combined chromatography–mass spectrometry system. To identify the eluting components and to determine their molecular mass, the mass spectrometry data are recorded by scanning a mass analyzer over a wide mass range, with the upper limit dictated by the highest expected molecular mass of the analyte and the lower limit by the background ions from the chromatography mobile phase. The mass spectrometry signal is plotted as an ion chromatogram or mass spectrum. Three different types of ion chromatograms can be retrieved from the mass spectrometry data: A *total ion current* (TIC) *chromatogram* is a plot of the sum of the signals due to all observed

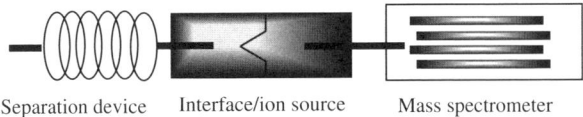

Separation device   Interface/ion source   Mass spectrometer

Figure 5.1. Conceptual diagram of the coupling of a separation device with a mass spectrometer.

ions in a defined mass range versus either the scan number or time. A peak in this chromatogram indicates the elution of a component. This trace is similar to a conventional UV chromatogram from an LC analysis and can be used to identify compounds provisionally on the basis of their retention times. A mass spectrum can be retrieved for any chosen scan number in the TIC chromatogram to provide qualitative identification of a compound. A *mass chromatogram*, also known by various other names, such as *ion-extraction chromatogram and reconstructed ion chromatogram*, is a plot of the ion current due to a characteristic ion versus either the scan number or time. This chromatogram is generated postacquisition from the TIC trace. It is a useful means to distinguish coeluting components and to identify a homologous series of compounds. For quantitative measurements, a *selected-ion monitoring* (SIM) *chromatogram* is acquired by recording the ion signal due to only one or more compound-specific ions. This trace is also useful to selectively detect a homologous series of compounds. For quantification, the area or height of the analyte peak is compared with that of the internal standard. The quantitative information can also be obtained from a mass chromatogram. It should, however, be remembered that although a mass chromatogram and an SIM chromatogram are identical in terms of selectivity, the latter provides much higher sensitivity, owing to the increased time spent monitoring each ion selected.

Signal acquisition from a chromatography system is challenging because the volume of data is enormous and the peak width is often very narrow. In addition, the partial pressure of the sample that enters the mass spectrometry ion source changes continuously. Therefore, a rapid scanning mass spectrometry system is very essential so that no useful data point will be missed during the scan. Acquisition of required data points ensures that there is no distortion in the peak intensities and that the chromatographic peak profile is not skewed (see Figure 5.2). It also ensures that complete information is obtained for all coeluting components in a narrow peak.

Data-Dependent Acquisition The specificity of mass spectrometry data is enhanced further when a spectrum is acquired in the tandem MS (MS/MS)

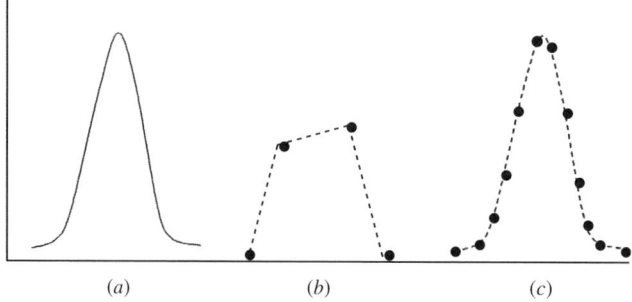

Figure 5.2. (*a*) Ideal chromatographic peak; (*b*) peak with insufficient data points; (*c*) peak with large and nearly sufficient data points. This peak profile nearly matches an ideal peak.

mode. With the current mass spectrometry data systems, it has become feasible to acquire MS/MS spectra during acquisition of the TIC trace. This procedure involves switching from the MS mode to the MS/MS mode by a technique known as *data-dependent acquisition* [2]. The data system performs a cyclic series of scans in which a full-scan spectrum is acquired first. The precursor ion is selected automatically, and its MS/MS spectrum is acquired. This cycle repeats several times as long as the preselected intensity remains above threshold. The MS/MS spectra of all components present in a single eluting LC peak can thus be acquired provided that their intensity is above the preselected threshold.

Peak Parking Another approach to acquiring complete information in a narrow peak is to extend the analysis time over peaks of interest. In this approach, termed *peak parking* or *variable-flow chromatography*, the column head pressure, and hence the flow rate, are reduced instantaneously [3,4]. With this approach, it is possible to perform higher-resolution narrow mass scans and acquire MS/MS scans on all coeluting components in a single narrow peak.

5.2.3. Characteristics of Mass Spectrometers

High scan speed, adequate mass range, reasonable mass resolution, high sensitivity, wide dynamic range, tolerance to high pressures, and low cost are some of the attributes that are sought in a mass spectrometry system. Quadrupole mass filters, quadrupole ion traps (QITs), and linear ion traps (LITs) all offer most of these desirable features. A QIT, however, is more compact, cheaper, and sensitive than a quadrupole mass filter. LIT is the most sensitive of these types of instruments. The tandem mass spectrometry facility is an added incentive for the coupling of chromatography with QIT and LIT. Magnetic sector-, Fourier transform (FT)–ion cyclotron resonance (ICR)-, and time-of-flight (TOF)-based mass spectrometry systems can also be coupled with chromatography. These analyzers possess certain unique features that are not available with quadrupole-based systems. Accurate mass measurement is performed routinely with these instruments. The fast spectrum-acquisition rate and multiplex detection capability of TOF–MS and FT–ICR–MS instruments are attractive features for GC/MS and LC/MS combinations.

5.3. CHROMATOGRAPHIC PROPERTIES

The primary function of chromatography is separation of a mixture of compounds into individual components. It is a dynamic separation system that is composed of two media, a stationary phase and a mobile phase. The stationary phase in most applications is a liquid supported on the surface of an inert solid support and is usually packed in a column. The mobile phase can be a gas, a liquid, or a supercritical fluid. Its purpose is to transport the sample through the column, a process known as *elution*. The components in a separation mixture distribute between these two phases to a different extent. A component that interacts strongly with the stationary phase moves slowly, and as a result it is separated from the fast-moving components, which have less affinity for the stationary phase. The emerging components are detected by a detector placed at the end of the column.

A chromatographic peak should be narrow and Gaussian in nature. In practice, the peaks are often broad and non-Gaussian; the more time the solute spends in a column, the broader the peak. The performance of a chromatographic system is described in terms of a number of parameters, including capacity factor, selectivity factor, plate height, plate number, and resolution.

The *capacity factor*, k', describes the migration rate of the solutes, and is defined as the ratio of the time a solute spends in the stationary phase versus the time it spends in the mobile phase. In practice, it can be measured from a given chromatogram (Figure 5.3) using the relation

$$k' = \frac{t_r - t_m}{t_m} \tag{5.1}$$

where t_r is the retention time of the solute and t_m is the dead time, the time taken by an unretained component to elute from the column. Ideal separation is obtained when k' lies between 1 and 5.

The *selectivity factor*, α, defines the extent of separation between two components A and B, and is given by the ratio of their capacity factors. Experimentally, it is determined by Eq. (5.2), where B refers to a late-eluting component (i.e., $t_B > t_A$ and $\alpha > 1$):

$$\alpha = \frac{k'_B}{k'_A} = \frac{(t_r)_B - t_m}{(t_r)_A - t_m} \tag{5.2}$$

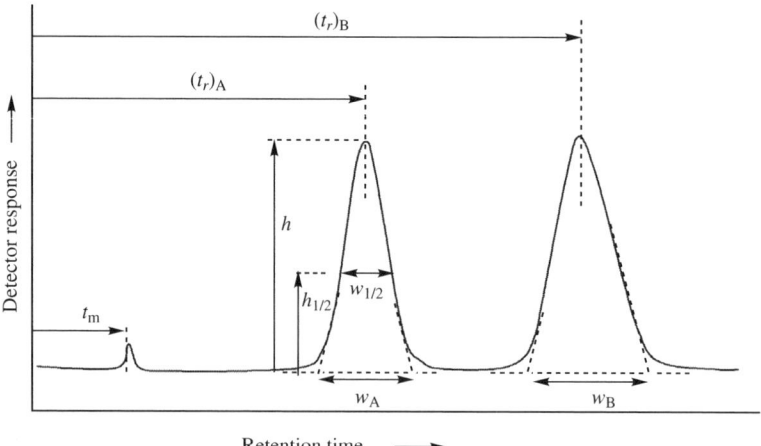

Figure 5.3. Typical chromatogram of a two-component (A and B) sample. The first little peak is of an unretained component.

The quantitative measures of column efficiency are the plate height and plate number. *The plate height, H*, is defined as the length of column that participates in one mass transfer equilibrium between the stationary phase and the mobile phase. The *plate number, N*, is the measure of the total number of plates in a column. The two terms are related to each other through the column length, L, by the expression $N = L/H$. The value of N can be measured from the chromatogram using the equation

$$N = 16 \left(\frac{t_r}{w}\right)^2 = 5.54 \left(\frac{t_r}{w_{1/2}}\right)^2 \tag{5.3}$$

where w is the peak width at the base and $w_{1/2}$ at half-peak height (see Figure 5.3).

▶ **Example 5.1** A 2-m-long glass capillary GC column was used for separation of a mixture of hydrocarbons. With the carrier gas flow rate of 50 mL min^{-1}, the width of a 7.5-min GC peak was found to be 0.3 min wide at the base. Calculate the effective number of theoretical plates.

Solution The retention time of this peak is 7.5 min and its width at the base is 0.3 min. Therefore, using Eq. (5.3), we obtain $N = (16) (7.5/0.3)^2 = 10{,}000$.

The separation of two components A and B is expressed in terms of the resolution, R, given by

$$R = \frac{2[(t_r)_B - (t_r)_A]}{w_A + w_B} \tag{5.4}$$

$$R = \frac{\sqrt{N}}{4} \frac{\alpha - 1}{\alpha} \frac{k'_B}{1 + k'_B} \tag{5.5}$$

Careful scrutiny of these two equations reveals that separation of the two target components can be optimized by manipulation of the k', α, and N terms. The first two terms are known as *thermodynamic effects*, and the third term is associated with kinetic features of the column. The k' term is optimized by increasing the temperature of the column (as in GC) and by changing the mobile-phase composition (as in LC). The options available to optimize α are a change in the mobile-phase composition, column temperature, and stationary-phase composition. N can be improved by increasing the length of the column and reducing H, which is accomplished by changing the mobile-phase flow rate and reducing the size of the solid support, the thickness of the liquid stationary phase, the viscosity of the mobile-phase solvent, the temperature of the column, and the diameter of the column. The use of capillary columns and columns packed with small particles is a common practice in GC/MS and LC/MS applications.

▶ **Example 5.2** Calculate the resolution between two components that elute at 10.2 and 11.0 min. Their base peak widths are 0.46 and 0.52 min, respectively.

Solution By using the mathematical relation given in Eq. (5.4), we obtain

$$R = \frac{(2)(11.0 - 10.2)}{0.46 + 0.52} = 1.63$$

5.4. GAS CHROMATOGRAPHY/MASS SPECTROMETRY

Gas chromatography/mass spectrometry was first used in the 1950s [5]. Since then it has acquired an enviable status as a modern analytical technique for the sensitive, high-resolution, and selective identification and quantification of a wide range of compounds. Its use is mentioned frequently in the fields of forensic drug analysis, structural characterization of biomolecules, toxicology, medicinal chemistry, clinical science, and environmental chemistry. GC/MS is considered a standard technique for positive identification of the presence of a target compound. Methods based on the GC/MS protocol are readily acceptable in medical–legal defense. GC/MS is, however, restricted to relatively volatile and thermally stable organic compounds. Some non volatile compounds can be made volatile at the temperatures used for separation after chemical derivatization. Coupling of GC with mass spectrometry has been reviewed [6].

5.4.1. Basic Principles of Gas Chromatography

In gas chromatography, a sample is vaporized in a heated injector block and deposited onto the head of a chromatographic column that contains a nonvolatile liquid stationary phase. The components of a mixture are separated on the basis of their varying affinity for, and solubility in, the stationary phase. Elution of the separated components is effected by the flow of an inert carrier gas (usually, helium). It is a usual practice to perform GC separations at elevated temperatures (150 to 300°C) to bring many not-so-volatile, but thermally stable compounds under its domain. In early applications, GC analysis was carried out on wide-bore (1 to 3 mm i.d.) packed columns. Modern analytical GC is practiced with more efficient, faster, and longer capillary columns. The inside wall of these fused-silica open-tubular (FSOT) columns is coated with a thin layer of a stationary phase, usually an organosiloxane polymer. Polydimethyl siloxane, $-[-O-Si(CH_3)_2-]_n-$, is a general-purpose nonpolar stationary phase. The replacement of the methyl groups with polar functional groups, such as phenyl, trifluoropropyl, or cyanopropyl, increases the polarity of this phase. Polyethylene glycol is another common polar stationary phase. GC is commonly practiced with flame ionization, electron capture, thermionic, and thermal conductivity detectors;

all of them lack the ability to provide the chemical composition or structure of the separated components.

5.4.2. Interfaces for Coupling Gas Chromatography with Mass Spectrometry

The first coupling of GC was demonstrated with a TOF mass spectrometer [5]. Gas chromatography and mass spectrometry are both highly compatible with respect to the sample size, and both deal with the gaseous samples. Therefore, the coupling GC with mass spectrometry is a straightforward proposition. The only incompatibility between the two systems is the pressure mismatch. For example, the separated components exit from the GC column at atmospheric pressure, whereas the mass spectrometry source operates at 10^{-6} to 10^{-5} torr. This mismatch is not a serious handicap in the coupling of capillary columns with mass spectrometry. The low gas flow rates (e.g., in the range 1 to 2 mL min^{-1}) of capillary columns can readily be accommodated by the vacuum system of modern mass spectrometers; no interface is necessary for columns up to 320 µm internal diameter (i.d.); the exit end of these columns is simply inserted into the ion source through a heated sheath (Figure 5.4a). The high-capacity pumping systems of modern mass spectrometers can effectively pump away the extra gas load from a capillary column.

Because of the greater gas load (e.g., 10 to 20 mL/min) of larger-diameter open-tubular and packed columns, their coupling requires an appropriate interface. One such device is an *open-split interface* [7]. It consists of a chamber in which a high flow of the purge gas is maintained at atmospheric pressure (see Figure 5.4b). The purge gas removes most of the carrier gas from the GC effluent and allows only a small fraction to enter the mass spectrometer through a capillary tube.

Another common interface for packed columns is a *jet separator*, shown in Figure 5.4c. It consists of a partially evacuated chamber in which a jet of GC effluent is sprayed through a small nozzle [8]. It operates on the principle that the fast-moving lighter-mass components effuse at a faster rate than do the slower heavier-mass species (solid circles in Figure 5.4c). Thus, the lighter carrier-gas molecules effuse faster at a wider angle, whereas the heavier sample components travel over a narrow angle around the central axis. The analyte-enriched middle portion is sampled into the ion source of the mass spectrometer through a skimmer placed opposite the emerging jet.

A *molecular-effusion interface* (also known as a *Watson–Biemann interface*) is made of a glass-frit tube that is placed in an evacuated chamber (Figure 5.4d). One end of this tube is connected to the GC column and the other to the ion source of a mass spectrometer [9]. The GC effluent is preferentially enriched in the sample molecules by the molecular effusion process; the lighter gas molecules effuse faster through the pores of the fritted tube and are pumped away, whereas the enriched sample molecules enter the ion source. An interface using a *silicon-rubber membrane* for enrichment of the solute molecules has also

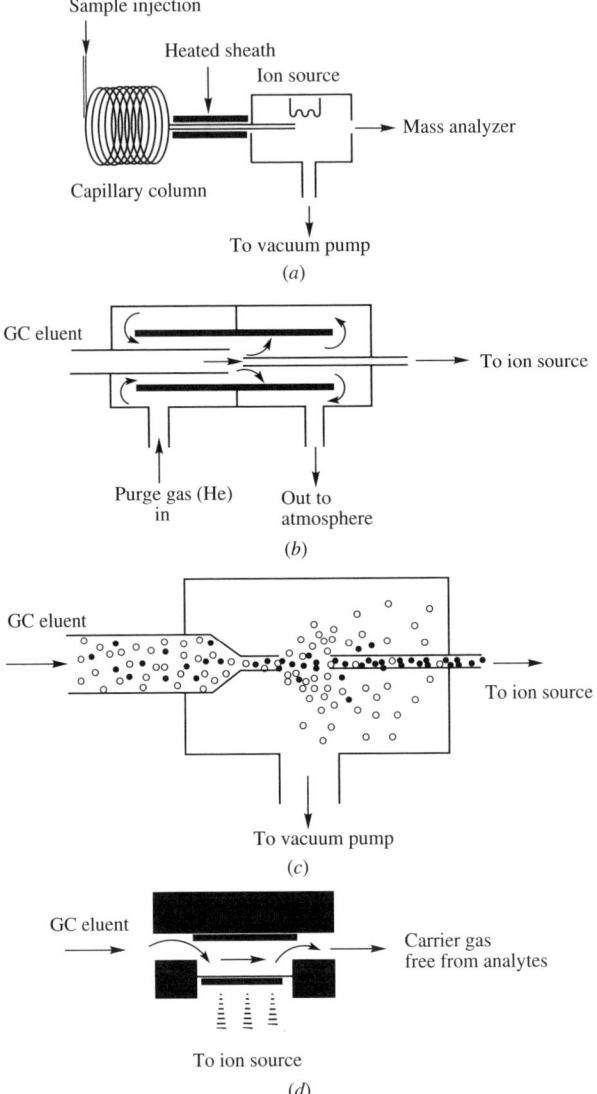

Figure 5.4. Various forms of GC/MS interfaces: (*a*) direct coupling of capillary columns with MS; (*b*) open-split interface; (*c*) jet-separator interface; (*d*) molecular-effusion interface.

found applications [10]. This membrane allows nonpolar organic molecules to pass through it selectively, and blocks the passage of the carrier gas.

Electron ionization (EI) and chemical ionization (CI) ion sources are both compatible with the gas-phase GC-eluting components. EI especially provides both molecular ion and fragment ion information that can be used independently or compared with a reference mass spectrum in a spectral library to identify known and unknown compounds. Commercially marketed libraries are available that contain over 100,000 spectra (see Section 6.4.6).

5.5. LIQUID CHROMATOGRAPHY/MASS SPECTROMETRY

More than 85% of compounds in nature fall into the category of being polar and thermally labile, which are not amenable to GC/MS. In the past, a strong need for a similar combination of HPLC and MS was felt for the analysis of such compounds. Although attempts to combine LC with MS were initiated over 30 years ago, a robust and dependable combination that could be adapted for a wide range of applications has emerged only after the discovery of atmospheric-pressure ionization techniques. Currently, the practice of LC/MS arguably has become the single most widely used analytical technique. A variety of applications has been envisioned for both qualitative and quantitative analysis of complex mixtures of biochemical, inorganic, and organic compounds. HPLC/MS has become the mainstay of proteomics and pharmaceutical laboratories.

5.5.1. Basic Principles of HPLC Separation

HPLC is practiced in the following formats:

- *Adsorption chromatography*, in which the stationary phase consists of a high-surface-area solid adsorbent; the solutes physically adsorb on the stationary phase, while the liquid mobile phase tries to dislodge them.
- *Partition chromatography* involves distribution of the solutes between a liquid stationary phase and a liquid mobile phase.
- *Ion-exchange chromatography*, used for ionic solutes, involves competition for solute ions between immobilized ionic sites on the stationary phase (e.g., ion-exchange resin) and liquid mobile phase.
- *Size-exclusion chromatography* separates solutes on the basis of their size. The stationary phase consists of a polymer matrix of various pore sizes. Solutes are separated according to their ability to penetrate the polymer matrix. The small solutes penetrate more and lag behind the larger solutes.
- *Affinity chromatography* involves specific interaction between one type of solute molecule and a second, complementary type of molecule that is immobilized on a solid support.

Of these, partition chromatography, especially the *reversed-phase (RP) mode*, is the most widely used technique for the separation of several classes of compounds. In RP–HPLC (see Table 5.1), the stationary phase is a nonpolar matrix and the mobile phase is a polar solvent (e.g., water mixed with a polar organic modifier such as methanol, isopropanol, or acetonitrile). Mobile phase and stationary phase both play prominent roles in the separation mechanism. A sample is applied onto the head of a column filled with an appropriate stationary phase. The

TABLE 5.1. Typical RP–HPLC Columns

Column	Dimensions (Length × i.d., in mm)	Flow Rate (μL min^{-1})	Injection Volume (μL)
Analytical	150 × 4.6	500–2000	10–500
Narrow-bore	150 × 2.1	50–500	5–50
Microbore	150 × 1.0	10–100	1–20
Capillary	300 × 0.3	1–10	0.05–2
Nanoscale	500 × 0.05	0.05–0.5	0.01–1

components of a mixture are partitioned differentially between the two phases and are separated when eluted by a stream of a liquid mobile phase. Two common modes of elution are *isocratic*, in which the solvent composition remains constant during elution, and *gradient*, in which the solvent composition is changed either in steps or in a continuous fashion during elution.

The basis of *RP–HPLC separation* is the hydrophobic interaction between the analyte and the nonpolar matrix of the stationary phase. A typical stationary phase is prepared by chemically bonding a long-chain alkyl group, such as *n*-octadecyl (C_{18}) to porous silica. The smaller alkyl groups, such as *n*-octyl, *n*-butyl, and *n*-ethyl, can also be used for specific applications. In practice, the nonpolar, hydrophobic solutes interact strongly with the stationary phase. In contrast, relatively hydrophilic compounds spend more time in the polar aqueous mobile phase and thus elute earlier. Increasing the strength of the eluting solvent enables elution of the more strongly retained hydrophobic solutes. The RP–HPLC separation is further fine-tuned by use of appropriate buffers and ion-pairing reagents in the mobile phase. Although a large number of buffer systems has been used in a conventional ultraviolet–visible detection system, volatile ion-pairing reagents work best for LC/MS operation.

Partition chromatography is also practiced in the *normal-phase mode*, in which the stationary phase is a polar matrix and the mobile phase at the start of the separation is a polar solvent. The polar solutes prefer to remain in the stationary phase and elute late.

5.5.2. Fast-Flow Liquid Chromatography

An important area of research in LC/MS is to reduce the analysis time. The driving force for the development of high-speed separations is the need for high-throughput analysis; such a development can especially speed up the drug-discovery process. Several imaginative attempts have been made to develop fast LC. One approach is to use short LC columns (typically, <50 mm long) that are packed with small particles (e.g., these with 3-μm diameter) [11,12]. A decrease in the analysis time by a factor of about 7.5 is realized when the column length is reduced from 250 mm to 33 mm. The small stationary-phase packing also shortens the analysis time because of the inverse relation between the optimal linear mobile-phase velocity and the particle diameter.

Another attempt to perform high-speed separation is to use ultrahigh or turbulent flows (over 30 mL/min) [13]. At turbulent flow rates, the solvent front profile is plug-like rather than the usual parabolic. Therefore, a considerable improvement in column efficiency is achieved. These prohibitive ultrahigh flow rates are, however, possible when columns are packed with particles over 50 µm.

High-speed chromatography can also be accomplished by the use of special packing materials. Three different types of stationary-phase packing—perfusion, nonporous, and monolithic—have provided fast flows through the LC columns. The perfusion packing [14] is made of polymer particles with a network of larger through-pores (e.g., 6000 to 8000 Å in diameter) and small diffusive pores (300 to 1500 Å). The through-pores allow 10 to 20-fold faster flow of a mobile phase through the packing, resulting in increased mass transfer rates and consequently, increased separation efficiency. The nonporous silica packing is made from small (e.g., 1.5 µm in diameter) silica or resin particles. With this type of packing, stagnant flow, one of the major causes of band broadening, is eliminated to increase mass transfer rates. Small particle packing, however, results in higher back pressures. Therefore, short columns 2 to 3 cm in length must be used for fast separations. Monolithic columns contain a continuous mass of porous silica rather than conventional particle packing. The silica skeleton of this packing contains two different types of pores, macropores and mesopores, giving lower pressure drops and higher mass transfer rates. Because of low-pressure drops, fast separations are easy to achieve with monolithic columns. Several examples of coupling of these columns with ESI–MS can be found in the literature [15,16].

Ultraperformance Liquid Chromatography (UPLC) This mode of LC is performed with columns that are packed with very small particles (ca. 1.7 µm in diameter) [17,18]. The benefits of these columns are high-resolution separation at exceedingly large theoretical plates. Small-particle packing creates operating pressures that are in the range 6000 to 15,000 psi and requires special pumping system that is capable of operating at such high pressures. Such a system has become commercially available [18].

5.6. INTERFACES FOR COUPLING LIQUID CHROMATOGRAPHY WITH MASS SPECTROMETRY

The coupling of LC with mass spectrometry is not as straightforward as a similar combination of GC with MS. There are several fundamental differences in the operating environment of HPLC and mass spectrometry. The first mismatch is the solvent flow rate. The separation in conventional wide-bore analytical columns is accomplished at liquid flow rates of 0.5 to 1.5 mL/min. Unlike GC/MS, this liquid produces a gas flow too large for safe operation of the mass spectrometry vacuum system (10^{-5} to 10^{-8} torr). For example, 1.0 mL of water will produce about 1.0×10^5 m^3 of gas load when introduced in a mass spectrometer at 10^{-5} torr pressure (see Example 5.3). Liquid flow rates below 10 µL/min can be accepted safely by a mass spectrometry system. Another problem is the incompatibility of common HPLC solvents and nonvolatile additives with mass spectrometry operation. The third problem in the early years of LC/MS efforts was unavailability of ionization methods that could be used for thermally labile and nonvolatile compounds. Over the years, a suite of different LC/MS interfaces has emerged; as of today, some of them are of historical importance only and are discussed here only for pedagogic interest. Advances in the liquid introduction devices have been reviewed [6,19].

▶ **Example 5.3** Calculate the volume of water vapors that are produced at 1×10^{-5} torr from 1 mL of water at 1 atm.

Solution The volume, V_1, of water vapors at 1 atm,

$$P_1 = (1 \text{ mL})(1 \text{ g/mL})(1 \text{ mol}/18.02 \text{ g})(22.414 \text{ L/mol})(10^{-3} \text{ m}^3/\text{L})$$
$$= 1.2438 \times 10^{-3} \text{ m}^3$$

The volume, V_2, of water vapors at 1×10^{-5} torr,

$$P_2, = \frac{P_1 V_1}{P_2} = \frac{760 \text{ torr} \times 1.2438 \times 10^{-3} \text{ m}^3}{1 \times 10^{-5} \text{ torr}} = 9.45 \times 10^4 \text{ m}^3$$

5.6.1. Moving-Belt Interface

The moving-belt interface was introduced in the 1970s [20–22]. The heart of this interface is a moving belt made of a 0.12-mm-diameter stainless-steel wire that serves to transport the column effluent into the mass spectrometry ion source (Figure 5.5). In later designs, the wire was replaced with a stainless-steel or polyimide ribbon to increase the sample surface area. The moving-belt interface is restricted primarily to compounds of appreciable volatility and thermal stability. The LC effluent is deposited in a uniform manner over the moving belt. The solvent is removed when the sample passes under an infrared (IR) heater and through two differentially pumped vacuum chambers. The volatile components are flash-evaporated into the ionization chamber for subsequent ionization by either EI or CI. Not so easily volatilized components can be ionized by focusing a FAB beam onto the end of the moving belt. Normal-phase LC solvents are highly compatible with this interface. With increasing aqueous content of the mobile phase, the formation of droplets on the belt creates a problem in the solvent evaporation. To certain extent, spray deposition of the LC effluent can overcome this droplet-formation problem. Depending on the aqueous content, solvent flow rates of 0.5 to 1.5 mL min^{-1} are readily acceptable.

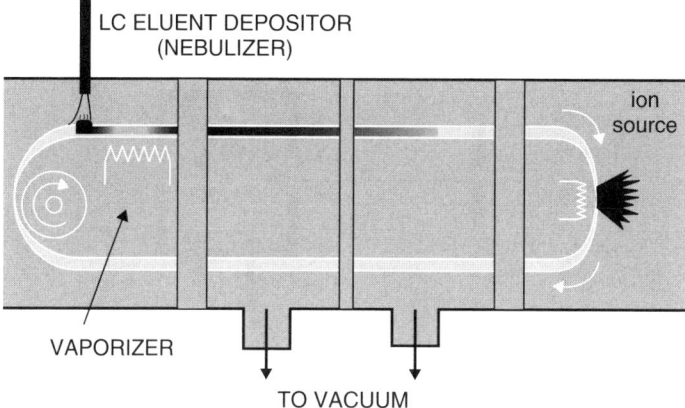

Figure 5.5. Moving-belt interface for LC/MS coupling. (Reproduced from ref. 6 by permission of Wiley-Interscience, copyright © 1997.)

5.6.2. Direct-Liquid Introduction Probe

Several designs of a direct-liquid introduction probe, popular in the early days of LC/MS, have been used to introduce LC effluent directly into the MS ion source [6,23]. One version contains a diaphragm with a pinhole at the end of a capillary tube through which the LC effluent is sprayed into the desolvation chamber [24]. The solute molecules enter the mass spectrometry ion source, where they are ionized via a solvent-mediated CI process. A flow splitter is needed to divert most of the LC solvent from conventional columns because only 10 to 50 $\mu L\,min^{-1}$ liquid flow rates can be accommodated without breakdown of the mass spectrometry vacuum. The microbore capillary columns are connected directly to the ion source. One of the practical operational difficulties of this probe is frequent clogging of the small orifice.

Electron Ionization for LC/MS Attempts have also been made to introduce LC effluents directly into the EI ion source [25]. Figure 5.6 shows a recent version of the direct LC/EI–MS interface. This design uses a nanoscale flow rate (0.8 $\mu L\,min^{-1}$) to nebulize the eluant inside the high vacuum of the ion source. The micronebulizer has a cone-shaped tip slightly bent sideways to allow aerosol droplets of the LC eluant to be directed asymmetrically toward a hot target surface. During the short journey, the droplets get converted to particles, and eventually, to vapor phase upon striking the hot target. The solvent vapors are removed efficiently and solute particles are ionized by EI.

5.6.3. Continuous-Flow Fast Atom Bombardment Interface

The *continuous-flow* (*CF*)–*FAB probe*, discussed in Section 2.11, has also achieved some success as an LC/MS interface for the analysis of nonvolatile and

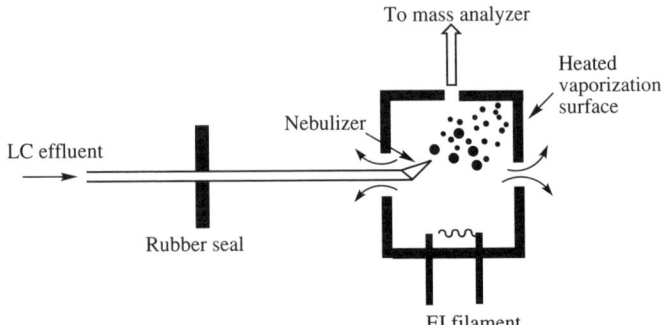

Figure 5.6. Direct nanoscale LC/EI–MS interface. (From ref. 25.)

thermally labile compounds [26–29]. The HPLC effluent, after passing through the central capillary of the CF–FAB probe, spreads over the probe tip, where it is bombarded with a high-energy atom or ion beam. In the alternative design, called *frit-probe interface*, a porous frit is used in place of the probe tip. Because stable operation is achieved with a maximum flow rate of only 1–10 µL/min, the CF–FAB interface is restricted to microbore and capillary columns. Applications of conventional wide-bore columns require postcolumn flow splitting (e.g., 100 : 1). Glycerol is the most common matrix additive, and it is either mixed with LC mobile-phase precolumn or added postcolumn through a coaxial fused-silica capillary [29]. In the later approach, the liquid matrix flows through the outer capillary, whereas the LC solvent is delivered through the inner capillary. The inner and outer capillaries both terminate at the probe tip, where mixing of the two liquid systems takes place. With precolumn mixing of the eluting buffer and glycerol, a possibility of peak tailing exists.

5.6.4. Thermospray Interface

The emergence of thermospray ionization heralded a first ideal interface for a wide range of molecules [30,31]. With the introduction of this interface, LC/MS was accepted as a routine analytical technique. A major beneficiary of this interface was the pharmaceutical industry, which used this system to characterize drugs and metabolites. The construction and basic principle of thermospray ion source was discussed in Section 2.14; briefly, it consists of a heated probe, a desolvation chamber, and an ion extraction skimmer. When passed through a resistively heated capillary, the HPLC effluent, emerges as a mist of fine droplets into a heated desolvation chamber. Ionization of the solute molecules occurs by direct evaporation of the preformed ions or solvent-mediated chemical ionization. Thus, unlike the interfaces discussed above, the thermospray system acts as an ion source as well as an interface. Thermospray is ideally suited to coupling with conventional wide-bore columns. It is, however, confined primarily to reversed-phase HPLC separations, and it is less compatible with nonvolatile solvents. Also, a significant concentration of volatile buffers (usually, millimole amounts of ammonium acetate) must be maintained in the mobile phase. The dissociation of this buffer produces gaseous ammonium and acetate ions, both might act as CI reagent ions for subsequent ionization of the separated components. A decrease in sensitivity is noted with high concentrations of organic solvents such as that used at the upper end of an RP–HPLC gradient analysis or in normal-phase HPLC. Thus, it is possible that the less-abundant components might escape detection during their elution at the upper end of an RP-HPLC gradient analysis. Under such circumstances, the filament-on mode of ionization is supplemented to generate plasma of solvent ions that will assist in solvent-initiated chemical ionization.

5.6.5. Particle–Beam Interface

A particle–beam interface is a useful device that can provide library-searchable EI spectra of LC-separated solutes [6,32–36]. This device uses a jet separator to free the solute molecules from volatile solvents and produces a high-velocity particle beam of gas-phase solute particles. The original device was termed MAGIC (monodisperse aerosol generating interface for chromatography) by its inventors [36].

As shown in the schematic of the particle–beam interface (Figure 5.7), three steps, nebulization, desolvation, and momentum separation, are involved in the conversion of analytes from LC effluents into gas-phase species. First, the LC effluent is nebulized pneumatically by forcing the liquid and helium carrier gas through a small orifice. Desolvation of the droplets in the mist occurs when they travel at very high velocity through the desolvation chamber. The temperature of this chamber is maintained a few degrees above ambient temperature to provide thermal energy just sufficient to compensate for the latent heat of vaporization of the solvent. The third step is momentum separation of the solute particles from

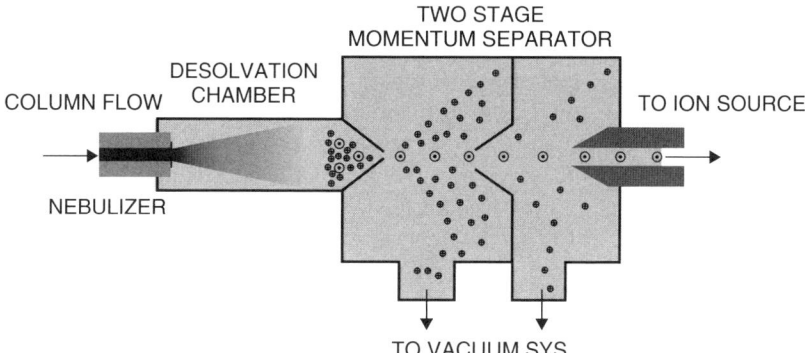

Figure 5.7. Particle–beam interface for LC/MS coupling. (Reproduced from ref. 6 by permission of Wiley-Interscience, copyright © 1997.)

the solvent stream. This process occurs when the mixture of solvent vapors, sample molecules, and helium gas is passed through a two-stage momentum (jet) separator that consists of two reduced-pressure chambers separated by skimmer orifices. Upon entering the first low-pressure region, the mixture expands into a supersonic jet; the high-momentum components, which are primarily heavier solute particles, remain confined to the central axis of the expanding jet, while the low-mass solvent and carrier gas molecules diffuse away at a wider angle and are removed efficiently by the pumping system. The high-momentum components enter the second stage of the momentum separator, where this process is repeated. Ultimately, a beam of uncharged particles, free from the solvent molecules, enters the mass spectrometry ion source. Optimal performance is achieved with normal-phase solvents at flow rates of 0.5 to 1.5 mL/min.

5.6.6. Electrospray Ionization Interface

With the development of an atmospheric-pressure ionization (API) source, coupling of LC with MS has become a routine matter. The ESI format of API is the most appropriate interface for the LC/MS combination because (1) of its potential for the analysis of a variety of nonvolatile and thermally labile molecules of low to very high molecular mass at unprecedented low detection sensitivity, (2) ionization occurs at atmospheric pressure, (3) of its compatibility with RP–LC solvents, and (4) a range of solvent flow can be accepted. As a consequence, the LC/ESI–MS combination has gained prominence in several areas of research, such as to sequence proteins; to identify mixtures of compounds, tryptic maps, and posttranslational modifications in proteins; to elucidate structure of metabolic products; to analyze drugs, pesticides, and toxins; and to screen combinatorial libraries. The development of LC/ESI–MS has also greatly advanced the science of quantification. Several reviews of LC/ESI–MS technology have appeared in the literature [37–40].

The composition and flow rate of the solvent are two variables that are paramount for optimum operation of the ESI system. The flow rate determines the size as well as the size distribution of the droplets formed during ESI. A conventional ESI source operates at a flow rate of 1 to 10 µL/min. At higher flow rates, the spray is not stable because of the formation of larger droplets, which lead to electrical breakdown. Similarly, a fluid with high surface tension, such as pure water, is difficult to electrospray, but many polar solvents commonly used in RP–HPLC (e.g., methanol, ethanol, isopropanol, and acetonitrile) are suitable for the electrospray operation. Nonpolar solvents are difficult to disperse; therefore, normal-phase HPLC is not easy to implement with the ESI process unless a polar solvent is admixed with the nonpolar mobile phase.

HPLC is performed with various size columns that range from 0.1 to 4.6 mm in i.d. The common types of columns and their characteristics are given in Table 5.1. The optimum mobile-phase flow rate is lower when the size of the column is reduced. For example, a decrease in the column i.d. from 4.6 mm to 320 µm reduces the solvent flow rate from 1 mL/min to 4.9 $\mu L\, min^{-1}$. The sensitivity, efficiency, sample-loading capacity, and sample size are the other important criteria in the selection of a specific-size column. Smaller-diameter columns provide increased efficiency of separation. Although the injection volume and loading capacity of capillary columns are much lower than they are for wide-bore columns, the former offers much higher sample peak concentration at the detector, to provide greater detection sensitivity. As an example, the sensitivity of microbore columns is greater than that of wide-bore columns by a factor of 100. A fourfold increase in sensitivity can be realized when the column diameter is reduced by one-half. Small columns also have the advantage of less solvent consumption and waste disposal. To accommodate a wide range of HPLC flow rates and columns, several designs of the ESI interface have emerged. With a proper ESI interface and flow splitting, a column of any dimension can be combined with mass spectrometry.

ESI Interface for Wide and Narrow-Bore Analytical Columns Because of their reasonably high efficiency and large sample loads, most applications of HPLC employ narrow-bore (2.1 mm i.d.) and wide-bore (standard analytical columns of 4.6 mm i.d.) columns. These columns can be coupled directly with a pneumatically assisted ESI source (also known as an ionspray source) and a heated source inlet (e.g., the megaflow ESI source), both of which can readily accept the operating solvent flow rates (0.2 to 1.0 $mL\,min^{-1}$) of these columns. The pneumatically assisted ESI probe uses a coaxial flow of sheath gas, which helps to extend the acceptable liquid flow [41]. The megaflow ESI source inlet employs either a heated capillary or a heated metal block to allow effective desolvation of the bigger drops (see Figures 2.17 and 2.18). Wide-bore columns with still higher flow rates (ca. 2.0 $mL\,min^{-1}$) must be coupled to mass spectrometry directly via this interface.

A conventional ESI source can also serve as an interface for narrow- and wide-bore columns, provided that postcolumn flow splitting is used. Often, flow splitting is essential to allow recovery of a substantial portion of the eluting components for other experiments. A 1.0-mL min^{-1} flow will require a 100:1 split. No degradation of the signal intensity owing to flow splitting is observed, because at low flow rates, the ESI source behaves as a concentration-sensitive detector.

The clogging of the sample orifice is one main concern in the use of high-flow LC/MS systems. Deposition of solids and liquids on the extraction lens of the interface also reduces detection sensitivity. To make the ESI interface more robust, several innovative designs have been used to sample the ion spray (see Figure 5.8). These designs include off-axis ion sampling, the "pepperpot" device, the cross-flow device, and orthogonal or Z-spray designs. In the off-axis design, the bulk of the solvent droplets and vapors strike the lens surface, whereas the ions are extracted by the electric field applied to the skimmer lenses. The pepperpot device uses chicane-type tunnels to prevent solvent from entering directly into the mass spectrometer. The cross-flow device uses a similar strategy. Three different orthogonal extraction schemes are shown in Figure 5.8, in which

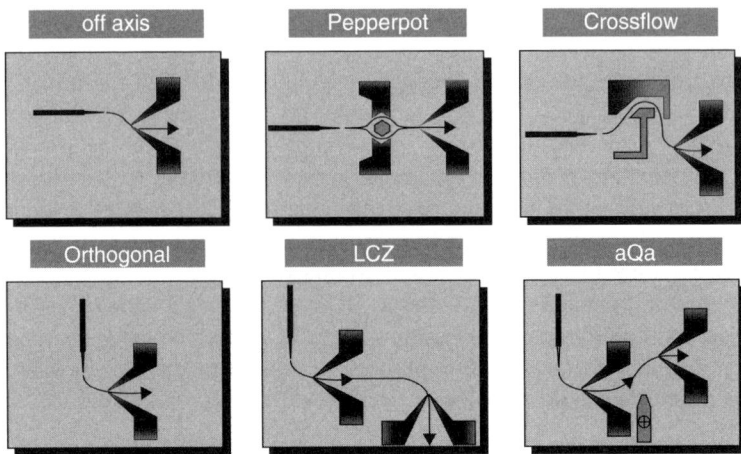

Figure 5.8. Innovative designs to enhance interface robustness. (Reproduced from ref. 6 by permission of Wiley-Interscience, copyright © 1997.)

ions are sampled into the mass spectrometer with one or two 90° extraction steps, and the neutrals are collected by a plate that is located in the probe axis. The aQa design combines orthogonal extraction and off-axis extraction concepts.

ESI Interface for Capillary-LC and Nano-LC Columns Currently, the applications of capillary and nano-LC are on the upswing; especially for many biochemical studies, where the sample amounts and volumes are both limited. For such samples, packed capillary columns of 50 to 300 μm i.d. are the ideal solutions. As pointed out above, the combined use of small-i.d. columns with an ES ion source has the advantage of optimal detection sensitivity because of its concentration-dependent response. Because these columns operate in the flow range nanoliters to microliters per minute, an ideal LC/MS system is realized when these columns are connected directly to nanospray or microspray sources [42,43]. The coupling of these columns to a conventional ES ion source can also be accomplished if an additional sheath liquid is added to increase the flow to a range that is acceptable by the source.

One of the limitations in the use of capillary columns is an irreproducible and unstable gradient in the nL/min flow range. A few years ago, the standard procedure to achieve reliable gradients was to use precolumn flow splitting [44]. Any conventional or micro-LC pumping system could serve to deliver low flow rates by using an appropriate homemade or commercial flow splitter. Currently, LC pumps for capillary columns have become commercially available that can couple capillary columns directly with a nanospray source [45].

Microbore columns (0.8 to 1 mm i.d.) offer a compromise between wide-bore and capillary columns. Chromatography separation with these columns is carried out at flow rates of 20 to 100 $\mu L\,min^{-1}$. HPLC hardware is available that can deliver a precise volume and solvent composition in this low flow range. These columns can be coupled to mass spectrometry via a pneumatically assisted ESI source with no precolumn or postcolumn flow splitting [46].

5.6.7. Atmospheric-Pressure Chemical Ionization Interface

Atmospheric-pressure chemical ionization (APCI) is another robust and dependable interface for the LC/MS combination. It is highly compatible for medium- and low-polarity small compounds [47]. The usefulness of LC/APCI–MS has been demonstrated for a wide range of applications in the bioanalytical, pharmaceutical, and environmental fields [48]. This interface shares many common features with ESI and thermospray. Similar to thermospray, APCI can also accommodate solvent flow rates in the range 0.5 to 2 mL min^{-1}. A main feature of this interface is a heated nebulizer inlet probe that helps to nebulize the LC effluent [49]. A high-voltage needle creates a corona discharge to generate reagent ions for subsequent solvent-initiated chemical ionization of the solute molecules. In many commercial designs, a common source housing is used for APCI and ESI. An APCI interface suitable for semimicro-LC is also available [50]. It consists of a stainless-steel capillary that is inserted into a stainless-steel block, and the vaporizer and nebulizer blocks are combined into one.

5.6.8. Atmospheric-Pressure Photoionization Interface

Atmospheric-pressure photoionization (APPI) is complementary to the ESI and APCI modes of ionization. Unlike ESI and APCI, it can be applied to the analysis of nonpolar molecules, and is relatively less susceptible to ion-suppression and salt-buffer effects. A heated nebulizer is also used to couple LC with APPI–MS [51]. The nebulizer is made of a long quartz tube, one end of which is connected to the LC column and the other to a mass spectrometer (Figure 5.9). The LC solvent enters the nebulizer at a flow rate of 200 µL min^{-1}. APPI is performed in the presence of an ionizable dopant (toluene or acetone), which is delivered to the nebulizer via a syringe pump. The high temperature (450°C) in the nebulizer causes the LC eluant and dopant to evaporate completely before they enter the ionization region. A unique feature of APPI is that ionization of the LC effluent can be suppressed by a proper choice of the photon energy, thus minimizing the background signal and improving the detection sensitivity. The sample ion signal is also enhanced because ionization occurs at atmospheric pressure in the presence of an excess of a dopant. Thus, LC/APPI–MS provides improved detection sensitivities for nonpolar molecules over those provided by LC/ESI–MS and APCI–MS. In another design, an orthogonal spray concept has been used for LC/APPI–MS coupling [52]. This design is equally effective at low LC flow rates (ca. 50 µL/min).

5.6.9. Coupling LC with TOF–MS

The duty cycle of most mass analyzers is long compared to the chromatography peak width. This mismatch might pose a problem in obtaining representative

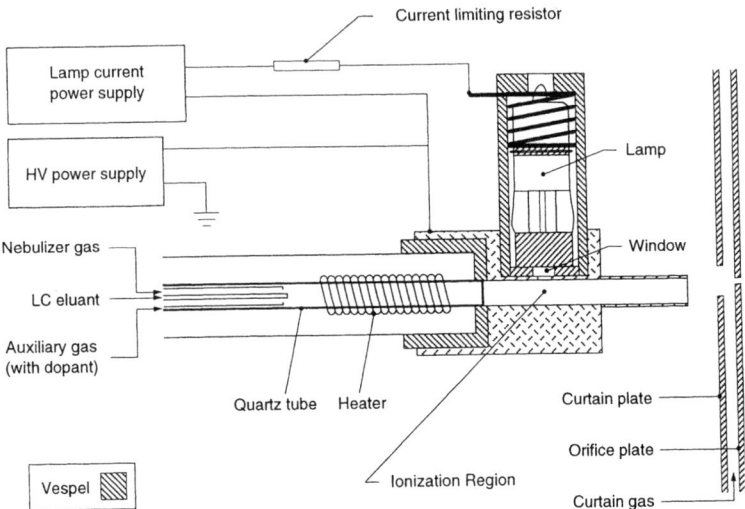

Figure 5.9. Schematic of the APPI ion source. Its main components are a heated nebulizer probe and photoionization lamp. (Reproduced from ref. 51 by permission of the American Chemical Society, Washington, DC, copyright © 2000.)

data with capillary LC systems (see Figure 5.2). The high scan speed of a TOF mass spectrometer is a highly attractive feature for an LC/MS combination. Also, a TOF instrument has multiplex detection capability. However, the coupling of TOF–MS with LC is not straightforward because of the pulse nature of the TOF operation. Nevertheless, some success has been achieved to couple LC with ESI and MALDI modes of ionization. The most common approach to couple ESI with TOF–MS makes use of the orthogonal ion-extraction concept, in which ESI-produced ions are stored between each duty cycle and are pushed into the flight tube by a high-voltage pulse.

5.6.10. Coupling LC with MALDI–MS

Several attempts have been made to combine LC online with MALDI [53]. In one approach, a *continuous-flow probe*, similar to that used for CF-FAB, has shown success in introducing LC effluents directly into the MALDI source [54,55]. The matrix solution, composed of 3-nitrobenzoyl alcohol, 0.1% TFA, 1-propanol, and ethylene glycol ($3:3:5:9$ by volume), is mixed postcolumn with the LC flow. A UV laser beam is focused onto the tip of the probe to ionize the emerging LC effluent.

A *pulsed sample introduction* (PSI) *interface* and laser-induced multiphoton ionization have been also used to couple LC with TOF instruments [56]. The interface consists of a heated capillary for aerosol generation and a high-temperature pulsed nozzle for sample vaporization. The LC effluent enters the heated capillary at a flow rate between 0.5 and 1.6 mL min^{-1}. A solenoid allows the sample vapor to enter the mass spectrometer in a pulse form.

Another design is based on the concept of aerosol formation [53]. In this interface, LC effluent premixed with a matrix solution flows through a stainless-steel tube into a pneumatic nebulizer and is sprayed directly into the ion source of a TOF mass spectrometer. Following the removal of the solvent, the spray is bombarded with a beam of photons from a UV laser.

A *rotating-wheel interface* has been devised to couple LC and CE with MS [57]. The LC effluent premixed with the MALDI matrix flows through a fused-silica capillary at the rate of 100 to 400 nL min^{-1}. The tip of the separation capillary is in contact with the rotating wheel. The solution deposited is dried rapidly in vacuum of the MALDI source, leaving a narrow trace of cocrystallized analyte–matrix mixture. The wheel is transported to the repeller, where laser desorption–ionization of the analyte–matrix mixture occurs. A similar system that uses a rotating ball coated with matrix for deposition of a single drop is also available for online LC/MALDI–MS [58]. The droplet deposition is controlled by a piezoelectric-actuated droplet generator.

5.7. MULTIDIMENSIONAL LC/MS

Although the separation efficiency of HPLC is high, optimal resolution of components of a very complex matrix is still a challenge with a single LC separation. Many coeluting components may escape detection because of mutual signal suppression in MS detection, and in the worst scenario they may have the same nominal mass. A solution to this problem is multidimensional chromatography, in which two orthogonal separation steps are combined; the fractions separated by the first separation step are presented to the second system using a column-switching technique. The second separation system must be faster and compatible with the MS ionization step. The most commonly used system comprises a combination of ion exchange and RP–HPLC [59]. For example, a cation-exchange column is used as a first-dimension separation of peptides in a complex protein digest. Several other fruitful two-dimensional systems include a combination of SEC with RP–HPLC [60], chromatofocusing with LC (discussed in Chapter 8 on page 297), affinity chromatography with LC [61,62] and CE [63], LC with CE [64], and capillary isoelectric focusing (CIEF) with LC [65].

A two-dimensional CIEF–RPLC/MS system that was developed to resolve a protein mixture is shown in Figure 5.10. It consists of a CIEF system that is outfitted with a microdialysis membrane-based cathodic cell, a 1-μL injection loop, a C_{18} trap, and a C_4 LC column [65]. A protein mixture is separated by the CIEF system and the focused bands are carried hydrodynamically past the cathodic cell into the injection loop, which is a part of a six-port microselection valve. By switching the position of the microselection valve, the fraction collected is transported from the injection loop to a C_{18} trap column, which is attached to a second six-port valve of a capillary LC system. After the protein fraction is trapped, both six-port valves are switched to their original position. The trapped protein fraction is eluted onto the C_4 analytical column for further fractionation. Meantime, another fraction from the CIEF step is collected

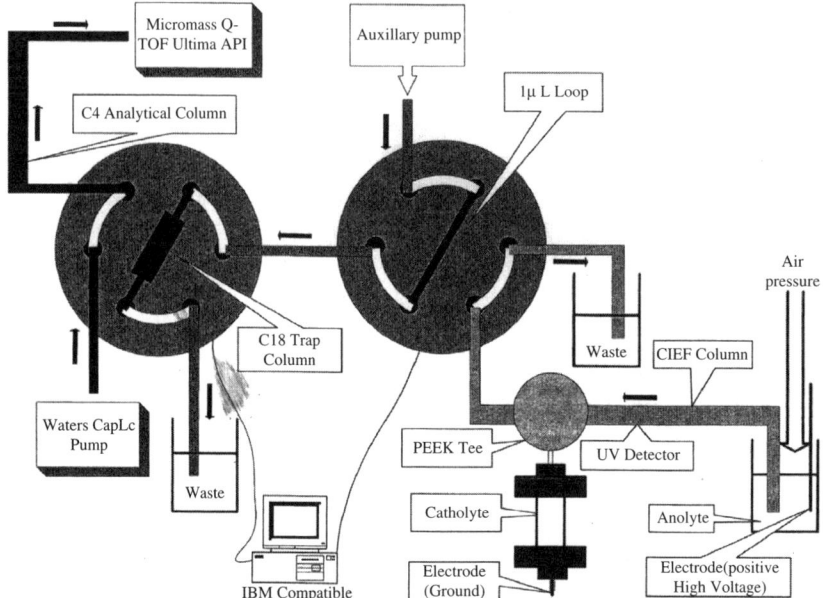

Figure 5.10. General view of the CIEF/RPLC–MS system. (Reproduced from ref. 65 by permission of the American Chemical Society, Washington, DC, copyright © 2004.)

by a 1-μL loop, and the entire sequence is repeated until all of the CIEF fractions have been analyzed by nano-ESI–MS. The arrangement effectively removes the CIEF ampholytes, which otherwise perturbs the ESI of the separated analytes.

5.8. CAPILLARY ELECTROPHORESIS/MASS SPECTROMETRY

Capillary electrophoresis (CE) is another high-resolution separation device that can benefit greatly from the highly structure-specific and sensitive detection capability of mass spectrometry. The basis of separations in CE is the electrically driven migration of ions in free solutions. It offers an orthogonal and complementary approach to HPLC separation. The hallmarks of CE are unparalleled resolving power (over 1 million theoretical plates), ultrahigh sensitivity (in the attomole range), low volumes (nanoliters), and a high speed of analysis. The reason for high separation efficiency is that the solvent flow profile inside the column is flat. Currently, CE is meeting the challenge of analyzing the sample volumes as low as from a single biological cell. Several areas of biochemical research have benefited from emergence of the CE/MS combination. Its applications include the analysis of peptides, proteins, enzymatic digests, noncovalent complexes, DNA adducts, pharmaceutical drugs and their metabolites, and compounds of environmental concern.

5.8.1. Basic Principles of Capillary Electrophoresis

Analytical selectivity in CE is achieved by the differential migration of electrically charged particles or ions in solution under the influence of an applied electric field. CE embodies the separation principle of traditional polyacrylamide gel electrophoresis (PAGE) and the online detection and automation strategies of chromatography. The separation efficiency of PAGE is poor because of the heat generation caused by high applied voltages. Jorgenson and Lukacs demonstrated that effective heat dissipation could be achieved in small-diameter open-tubular fused-silica capillaries to provide highly efficient separations [66]. CE is performed in a number of formats, described below.

Capillary Zone Electrophoresis Capillary zone electrophoresis (CZE, referred in this book simply as CE) is the most common and simple mode of CE. In this setup, a 50- to 100-cm-long narrow-bore (25 to 75 μm i.d.) capillary is filled with a buffer and the ends are immersed in separate buffer reservoirs (Figure 5.11). A very high potential (200 to 500 V/cm) is applied between the two ends. The electrostatic force drives the cationic analytes toward the cathode and anionic analytes toward the anode; the neutral species are unaffected. The rate of electrophoretic mobility is proportional to the charge-to-size ratio of the analyte; highly charged small ions move at a faster rate than do the low-charged large ions. The analyte species also migrate under the influence of electroosmosis, which is produced as a consequence of the surface charge on the inside wall of the silica capillary. Above pH 2, the ionized negatively charged silanol groups (SiO^-) on the capillary wall attract the positively charged buffer ions from the solution to produce an inner static positively charged double layer and an outer diffuse layer. The resulting zeta potential, ξ, forces movement of the bulk liquid (in the low-nL/min range) toward the cathode. Because of the overwhelming nature of this flow, all ions, regardless of their charge, migrate toward the cathode. The migration velocity of the analyte is thus the sum of the electrophoretic mobility,

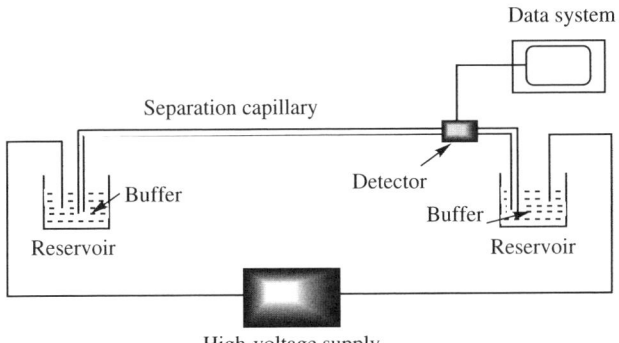

Figure 5.11. Schematic representation of a CE apparatus. (Reproduced from C. Dass, *Principles and Practice of Biological Mass Spectrometry*, Wiley-Interscience, 2001.)

μ_e, and electroosmotic mobility, μ_{EO} (i.e., $\mu = \mu_e + \mu_{EO}$). The cationic species move faster than the neutral species, which in turn move faster than the anionic species. Thus, the simultaneous detection of all ionic solutes is achieved. The migration time, t, is given by

$$t = \frac{l}{\mu E} = \frac{lL}{\mu V} \tag{5.6}$$

where l the length of the capillary from the anode end to the detector, L the total capillary length, E the electric field, and V the applied voltage. The solvent-flow profile within the capillary is flat and thus yields very narrow peaks.

Miceller Electrokinetic Chromatography (MEKC) This mode of CE is a hybrid of electrophoresis and chromatography, and it is one of the two CE techniques that can be applied to neutral and ionic species (the other being CEC). In this mode, a surfactant, at concentrations above the critical micelle concentration, is added to the separation buffer. The micelles have exposed polar head groups and buried hydrophobic tails, and depending on the charge, move with or against the EOF. For example, an SDS micelle is negatively charged and thus tends to move toward the anode. The EOF, however, pushes it toward the cathode, but at a slower rate than the net flow of the buffer. Similar to reversed-phase HPLC packing, the micelles act as a pseudo-stationary phase, in which neutral molecules can partition to different extents according to their hydrophobicity. When in the SDS micelle, the migration of a neutral molecule will be slowed. The rate of migration of ions is also affected by their varying degrees of electrostatic attraction for the polar head groups of the micelles. Anions will be repelled by the negatively charged micelle and thus move along with the EOF. Cations elute last, owing to strong electrostatic attraction with the negatively charged micelle.

Capillary Gel Electrophoresis (CGE) This mode of CE is used to separate macromolecules on the basis of their size difference in porous gels (e.g., polyacrylamide gel), which behave as molecular sieves. The capillary is filled with a gel and a suitable buffer. The macromolecules, such as DNA and proteins, migrate in the gel and buffer-filled capillary, and separation occurs by the combined effect of molecular sieving and electrophoretic mobility.

Capillary Isoelectric Focusing CIEF separates amphiprotic species, such as proteins or peptides, on the basis of differences in the isoelectric point (pI). The capillary is filled with a mixture of carrier ampholytes and solute, and its cathodic end is placed in a basic buffer and the anodic end in an acidic solution. When an electric potential is applied, the ampholytes migrate until they become uncharged; thus, a pH gradient is formed throughout the length of the capillary. The solutes migrate similarly until they reach the region of identical pI. This step is known as focusing. After focusing, the ampholyte and solute bands are mobilized past the detector.

Capillary Isotachophoresis CIT is a moving-boundary technique. The sample is sandwiched between a leading electrolyte with mobility higher than any of the sample components, and a terminating electrolyte with mobility lower than any of the sample components. Upon application of the electrical field, the sample ions are separated into bands according to their electrophoretic mobility. Once the band is formed, all ions that are separated migrate at the same velocity.

Capillary Electrochromatography CEC, also a hybrid of CE and LC, combines the best aspects of both modes of separation [67]. In this approach, a CE capillary is packed with a microparticulate LC stationary phase or a monolithic continuous-bed sorbent, and the solvent is driven by electroosmosis generated at the solid–liquid interface within the capillary. Separation of analytes is the result of distribution of solutes between the liquid stationary phase and the mobile phase. This arrangement has the advantage that the leading edge of solvent flow inside the capillary has a planar profile rather the parabolic laminar profile that exists for pump-driven solvents in conventional LC columns. The result is an improved separation efficiency compared to the efficiency that can be achieved with HPLC, and the advantage of the high sample capacity of HPLC is retained.

5.8.2. Interfaces for Coupling Capillary Electrophoresis with Mass Spectrometry

Almost all operational modes of capillary electrophoresis have been combined with mass spectrometry via CF–FAB [68–70], MALDI [57,71], and ESI [72–76]. Of these three technologies, ESI has attained an enviable status as the interface of choice to couple CE with mass spectrometry. A CE/MS interface must address the problem of the low liquid flow rates of the CE process. Other considerations are buffer compositions and electrical contact between the separation buffer and the interface. As with LC/MS systems, only volatile buffers are tolerated in a CE/MS operation. Also, high salt concentrations must be avoided for stable operation of a mass spectrometry system.

Electrospray Ionization Interface The ESI source is the preferred mode of CE/MS coupling. A general view of the CE/ESI–MS coupling is shown in Figure 5.12. The inlet of the CE capillary and the separation anode are immersed in the buffer vial. The other end of the capillary is inserted into the ESI needle, which is connected to the electrospray voltage supply. A standard ESI source is ideally suited for this purpose. There is, however, a mismatch in the flow rates of the two systems. Another major concern is the electrical contact with the CE capillary. The first successful CE/MS combination was described in 1987 [73]. In this prototype design, the cathode end of the separation capillary terminated within a stainless-steel capillary, which acted as the CE cathode and the ESI needle. Currently, three main categories of CE/ESI–MS interfaces exist: sheathless interface, sheath-flow interface, and liquid-junction interface (see Figure 5.13).

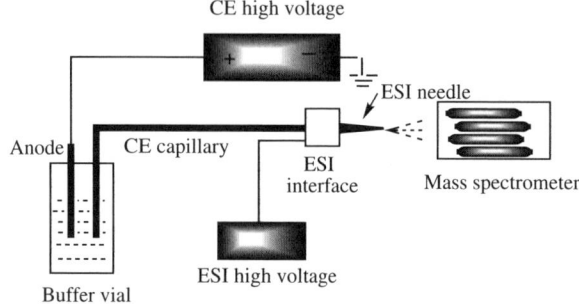

Figure 5.12. General view of CE/ESI–MS coupling.

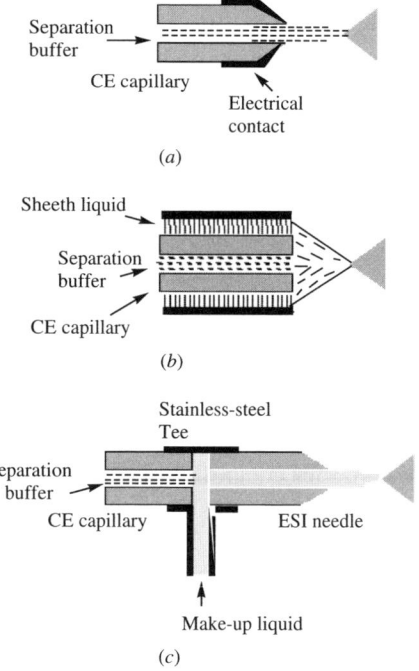

Figure 5.13. Conceptual views of (a) sheathless, (b) sheath-flow, and liquid-junction interfaces. (From ref. 72.)

1. *Sheathless interface.* In the sheathless approach (Figure 5.13a), the CE capillary is coupled directly to the ESI source. The electrical contact for the CE circuit at the capillary terminus, as well as for the ESI electrical field, has been established by coating the capillary tip with a conducting metal [74], by inserting a gold electrode into the outlet of the CE capillary, or by inserting a platinum wire electrode through a hole near the

tapered capillary [75]. In another version, the CE capillary is inserted into a stainless-steel liner, whose tip is ground to a sharp end. A narrow gap (ca. 10 μm) between the two capillaries helps to establish electrical contact via a liquid film that builds up around the outer surface of the CE capillary tip [76].

The coupling CE through the sheathless concept is optimally achieved via a nano-ESI source because both operate at nL/min flow rates. In one version, the CE column is connected to the sprayer tip via a microdialysis device [77,78]. In another simple version, a single piece of tapered CE column with a 20-μm-i.d. tip has been used for separation as well as microspraying [78]. This combination results in high ionization efficiency and low detection limit. The detection limit of 0.1 to 5 fmol for peptide standards has been reported.

The sheathless design has the advantages of high detection sensitivity, low flow rates, long-term stability, and the absence of interference from the sheath liquids. A major concern of this design is to find an optimum buffer combination for CE separation and ESI operation.

2. *Sheath–flow interface.* The sheath–flow design is the most common approach for CE/MS coupling (Figure 5.13b) [79,80]. It is versatile and easy to implement. In this design, the separation capillary is inserted into a narrow metal capillary through which the makeup liquid is introduced coaxially. Optionally, a third concentric capillary might be used to add a sheath gas to assist nebulization. The sheath liquid is required for stable operation of the standard ESI source, and to establish the electrical contact at the cathode end of the capillary. A wide range of buffer systems can be used because the sheath liquid and the separation buffer can be optimized independently for desired efficiency in CE separation and for stable electrospray operation. For positive-ion analysis, the sheath liquid consists of a 1 : 1 (v : v) mixture of water–methanol that contains 0.1% of either formic or acetic acid. The sheath–flow interface is more robust and reproducible and provides enhanced signal-to-noise ratios and improved separations. Its limitation is that the added sheath liquid might degrade the detection sensitivity.

3. *Liquid–junction interface.* In this type of interface, a stainless-steel tee is used to introduce the makeup liquid (Figure 5.13c) [81,82]. The cathode end of the CE capillary and the ESI needle are introduced from opposite ends of the tee, with a narrow gap (10 to 25 μm) between the two terminals. The electrical contact between the CE capillary and the ESI needle is established through the makeup liquid that surrounds the junction of the two capillaries. The liquid–junction interface is easy to assemble and operate. The additional liquid flow often degrades the detection sensitivity. CE separation is also compromised at the liquid junction.

Some more interfaces are available that combine the versatility of the sheath–liquid design with the sensitivity of the sheathless format. In one design, two

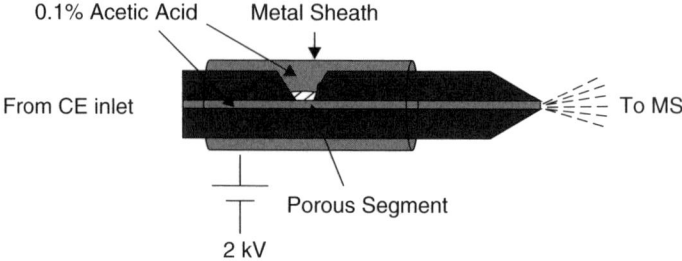

Figure 5.14. CE/ESI–MS interface that uses a porous plug capillary for electrical contact. (Reproduced from ref. 85 by permission of the American Chemical Society, Washington, DC, copyright © 2003.)

capillaries, one for CE separation and the other for makeup liquid, are inserted into a nano-ESI tip to provide a low-flow interface [83]. In another design, a split-flow approach is used to provide electrical connection to the CE capillary outlet [84]. A part of the CE buffer is diverted out of the capillary through a small opening near the capillary outlet. The electrical contact is established by inserting this portion of the capillary into the existing ESI needle and filling the needle with a background electrolyte. An improved version of this design incorporates a porous plug in place of a small opening to divert the flow [85]. This concept is illustrated in Figure 5.14. The porous plug is created by thinning a small section of the CE capillary by drilling a well into it and etching the remaining thickness of the wall. Both of these designs have the advantage of zero dead volume.

Continuous-Flow Fast Atom Bombardment Interface Although currently not a popular approach, the coupling of CE mass spectrometry was once achieved via a CF–FAB probe. Makeup flow is required in this coupling because of the mismatch of the low flow rates of the CE solution with the liquid flow rates of stable CF–FAB operation. The sheath–flow and liquid–junction designs discussed above have been used successfully for this purpose [69–71]. The sheath–flow design has the advantages that the composition and the flow rates of the CE effluents and of the FAB matrix solution can be optimized independently, and that the separation efficiency is higher.

Coupling CE with MALDI–MS Similar to LC–MALDI coupling, coupling of CE with MALDI is not a common practice. In one approach, known as continuous-flow laser vaporization/ionization, direct introduction of CE eluant into a MALDI source has produced good results [71]. The matrix is mixed with the CE buffer, and the emerging liquid is bombarded with a laser beam. The energy absorbed by the matrix from the laser (e.g., $CuCl_2$ at 248-nm excimer laser) is used to vaporize and ionize the CE eluant. The rotating-wheel approach discussed in Section 5.6.10 for online LC/MALDI–MS coupling has also been adapted for a CE–MALDI interface [57].

Practical Considerations For the success of online CE/MS, the ionization mode must be compatible with the electrical charge of the eluting components. In practice, CE separation works best for anions, whereas the optimum mass spectrometry response is obtained for cations. The postcolumn addition of an appropriate sheath buffer minimizes this mismatch. The intrinsic narrow time width (<1 s) of CE peaks also poses a practical problem, a solution for which is ultrarapid scanning of the spectral data as is done in TOF–MS or use the concept of *programmable field-strength gradient* (also known as *voltage stepping*). In the latter approach, similar to peak parking, the CE voltage is reduced just prior to elution of the analyte peak to increase the elution time of the analyte peak [86]. Another practical problem is poor concentration sensitivity of CE, which can be alleviated with a preconcentration step. Online sample stacking with isotachophoresis has been adapted successfully for this purpose [87,88]. Another approach to improving the sample concentration is to use an online RP–HPLC packing cartridge [89] or an impregnated membrane [90]. In the chromatography preconcentration step, a C_{18} packing column is connected between the transfer capillary and the separation capillary via a Teflon sleeve. The analytes are first trapped by the C_{18} packing and then eluted into the CE column. A 110-fold enrichment in the sample concentration has been achieved with this setup. The concept and setup of the impregnated membrane preconcentration step are similar. In this case, in place of a packed precolumn, a membrane impregnated with C_{18} or polymer (e.g., styrene–divinyl benzene) adsorptive material is used [90].

Coupling Other Modes of CE with MS Because of practical difficulties, the coupling of other modes of CE with MS is not very common. However, some attempts have been made to couple CIEF with ESI–MS [80,91], CEC–MS [92], CGE–MS [93], and MEKC–MS [83,94,95]. A simple MEKC–MS interface is described (Figure 5.15). This interface comprises two parallel capillaries, one for separation and the other for makeup liquid. Both have the dimensions 50 μm i.d. × 155 μm o.d. and have tapered ends. The capillaries are housed in a larger capillary (530 μm i.d. × 690 μm o.d.) with a beveled edge. The larger capillary serves as a nano-ESI sprayer. A makeup liquid allows control of the composition of the spray solution.

5.9. AFFINITY CHROMATOGRAPHY/MASS SPECTROMETRY

Although an LC/MS combination provides high-resolution separation and the specific detection of analytes in a complex mixture, in many situations it is necessary to provide purification, concentration, and isolation of desired components (e.g., in a biological matrix). This step is performed by affinity chromatography. Practical examples are the selective capture of a specific antigen by the corresponding antibody and the isolation of phosphorylated proteins and peptides [96,97]. An

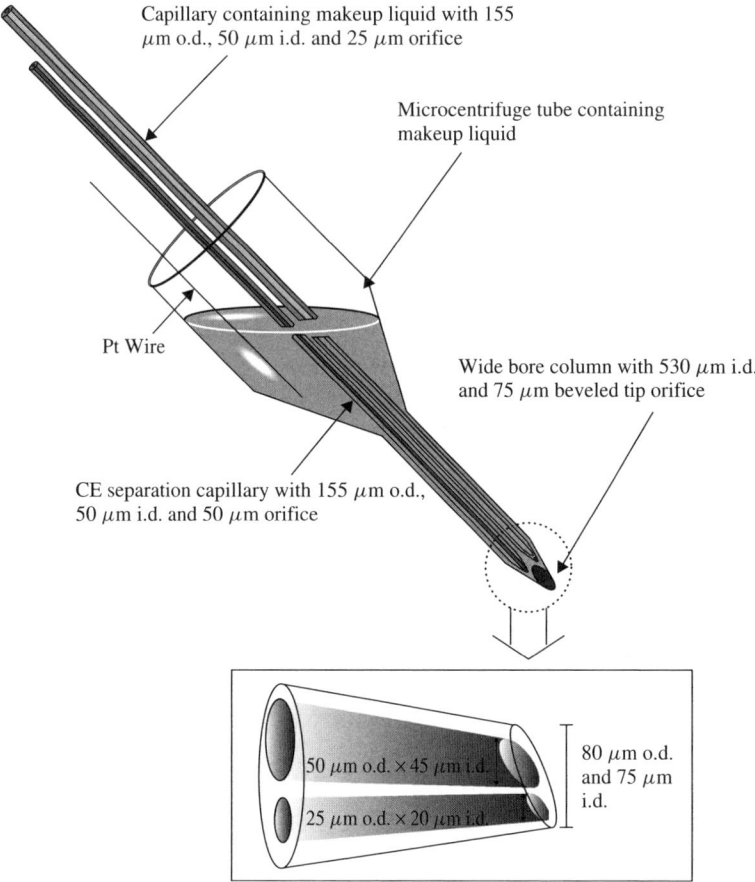

Figure 5.15. Schematic of the low-makeup beveled CE/ESI–MS interface. (Reproduced from ref. 95 by permission of the American Chemical Society, Washington, DC, copyright © 2004.)

example of the coupling of affinity chromatography to mass spectrometry via an RP–LC system is shown in Figure 5.16. This setup is used to isolate phosphopeptides selectively from a protein digest. It consists of a nanoflow LC system, a six-port switching valve, a dual TiO_2/C_{18} precolumn, and a C_{18} analytical column. The two-dimensional separation of phosphopeptides is achieved by loading the sample onto the dual precolumn. During this step, phosphopeptides are trapped by TiO_2 and nonphosphopeptides by C_{18} packing. Next, the trapped nonphosphopeptides are separated on the C_{18} analytical column with gradient elution. After this step, the TiO_2-trapped phosphopeptides are transferred to the C_{18} precolumn, where they are concentrated further. A second gradient elution separates phosphopeptides onto the C_{18} analytical column.

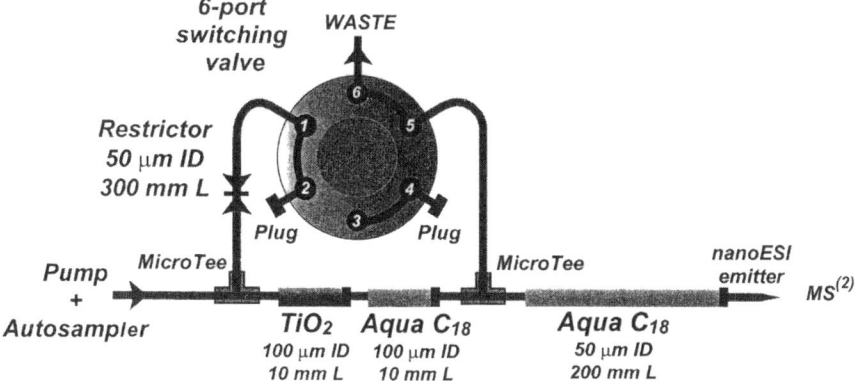

Figure 5.16. Two-dimensional LC/MS setup consisting of a six-port switching valve, a dual TiO_2/C_{18} precolumn, and a C_{18} analytical column. (Reproduced from ref. 98 by permission of the American Chemical Society, Washington, DC, copyright © 2004.)

5.10. SUPERCRITICAL-FLUID CHROMATOGRAPHY/MASS SPECTROMETRY

Supercritical-fluid chromatography (SFC) is a hybrid of GC and LC in which the mobile phase is a gas (e.g., CO_2) at a pressure above its critical pressure. It is applicable to nonvolatile and thermally labile compounds of low to moderate polarity. The coupling of SFC with mass spectrometry exploits the liquidlike properties of supercritical fluids, and the ease of coupling of GC with mass spectrometry [99–101]. Packed and capillary column chromatography columns have both been utilized. Although all major types of LC interfaces, including thermospray, particle–beam, and moving belt, have been coupled with SFC in the past, the current applications rely primarily on APCI and ESI interfaces [100,101]. A key feature of any SFC–MS interface is a heated transfer line, which is required to compensate for cooling caused by expansion of the supercritical fluid.

5.11. COUPLING PLANAR CHROMATOGRAPHY WITH MASS SPECTROMETRY

The most common format of planar chromatography is TLC, in which separation occurs on a flat glass plate that has been coated with a thin layer of adsorbent called the stationary phase. Silica gel is the most widely used stationary phase. The sample is applied as a tiny spot about 1 cm above the lower edge of the TLC plate, and elution occurs by immersing this end of the plate in the mobile phase that is contained in an enclosed jar. The solvent rises past the sample spot by the capillary action to separate the components of the sample. The separated components are detected by staining the plate with a reagent or by incorporating a fluorescent dye into the stationary phase. TLC has the advantage of speed of analysis and at a cost that is much lower than that of LC methods; the resolution is, however, low. It can handle only simple mixtures with only four to seven components.

A more refined version of TLC is *overpressured-layer chromatography* (OPLC), in which the mobile phase is forced through the TLC sorbent material [102]. A TLC plate is covered with a sheet of Teflon membrane and is subjected to high pressure (ca. 50 bar). Separation is effected by pumping the mobile phase through the sorbent layer. Another forced-flow TLC system is *rotational planer chromatography* (RPC), in which the TLC plate is rotated at a high angular velocity [103]. The centrifugal force generated forces the mobile phase to move in a circular profile. The components are separated in annular rings rather than as linear spots.

Because coupling greatly increases the information content of the component separated, several efforts have been made to combine TLC with mass spectrometry [104–109]. The MALDI mode of ionization constitutes an ideal system for analysis of the TLC-separated components. In a simple version, MALDI can be performed directly on a TLC plate or after transfer of the sample onto a polymer membrane. The TLC plate is sprayed with a matrix solution, and the plate is attached to the MALDI target with a double-sided adhesive tape [104–106]. Staining the plate with a suitable reagent (e.g., orcinol–H_2SO_4) can improve MALDI surface sampling of the components separated. A more improved system uses a hybrid TLC–MALDI plate that contains two side-by-side layers: one of silica adsorbent for separation of the mixture and the other of MALDI matrix for detection (Figure 5.17) [107]. The MALDI layer is formed by scraping the silica from a conventional TLC plate and spraying the matrix solution that was

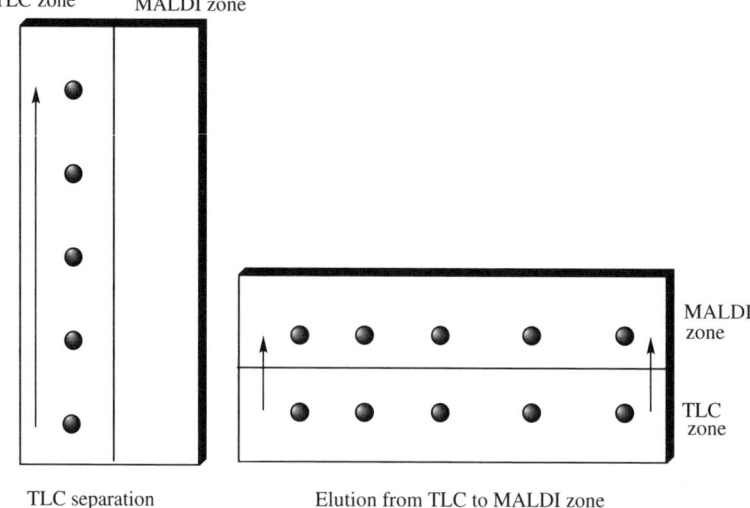

Figure 5.17. Hybrid TLC–MALDI plate for coupling with TOF–MS. (From ref. 107.)

prepared in a toluene–ethyl ether (60:40, v:v) solvent mixture on the exposed TLC backing. A mixture of compounds is separated in the TLC layer. Next, the plate is rotated 90° and the separated components are transported to the MALDI layer by perpendicular elution of the TLC plate.

The continuous-flow profile of the mobile phase in OPLC and RPC has provided an opportunity to couple them to mass spectrometry through an API interface. In an OPLC/ESI–MS system, the coupling is achieved by connecting the OPLC outlet with the inlet of the ESI probe, which can accept a solvent flow rate of 60 to 120 µL min^{-1} [108]. Coupling of a commercial CycloGraph RPC system with APCI mass spectrometry has also been demonstrated [109]. Details of this coupling are shown in Figure 5.18. The self-aspiration of the heated APCI nebulizer delivers about 20 µL min^{-1} of the total 3.0 mL min^{-1} RPC flow to the mass spectrometer. The remainder of the flow is diverted to a fractional collector.

OVERVIEW

The online coupling of a separation device with mass spectrometry is an analytical approach that can help in the analysis of real-world samples such as environmental samples and biological tissues and fluids. This online marriage between two stand-alone analytical techniques can provide an unequivocal characterization of individual components of such complex samples and greatly increases the information content of those components. Several issues that must be addressed to achieve an ideal combination. A major concern is the pressure mismatch. The solvent incompatibility also becomes an issue in the coupling of LC and CE with mass spectrometry. Thus, online coupling requires an interface that can transport the separated components into the ion source without affecting their resolution or the performance of a mass spectrometer. Not all types of mass spectrometers

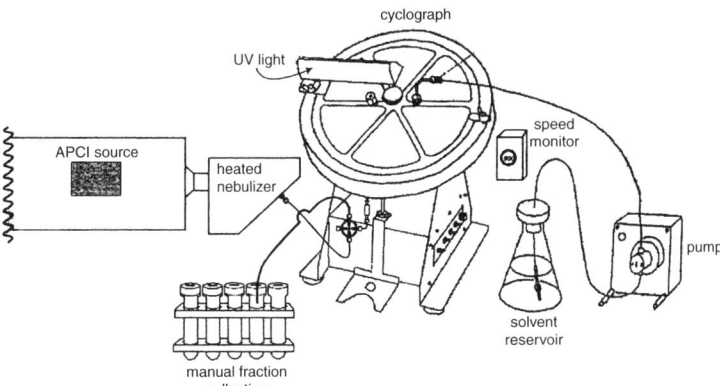

Figure 5.18. Schematic representation of the coupling of the CycloGraph RPC system with APCI–MS. (Reproduced from ref. 109 by permission of the American Chemical Society, Washington, DC, copyright © 2004.)

are suitable for the online coupling of a separation device. Only those that have a high scan speed and can tolerate high pressures (e.g., quadrupole mass filters, quadrupole ion traps, linear ion traps, TOF instruments) are considered ideal for this combination.

GC is an ideal separation technique for small thermally stable volatile molecules. Capillary columns provide high-resolution separation of complex mixtures. These columns, owing to their low gas flow requirements, can be coupled directly to mass spectrometry. A variety of interfaces, such as an open-split interface, a jet-separator interface, and a molecular-effusion interface, are used to handle larger carrier gas flows of packed GC columns.

LC and CE, on the other hand, are suitable for separation of mixtures that contain nonvolatile and thermally labile compounds. The coupling of these liquid-phase separation techniques with mass spectrometry is not very straightforward. Development of API techniques, especially ESI, has provided an ideal opportunity to couple LC and CE with mass spectrometry. ESI acts as an ion source and as an interface. Different types of ESI interfaces are available that can accept a wide range of solvent flow rates that are employed in LC separation. As a consequence, all types of LC columns, from smaller-bore capillaries to wide-bore analytical columns, have been combined with ESI–MS. Several other interfaces have also achieved some degree of success in combining LC with mass spectrometry. These include a moving-belt interface, a direct-liquid introduction probe, a CF–FAB probe, a thermospray interface, a particle–beam interface, and an atmospheric-pressure photoionization interface. Although not as common as LC/ESI–MS technology, LC has also been coupled with MALDI–MS.

Multidimensional chromatography, in which two orthogonal separation steps are combined (e.g., a combination of ion exchange and RP–HPLC) provides a solution to very complex sample mixture.

CE provides a complementary approach to HPLC separation. It is performed in several different formats, including capillary zone electrophoresis, miceller electrokinetic chromatography, capillary gel electrophoresis, capillary isoelectric focusing, isotachophoresis, and capillary electrochromatography. Of these formats, capillary zone electrophoresis is the most popular separation technique. The most successful coupling of CE with mass spectrometry is achieved via an ESI interface. The three most practical designs are sheathless interface, sheath–flow interface, and liquid–junction interface.

The coupling of other modes of chromatography, such as affinity chromatography, supercritical-fluid chromatography, and planar chromatography, has also been demonstrated successfully.

EXERCISES

5.1. Why is the sensitivity of the SIM chromatogram greater than that obtained with a mass chromatogram?

5.2. Why is it important to use a rapid-scanning mass spectrometry system for data collection from a high-resolution separation device?

5.3. In what modes is LC performed?

5.4. What are the challenges when coupling LC with mass spectrometry?

5.5. Describe the basis of separation in capillary electrophoresis.

5.6. Why is micellar electrokinetic chromatography able to resolve neutral molecules, whereas capillary zone electrophoresis cannot?

5.7. What is the order of elution of ionic and neutral compounds from a capillary zone electrophoresis column and from a micellar electrokinetic chromatography column?

5.8. How is the electrical contact between the separation buffer and the interface established in a CE/ESI–MS combination?

ADDITIONAL READING

1. R. L. Grob, *Modern Practice of Gas Chromatography*, Wiley, New York, 1995.
2. V. R. Meyer, *Practical High-Performance Liquid Chromatography*, Wiley, New York, 1999.
3. E. D. Katz, ed., *High-Performance Liquid Chromatography: Principles and Methods in Biotechnology*, Wiley, New York, 1996.
4. P. Camilleri, ed., *Capillary Electrophoresis: Theory and Practice*, CRC Press, Boca Raton, FL, 1998.
5. P. G. Righetti, ed., *Capillary Electrophoresis in Analytical Biotechnology*, CRC Press, Boca Raton, FL, 1998.
6. F. W. McLafferty, *Chemtech* **22**, 182–189 (1992).
7. F. G. Kitson, B. S. Larsen, and C. N. McEwen, *A Practical Guide to Gas Chromatography and Mass Spectrometry*, Academic Press, San Diego, CA, 1996.
8. R. Willoughby, E. Sheehan, and S. Mitrovich, *A Global View of LC/MS*, Global View Publishing, Pittsburgh, PA, 1995.
9. M. C. McMaster, *LC/MS: A Practical User's Guide*, Wiley, Hoboken, NJ, 2005.

REFERENCES

1. K. B. Tomer, Separations combined with mass spectrometry, *Chem. Rev.* **101**, 297–328 (2001).
2. C. L. Gatlin et al., Automated identification of amino acid sequence variations in proteins by HPLC/microspray tandem mass spectrometry, *Anal. Chem.* **72**, 757–763 (2000).
3. M. T. Davis and T. D. Lee, Variable flow liquid chromatography–tandem mass spectrometry and computer analysis of complex protein digest mixtures, *J. Am. Soc. Mass Spectrom.* **8**, 1059–1069 (1997).
4. M. T. Davis and T. D. Lee, Rapid protein identification using a microscale electrospray LC/MS system on an ion trap mass spectrometer, *J. Am. Soc. Mass Spectrom.* **8**, 1059–1069 (1997).

5. R. S. Gohlke, Time-of-flight mass spectrometry and gas–liquid partition chromatography, *Anal. Chem.* **31**, 535–541 (1959).
6. J. Abian, The coupling of gas and liquid chromatography with mass spectrometry, *J. Mass Spectrom.* **34**, 157–168 (1999).
7. R. F. Arendale, R. F. Severson, and O.T. Chortyk, Open-split interface for GC-MS, *Anal. Chem.* **56**, 1533–1537 (1984).
8. R. Ryhage, MS as a detector for GC, *Anal. Chem.* **36**, 759–764 (1964).
9. J. T. Watson and K. Biemann, High-resolution MS of GC effluents, *Anal. Chem.* **37**, 844–851 (1965).
10. D. R. Black, R. A. Flath, and R. Teranishi, Membrane molecular separators for GC-MS interfaces, *J. Chromatogr. Sci.* **7**, 284–289 (1969).
11. D. A. Volmer, Multiresidue determination of sulfonamide antibiotics in milk by short-column liquid chromatography coupled with electrospray ionization tandem mass spectrometry, *Rapid Commun. Mass Spectrom.* **10**, 1615–1620 (1996).
12. H. Zhang, K. Heinig, and J. Henion, Atmospheric pressure ionization time-of-flight mass spectrometry coupled with fast liquid chromatography for quantitation and accurate mass measurement of five pharmaceutical drugs in human plasma, *J. Mass Spectrom.* **35**, 423–431 (2000).
13. J. Ayrton, G. J. Dear, W. J. Leavens, D. N. Mallett, and R. S. Plumb, The use of turbulent flow chromatography/mass spectrometry for the rapid, direct analysis of a novel pharmaceutical compound in plasma, *Rapid Commun. Mass Spectrom.* **11**, 1953–1958 (1997).
14. D. B. Kassel, B. Shushan, T. Sakuma, and J.-P. Salzmann, Evaluation of packed capillary perfusion column HPLC/MS/MS for the rapid mapping and sequencing of enzymic digests, *Anal. Chem.* **66**, 236–243 (1994).
15. Y. Hsieh, G. Wang, Y. Wang, S. Chackalamannil, and W. A. Korfmacher, Direct plasma analysis of drug compounds using monolithic column liquid chromatography and tandem mass spectrometry, *Anal. Chem.* **75**, 1812–1818 (2003).
16. A. R. Ivanov, L. Zang, and B. L. Karger, Low attomole electrospray ionization–MS and –MS/MS analysis of protein tryptic digests using 20-mm-i.d. polystyrene–divinylbenzene monolithic capillary columns, *Anal. Chem.* **75**, 5306–5316 (2003).
17. M. E. Swartz, UPLC: An introduction and review, *J. Liq. Chromatogr.* **28**, 1253–1263 (2005).
18. Y. Wu, W. B. Hobbins, and J. R. Engen, Ultra performance liquid chromatography (UPLC) further improves hydrogen/deuterium exchange mass spectrometry, *J. Am. Soc. Mass Spectrom.* **17**, 163–167 (2006).
19. W. M. A. Niessen, Advances in instrumentation in liquid chromatography–mass spectrometry and related liquid-introduction techniques. *J. Chromatogr. A* **794**, 407–435 (1998).
20. P. Arpino, Combined liquid chromatography mass spectrometry I: Coupling by means of a moving belt interface, *Mass Spectrom. Rev.* **8**, 35–55 (1989).
21. W. H. McFadden, H. L. Schwartz, and D. C. Bradford, Direct analysis of liquid chromatographic effluents, *J. Chromatogr.* **122**, 389–396 (1976).
22. M. J. Hayes, E. P. Lankmayer, P. Vouros, and B. L. Karger, Moving belt interface with spray deposition for liquid chromatography/mass spectrometry, *Anal. Chem.* **55**, 1745–1752 (1983).

23. M. A. Baldwin and F. W. McLafferty, Liquid chromatography–mass spectrometry interface, I: Direct introduction of liquid solutions into a chemical ionization mass spectrometer, *Org. Mass Spectrom.* **7**, 1111–1112 (1973).
24. A. Melera, Design, operation and applications of a novel LCMS CI interface, *Adv. Mass Spectrom.* **8B**, 1597 (1980).
25. A. Cappiello, G. Famiglini, and P. Plama, Electron ionization for LC/MS, *Anal. Chem.* **75**, 486A–503A (2003).
26. Y. Ito, T. Takeuchi, D. Ishii, and M. Goto, Direct coupling of micro high-performance liquid chromatography with fast atom bombardment mass spectrometry, *J. Chromatogr.* **346**, 161–166 (1985).
27. R. M. Caprioli, W. T. Moore, B. DaGue, and M. Martin, Microbore high-performance liquid chromatography–mass spectrometry for the analysis of proteolytic digests by continuous-flow fast-atom bombardment mass spectrometry, *J. Chromatogr.* **443**, 355–362 (1988).
28. M. A. Moseley, L. J. Deterding, K. B. Tomer, and J. W. Jorgenson, Nanoscale packed-capillary liquid chromatography coupled with mass spectrometry using a coaxial continuous-flow fast atom bombardment interface, *Anal. Chem.* **63**, 1467–1473 (1991).
29. J. S. M. deWit, M. A. Moseley, L. J. Deterding, K. B. Tomer, and J. W. Jorgenson, Design of a coaxial continuous flow fast atom bombardment probe, *Rapid Commun. Mass Spectrom.* **2**, 100–104 (1988).
30. C. R. Blakley and M. L. Vestal, Thermospray interface for liquid chromatography/mass spectrometry, *Anal. Chem.* **55**, 750–754 (1983).
31. P. Arpino, Combined liquid chromatography mass spectrometry, II: Techniques and mechanisms of thermospray, *Mass Spectrom. Rev.* **9**, 631–669 (1990).
32. C. S. Creaser and J. W. Stygall, Particle beam liquid chromatography–mass spectrometry: instrumentation and applications—a review, *Analyst* **118**, 1467–1480 (1993).
33. A. Cappiello, Is particle beam an up-to-date LC–MS interface? State of the art and perspectives, *Mass Spectrom. Rev.* **15**, 283–296 (1996).
34. A. Cappiello, G. Famiglini, A. Lombardozzi, A. Massari, and G. G. Vadala, Electron capture ionization of explosives with a microflow rate particle beam interface, *J. Am. Soc. Mass Spectrom.* **7**, 753–758 (1996).
35. R. C. Willoughby and R. F. Browner, Monodisperse aerosol generation interface for combining liquid chromatography with mass spectroscopy, *Anal. Chem.* **56**, 2626–2631 (1984).
36. A. Cappiello and G. Famiglini, Capillary-scale particle–beam liquid chromatography/mass spectrometry interface: can electron ionization sustain the competition? *J. Am. Soc. Mass Spectrom.* **9**, 993–1001 (1998).
37. K. B. Tomer, M. A. Moseley, L. J. Deterding, and C. E. Parker, Capillary liquid chromatography/mass spectrometry, *Mass Spectrom. Rev.* **13**, 431–457 (1996).
38. J. Abian, A. J. Osterkamp, and E. Gelpi, Comparison of conventional, narrow-bore, and capillary liquid chromatography/mass spectrometry for electrospray ionization mass spectrometry, *J. Mass Spectrom.* **34**, 244–254 (1999).
39. C. Dass, High-performance liquid chromatography–electrospray ionization mass spectrometry, in F. Settle, ed., *Handbook of Instrumental Techniques for Analytical Chemistry*, Prentice Hall, Upper Saddle River, NJ, 1998, pp. 647–664.

40. C. Dass, Recent developments and applications of high-performance liquid chromatography–electrospray ionization mass spectrometry, *Curr. Org. Chem.* **3**, 193–209 (1999).
41. T. R. Covey, E. C. Huang, and J. D. Henion, Structural characterization of protein tryptic peptides via liquid chromatography/mass spectrometry and collision-induced dissociation of their doubly charged molecular ions, *Anal. Chem.* **63**, 1193–1200 (1991).
42. M. Raida, Capillary and nano-LC coupled to mass spectrometry, in J. Silberring and R. Ekman, eds., *Mass Spectrometry and Hyphenated Techniques in Neuropeptides Research*, Wiley-Interscience, New York, 2002, pp. 109–134.
43. M. R. Emmett, F. M. White, C. L. Hendrickson, S. D.-H. Shi, and A. G. Marshall, Applications of micro-electrospray liquid chromatography to FT-ICR MS to enable high-sensitivity biological analysis, *J. Am. Soc. Mass Spectrom.* **9**, 333–340 (1998).
44. J. C. Le, J. Hui, V. Katta, and M. F. Rohde, Development and evaluation of on-line nanolitter flow analysis of protein digests by pneumatic-splitter electrospray liquid chromatography–mass spectrometry, *J. Am. Soc. Mass Spectrom.* **8**, 703–712 (1997).
45. M. T. Davis and T. D. Lee, Low flow high-performance liquid chromatography solvent delivery system designed for tandem capillary liquid chromatography-mass spectrometry, *J. Am. Soc. Mass Spectrom.* **6**, 571–577 (1995).
46. E. C. Huang and J. D. Henion, LC/MS and LC/MS/MS determination of protein tryptic digests, *J. Am. Soc. Mass Spectrom.* **1**, 158–165 (1990).
47. T. R. Covey, E. D. Lee, A. P. Bruins, and J. D. Henion, Liquid chromatography/mass spectrometry, *Anal. Chem.* 58, 1451A-1463A (1986).
48. E. Gelpi, Biomedical and biochemical applications of liquid chromatography–mass spectrometry, *J. Chromatogr. A* **703**, 59–80 (1995).
49. M. S. Sakairi and H. Kambara, Characteristics of a liquid chromatograph/atmospheric pressure ionization mass spectrometer, *Anal. Chem.* **60**, 774–780 (1988).
50. T. Nabeshima, Y. Takada, and M. S. Sakairi, Atmospheric-pressure chemical ionization interface for semimicro liquid chromatography/mass spectrometry, *Rapid Commun. Mass Spectrom.* **11**, 715–718 (1997).
51. D. B. Robb, T. R. Covey, and A. P. Bruins, Atmospheric pressure photoionization: an ionization method for liquid chromatography–mass spectrometry, *Anal. Chem.* **72**, 3653–3659 (2000).
52. K. A. Hanold, S. M. Fischer, P. H. Cormia, C. E. Miller, and J. A. Syage, Atmospheric pressure photoionization, 1: General properties for LC/MS, *Anal. Chem.* **76**, 2842–2851 (2004).
53. K. K. Murray, Coupling matrix-assisted laser desorption/ionization to liquid separations, *Mass Spectrom. Rev.* **16**, 283–299 (1997).
54. D. Nagra and L. Li, Liquid chromatography–time-of-flight mass spectrometry with continuous-flow matrix-assisted laser desorption ionization, *J. Chromatogr. A* **711**, 235–245 (1995).
55. S. J. Lawson and K. K. Murray, Continuous flow infrared matrix-assisted laser desorption/ionization with a solvent matrix, *Rapid Commun. Mass Spectrom.* **14**, 129–134 (2000).

56. A. P. L. Wang, X. Guo, and L. Li, Liquid chromatography/time-of-flight mass spectrometry with a pulsed sample introduction interface, *Anal. Chem.* **66**, 3664–3675 (1994).
57. J. Preisler, F. Foret, and B. L. Karger, On-line MALDI–TOF MS using a continuous vacuum deposition, *Anal. Chem.* **70**, 5278–5287 (1998).
58. X. Zhang, D. A. Narcisse, K. K. Murray, On-line single droplet deposition for MALDI mass spectrometry, *J. Am. Soc. Mass Spectrom.* **15**, 1471–1477 (2004).
59. C. Delahunty and J. R. Yates III, Protein identification using 2D-LC-MS/MS, *Methods* **35**, 248–255 (2005).
60. G. J. Opiteck, J. W. Jorgenson, and R. J. Anderegg, Two-dimensional SEC/RPLC coupled to mass spectrometry for the analysis of peptides, *Anal. Chem.* **67**, 2283–2291 (1997).
61. M. L. Nedved, S. Habibi–Goudarzi, B. Ganem, and J. D. Henion, Characterization of benzodiazepine "combinatorial" chemical libraries by on-line immunoaffinity extraction, coupled column HPLC-ion spray mass spectrometry–tandem mass spectrometry, *Anal. Chem.* **68**, 4228–4236 (1996).
62. A. de Jong, Contribution of mass spectrometry to contemporary immunology, *Mass Spectrom. Rev.* **17**, 311–315 (1998).
63. P. Cao and J. T. Stults, Mapping the phosphorylation sites of proteins using on-line immobilized metal affinity chromatography/capillary electrophoresis/electrospray ionization multiple stage tandem mass spectrometry, *Rapid Commun. Mass Spectrom.* 14, 1600–1606 (2000).
64. K. C. Lewis, G. J. Opiteck, J. W. Jorgenson, and D. M. Sheeley, Comprehensive online RPLC-CZE-MS of peptides, *J. Am. Soc. Mass Spectrom.* **8**, 495–500 (1997).
65. F. Zhou and M. V. Johnston, Protein characterization by on-line capillary isoelectric focusing and reversed-phase liquid chromatography and mass spectrometry, *Anal. Chem.* **76**, 2734–2740 (2004).
66. K. D. Lukacs and J. W. Jorgenson, Capillary zone electrophoresis: effect of physical parameters on separation efficiency and quantitation, *J. High Result. Chromatogr.* **8**, 407–411 (1985).
67. K. K. Ugner, A critical appraisal of capillary electrochromatography, *Anal. Chem.* **74**, 200A–207A (2002).
68. M. A. Moseley, L. J. Deterding, K. B. Tomer, and J. W. Jorgenson, Capillary-zone electrophoresis/fast-atom bombardment mass spectrometry: design of an on-line coaxial continuous-flow interface, *Rapid Commun. Mass Spectrom.* **3**, 87–93 (1989).
69. R. M. Caprioli, W. T. Moore, M. Martin, B. DaGue, K. Wilson, and S. Moring, Coupling capillary zone electrophoresis and continuous-flow fast atom bombardment mass spectrometry for the analysis of peptide mixtures, *J. Chromatogr.* **480**, 247–257 (1989).
70. N. J. Reinhoud, W. M. A. Niessen, U. R. Tjaden, L. G. Gramberg, E. R. Verheij, and J. van der Greef, Performance of a liquid-junction interface for capillary electrophoresis mass spectrometry using continuous-flow fast-atom bombardment, *Rapid Commun. Mass Spectrom.* **3**, 348–357 (1989).
71. J. Preisler, P. Hu, T. Rejtar, and B. L. Karger, Capillary electrophoresis on-line MALDI–TOF MS using a continuous vacuum deposition, *Anal. Chem.* **72**, 4785–5795 (2000).

72. M. Wetterhall, T. Johnson, and J. Bergquist, Capillary electrophoresis coupled to mass spectrometry for peptide and protein analysis, in J. Silberring and R. Ekman eds., *Mass Spectrometry and Hyphenated Techniques in Neuropeptides Research*, Wiley-Interscience, New York, 2002, pp. 135–154.
73. J. A. Olivares, N. T. Nguyen, C. R. Yonker, and R. D. Smith, On-line mass spectrometric detection for capillary zone electrophoresis, *Anal. Chem.* **59**, 1230–1232 (1987).
74. R. D. Smith, J. A. Olivares, N. T. Nguyen, and H. R. Udseth, Capillary zone electrophoresis–mass spectrometry using an electrospray ionization interface, *Anal. Chem.* **60**, 436–441 (1988).
75. P. Cao and M. Moini, A novel sheathless interface for capillary electrophoresis/electrospray ionization mass spectrometry through in-capillary electrode, *J. Am. Soc. Mass Spectrom.* **8**, 561–564 (1997).
76. M. A. Petersson, G. Hulthe, and E. Fogelqvist, New sheathless interface for coupling capillary electrophoresis to electrospray mass spectrometry evaluated by the analysis of fatty acids and prostaglandins, *J. Chromatogr. A* **854**, 141–154 (1999).
77. E. Rohde, A. J. Tomlinson, D. H. Jonson, and S. Naylor, Comparison of protein mixtures in aqueous humor by membrane preconcentration–capillary electrophoresis–mass spectrometry, *Electrophoresis* **19**, 2361–2370 1998).
78. J. F. Kelly, L. Ramaley, and P. Thibault, Capillary zone electrophoresis–electrospray mass spectrometry at submicroliter flow rates: practical considerations and analytical performance, *Anal. Chem.* **69**, 51–60 (1997).
79. R. D. Smith, C. J. Barinaga, and H. R. Udseth, Improved electrospray ionization interface for capillary zone electrophoresis–mass spectrometry, *Anal. Chem.* **60**, 1948–1952 (1988).
80. H. R. Udseth, J. A. Loo, and R. D. Smith, Capillary isotachophoresis/mass spectrometry, *Anal. Chem.* **61**, 228–232 (1989).
81. E. D. Lee, W. Muck, J. D. Henion, and T. R. Covey, Liquid junction coupling for capillary zone electrophoresis/ion spray mass spectrometry, *Biomed. Environ. Mass Spectrom.* **18**, 844–850 (1989).
82. T. Wachs, R. L. Sheppard, and J. Henion, Design and applications of a self-aligning liquid junction–electrospray interface for capillary electrophoresis–mass spectrometry, *J. Chromatogr. B* **685**, 335–342 (1996).
83. Y.-R. Chen, M.-C. Tseng, Y.-Z. Chang, and G.-R. Her, A low-flow CE/ESI MS interface for capillary-zone electrophoresis, large-volume sample stacking, and micellar electrokinetic chromatography, *Anal. Chem.* **75**, 503–508 (2003).
84. M. Moini, Design and performance of a universal sheathless capillary electrophoresis to mass spectrometry interface using split-flow technique, *Anal. Chem.* **73**, 3497–3501 (2001).
85. J. T. Witt and M. Moini, Capillary electrophoresis to mass spectrometry interface using a porous junction, *Anal. Chem.* **75**, 2188–2191 (2003).
86. D. R. Goodlett, J. H. Wahl, H. R. Udseth, and R. D. Smith, Reduced elution speed detection for capillary electrophoresis/mass spectrometry, *J. Microcolumn. Sep.* **5**, 57–62 (1993).
87. N. J. Reinhoud, A. P. Tinke, U. R. Tjaden, W. M. A. Niessen, and J. van der Greef, Capillary isotachophoretic analyte focusing for capillary electrophoresis with mass

spectrometric detection using electrospray ionization, *J. Chromatogr.* **627**, 263–271 (1992).

88. T. J. Thompson, F. Foret, P. Vouros, and B. L. Karger, Capillary electrophoresis/electrospray ionization mass spectrometry: improvement of protein detection limits using on-column transient isotachophoretic sample preconcentration, *Anal. Chem.* **65**, 900–906(1993).

89. D. Figeys, A. Durect, and R. Aebersold, Identification of proteins by capillary electrophoresis–tandem mass spectrometry: evaluation of an online solid-phase extraction device, *J. Chromategr. A*, **763**, 295–306 (1997).

90. Q. Yang, A. J. Tomlinson, and S. Naylor, Membrane preconcentration CE, *Anal. Chem.* **71**, 183A–189A (1999).

91. J. Ding and P. Vouros, Advances in CE/MS, *Anal. Chem.* **71**, 378A–385A (1999).

92. J. Ding, T. Barlow, A. Dipple, and P. Vouros, Separation and identification of positively charged and neutral nucleoside adducts by capillary electrochromatography–microelectrospray mass spectrometry, *J. Am. Soc. Mass Spectrom.* **9**, 823–829 (1998).

93. A. Harsch and P. Vouros, Interfacing of CE in a PVP matrix to ion trap mass spectrometry: analysis of isomeric and structurally related (N-acetylamino)fluorene-modified oligonucleotides, *Anal. Chem.* **70**, 3021–3027 (1998).

94. M. H. Lamoree, U. R. Tjaden, and J. van der Greef, Online coupling of micellar electrokinetic chromatography to electrospray mass spectrometry, *J. Chromatogr.* **712**, 219–225 (1995).

95. Y.-R. Chen, M.-C. Tseng, Y.-Z. Chang, and G.-R. Her, A low-makeup beveled tip capillary electrophoresis/ESI MS interface for micellar electrokinetic chromatography and nonvolatile buffer capillary electrophoresis, *Anal Chem.* **76**, 6306–6312 (2004).

96. J. Cai and J. Henion, Quantitative capillary electrophoresis–ion spray mass spectrometry on a benchtop ion trap for the determination of isoquinoline alkaloids, *Anal. Chem.* **66**, 2103–2109 (1994).

97. J. W. Seu, R. Bersen III and N. H. H. Heegaard, On-line immunoaffinity-LC-MS for identification of amyloid disease markers in biological fluids, *Anal. Chem.* **75**, 1196–1202 (2003).

98. M. W. H. Pinske, M. P. Uitto, M. J. Hillhorst, B. Ooms, and J. R. Albert, Selective isolation at the femtomole level of phosphopeptides from proteolytic digests using 2D-nanoLC-ESI-MS/MS and titanium oxide precolumns, *Anal. Chem.* **76**, 3935–3943 (2004).

99. P. Arpino and P. Haas, Recent developments in supercritical fluid chromatography–mass spectrometry coupling, *J. Chromatogr. A* **703**, 479–488 (1995).

100. M. C. Ventura, W. P. Ferrell, C. M. Aurigemma, and M. J. Greig, Packed column supercritical fluid chromatography/mass spectrometry for high-throughput analysis, 2, *Anal. Chem.* **71**, 4223–4231 (1999).

101. P. J. R. Sjoberg and K. E. Markides, Capillary column supercritical fluid chromatography–atmospheric pressure ionization mass spectrometry: interface performance of atmospheric pressure chemical ionization and electrospray ionization, *J. Chromatogr. A* **855**, 317–327 (1999).

102. D. Nurok, Forced-flow techniques in planar chromatography, *Anal. Chem.* **72**, 634A–641A (2000).

103. Sz. Nyiredy, Progress in forced-flow planar chromatography, *J. Chromatogr. A* **1000**, 985–999 (2003).
104. J. Guitard, X. L. Hronowskiy, and C. E. Costello, Direct matrix-assisted laser desorption/ionization mass spectrometric analysis of glycosphingolipids on thin layer chromatographic plates and transfer membranes, *Rapid Commun. Mass Spectrom.* **13**, 1838–1849 (1999).
105. V. B. Ivleva et al., Coupling of TLC with vibrational cooling MALDI FTMS for the analysis of gangliosides, *Anal. Chem.* **76**, 6484–6491 (2004).
106. K. Dreisewerd et al., Analysis of native milk oligosaccharides directly from TLC plates by MALDI-orthogonal-time-of-flight mass spectrometry with glycerol matrix, *J. Am. Soc. Mass Spectrom* **17**, 139–150 (2006).
107. J. T. Mehl and D. M. Hercules, Direct TLC-MALDI coupling using hybrid plate, *Anal. Chem.* **72**, 68–73 (2000).
108. W. Chai, C. Letaux, A. M. Lawson, and M. S. Still, On-line overpressure thin layer chromatographic separation and electrospray mass spectrometry detection of glycolipids, *Anal. Chem.* **75**, 118–125 (2003).
109. G. J. Van Berkel, J. J. Llave, F. De Apadoca, and M. J. Ford, Rotational planer chromatography coupled on-line with atmospheric pressure chemical ionization mass spectrometry, *Anal. Chem.* **76**, 479–482 (2004).

PART II

ORGANIC AND INORGANIC MASS SPECTROMETRY

CHAPTER 6

ORGANIC MASS SPECTROMETRY

Since the 1960s, mass spectrometry has played a pivotal role in the field of structure elucidation and identification of organic compounds. Over the years, a wealth of knowledge has been gained on reactions of gas-phase ions from the use of a variety of mass spectrometric techniques. Mass spectrometry can be used to identify unknown compounds or to perform de nova structure determination. The former is relatively easy if one knows accurate mass and a reference spectrum. The latter is much more difficult and requires detailed knowledge of the rules for interpretation of a mass spectrum. Providing this knowledge is the focus of this chapter.

A highly practical approach for ionization of thermally stable and relatively volatile organic compounds is electron ionization (EI). An EI mass spectrum in most cases is a fingerprint signature of the target compound. Three types of vital information can be gleaned from this spectrum which assist in determination of the structure of an organic compound: (1) the molecular mass, (2) the elemental composition, and (3) the compound-specific fragment ions. Although a mass spectrum contains highly useful information for the identification of unknown compounds, it is not a magic bullet. Not all compounds can be identified on the basis of information from a mass spectrum only. Often, it becomes essential to employ other complementary techniques, such as nuclear magnetic resonance (NMR) and infrared (IR) spectroscopy, along with mass spectrometry, provided that one has a sufficient amount of material.

Fundamentals of Contemporary Mass Spectrometry, by Chhabil Dass
Copyright © 2007 John Wiley & Sons, Inc.

6.1. DETERMINATION OF MOLECULAR MASS

One of the most important pieces of information required to elucidate the molecular structure of an unknown organic compound is its molecular mass, which provides a window within which the elemental composition and the final structure of the compound must fit. Therefore, the first essential step to identifying a compound is to measure its molecular mass by determining the m/z value of the molecular ion. Molecular mass measurements can be performed at either low or high resolution. A low-resolution measurement provides information about the nominal mass of the analyte, and its elemental composition can be also determined for low-molecular-weight compounds from the isotopic pattern. From a high-resolution mass spectrum, the accurate molecular mass can be determined, from which it is also feasible to deduce the elemental composition. Chemists who work with synthetic compounds and natural products rely heavily on the exact mass measurement data for structural assignment. This value is acceptable in lieu of the combustion or other elemental analysis data. An acceptable value of the measured mass should be within 5 ppm of the accurate mass [1]. As shown below, the mass measurement error is reported either in parts per million (ppm) or in millimass units (mmu).

$$\text{ppm} = \frac{(\text{accurate mass} - \text{measured mass})}{\text{accurate mass}} \times 10^6$$

$$\text{mmu} = (\text{accurate mass} - \text{measured mass}) \times 10^3$$

6.1.1. Molecular Mass Measurements at Low-Mass Resolving Power

A low-resolution mass measurement is a simple procedure and can be performed with most of the mass spectrometric systems discussed in Chapter 3. The instrument is set up at a resolving power (RP) above 1000. At this low resolving power, the molecular ions of most organic compounds that differ by a unit mass are well separated. The mass spectrometer is first mass-calibrated with an external calibration procedure. During the calibration scan, the computer stores the peak centroid (the center of gravity) time and the area of each peak. The mass of the ion is related exponentially to the peak centroid time:

$$M = M_{\max} e^{-t/\tau} \qquad (6.1)$$

where $M_{\max}$ is the maximum mass value in the scan, t the elapsed time from the start of the scan, and τ a time constant. The reference compound may be removed, and the full-scan spectrum of the target compound is acquired.

6.1.2. Molecular Mass Measurements at High-Mass Resolving Power

For accurate mass measurements, it is absolutely essential that the ion of interest is completely resolved from all other neighboring ions. Any unresolved interfering ion would otherwise shift the position of the peak of interest and introduce an error in mass measurement. High resolving power is thus essential for accurate mass measurements (e.g., a resolving power of ca. 10,000). The need for higher resolving power is not as critical for small molecules but becomes more important as the m/z value of the analyte ion increases. For example, according to Eq. (3.1), a resolving power of 770 would mass-resolve $C_2H_4^{+\bullet}$ (mass = 28.031300) and $CO^{+\bullet}$ (mass = 27.994915). On the other hand, a resolving power of 5500 is required to separate $C_{13}H_{16}-C_2H_4^{+\bullet}$ and $C_{13}H_{16}-CO^{+\bullet}$ ions, although the mass difference in both of these examples is 0.036385 u. Also, for low-mass compounds, the possible elemental compositions are few but become exceedingly large as the mass increases.

As discussed in Chapter 3, high-resolution accurate mass measurements were traditionally performed with double-focusing magnetic-sector mass spectrometers. Fourier transform ion cyclotron (FT–ICR) and reflectron-time-of-flight (TOF) mass spectrometers now play a more important role in high-resolution accurate mass measurements, and magnet-sector instruments are now a small fraction of the total mass spectrometers used in compound identification. High-resolution accurate mass measurements have also been made when quadrupole mass filters are operated in the multiple-pass mode and in higher-stability regions [2], but this approach is seldom used on a routine basis. Similarly, by scanning the electric fields very slowly, quadrupole ion traps can also function as high-resolution instruments [3,4], but they give poor accurate mass measurement. Orbitrap, a recent addition to the family of mass analyzers, has also emerged as a possible tool for accurate mass measurements.

One of the following procedures can be used to acquire high-resolution mass measurement data with magnetic-sector instruments.

Full-Scan Mode One of the simplest ways to obtain an accurate mass value is to record the spectrum at high resolving power in the presence of a reference compound. The instrument is set to a mass resolving power of ca. 10,000, and the magnet is scanned slowly (3 to 5 s/decade; a decade means 10 to 100 or 100 to 1000) to encompass the complete mass range of interest. Accurate masses of all ions present in the spectrum can be determined in one injection of the sample by this technique. The instrument is first mass-calibrated by acquiring the spectrum of the reference compound alone. From the known mass reference ions, the masses of the sample ions are determined with an accuracy of less than 5 ppm. For Mattauch–Herzog geometry double-focusing mass spectrometers, ions in a wide mass range are detected simultaneously by a focal plane detector such as a diode array detector. From the precise position of each ion, the accurate mass is measured with respect to the ions of known mass. Although these instruments played an important role in accurate mass measurements in the 1960s and 1970s, they are seldom used today.

Peak-Matching Mode Peak matching is a more accurate mass measurement technique (<0.3 ppm precision). Instead of acquiring the complete spectrum, the mass of only one ion is determined at a time. In a typical manual procedure, the magnetic field is held constant and the accelerating voltage is changed in such a manner that a reference ion and the analyte ion are made to follow the same trajectory (when the two peaks are overlaid on a storage oscilloscope, they must have the same radius of trajectory; they are said to be *peak matched*). The lower of the two masses is brought to focus first, and by changing the accelerating voltage to within a precision of less than 1 ppm, the higher mass ion is superimposed on it. The unknown mass is calculated from that ratio using

$$m_2 = m_1 \frac{V_1}{V_2} \tag{6.2}$$

The mass of the reference ion must be within 2% of the unknown mass to give the highest accuracy. Current generation of instruments can perform peak-matching experiments automatically, often by slow scanning of the accelerating voltage. Computer programs are available to perform signal averaging, smoothing, and peak centroiding to provide improved mass accuracy [5].

6.1.3. Molecular Mass Measurements by ESI and MALDI

Accurate mass measurements with ESI are performed in combination with an FT–ICR instrument [6,7]. This instrument currently yields the highest resolving power and the most accurate mass measurement of any mass spectrometry instrument. The space-charge effects and nonlinearity of the trapping electric fields limit the current accuracy in mass measurement. Use of an internal calibrant can minimize this effect. By bracketing the unknown mass with two references, an accuracy of less than 3 ppm for the elemental analysis can be achieved [7]. ESI has also been combined with quadrupoles, provided that the compound is pure [2] or online with HPLC [9].

Accurate mass measurement data with MALDI are generally acquired by coupling it with DE–TOF–MS [10–13]. In these instruments, the linear relation between m/q and t^2 no longer exists. The generally accepted mass calibration equation is $m/q = at^2 + b$ [Eq. (3.13)]. It is assumed in this equation that all ions that exit the source have the same kinetic energy; that assumption may not be true for MALDI-formed ions. A third constant term is usually included into this equation to account for the initial velocity spread [11]. Two types of calibrations are performed in MALDI-TOF-MS experiments: internal mass calibration and external mass calibration. In the former, calibration standards are mixed with the sample-matrix preparation, whereas in the external calibration, the calibrant is spotted at a separate place on the MALDI plate. An accuracy of ±5 ppm can be attained routinely with TOF–MS by using two reference mass ions as internal calibrant. A higher-order calibration function can improve the mass measurement accuracy further. Increasing the number of terms in the calibration equation, however, requires knowledge of more instrument parameters. Alternatively, two-step calibration is recommended [12]. The first step involves external calibration, in which high-order polynomial function is used to establish the relation between m/q and t^2. This calibration is performed in the entire mass range of interest by fitting the square of the flight times observed for reference ions of a polymeric mixture. The second-step calibration uses two internal reference masses to account for a first-order correction for sample position-dependent errors. An average mass error of about 1 ppm has been achieved with this procedure.

6.1.4. Mass Calibration Standards

A reference compound is needed to determine the molecular mass. The mass of an unknown is computed by comparing its signal on the mass axis with that of the known mass reference peaks. A reference compound is also required to calibrate the data system and to tune and performance-check the instrument. A calibration standard has the following desirable characteristics: (1) it should yield a sufficient number of regularly spaced abundant ions across the entire scan range; (2) those reference ions should have negative mass defects to prevent overlap with the usual compounds containing C, H, N, and O; and (3) it should be readily available, chemically inert, and sufficiently volatile.

Some of the commonly used calibration standards in EI and CI are listed in Table 6.1, and a complete list of the ions produced along with their masses and relative abundances are given in Appendix D. The most widely used mass calibrants in EI-MS are perfluorokerosene (PFK) and perfluorotributylamine [PFTBA, $(C_4F_9)_3N$; also known as heptacosa]. Both compounds perform well for compounds up to 900 u. For high-mass (up to 3000 u) analysis, triazines and Ultramark (a mixture of fluorinated phosphazenes) are suitable as reference mass markers. For ESI measurements, CsI, poly(ethylene glycol) (PEG), poly(ethylene glycol) monomethyl ether) (PEGMME), poly(ethylene glycol) bis(carboxymethyl ether), and poly(propylene glycol) (PPG) are the mass calibrants. For MALDI mass measurements, several reference compounds have been used as reference mass markers, including α-CHCA matrix (dimer + H^+), 4-hydroxy-3-methoxycinnamic acid (trimer + Na^+), angiotensin I, angiotensin II, bradikinin, desArg1-bradikinin, substance P, gramicidin, and autodigestion products of trypsin (see Appendix D).

6.2. MOLECULAR FORMULA FROM ACCURATE MASS VALUES

Another piece of information that is critical to structure elucidation of an unknown compound is its molecular formula or elemental composition. One way to obtain this information is the accurate mass of the molecular ion. With that knowledge, the elemental composition can be assigned to compounds with molecular mass up to about 400 u.

Fortunately, the atomic masses of isotopes are nonintegers (see Table 6.2 and Appendix C). This feature imparts a unique value to the molecular mass of each compound or each formula. A nominal mass may have several combinations of elemental compositions, but the accurate mass can match one composition only. For example, a few of the molecular formulas that can be assigned to a 70-u

TABLE 6.1. Mass Calibration Standards for EI, CI, FAB, ESI, and MALDI[a]

Compound	Structure	Ionization Mode	Mass Range (Da)
Perfluorokerosene (PFK)	$CF_3(CF_2)_nCF_3$	EI, CI	0–900
Perfluorotributylamine	$(C_4F_9)_3N$	EI, CI	0–600
Ultramark F series (Fomblin)	$CF_3O(CFCF_3CF_2O)_m-(CF_2O)_nCF_3$	EI, CI	Up to 3500
Ultramark 1621 (perfluoroalkoxycyclo triphosphazines)		EI, CI	Up to 3000
		ESI	Up to 6000
Tris(perfluoroalkyl)-s-triazine		EI	Up to 1500
Glycerol	$CH_2(OH)CH(OH)-CH_2OH$	FAB	20-1200
Cesium iodide	CsI	FAB	130-30,000
		ESI	Up to 3000
CsI/glycerol		FAB	90-3500
CsI/NaI/RbI		FAB	20-1000
LiI/NaI		FAB	0-1000
Poly(ethylene glycol)	$H(OCH_2CH_2)_nOH$	FAB, ESI	50-2000
Poly(propylene glycol)	$H(OCH(CH_3)CH_2)_nOH$	FAB, ESI	50-2000
Poly(ethylene glycol monomethyl ether)		ESI	50-2000
Myoglobin		ESI	600-2500
Lysozome		ESI	1000-2100
Polypropylene glycol sulfate		ESI (negative)	300-1700
Sodium trifluoroacetate	CF_3COONa	ESI	100-4000
Tetraethylammonium iodide	$(C_2H_5)_4NI$	ESI (negative)	Up to 6000
Water clusters	$H(H_2O)_n$	ESI	Up to 4000
Cesium tridecafluoroheptanoate	$C_7F_{13}O_2Cs$	ESI	Up to 10,000

[a] MALDI standards: CHCA dimer (379.0930 Da); gramicidin-S (1141.5 Da); ACTH (18–39; 2465.7 Da); angiotensin I (1295.6775 Da); angiotensin II (1045.5345 Da); bradikinin (1060.5692 Da); bovine insulin (5733.5 Da); horse heart cytochrome c (12,360 Da); horse heart myoglobin (16,951 Da); bovine trypsinogen (23,957 Da); bovine serum albumin (66,431 Da); trypsin autodigestion products.

nominal mass are $C_3H_2O_2$ ($m = 70.0054$ u), C_3H_4NO ($m = 70.0293$ u), C_4H_6O ($m = 70.0419$ u), $C_3H_6N_2$ ($m = 70.0532$ u), C_4H_8N ($m = 70.0657$ u), and C_5H_{10} ($m = 70.0783$ u); each has a unique exact mass value. However, C_4H_6O is the only possible molecular formula for an ion with a mass of 70.0419 u. The other possibilities can be ruled out readily when the accurate mass of the ion is known within 11.3 mmu. Unfortunately, there are more possible combinations of molecular formulas with an increase in the number of atoms in a molecule. Thus, the usefulness of the accurate mass information for the unequivocal identification of a molecular formula also gains significance with the increasing mass of the molecule. For example, a nominal mass of 40 u has only four possible molecular formulas, whereas mass 70 and 100 u have 11 and 23 different formulas, respectively. A list of the molecular formulas for various masses has been compiled [14]. Algorithms and computer programs are also available for this purpose. Online facilities are listed in Appendix E.

TABLE 6.2. Exact Masses of Isotopes of Some Common Elements

Element	Average Mass	Nuclide	Isotopic Mass
Hydrogen	1.00794	^{1}H	1.007825
		^{2}H (D)	2.014102
Carbon	12.0110	^{12}C	12.000000
		^{13}C	13.003355
Nitrogen	14.00674	^{14}N	14.003074
		^{15}N	15.000109
Oxygen	15.9994	^{16}O	15.994915
		^{17}O	16.999131
		^{18}O	17.999160
Fluorine	18.9984	^{19}F	18.998403
Silicon	28.0855	^{28}Si	27.976927
		^{29}Si	28.976495
		^{30}Si	29.973770
Phosphorus	30.9738	^{31}P	30.973762
Sulfur	32.0660	^{32}S	31.972072
		^{33}S	32.971459
		^{34}S	33.967868
Chlorine	35.4527	^{35}Cl	34.968853
		^{37}Cl	36.965903
Bromine	79.9094	^{79}Br	78.918336
		^{81}Br	80.916289
Iodine	126.9045	^{127}I	126.904477

6.3. MOLECULAR FORMULA FROM ISOTOPIC PEAKS

Another useful resource for determining elemental compositions is a low-resolution mass spectrum. This magical trickery is the result of another unique gift to us: the fact that most elements exist in several distinct isotopic forms, each with fixed relative abundances (RAs). For example, natural hydrogen is a mixture of 99.985% ^{1}H and 0.015% ^{2}H, and natural carbon is a mixture of 98.90% ^{12}C and 1.10% ^{13}C. By definition, all isotopes of an element contain the same number of protons but different numbers of neutrons; they have the same atomic number but a different mass number (e.g., ^{12}C and ^{13}C are isotopes, and so are ^{1}H and ^{2}H). The relative isotopic abundances of commonly present elements in organic compounds are listed in Table 6.3. According to McLafferty's scheme, those elements can be classified into three categories: A, A + 1, and A + 2 [Additional Reading 1]. By convention, A-type elements are those that exist in one isotopic form only (e.g., fluorine, phosphorus, and iodine). For the sake of calculations, hydrogen is also included in this category because the abundance of its A + 1 isotope, ^{2}H, is negligibly small. Carbon and nitrogen are considered A + 1 elements because their second isotope is not

TABLE 6.3. Relative Abundances of Some Common Elements

Element	A Isotope	RA[a]	A + 1 Isotope	RA	A + 2 Isotope	RA	Element Type
Hydrogen	^{1}H	100	^{2}H	0.015	—	—	A
Carbon	^{12}C	100	^{13}C	1.11	—	—	A + 1
Nitrogen	^{14}N	100	^{152}N	0.37	—	—	A + 1
Oxygen	^{16}O	100	^{17}O	0.04	^{18}O	0.2	A + 2
Fluorine	^{19}F	100	—	—	—	—	A
Silicon	^{28}Si	100	^{29}Si	5.1	^{30}Si	3.4	A + 2
Phosphorus	^{31}P	100	—	—	—	—	A
Sulfur	^{32}S	100	^{33}S	0.79	^{34}S	4.4	A + 2
Chlorine	^{35}Cl	100	—	—	^{37}Cl	32.0	A + 2
Bromine	^{79}Br	100	—	—	^{81}Br	97.3	A + 2
Iodine	^{127}I	100	—	—	—	—	A

[a] RA, relative abundance in %.

negligible and it is 1 u higher than the most abundant isotope. The A + 2 elements include oxygen, silicon, sulfur, chlorine, and bromine; their heavier isotope is 2 u higher than the lower mass isotope (i.e., the primary isotope). Oxygen, silicon, and sulfur also have isotopes that are 1 u higher than the most abundant isotope.

A striking feature of a mass spectrum is the existence of satellite peaks. Those peaks are isotopically shifted lines that appear at masses one or more units higher than the main peak M; the mass of M is calculated using the atomic masses of the most abundant isotopic species (i.e., the primary isotope). The satellite peaks, designated as M + 1, M + 2, and so on, reflect the differences in the natural abundances of the isotopes. From the elemental composition of a molecular ion or fragment ion, the abundances of its satellite peaks (i.e., the isotopic pattern) can be predicted (see Example 6.1). Consider a compound of the general formula $C_xH_yN_zO_n$; the abundance of the [M + 1] ion relative to [M] = 100% is given by

$$\%[M+1] = 100 \left(\frac{[M+1]}{[M]} \right) = x(1.11) + y(0.015) + z(0.37) + n(0.037) \quad (6.3)$$

If S and Si atoms are also present, their contribution to %[M + 1] is calculated by multiplying the number of S atoms with 0.8 and of Si atoms by 5.1, and adding these terms to Eq. (6.3).

▶ **Example 6.1** Calculate the %[M + 1] and %[M + 2] for butyl propionate ($C_7H_{14}O_2$) and 2,3-dihydronaphthalene ($C_{10}H_{10}$) ions.

Solution

%[M + 1] of $C_7H_{14}O_2^+$: = $(7 \times 1.11) + (14 \times 0.015) + (2 \times 0.04) = 8.06$

%[M + 2] of $C_7H_{14}O_2^+$: = contribution of seven carbons + $2 \times \%^{18}O$

$= 0.29 + (2 \times 0.20) = 0.69$

%[M + 1] of $C_{10}H_{10}^+$: = $(10 \times 1.11) + (10 \times 0.015) = 11.25$

%[M + 2] of $C_{10}H_{10}^+$: = contribution of 10 carbons = 0.60

(*Note:* The contribution of ^{13}C to %[M + 1] is from Table 6.5.)

The effect of the A + 2 elements is more striking for Cl and Br because the A + 2 isotope is highly abundant (unlike that of O or S). For example, the contribution of one chlorine atom to the abundance of the M + 2 peak is 32.5%, and that of one bromine atom is 98.0%. In other words, for a compound that contains one chlorine atom, its [M + 2] will be approximately one-third of [M]. The contribution of more than one A + 2 element is calculated from expansion of the binomial $(a + b)^n$, where a is the natural abundance of the light isotope, b that of the heavy isotope, and n the number of atoms of that element present. If two different A + 2 elements are present (e.g., Cl and Br), the isotopic pattern is calculated by the expansion of the product given in

$$(a + b)^n (c + d)^m \tag{6.4}$$

where c and d are natural abundances of the light and heavy isotopes of the second element and m is the number of that element. The isotopic patterns calculated for the most frequently encountered A + 2 elements are presented in Table 6.4. For a visual impact of this pattern, the data are presented pictorially in Figure 6.1. An example of a spectrum of a chlorocompound (dichloromethane) is shown in Figure 6.2. The isotopic pattern of two chlorine atoms can be seen for m/z 84, 86, and 88, and that for one chlorine atom for m/z 49 and 51.

▶ **Example 6.2** Write all possible isotopic compositions for the [M], [M + 1], and [M + 2] ions of the formula C_8H_9OCl.

Solution

$[M](156) = {}^{12}C_8^1 H_9^{16}O^{35}Cl$

$[M+1](157) = {}^{12}C_7^{13}C_1^1 H_9^{16}O^{35}Cl, {}^{12}C_8^1 H_8^2 H_1^{16}O^{35}Cl, {}^{12}C_8^1 H_9^{17}O^{35}Cl$

$[M+2](158) = {}^{12}C_6^{13}C_2^1 H_9^{16}O^{35}Cl, {}^{12}C_7^{13}C_1^1 H_8^2 H_1^{16}O^{35}Cl, {}^{12}C_7^{13}C_1^1 H_9^{17}O^{35}Cl,$

${}^{12}C_8^1 H_8^2 H_1^{17}O^{35}Cl, {}^{12}C_8^1 H_9^{18}O^{35}Cl,$ and ${}^{12}C_8^1 H_9^{16}O^{37}Cl$

TABLE 6.4. Abundances of Isotopic Peaks (percent) for Combinations of Chlorine and Bromine

Element	[M]	[M + 2]	[M + 4]	[M + 6]	[M + 8]	[M + 10]
Cl	100	32.5	—	—	—	—
Cl_2	100	65.3	10.6	—	—	—
Cl_3	100	97.8	31.9	3.47	—	—
Cl_4	78	100	48	10	1	—
Cl_5	63	100	64	20	3	0.4
Br	100	97.9	—	—	—	—
Br_2	51	100	49	—	—	—
Br_3	34	100	97	32	—	—
ClBr	77	100	24	—	—	—
Cl_2Br	62	100	45	6	—	—
$ClBr_2$	44	100	70	13	—	—
Cl_2Br_2	39	100	89	31	4	—

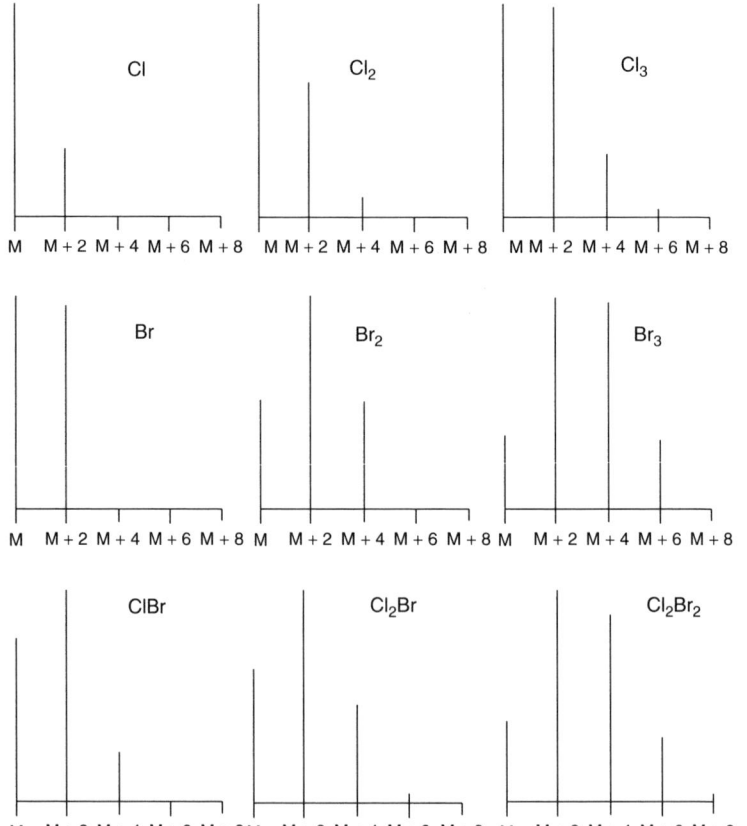

Figure 6.1. Abundances of ions from chlorine- and bromine-containing compounds.

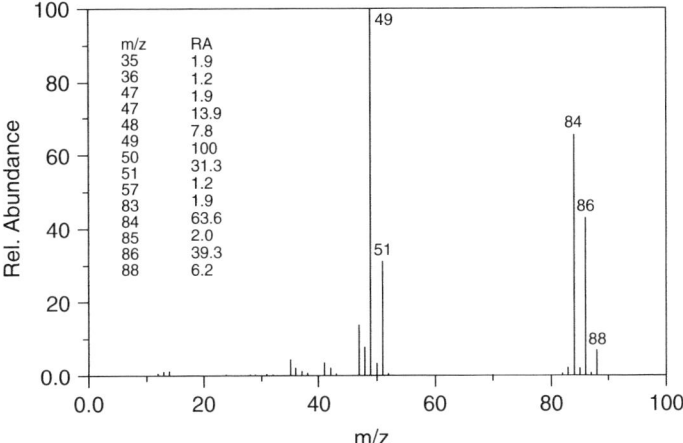

Figure 6.2. EI mass spectrum of dichloromethane.

Compounds with a large number of carbon atoms also show measurable contributions to [M + 2] because of the probability that the molecular formula might contain two ^{13}C atoms. This contribution can be calculated from the formula

$$\%[M+2] = 100\left(\frac{[M+2]}{[M]}\right) = \frac{[x(1.08)]^2}{200} \qquad (6.5)$$

or from the data shown in Table 6.5.

▶ **Example 6.3** Calculate the isotopic pattern due to (a) two Cl atoms, (b) two Br atoms, and (c) one Cl and two Br atoms.

Solution

(a) $a \sim 3, b \sim 1$, and $n = 2$, and the binomial expansion is $(a+b)^2 = a^2 + 2ab + b^2$; therefore, [M] : [M + 2] : [M + 4] = 9:6:1.

(b) $a \sim 1, b \sim 1$, and $n = 2$; therefore, [M] : [M + 2] : [M + 4] = 1:2:1.

(c) $a \sim 3, b \sim 1$, and $n = 1$, and $c \sim 1, d \sim 1$, and $m = 2$, and the binomial expansion is $(a+b)(c+d)^2 = ac^2 + 2acd + ad^2 + bc^2 + 2bcd + bd^2$. Collecting all M ($^{35}Cl^{79}Br^{79}Br$; ac^2), M + 2 ($^{35}Cl^{79}Br^{81}Br = 2acd$; $^{37}Cl^{79}Br^{79}Br = bc^2$), M + 4 ($^{35}Cl^{81}Br^{81}Br = ad^2$; $^{37}Cl^{79}Br^{81}Br$; $2bcd$), and M + 6 ($^{37}Cl^{81}Br^{81}Br = bd^2$) terms together gives isotope abundance ratios of 3 : (6 + 1) : (3 + 2) : 1 = 3:7:5:1.

TABLE 6.5. Isotopic Contribution of Carbon Atoms

Number of Carbon Atoms	Contribution to: [M + 1]	Contribution to: [M + 2]	Number of Carbon Atoms	Contribution to: [M + 1]	Contribution to: [M + 2]
1	1.1	0.01	16	17.6	1.5
2	2.2	0.02	17	18.7	1.7
3	3.3	0.05	18	19.8	1.9
4	4.4	0.10	19	20.9	2.2
5	5.5	0.15	20	22.0	2.4
6	6.6	0.22	21	23.1	2.6
7	7.7	0.29	22	24.2	2.9
8	8.8	0.38	23	25.3	3.2
9	9.9	0.49	24	26.4	3.4
10	11.0	0.60	25	27.5	3.7
11	12.1	0.73	26	28.6	4.1
12	13.2	0.86	27	29.7	4.4
13	14.3	1.01	28	30.8	4.7
14	15.4	1.18	29	31.9	5.0
15	16.5	1.35	30	33.0	5.4

Computational methods are available to calculate the isotope distribution for an ion of a specific elemental composition and known isotopic abundances [15,16]. Online facilities for this purpose are also available at several Web sites [17,18].

As a corollary, from knowledge of the relative abundances of satellite peaks, the molecular formula of a compound can be derived. One should, however, exercise the judgment that a relative experimental error of 10% in the abundance values measured may exist. The following steps provide a helpful guideline:

1. Recognize the m/z value of the [M] ion of interest. It is usually represented by the largest peak in the cluster, except when chlorine and bromine atoms or when a large number of carbon atoms (ca. 100) are present.
2. Look for the presence of the A + 2 elements. Because the isotopic pattern (each peak is unusual and is separated by 2; e.g., M, M + 2, M + 4, etc.), the presence of Cl and Br and to some extent of Si and S is easy to discern. Use the data presented in Table 6.4 and Figure 6.1 to assign the number of these elements. However, remember the cardinal rule: If the %[M + 2] is less than 3, then Cl, Br, S, and Si are absent.
3. Next, calculate the number of carbon atoms from the [M + 1] value by using

$$\text{number of carbon atoms}(x) = \frac{[M+1]}{[M](0.011)} \quad (6.6)$$

This equation is derived from Eq. (6.3) with the assumption that the contributions of hydrogen and oxygen to [M + 1] are negligible and that the compound contains no nitrogen, silicon, or sulfur atoms. When the number of nitrogen atoms are known (see Section 6.4.3), this calculation can be refined by subtracting 0.0037 times the abundance of the M ion from the [M + 1] value for each nitrogen atom. Similarly, subtract 5.1 and 0.8 to account for the presence of each silicon and sulfur atom, respectively, from the [M + 1] value.

4. After calculating the number of carbon atoms, estimate the number of oxygen atoms from the %[M + 2] value with the knowledge that each oxygen atom contributes 0.20% to the abundance of the M + 2 peak. Before this calculation is performed, the contribution to the %[M + 2] value of two ^{13}C atoms must be accounted for.

5. Assign the A-type elements. The most common of these elements is hydrogen, the number of which is estimated from the m/z value of the molecular ion minus the sum of the masses of all other atoms so far identified. The presence of F and I is usually very obvious from the characteristic loss of F (19 u) or HF (20 u) and I (127 u) or HI (or 128 u) from the molecular ion, respectively.

▶ **Example 6.4** The following ions are observed in the EI mass spectra of the molecular ion region. Determine the elemental composition in each example.

(a)

m/z	RA
121	0.71
122	78
123	6.0
124	0.54

(b)

m/z	RA
112	—
113	100
114	6.1
115	32.8
116	1.98

Solution

(a) The most abundant ion of m/z 122 appears to be the molecular ion [M]. First normalize all abundance values to [M] = 100%. Therefore, new normalized %RA values are [M − 1](121) = 0.91, [M] (122) = 100, [M + 1](123) = 7.7, and [M + 2](124) = 0.69. Because M is an even mass, it contains either no nitrogen or even-numbered nitrogens. Also, Si, S, Cl, and Br atoms are absent because %[M + 2] is less than 3. However, oxygen might be present. Calculate the number (x) of carbon atoms using Eq. (6.6). The value of x calculated is 7.0. Therefore, a maximum of seven carbon atoms could be present. Next, calculate the number of oxygen atoms. But before that, subtract the contribution of ^{13}C due to seven carbons (see Table 6.5) on the [M + 2] value (i.e., new [M + 2] = 0.69 − 0.29 = 0.40%). This value gives the number of oxygen atoms = 0.42/0.20 = 2. This brings the contribution of seven carbons and two oxygens to the m/z value = (7 × 12) + (2 × 16) = 116.

The presence of nitrogen atoms can now be ruled out because it will increase the value of m/z beyond 122. Finally, the number of hydrogens = $122 - 116 = 6$. Thus, the elemental composition is $C_7H_6O_2$.

(b) The arguments used in part (a) are also applied here. The m/z 113 ion is assigned as [M]. It is odd-numbered; therefore, it contains at least one nitrogen. The [M + 2] indicates the presence of one chlorine. The number of carbon atoms calculated is five. Thus, the elemental composition is C_5H_4NCl.

6.4. GENERAL GUIDELINES FOR INTERPRETATION OF A MASS SPECTRUM

Besides knowing how to find the elemental composition (Section 6.3), interpretation of the spectrum of an unknown compound requires an understanding of concepts described below [Additional Reading 1].

6.4.1. Odd- and Even-Electron Ions

As mentioned in Section 1.4, the removal of an electron from a neutral molecule during electron ionization generates an *odd-electron* (OE) *ion*, by definition, an ion that contains an unpaired electron. It is denoted by placing +• beside the molecular formula (e.g., $C_6H_6^{+\bullet}$). An $OE^{+\bullet}$ ion may isomerize to a more stable *distonic ion*, an ion in which the charge and radical sites are located on different atoms [19]:

$$CH_3CH_2CH_2CH=O^{+\bullet} \longrightarrow CH_2CH_2CH_2CH=OH^+ \quad (6.7)$$
$$\text{OE ion} \qquad\qquad \text{OE ion distonic ion}$$

In contrast, *even-electron (EE) neutrals and ions* are those species in which the outer-shall electrons are fully paired. When an OE^+ ion fragments, in most cases, the charged product is an EE^+ ion and the neutral is an OE species (a radical):

$$OE^{+\bullet} \longrightarrow OE^{\bullet} + EE^+ \quad \text{or}$$
$$OE^{+\bullet} \longrightarrow OE^{+\bullet} + EE \quad \text{(rearrangement reaction)} \quad (6.8)$$

An OE^+ ion can also fragment via a rearrangement reaction to expel an EE neutral species and form an OE^+ fragment ion [reaction (6.8)]. EE^+ ions are more stable than OE^+ ions. The fragmentation of EE^+ ions follows different rules [20]. Most of the EE^+ ions fragment to other EE^+ ions and a neutral molecule:

$$EE^+ \longrightarrow EE^+ + EE \quad \text{or} \quad EE^+ \longrightarrow OE^{+\bullet} + OE^{\bullet} \quad (6.9)$$

In some cases, they fragment to $OE^{+\bullet}$ ions and an OE radical [reaction (6.9)].

6.4.2. Recognizing the Molecular Ion

The recognition of a molecular ion in the mass spectrum is the most critical step to identifying an organic compound. This information provides the molecular mass and the elemental composition of the compound, in the absence of which it may be a losing battle to identify an unknown. At times, however, it is not easy to identify the molecular ion correctly because of the interference of background ions and due to the fact that many labile molecules with respect to ionization fail to yield the molecular ion at 70-eV ionization. The correct molecular ion must meet the following criteria:

1. It must be the most abundant ion in the high-mass cluster. In most cases this statement is true, but there are a few exceptions, especially when the compound contains more than two Cl atoms or one Br atom. For example, in the spectrum of CH_2Br_2, the most abundant ion in the molecular ion region is at m/z 174, although its molecular mass is 172 u. Also, for a few compounds, the $(M-1)^+$ ion is more abundant than the $M^{+\bullet}$ ion.
2. It must be an $OE^{+\bullet}$ ion; that is, it must be at an even m/z value, except when the compound contains an odd number of nitrogens (see Section 6.4.3).
3. It must provide a logical explanation for the neutral that was expelled. For example, the presence of $(M-1)^+$, $(M-2)^{+\bullet}$, and even $(M-3)^+$ ions due to the losses of H, H_2, and $H + H_2$, respectively, from the molecular ion is reasonable and logical. So is the loss of 15 (CH_3), 16 u (CH_4), 17 u (OH or NH_3), 18 u (H_2O), 19 u (F), 20 u (HF), 27 u (HCN), 28 u (CO, C_2H_4, or CH_2N), 29 u (CHO, C_2H_5, or CH_2N), 31 u (OCH_3), or 32 u (CH_3OH). Losses of 4 to 14 and 21 to 25 u are highly unexpected. A detailed list of the neutral losses is given in Table 6.6.
4. The $M + 1$ peak must not be too large and must accommodate a reasonable number of carbon atoms; otherwise, the suspected $M + 1$ peak is the molecular ion peak. For example, m/z 147 is the base peak in the EI spectrum of cinnamic acid, whereas its molecular ion peak (m/z 148) is of 70% relative abundance (see Figure 6.3).

According to common wisdom, if the molecular ion is in doubt, a CI or ESI spectrum should be obtained. These techniques produce abundant EE or protonated molecule ions (MH^+), from which the molecular ion can readily be discerned.

6.4.3. Nitrogen Rule

It is a fortunate coincidence that the most abundant isotope of all elements has either even mass and even valence, or both are odd, except for nitrogen, which has even mass but odd valence. This property leads to what is known as the

TABLE 6.6. Neutrals Lost

Neutral Lost	Formula of the Neutral	Precursor Molecule
1	H	Aldehydes, Ar–CH_3, alkynes
2	H_2	—
15	CH_3	Alkanes, methyl derivatives
16	O	$ArNO_2$, amine oxides, sulfoxides
	NH_2	Carboxamides, sulfonamides
17	OH	Carboxylic acids, oximes
	NH_3	—
18	H_2O	Alcohols, aldehydes, ketones, ethers
19	F	Fluoroalkanes
20	HF	Fluoroalkanes
26	CH≡CH	Aromatic hydrocarbons
27	C_2H_3	Alkenes, cycloalkanes
	HCN	Alkyl cyanides, nitrogen heterocycles
28	$CH_2=CH_2$	Ketones, cyclohexenes
	CO	Phenols, Ar–CO, R–CO, cyclic ketones
29	C_2H_5	Alkanes, ethyl derivatives
	CHO	Aromatic aldehydes
30	C_2H_6	Branched alkanes
	NO	Nitroamines, nitroesters
31	OCH_3	Methoxy derivatives, methyl esters
	CH_5N	—
32	CH_3OH	Ethers, esters
	S	Sulfides
34	H_2S	Thiols, methyl sulfides
35	Cl	Alkyl chlorides
36	HCl	Alkyl chlorides
40	$CH_3C≡CH$	Aromatic hydrocarbons
41	C_3H_5	Esters
42	$CH_3CH=CH_2$	Ketones, cyclohexenes
	$CH_2=C=O$	—
43	C_3H_7	Alkanes, propyl derivatives
44	C_3H_8	Alkanes
	CO_2	Esters, acids, anhydrides
	$CONH_2$	Amides
45	OC_2H_5	Ethoxy derivatives, ethyl esters
	COOH	Acids
46	[H_2O + $CH_2=CH_2$]	—
	NO_2	Nitro compounds
47	CH_2SH	—
48	CH_4S	—
56	C_4H_8	—
80	HBr	Alkyl bromides

GENERAL GUIDELINES FOR INTERPRETATION OF A MASS SPECTRUM 213

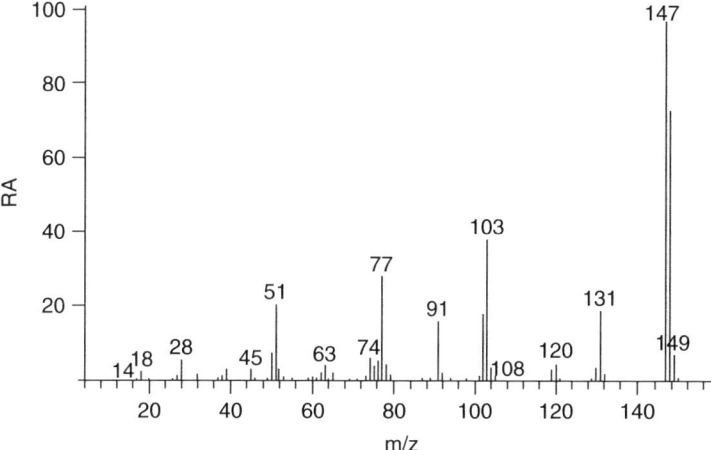

Figure 6.3. EI mass spectrum of cinnamic acid that was acquired on a GC/MSD instrument.

nitrogen rule: If a molecule has an odd number of nitrogen atoms, its molecular mass will be an odd number; and if it contains no nitrogens or even-number nitrogens, its molecular mass will be an even number. The consequence of this rule is that the molecular mass of an $OE^{+\bullet}$ ion will be an even number (and of EE^+ ions will be an odd number) if it contains no or even-numbered nitrogens. For example, the molecular mass of benzene (C_6H_6, 78 u), methyl phenyl ketone (C_8H_8O, 120 u), and *o*-nitroaniline ($C_6H_6N_2O_2$, 138 u) are all even-numbered, and that of ammonia (NH_3, 17 u) and pyridine (C_5H_5N, 79 u) are odd-numbered.

The nitrogen rule has two important applications: (1) it can be used to identify whether an ion is OE or EE (see Example 6.5), and (2) it can provide some idea of the presence or absence of nitrogen atoms in a compound (see Example 6.6).

▶ **Example 6.5** Determine whether ions of the formula (a) C_7H_8N, (b) $C_7H_6O_2$, and (c) C_4H_9 are $OE^{+\bullet}$ or EE^+ ions.

Solution

(a) The nominal mass of C_7H_8N is even, $(7 \times 12) + (8 \times 1) + (1 \times 14) = 106$, but it has one (odd number) nitrogen; therefore, it is an EE^+ ion.

(b) The nominal mass of $C_7H_6O_2$ is even, $(7 \times 12) + (6 \times 1) + (2 \times 16) = 122$, and it contains no nitrogen atoms; therefore, it is an $OE^{+\bullet}$ ion.

(c) The nominal mass of C_4H_9 is odd, $(4 \times 12) + (9 \times 1) = 57$, and it contains no nitrogen atoms; therefore, it is an EE^+ ion.

▶ **Example 6.6** Determine whether or not the molecular ions formed at (a) *m/z* 108 and (b) *m/z* 101 contain nitrogen. These ions contain seven and five carbons, respectively.

Solution

(a) Remember that all molecular ions are $OE^{+\bullet}$ ions. Also, because *m/z* 108 is even, it has either no nitrogens or an even number of nitrogens. Because it has seven carbons, the presence of any nitrogen atom is ruled out (the presence of two nitrogens will exceed the *m/z* value).

(b) The value *m/z* 101 is odd; it has an odd number of nitrogens. Because it has five carbons, it can have a maximum of only one nitrogen (the presence of three nitrogens will exceed the *m/z* value).

6.4.4. Value of the Rings Plus Double Bonds

The value of the rings plus double bonds (R + DB value) is a measure of the degree of unsaturation in a molecule. It is also a convenient means of ascertaining whether an ion is an OE or an EE species. *The R + DB value will be an integer number for $OE^{+\bullet}$ ions and an odd multiple of 0.5 for EE^+ ions.* If the elemental composition of an ion is known, the value of R + DB can be calculated. For a molecule of the general formula $C_x H_y N_z O_n$, R + DB is given by

$$R + DB = x - \frac{y}{2} + \frac{z}{2} + 1 \qquad (6.10)$$

This formula is based on the fact that each ring or double bond in a hydrocarbon decreases the number of hydrogen atoms by two units, and each nitrogen atom increases the number of hydrogen atoms by one unit. If elements other than C, H, N, and O are present, they are counted with one of these elements by matching their valences. For example, all tetravalent elements are counted with C, the trivalent elements with N, the bivalent elements with O, and the monovalent elements with H. Also, remember that a triple bond is treated as two double bonds.

▶ **Example 6.7** Calculate the value of R + DB for (a) methyl phenyl ketone (C_8H_8O), (b) aniline (C_6H_7N), and (c) 1-chloronaphthalene ($C_{10}H_7Cl$).

Solution

(a) R + DB for $C_8H_8O = 8 - 8/2 + 1 = 5$.
(b) R + DB for $C_6H_7N = 6 - 7/2 + 1/2 + 1 = 4$.
(c) R + DB for $C_{10}H_7Cl = 10 - 8/2 + 1 = 7$.

▶ **Example 6.8** Determine on the basis of the R + DB value whether the ions in Example 6.5 are $OE^{+\bullet}$ or EE^+ ions. Also, decide which of these elemental compositions would represent a possible molecular ion.

Solution

(a) R + DB for $C_7H_8N = 7 - 8/2 + 1/2 + 1 = 4.5$; it is EE^+ ion.
(b) R + DB for $C_7H_6O_2 = 7 - 6/2 + 1 = 5$; it is $OE^{+\bullet}$ ion.
(c) R + DB for $C_4H_9 = 4 - 9/2 + 1 = 0.5$; it is EE^+ ion.

Because only an OE^+ ion can be a molecular ion, the formula $C_7H_6O_2$ is a possible molecular ion.

6.4.5. Systematic Steps in Interpreting a Mass Spectrum

The following general steps must be followed to interpret the mass spectrum of an unknown compound [21; Additional Reading 1]:

1. Recognize the molecular ion (Section 6.4.2).
2. Apply the nitrogen rule (Section 6.4.3).
3. From the isotope pattern, determine the elemental composition (Section 6.3). Does it fit with the mass of the molecular ion? If the molecular ion is of low abundance and fails to yield a reasonable elemental composition, calculate the composition of the next abundant fragment ion and add to it the composition of the neutral lost to arrive at the composition of the target molecule (see Table 6.6 for the formulas of neutral losses).
4. Calculate the R + DB value (Section 6.4.4). In addition to ascertaining whether an ion is OE or EE, this information could provide a major clue to the structure (e.g., whether it is a saturated or an unsaturated or contains a ring).
5. Make a judgment of the abundance of the $M^{+\bullet}$ ion. It reflects the stability of the molecule. The $M^{+\bullet}$ ion signal increases with increased unsaturation and number of rings but decreases substantially with chain length and chain branching; aromatic compounds yield a prominent $M^{+\bullet}$ ion. In contrast, the molecules that have a facile cleavage site produce low abundant $M^{+\bullet}$.
6. Pay attention to the general appearance of the spectrum; long-chain aliphatic compounds show a characteristic hydrocarbon pattern, whereas aromatic compounds produce a few but highly abundant ions.
7. Recognize characteristic losses of neutral species from the molecular ion.
8. Look for a low-mass ion series. A series of ions of *m/z* 29, 43, 57, 71, ... is indicative of alkanes and aliphatic ketones. Alkenes and cycloalkanes will produce a series of ions of *m/z* 27, 41, 55, 69, ... Similarly, ions of *m/z* 30,

44, 58, 72, ... are produced from long-chain amines, and ions of m/z 31, 45, 59, 73, ... are the results of the fragmentation of long-chain alcohols and ethers. An ion series that contains m/z 39, 51, 63, and 77 is the signature of the phenyl group–containing compounds, whereas the ions of m/z 39, 51, 65, and 91 indicate the presence of the benzylic group.

9. Look for abundant $OE^{+\bullet}$ ions in the spectrum (e.g., m/z 60, 74, 92, ...); these ions are formed by the loss of a stable neutral molecular species via specific rearrangement reactions (e.g., McLafferty rearrangement, see Section 6.5.2) or nonspecific reactions (e.g., loss of water), and can provide a clue to the presence of certain functional groups.

10. Look for even-mass ions at the low-mass end of the spectrum [e.g., m/z 30 ($CH_2=N^+ H_2$)], the absence of which indicates that the molecule is of even mass.

11. Also, look for other unique structure-specific ions (e.g., m/z 77, 91, 93, 105, 127, ... see Section 6.5.5 and Tables 6.7 and 6.8).

12. Deduce the structure by applying fragmentation rules. You should be able to rationalize most of the peaks in the spectrum by reasonable ion-chemistry processes (sometimes called *arrow pushing*).

6.4.6. Mass Spectral Compilations

An easier approach to identifying an organic compound is to match the spectrum to the existing collection of spectra. There is a caveat that isomeric compounds might yield identical spectra. In such situations, identification of the compound might not be unequivocal. The following collections of libraries are useful resources for this purpose:

- The *Wiley Registry of Mass Spectral Data*, the most comprehensive library of mass spectral data. The 7th edition contains over 338,000 EI mass spectra. It is marketed by John Wiley & Sons and Palisade Corporation.
- The *NIST/EPA/NIH Mass Spectral Library*, another useful resource for mass spectral data. The 2005 version (NIST05) contains 190,825 spectra. A smaller collection can be accessed over the Internet (http://webbook.nist.gov/chemistry).
- The *Eight Peak Index of Mass Spectra*, published by the Mass Spectrometry Data Center of the Royal Society of Chemistry, contains a collection of over 80,000 spectra.
- The National Institute of Advanced Industrial Science and Technology in Japan also maintains a small collection of EI spectra that can be accessed online at http://aist.go.ip/RIODB/SDBS/cgi-bin/cre_index.cgi.

6.5. FRAGMENTATION PROCESSES

Fragmentation reactions of radical cations are well documented [Additional Reading 1]. Ionization occurs by the removal of an electron from either a sigma (σ),

TABLE 6.7. Selected Common Fragment Ions with Their Formulas

m/z	Formula	m/z	Formula
15	CH_3	65	C_5H_5
16	O	69	C_5H_9, $CH_3CH=CHC=O$,
17	OH		$CH_2=C(CH_3)C=O$
18	H_2O, NH_4	71	C_5H_{11}, $C_3H_7C=O$
19	F, H_3O	74	$[CH_2COOCH_3 + H]$
26	CN, C_2H_2	77	⌬$^+$ (C_6H_5)
27	C_2H_3		
28	CO, C_2H_4		
29	C_2H_5, CHO	85	C_6H_{13}, $C_4H_9C=O$
30	CH_2NH_2, NO	88	$[CH_2COOC_2H_5 + H]$
31	CH_2OH, OCH_3	91	C_7H_7 (tropolium ion)
32	O_2	92	⌬—CH_2 + H
33	SH		
34	H_2S		
35	Cl	93	⌬—O
36	HCl		
39	C_3H_3		
43	C_3H_7, $CH_3C=O$, C_2H_5N	94	⌬—O + H
44	$[CH_2CHO + H]$, CH_3CHNH_2,		
	CO_2, $NH_2C=O$, $(CH_3)_2N$		
45	CH_3CHOH, CH_2CH_2OH,	102	$[CH_2COOC_3H_7 + H]$
	CH_2OCH_3, COOH	105	⌬—CO, ⌬—CH_2CH_2,
47	CH_2SH, CH_3S		
49	CH_2Cl		⌬—$CHCH_3$
53	C_4H_5		
55	C_4H_7, $CH_2=CHC=O$		
56	C_4H_8		
57	C_4H_9, $C_2H_5C=O$	107	⌬—CH_2O, ⌬—OH / CH_2
58	$[CH_3\overset{O}{\underset{\|}{C}}CH_2 + H]$, $C_2H_5CHNH_2$,		
	$(CH_3)_2NCH_2$, $C_2H_5NHCH_2$	108	⌬—CH_2OH
59	$(CH_3)_2COH$, $CH_2OC_2H_5$, $COOCH_3$,	122	$[C_6H_5COO + H]$
	CH_3OCHCH_3, CH_3CHCH_2OH,	127	I
	C_2H_5CHOH, $[NH_2COCH_2 + H]$	128	HI
60	$[CH_2COOH + H]$		
61	$CH_3COO + 2H$, CH_2CH_2SH,		
	CH_2SCH_3		

TABLE 6.8. Structure-Specific Ions

m/z	Structure	Fragmentation Reaction	Precursor Structure
30	CH_2NH_2	α-Cleavage	Primary amines
31	CH_2OH, OCH_3	α-Cleavage	Primary alcohols
44	$CH_2C(OH)H$	McLafferty rearrangement	Aldehydes
45	CH_2OCH_3	α-Cleavage	Methyl ethers
	COOH	α-Cleavage	Carboxylic acids
47	CH_2SH	α-Cleavage	Thiols
49	CH_2Cl	α-Cleavage	Alkyl chlorides
58	$CH_2C(OH)CH_3$	McLafferty rearrangement	Methyl ketones
59	$COOCH_3$	α-Cleavage	Methyl esters
	$CH_2C(OH)NH_2$	McLafferty rearrangement	Amides
60	$CH_2C(OH)OH$	McLafferty rearrangement	Carboxylic acid
61	$CH_3COO + 2H$	Double H-rearrangement	Esters of long-chain alcohols
	CH_2SCH_3	α-Cleavage	Methyl thioethers
72	$CH_2C(OH)C_2H_5$	McLafferty rearrangement	Ethyl ketones
74	$CH_2C(OH)OCH_3$	McLafferty rearrangement	Methyl esters
77	C_6H_5	α-Cleavage	Benzene ring–containing compounds
88	$CH_2C(OH)OC_2H_5$	McLafferty rearrangement	Ethyl esters
91	C_7H_7 (tropolium ion)	α-Cleavage	Phenyl alkanes
92	C_6H_5-CH_2 + H	McLafferty rearrangement	Phenyl alkanes
93	C_6H_5-O	α-Cleavage	Phenyl ethers
94	C_6H_5-O + H	McLafferty rearrangement	Phenyl ethers
105	C_6H_5-CO	α-Cleavage	Phenyl ketones
107	C_6H_5-CH_2O	α-Cleavage	Benzyl ethers
108	C_6H_5-CH_2OH	H-rearrangement	Benzyl esters
120	$CH_2C(OH)C_6H_5$	McLafferty rearrangement	Phenyl ketones
122	$[C_6H_5COO + H]$	McLafferty rearrangement	Esters of benzoic acids
127	I	Inductive effect	Iodocompounds
128	HI	H-rearrangement	Iodocompounds
136	$CH_2C(OH)OC_6H_5$	McLafferty rearrangement	Phenyl esters

π, or nonbonding (*n*) orbital:

$$\sigma\text{-electron ionization:} \quad CH_3CH_2-CH_3 \xrightarrow{e^-} CH_3CH_2 + {}^\bullet CH_3$$
$$\pi\text{-electron ionization:} \quad CH_3CH=CH_2 \xrightarrow{e^-} CH_3CH \overset{+\bullet}{-} CH_2 \quad (6.11)$$
$$n\text{-electron ionization:} \quad CH_3\ddot{N}H_2 \xrightarrow{e^-} CH_3\overset{+\bullet}{N}H_2$$

The most-favored site for the loss of an electron in the molecule is the one that has the lower ionization energy (IE). The order of this preference is non-bonding electron > conjugated π-electron > π-electron > σ-electron. Although the charge cannot be localized to one specific site, for the purpose of rationalizing fragmentation reactions, the location of the charge follows the same order. With an unspecified location of the charge and the radical, the $OE^{+\bullet}$ ion is represented by placing the symbol +• at the end of the molecular formula (e.g., $CH_3CH_2CH_3{}^{+\bullet}$), but with a specified location, + and • are both placed at the appropriate sites. Both of these sites represent an electron deficiency and thus can induce fragmentation in the active ion. Fragmentation reactions are of two types: simple bond-cleavage reactions and rearrangement reactions.

6.5.1. Simple Bond-Cleavage Reactions

Simple bond cleavage is the dominating fragmentation process in the EI mass spectra of organic compounds. By definition, these reactions are those fragmentations in which a direct cleavage of bonds occurs (i.e., there is no formation of new bonds between previously unconnected atoms). Often, these reactions occur via several competitive pathways:

$$ABCD^{+\bullet} \rightarrow ABC^+ + {}^\bullet D \text{ (or } {}^\bullet ABC + D^+)$$
$$ABCD^{+\bullet} \longrightarrow AB^+ + {}^\bullet CD \text{ (or } {}^\bullet AB + CD^+) \quad (6.12)$$
$$ABCD^{+\bullet} \longrightarrow A^+ + {}^\bullet BCD \text{ (or } {}^\bullet A + BCD^+)$$

Retention of the charge is governed by *Stevenson's rule*, which states that a fragment that retains the charge is the one with the lowest ionization energy [22,23].

Sigma-Bond (σ) Cleavage: Upon removal of a σ-electron, the C–C bond becomes elongated and much weaker, and ultimately fragments at that site; formation of a charged moiety and a neutral fragment is the consequence:

$$(CH_3)_3C-CH_2CH_3 + e^- \longrightarrow (CH_3)_3C + {}^\bullet CH_2CH_3 + 2e^- \longrightarrow$$
$$(CH_3)_3C^+ + {}^\bullet C_2H_5 \quad (6.13)$$

This process is known as *sigma-bond cleavage* or simply *σ-cleavage*. The favored charged fragment will be the one that can better stabilize the positive charge. For example, secondary and tertiary carbocations are much more stable than primary carbocations. Sigma cleavage is the most common fragmentation encountered in saturated alkanes. Ionization and fragmentation are favored at more substituted carbon atom (i.e., at the branching site), and when there is a choice, the larger group is apparently lost preferentially.

▶ **Example 6.9** Show the preference of the loss of neutral species from 3-methylhexane$^{+\bullet}$ (**1**).

Solution Because C_3H_7 is a larger alkyl group than C_2H_5 and CH_3, the preference of the neutral lost is $^\bullet C_3H_7 > {}^\bullet C_2H_5 > {}^\bullet CH_3 > H$.

$$C_3H_7 - \overset{\overset{\displaystyle H}{|}}{\underset{\underset{\displaystyle CH_3}{|}}{C}} - C_2H_5$$

(**1**)

Radical Site–Initiated Fragmentation The cleavage of a single σ-bond remote from the site of ionization is also a common phenomenon of radical cations. This fragmentation is initiated by the radical site in a variety of heteroatom- and unsaturated functional group–containing compounds (e.g., ethers, alcohols, ketones, esters, amines, alkenes). The driving force for this reaction is the radical character of the ion, which has a strong tendency for electron pairing. As in

$$R_1 \overset{\frown}{\underset{\beta}{-}} CH_2 \overset{\bullet +}{\underset{\alpha}{-}} X - R_2 \quad \overset{\alpha}{\longrightarrow} \quad {}^\bullet R_1 + CH_2 \overset{+}{=} XR_2 \quad \text{charge retention} \qquad (6.14)$$
$$\overset{\alpha}{\longrightarrow} \quad \overset{+}{R_1} + {}^\bullet CH_2 XR_2 \quad \text{charge migration}$$

the charge and the odd electron are localized on the heteroatom (X) of the molecule. This odd-electron migrates to form a new bond with the carbon that is adjacent to (α to) the atom containing the odd electron, and concurrently abstracts an electron from another bond to this α-carbon atom, leading to cleavage of that bond; a strongly stabilized EE charged fragment and a neutral radical are formed. This homolytic bond cleavage is commonly known as α-*cleavage*. The movement of a single electron is denoted by a "fishhook" half-arrow. In some compounds, the α-cleavage may also involve charge migration provided that the IE of R_1 is less than that of $^\bullet CH_2XR_2$ (Stevenson's rule). In such a case, the charged fragment is R_1.

In ethers, a resonance-stabilized oxonium ion is formed [reaction (6.15)], whereas in ketones, the acylium ion is the charged fragment [reaction (6.16)]. In ketones, cleavage of bonds on either side of the carbonyl carbon can occur, but the abundant ion is the one that results from the loss of the larger alkyl radical. The α-cleavage reactions of alcohols and amines are depicted in reactions (6.17) and (6.18), respectively, and for these compounds, the loss of the larger alkyl group is apparently more favored.

$$CH_3-CH_2-\overset{\bullet+}{O}-C_2H_5 \xrightarrow{\alpha} {}^{\bullet}CH_3 + CH_2=\overset{+}{O}C_2H_5 \longleftrightarrow {}^{+}CH_2-OC_2H_5 \quad (6.15)$$

$$CH_3-\underset{\underset{\displaystyle O}{\|}}{C}-C_2H_5 \xrightarrow{\alpha} CH_3C\equiv\overset{+}{O} \longleftrightarrow CH_3\overset{+}{C}=O + {}^{\bullet}C_2H_5 \quad (6.16)$$

$$CH_3-CH_2-\overset{\bullet+}{O}H \xrightarrow{\alpha} {}^{\bullet}CH_3 + CH_2=\overset{+}{O}H \quad (6.17)$$

$$CH_3-CH_2-\overset{\bullet+}{N}H_2 \xrightarrow{\alpha} {}^{\bullet}CH_3 + CH_2=\overset{+}{N}H_2 \quad (6.18)$$

A double bond in alkenes is also known to initiate α-cleavage reactions. Upon ionization, the radical can reside on either carbon atom of the double bond, leading to formation of a resonance-stabilized allyl cation:

$$CH_3-CH_2-CH\overset{\bullet+}{-}CH_2 \xrightarrow{\alpha} {}^{\bullet}CH_3 + CH_2=CH-\overset{+}{C}H_2 \longleftrightarrow \overset{+}{C}H_2-CH=CH_2 \quad (6.19)$$

An aromatic compound with a carbon atom attached to the ring also participates in an α-cleavage. In this case the bond next to this benzylic carbon is cleaved if the ring is only monosubstituted, and the benzylic cation or its rearranged and resonance-stabilized tropylium cation (**2**) of m/z 91 is the charged product:

$$(6.20)$$

m/z 91   (**2**)

A typical example of the spectrum of an aromatic hydrocarbon is shown in Figure 6.4.

The proclivity for the α-cleavage reaction of a radical site parallels its electron-donating tendency and the charge-stabilizing ability of the heteroatom. Thus, compared to other fragmentations, the preference for α-cleavage is N > S, O, π, R > Cl, Br > H. This order is clearly manifested in the fragmentation of bifunctional molecules, such as $HO-CH_2-CH_2-NH_2$, in which the most-favored α-cleavage product is $CH_2=\overset{+}{N}H_2$ (its RA is 57% versus <3% for the oxygen-directed product $CH_2=\overset{+}{O}H$).

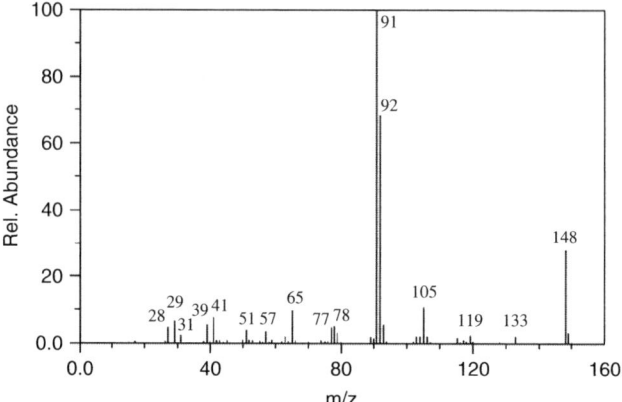

Figure 6.4. EI mass spectrum of 1-phenylbutane that gives a McLafferty-type rearrangement product at m/z 92.

▶ **Example 6.10** Which ion will be the dominant α-cleavage product in the mass spectrum of $HO-CH_2-CH_2-SH$?

Solution Because sulfur is a better electron donor than oxygen, $CH_2=\overset{+}{S}H$ will be the dominant α-cleavage product.

Charge Site–Initiated Fragmentation The charge site can also initiate the cleavage of a σ-bond in radical cations. In this case, because of the inductive effect of the charge site, both electrons of the bond that cleaves migrate together (denoted by a conventional full arrow) to the charge-bearing atom, and the bond cleaved is the one that is adjacent to the charge-bearing atom. This process, termed *inductive (i)-cleavage*, is an example of heterolytic bond cleavage, and it requires migration of the charge from the original site to a new site.

$$CH_3-CH_2-\overset{\bullet+}{O}-C_2H_5 \xrightarrow{i} CH_3-\overset{+}{C}H_2 + {}^{\bullet}OC_2H_5 \quad (6.21)$$

$$CH_3-CH_2-\overset{\bullet+}{Br} \xrightarrow{i} CH_3-\overset{+}{C}H_2 + {}^{\bullet}Br \quad (6.22)$$

As shown by the above reactions, the products of the fragmentation are an EE^+ ion and a neutral radical. The preference for the i-cleavage is reverse of the α-cleavage, that is, halogens exhibit a greater tendency for inductive cleavage than O, S, or N atoms.

Certain even-electron ions that are the primary products of an initial α-cleavage of radical cations can also undergo further fragmentation via the inductive effect. Consider the reaction

$$CH_3-CH_2-\overset{\bullet+}{O}-C_2H_5 \xrightarrow{\alpha} CH_2=\overset{+}{O}-C_2H_5 \xrightarrow{i} CH_2=O + \overset{+}{C}_2H_5 \quad (6.23)$$

The primary EE$^+$ ion product CH$_2$=O$^+$–C$_2$H$_5$ of the initial α-cleavage undergoes further inductive cleavage to produce the $^+$C$_2$H$_5$ EE$^+$ ion. According to the even-electron ion fragmentation rule (Section 6.4.1), both products are even-electron species. Although it does not belong here, there is another class of fragmentation process, known as *charge-remote fragmentation* (CRF), in which fragmentation is initiated remote from the charge site [24,25]. The CRF is not common for radical cations, but can occur for stable EE$^+$ ions such as those produced by CI, FAB, and ESI of fatty acids and peptides.

6.5.2. Rearrangement Reactions

Rearrangement reactions are those fragmentation reactions in which cleavage of the target ion proceeds through an intermediate structure; these two steps can occur in a concerted fashion or in a step-wise manner. This process involves the formation of new bonds prior to cleavage:

$$\text{ABCD}^{+\bullet} \longrightarrow \text{BCDA}^{+\bullet} \longrightarrow \text{BC}^{+\bullet} + \text{AD} (\text{or } \text{BC} + \text{AD}^{+\bullet}) \quad (6.24)$$

Although a variety of rearrangement reactions of radical cations are known, only those that are of high diagnostic value in structure elucidation are discussed here.

McLafferty Rearrangement One of the well-recognized and most-studied rearrangement reactions of gas-phase ions is the McLafferty rearrangement, also known as γ-hydrogen rearrangement [26; Additional Reading 1]. This reaction occurs in radical cations of a variety of compounds that have an unsaturated functionality, such as in ketones, aldehydes, carboxylic acids, esters, amides, olefins, phenylalkanes, and many more. The reaction involves the transfer of a γ-hydrogen to an unsaturated functional group, followed (or accompanied) by the cleavage of the α, β-bond in the intermediate to result in the expulsion of an alkene and the formation of an OE$^{+\bullet}$ ion:

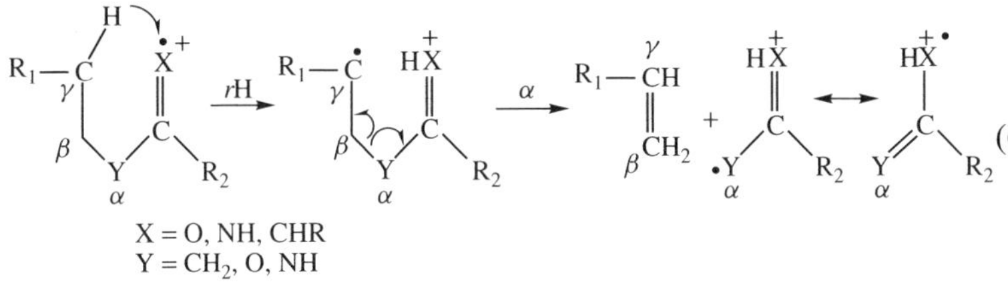

$$X = O, NH, CHR$$
$$Y = CH_2, O, NH$$

(6.25)

where *r*H indicates a rearranged hydrogen. Requirements for this reaction to occur are (1) the presence in a molecule of a γ-hydrogen, (2) a group that can accept this hydrogen, (3) the possibility of forming a sterically favorable six-membered transition state, and (4) that the α, β-bond must be a single bond. The characteristic odd-electron charged products of this reaction easily stand out in the EI mass spectra because they appear at even *m/z* values for zero or even-numbered nitrogen compounds, whereas most other ions in the spectra are at odd m/z values. Their presence can provide structurally meaningful information.

A typical example of the McLafferty rearrangement is the loss of propene from 2-hexanone:

Ketones:

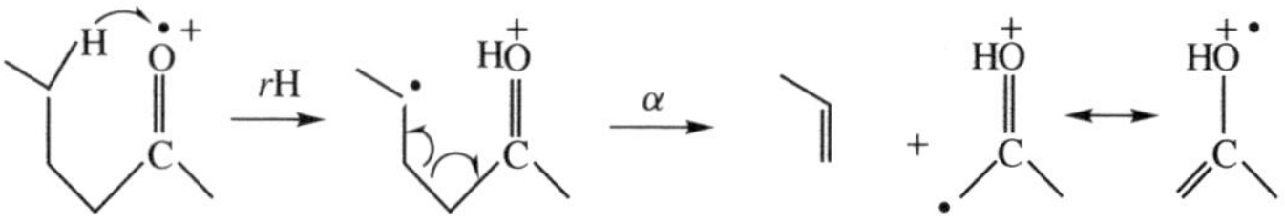

$$\text{(scheme)} \tag{6.26}$$

The impetus for this reaction is the formation of a stable enol radical cation and the elimination of a stable alkene. Other examples of facile McLafferty rearrangement follow:

Esters:

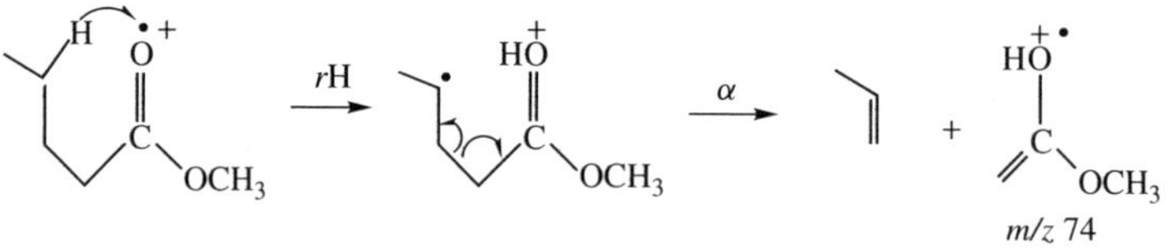

$$\text{(scheme, } m/z\ 74\text{)} \tag{6.27}$$

Phenyl alkanes:

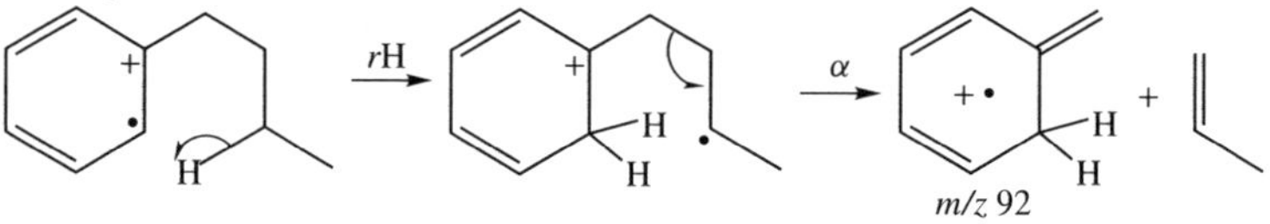

$$\text{(scheme, } m/z\ 92\text{)} \tag{6.28}$$

Phenyl ethers:

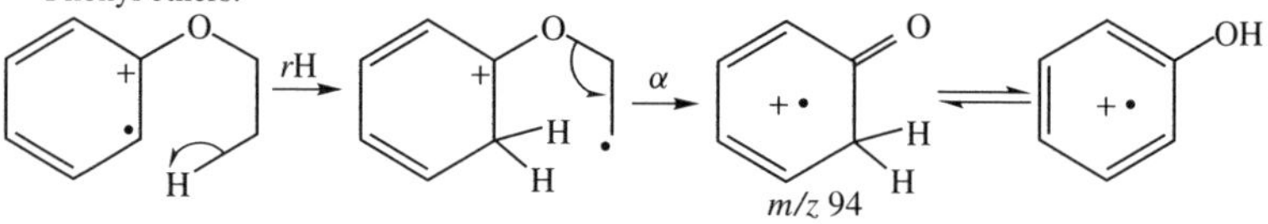

$$\text{(scheme, } m/z\ 94\text{)} \tag{6.29}$$

As shown in reaction (6.28), the formation of m/z 92 is a diagnostic reaction of alkyl benzenes that have at least three carbon atoms in the alkyl chain (see Figure 6.4). In this example of the McLafferty rearrangement, the benzene ring functions as an efficient γ-hydrogen acceptor.

▶ **Example 6.11** Predict the ionic product of the γ-hydrogen rearrangement of ionized ethyl phenyl ether.

Solution The reaction produces a characteristic ion at m/z 94 [reaction (6.29)].

Hydrogen Rearrangement to a Saturated Heteroatom
The elimination of a molecule of water, an acid, or ketene from certain radical cations can be explained by an intramolecular transfer of a hydrogen atom to a saturated heteroatom. A prototypical example is the loss of water from primary alcohols with a chain length of at least four carbon atoms:

(6.30)

This reaction is exemplified in Figure 6.5, which is the spectrum of 1-hexanol (MW = 102 u). This alcohol fails to yield the $M^{+\bullet}$ ion but exhibits the loss of H_2O to produce an ion at m/z 84. The water loss is, however, nonspecific and occurs via a six-membered ring (other rings are also possible). For example, the loss of water from cyclohexanol occurs via 1,2-, 1,3-, and 1,4-processes. Another example of nonspecific water loss is from tetralol. The loss of HCl from alkyl chlorides also requires H-transfer to the ionized Cl atom:

(6.31)

Phenyl and benzyl acetates both show the loss of ketene via H-transfer to an ether O atom:

(6.32)

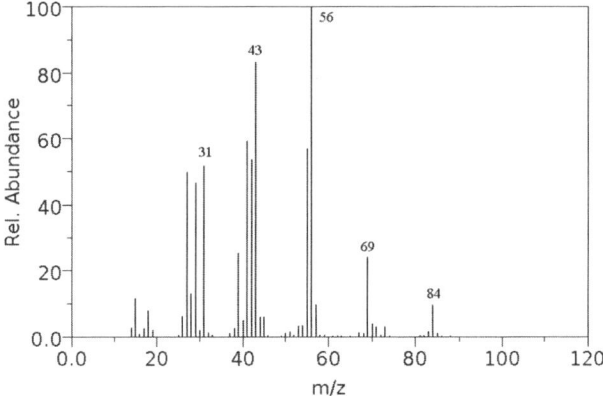

Figure 6.5. EI mass spectrum of 1-hexanol (MW = 102 u). The $M^{+\bullet}$ ion is missing, but the loss of H_2O from it is conspicuous.

Ortho Rearrangement The *ortho* effect is observed in those bis-substituted aromatic compounds that have the two appropriate substituents in the *ortho* positions. Similar to the McLafferty rearrangement, this reaction also requires the transfer of a hydrogen atom from one of the substituents to the acceptor group of the other substituents through a six-membered transition state from which a neutral molecule is eliminated via an inductive effect: for example HX in the reaction

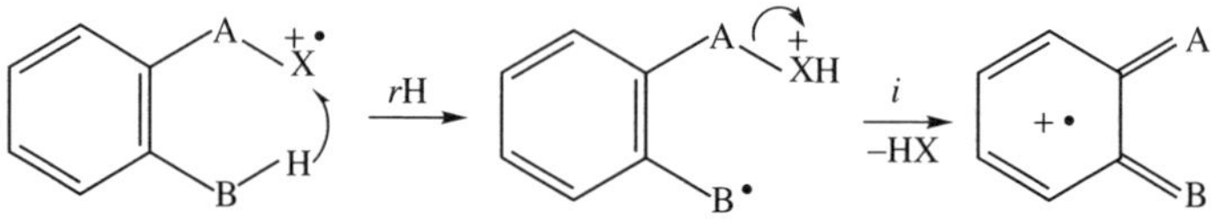

$$\text{(structure)} \xrightarrow{rH} \text{(structure)} \xrightarrow{i}_{-HX} \text{(structure)} \quad (6.33)$$

This reaction is useful in differentiating *ortho* from *meta* and *para* isomers. Two instructive examples of the *ortho* effect are shown in reaction (6.34) and in Example 6.11:

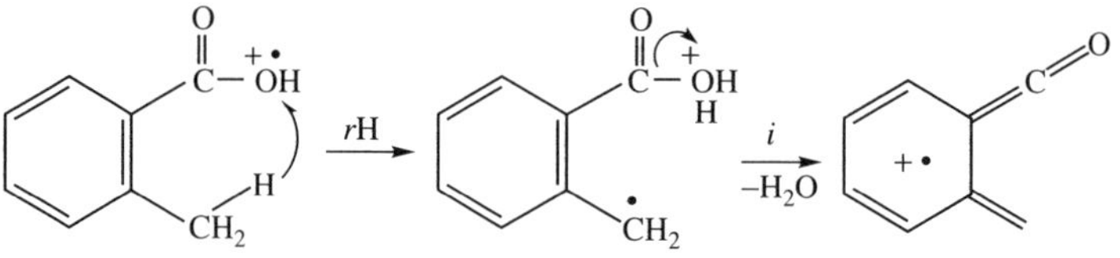

$$\text{(structure)} \xrightarrow{rH} \text{(structure)} \xrightarrow{i}_{-H_2O} \text{(structure)} \quad (6.34)$$

▶ **Example 6.12** Predict the products of the *ortho* rearrangement in ionized methyl 2-hydroxybenzoate.

Solution The reaction exhibits the loss of methanol to produce a characteristic ion at m/z 120:

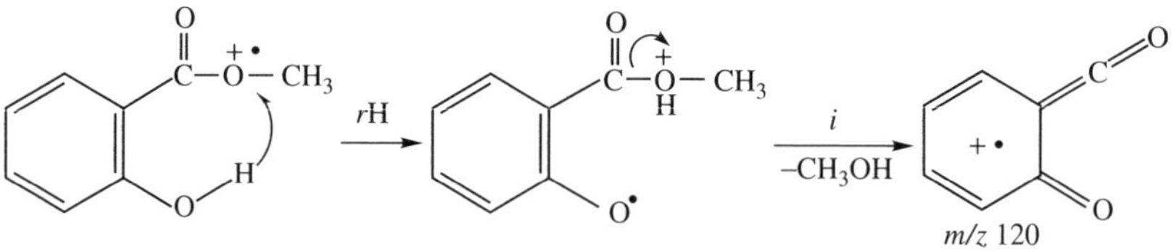

$$\text{(structure)} \xrightarrow{rH} \text{(structure)} \xrightarrow{i}_{-CH_3OH} \text{(structure, } m/z\ 120)$$

(6.35)

Double-Hydrogen Rearrangement The double-hydrogen rearrangement, also known as the *McLafferty + 1 rearrangement*, occurs primarily in ionized esters, thioesters, and amides that are derived from alcohols, thiols, and amines, respectively, and have an aliphatic chain at least three carbon atoms long. As illustrated in the spectrum of butylacetate (Figure 6.6), the intermediate product formed after the initial H-transfer reaction of the normal McLafferty reaction still has the propensity to undergo another γ-hydrogen transfer. Cleavage of the

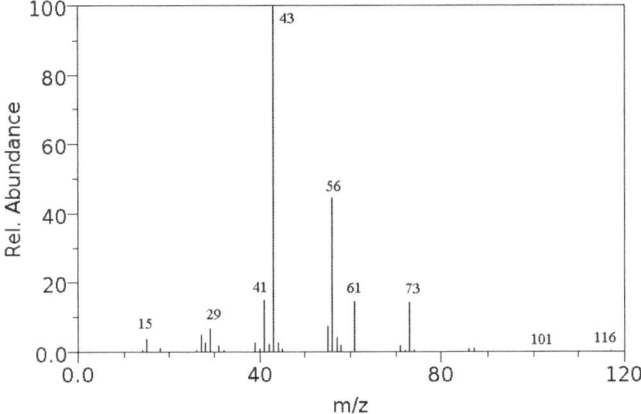

Figure 6.6. EI mass spectrum of butyl acetate (MM = 116 u). The M$^{+\bullet}$ ion is barely visible, but a McLafferty + 1 product is prominent at m/z 61.

α, β-bond, following this double H-transfer reaction, produces an EE$^+$ ion with one mass unit higher than the usual McLafferty product, and the neutral expelled is an alkenyl radical:

(6.36)

The McLafferty + 1 reaction also contains structurally diagnostic information. For example, esters of long-chain alcohols can be recognized by the peaks at m/z 61, 75, 89, and so on, resulting from the loss of an alkenyl radical.

6.5.3. Fragmentation of Cyclic Structures

The fragmentation of ionized cyclic compounds proceeds through rupture of the cyclic skeleton. This fragmentation requires the cleavage of at least two σ-bonds in the ring. In addition, cyclic compounds also participate in H-rearrangement reactions. There is no change in the mass of the ion when only one σ-bond cleaves in the ring, but with two- and three-ring bond cleavages, OE and EE ions, respectively, are formed. Also, loss of the ring substituent is favored over cleavages of the ring bonds. An example is the preferential loss of the methyl group from methylcyclohexane. A few examples of these reactions is enumerated here.

Cycloreversion Reactions A cycloreversion reaction is the reverse of a cycloaddition reaction and leads to the formation of the starting reactants through the cleavage of two bonds in the ring [18]. A typical example is the formation of $C_2H_4^{+\bullet}$ and neutral C_2H_4 from the cyclobutane radical cation. As shown in reaction (6.37), this reaction proceeds through the intermediacy of a distonic ion. The radical cations of a variety of other four-membered cyclic compounds, such as cyclobutanones (**3**), diketene (**4**), oxetane (**5**), cyclobutylamine (**6**), and thiocyclobutane (**7**), are known to participate in cycloreversion reactions [27].

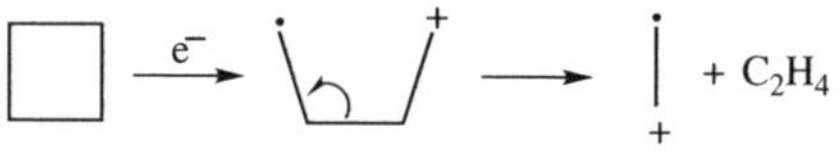

$$\square \xrightarrow{e^-} \overset{+}{\diagdown\diagup} \longrightarrow \underset{+}{|} + C_2H_4 \qquad (6.37)$$

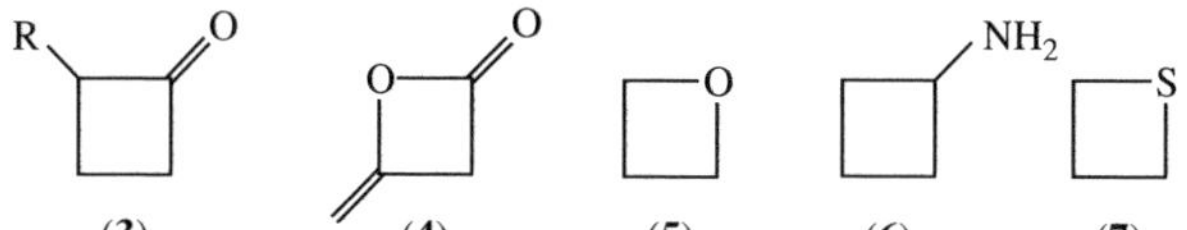

(3) (4) (5) (6) (7)

Retro-Diels–Alder Reaction Cyclohexene (Figure 6.7) and its derivatives undergo a retro-Diels–Alder (RDA) reaction upon EI, which is a special case of the cycloreversion reaction [27,28]. This reaction is the reverse of a well-known [4 + 2] cycloaddition reaction between a conjugated diene and an olefin, named the *Diels–Alder reaction*, after its discoverers. The impetus for the RDA reaction is the ring double bond, which upon ionization creates a radical site and a charge site:

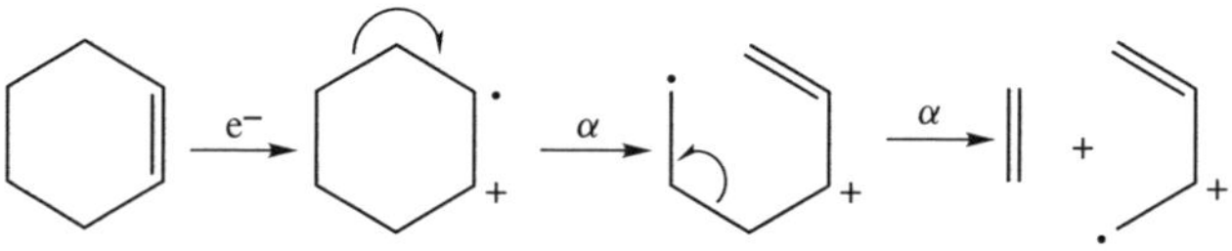

$$(6.38)$$

The radical site initiates two successive α-cleavage reactions that lead to the loss of an ethene molecule. In most reactions the charged fragment is the diene radical cation. The charge site is, however, dictated by the IE values of the corresponding neutrals. According to Stevenson's rule, the charge is retained on the fragment with the lower IE value. For example, in contrast to the reaction of the cyclohexene radical cation, the RDA product of 4-phenylcyclohexene$^{+\bullet}$ is ionized styrene:

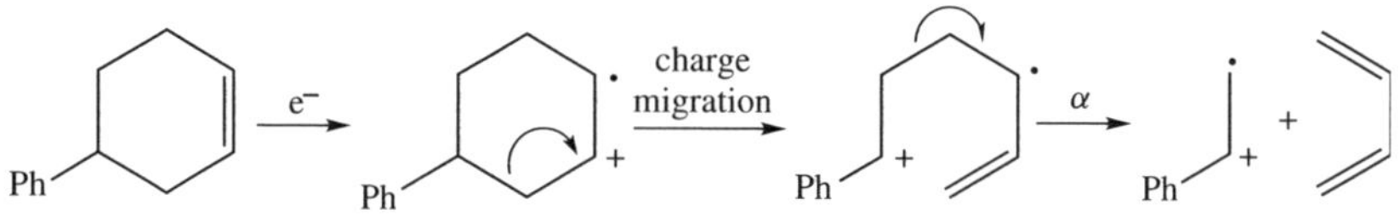

$$(6.39)$$

This process is quite useful in the structure determination of various natural products, such as cyclic isoprenoids.

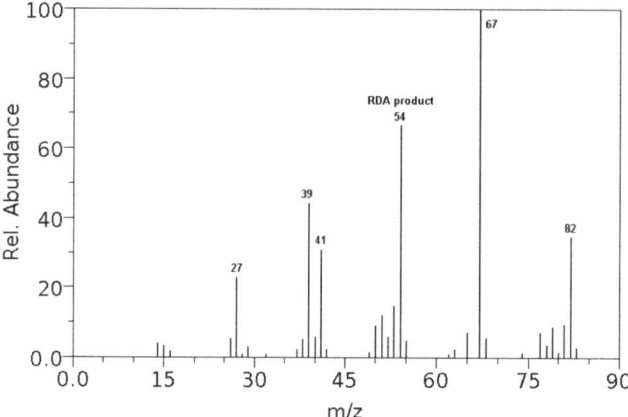

Figure 6.7. EI mass spectrum of cyclohexene.

Fragmentation of Cycloalkanes Similar to cyclobutane, higher-member cycloalkanes also exhibit the loss of an ethene molecule. As shown in reaction (6.40) for cyclohexane, this reaction requires cleavage of two bonds in the ring. Similarly, the origin of a series of ions at even-numbered m/z values in the spectrum of cyclodecane can be explained by the consecutive losses of alkene molecules from the initially formed distonic ion.

(6.40)

The H-transfer to the radical site in the distonic ion, preferably via a six-membered transition state, followed by the loss of alkyl radicals and eventual formation of an allyl cation, also competes with the ethene loss [reaction (6.40)]. In alkyl-substituted cycloalkanes, loss of the alkyl side chain overwhelms other fragmentation processes due to branching at that site. For example, m/z 83 is the base peak in the spectra of methyl- and n-butylcylohexanes.

Fragmentation of Heteroatom-Containing Cyclic Ions The possible fragmentations of heteroatom-containing cyclic ions are shown in the following reaction which shows that the charged fragments are formed via three different reaction pathways:

1. Cleavage of two σ-bonds:

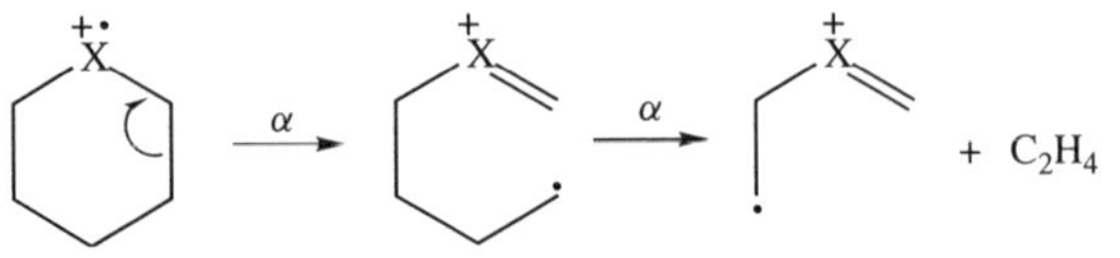

$$\text{cyclic } X^{+\bullet} \xrightarrow{\alpha} \cdot \xrightarrow{\alpha} + C_2H_4 \qquad (6.41)$$

2. Inductive effect:

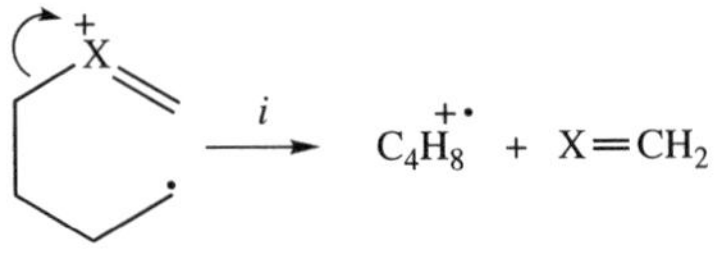

$$\xrightarrow{i} C_4H_8^{+\bullet} + X=CH_2$$

3. H-transfer reaction

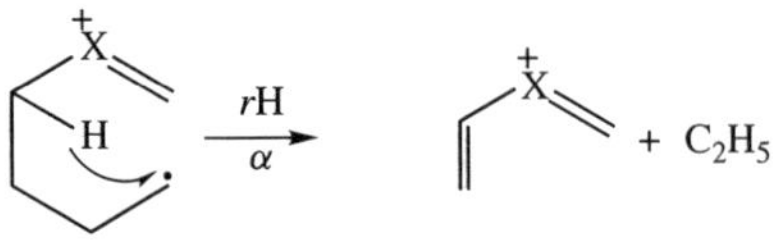

$$\xrightarrow[\alpha]{rH} + C_2H_5$$

The specificity of a particular reaction is dictated by the stability of the product. For example, the $C_4H_8^{+\bullet}$ ion formed by the inductive effect is the major product in the EI spectrum of tetrahydopyran (**8**):

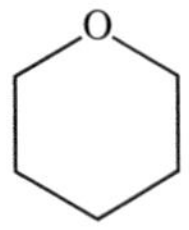

(**8**)

and the double σ-bond cleavage product is of minor significant. Ionized piperidine exhibits the loss of $^\bullet CH_3$ and $^\bullet C_2H_5$ due to H-transfer reactions and of C_2H_4 due to a double α-cleavage, to produce ions at m/z 70, 56, and 57, respectively [reaction (6.42); Figure 6.8] [29]:

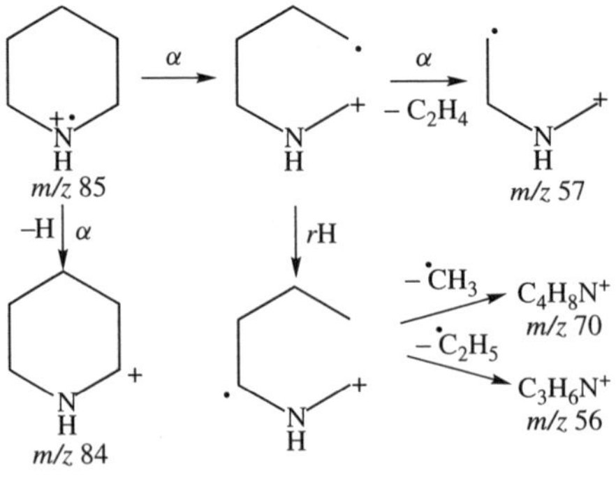

$$\begin{array}{c} m/z\ 85 \xrightarrow{\alpha} \cdot \xrightarrow{\alpha,\ -C_2H_4} m/z\ 57 \\ -H\Big|\alpha \qquad \Big| rH \\ m/z\ 84 \qquad \xrightarrow[-\dot{C}_2H_5]{-\dot{C}H_3} \begin{array}{c} C_4H_8N^+\ m/z\ 70 \\ C_3H_6N^+\ m/z\ 56 \end{array} \end{array} \qquad (6.42)$$

An abundant M − 1 ion (m/z 84) also results, due to loss of the α-H.

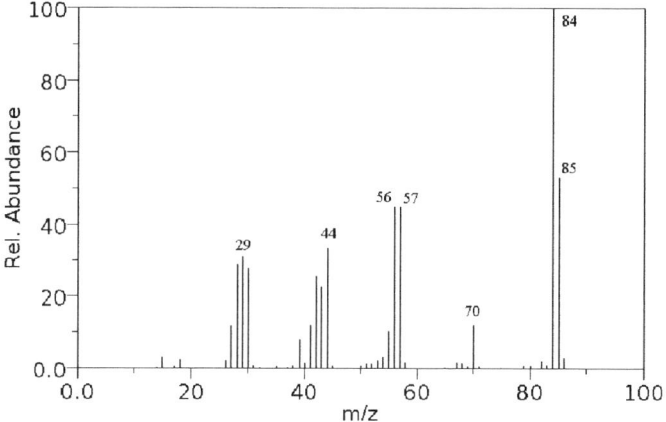

Figure 6.8. EI mass spectrum of piperidine (MW = 85 u).

Similar fragmentation processes also operate in cycloketones. As exemplified for ionized cyclohexanone (Figure 6.9), α-cleavage of the C–C bond adjacent to the carbonyl group to form a stable distonic ion is the first step of fragmentation, followed by the loss of either CO (inductive effect) or C_2H_4 (α-cleavage):

(6.43)

The base peak in the mass spectrum of cyclohexanone is, however, m/z 55, which arises by the loss of a propyl radical following the six-center H-transfer reaction in the initially formed distonic ion. Loss of both CO and C_2H_4 is observed in fragmentation of the cyclobutanone radical cation [30]. Similar fragmentation pathways are exhibited by ionized cycloalkylamines.

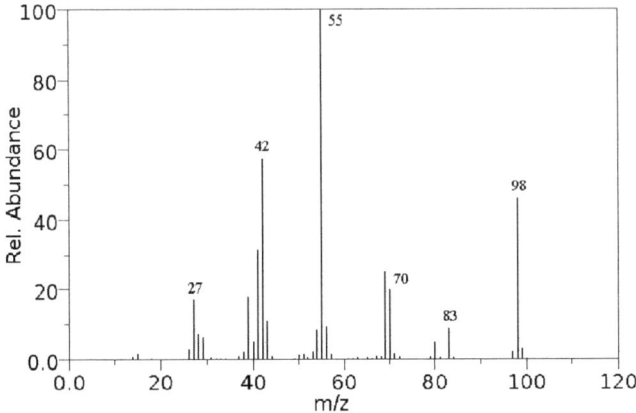

Figure 6.9. EI mass spectrum of cyclohexanone (MW = 98 u).

6.5.4. Differentiation of Isomeric Structures

Unlike NMR spectroscopy, mass spectrometry is poorly disposed in its ability to differentiate isomeric compounds. Several examples are, however, known that demonstrate the ability of mass spectrometry in unambiguous identification of isomers. Some representative examples are discussed here:

1. 1-Butanol, 2-butanol, and 2-methyl-2-propanol can be distinguished on the basis of their propensity for α-cleavage reaction. Ionized 1-butanol yields mainly an ion of m/z 31 ($CH_2=\overset{+}{O}H$), 2-butanol an ion of m/z 45 ($CH_3CH=\overset{+}{O}H$), and 2-methyl-2-propanol an ion of m/z 59 [$(CH_3)_2C=\overset{+}{O}H$]:

$$
\begin{array}{c}
C_2H_5CH_2\frown CH_2\!-\!\overset{\cdot+}{O}H \xrightarrow{\alpha} {}^\cdot C_3H_7 + CH_2=\overset{+}{O}H \\
m/z\ 31 \\[4pt]
CH_3CH_2\frown \overset{\overset{\cdot+}{O}H}{C}H\!-\!CH_3 \xrightarrow{\alpha} {}^\cdot C_2H_5 + CH_3CH=\overset{+}{O}H \\
m/z\ 45 \\[4pt]
CH_3\frown \underset{\underset{CH_3}{|}}{\overset{\overset{\cdot+}{O}H}{C}}\!-\!CH_3 \xrightarrow{\alpha} {}^\cdot CH_3 + \underset{\underset{CH_3}{|}}{CH_3C}=\overset{+}{O}H \\
m/z\ 59
\end{array}
\tag{6.44}
$$

Similarly, isomeric primary, secondary, and tertiary amines are differentiated on the basis of their distinct α-cleavage products.

2. As shown in Figure 6.10, the radical cations of 3-pentanone and 3-methyl-2-butanone yield distinct α-cleavage products:

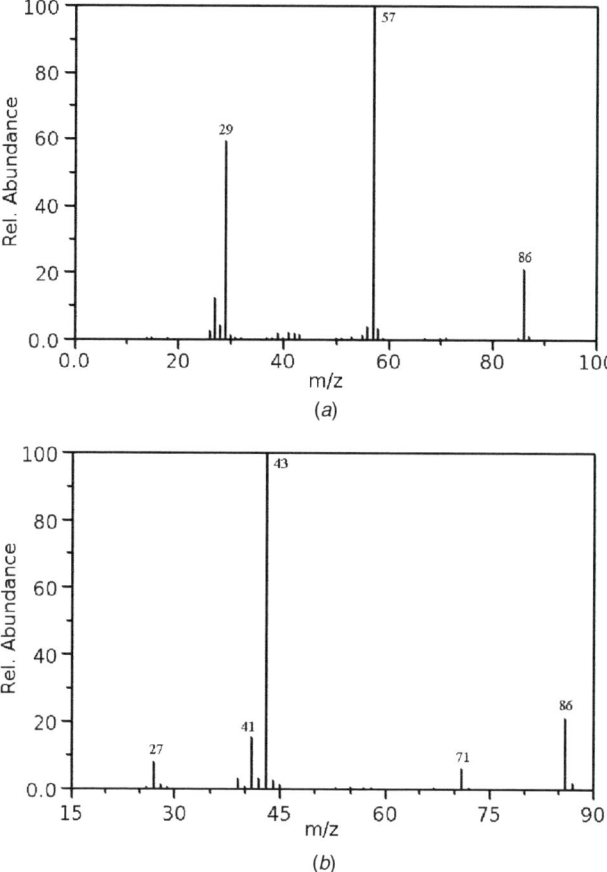

Figure 6.10. EI mass spectra of (*a*) 3-pentanone and (*b*) 3-methyl-2-butanone.

The former produces m/z 57 ($C_2H_5CO^+$), whereas the latter yields m/z 43 (CH_3CO^+).

3. The McLafferty rearrangement can also provide isomer-specific data. For example, this rearrangement is able to distinguish 4-methyl-2-pentanone

234 ORGANIC MASS SPECTROMETRY

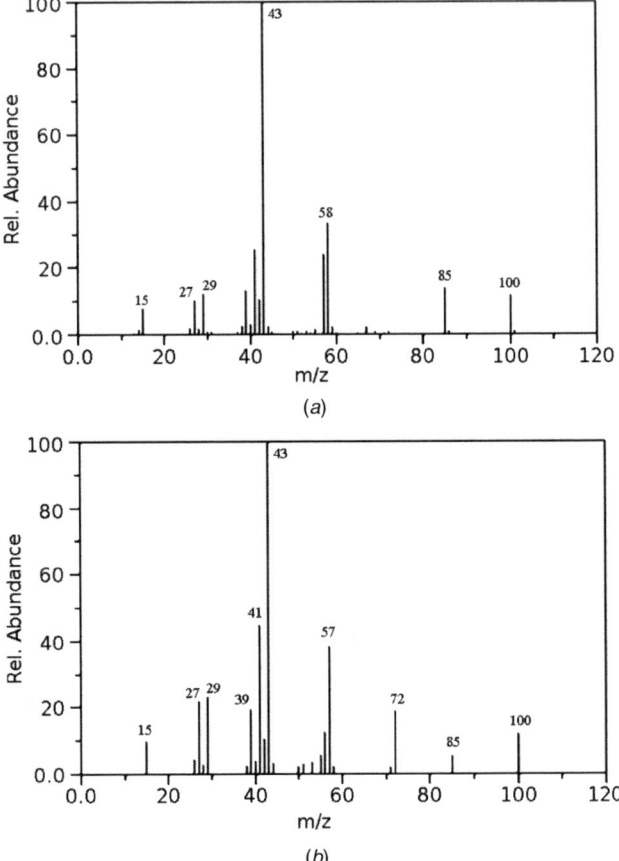

Figure 6.11. EI mass spectra of (*a*) 4-methyl-2-pentanone and (*b*) 3-methyl-2-pentanone. These two compounds can be distinguished on the basis of the McLafferty rearrangement products m/z 58 and 72, respectively.

and 3-methyl-2-pentanone; the former structure leads to the formation of an ion of m/z 58 ($C_2H_6CO^{+\bullet}$), whereas the latter produces an ion of m/z 72 ($C_3H_8CO^{+\bullet}$) [reaction (6.46); Figure 6.11]:

(6.46)

4. The RDA fragmentation is another diagnostic reaction that can be applied to differentiate isomeric compounds. For example, on the basis of distinct RDA products, 1-tetralol$^{+\bullet}$ and 2-tetralol$^{+\bullet}$ can be distinguished. The RDA fragmentation of 1-tetralol$^{+\bullet}$ results in the loss of an ethene molecule, whereas 2-tetralol$^{+\bullet}$ undergoes the loss of C_2H_3OH (possibly vinyl alcohol):

$$\text{1-tetralol}^{+\bullet} \xrightarrow{RDA} \text{product}^{+\bullet} + C_2H_4$$

$$\text{2-tetralol}^{+\bullet} \xrightarrow{RDA} \text{product}^{+\bullet} + C_2H_3OH \tag{6.47}$$

5. Many isomeric bis-substituted benzenes can be distinguished on the basis of the *ortho* effect. Some examples were presented in Section 6.5.2.

6.5.5. Structurally Diagnostic Fragment Ions

Often, the presence of certain ions in the spectrum can provide a helpful hint for the successful identification of a compound. Some typical diagnostic ions are listed in Table 6.8, along with their structures, the type of reaction that leads to their formation, and the precursor molecule. The ion of m/z 30, formed via the α-cleavage reaction, is characteristic of primary amines [reaction (6.18)]. Similarly, the ion of m/z 31 provides the identity of primary alcohols [reactions (6.17) and (6.44)]. Aldehydes, with no substitution at the α-carbon, can be identified by the characteristic McLafferty rearrangement product ion of m/z 44 [$CH_2C(OH)H^{+\bullet}$]. The McLafferty rearrangement also yields structure-specific ions of m/z 58 [$CH_2C(OH)CH_3^{+\bullet}$] [Eq. (6.46)], 59 [$CH_2C(OH)NH_2^{+\bullet}$], and 120 [$CH_2C(OH)C_6H_5^{+\bullet}$] for methyl ketones, amides, and phenyl ketones, respectively. Similarly, carboxylic acids that are unsubstituted at the α-carbon can be identified by the presence of an OE$^{+\bullet}$ ion at m/z 60 [$CH_2C(OH)_2^{+\bullet}$], also formed by the McLafferty rearrangement. Analogous ions of m/z 74 [$CH_2C(OH)OCH_3^{+\bullet}$; reaction (6.27)], 88 [$CH_2C(OH)OC_2H_5^{+\bullet}$], or 136 [$CH_2C(OH)OC_6H_5^{+\bullet}$] would provide an indication for methyl, ethyl, or phenyl esters, respectively. The m/z 60 ion is also present in the spectra of esters of acetic acid [reaction (6.36)]. Another ion diagnostic of esters of acetic acid, which otherwise fail to yield any detectable molecular ion, is of m/z 61 [$CH_3C(OH)_2^{+}$]. This ion is formed by the McLafferty + 1 rearrangement [reaction (6.36)].

The ion of m/z 77 ($C_6H_5^+$) usually points to the presence of a benzene ring. The ion of m/z 91 ($C_6H_5CH_2^+$; formed by α-cleavage) is the signature of phenyl alkanes [reaction (6.20)]. If m/z 92 ($C_7H_8^{+\bullet}$) is also present along with m/z 91, it provides additional information that the substituent hydrocarbon chain is at least three carbons long [reaction (6.28)]. Phenyl ethers can be identified by the presence of an ion of m/z 93 ($C_6H_5O^+$):

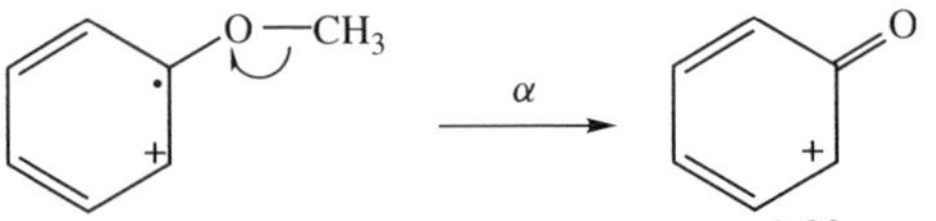

$$\text{(6.48)}$$

Ethyl or longer-chain phenyl ethers, in addition, would produce an ion of m/z 94 ($C_6H_6O^{+\bullet}$) via McLafferty-type rearrangement [reaction (6.29)]. The ion of m/z 93 ($C_6H_7N^+$) is also produced from ionized 2-alkyl and 4-alkylpyridines via the McLafferty rearrangement. The m/z 105 ion ($C_6H_5CO^+$), which is produced via α-cleavage, is evidence for the presence of phenyl ketones and esters and amides of benzoic acids. Esters of benzoic acids will also produce a characteristic ion of m/z 122 as a result of the McLafferty rearrangement:

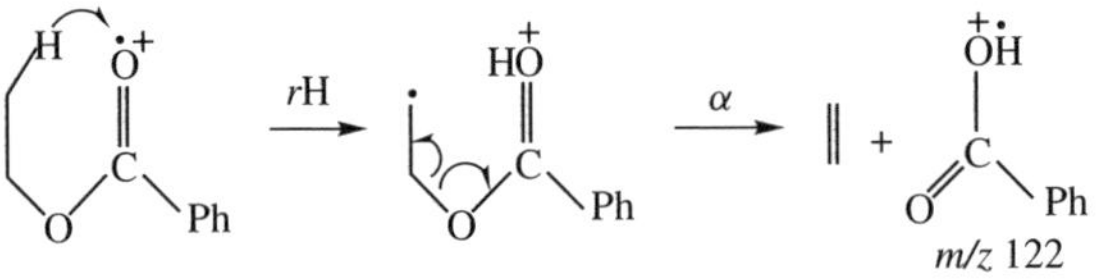

$$\text{(6.49)}$$

Benzyl esters can be recognized on the basis of an ion of m/z 108 that is formed by the elimination of a ketene molecule [reaction (6.32)]. Ions at m/z 127 and 128 are indications of iodo compounds. A strong signal at m/z 149 (**9**) is always present in the spectra of esters of phthalic acids and in samples that might be contaminated with plasticizers.

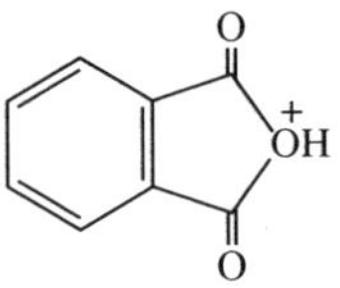

(**9**) (m/z 149)

▶ **Example 6.13** Identify the compound whose spectrum is shown in Figure 6.12.

Solution From the high-mass cluster it is obvious that the unknown contains a bromine atom. The difference in mass between m/z 122 and 43 supports this observation. The elemental composition of m/z 43 is $C_3H_7^+$, which could be an isopropyl or propyl cation. The unknown is either 1- or 2-bromopropane, both of which give nearly identical mass spectra.

FRAGMENTATION PROCESSES   **237**

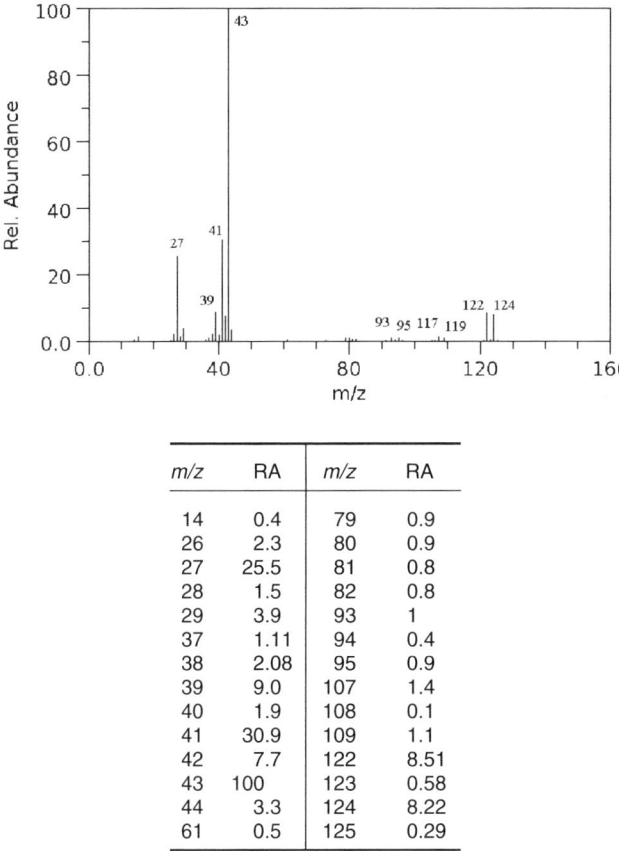

Figure 6.12. Mass spectra of an unknown for Example 6.13.

▶ **Example 6.14**   Identify the compound whose spectrum is shown in Figure 6.13.

Solution   The *m/z* 122 ion is identified as the molecular ion. The even mass indicates the presence of no nitrogen or even-numbered nitrogens. First, assuming no nitrogen, and calculating the number of carbon atoms, we arrive at seven carbons. The abundance of the [M + 2] ion indicates the presence of two oxygen atoms. The masses of these atoms account for 116 u of the molecular mass. Thus, the presence of any nitrogen can be ruled out. The difference of 6 u between 116 and 122 u indicates that the molecule has six hydrogens. Thus, the calculated elemental composition is $C_7H_6O_2$. The compound is highly unsaturated, as indicated by the R + DB value of 5. High abundance of the molecular ion suggests that the compound is highly stable. The presence of a benzene ring is indicated by *m/z* 77

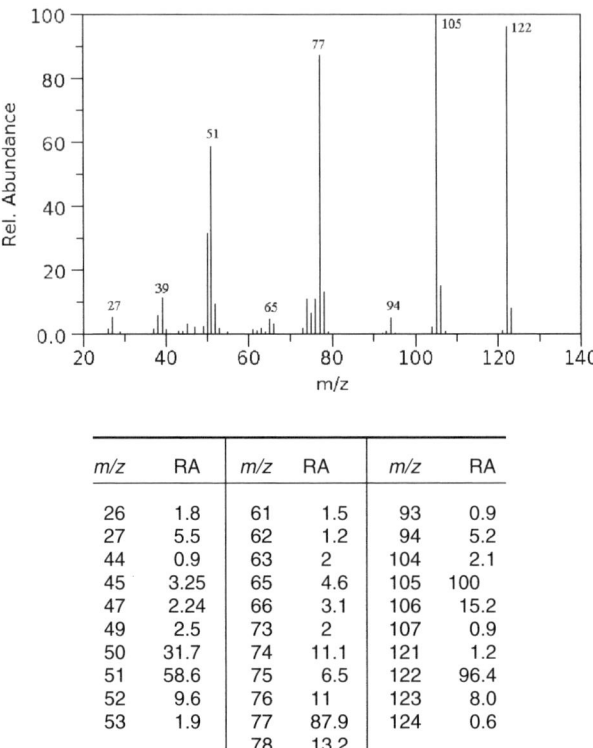

Figure 6.13. Mass spectra of an unknown for Example 6.14.

and confirmed by other diagnostic ions at m/z 39, 51, and 65. The base peak represents an ion of m/z 105, which could be a $C_6H_5CO^+$ ion; the other possibility, $C_6H_5CH_2CH_2^+$, can be ruled out on the basis of the number of hydrogen atoms. The loss of 17 u from the molecular ion also supports the $C_6H_5CO^+$ structure. The mass difference of 45 between 77 and 122 u indicates the presence of a COOH group. The unknown is benzoic acid.

6.6. FRAGMENTATION REACTIONS OF SPECIFIC CLASSES OF COMPOUNDS

This section summarizes the characteristic fragmentation reactions of most common classes of organic compounds.

6.6.1. Hydrocarbons

Straight-Chain Alkanes Mass spectra of straight-chain hydrocarbons are characterized by low-abundant $M^{+\bullet}$ ions. These compounds undergo principal ionization at the C—C σ-bond and fragment all along the hydrocarbon chain by

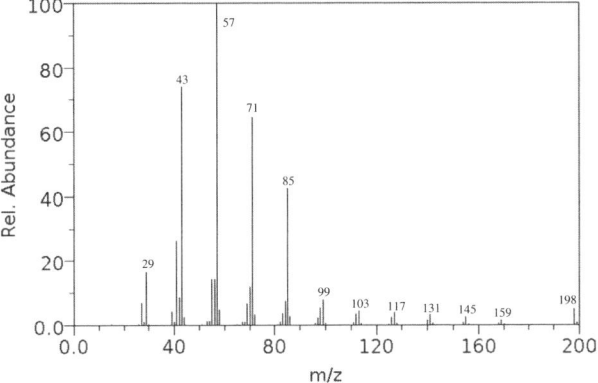

Figure 6.14. EI mass spectrum of tetradecane ($C_{14}H_{30}$; MW = 198).

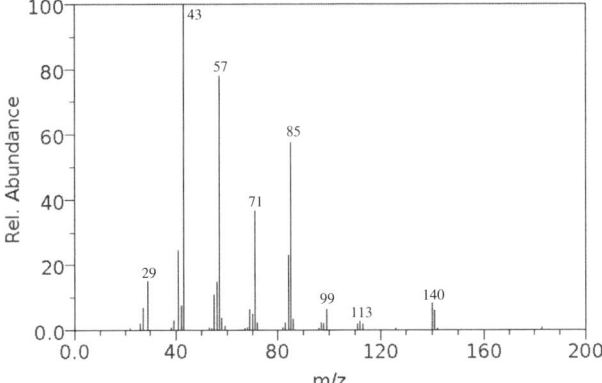

Figure 6.15. EI mass spectrum of 5-methyltridecane ($C_{14}H_{30}$; MW = 198).

cleavage of the C—C bond to produce a cluster of ions, each cluster separated by 14 u (see Figure 6.14 for the spectrum of tetradecane). The $C_nH_{2n+1}^+$ ion series is the most abundant (i.e., ions of m/z 29, 43, 57, 71, ...) of these cluster ions. Another unique feature of the spectra of linear alkanes is a smooth distribution of clusters at the low-mass end, where C_3 and C_4 ions are the most abundant. The high-mass region of the spectra is populated with low-abundance ions.

Branched-Chain Alkanes Branching reduces the $M^{+\bullet}$ ion abundance. Also, the smooth bell-shaped distribution is interrupted by preferred cleavage at the branching point, and loss of the largest group is the dominant reaction (see Figure 6.15 for the spectrum of 5-methyltridecane and compare this spectrum with the spectrum of isomeric tetradecane in Figure 6.14). In Figure 6.15, the abundance of $M^{+\bullet}$ is lower than that of the $M^{+\bullet}$ of the straight-chain isomer (Figure 6.14), and the branch point is indicated by the prominence of m/z 85 [$CH_3(CH_2)_3CHCH_3^+$] and 141 [$CH_3(CH_2)_7CHCH_3^+$].

Cycloalkanes The $M^{+\bullet}$ ions of cycloalkanes are more stable than their acyclic counterparts because the ions can accommodate a ring C—C bond cleavage without changing m/z. A major fragmentation of cycloalkanes is the loss of an ethene molecule and formation of even-numbered $OE^{+\bullet}$ ions. This reaction requires the cleavage of two bonds in the ring [reaction (6.40)]. The presence of these ions in the spectrum is a useful means of identifying a cycloalkane. Another important fragmentation pathway of cycloalkanes is H-transfer in the initially formed distonic ion, followed by α-cleavage to produce the $C_nH_{2n-1}^+$ ion series (i.e., ions of m/z 27, 41, 55, 69, ...). With substituted cycloalkanes, loss of the side chain is the dominant reaction.

Alkenes Ionization occurs at the site of the double bond, and the relative abundance of $M^{+\bullet}$ ion is higher than that formed in the corresponding saturated alkanes. The double bond is, however, mobile, and leads to structurally insensitive spectra for various isomeric alkenes. The allylic cleavage initiated by the radical site is a favored fragmentation of alkene radical cations [reaction (6.19)] to produce a dominant $C_nH_{2n-1}^+$ ion series (i.e., ions of m/z 27, 41, 55, 69, ...). The McLafferty rearrangement can occur prior to double-bond migration for certain alkenes and is structurally diagnostic.

Cycloalkenes These compounds yield even more distinct $M^{+\bullet}$ ions than do their saturated cycloalkane analogs. The RDA reaction is the signature of the cyclohexene-containing ring system [reactions (6.38) and (6.39)].

Aromatic Hydrocarbons Fragmentation of the benzene ring requires significantly high activation energy levels. Therefore, the $M^{+\bullet}$ ion is overwhelmingly abundant in the spectra of most aromatic hydrocarbons, and benzene ring–related ions (e.g., m/z 39, 51, 63, and 77) are of low abundance. The α-cleavage of the benzylic bond in ionized alkyl benzenes generates the structurally diagnostic m/z 91 [reaction 6.20)], the preponderance of which is due to a resonantly stabilized tropylium cation (**2**). Subsequent expulsion of the C_2H_2 moiety from a tropylium ion that has internal energy forms another characteristic ion of m/z 65. With a larger alkyl group, a series of ions at m/z 91, 105, 119, ... are observed. Another signature fragment of phenyl alkanes with an alkyl chain ≥ three carbons long is a McLafferty-type rearrangement that produces an $OE^{+\bullet}$ ion at m/z 92 [reaction (6.28) and Figure 6.4].

6.6.2. Alcohols

Aliphatic Alcohols The most common site of ionization in aliphatic alcohols is the oxygen atom of the OH group. The resulting $M^{+\bullet}$ ion is usually of low abundance for primary and secondary alcohols and is absent in most tertiary alcohols. The spectra of alcohols are dominated by α-cleavage reactions [reaction (6.17)]; the resonance-stabilized oxynium ion ($RCH=\overset{+}{O}H$; e.g., of m/z 31, 45, 59, ...) is the major product. The m/z 31 ($CH_2=\overset{+}{O}H$) is the signature of primary alcohols. As shown in reaction (6.44), it is often feasible to distinguish the spectra of isomeric alcohols. Another diagnostic reaction of primary alcohols is the loss of H_2O [reaction (6.30)]. Subsequent loss of an alkene molecule also occurs. These even-numbered $OE^{+\bullet}$ ions are conspicuous at [M − 18], [M − 46], [M − 74], ... in the spectra of primary alcohols (Figure 6.5). In addition, the spectra of long-chain alcohols show the features of hydrocarbon fragmentation.

Cyclic Alcohols Many of the fragmentation reactions of cyclic alcohols are similar to those of cycloalkanes. For example, α-cleavage of the ring bond is followed by the loss of an alkene molecule to produce even-numbered $OE^{+\bullet}$ ions. Hydrogen rearrangement, the subsequent loss of alkyl radicals, and H_2O loss also compete with this reaction.

Aromatic Alcohols The molecular ion of aromatic alcohols and phenols is relatively abundant. Unsubstituted benzyl alcohols produce the characteristic benzylic ion (m/z 91) via α-cleavage. Benzyl alcohols also undergo McLafferty-type rearrangement involving the phenyl ring. The loss of H_2O is also observed, especially in *ortho*-substituted phenols and benzyl alcohols. Phenols also exhibit the loss of CO and HCO molecules.

6.6.3. Ethers

Aliphatic Ethers The abundance of the molecular ion from aliphatic ethers is usually low but higher than that of the corresponding alcohols. These compounds exhibit radical site− and charge site−initiated fragmentation reactions. Radical site−initiated reactions produce resonance-stabilized oxynium ions (e.g., of m/z 45, 59, 73, ...). As exemplified in the reaction

$$CH_3CH=\overset{+}{O}-CH_2 \xrightarrow[\alpha]{rH} CH_3CH=\overset{+}{O}H + C_2H_4 \quad (6.50)$$
$$\underset{H-CH_2}{} \quad\quad m/z\ 45$$

these EE^+ ions can fragment further via a β-hydrogen transfer and the loss of an alkene molecule. Competing with α-cleavage [reaction (6.15)] is *i*-cleavage to form alkyl and related carbenium ions with the loss of an alkoxy radical [reaction (6.21)]. This process is more facile for ethers than for alcohols, presumably because •OR is more stable than •OH. The *i*-cleavage results in charge migration to the radical site and cleavage of the C−O bond [reaction (6.21)].

Aromatic Ethers In parallel with other aromatic compounds, the molecular ions of aromatic ethers are reasonably abundant. Cleavage of the bond β to the ring, as shown in reaction (6.48) for methyl phenyl ether, is typical of aromatic ethers, and forms a structure-specific ion of m/z 93. Benzyl ethers, on the other hand, produce m/z 91. Another typical reaction is hydrogen transfer, followed by the loss of an aldehyde from methyl phenyl ether and an alkene from phenyl and benzyl ethers with a two-carbon or larger alkyl chain [reaction (6.29)].

6.6.4. Aldehydes and Ketones

Ionization of aldehydes and ketones takes place by the removal of an electron from the carbonyl oxygen to yield a relatively (when compared to alkanes and alkenes) intense molecular ion signal, especially for ketones. An abundant $[M - 1^+]$ ion to form an acylium ion is the hallmark of aldehydes. In aromatic aldehydes, this ion might be even more abundant than the $M^{+\bullet}$ ion. Aliphatic aldehydes and ketones both show characteristic ion series at m/z 29, 43, 57, 71, ... that are due to acyl ions ($C_nH_{2n+1}CO^+$) as well as alkyl ions ($C_nH_{2n+1}^+$). The α-cleavage plays a major role in the fragmentation of ketones; loss of the larger alkyl chain is favored [reaction (6.16)]. Charge site–initiated fragmentations are also important for aldehydes and ketones. When an alkyl chain attached to the carbonyl group is $\geq C_3$, the McLafferty rearrangement is also observed [reaction (6.26)]. As shown in Section 6.5.4, the α-cleavage and McLafferty rearrangement reactions can both provide isomer-specific spectra for some ketones.

As with aromatic ethers, the radical site–initiated reaction in phenyl alkyl ketones cleaves the bond β to the ring to form a structurally diagnostic ion of m/z 105 ($C_6H_5CO^+$). Further loss of CO from this ion produces the phenyl cation of m/z 77 (see Figure 1.3). Similar to aliphatic ketones, the aromatic ketones also undergo McLafferty rearrangement when the alkyl chain is $\geq C_3$.

As mentioned earlier [see reaction (6.43)], cyclic ketones undergo three distinct fragmentation pathways: elimination of an alkene via cleavage of two σ-bonds in the ring, charge site–initiated reactions, and H-transfer reactions.

6.6.5. Carboxylic Acids

Aliphatic acids yields low-abundance molecular ions. A prominent signal at m/z 60 [$CH_2C(OH)_2^{+\bullet}$], formed by the McLafferty rearrangement,

$$\text{(6.51)}$$

R = H, m/z 60
R = CH_3, m/z 74

is diagnostic of the straight-chain carboxylic acids, which have four carbons or more. Substitution at the α-carbon commensurately increases the mass of that McLafferty ion. For example, in the spectrum of 2-methylpentanoic acid, the McLafferty ion appears at m/z 74, whereas in hexanoic acid, the McLafferty product is of m/z 60 [reaction (6.51)]. $[M^{+\bullet} - OH]$ and $[M^{+\bullet} - COOH]$ are other fragmentation reactions that are typical of carboxylic acids. The cleavage at each C—C bond in the alkyl chain of long-chain acids produces two types of ion series, one in which the charge is retained by the oxygen-containing fragments (i.e., $C_nH_{2n}COOH^+$) and the other, characteristic of the hydrocarbon chain (i.e., $C_nH_{2n+1}^+$).

The molecular ion of aromatic acids is more stable and hence more abundant. The sequential losses of the OH and CO moieties are observed from that ion. In the presence of an *ortho* group with a labile hydrogen, a prominent loss of H_2O can also occur [reaction (6.34)], followed by the loss of CO.

6.6.6. Esters

Molecular ions of appreciable abundance are observed in the EI spectra of aliphatic esters. Ionization can occur at either oxygen atom. Similar to aliphatic acids, straight-chain methyl esters also produce the structurally diagnostic McLafferty ion at m/z 74 [reaction (6.27)]. The m/z value of the McLafferty ion can identify the alcohol moiety as well as the substituent at the α-carbon. For example, in either ethyl esters or when a methyl group group is substituted at the α-carbon, the McLafferty ion will appear at m/z 88 (Figure 6.16). In addition to the McLafferty rearrangement, esters (R—COOCH$_2$R′) participate in radical site– and charge site–initiated fragmentation reactions to produce RCO^+, $R'CH_2OCO^+$, R^+, $R'CH_2O^+$, $RCO_2CH_2^+$, and $CH_2R'^+$ ions, as exemplified in reaction (6.52) for ethyl butanoate (Figure 6.16). Charge site–initiated products [e.g., $C_3H_7^+$ and $C_2H_5O^+$ in reaction (6.52)]

$$\begin{array}{c}
\overset{+\bullet}{\underset{\|}{O}} \\
C_3H_7-C-O-C_2H_5
\end{array} \xrightarrow{\alpha} \begin{array}{cc} C_3H_7C\equiv\overset{+}{O} & \text{and} \quad C_2H_5OC\equiv\overset{+}{O} \\ m/z\ 71\ \diagdown -CO & m/z\ 73\ \diagdown -CO \end{array}$$

$$\xrightarrow{i} \quad C_3H_7^+ \quad \text{and} \quad C_2H_5\overset{+}{O}$$
$$\qquad m/z\ 43 \qquad\qquad m/z\ 45 \qquad\qquad (6.52)$$

$$\begin{array}{c}
O \\
\underset{\|}{}\ \overset{+\bullet}{} \\
C_3H_7-C-O-C_2H_5
\end{array} \xrightarrow{\alpha} \begin{array}{cc} C_3H_7CO\overset{+}{O}CH_2 & \text{and} \quad C_2H_5OCO \\ m/z\ 101 & m/z\ 73 \end{array}$$

$$\xrightarrow{i} \quad C_3H_7\overset{+}{C}O \quad \text{and} \quad C_2H_5^+$$
$$\qquad m/z\ 71 \qquad\qquad m/z\ 29$$

can also arise by the loss of CO from the initial α-cleavage products. In most esters, fragmentation, however, occurs preferentially at the C—O bond. Methyl esters of straight-chain fatty acids fragment to give a series of ions of the general formula $CH_3OCO(CH_2)_n^+$ by cleavage of the hydrocarbon chain. Esters of long-chain alcohols also exhibit McLafferty + 1 rearrangement to produce an ion of structurally diagnostic value, the mass of which depends on the alkyl chain attached to the carbonyl group [e.g., 61, for a methyl group and 75 for an ethyl group; reaction (6.36)].

The general fragmentations discussed for aliphatic esters also operate in ionized aromatic esters, but the charge resides on the aryl moiety. For example, in

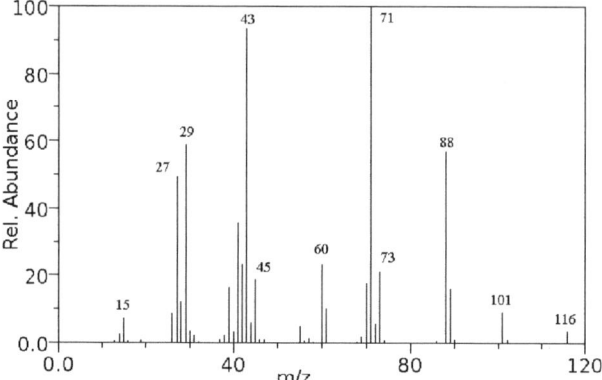

Figure 6.16. EI mass spectrum of ethyl butanoate (MW = 116); the McLafferty ion appears at m/z 88. Formation of other prominent ions is explained in reaction (6.52).

methyl benzoate, the $[M - 31^+]$ peak is the base peak. With substitution at the *ortho* position in the benzene ring, loss of an ROH molecule is a favored diagnostic reaction (see Example 6.12). When the benzene ring is part of the alcohol (e.g., as in benzyl and phenyl acetates), H-rearrangement leads to the elimination of a ketene molecule [reaction (6.32)].

6.6.7. Nitrogen-Containing Compounds

Amines The molecular ion of alkyl amines is of an odd m/z. Although the removal of an electron from the N atom in amines is a low-energy process, the resulting molecular ion is either of low abundance or absent because fragmentation [α-cleavage reaction; reactions (6.53 and 6.54)],

$$CH_3(CH_2)_n CRR'NH_2^{+\bullet} \longrightarrow CH_3(CH_2)_n^\bullet + CRR'{=}NH_2^+ \quad (6.53)$$

$$CH_3(CH_2)_n CH_2 NRR'^{+\bullet} \longrightarrow CH_3(CH_2)_n^\bullet + H_2C{=}NRR'^+ \quad (6.54)$$

directed by the nitrogen, is facile. For primary amines, the α-cleavage product (e.g., immonium ion, $CRR'{=}N^+H_2$) is invariably the most abundant ion; often, it is the base peak [reaction (6.53)]. Primary *n*-alkylamines yield a homologous series of ions of the general formula $C_n H_{2n+2} N^+$ (i.e., ions of m/z 30, 44, 58,...) that are formed by successive cleavage of the C—C bonds in the alkyl chain, but α-cleavage is the most dominant [29]. Another mechanism responsible for formation of these even-electron ions is ring formation:

$$\text{(6.55)}$$

A characteristic fragmentation of secondary and tertiary amines is the β-hydrogen transfer reaction

$$CH_3CH=\overset{+}{N}H-CH_2\text{...}\underset{\alpha}{\overset{rH}{\longrightarrow}} CH_3CH=\overset{+}{N}H_2 + C_2H_4 \quad (6.56)$$

As with ethers, this reaction occurs in the immonium ion that is formed by primary α-cleavage of amines and is the source of $C_nH_{2n+2}N^+$ ions.

Compared to aliphatic amines, the abundance of the molecular ion of alkyl aryl amines is higher. Fragmentation of the molecular ion is controlled by the N atom and leads to cleavage of the neighboring C–C bond:

$$\text{Ph-}\overset{+\cdot}{N}H-CH_2-CH_2R \xrightarrow{\alpha} \text{Ph-}\overset{+}{N}H=CH_2 + {}^{\cdot}CH_2R \quad (6.57)$$

Cyclic Amines Similar to cycloketones [reaction (6.43)], the first step in the fragmentation of ionized cyclic amines is α-cleavage of the C–C bond next to the C—N bond to produce a distonic ion [20]. Subsequent fragmentation proceeds via the loss of an alkene molecule or via an H-transfer followed by the loss of an alkyl radical. A typical example of the latter fragmentation is the loss of an ethyl radical from the cyclopentylamine radical cation. Ionized cyclic amines also exhibit an abundant (M − 1) ion from loss of the α-H (see Figure 6.8). As exemplified for piperidine, cyclic amines also participate in these generic fragmentation reactions of heterocyclic compounds [reaction (6.42)].

Amides In alkyl amides, ionization can occur at either of the heteroatoms (i.e., at either an oxygen or a nitrogen atom). A dominant reaction of ionized amides is α-cleavage initiated at either radical site. Both of these reactions, however, lead to the same product ($CONH_2^+$; m/z 44). Secondary amides will produce an ion series at m/z 44, 58, 72,... via α-cleavage. With the availability of the γ-hydrogen in the acyl moiety, ionized amides also undergo the McLafferty rearrangement. This reaction is of greater importance in primary amides and produces a characteristic ion of m/z 59. Aromatic amides show a prominent acylium ion ($ArC\equiv O^+$) peak via α-cleavage.

Nitriles The molecular ion is either of low abundance or entirely absent in the EI spectra of aliphatic nitriles. The molecular ion can, however, be recognized by the presence of an abundant (M − 1)$^+$ ion. Ionized aliphatic nitriles show a preference for the McLafferty rearrangement, where the rearranging H migrates to a triple bond, over an α-cleavage. The resulting m/z 41 ($CH_2=CNH^{+\bullet}$) is, however, of little diagnostic value because an ion ($C_3H_5^+$) of the same m/z value is produced due to fragmentation of the alkyl chain. In addition to the usual hydrocarbon pattern, a series of ions [$(CH_2)_nCN^+$] of m/z 40, 54, 68, 82,..., formed by a simple cleavage of the C–C bonds in the alkyl chain, can also be seen.

6.6.8. Sulfur-Containing Compounds

Sulfur-containing compounds (thiols and sulfides) are easily recognized from the M + 2 isotopic peak; each sulfur contributes 4.4% to the abundance of the M + 2 ion. The fragmentation patterns of thiols (mercaptans) and sulfides (thioethers) parallel the corresponding alcohols and ethers. For example, similar to the alcohol series at m/z 31, 45, 59, ..., the α-cleavage in thiols produces a series of ions at m/z 47, 61, 75, 89, ..., and each ion has a satellite peak 2 u higher, due to ^{34}S. In addition, thiols exhibit a characteristic loss of H_2S, followed by the elimination of alkene moieties to produce peaks at $(M - 34)^{+\bullet}$, $(M - 34 - C_2H_4)^{+\bullet}$, and so on. In contrast, secondary thiols show a characteristic peak at $(M - SH)^+$. Aromatic thiols also behave similarly to phenols under EI conditions. In addition, they show ions at $(M - S)^{+\bullet}$, $(M - SH)^+$, and $(M - C_2H_4)^{+\bullet}$.

α-Cleavage is a prominent reaction in sulfides and produces the $RS^+=CH_2$ ion; as expected, the emphasis is on loss of the largest group. Similar to the reaction shown in reaction (6.50) for ethers, this primary ion also undergoes H-rearrangement and subsequent loss of an alkene molecule.

6.6.9. Halogen-Containing Compounds

Identification of chloro and bromo compounds is a relatively simple matter because of the unique isotopic pattern. The presence of fluoro and iodo compounds, although not easy, can be inferred from the conspicuously low [M + 1]/[M] ratio, which is due to the fact that F and I are monoisotopic. The molecular ion in aliphatic chlorides is visible only in lower monochlorides. With an increase in the number of chlorine atoms, the abundance of the molecular ion decreases further. The i-cleavage to expel a halogen atom often produces an abundant ion (e.g., the base peak in the mass spectrum of t-butyl chloride is $C_4H_9^+$). The α-cleavage is of low consequence in alkyl chlorides, but the loss of an alkyl radical can be prominent when the alkyl chain is longer than four carbons; the product is a five-membered ring halonium ion:

$$\underset{}{\overset{R}{\underset{Cl}{\bigcirc}}^{+\bullet}} \quad \xrightarrow{rd} \quad {}^{\bullet}R \quad + \quad \underset{Cl}{\bigcirc} \quad (6.58)$$

The fragmentation reactions of alkyl bromides and iodides are similar to those of the corresponding chloro compounds. Losses of a halogen radical and an HX molecule are also observed, but to a different extent. The former is more common in bromo and iodo compounds, whereas the latter is common in chloro and fluoro compounds.

Thus, if one understands the basic ion chemistry and the fragmentation directed by functional groups in simple molecules, one is reasonably prepared to interpret the spectra of more complex polyfunctional compounds. Further the fragmentation of EE ions is also seen in the product-ion spectra of $[M + H]^+$ ions produced by CI, ESI, and FAB.

6.7. THEORY OF ION DISSOCIATION

Under the operating conditions of an EI source, the fragmentation of ions is strictly unimolecular. The unimolecular dissociation of gas-phase species can be explained elegantly by two statistical theories that were developed almost simultaneously in the early 1950s: the quasiequilibrium theory (QET), which deals mainly with ionic species [31–33], and the Rice–Ramsperger–Kassel–Marcus (RRKM) theory, applicable to neutral molecules [31,32,34]. Both of these theories are based on the assumption that the time of ion dissociation is long compared to the time required for ionization and excitation of the molecule, but slower than the rate of distribution of the excitation energy among the various possible energy states of the ion.

Ionization of a molecule by the removal of an electron is a very fast process (ca. 10^{-16} s) compared to the time, 10^{-14} s, of one vibration. Thus, ionization is essentially a vertical (Frank–Condon) process. In EI, ions formed have a broad energy distribution, with each ion having a specific amount of energy, known as the *internal energy* (E). The distribution of energy can be represented schematically by a Wahrhaftig diagram (Figure 6.17), a plot of the probability

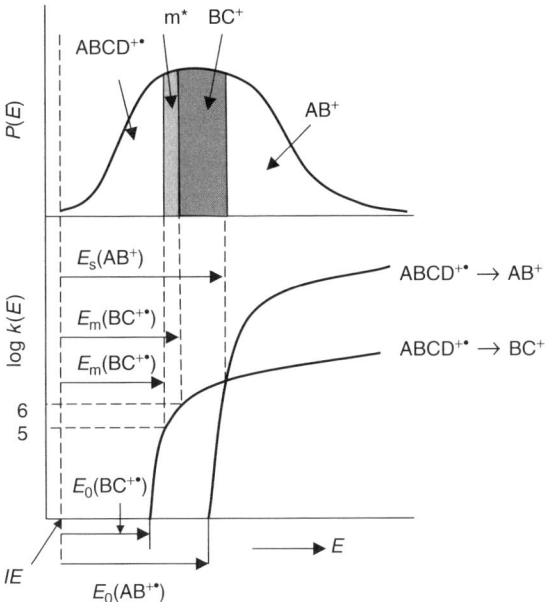

Figure 6.17. Warhaftig diagram: The upper portion represents the probability of a particular ion at a specified energy. The lower portion represents the rate constant of a unimolecular fragmentation of the ABCD$^{+\bullet}$ radical cation as a function of internal energy; the steep curve is for a direct cleavage reaction, and the shallow curve is for a rearrangement reaction. (Redrawn from F. W. McLafferty and F. Tureček, *Interpretation of Mass Spectra*, 4th ed.)

function $P(E)$ (i.e., the proportion of ions with a certain energy) versus E. The true distribution is difficult to determine, and an estimate can be made from the photoelectron spectrum of the molecule. The energy imparted to the ion in excess of IE is distributed statistically among the various internal degrees of freedom; that process implies that a molecule upon ionization does not dissociate immediately but survives for the time required for conversion of the internal energy into the vibrational and rotational energies of the ground electronic state of the ion. When a particular vibrational mode contains more energy than a critical energy (E_0, also called *activation energy*), the ion dissociates through that vibrational channel. As a consequence, decomposition of the ion is independent of the mode of ionization and depends only on its internal energy and the rate constant function $k(E)$; these relationships imply that the rate constant of a particular reaction is a function of the internal energy of the ion. This function can be calculated by the RRKM expression

$$k(E) = \frac{\sigma N*(E - E_0)}{h\rho(E)} \tag{6.59}$$

in which σ is the reaction degeneracy factor, $N*(E - E_0)$ is the number of internal energy states in the transition state in the energy interval $(E - E_0)$, h is the Planck's constant, and $\rho(E)$ is the density of states.

As discussed in Section 6.5, fragmentation can occur via either direct cleavage (e.g., $ABCD^{+\bullet} \rightarrow AB^+ + {}^{\bullet}CD$) or a rearrangement reaction (e.g., $ABCD^{+\bullet} \rightarrow BCDA^{+\bullet} \rightarrow BC^{+\bullet} + AD$). The internal energy and structure of the ion are the determining factors as to which reaction will occur and to what extent. Each reaction proceeds via a transition state of certain critical energy (E_0). The plot of log k versus internal energy for the two competitive reactions is shown in the lower portion of Figure 6.17. The time scale in this plot corresponds to ions formed in magnetic-sector instruments, although it would be similar to ions mass-analyzed in a time-of-flight or linear quadrupole. The plot shows that both thermodynamic and kinetic effects play a role in determining the abundance of a specific fragmentation product. The transition-state geometry of the direct cleavage reaction is "loose" because there is no obstruction to vibrational and rotational motions, and thus it has favorable entropy:

$$\begin{array}{c}
\text{A—B—C—D}^{+\bullet} \diagup \diagdown \begin{array}{c} \text{A------D}^{+\bullet} \\ \text{B==C} \\ \text{tight complex} \end{array} \longrightarrow \text{A—D} + \text{B=C}^{+\bullet} \\ \\ \begin{array}{c} \text{A—B}^+\text{----C}^{\bullet}\text{—D} \\ \text{loose complex} \end{array} \longrightarrow \text{A—B}^+ + {}^{\bullet}\text{C—D}
\end{array} \tag{6.60}$$

In contrast, a rearrangement reaction has a "tight" transition state because one or more rotations are stopped in this more constrained cyclic geometry.

Thus, a rearrangement reaction has low activation energy because a part of the energy required for bond cleavage (e.g., A—B and C—D) is compensated for by the energy that is released in formation of new bonds (e.g., A—D), but the reaction is less likely to occur, owing to the lower entropy of the intermediate transition state. The $k(E)$ function versus energy curve also rises faster in a direct reaction. As a consequence, at lower internal energies, the product of the rearrangement reaction (i.e., $BC^{+\bullet}$) dominates the spectrum, whereas at higher internal energies, the direct-cleavage reaction has a more favorable rate constant and its dissociation product AB^+ is formed more readily. Figure 6.17 also shows that the fragmentation products are rarely observed at the precise value of E_0 but at an energy (E_s) that is in excess of E_0. This extra energy ($E_0 - E_s$) is known as the *kinetic shift*; it is the additional amount of energy required by the dissociating ion to produce a detectable number of product ions. For most ions (<200 m/z), this energy is small (<0.01 eV). Ions with a k value <10^6 s^{-1} are stable, with a lifetime of 10^{-5} s or greater, and reach the detector intact. Ions with a k value >10^6 s^{-1} will decompose in the ion source. There is a small population of ions with k in the range 10^5 to 10^6 s^{-1} that are kinetically unstable, yet do not dissociate in the ion source, due to an unfavorable kinetic function. They do fragment, but only after exiting the ion source. These ions, termed *metastable ions*, can be detected as low-abundance ions in sector instruments or by MS/MS techniques. Their existence is validation of this kinetic picture, whereby ions decompose over a range of times depending on their internal energy.

A pictorial description of the RRKM expression is presented in Figure 6.18, which shows a potential energy curve for two competitive reactions: one is a

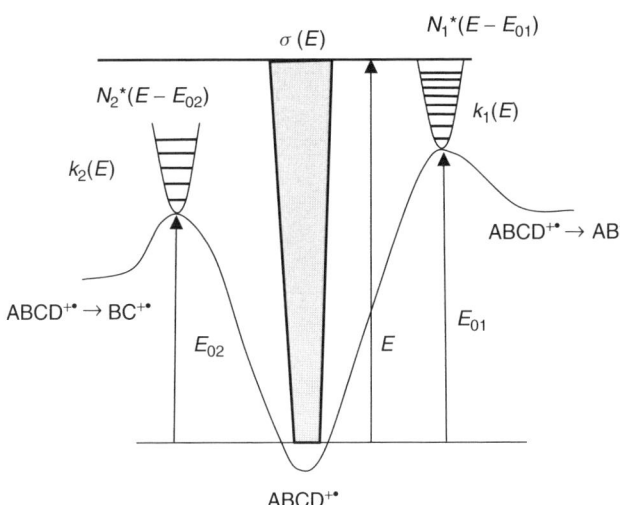

Figure 6.18. Competitive dissociation kinetics of rearrangement (left side) and direct cleavage (right side) reactions. The middle portion (shaded area) shows the increase in density of states with energy. (From ref. 35.)

direct cleavage of $ABCD^{+\bullet}$ with the rate constant function $k_1(E)$ and the critical energy E_{01}, and the other is a rearrangement reaction (i.e, isomerization of $ABCD^{+\bullet}$ to $BCDA^{+\bullet}$, followed by its dissociation) with the rate constant function $k_2(E)$ and the critical energy E_{02} [35]. The tight transition state for the rearrangement reaction has a smaller number of quantum states than that for the direct cleavage reaction. It is evident that ion fragmentation occurs only when $E \geq E_0$. When the internal energy E_{01} is less than that of the isomerization barrier, only the product from direct cleavage will be seen in the spectrum. When the isomerization barrier is below E_{01}, at an energy interval $E_{01} - E_{02}$, only the products from the rearrangement reaction are formed. At higher values of E, the $k_1(E)$ function for the direct cleavage reaction increases much faster than the $k_2(E)$. Therefore, products of the direct cleavage reaction will dominate at higher internal energies.

6.8. STRUCTURE DETERMINATION OF GAS-PHASE ORGANIC IONS

As shown in Figure 6.19, a molecular ion can be classified as a dissociating, nondissociating, or stable ion:

- The internal energy of *dissociating ions* is above the threshold for the first dissociation channel. Because their lifetimes are very short, they dissociate in the ion source.

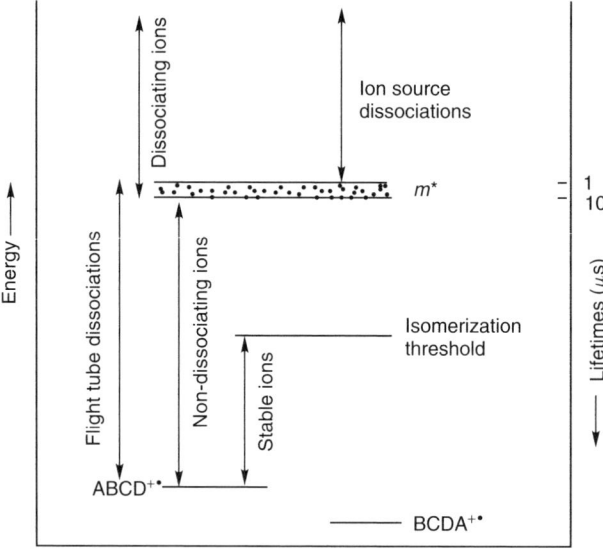

Figure 6.19. Classification of ions on the basis of their energy. The m^* are metastable ions.

- Ions whose internal energy is below the first dissociation channel are called *nondissociating ions*; they arrive at the detector intact. They, however, may or may not retain the structure of the ions formed initially.
- *Stable ions* are nondissociating ions that have energies below the threshold for isomerization. They retain the structure of the initially formed ions and are a true representation of the structure of the neutral precursor.

Unequivocal identification of the structure of gas-phase ions is not a simple task, owing primarily to the possibility that the ion formed initially may not retain the structure of its neutral precursor. As a consequence, the mass spectrum may not be a true representation of the ions formed initially. In many cases, ions undergo isomerization to a more stable structure before fragmentation can occur. In several examples in the literature, the solution-phase stability is reversed for gas-phase ions. Ketones are more stable in solution than are their enol tautomers, but the reverse is true for gas-phase radical cations.

To this end, the specialized techniques described below have been developed to determine the structure of gas-phase ions [36]. In most of these methods, the measurements are compared against a reference compound, with the premise that if the unknown ion and the ion of a known structure give identical mass spectral data, the two ions possess the same structure. The reader is, however, reminded that ions sampled by these methods have different lifetimes and energy content. Therefore, the conclusions drawn from different methods may not be identical, but rather, depend on the type of ions sampled, their energy content, and the isomerization threshold.

Determination of Heat of Formation From measurements of the adiabatic IE and appearance energy (AE), the heat of formation (ΔH_f) of an ion can be calculated [37]. This method provides data on the ions that are formed at threshold. Therefore, the structure deduced may be directly related to the neutral precursor. From the ΔH_f values, a potential energy diagram can be constructed to help understand the mechanism of isomerization of ions and their dissociation pathways. From the values of IE and AE measured, the ΔH_f of the molecular ion (say, $ABCD^{+\bullet}$) and its ionized product (say, AB^+) can be calculated using Eqs. (6.61) and (6.62), respectively:

$$\Delta H_f(ABCD^{+\bullet}) = IE(ABCD) + \Delta H_f(ABCD) \tag{6.61}$$

$$\Delta H_f(AB^+) = AE(AB^+) + \Delta H_f(ABCD) - \Delta H_f(^\bullet CD) - E_{exc} \tag{6.62}$$

where E_{exc} is the sum of the kinetic shift (E_{kin}) and reverse activation energy (E_r) and can be measured experimentally from the kinetic energy released in a metastable fragmentation. These relationships are depicted in Figure 6.20.

The ΔH_f values of the neutrals can be found in compilations of the experimental values [38] or can be calculated from the group additive principle [39]. IE and AE are measured with EI or photoionization techniques, and their values are readily available at the NIST Web site (address given in Section 6.4.6).

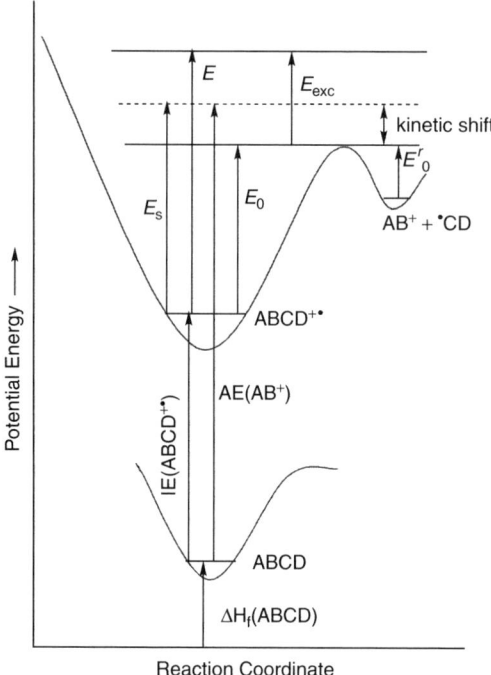

Figure 6.20. Thermochemistry of $ABCD^+$ and its fragment ions. (Redrawn from ref. 37.)

Metastable-Ion Spectra As shown in Figures 6.17 and 6.19, the metastable ions are characterized as having a narrow range of internal energies just above the threshold for the lowest dissociation. Fragmentation of these ions produces relatively broad peaks in the spectrum, owing to conversion of the internal energy of the dissociating ions into the translational energy of the products. The products result from the lowest-energy pathways. This method of ion structure determination is based on the concept that if the two ion structures in question exhibit the same fragmentation reactions and their respective products have the same abundance ratios, they have identical structures or are composed of a mixture having identical structures in approximately the same ratio [40]. The spectrum of the unknown is compared with a reference structure, and if these two spectra are nearly identical, one may conclude that the structures of the two ions are the same.

Kinetic-Energy Release (KER) Measurements The shape and width of the metastable peaks can also provide structure-specific information. The width of a peak provides a direct measure of the magnitude of the KER [37]. Metastable peaks usually have a Gaussian, flat-topped, or dish-topped shape. A Gaussian peak indicates a small energy release, whereas a large value of KER gives rise to a flat-topped or dish-topped peak. These measurements are usually made on magnet-sector instruments [37].

Isotope-Labeling Experiments Isotope labeling can establish which atoms are involved in the dissociation of an ion. In conjunction with metastable ion and KER measurements, isotope labeling is a useful procedure to pinpoint the mechanism of an ion-fragmentation pathway. A complication can arise, however, owing to the randomization or scrambling of the isotope label. Even in such cases, fruitful conclusions can be drawn. For example, complete scrambling indicates that participating sites are equivalent. Isotope labeling is also helpful to resolve ambiguities in ion–molecule reactions.

Collision-Induced Dissociation (CID) Spectra This technique has contributed a large volume of data on the structure determination of gas-phase ions and deals with nondissociating ions [41]. These high-translational energy ions collide with neutral gas atoms to undergo collision activation and subsequent dissociation (see Section 4.3.1). The CID spectrum is believed to be characteristic of the ions formed initially, except in cases where the isomerization barrier is very low. In such an event, the CID spectra will be representative of the mixture of interconverting ions. When the isomerization barrier is low, ions can be generated having energies below that barrier using a low-energy charge-exchange process [42]. Often, a reference ion (of known structure) is studied, and if its product-ion spectrum is nearly identical to that of the unknown, one may conclude that the structure(s) of the ions is(are) the same [43]. High-energy collisions are preferred in this exercise because they are less dependent on internal energy.

Charge-Permutation Reactions Several examples are known where charge-permutation reactions (see Section 4.4), especially charge-stripping reactions, have provided isomer-specific data [44]. Sampled ions have energies close to the ground state. Therefore, this technique yields distinct spectra even when the isomerization barrier between the two interconverting ions is very low. The population of the ions that are sampled is, however, small; thus, the acquisition and averaging of several scans are necessary.

Ion–Molecule Reactions In this approach, the structure of the target ion is determined by reacting it with a neutral molecule, and the products of the reaction are compared with those that result in a similar reaction with an ion of known structure. Reactions are conducted normally in the ion cyclotron resonance cell at a low pressure (ca. 10^{-7} torr) or in a quadrupole ion trap at a moderate pressure (ca. 10^{-3} torr). Sampled ions have a lifetime in the millisecond range and thus have survived fragmentation. Because those ion–molecule reactions that are detectable in mass spectrometry are usually exothermic, the intermediate adduct is rarely observed. Therefore, the structural features of the target ion or its adduct are derived from either the fragmentation pattern or from isotope-labeling experiments.

Ion–molecule reactions have also been investigated at high pressures in the CI source of a tandem-sector mass spectrometer [43,44]. In this technique it is feasible to "freeze out" the ion–molecule adduct by collisional stabilization with a neutral bath gas and to study the adduct by the usual CID approach.

Field Ionization Kinetics (FIK) Although limited to select research laboratories, FIK can also reveal the mechanism of ion dissociation [45]. This technique studies ions with lifetimes in the range 10^{-11} to 10^{-9} s. Ions are generated under the influence of very high potential by field ionization (see Section 2.6). The data are used to obtain the relation of the rate constant function $k(t)$ to the reaction time t. In combination with isotope labeling, FIK has been successful in identifying the fragmentation mechanism of a variety of gas-phase ions.

Ab initio Calculations The structure of gas-phase ions can also be determined using molecular orbital calculations [46]. The ab initio calculations permit an estimation of the ΔH_f values and equilibrium geometries of ions, from which the potential energy diagrams can be constructed. These diagrams reveal the mechanistic pathways for isomerization and decomposition reactions and estimates of the energy barriers involved in these reactions.

OVERVIEW

Mass spectrometry has an important role in the structure elucidation and identification of organic compounds. This task is accomplished by measuring the molecular mass and the elemental composition of the compound, and knowing the compound-specific fragment ions from the EI mass spectrum. The elemental composition can be determined from the isotopic pattern of the molecular ion or from its accurate molecular mass. The feat of determining the elemental compositions from a low-resolution mass spectrum is like puzzle solving, and hence satisfying when successful. This feat is accomplished from knowledge of the relative abundances of satellite peaks (i.e., M, M + 1, M + 2, etc.; Section 6.3).

Systematic steps for interpretation of the spectrum of an unknown compound are enumerated in Section 6.4.5. It requires an understanding of how to calculate the elemental composition and the ring + double-bond values (Section 6.4.4), of the nitrogen rule (Section 6.4.3), and of the fragmentation rules (Section 6.5). First, the molecular ion is recognized from the high-mass ion cluster. Next, the elemental composition and ring + double-bond value are calculated. Characteristic losses of neutral species from the molecular ion and unique structure-specific fragment ions are recognized. With this knowledge, the structure of the target compound is identified. Once the structure is assigned, the mass spectrum is rationalized to ensure that the fragment ions are consistent with the structure proposed and with the rules of fragmentation.

Odd-electron ions fragment via reasonably well understood fragmentation processes. Fragmentation reactions are classified either as simple cleavage reactions or as rearrangement reactions. In the former, direct cleavage of bonds occurs. Typical examples are sigma-bond cleavage, radical site–initiated fragmentation, and charge site–initiated fragmentation. Sigma-bond cleavage, the most common fragmentation encountered in saturated alkanes, involves cleavage of the C–C single bond. In radical site–initiated fragmentation, there is cleavage of a single σ-bond remote from the site of ionization. This reaction is a common occurrence in a variety of heteroatom- and unsaturated functional group–containing compounds such as ethers, alcohols, ketones, esters, amines, and alkenes. Charge site–initiated fragmentation, favored in halogen-containing compounds, is caused by the inductive effect of the charge site and the bond cleaved is the one adjacent to the charge-bearing atom.

In rearrangement reactions, the target ion first isomerizes to an intermediate structure prior to cleavage. The most frequently observed rearrangement reaction, the McLafferty rearrangement, occurs in ketones, aldehydes, carboxylic acids, esters, amides, and olefins. It involves the transfer of a γ-hydrogen to an unsaturated functional group, followed by cleavage of the α, β-bond in the intermediate structure and the expulsion of an alkene. The $OE^{+\bullet}$ ions thus formed are diagnostic of a compound undergoing McLafferty rearrangement. Other common rearrangement reactions are hydrogen rearrangement to a saturated heteroatom (e.g., oxygen, halogen), *ortho* rearrangement (in benzene ring–containing aromatic compounds), and double-hydrogen rearrangement, each provide ionic product diagnostic of the target compound.

Cycloreversion reactions, especially the retro-Diels–Alder reaction, also yield diagnostic product ions. Fragmentation reactions of each class of compounds (e.g., alkanes, ketones, alcohols, esters) are also discussed (Section 6.6). Upon EI, each class of molecules fragments in a typical fashion that can identify the precursor structure.

A discussion on the unimolecular dissociation of gas-phase species in terms of two statistical theories, the quasi-equilibrium theory and the Rice–Ramsperger–Kassel–Marcus theory, is also included. The formation of ionic products is related to the internal energy of the decomposing ion and the rate constant function of the fragmentation channel. At lower internal energies, the product of the rearrangement reaction dominates the mass spectrum, whereas at higher internal energies, the direct-cleavage reaction has a more favorable rate constant, and its dissociation product is seen prominently in the spectrum.

Isomerization of ionic structures is frequently observed. Several mass spectrometry techniques have been developed to identify the structure of the ion formed initially or its isomerized form. These methods are based on determination of the heat of formation, study of metastable ion spectra and CID spectra, experiments involving kinetic-energy-release measurements, isotope labeling, charge-permutation reactions, ion–molecule reactions, field ionization kinetics, and ab initio calculations.

EXERCISES

6.1. The molecular mass of benzoic acid was measured as to 122.0343 u. Find the mass measurement error in ppm.

6.2. The molecular mass of naphthalene was measured with an error of 2 mmu. What would this mass measurement error be in ppm?

6.3. The accurate mass of an unknown organic compound was determined by peak matching against the CF_3^+ ion of PFK. The accelerating voltage was reduced by 93.099% of the value required to focus CF_3^+ ion. Which one of the following is the correct elemental composition of the unknown: $C_4H_{10}O$, $C_4H_6D_2O$, C_5H_{14}, or $C_3H_{10}N_2$?

6.4. Calculate the isotopic pattern that would be expected in the molecular ion region for (a) p-chlorobenzophenone ($C_{13}H_9OCl$), (b) tetramethylsilane [$(CH_3)_4Si$], and (c) methionine (C_5H_9NOS).

6.5. Write possible isotopic compositions for ions in the molecular ion cluster for o-bromobenzophenone ($C_{13}H_9OBr$).

6.6. Calculate the isotopic pattern in the molecular ion region for a compound that contains two chlorine and one bromine atoms.

6.7. Calculate the elemental composition from the following spectral data:

Ion Type	m/z	Relative Abundance
M	115	100
M + 1	116	7.1
M + 2	117	0.42

6.8. Determine the elemental composition from the following spectral data:

Ion Type	m/z	Relative Abundance
M	192	100
M + 1	193	12.3
M + 2	194	1.31
M + 3	195	0.01

6.9. Predict whether the following structures would produce an odd- or an even-electron ion: (a) C_3H_8S, (b) C_7H_7NO, and (c) C_5H_{12}.

6.10. A compound contains nine carbons, and its molecular ion is observed at m/z 134. How many nitrogens, if any, are there in its structure?

6.11. Calculate the R + DB values for the compounds (a) $C_{13}H_9OCl$, (b) C_3H_7COOH, and (c) $C_3H_7CONH_3$, and decide whether any of these formulas could represent a molecular ion.

6.12. Where will the σ-bond cleavage preferentially occur in 2-methylbutane?

6.13. Predict the dominant α-cleavage product in the mass spectrum of (a) $HO-CH_2-CH_2-Cl$ and (b) $C_6H_4-CH_2-CH_2-NH_2$.

6.14. Show the ionic product of the retro-Diels–Alder reaction in 3-chlorocyclohexene.

6.15. Predict the order of relative abundances of the α-cleavage products in the EI mass spectrum of 3-methyl-3-hexanol.

6.16. Predict the ions that would be expected in the spectrum of 3-hexanone (i.e., ethyl propyl ketone; $C_2H_5COC_3H_7$). Also, show the mechanism of their formation.

6.17. Predict the ions that would be expected from the McLaffery rearrangement in the EI mass spectrum of (**a**) *n*-butyramide ($C_3H_7CONH_2$), (**b**) ethyl heptanoate ($C_6H_{13}COOC_2H_5$), and (**c**) 2−*n*-propylpyridine.

6.18. Rationalize the formation of the following ions through a possible mechanism: (a) *m/z* 30 in primary amines (e.g., in $C_2H_5NH_2$), (b) *m/z* 92 from butylbenzene [$C_6H_5-(CH_2)_3CH_3$], and (c) *m/z* 58 from butyl methyl ketone ($CH_3-CO-C_4H_9$)

6.19. Explain the formation of *m/z* 90 from *o*-methylphenol ($CH_3-C_6H_4OH$) and *m/z* 118 from 2-methylmethylbenzoate ($CH_3-C_6H_4COOCH_3$).

6.20. Two ions in the spectrum of cyclopentanone are *m/z* 55 and 56. Write reasonable mechanisms for their formation.

6.21. Distinguish the following pairs of isomeric compounds: (**a**) 4-phenylcyclohaxene and 3-phenylcyclohaxene, and (**b**) (1-methyl-2-propyl) benzene and (2-methylpropyl) benzene. Justify your answers through possible fragmentations.

6.22. Explain with a reasonable mechanism the presence of *m/z* 57 and 75 in the spectrum of the propyl ester of propionic acid ($C_2H_5COOC_3H_7$).

6.23. The ion of *m/z* 56 is the base peak in the spectrum of cyclohexylamine. Postulate the mechanism for its formation.

6.24. In the spectrum of 1-chloropropane, *m/z* 42 is the base peak, whereas the relative abundance of this ion is only about 20% in the spectrum of 2-chloropropane. Instead, *m/z* 43 is the base peak in this spectrum. Provide the rational for this difference.

6.25. Determine the structure of the unknown from the EI mass spectrum of Figure 6.21.

6.26. Identify the unknown compound whose EI mass spectrum is shown in Figure 6.22.

6.27. Postulate the structure of the unknown shown in Figure 6.23.

6.28. Provide a characteristic EI mass spectral feature that will distinguish 3-methyl-2-butanone and 2-pentanone.

6.29. Postulate the product of the double-hydrogen rearrangement in octyl acetate.

6.30. Provide the geneses of *m/z* 105 (100%), 122 (27%), and 123 (47%) in the EI mass spectrum of propyl benzoate.

6.31. What are the basic requirements for the McLafferty rearrangement?

258 ORGANIC MASS SPECTROMETRY

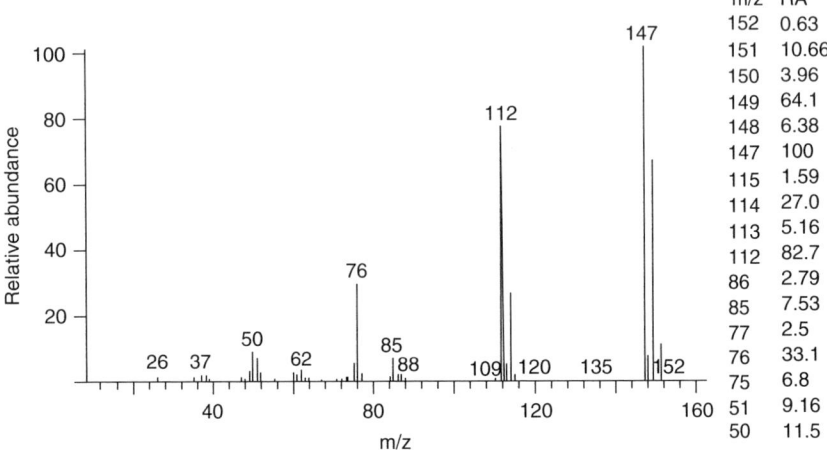

Figure 6.21. EI mass spectrum of an unknown of Exercise 6.25.

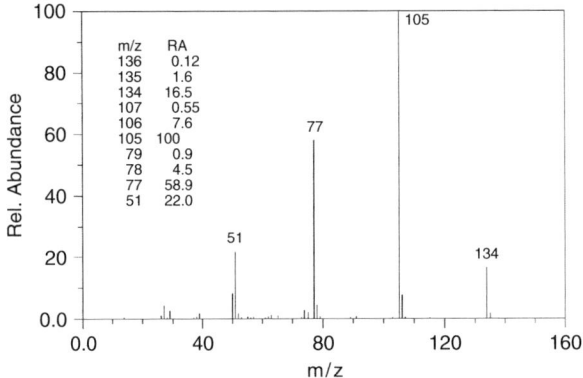

Figure 6.22. EI mass spectrum of an unknown of Exercise 6.26.

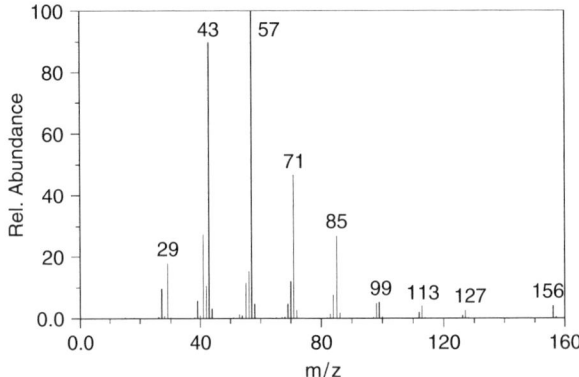

Figure 6.23. EI mass spectrum of an unknown of Exercise 6.27.

ADDITIONAL READING

1. F. W. McLafferty and F. Tureček, *Interpretation of Mass Spectra*, 4th ed., University Science Books, Sausalito, CA, 1993.
2. T. A. Lee, *A Beginner's Guide to Mass Spectral Interpretation*, Wiley, New York, 1998.
3. R. M. Smith, *Understanding Mass Spectra: A Basic Approach*, Wiley-Interscience, New York, 1999.
4. N. M. M. Nibbering, vol. ed., M. L. Gross and R. M. Caprioli, ser. eds., *The Encyclopedia of Mass Spectrometry*, Vol. 4, *Fundamentals of and Applications to Organic (and Organometallic Compounds)*, Elsevier, Amsterdam, The Netherlands, 2005.

REFERENCES

1. M. L. Gross, Accurate masses for structure confirmation, *J. Am Soc. Mass Spectrom.* **5**, 57 (1994).
2. M. H. Amad and R. S. Houk, High-resolution mass spectrometry with a multiple pass quadrupole mass analyzer, *Anal. Chem.* **70**, 4885–4889 (1998).
3. J. D. Williams, K. A. Cox, R. G. Cooks, and J. C. Schwartz, High mass-resolution using a quadrupole ion-trap mass spectrometer, *Rapid Commun. Mass Spectrom.* **5**, 327–329 (1991).
4. F. A. Londry, G. J. Wells, and R. E. March, Enhanced mass resolution in a quadrupole ion trap, *Rapid Commun. Mass Spectrom.* **7**, 43–45 (1993).
5. C. G. Hammar, G. Pettersson, and P. T. Carpenter, Computerized mass fragmetography and peak matching, *Biomed. Mass Spectrom.* **1**, 397–411 (1974).
6. J. Wu, S. T. Fannin, M. A. Franklin, T. F. Molinski, and C. B. Lebrilla, Exact mass determination for elemental analysis of ions produced by matrix-assisted laser desorption, *Anal. Chem.* **67**, 3788–3792 (1995).
7. R. D. Burton, K. P. Matuszak, C. H. Watson, and J. R. Eyler, Exact mass measurements using a 7 tesla Fourier transform ion cyclotron resonance mass spectrometer in a good laboratory practices–regulated environment, *J. Am Soc. Mass Spectrom.* **10**, 1291–1297 (1999).
8. N. Huang, M. M. Siegal, G. H. Kruppa, and F. H. Laukien, Automation of a Fourier transform ion cyclotron resonance mass spectrometer for acquisition, analysis, and e-mailing of high-resolution exact-mass electrospray ionization mass spectral data, *J. Am Soc. Mass Spectrom.* **10**, 1166–1173 (1999).
9. T. Storm, C. Hartig, T. Reemtsma, and M. Jekel, Exact mass measurements on-line with high-performance liquid chromatography on a quadrupole mass spectrometer, *Anal. Chem.* **73**, 589–595 (2001).
10. D. H. Russell and R. D. Edmondson, High-resolution mass spectrometry and accurate mass measurements with emphasis on the characterization of peptides and proteins by matrix-assisted laser desorption/ionization time-of-flight mass spectrometry, *J. Mass Spectrom.* **32**, 263–276 (1997).
11. C. C. Vera, R. A. Zubarev, H. Ehring, P. Hakansson, and B. U. R. Sundquvist, A three-point calibration procedure for matrix-assisted laser desorption/ionization mass spectrometry utilizing multiply charged ions and their mean initial velocities, *Rapid Commun. Mass Spectrom.* **10**, 1429–1432 (1996).

12. T. Fukai, J. Kuroda, and T. Nomura, Accurate mass measurement of low molecular weight compounds by matrix-assisted laser desorption/ionization time-of-flight mass spectrometry, *J. Am Soc. Mass Spectrom.* **11**, 458–463 (2000).
13. J. Gobom, M. Mueller, et al., A calibration method that simplifies and improves accurate determination of peptide molecular masses by MALDI-TOF MS, *Anal. Chem.* **74**, 3915–3923 (2002).
14. J. H. Beynon and A. E. Williams, *Mass and Abundance Tables for Use in Mass Spectrometry*, Elsevier, Amsterdam, The Netherlands, 1963.
15. C. S. Hsu, Diophantine approach to isotopic abundance calculations, *Anal. Chem.* **56**, 1356–1361 (1984).
16. A. L. Rockwood and S. L. Van Orden, Ultrahigh-speed calculation of isotope distributions, *Anal. Chem.* **68**, 2027–2030 (1996).
17. http://www.sisweb.com/mstools.htm.
18. http://www.shef.ac.uk/~chem/chemuter/isotopes.html.
19. S. Hammerum, Distonic radical cations in gaseous and condensed phase, *Mass Spectrom. Rev.* **7**, 123–202 (1988).
20. M. Karni and A. Mandelbaum, The "even-electron rule," *Org. Mass Spectrom.* **15**, 53–64 (1980).
21. J. T. Watson and O. D. Sparkman, Rules for mass spectral interpretation: The standard interpretation procedure, in N. M. M. Nibbering, vol. ed., M. L. Gross and R. M. Caprioli, ser. eds., *The Encyclopedia of Mass Spectrometry*, Vol. 4, *Fundamentals of and Applications to Organic (and Organometallic Compounds)*, Elsevier, Amsterdam, The Netherlands, 2005, pp. 719–730.
22. D. P. Stevenson, Ionization and dissociation by electronic impact: ionization potentials and energies of formation of *sec*-propyl and *ter*-butyl radicals—some limitations of the method, *Disc. Faraday Soc.*, 35–45 (1951).
23. M. L. Gross and D. Giblin, Stevenson's rule, in N. M. M. Nibbering, vol. ed., M. L. Gross and R. M. Caprioli, ser. eds., *The Encyclopedia of Mass Spectrometry*, Vol. 4, *Fundamentals of and Applications to Organic (and Organometallic Compounds)*, Elsevier, Amsterdam, The Netherlands, 2005, pp. 59–64.
24. C. Cheng and M. L. Gross, Applications and mechanism of charge-remote fragmentation: applications and mechanism, *Mass Spectrom. Rev.* **19**, 398–420 (2000).
25. M. L. Gross, Charge-remote fragmentation: applications and mechanism, in N. M. M. Nibbering, vol. ed., M. L. Gross and R. M. Caprioli, ser. eds., *The Encyclopedia of Mass Spectrometry*, Vol. 4, *Fundamentals of and Applications to Organic (and Organometallic Compounds)*, Elsevier, Amsterdam, The Netherlands, 2005, pp. 361–370.
26. D. G. I. Kingston, J. T. Bursey, and M. M. Bursey, Intramolecular hydrogen transfer in mass spectra II. McLafferty rearrangement and related reactions, *Chem. Rev.* **2**, 215–242 (1974).
27. C. Dass, Cycloreversion reactions, in N. M. M. Nibbering, vol. ed., M. L. Gross and R. M. Caprioli, ser. eds., *The Encyclopedia of Mass Spectrometry*, Vol. 4, *Fundamentals of and Applications to Organic (and Organometallic Compounds)*, Elsevier, Amsterdam, The Netherlands, 2005, pp. 378–385.
28. F. Tureček and V. Hanus, Retro-Diels–Alder reactions in mass spectrometry, *Mass Spectrom. Rev.* **3**, 85–152 (1984).

29. H.-E. Audier, Alkylamine radical cations, in N. M. M. Nibbering, vol. ed., M. L. Gross and R. M. Caprioli, ser. eds., *The Encyclopedia of Mass Spectrometry*, Vol. 4, *Fundamentals of and Applications to Organic (and Organometallic Compounds)*, Elsevier, Amsterdam, The Netherlands, 2005, pp. 126–134.
30. C. Dass and M. L. Gross, [2 + 1] Cycloaddition of ketene radical cation and ethylene, *J. Am. Chem. Soc.* **106**, 5775–5780 (1984).
31. T. Baer and W. L. Hase, *Unimolecular Reaction Dynamics: Theory and Experiments*, Oxford University Press, New York, 1996.
32. T. Baer and B. Sztaray, The statistical theory for unimolecular decay of organic and organometallic ions, in N. M. M. Nibbering, vol. ed., M. L. Gross and R. M. Caprioli, ser. eds., *The Encyclopedia of Mass Spectrometry*, Vol. 4, *Fundamentals of and Applications to Organic (and Organometallic Compounds)*, Elsevier, Amsterdam, The Netherlands, 2005, pp. 9–18.
33. H. M. Rosenstock, M. B. Wallenstein, A. L. Wahrhaftig, and H. Eyring, Absolute rate theory for isolated systems and mass spectra of the polyatomic molecules, *Proc. Natl. Acad. Sci. USA* **38**, 667–668 (1952).
34. R. A. Marcus and O. K. Rice, The kinetics of the recombination of methyl radicals and iodine atoms, *J. Phys. Colloid. Chem.* **55**, 894–908 (1951).
35. F. Tureček, Thermochemical and kinetic factors controlling fragmentation, in N. M. M. Nibbering, vol. ed., M. L. Gross and R. M. Caprioli, ser. eds., *The Encyclopedia of Mass Spectrometry*, Vol. 4, *Fundamentals of and Applications to Organic (and Organometallic Compounds)*, Elsevier, Amsterdam, The Netherlands, 2005, pp. 1–9.
36. J. L. Holmes, Structure and mechanism in organic ions, in N. M. M. Nibbering, vol. ed., M. L. Gross and R. M. Caprioli, ser. eds., *The Encyclopedia of Mass Spectrometry*, Vol. 4, *Fundamentals of and Applications to Organic (and Organometallic Compounds)*, Elsevier, Amsterdam, The Netherlands, 2005, pp. 287–297.
37. R. G. Cooks, J. H. Beynon, R. M. Caprioli, and G. R. Lester, *Metastable Ions*, Elsevier, Amsterdam, The Netherlands, 1973, pp. 1–296.
38. http://webbook.nist.gov/
39. S. W. Benson, *Thermochemical Kinetics*, Wiley, New York, 1976.
40. T. W. Shannon and F. W. McLafferty, Identification of gaseous organic ions by the use of metastable peaks, *J. Am. Chem. Soc.* **88,** 5021–5022 (1966).
41. W. F. Haddon and F. W. McLafferty, Metastable ion characteristics, VII: Collision-induced metastables, *J. Am. Chem. Soc.* **90**, 4745–4746 (1968).
42. C. Dass and M. L. Gross, Electrocyclic ring opening of 1-phenylcyclobutene and 3-phenylcyclobutene radical cations, *J. Am. Chem. Soc.* **105,** 5724–5729 (1984).
43. C. Dass, Dimerization of ketene radical cation in the gas phase, *Rapid Commun. Mass Spectrom.* **7**, 95–98 (1993).
44. C. Dass, D. A. Peake, and M. L. Gross, Structures of gas phase C_5H_8 radical cations: a collisional ionization study, *Org. Mass Spectrom.* **21**, 741–746 (1986).
45. N. M. M. Nibbering, Mechanistic studies by field ionization kinetics, *Mass Spectrom. Rev.* **3**, 445–477 (1984).
46. L. Radom, Chemistry by computer: a theoretical approach to gas phase ion chemistry, *Org. Mass Spectrom.* **26**, 359–373 (1990).

CHAPTER 7

INORGANIC MASS SPECTROMETRY

Besides its applications to organic and biochemical compounds, mass spectrometry is actively applied in the analysis of inorganic materials [1,2]. Several scientific disciplines, including materials science, nuclear science, agricultural science, cosmochemistry, environmental science, planetary science, geology, and meteorology, have benefited from developments in mass spectrometry. Many volatile inorganic materials can be analyzed by conventional electron ionization (EI)–MS. Less volatile materials are accessible to desorption ionization techniques, notably fast atom bombardment, secondary-ion mass spectrometry (SIMS), and laser desorption/ionization (LDI). We discussed these techniques in Chapter 2. A large number of inorganic materials are refractory in nature and require special ionization methods. This chapter is devoted specifically to those methods.

7.1. IONIZATION OF INORGANIC COMPOUNDS

Several methods are available to ionize inorganic materials, some for volatile compounds and others for nonvolatile and refractive materials:

- Electron ionization: used for volatile compounds (we discuss this method in detail in Section 2.3).
- Thermal ionization
- Spark-source ionization
- Glow discharge ionization
- Inductively coupled plasma ionization

Fundamentals of Contemporary Mass Spectrometry, by Chhabil Dass
Copyright © 2007 John Wiley & Sons, Inc.

- Secondary-ion mass spectrometry
- Resonance ionization

7.2. THERMAL IONIZATION MASS SPECTROMETRY

Thermal ionization mass spectrometry (TIMS) is used primarily for elemental analysis and is especially in demand for the precise and accurate measurement of isotopic ratios [1–3]. It is applicable to elements of low first ionization energy (IE) and to nonmetals with high electron affinity (EA). It works on the principle of thermal ion evaporation from a heated surface [3–5]. Ionization is a two-step process: evaporation of the sample from a heated metal filament and removal of an electron from the evaporated analyte. A typical thermal ionization (TI) source consists of two heated filaments; one is called the sample filament and the other the ionization filament. Both filaments are constructed from high-purity W, Pt, Ta, or Re. The sample solution is deposited on the sample filament, and is evaporated in the ion source vacuum by heating the filament. The evaporated species settles on the second filament, where ionization occurs by the loss of an electron. A double-filament assembly TI source is shown in Figure 7.1. The single- and triple-filament ion source assemblies are also popular. In a single-filament ion source, conversion of an analyte to the gas phase and subsequent ionization take place on the same filament. In the triple-filament ion source assembly, two sample filaments are used for comparative analysis of two different samples. Multifilament systems have the advantage that evaporation and ionization processes can be controlled independently to improve ionization efficiency. For example, an intense and longer-lasting sample beam can be generated by heating the sample filament at a lower temperature and the ionization filament at a higher temperature

The basis of thermal ionization is that when a neutral analyte species approaches a hot metallic surface, its Fermi levels are nearly equal to those of the metal filament, with the consequence that an electron can tunnel from the analyte

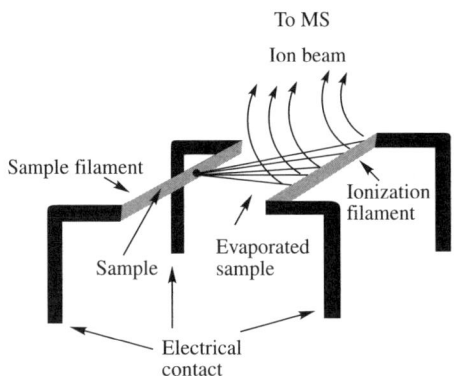

Figure 7.1. Double-filament assembly thermal ionization source.

species to the filament. The result is the formation of positive ions of the analyte. Negative ions are formed by the reverse process (i.e., the transfer of an electron from the metal filament to the analyte). The ionization efficiency E^+ for positive ions is expressed by the Saha–Langmuir equation,

$$E^+ = \frac{N^+}{N_0} = A \, \exp\left(\frac{\phi - I}{kT}\right) \quad (7.1)$$

where N^+ and N_0 are the numbers of positive ions and neutral atoms that are ejected from the filament, respectively; A is a constant; ϕ (in eV) is the work function of the filament, I (in eV) is the first ionization energy (IE) of the element being analyzed; k is the Boltzmann constant; and T (in kelvin) is the absolute temperature of the filament. Equation. (7.2) is the corresponding expression that describes the efficiency of negative-ion formation, where N^- is the number of negative ions, B is a constant, and EA is the electron affinity of the ionized species (in eV):

$$E^- = \frac{N^-}{N_0} = B \, \exp\left(\frac{\text{EA} - \phi}{kT}\right) \quad (7.2)$$

It is obvious from Eq. (7.1) that the ionization efficiency is higher for elements with low IE values, such as alkali metals. The ion production is also higher at higher values of the filament work function and heating temperature. Negative ionization is employed for nonmetals, metalloids, and transition metal oxides. Equation. (7.2) implies that for efficient ion formation in the negative-ion mode, filaments with a low work function must be used. Thus, TI can be made a highly selective process by a proper choice of certain operating parameters. Also, ions formed by this process have a low energy distribution (0.1 to 0.2 eV). Therefore, TI can be optimally coupled to a low-resolution mass spectrometer such as a single-sector magnetic field instrument. One major shortcoming of this technique is that multielement analysis is not easy to perform.

7.3. SPARK-SOURCE MASS SPECTROMETRY

Spark-source ionization, developed by Dempster [6], has long served as a useful device for the simultaneous analysis of a wide range of elements [7]. It is applicable primarily to solid-state samples. The ions are formed in a high-intensity spark plasma that can be generated between two electrically conducting electrodes. Two types of ion sources are known, the low-voltage direct current (dc) arc source and the high-voltage radio-frequency (rf) spark source; the latter version is more popular with commercial instruments. In this source (shown in Figure 7.2), an rf voltage is applied in pulses to create a spark across a narrow gap between the two pin-shaped electrodes. Typically, depending on the nature of the sample, a series of rf pulses of 20 to 100 ms duration, each

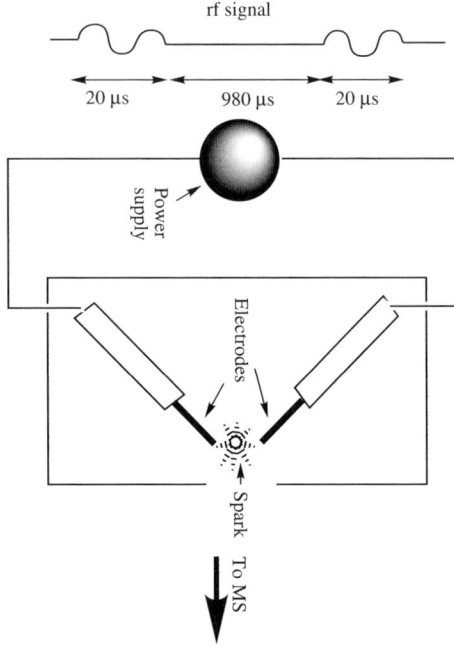

Figure 7.2. Schematic of a radio-frequency spark source.

consisting of 1 MHz rf voltage of 10 to 100 kV, are used. The sample material undergoes evaporation, atomization, and ionization by the EI process in the spark plasma to form positive or negative ions. Electrically conducting or low-resistance semiconductor samples are mounted as one of the electrodes; nonconducting samples (e.g., many oxides, ceramics) are mixed in powder form with high-purity graphite powder. A cylindrical electrode is cast from this mixture with a mold and hydraulic press. Alternatively, the mixed powder can be placed in the cup-shaped electrode of a conducting material.

Because the ions produced have a wide energy distribution (2 to 3 keV), the spark source best works with high-resolution mass analyzers such as the Mattauch–Herzog geometry double-focusing instrument (see Section 3.3.3), to provide a reasonable mass resolution for multielement analysis. A photographic plate detects all of the separated ion beams simultaneously at the focal plane of this double-focusing instrument. This arrangement forms the basis of simultaneous analysis of multielement samples at trace levels by spark-source mass spectrometry. Instead of a photographic plate, multi-ion counting with individual separate channeltrons is also used for the simultaneous analysis of multielement samples [8]. Ions formed are mostly singly charged ions with some contribution from multiply charged ions. Matrix-related ions are also formed. A spark source is a useful device for trace-level detection of elements in a variety of matrices, provided that high accuracy is not a requirement.

7.4. GLOW DISCHARGE IONIZATION MASS SPECTROMETRY

Glow discharge ionization is another suitable mode of ionization for solid-state samples [1,2,9,10]. Glow discharge is a versatile ion source that performs the functions of both sample atomization and ionization. Atomization occurs by the impact of high-energy ions, and the atomized species are ionized in the plasma. A glow discharge is a type of plasma that consists of a partially ionized high-density gas with nearly equal concentrations of positive and negative ions. A plasma also contains a small population of excited atomized species, whose relaxation produces a low-intensity glow: hence the name glow discharge. A typical dc glow discharge ion source consists of two electrodes that are fitted inside a cell that is filled with an inert gas (usually, Ar) at a low pressure of 0.1 to 10 torr (see Figure 7.3). A potential difference of approximately 1 kV is applied between the electrodes to convert the buffer gas into positive ions (e.g., Ar^+) and electrons. The potential difference accelerates the Ar^+ ions toward the cathode, which is impregnated with the sample material. The impact of Ar^+ ions on the cathode sputters neutral sample atoms for their subsequent ionization in the self-sustaining plasma by one of two processes: electron ionization [Eq. (7.3)] or Penning ionization with metastable argon atoms (Ar^*) [Eq. (7.4)]:

$$M + e^- \longrightarrow M^+ + 2e^- \tag{7.3}$$

$$M + Ar^* \longrightarrow M^+ + Ar + e^- \tag{7.4}$$

The cathode is usually made of conducting samples in the form of a pin, disk, or hollow cylinder. The nonconducting samples are pelleted with high-purity conducting binders such as graphite or copper. Argon is invariably used as a glow discharge gas because of its high sputtering efficiency and a number of useful metastable states.

Some designs of glow discharge ion sources use pulsed dc glow discharge and rf glow discharge. The pulsed dc source, in which voltage and current are

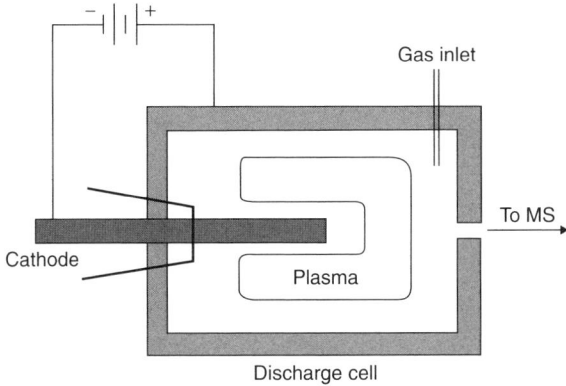

Figure 7.3. Schematic of a glow discharge cell.

applied in short-duration pulses, has the advantage of high yields of the sputtered sample atoms, owing to higher peak voltages and currents, and hence of high sensitivity of the analysis. The rf source has the advantage that it can directly analyze nonconducting samples.

Quadrupole and magnetic-sector instruments are the common type of mass analyzers for the detection of the GD-formed ions. The use of quadrupole ion traps, FT–ICR, and TOF instruments has also been explored [11–13]. Specifically, quadrupole ion traps and FT–ICR mass spectrometers are of particular interest because they can accumulate and store ions for a desired length of time for subsequent CID or ion–molecule reactions that will eliminate isobaric interferences. The fast-scan facility of a TOF mass analyzer has the advantage that it can be used to monitor even short-lived transient signals.

7.5. INDUCTIVELY COUPLED PLASMA MASS SPECTROMETRY

Inductively coupled plasma (ICP) *ionization* has currently assumed a more prominent role in the field of elemental and isotopic analysis [1,2,14]. It is applicable to solid-state as well as to solution-phase samples. A *plasma* is defined as a form of matter that contains a significant concentration of ions and electrons. The heart of this technique is a plasma torch, first developed as an efficient source for optical emission spectroscopy (OES) [15,16]. Multielement analysis with OES has, however, some serious shortcomings, such as complicated spectra, spectral interferences, high background levels, and inadequate detection of some rare-earth and heavy elements. The high ionization efficiency (>90%) of ICP for most elements is an attractive feature for its coupling to mass spectrometry.

7.5.1. Inductively Coupled Plasma Ion Source

Figure 7.4 shows a typical ICP torch that can be used as an ion source for mass spectrometry analysis [15,16]. It consists of an assembly of three concentric quartz tubes. The outermost tube, approximately 100 mm in length and 18 mm in i.d., is the plasma torch. The innermost tube (1.5 mm i.d) serves as a sample introduction capillary. The middle tube has a 14-mm i.d. Three different flows of argon gas are maintained separately through these three tubes. The first steam of argon flows at the rate of 12 to 15 L min^{-1} between the walls of the outer and middle tubes to cool the inner surface of the outer tube. An auxiliary flow of up to 1 L min^{-1} flows through the middle tube. Finally, a flow of 0.5 to 1 L min^{-1} is used as a sample carrier through the innermost tube. A water-cooled induction coil surrounds the top of the outer tube. The coil is powered by an rf generator of 0.5 to 2 kW and 27 MHz frequency. Initially, the plasma is seeded by a spark from a Tesla coil. The flux of electrons from the coil initiates ionization of the argon. The fluctuating magnetic field produced by the induction coil forces the ions and electrons to move in closed annular paths. Collisions of these ions with the neutral species that are encountered in their flight path create

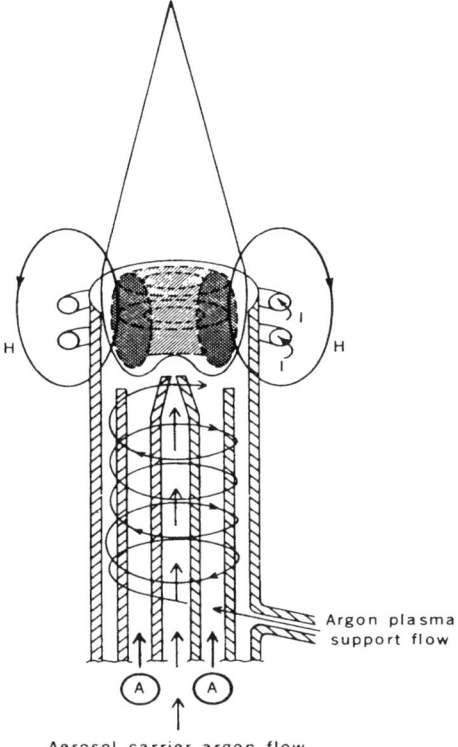

Figure 7.4. Schematic of a typical ICP torch. (Reproduced from ref. 15 by permission of the International Union of Pure and Applied Chemistry, Research Triangle Park, NC, copyright © 1977.)

resistance to their motion and generate Joule or ohmic heating. The heat is able to sustain the plasma. The temperature of the central part (10 mm beyond the rf coil) of the plasma approaches 8000 K, where efficient ionization of most elements occurs. The high temperature mandates thermal isolation of the outer quartz tube, achieved by a tangential flow of argon around the walls of the tube. The sample is carried by argon into the plasma, where the dissolved sample is vaporized, atomized, and ionized. The ionized species are sampled by a mass spectrometer for their m/z measurement.

7.5.2. Coupling an ICP Source with Mass Spectrometry

The ICP source operates at atmospheric pressure, whereas a mass spectrometer requires a pressure of $<10^{-6}$ torr. Therefore, the coupling of an extremely hot atmospheric pressure ICP torch with a mass spectrometer that operates at high vacuum is a challenge. A suitable interface is required to overcome this challenge. ICP–MS coupling was first attempted in the early 1980s [17,18]. Conceptually,

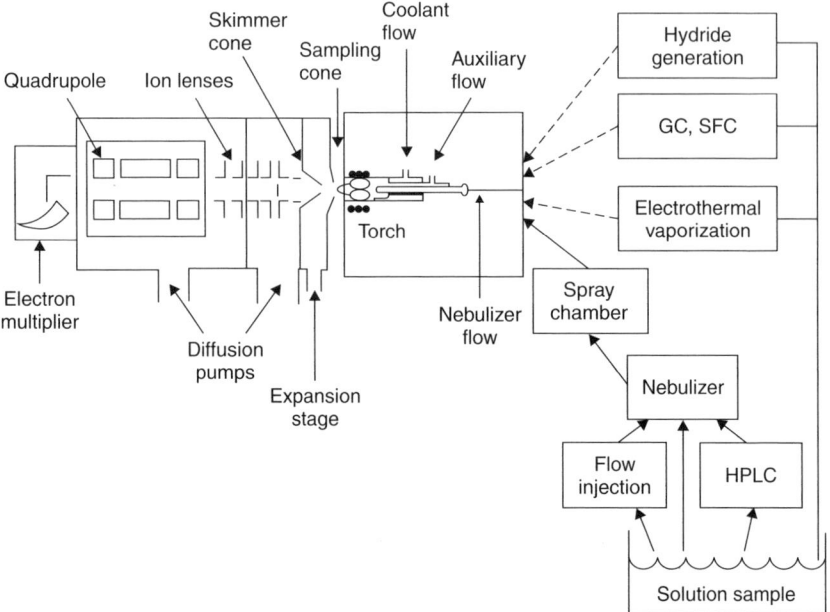

Figure 7.5. Schematic of an ICP–MS system. Various gaseous (dashed lines) and liquid (solid lines) sample introduction devices are also shown. (Reproduced from ref. 19 by permission of the American Chemical Society, Washington, DC, copyright © 1993.)

the interface used for this purpose is similar to that employed in the coupling of APCI and ESI ion sources with mass spectrometry (see Sections 2.15 and 2.17). As shown in Figure 7.5, the interface consists of differentially pumped chambers that are separated by two metal cones [19]. The fist cone is a water-cooled sampling cone with an orifice of 1 mm, and the second, called a skimmer cone, has a still smaller aperture. The region between the two cones is maintained at a pressure of 1 torr, where the hot plasma transmitted through the sampling cone undergoes rapid expansion. A fraction of this gas (rich in sample ions) is sampled by the skimmer cone into a region maintained at still lower pressure. After passing through a lens assembly, the ions are accelerated into the mass analyzer for m/z analysis. Although magnetic-sector, FT–ICR, and TOF instruments have been coupled with an ICP source, a quadrupole remains the preferred type of mass spectrometer for this combination because of its obvious ability to readily accommodate atmospheric-pressure devices.

7.5.3. Sample Introduction Systems for an ICP Source

Various sample introduction devices are available for ICP–MS analysis; some are suitable for solution-phase samples and others for solid samples. The most common device for solution-phase samples is a pneumatic nebulizer. Two popular designs are the concentric tube type and the cross-flow type. In the *concentric tube design*, the argon carrier gas flows along a concentric sample capillary into which the sample solution is introduced with a peristaltic pump. The high-velocity argon gas dissociates the emerging sample liquid flow into fine droplets which are carried into the plasma for atomization and ionization. In the *cross-flow design*, carrier gas flows at right angles across the tip of the sample capillary to convert the sample solution into a mist of fine droplets. *An ultrasonic nebulizer* can also be employed to carry the sample into the ICP plasma. In this device, the sample solution is pumped over a vibrating piezoelectric quartz crystal. Rapid oscillations of the quartz crystal break the liquid into a fine mist. With this arrangement, a more dense and homogeneous aerosol is produced. Other sample introduction techniques used for special applications are electrothermal vaporization, laser ablation, spark and arc ablation, and hydride generation [2]. An electrothermal vaporizer is an enclosed chamber, in which sample is placed (liquid or solid) on a carbon rod or tantalum filament and heated by an electric current. The evaporated sample is swept by a flow of argon. Hydride generation is a special technique used for toxic elements such as As, Sb, Sn, Se, Bi, and Pb. The acidified solutions of those elements are treated with $NaBH_4$ in a flask to produce volatile hydrides, which are swept by argon into the ICP plasma. Gas chromatography, ion chromatography, HPLC, and capillary electrophoresis have also been coupled to introduce samples into ICP plasma [2].

7.5.4. Spectral Interferences

Detection limits for some elements are severely compromised by isobaric spectral interferences which might arise either from the analyte itself or due to isotopes of other elements that are present in the sample or from polyatomic ions, oxides, and hydroxides that are generated from the plasma gas, the matrix, the solvent used in dissolving the sample. A few of the spectral interferences are listed in Table 7.1. Following are some of the ways to reduce these spectral interferences:

1. *Use of isotopes that are free of isobaric overlap.* The isobaric elemental interferences can be eliminated by choosing as the analyte a less abundant isotope of the target element. For example, $^{40}Ar^+$ interferes in the analysis of calcium because of its overlap with $^{40}Ca^+$. This interference can be eliminated when measurements are done with the second-most abundant isotope, $^{44}Ca^+$.
2. *Use of a cool plasma.* Several polyatomic species (e.g., $^{40}Ar^{2+}$, $^{38}ArH^+$, $^{16}O_2^+$, $^{16}OH^+$, and $^{14}N^{16}OH^+$) are formed in the hot plasma, which is generated at 1000 to 1400 W rf power and a nebulizing gas flow of 0.8 to 1.0 L/min. However, by generating the plasma at 500 to 800 W rf power and a nebulizing gas flow of 1.5 to 1.8 L/min, the argon- and matrix-based polyatomic species that are prevalent in the hot plasma can be minimized, and significantly improved detection limits are realized [20]. One major shortcoming of the cool plasma is that not enough energy is available in

TABLE 7.1. Common Isobaric Spectral Interferences in ICP–MS

Analyte	m/z	Interference
$^{24}Mg^+$	24	$^{12}C^{12}C^+$
$^{28}Si^+$	28	$^{14}N^{14}N^+$, $^{12}C^{16}O^+$
$^{31}P^+$	31	$^{14}N^{16}OH^+$
$^{32}S^+$	32	$^{16}O^{16}O^+$
$^{39}K^+$	39	$^{38}ArH^+$, $^{37}ClH_2^+$
$^{40}Ca^+$	40	$^{40}Ar^+$
$^{48}Ti^+$	48	$^{16}O_3^+$, $^{14}N^{16}O^{18}O^+$, $^{32}S^{16}$, $^{29}S^{19}F^+$
$^{51}V^+$	51	$^{35}Cl^{16}O^+$, $^{34}S^{16}OH^+$
$^{52}Cr^+$	52	$^{40}Ar^{12}C^+$, $^{35}Cl^{16}OH^+$, $^{34}S^{18}O^+$
$^{56}Fe^+$	56	$^{40}Ar^{16}O^+$, $^{40}Ca^{16}O^+$
$^{58}Ni^+$	58	$^{42}Ca^{16}O^+$
$^{59}Co^+$	59	$^{43}Ca^{16}O^+$, $^{42}Ca^{16}OH^+$
$^{63}Cu^+$	63	$^{31}P^{16}O_2^+$, $^{46}Ca^{16}OH^+$
$^{64}Zn^+$	64	$^{32}S^{16}O_2^+$, $^{31}P^{16}O_2H^+$, $^{32}S_2^+$, $^{48}Ca^{16}O^+$
$^{69}Ga^+$	69	$^{37}Cl^{16}O_2^+$
$^{74}Ge^+$	74	$^{37}Cl^{37}Cl^+$
$^{75}As^+$	75	$^{40}Ar^{35}Cl^+$, $^{40}Ar^{19}F^{16}O^+$
$^{80}Se^+$	80	$^{40}Ar^{40}Ar^+$

the plasma to atomize those elements that form strong bonds with oxygen (e.g., vanadium oxide and hydroxide) or fluorine. Also, elements with a high ionization energy cannot be ionized in the cool plasma.

3. *Use of a collision cell.* Another novel approach to eliminating isobaric spectral interferences is to use a collision cell [21]. A collision cell approach is similar to the fragmentation scheme used in tandem mass spectrometry. An rf-only multipole collision cell is positioned between the ion extraction optics and the mass analyzer and is filled with hydrogen or helium as a collision gas. Polyatomic interfering species, such as $^{40}Ar^+$, $^{40}Ar^{16}O^+$, and $^{38}ArH^+$, are converted to neutral species or noninterfering ionic species, and thus are shifted away from the m/z of the analyte.

4. *Use of a dynamic reaction cell (DRC).* In the DRC approach (Figure 7.6), a wide-bandpass quadrupole filter is used as a reaction cell [22]. A highly reactive gas (e.g., NH_3) is used to convert the interfering species into either neutral species or ionic species with masses that differ from the analyte. These ionic species can be filtered out by adjusting the electrical fields of the cell. As an example, in the analysis of $^{56}Fe^+$, the isobaric species $^{40}Ar^{16}O^+$ interferes. The ion–molecule reaction of ammonia transforms the $^{40}Ar^{16}O^+$ into noninterfering argon and oxygen atoms:

$$ArO^+ + NH_3 \longrightarrow Ar + O + NH_3^+ \qquad (7.5)$$

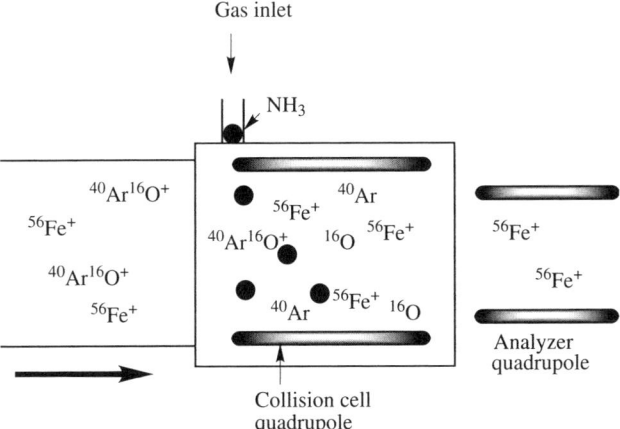

Figure 7.6. Schematic of a dynamic reaction cell for removal of interferences in an ICP-MS analysis. An example of the removal of $^{40}Ar^{16}O^+$ interference is shown here.

This approach has been highly successful in the ultratrace analysis of materials used in the semiconductor industry [21].

7.5.5. Laser Ablation–ICP–MS

Current applications of LA–ICP–MS are on the rise [1,2,23,24]. A desirable feature of this technique is its applicability to solid samples; conducting and nonconducting solids can both be analyzed without wet chemical sample preparation, thus avoiding any sample contamination during time-consuming dilution and digestion steps. The principle of LA–ICP–MS is illustrated in Figure 7.7. In essence, the sample is bombarded with a pulsed, high-energy, focused laser beam in an inert-gas atmosphere to create a plume of the sample particles. Almost all types of laser sources have been used to ablate solid samples. The ablation chamber is flushed with an inert carrier gas to transport the sample plume to the ICP torch, where the sample particles are vaporized, atomized, and ionized in the plasma. The ions are transported to the mass spectrometry system for their detection. LA–ICP–MS has been used to determine the composition of refractory ceramic materials, the trace element composition of inorganic materials, the fingerprinting of archaeological and forensic objects, and the origin of the masterpiece paintings. Other applications include analysis of meteorites, minerals, optical materials, cement, bones, and gemstones.

7.6. RESONANCE IONIZATION MASS SPECTROMETRY

Resonance ionization (RI) is a selective ionization approach that has become widely recognized for elemental analysis with the availability of tunable lasers

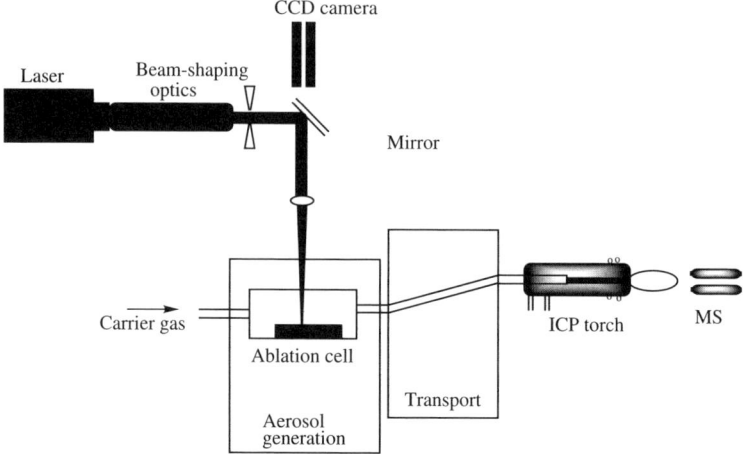

Figure 7.7. Instrumental configuration of a laser ablation–ICP–MS system. Various gaseous (dotted lines) and liquid (solid lines) sample introduction devices are also shown. (Redrawn from ref. 23.)

[25–28]. Almost every element in the periodic table can be selectively ionized with nearly 100% efficiency by this technique. The basic principle of RI–MS was explained in Chapter 2 (Section 2.5) for molecular species. The same principle applies to atomic species. In short, a photon of a well-defined wavelength from a tunable laser is absorbed by a gas-phase atom to excite an electron from its ground state to a higher-energy excited state. Under normal circumstances, the excited electron returns to the ground state by emission of a photon of characteristic frequency or by collisional deactivation. There is, however, a probability that this excited electron is pushed to higher energy levels, ultimately to be ejected from the atom if one or more photons can interact with the excited atom. The positive ions thus formed are accelerated into a mass spectrometer and their m/z is measured.

Several excitation and ionization schemes of atoms are available [25]. As shown in Figure 7.8, these schemes include the use of one-color photons (i.e., those of the same frequency, ω_1) or of two-color photons (i.e., those of different frequencies, ω_1 and ω_2). Schemes 1 and 2 employ photons of the same frequency, whereas in schemes 3, 4, and 5, photons of different frequencies are used. With these five schemes, all of the elements except He, F, Ne, and probably Ar can be detected by RI–MS. This technique has the potential of one-atom detection [25]. The main strength of RI–MS is its ability to detect one element selectively in the presence of a large number of other elements. Isobaric interferences that are often encountered in some of the other techniques discussed in this chapter are eliminated. With the facility of high selectivity, the multielement capability is, however, lost.

One basic requirement for the RI–MS technique is that the sample be converted into the gas phase. One such approach is thermal atomization, in which the

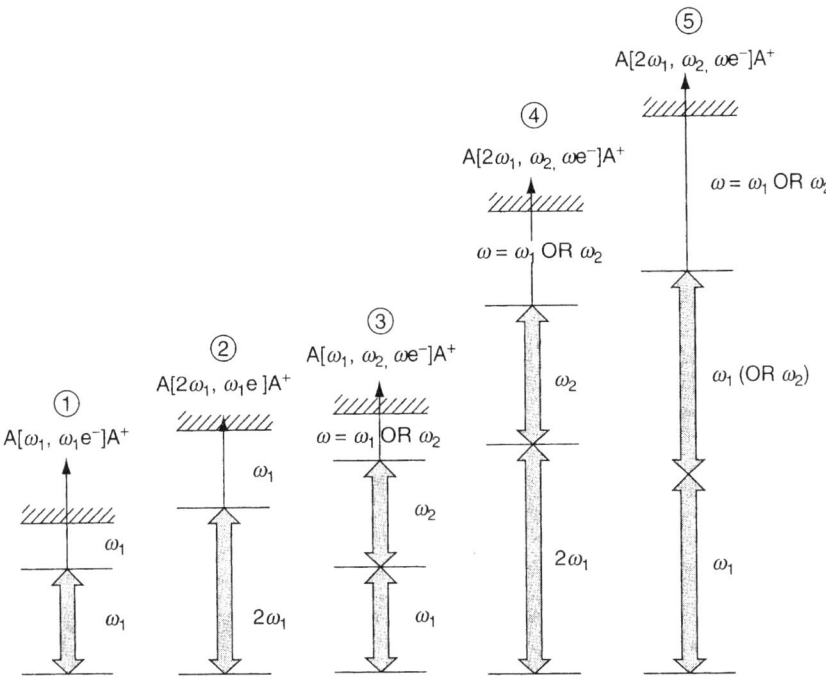

Figure 7.8. Classification of various resonance ionization schemes for the ionization of atomic species. ω_1 and ω_2 represent photons of frequency 1 and frequency 2, respectively, and $2\omega_1$, frequency-doubled photons. (Reproduced from ref. 25 by permission of the American Physical Society, College Park, MD, copyright © 1979.)

chemically isolated analyte is deposited electrolytically on a high-purity rhenium filament, with atomization occurring by heating the filament to >1000°C. Other common atomization techniques, such as laser and spark ablation, glow discharge, graphite furnace, and ion bombardment, have also been combined with RI–MS. Magnetic-sector, quadrupole, and TOF mass analyzers have all been coupled successfully with RI.

7.7. ISOTOPE RATIO MASS SPECTROMETRY

Isotope ratio mass spectrometry (IR–MS) is a special branch of mass spectrometry in which abundance ratios of stable isotopes of elements are determined with high accuracy [Additional Reading 5]. The rates at which heavier isotopes participate in chemical and physical processes are slightly different from those for lighter isotopes. The difference in rates leads to a subtle variation in the natural abundances of isotopes, owing to a variety of fractionation processes. The isotope ratio measurements can be highly informative regarding a range of natural processes. Information about precise isotope ratios is very important in the nuclear, geological, agricultural, environmental, and health industries.

Isotope ratios in a sample are measured and expressed relative to internationally accepted standard samples of accurately known isotope ratios. For example, PDB (Pee Dee Belemnite; $CaCO_3$) is used as a standard for carbon; its $^{13}C/^{12}C$ ratio is 0.0112372 [29]. Similarly, Vienna SMOW (Standard Mean Ocean Water) is the standard for oxygen and hydrogen, and air is the standard for nitrogen. Following are the commonly accepted definitions for reporting isotope ratios:

- *Parts per thousand or per mil (‰)*: the isotope ratio of a heavier isotope to a lighter isotope is expressed as parts per thousand difference from the appropriate standard, denoted by the symbol δ:

$$\delta = \left(\frac{R_{\text{sample}}}{R_{\text{standard}}} - 1\right) \times 1000 \tag{7.6}$$

where R stands for the isotope abundance ratio (e.g., $^{13}C/^{12}C$).

- *Atom percent* (atom %): the percent abundance of the heavier isotope in a sample is expressed as

$$\text{atom}\% = \frac{n_a}{\sum n_i} \times 100 \tag{7.7}$$

where n_a refers to the abundance of the relevant isotope and n_i represents all stable isotopes.

- *Atom % excess*: the atom % of the heavier isotope in a sample in excess of its natural abundance in a standard.

▶ **Example 7.1** $(D/H)_{\text{SMOW}} = 155.76 \times 10^{-6}$. If D/H for a water sample was found to be 150.00×10^{-6}, express this ratio in terms of the δ(D) value.

Solution

$$\delta = \left(\frac{R_{\text{sample}}}{R_{\text{standard}}} - 1\right) \times 1000 = \left(\frac{150.00 \times 10^{-6}}{155.76 \times 10^{-6}} - 1\right) \times 1000 = -36.98$$

▶ **Example 7.2** The PDB standard is no longer available. It has been replaced by new carbonate reference materials (e.g., NBS-18, NBS-19, NBS-20, and NBS-21). The $\delta(^{13}C)$ value of the NBS-19 is reported as +1.95 against PDB. Calculate the $^{13}C/^{12}C$ ratio for NBS-19.

Solution By rearranging Eq. (7.6) in terms of R_{sample}, we get

$$R_{\text{NBS}-19} = \left(\frac{\delta}{1000} + 1\right) R_{\text{PDB}} = \left(\frac{1.95}{1000} + 1\right)(0.0112372) = 0.0112591$$

7.7.1. Isotope Ratio MS Systems

Isotope ratio mass spectrometers are highly specialized instruments that can provide a precision of seven significant figures. All of the ionization techniques discussed above can be used for isotope ratio measurements. Solid samples are ionized conveniently by thermal ionization or LA–ICP. Gaseous samples are analyzed by electron ionization. The determination of isotope ratios of $^2H/^1H$, $^{13}C/^{12}C$, N^{15}/N^{14}, $^{18}O/^{16}O$, and $^{34}S/^{32}S$ generally requires gas-phase samples. For these ratio measurements, the original samples are first converted into an appropriate elementary gas such as H_2, CO_2, N_2, or SO_2. The two most common forms of sample introduction devices for those gaseous samples are dual viscous flow and continuous-flow inlet systems. The *dual viscous flow inlet system* uses two capillaries, one for the sample and the other for the standard [30]. The capillaries have a crimp just prior to the entry into the ion source to maintain a viscous flow of the gases into the source. The dimensions of the capillary are chosen such that fractionation of the sample does not occur (i.e., the isotopic composition is not altered during transport). Typically, only 10% of the total sample is consumed. Also, this arrangement allows long measurements times at stable conditions to yield high precision. In a *continuous-flow inlet system*, a carrier gas flows continuously into the ion source, and the sample is injected into this flow [29,31–33]. The entire sample is consumed in a relatively short analysis time to provide signal enhancement. The precision of the analysis is, however, lower than that obtained in the dual viscous flow design. Gas chromatography (GC) can be coupled through the continuous-flow inlet system [29,31]. This online coupling can be used to measure $^2H/^1H$, $^{13}C/^{12}C$, N^{15}/N^{14}, and $^{18}O/^{16}O$ isotopic ratios in organic compounds. Combustion and reduction furnaces are also placed between GC and MS. The combustion furnace converts organic compounds to CO_2, H_2O, and a variety of nitrogen oxides. The reduction furnace helps to convert nitrogen oxides to N_2. For isotopic analysis in nonvolatile and thermally labile compounds, LC can be coupled to an isotope ratio mass spectrometer [32].

Special mass spectrometry systems are built for isotope ratio measurements. Most isotope ratio mass spectrometers consist of a single-focusing magnetic sector instrument and a multiple Faraday cup detection system. Because the Faraday cup exhibits a stable response, it is an ideal detector for isotope ratio measurements. Simultaneous collection of relevant ion beams from all isotopes provides high-precision isotopic measurements. A three-Faraday cup detection system is shown in Figure 7.9. The multicup assembly is placed at the focal plane of the mass analyzer and can be used for the simultaneous detection of each isotopic form of the analyte species (e.g., m/z 44, 45, and 46 from CO_2). Commercial instruments with up to nine collectors are available.

7.7.2. Applications of Isotope Ratio MS

One of the important applications of IRMS is to measure the natural variations of isotopes. These variations are used as tracers of natural processes in geology, hydrology, glaciology, cosmochemistry, and atmospheric chemistry [29]. A study

278 INORGANIC MASS SPECTROMETRY

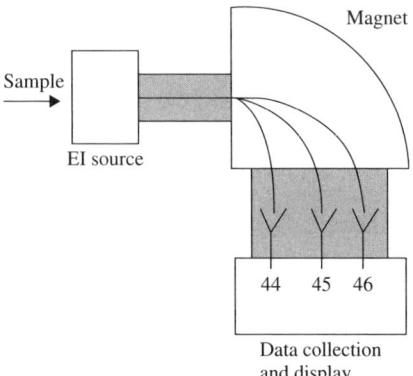

Figure 7.9. Isotope ratio mass spectrometer with a three-cup detection system.

of fossils can shed light on how life has evolved on Earth. By measuring the bulk isotope composition, it is possible to detect food adulteration. Investigations of archaeological artifacts, global climatic changes, organic composition of meteorites, athletic doping, and metabolism of pharmaceutical drugs are some of the other applications of IR–MS. The nuclear industry has benefited extensively from IR–MS. IR–MS has been able to settle the debate as to whether dinosaurs were warm-blooded, high-metabolism homeotherms (animals that maintain their body temperature within a limited range, e.g., mammals and birds) or cold-blooded, low-metabolism heterotherms (animals whose body temperature varies with environment; e.g., the Kodo dragon). Through the oxygen isotopic composition of bone phosphate, it has been established that dinosaurs are homeothermic [34,35].

7.8. ACCELERATOR MASS SPECTROMETRY

Accelerator mass spectrometry is an ultrasensitive technique that can measure isotope ratios at the 10^{-15} level. The concept of AMS was proposed by Alvarez and Cornog in 1939 [36]. Its applications to the determination of isotope ratios of radioisotopes were stipulated independently by Muller [37] and Bennet et al. [38] in 1977. AMS has made a tremendous impact in radiocarbon dating. A normal mass spectrometer cannot be used for this purpose. First, the amount of the radionuclide ^{14}C in the samples is less than the detection limits of even the most sensitive instruments. Second, the interference of a large amount of isobaric ^{14}N that is always present in the mass spectrometry systems precludes the accurate quantitation of ^{14}C. Two novel concepts have been incorporated in AMS to overcome these shortcomings. First, ions are accelerated to MeV energies so that the sensitive detection of ultratrace amounts of radionuclides can be performed by high-energy ion counting with gas-filled ionization chamber detectors. The high-energy ion-counting system also reduces detector background. Second, the interference of ^{14}N and other isobaric species is eliminated through the generation

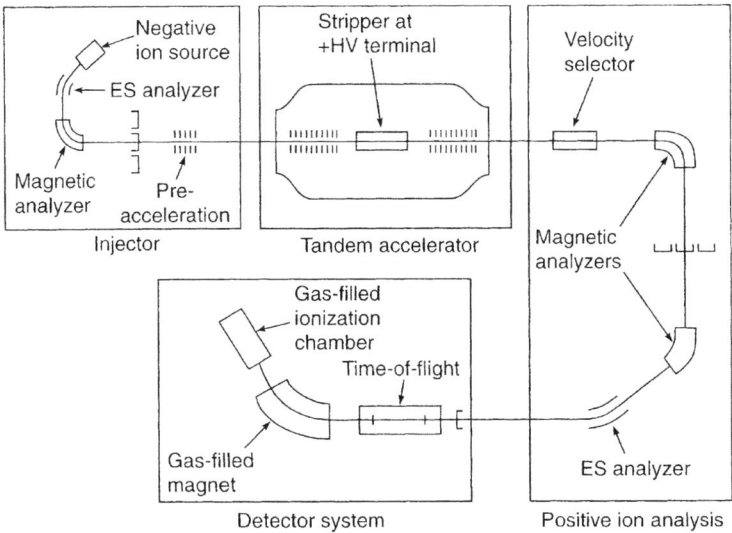

Figure 7.10. Schematic diagram of an AMS instrument. (Reproduced from ref. 39 by permission of the American Association for the Advancement of Science, Danvers, MA, copyright © 1987.)

of negative ions and subsequent high-energy acceleration; under these conditions, $^{14}N^-$ ions do not survive and fail to reach the detector.

A typical AMS instrument is shown in Figure 7.10. It consists of five main units: an ion source, an ion-extraction and acceleration system, an MeV accelerator, a mass analyzer, and an ion-detection system. In the ion source, negative ions are formed by bombarding the sample with a Cs^+ ion beam. The sputtered ions are extracted at 10 to 20 keV into a two-stage mass filter for the separation of isotopes. The mass-selected negative ions (e.g., $^{14}C^-$, $^{14}N^-$, $^{12}CH_2^-$, $^{13}CH^-$, $^7Li_2^-$, etc., all of mass 14 u) enter a dual-stage tandem accelerator; the first stage is used for negative-ion acceleration and the second stage for positive-ion acceleration. The first stage accelerates negative ions to MeV energies into a thin foil or a gas absorber stripper that is held at 2 to 10 MV positive potential. In this device the ions are stripped of several electrons, and are converted to multiply charged positive ions which undergo acceleration in the second part of the tandem accelerator. The stripping processes are sufficiently energetic to dissociate most multiply charged molecular ions rapidly without affecting any atomic ions with a +3 or higher charge. These stable species enter a multisector magnetic analyzer, where they are mass-resolved from molecular fragments. The mass-resolved atomic ions finally enter the detection assembly and are detected by a gas-filled ionization chamber. The detection assembly also contains a TOF mass analyzer for additional positive-ion mass discrimination. The ion current of each stable isotope is measured by alternatively selecting them at a removable Faraday cup and of the radioisotope by the gas-filled detector to provide isotope ratios with an abundance sensitivity in the range 1×10^{-15}.

Most applications of AMS are in the measurement of rare radionuclides (e.g., ^{10}Be, ^{14}C, ^{26}Al, ^{36}Cl, ^{129}I) with half-lives from 5000 to 16×10^{-15} [39]. AMS has been applied in biochemical research to determine ^{14}C, ^{41}Ca, and ^{26}Al in the femtogram to gram range [40] and in environmental science to determine ^{10}Be, ^{14}C, ^{26}Al, and ^{129}I [41]. A much publicized application of AMS is the determination of the age of the Shroud of Turin, a piece of cloth believed to be used to wrap Christ's body after the Crucifixion [42]. It was concluded through experiments conducted by three independent laboratories that the shroud is not authentic and belongs to the medieval age (A. D. 1260–1390).

For the direct analysis of ^{14}C in biological species, an interface is available [43]. The liquid samples are deposited into a bed of CuO powder which is held on a refractory support in an enclosed chamber. The CuO matrix is heated locally by an infrared laser, and the CO_2 evolved is swept away by a flow of He into the ion source.

7.9. ISOTOPE DILUTION MASS SPECTROMETRY

Isotope dilution mass spectrometry (ID–MS) is widely accepted as a quantification procedure of proven accuracy in elemental analysis and isotope ratio measurements [4]. Several areas of research in nuclear science, geochronology, medicinal chemistry, environmental science, and agricultural science have benefited from this technique. ID–MS is applicable to all elements that have at least two stable isotopes. Monoisotopic elements can be analyzed only if they have a long-lived natural or artificial radioisotope. For example, iodine and thorium have been determined with spikes of the long-lived isotopes ^{129}I and ^{230}Th, respectively [44]. TI–MS and ICP–MS are the methods of choice for accurate ID–MS analysis. ICP–MS has the advantage that several elements can be analyzed simultaneously under the same experimental conditions. Other ionization techniques discussed in this chapter have also been coupled with ID-MS.

In principle, this procedure involves the addition to the sample of a known amount of a less abundant enriched isotope of the same element, called a spike. Both isotopes must be in the same chemical form. The mixture is isotopically equilibrated, and the altered ratio of the unknown number of the analyte atoms to the known number of the spike atoms is measured with mass spectrometry. Often, chemical processing and separation are required to provide a suitable form for mass spectrometric analysis and to eliminate any isobaric interferences.

The principle of ID–MS is depicted in Figure 7.11, a bar graph spectrum of the sample, which has two isotopes, to which a known amount of spike, the enriched lighter isotope, was added. The measured isotope ratio, R, is given by

$$R = \frac{N_x A_x + N_s A_s}{N_x B_x + N_s B_s} \tag{7.8}$$

where N_x and N_s are the number of atoms of the element in the unknown sample and spike, respectively; A_x and B_x are the isotope abundances in percent

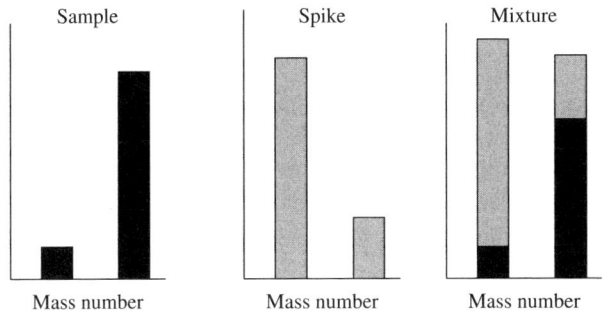

Figure 7.11. Bar graph spectra of the sample, spike, and mixture.

of isotopes A and B in the sample; A_s and B_s are the corresponding abundances in the spike. Solving the equation for N_x gives

$$N_x = N_s \frac{A_s - RB_s}{RB_x - A_x} \qquad (7.9)$$

From the known weights of the sample and spike, W_x and W_s, the concentration C_x of the element in the sample can be calculated from

$$C_x = \frac{C_s W_s}{W_x} \frac{A_s - RB_s}{RB_x - A_x} \qquad (7.10)$$

The detection limit of a linear calibration is defined as three times the standard uncertainty of the concentration of the blank. This definition is, however, not an ideal representation of the detection limit in ID–MS. A formulation for the detection limit for ID–MS is available [45], where the ID–MS detection limit is described as a function of the enrichment of the isotopic spike and of the uncertainties in the measurement of the spiked isotope. It states that when the spike is not enriched isotopically, the detection limit is infinite and unusable. When the spike is enriched in either isotope, the ID–MS measurement uncertainties approach the linear calibration detection limit.

OVERVIEW

An analysis of inorganic materials is required in several scientific disciplines, such as materials science, nuclear science, agriculture science, cosmochemistry, environmental science, planetary science, geology, and meteorology. In addition to EI, SIMS, FAB, and LDI, several specialized ionization techniques are available to analyze refractory inorganic materials: thermal ionization, spark-source ionization, glow discharge ionization, inductively coupled plasma ionization, and resonance ionization. Thermal ionization occurs when a sample is placed on a heated filament. In spark-source ionization, the ions are formed in high-intensity spark plasma struck between two electrically conducting electrodes. One of the electrodes contains, or is made of, the sample material. In glow discharge ionization, the sample atoms are sputtered by Ar^+ ion bombardment and are ionized by electrons and metastable species of the plasma. In ICP ionization, a plasma torch is used to ionize atomized analyte species. In this torch, the plasma is created by fast-moving Ar^+ ions and electrons in a fluctuating magnetic field produced by the rf-powered induction coil. The fast-moving ions collide with the neutral species to produce ohmic heating, which can maintain the plasma to temperatures as high as 8000 K required for efficient ionization of most elements. The sample is introduced into the plasma in solution form with a pneumatic nebulizer. For solid samples, laser ablation–ICP is the preferred approach, where the sample is bombarded with a pulsed high-energy focused laser beam in an inert-gas atmosphere. The plume of the sample particles is swept away to the ICP torch by Ar. Resonance ionization is a highly selective mode of ionization in which gas-phase atomized species are irradiated with photons of a well-defined wavelength from a tunable laser to eject an electron from the atom.

The precise measurement of isotope ratios has influenced many fields, including the nuclear, geological, agricultural, environmental, and healthcare industries. Isotope ratio measurements are performed with specially built mass spectrometry systems; the most common ones are made of a single-focusing magnetic-sector mass spectrometer and a multiple-Faraday-cup detection system. Very low isotope ratios are measured with accelerator mass spectrometry, which is especially applicable to the measurement of isotope ratios of radioisotopes. Such measurements are performed with specially built mass spectrometry systems that can accelerate ions to MeV energies. This field has made significant contributions to radiocarbon dating.

A widely accepted quantification procedure for accurate elemental analysis and isotope ratio measurements is isotope dilution mass spectrometry. In this procedure the sample is spiked with a known amount of a less-abundant enriched isotope of the same element, and the ratio of the unknown number of analyte atoms to the known number of the spike atoms is measured using mass spectrometry.

EXERCISES

7.1. Which situation will produce more positive-ion current in a thermal ionization source: ionization of sodium salt on (**a**) rhenium ($\phi = 4.98$ eV) filament heated to 2200°C or (**b**) tungsten ($\phi = 4.58$ eV) filament heated to 2100°C?

7.2. What procedures are followed to eliminate isobaric spectral interferences in ICP–MS?

7.3. How are glow discharge plasma and inductively coupled plasma generated?

7.4. What makes resonance ionization a highly selective ionization technique?

7.5. $(^{18}O/^{16}O)_{SMOW} = 2005.20 \times 10^{-6}$. If $(^{18}O/^{16}O)$ for a water sample were found to be 2025.00×10^{-6}, what would the $\delta(^{18}O)$ value for the water sample be?

7.6. The $\delta(^{18}O)$ value of a water sample is reported as -30.02 against the SMOW standard. Calculate the $^{18}O/^{16}O$ ratio for the water sample.

ADDITIONAL READING

1. S. K. Aggarwal and H. C. Jain, eds., *Introduction to Mass Spectrometry*, Indian Society for Mass Spectrometry, Trombay, India, 1997.
2. J. R. De Laeter, *Applications of Inorganic Mass Spectrometry*, Wiley-Interscience, New York, 2001.
3. C. M. Barshick, D. C. Douglas, and D. H. Smith, eds., *Inorganic Mass Spectrometry*, Marcel Dekker, New York, 2000.
4. F. Adams, R. Gijbels, and R. Van Grieken, eds., *Inorganic Mass Spectrometry*, Wiley, New York, 1988.
5. I. T. Platzner, *Modern Isotope Ratio Mass Spectrometry*, Wiley, New York, 1997.

REFERENCES

1. J. S. Becker and H.-J. Dietze, Inorganic trace analysis by mass spectrometry, *Spectrochim. Acta B*, 1475–1506 (1998).
2. J. S. Becker and H.-J. Dietze, Inorganic mass spectrometry methods for trace, ultratrace, isotope, and surface analysis, *Int. J. Mass Spectrom.* **197**, 1–35 (2000).
3. K. G. Heumann, S. Eisenhut, S. Gallus, E. H. Hebeda, R. Nusko, A. Vengosh, and T. Walczyk, Recent developments in thermal ionization mass spectrometry techniques for isotope analysis, *Analyst* **120**, 1291–1299 (1995).
4. K. G. Heumann, Isotope dilution mass spectrometry (IDMS) of the elements, *Mass Spectrom. Rev.* **11**, 41–67 (1992).
5. M. G. Inghram and W. A. Chupka, Surface ionization source using multiple filaments, *Rev. Sci. Instrum.* **24**, 518–520 (1953).
6. A. J. Dempster, Ion sources for mass spectroscopy, *Rev. Sci. Instrum.* **7**, 46–49 (1936).
7. R. Gijbels, Elemental analysis of high-purity solids by mass spectrometry, *Talanta* **37**, 363–373 (1990).
8. K. P. Jochum, Trace element analysis of geological samples by modern spark source mass spectrometry using multi-ion counting, *Spectrosc. Eur.* **9**, 22–27 (1997).
9. W. W. Harrison, C. M. Barshick, J. A. Klinger, P. H. Ratliff, and Y. Mei, Glow discharge techniques in analytical chemistry, *Anal. Chem.* **62**, 943A–949A (1990).
10. F. L. King and W. W. Harrison, Glow discharge mass spectrometry: an introduction to the technique and its utility, *Mass Spectrom. Rev.* **9**, 285–317 (1990).
11. D. C. Duckworth, C. M. Barshick, D. H. Smith, and S. A. McLuckey, Dynamic range extension in glow discharge quadrupole ion trap mass spectrometry, *Anal. Chem.* **66**, 92–98 (1994).

12. K. E. Milgram et al., High-resolution inductively coupled plasma fourier transform ion cyclotron resonance mass spectrometry, *Anal. Chem.* **69**, 3714–3721 (1997).
13. D. M. McClenathan, S. J. Ray, W. C. Wetzel, and G. M. Hieftje, Plasma source TOFMS, *Anal. Chem.* **76**, 159A–166A (2004).
14. D. Beauchemin, Inductively coupled plasma mass spectrometry, *Anal. Chem.* **78**, 4111–4136 (2006).
15. V. A. Fassel, Current and potential applications of ICP–AES in the exploration, mining and processing of materials, *Pure & Appld. Chem.* **49**, 1533–1545 (1977).
16. V. A. Fassel, Quantitative and elemental analysis by plasma emission spectroscopy, *Science* **202**, 185–191 (1978).
17. R. S. Houk, V. A. Fassel, G. D. Flesch, H. Svec, A. L. Gray, and C. E. Taylor, Inductively coupled argon plasma as an ion source for mass spectrometry determination of trace elements, *Anal. Chem.* **52**, 2283–2289 (1980).
18. A. L. Gray and A. R. Date, Inductively coupled plasma source mass spectrometry using continuous flow ion extraction, *Analyst* **108**, 1033–1050 (1983).
19. N. P. Vela, L. K. Olsen, and J. A. Caruso, Elemental speciation with plasma mass spectrometry, *Anal. Chem.* **65**, 585A–597A (1993).
20. K. Sakata and K. Kawabata, Reduction of fundamental polyatomic ions in inductively coupled plasma mass spectrometry, *Spectrochim. Acta* **49B**, 1027–1038 (1994).
21. T. J. Rowen and R. S. Houk, Attenuation of polyatomic ion interferences in inductively coupled plasma mass spectrometry by gas-phase collisions, *Appl. Spectrosc.* **43**, 976–980 (1989).
22. K. Kawabata and Y. Kishi, Dynamic reaction cell ICPMS for trace metal analysis of semiconductor materials, *Anal. Chem.* **75**, 423A–428A (2003).
23. B. Hattendorf, C. Latkoczy, and D. Günther, Laser ablation ICPMS, *Anal. Chem.* **75**, 341A–347A (2003).
24. D. B. Aeschliman, S. J. Bajic, J. Stanley, D. Baldwin, and R. S. Houk, Multivariate pattern matching of trace elements in solids by laser ablation inductively coupled plasma-mass spectrometry: source attribution and preliminary diagnosis of fractionation, *Anal. Chem.* **76**, 3119–3125 (2004).
25. G. S. Hurst, M. G. Pyne, S. D. Kramer, and J. P. Young, Resonance ionization spectrometry and one atom detection, *Rev. Mod. Phys.* **51**, 767–819 (1979).
26. G. S. Hurst and M. G. Pyne, *Principles of Resonance Ionization Spectroscopy*, Adam Hilger, Philadelphia, PA, 1988.
27. D. H. Smith, J. P. Young, and R. W. Shaw, Elemental resonance ionization mass spectrometry: a review, *Mass Spectrom. Rev.* **8**, 345–378 (1989).
28. G. S. Hurst, Counting the atoms: some applications in chemistry, *J. Chem. Educ.* **59**, 895–899 (1982).
29. W. A. Brand, High precision isotope ratio monitoring techniques in mass spectrometry, *J. Mass Spectrom.* **31**, 225–235 (1996).
30. C. R. McKinny, J. M. McCrea, S. Epstein, H. A. Allen, and H. C. Urey, Improvements in mass spectrometers for the measurement of small differences in isotope abundance ratios, *Rev. Sci. Instrum.* **21**, 724–730 (1950).
31. D. E. Mathews and J. M. Hayes, Isotope-ratio-monitoring gas chromatography–mass spectrometry, *Anal. Chem.* **50**, 1465–1473 (1978).

32. R. J. Caimi and J. T. Brenna, High-precision liquid chromatography–combustion isotope ratio mass spectrometry, *Anal. Chem.* **65**, 3497–3500 (1993)
33. R. J. Caimi and J. T. Brenna, High-sensitivity liquid chromatography–combustion isotope ratio mass spectrometry of fat-soluble vitamins, *J. Mass Spectrom.* **30**, 466–472 (1995).
34. R. E. Barrick and W. J. Showers, *Science* **265**, 222–224 (1994).
35. W. J. Showers, R. E. Barrick, and B. Genna, Isotopic analysis of dinosaurs bones, *Anal. Chem.* **74**, 142A–150A (2002).
36. L. W. Alvarez and R. Cornog, Helium and hydrogen of mass 3, *Phys. Rev.* **56**, 379 (1939).
37. R. A. Muller, Radiocarbon dating with a cyclotron, *Science* **196**, 489–494 (1977).
38. C. L. Bennet, R. P. Beukens, M. R. Glover, H. E. Gove, R. B. Liebert, A. E. Litherland, K. H. Purser, and W. E. Soundheim, Radiocarbon dating using electrostatic accelerators: negative ions provide the key, *Science* **198**, 508–510 (1977).
39. D. Elmore and F. M. Phillips, Accelerator mass spectrometry for measurement of long-lived radioisotopes, *Science* **236**, 543–550 (1987).
40. J. S. Vogel and K. W. Turteltaub, Accelerator mass spectrometry in biomedical research, *Nucl. Instrum. Methods B* **92**, 445–453 (1994).
41. J. Rucklidge, Accelerator mass spectrometry in environmental geoscience, *Analyst* **120**, 1283–1290 (1995).
42. P. E. Damon et al., Radiocarbon dating of the Shroud of Turin, *Science* **337**, 611–615 (1989).
43. R. G. Liberman, S. R. Tannenbaum, et al., An interface for direct analysis of ^{14}C in non-volatile samples by AMS, *Anal. Chem.* **76**, 328–334 (2004).
44. J. D. Fassett and P. J. Paulsen, Isotope dilution mass spectrometry for accurate elemental analysis, *Anal. Chem.* **61**, 643A–649A (1989).
45. L. L. Yu, J. D. Fassett, and W. F. Guthrie, Detection limit of isotope dilution mass spectrometry, *Anal. Chem.* **74**, 3887–3891 (2002).

PART III

BIOLOGICAL MASS SPECTROMETRY

CHAPTER 8

PROTEINS AND PEPTIDES: STRUCTURE DETERMINATION

Proteins are complex macromolecules: and arguably, the most important of all biological compounds. The word *protein* has its origin in the Greek word *proteios*, which means "of first importance." Proteins are the ultimate functional macromolecules in a cellular system and play crucial roles in virtually all biological processes. Just about every cell and organ of our body uses proteins to create the chemistry of life. The important functions of proteins include as enzymes, receptors, antibodies (for immune protection), and hormones, to control the expression of genes; as a means of transport and storage of small molecules; and many more. They are the major components of muscles, bones, skin, and hair.

Proteins and peptides are polymers of amino acids. There are 20 naturally occurring amino acids (see Figure 8.1a for the general formula of an amino acid). All amino acids contain a central carbon atom. A hydrogen atom, an amino group, a carboxylic acid group, and an R group are attached to this carbon atom. The R group imparts a distinct character to an amino acid. All naturally occurring amino acids except glycine contain a chiral carbon and exist in the L-form (Figure 8.1b). Figure 8.2 depicts the detailed structures of all of the 20 amino acids, together with their three-letter abbreviations and one-letter code. On the basis of a distinct R group, all amino acids are categorized into six distinct groups: whether they are hydrophilic or hydrophobic and whether they neutral, acidic, or basic.

In a protein or a peptide, a long chain of amino acids is held together in a head-to-tail arrangement; the α-carboxylic group of one amino acid is joined by the α-amino group of another amino acid via an amide linkage, also called the *peptide*

Fundamentals of Contemporary Mass Spectrometry, by Chhabil Dass
Copyright © 2007 John Wiley & Sons, Inc.

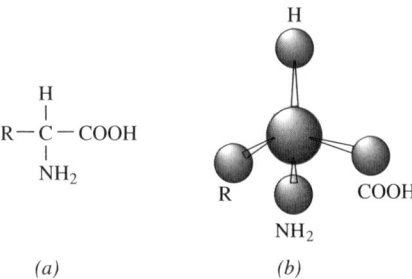

Figure 8.1. (*a*) General structure and (*b*) L-configuration of an amino acid.

bond. The amino acids in the peptide chain are called *residues*. The residue that has a free amino group is called the *N-terminus*, and a free α-carboxylic acid group is the *C-terminus*. By convention, the N-terminus is placed at the left side and the C-terminus at the right side of a peptide or protein. Although there is no rigid rule, a chain length longer than 30 to 40 amino acids is called a *protein*, whereas a *peptide* is composed of fewer than 30 to 40 amino acids.

The synthesis of proteins in the cell is controlled by our genes (DNA). Within the genes, this information is stored in the form of a sequence of three specific bases, known as a *codon*, that directs which amino acid is to be added next to the growing chain of a protein. The synthesis starts with the transcription of the genetic information from the protein gene to the mRNA, followed by the translation of this information into protein synthesis. This relation, along with newly coined "omics" terms, is shown in Figure 8.3. The nascent proteins are transported to various intracellular and extracellular locations. During their transport through the endoplasmic reticulum, the Golgi complex, and secretary vesicles, proteins undergo a variety of posttranslational modifications (PTMs) including acylation (acetyl, formyl, and myristyl), carboxylation, glycosylation, lipidation, amidation, oxidation, nitration, nitrosylation, phosphorylation, and sulfation (see Table 9.1 for more PTMs). Consensus sequences have been defined for many posttranslational modifications. Proteins also undergo proteolytic cleavage before they are rendered fully functional. In addition, a protein that participates in a reaction may change its molecular form as a result of a physiological condition.

8.1. STRUCTURE OF PROTEINS

Proteins are organized into the following four levels of structures [1]:

1. The *primary structure:* This structure is a fundamental property of a protein and describes the linear sequence of the constituent amino acids. The primary structure represents what is encoded directly by an organism's *genome* (the elaborate structure of DNA in a cell). In addition, this structure includes any covalent modifications of certain amino acid residues (e.g., glycosylation, phosphorylation, acylation, amidation, lipidation, and sulfation). Each class of protein in the body

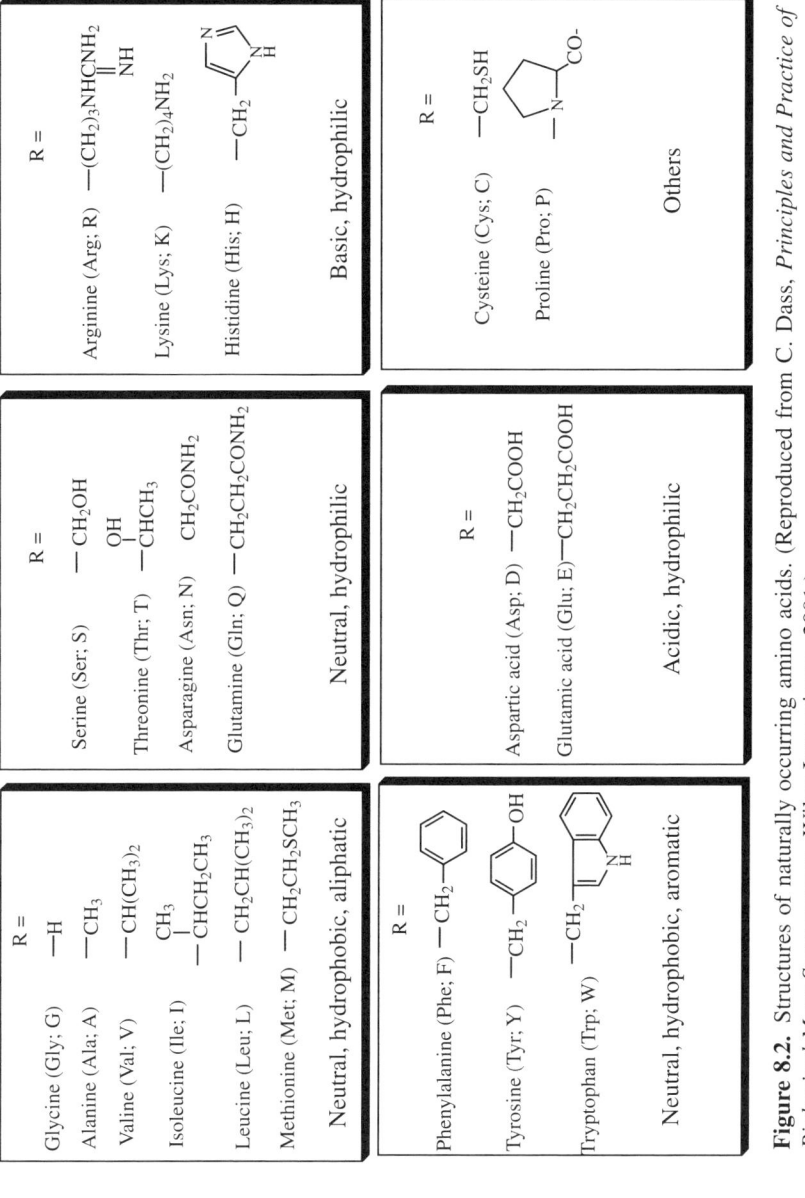

Figure 8.2. Structures of naturally occurring amino acids. (Reproduced from C. Dass, *Principles and Practice of Biological Mass Spectrometry*, Wiley-Interscience, 2001.)

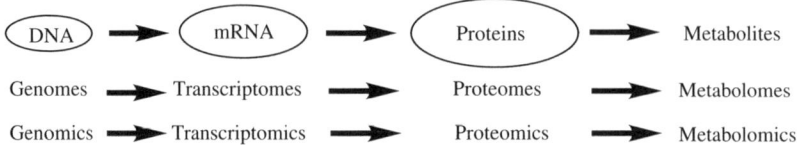

Figure 8.3. From DNA to proteins; a single DNA species is converted to many more proteins.

has its own unique amino acid sequence, which defines its biological function and chemical behavior. The peptide backbone is flexible because it can rotate about C_α–N and C_α–C bonds, with the angles of rotation being termed ϕ and ψ, respectively.

2. The *secondary structure:* represents the folding of short segments of proteins into defined structural patterns, such as α-helix, β-sheets, and turns (e.g., β-, γ-, and types I, II, and III). The orderly alignment of these protein chains is maintained as a result of hydrogen bonds between the backbone −C=O and H−N−groups; the helices are stabilized by intrachain hydrogen bonds, whereas the β-sheets are formed via hydrogen bonding between two or more polypeptide chains. A turn provides polypeptides the ability to bend and reorient to a required compact, globular structure. Proteins can also fold in the form of a random coil. In some proteins, all types of secondary structures are manifested.

3. The *tertiary structure:* describes how the secondary structural elements of a single protein chain interact with each other to fold into a three-dimensional conformation of protein molecules. Tertiary structures are stabilized via disulfide bridges, hydrogen bonding, salt bridges, and hydrophobic interactions.

4. The *quaternary structure:* determines how individual polypeptide chains (called *subunits*) interact to form a multimeric complex. Subunits are held together through polar and hydrophobic interactions.

Knowledge of the amino acid sequence of a protein is paramount to an understanding of biological events and the molecular basis of its biological activity, and to predicting its three-dimensional structure. The identification of the exact molecular form of a protein is also important from the viewpoint of human health. An aberration in a protein structure can lead to a particular disease. Furthermore, a disease can cause an aberration in a protein structure. In addition, the amino acid sequence of a protein is useful to unravel evolutionary events. All proteins from a common ancestor have greater sequence homology.

8.2. DETERMINATION OF THE SEQUENCE OF A PROTEIN

Traditionally, proteins have been identified by using Western blotting in combination with antibodies and photoaffinity labeling. Although rapid and sensitive, the use of antibodies, requires the ready availability of an extensive library of suitable antibodies. In addition, the cross-reactivity of antibodies with compounds of a similar sequence might create uncertainty. Similarly, photoaffinity labeling, followed by autoradiography, is also very sensitive but requires the use of hazardous radioactive labels, and its specificity is also questionable. Thus, subsequent sequencing of any provisionally identified proteins is obligatory for unequivocal characterization of the protein. The commonly applied techniques for generating sequence-specific information are:

1. *cDNA sequencing*. From the mRNA that codes for a protein, cDNA is made and cloned, from the sequence of which the corresponding amino acid sequence of the protein is derived. Although this procedure requires less effort and is more sensitive than the Edman procedure, errors may occur in a DNA-derived sequence because a protein expressed by recombinant DNA techniques is not necessarily identical to the native protein. Information about PTMs also remains elusive.

2. *Edman sequencing*. This procedure involves the sequential degradation of a protein from the N-terminus and the identification of the amino acids released one at a time by high-performance liquid chromatography (HPLC) [2,3]. Phenylisothiocyanate (PITC) is coupled to the N-terminus of the protein. The N-terminal amino acid is removed as a phenylthiohydantoin (PTH) derivative, which is identified by HPLC analysis. The degradation cycle is repeated continuously to identify subsequent amino acids. The Edman technique has several limitations. It is time consuming and less sensitive. The sensitivity decreases and interference increases with each step of the Edman cycle. Therefore, complete sequencing of polypeptides is seldom achieved. In addition, it is not applicable to peptides or proteins with a blocked N-terminus, and it does not allow direct identification of many PTMs.

3. *Mass spectrometric sequencing*. Mass spectrometry sequencing involves (a) digestion of proteins into smaller peptides, (b) characterization of peptides in the digest by mass measurement (MS analysis) or sequence ion analysis (MS/MS analysis), and (c) a computer protein database search [4,5]. It should be emphasized here that MS analysis and database searching leads to a probability of identity, not an absolute identification. Mass spectrometry–based methods have the advantage of short analysis times, high absolute sensitivity, straightforward sample preparation steps, and optimum specificity. Often, subfemtomole amounts will suffice to identify a protein. The following sections provide a detailed discussion of these three basic steps.

Mass spectrometry currently has assumed a central role in protein sequencing. This development has been possible with the introduction of two highly sensitive ionization techniques: electrospray ionization (ESI) and matrix-assisted laser desorption and ionization (MALDI) and the advent of improved instrumentation capable of high-mass and high-sensitivity detection. Currently, biopolymers with a molecular mass over 100,000 Da are analyzed routinely. In the past, fast atom bombardment (FAB) [6,7] and ^{252}Cf plasma desorption (PD) ionization [8] also played a limited role in protein sequencing. Mass spectrometry now has assumed an ever-increasing role in solving research problems in biological and medical sciences. The complete sequencing of peptides; determination of the molecular mass of intact proteins, peptides, and protein–protein complexes; verification of the primary structure of proteins predicted from the cDNA sequence; characterization of natural mutants; and identification of PTMs are now accomplished routinely by mass spectrometry techniques.

8.3. GENERAL PROTOCOL FOR AMINO ACID SEQUENCE DETERMINATION OF PROTEINS

Figure 8.4 illustrates a general protocol that can be tailored to identify a protein in a specific situation. The essential steps of this protocol are:

1. Homogenization of the biological sample (target tissue, biofluid, cell culture, or microorganism).
2. Affinity enrichment of the target proteins.
3. Separation of the target protein(s) using one- or two-dimensional sodium dodecyl sulfate (SDS)–polyacrylamide gel electrophoresis (PAGE) or HPLC techniques.
4. Measurement of the molecular mass of intact proteins.
5. Reduction–alkylation of proteins. The molecular mass measurement of the target protein before and after this step provides information on the number of cysteine residues and disulfide bonds.
6. Site-specific cleavage of reduced and alkylated proteins into smaller, more manageable peptides by chemical degradation or proteolysis.
7. Mass measurement of each cleaved peptide.
8. Sequence ion analysis of each cleaved peptide fragment by tandem mass spectrometry (MS/MS).

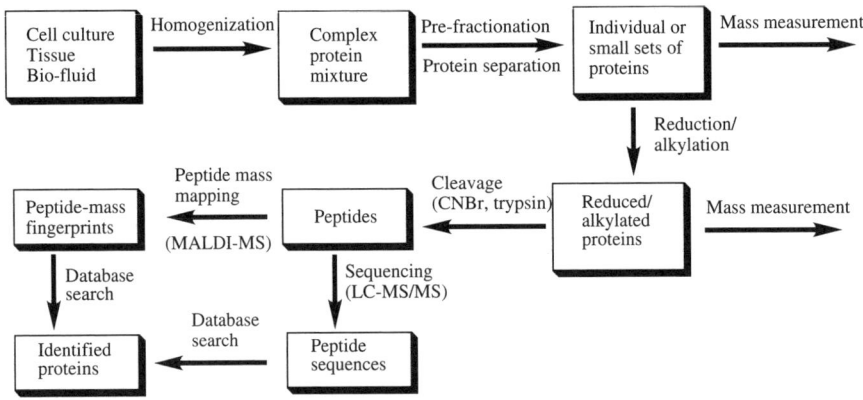

Figure 8.4. General protocol for the identification of proteins.

9. Identification of proteins by correlating the mass spectrometry data with entries in a protein or DNA database. Alternatively, the amino acid sequence of each cleaved peptide is determined by de novo methods.

8.3.1. Homogenization and Subcellular Fractionation

The process of protein identification begins with protein extraction and separation. Different biological samples, such as whole organ, tissue, fluid, or cell culture, are first processed to extract proteins. The tissue sample is first minced in the presence of a suitable lysis buffer, in which an appropriate denaturing agent, a reducing agent, and a cocktail of enzyme inhibitors are present. The minced tissue sample is homogenized in a blender.

Often, low-abundance proteins in a whole-cell lysate remain undetected. To enhance their detection, either preferential enrichment or subcellular fractionation into distinct organelles (e.g., nuclear membrane, mitochondria, endoplasmic reticulum, microsomes, and cytoplasm) is necessary [9]. This task is accomplished via differential centrifugation and density-gradient centrifugation by exploiting distinct physical properties, such as mass and density, of the organelles. The homogenized tissue sample is subjected to a series of centrifugation steps to yield subcellular fractions. Subcellular fractionation clarifies proteomic changes better than the analysis of whole-organ lysates.

8.3.2. Enrichment and Purification of Proteins

Use of mass spectrometry–based characterization of proteins often encounters difficulty in identifying low-copy-number proteins in biological mixtures because the dynamic range of ESI and MALDI techniques is about 10^4, whereas there is a wide dynamic range ($>10^{10}$) in the abundance of proteins in biological samples [10]. Thus, low-abundance proteins usually escape detection, owing to suppression of their ionization by high-abundance proteins. Further complications arise because there is a wide diversity in the PTMs of proteins. Therefore, preferential enrichment and purification of target proteins are essential steps in the identification of proteins in biological samples. Ammonium sulfate precipitation and affinity-based enrichment with antibodies or metal ion affinity columns are the proven techniques for this purpose. Commercial immunodepletion columns are available to enrich the plasma or serum samples for low-abundance proteins through removal of the topmost abundant proteins. As an example, Agilent's MARS (Multiple Affinity Removal System) technology is applied to remove from human plasma six of highly abundant proteins (albumin, IgG, IgA, transferrin, haptoglobin, and antitrypsin) [11]. GenWay Biotech's LC column can remove the 12 high-abundance proteins that represent 96% of the total mass. These columns use polyclonal antibodies to remove desired proteins.

Free-Flow Electrophoresis of Protein Samples To reduce the dynamic range of proteins to be analyzed, and subsequently to maximize the number of proteins identified, a prefractionation step is often incorporated into the separation scheme. A current procedure of interest uses free-flow electrophoresis (FFE), which is a solution-phase isoelectric focusing (IEF)–based approach [12]. The apparatus consists of a separation chamber made of two narrowly spaced parallel plates and contains two parallel electrodes to generate an electrical field perpendicular to the flow of the separation medium. A stabilizing solution flows continuously upward through this chamber to create a laminar flow of a pH gradient. The sample is injected near the bottom of the chamber, and proteins are separated into various fractions that span a linear pH gradient of 3 to 11. The proteins separated exit at the top of the chamber for collection into individual fractions.

Two-Dimensional Gel Electrophoresis 2-DE is the most powerful and robust technique available to separate a complex mixture of proteins from biological specimens [13–15]. It is a two-step process in which each dimension makes use of complementary properties of proteins for their separation. The first dimension is usually performed in the IEF mode, and the second in the conventional, size-based separation mode. The IEF mode separates proteins on the basis of their inherent charge or isoelectric point (pI values). In the second dimension, proteins are separated on the basis of molecular size, typically by SDS–PAGE. With this orthogonal separation scheme, a mixture that contains thousands of proteins can be separated in a single experiment to provide a two-dimensional image. The following key steps are used in a typical 2-DE separation of proteins:

1. For the first-dimension separation, a protein sample containing an appropriate cocktail of carrier ampholyte is applied to a standard immobilized pH gradient gel (IPG) strip and left overnight to rehydrate the strip. These strips are available in several narrow pH ranges for improved separation and are specially well suited for FFE-separated subfractions. Application of the appropriate electrical potential to the rehydrated strip results in separation of the proteins according to their intrinsic charge.

2. The focused strips are equilibrated with SDS–Tris buffer that also contains a reducing agent (e.g., 2-mercaptoethanol or dithiothreitol) and an alkylating agent (e.g., iodoacetamide) prior to second-dimension separation.

3. The IPG strips after reduction and alkylation are placed at the top of the SDS polyacrylamide gel slab.

4. The proteins separated are visualized by staining with colloidal Coomassie Blue (CCB). Other staining protocols use silver staining or SYPRO Ruby fluorescent dye (Molecular Probes, Eugene, Oregon or Invitrogen). SYPRO Ruby dye is the most sensitive of all staining agents, and in addition, it is well compatible with in-gel digestion. The stained images are captured with a fluorescence imager that operates at the excitation wavelength of 457 nm and the emission wavelength of 610 nm. Other suitable fluorescent dyes are Deep Purple (Amersham) and cyanine dyes (CyDye; Amersham).

5. The protein spots are picked up with an automatic spot picker for digestion and subsequent mass spectrometry analysis. Alternatively, the separated protein spots are transferred to an inert support (*Western blotting*). Commonly used inert supports are polyvinylidene difluoride (PVDF) and nitrocellulose. This process cleans and concentrates the separated protein before its further analysis.

2-DE technology allows parallel processing of over 1000 proteins expressed by an organism or cell type. It provides reproducible comparative image analysis of separated proteins. It also serves as a medium to store the separated proteins. Despite its popularity, 2-DE has several limitations. First, it is labor intensive and cannot be automated. It is not effective for very small and very large proteins. Similarly, highly acidic and basic proteins are difficult to separate. The limited solubility of hydrophobic and membrane proteins also pose a problem. In addition, it is encumbered by limited dynamic range and poor sensitivity toward high-abundance proteins.

Two-Dimensional Liquid Chromatography To overcome the limitations of 2-DE technology, several liquid phase–based approaches have become available. These methods include combinations of reversed-phase (RP)–HPLC, ion exchange, size exclusion, chromatofocusing (CF), IEF, and capillary electrophoresis to provide a two-dimensional separation strategy for complex mixtures of proteins and peptides.

In a currently popular two-dimensional combination, *chromatofocusing* is used as a first-dimension separation device and RP–HPLC as the second dimension [16]. Other combinations were discussed in Section 5.7. CF is a column-based liquid-phase separation technique in which proteins are fractionated on the basis of differences in their pI values. A weak ion-exchange column of suitable buffering capacity is preequilibrated with a starting buffer at high pH. A descending pH gradient is established in situ when the column is eluted with another, lower-pH buffer. Proteins with the lowest pI values are retained longer, whereas the highest-pI proteins move faster. Thus, proteins are selectively eluted at pH values proximal to their pI. Each fraction is loaded onto a RP–HPLC column, where the basis of separation is protein hydrophobicity.

8.4. MOLECULAR MASS MEASUREMENT OF PROTEINS

The mass measurement of an intact protein is the first important step in the characterization of that protein. It provides a frame within which the final structure must fit. The molecular mass information can be used for the following applications:

- To verify the correctness of the translated sequence and to identify point mutations.
- To identify a protein in a biological sample using a database search.
- To determine whether a protein has undergone PTMs. The molecular mass of the intact protein will increase by an amount equal to the mass of the

posttranslational moiety (e.g., by 80 Da for each phospho group). To determine whether a protein contains sugar residues, the molecular mass of the protein is measured before and after treating it with N-glycosidase F (PNGase F) and endo-β-N-acetylglucosaminidase H (endo H).
- To determine the number of cysteine residues and disulfide bonds in a protein (see Section 9.2.1 for details).

Traditionally, the molecular mass of proteins has been estimated with SDS–PAGE, in which the migration pattern of a protein is compared to a set of known mass proteins. The accuracy of this procedure is, however, very poor (5 to 20%) and mitigates against its use in the applications pointed out above. More accurate mass values are required to distinguish a mutation between Asp and Asn, Asn and Ile/Leu, Glu and Gln, and Lys and Glu/Gln. Currently, ESI and MALDI, combined with Q–TOF, orbitrap, and FT–ICR, have become a standard protocol to determine the molecular mass of intact proteins with an accuracy within ±0.01%.

The basis of molecular mass measurement by ESI is the formation of multiply charged ions of the analyte. For high-mass biopolymers, an envelope of a series of ions, each differing by one charge from its neighbor, is formed. By using Eqs. (2.31) and (2.32), the charge state of each multiply charged ion and the molecular mass of the protein can be determined accurately within ±0.001%. In contrast, MALDI produces a simpler spectrum that contains mainly the singly protonated ion.

Procedures have been developed to measure the molecular mass of proteins separated by gel electrophoresis directly by MALDI–MS [17–19]. In one option, blot membranes are used as substrates for laser irradiation. Soaking the membrane in a saturated matrix solution allows incorporation of the matrix. MALDI mass spectra are obtained by taping the cut membranes onto the metal sample target. The direct mass measurement of proteins from a gel without resorting to electroblotting is also feasible [19]. To avoid cracking of the gels in the vacuum of a mass spectrometer, specially prepared ultrathin gels with a lower percentage of acrylamide are used. IEF gels have no such cracking problem. The matrix is spotted onto the protein spot, or gels are soaked in the matrix solution.

8.5. PEPTIDE MASS MAPPING

In peptide mapping, proteins are characterized on the basis of their distinct proteolysis pattern. The rationale is that proteins that yield identical peptide maps have an identical amino acid sequence. Peptide mapping by mass spectrometry has been termed *peptide mass fingerprinting* [20]. In this technique, the purified protein is cleaved into smaller discrete segments by digestion with either a proteolytic enzyme or a chemical reagent, and the molecular mass of each resulting peptide is measured by a suitable mass spectrometry technique to create a distinct peptide mass map or fingerprint. Peptide mass mapping is a powerful technique to determine the primary sequence of a protein. Other uses of peptide mapping technique in protein chemistry are (1) to assess the quality of genetically engineered proteins, (2) to confirm the sequence of a native protein for which the cDNA sequence is known, (3) to evaluate the higher-order structures (discussed in Chapter 10), and (4) to elucidate noncovalent interactions of proteins. Peptide mapping is equally applicable to proteins that contain posttranslationally modified residues.

8.5.1. Reduction and Carboxymethylation

Because cysteine residues are prone to autooxidation, proteins that contain disulfide bonds or free sulfhydryl groups must be subjected to reduction–alkylation process prior to cleavage. Alkylation converts sulfhydryl groups to stable derivatives. Also, during the reduction–alkylation procedure, the three-dimensional structure of proteins is disrupted to allow more cleavage sites to be accessible. Proteins are first reduced by treatment with 2-mercaptoethanol or dithiothreitol to convert disulfide bonds to free sulfhydryl group:

$$\begin{matrix} \sim S \\ | \\ \sim S \end{matrix} \xrightleftharpoons{RSH} \begin{matrix} \sim S \\ \\ \sim SH \end{matrix} \xrightleftharpoons{RSH} \begin{matrix} \sim SH \\ \\ \sim SH \end{matrix} + RS-SR \quad (8.1)$$

which are subsequently converted to an S-carboxymethyl derivative by reacting with iodoacetic acid:

$$\begin{matrix} \sim SH \\ \\ \sim SH \end{matrix} + 2ICH_2COOH \longrightarrow \begin{matrix} \sim S-CH_2COOH \\ \\ \sim S-CH_2COOH \end{matrix} + 2HI \quad (8.2)$$

The reduction is carried out at alkaline pH (ca. 8.0) to generate a reactive intermediate (thiolate anion). 4-Vinylpyridine, iodoacetamide, or acrylamide can also be used in place of iodoacetic acid.

▶ **Example 8.1** A protein with two sulfide bonds and three cysteine residues is first reduced with 2-mercaptoethanol and then alkylated with iodoacetic acid. By how much will the mass of the protein increase?

Solution Reduction of a protein converts each sulfide bond to two sulfhydryl groups. Therefore, after reduction there are a total of $2 \times 2 + 3 = 7$ sulfhydryl groups in the protein. Because alkylation of each sulfhydryl group by iodoacetic acid increases the mass by 58 Da, the mass of the protein increases by $7 \times 58 = 406$ Da.

8.5.2. Cleavage of Proteins

The objective of cleaving a protein at specific sites is to generate peptides of suitable size for mass spectrometry analysis. Two methods are commonly used to cleave proteins: (1) hydrolysis with a suitable chemical reagent and (2) digestion

TABLE 8.1. Typical Protein-Cleaving Agents

Cleaving Agent	Specificity	Digestion Conditions [buffer; pH; temperature (°C)]
Chemical agents		
Cyanogen bromide	Met-X	70% TFA
N-Chlorosuccinimide	Trp-X	50% acetic acid
N-Bromosuccinimide	Trp-X	50% acetic acid
Highly specific proteases		
Trypsin	Arg-X, Lys-X	50 mM NH_4HCO_3; 8.5; 37
Endoproteinase Glu-C	Glu-X	50 mM NH_4HCO_3; 7.6; 37
Endoproteinase Arg-C	Arg-X	50 mM NH_4HCO_3; 8.0; 37
Endoproteinase Lys-C	Lys-X	50 mM NH_4HCO_3; 8.5; 37
Endoproteinase Asp-N	X-Asp	50 mM NH_4HCO_3; 7.6; 37
Nonspecific proteases		
Chymotrypsin	Phe-X, Tyr-X, Trp-X, Leu-X	50 mM NH_4HCO_3; 8.5; 37
Thermolysin	X-Phe, X-Leu, X-Ile, X-Met, X-Val, X-Ala	50 mM NH_4HCO_3; 8.5; 37
Pepsin	Phe-X, Tyr-X, Trp-X, Leu-X, Met-X	0.01 M HCl; 2.0; 37
Elastasase	Broad specificity	50 mM NH_4HCO_3; –; 37

with an endoprotease. A list of these cleaving reagents, along with the conditions of the reaction, is given in Table 8.1.

Degradation with Chemical Reagents A frequently used procedure is to react proteins at acidic pH with cyanogen bromide (CNBr), which cleaves the amide bond on the C-terminal side of methionine. This reaction converts methionine to homoserine, which under the acidic environment exists in the lactone form:

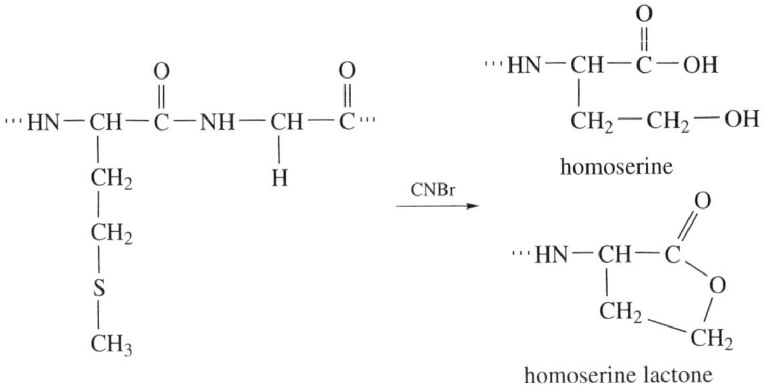

$$\cdots HN-CH(CH_2CH_2SCH_3)-C(=O)-NH-CH(H)-C(=O)\cdots \xrightarrow{CNBr} \cdots HN-CH(CH_2CH_2OH)-C(=O)-OH \quad \text{homoserine}$$

$$\cdots HN-CH-C(=O)-O-CH_2-CH_2 \quad \text{(ring)} \quad \text{homoserine lactone}$$

(8.3)

the mass of which is 48.1 Da less than the methionine residue. The peptides generated are large in size but few in number. N-Chloro- and N-bromosuccinimide are the other two chemical reagents that have found some utility to cleave a protein. These reagents have specificity for the amide bond C-terminal to Trp residue.

Digestion with Proteolytic Agents Endoproteases that have been used to produce reproducible peptide maps of proteins are listed in Table 8.1. Of these enzymes, trypsin is more or less a universal choice for most applications. It cleaves an amide bond on the C-terminal side of lysine and arginine residues and produces smaller peptides that contain a basic residue at the C-terminus, except for the C-terminus of the protein. This aspect is especially useful for the ESI process and for subsequent sequencing by tandem mass spectrometry. If needed, larger and overlapping peptides are generated by digesting the protein with endoproteinase Lys-C, endoproteinase Arg-C, endoproteinase Glu-C (also known as *Staphylococcus aureus* V8 protease), or endoproteinase Asp-N. The first three enzymes have specificities to the amide bond C-terminal to lysine, arginine, and glutamic acid residues, respectively. In contrast, endoproteinase Asp-N is specific to the bond N-terminal to aspartic acid. Another enzyme that cleaves the N-terminal side of the amide bond is thermolysin, which has broad specificity for leucine, isoleucine, methionine, phenylalanine, and tryptophan.

Often, digestion is performed in solution at the conditions of pH, temperature, and buffer specified (see Table 8.1), and in a denaturing environment to ensure complete endpoint digestion [21]. Care should also be taken to use volatile buffers; otherwise, the ionic components of the buffer will interfere with the mass spectrometry analysis. Ammonium carbonate and ammonium bicarbonate can both be used to adjust the pH of the digestion mixture to between 7.0 and 8.5. The volatile salts can be removed by lyophilization. In the event a nonvolatile buffer is used or any other nonvolatile components or salts are present, a solid-phase extraction (SPE) step might be required. Procedures are also available that use either immobilized trypsin packed into a small-diameter PEEK (polyether ether ketone) column [22] or immobilized on MALDI targets [23]. In the latter procedure, known as *on-probe digestion*, the proteolytic enzyme is attached covalently to the activated MALDI probe, and the protein is digested directly on the probe by the surface-tethered enzyme. The digestion is terminated after a period of time by adding the matrix to the probe.

Digestion of Gel-Separated Proteins Three digestion methods are used for gel-separated proteins: (1) the intact protein eluted from the gel and digested with a protease [24], (2) in-gel digestion of proteins [25], and (3) digestion of electroblotted proteins [26]. In-gel digestion requires the removal of SDS and other contaminants prior to mass spectrometry analysis of the proteolytic digests. One major advantage of immobilization of proteins onto the membrane is that most of the contaminants and SDS can be washed away prior to digestion. PVDF is the membrane of choice because it has a high-binding affinity and provides a cleaner chemical background for subsequent mass spectrometry analysis [27]. Carboxymethyl cellulose and Immobilon–CD membranes also exhibit a high

302 PROTEINS AND PEPTIDES: STRUCTURE DETERMINATION

recovery of blotted proteins. In a typical procedure, the gel-separated protein is electroblotted onto a PVDF membrane, visualized, and destained. Next, the spot is excised and subjected to the cleavage step. The cleaved peptides are extracted and analyzed by MALDI or ESI.

▶ **Example 8.2** A peptide of sequence Tyr–Phe–Ser–Asp–Phe–Leu–Arg–Trp–Val–Gly–Met–Glu–Ala–Asp–Phe–Gly–Thr–Lys–Ser–Asn–Ala–Leu is treated first with endoproteinase Asp–N and then with trypsin. How many peptides will be generated by this treatment, and what will be each peptide's molecular weight?

Solution Cleavage of the peptide with Asp–N will produce a total of three peptides: Tyr–Phe–Ser, Asp–Phe–Leu–Arg–Trp–Val–Gly–Met–Glu–Ala, and Asp–Phe–Gly–Thr–Lys–Ser–Asn–Ala–Leu. Further treatment of this mixture with trypsin will generate the following five peptides:

$$\text{Tyr–Phe–Ser} = \text{H} + \text{residue masses of Tyr, Phe, and Ser} + \text{OH}$$

$$= 1 + 163 + 147 + 87 + 17 = 415 \text{ Da}$$

Asp–Phe–Leu–Arg = 549 Da   Trp–Val–Gly–Met–Glu–Ala = 691 Da
Asp–Phe–Gly–Thr–Lys = 566 Da   Ser–Asn–Ala–Leu = 403 Da

8.5.3. Mass Spectrometric Analysis of Peptide Maps

Traditional methods to generate peptide maps involve fractionation of complex mixtures of peptides in a protein digest either with one-dimensional SDS–PAGE or RP–HPLC [28,29]. The mass spectrometry peptide-mapping protocol, in principle, is similar to these techniques, but it provides an added dimension of structure-specific data (i.e., the molecular mass). MALDI–MS [30,31], ESI-MS [32], LC/ESI–MS [33], and CE/ESI–MS [34] have currently replaced the traditional biochemical approaches. MALDI allows the direct analysis of unfractionated protein digests. The commonly used matrices are sinapinic acid, α-cyano-4-hydroxy cinnamic acid (α-CHCA), and 2,5-dihydroxybenzoic acid (DHB).

Although direct analysis of the unfractionated protein digests is feasible with ESI, it is the online combination of ESI with capillary LC that has made a significant impact [33,35,36]. With this technology, the separation of the peptides in complex enzyme digests and the determination of their molecular mass can be accomplished simultaneously. When LC is connected to a tandem MS instrument, it is feasible to obtain on-the-fly amino acid sequence for each peptide. The tryptic peptides produce doubly charged ions (except the C-terminal fragment), which have a high proclivity for peptide-bond fragmentation. With *data-dependent analysis*, molecular mass and sequence information are both obtained with the same

HPLC analysis. The use of nano-ES allows small sample amounts to be analyzed [37].

Another technique that is compatible with the analysis of peptide maps from small amounts of proteins is an online combination of CE and ESI–MS [34]. CE provides high-resolution separation that complements RP–HPLC; separation efficiencies of over 1 million theoretical plates with detection levels of <300 amol have been reported. The concentration of large analyte volumes into small volumes, a requirement for CE analysis, improves detection sensitivity considerably. This step can be accomplished by including either a small length of RP–HPLC packing within the main CE column or a polymeric membrane impregnated with the chromatographic stationary phase [38]. A downside of CE is the relatively high concentration limits of detection.

8.6. PROTEOMICS

The increasing availability of genomic sequences has ushered in a new field of biomedical research, known as *proteomics*, which deals with systematic characterization of gene products, the *proteome*. Proteomics is an experimental approach that encrypts the information contained in genomic sequences in terms of the structure, function, and control of biological processes [4,5]. Currently, the protein content of biological samples is analyzed exclusively by proteomics. The concept of the proteome has been defined as the entire protein complement expressed at a given time by a genome or by a cell or tissue type under a given condition (i.e., *proteome* is *prote*in + gen*ome*) [39]. The field of proteome analysis has gained significant importance simply because the data from genetic and genomic techniques by themselves are insufficient to describe biological processes; the genome is a static system and essentially identical in every cell of an organism, whereas the proteome is highly dynamic—meaning that the types of expressed proteins, their abundance, the types of PTMs, and subcellular location are greatly dependent on the environment and physiological state of the cell (i.e., whether it is healthy or has been subjected to insult). Thus, a single gene can be encoded into several protein isoforms. Proteome analysis thus assumes greater importance because it provides a more detailed picture of what goes on in the cell. It presents a unique opportunity to identify proteins as potential targets of therapeutic agents and biomarkers of disease and insult due to external agents (e.g., drugs and toxins) at much reduced time and cost. Mass spectrometry is playing an ever-increasing role in current proteome analysis because of its integration with high-resolution separation techniques and protein databases [4,5].

Within the broad umbrella of proteome analysis, there are a number of more tightly focused subfields, such as (1) *characterization proteomics*, which provides a survey of proteins presents in a cell, tissue, or biofluid, (2) *differential proteomics*, which provides identification of differentially expressed proteins in different physiological states, such as in healthy versus diseased subjects; and

(3) *functional proteomics*, which provides identification of a group of proteins involved in a specific function.

8.6.1. Strategies for Proteomics

Presently, mass spectrometry–based proteomics can be classified in two broad categories (Figure 8.5):

1. *Bottom-up proteomics*. This strategy relies on peptide-level information, such as mass or sequence, to identify a protein. Currently, this strategy dominates proteomics research. In this approach, the protein is digested, typically with trypsin, and the digest is analyzed with an appropriate mass spectrometry system to obtain the peptide masses and sequences. The database search with mass spectrometry data characterizes the protein. More details of the mass spectrometry platforms used in this approach are provided in the following sections.

2. *Top-down proteomics*. This strategy deals with intact protein molecules; no proteolytic cleavage is performed [41]. It involves the accurate molecular mass measurement of the intact protein using high-resolution mass spectrometry within ±2 Da, followed by a molecular mass database search. The identity of

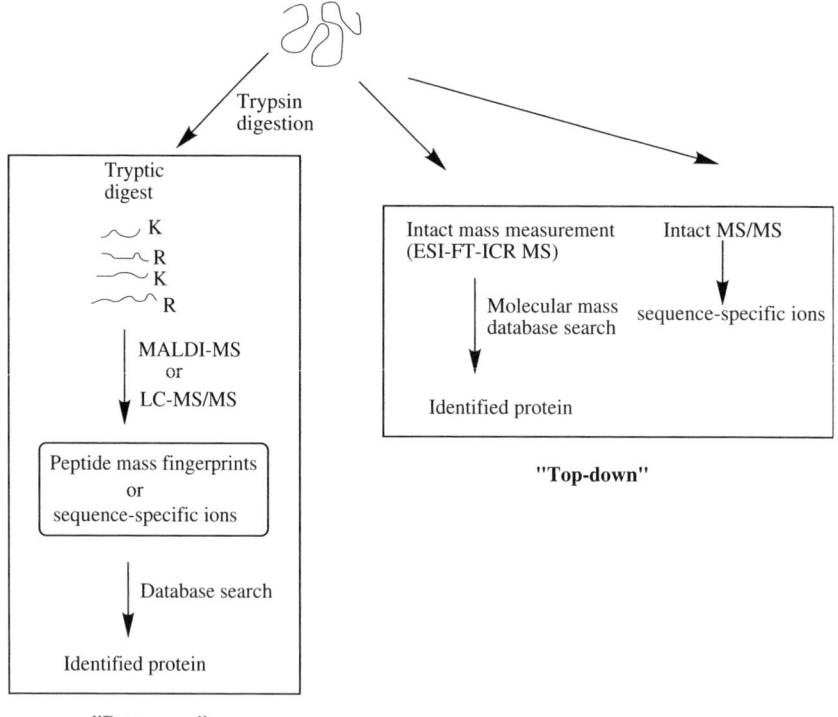

Figure 8.5. Bottom-up and top-down proteomics.

the protein can be further confirmed by obtaining the tandem mass spectrometry sequence-specific fragment ions of the intact protein. The top-down approach can also reveal the identity of any PTMs. Because of the current limitation of fragmentation techniques (see Section 4.3), this approach is restricted to small proteins such as cytochrome c. The combination of ESI with either Q–FT–ICR or LIT–FT–ICR MS is considered an ideal mass spectrometry system for top-down proteomics.

The field of proteome analysis is challenging because of the complexity of any given proteome, the broad dynamic range of protein abundance, the existence of multiple protein forms due to PTMs, proteolysis and splice variants, and the difficulty of quantification of absolute protein content. Some of these challenges can be met by the use of complementary approaches to proteomics. In general, four basic steps are involved in proteome analysis: (1) sample preparation, (2) separation and purification, (3) mass spectrometry analysis, and (4) database search.

Sample Preparation Sample preparation is one of the critical steps in proteome analysis. As discussed in Section 8.3, sample preparation from tissues, organs, and cell cultures requires homogenization or subcellular fractionation into distinct organelles prior to protein profiling by mass spectrometry. Subcellular fractionation allows deeper proteomic coverage. Biological fluids (blood, plasma, CSF), in addition, require depletion of major blood proteins, such as albumin, to preferentially enrich the low-abundance proteins. This task is typically accomplished by immunodepletion. The high-abundance proteins are retained in the column (bound fraction), and the low-abundance proteins are collected in the flow-through. Removal of the high-abundance proteins increases the sample capacity and improves the detection of low-abundance proteins. A typical scheme for sample preparation is shown in Figure 8.6.

Separation and Purification A high degree of purification of the homogenized subcellular fractionations into individual proteins (rarely achieved even by two-dimensional-PAGE for all proteins) or a simple mixture is an essential requirement for unambiguous protein identification. This aspect was also discussed in Section 8.3. The use of several orthogonal techniques (e.g., 2-DE or LC) becomes essential to reduce the complexity of the sample and improve chances of detection of low-abundance proteins.

Mass Spectrometry Analysis The following mass spectrometry approaches are commonly used in proteomics for protein identification (see Figure 8.7):

1. *Peptide-mass fingerprinting.* In this approach, also known as *peptide-mass mapping*, the protein is first subjected to enzymatic digestion to generate a set of peptides that are unique to this protein (Section 8.4) [20,42–47]. The molecular mass of each fragment is determined accurately (within ±0.5 Da) using MALDI–MS or ESI–MS. Correlation of these masses with the theoretical peptide

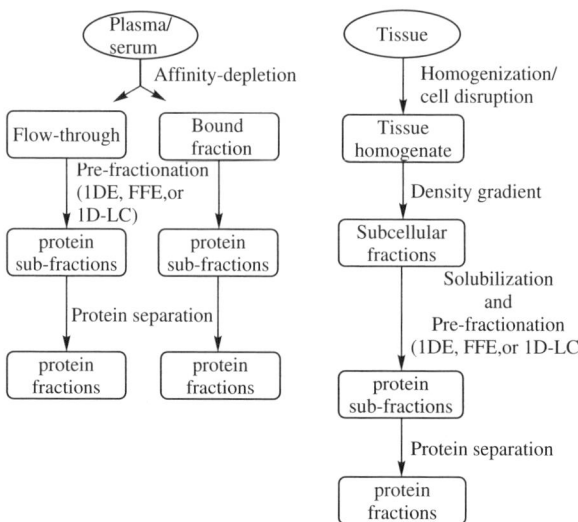

Figure 8.6. Sample preparation and separation schemes for proteins from biofluids and tissues.

masses in the database will identify the protein correctly. At least four or five peptide masses are sufficient to identify a protein.

2. *Peptide sequence tags.* Another popular approach (Figure 8.7 lower part) is based on a correlation of the MS/MS fragmentation patterns with the amino acid sequences in the protein databases [48–51]. The use of fragmentation patterns is far more discriminating than peptide mass fingerprinting data, and a high probability assignment of proteins is possible with an MS/MS spectrum of just one peptide. It should be emphasized, however, that identifications based on only one peptide can often be misleading. In this approach, the protein is first enzymatically cleaved into smaller segments. Peptides from this protein digest are sequenced by MS/MS, and the observed pattern of fragment ions are matched with the predicted patterns of fragment ions in translated genomic databases.

3. *Accurate-mass tags.* A less common approach is the use of accurate-mass tags (AMT) [52,53]. This method is a nontandem MS-based approach, the underlying basis of which is that when the mass of a peptide fragment can be measured with less than 1-ppm accuracy, the mass of a single peptide becomes a unique identifier of the protein. Such a high degree of mass accuracy is attained with FT–ICR instruments. Further improvement in this approach is to combine the HPLC retention time [54] or the pI values collected from IEF analysis [55].

Based on these approaches, several mass spectrometry platforms have been developed. At present, however, no single platform is available that can provide a complete study of the entire proteome of a specified biological sample. The use of multiple complementary approaches is recommended for complex samples. These approaches are described below.

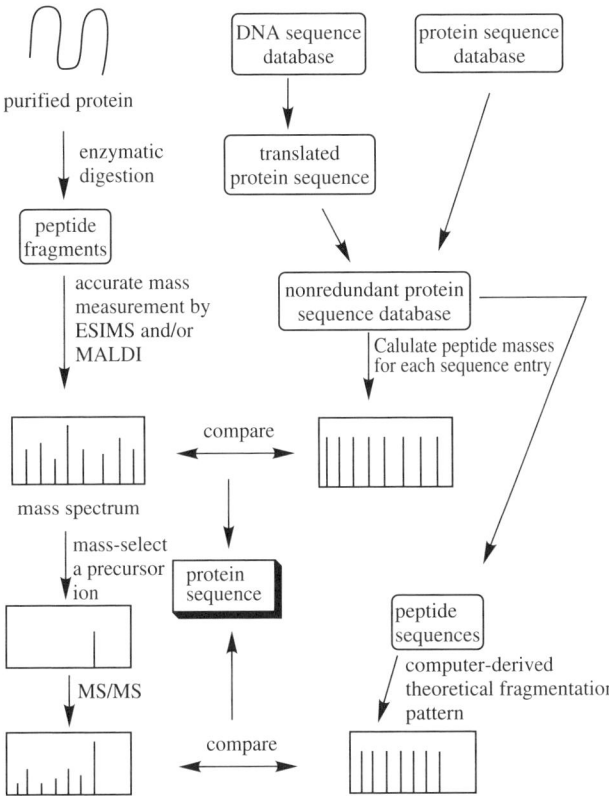

Figure 8.7. Mass spectrometry–based proteomics approach. (Redrawn from C. Dass, *Principles and Practice of Biological Mass Spectrometry*, Wiley-Interscience, 2001.)

1. *Peptide-mass fingerprinting with 2-DE–MALDI–TOF–MS.* This is the most notable platform for proteome analysis [44,56,57]. The complex mixture of proteins is separated by 2-DE, the protein spots are cut from the gel, and proteins are subjected to in-gel digestion with trypsin. Salts are removed from the tryptic digest with a C-18 ZipTip. The peptides are eluted with the matrix solution and are deposited directly onto a 96-well MALDI plate. Mass measurement by TOF–MS provides peptide-mass fingerprints for subsequent protein identification with database search.

2. *Peptide sequence tags with MALDI–MS/MS.* The gel-separated protein spots are digested as above and are deposited onto a large-format MALDI plate either directly or after separation by RP–HPLC. The matrix is added, and the spotted samples are analyzed with MALDI–MS/MS on a TOF/TOF tandem instrument [58]. MALDI–ion traps and MALDI–Q–TOF are also suitable for this purpose.

3. *Peptide sequence tags with multidimensional protein identification technology (MudPIT).* MudPIT is a gel-free "*shotgun*" *proteomic approach*, which does

not require any upstream protein separation [59–62]. Instead, the homogenized protein sample is tryptic-digested to provide a complex peptide digest that might contain over 10,000 peptides. This complex mixture is analyzed directly with multidimensional liquid chromatography online with ESI tandem mass spectrometry (2D–LC–ESI–MS/MS). The amino acid sequence ion is correlated with the calculated fragmentation pattern obtained from *in silico* tryptic digest of all proteins derived from the genome in a given database. In this two-dimensional LC system, strong cation-exchange (SCX) chromatography that acts as a first-dimension separation step is coupled to RP–HPLC. An off-line combination of SCX chromatography and RP–HPLC can also be employed instead; in this arrangement, the peptide mixture is fractionated by SCX, and each fraction is subjected individually to RP–HPLC separation online with ESI–MS/MS. Off-line separation by IEF, followed by online LC/ESI–MS/MS, is another possible option [62]. An off-line combination of RP–HPLC and RP–HPLC has also been used for proteome analysis of tissue extracts [63]. These various arrangements permit greater peak capacity, flexibility, and dynamic range of detection. The tandem mass spectrometry instrumentation includes a quadrupole ion trap (QIT), linear ion trap (LIT), or hybrid systems (Q–TOF, LIT–FTICR). LC–ESI–MS/MS is performed in the data-dependent analysis mode, which involves switching to the MS/MS mode when a doubly or triply charged ion is detected in the normal MS scan.

Database Search A final step in the proteome analysis is a database search with the acquired MS and MS/MS data. This new paradigm has become feasible, owing to integration of the mass spectrometry data with database searching. A variety of databases can be searched, including those for protein, DNA, and expressed-sequence tag (EST) databases (Table 8.2). This general approach is based on the hypothesis that the primary structure of a protein is a direct product of genome transcription (Figure 8.3), and therefore the protein's structure can be deduced from knowledge of the organism's genome.

In addition to the sequence entries (core data) of proteins, databases also contain annotation information, which provides the link to other databases and the literature. Links to information about functions of the protein identified, any PTMs, domain and sites, higher-order structures, homology to other proteins, associated diseases, sequence conflicts, and variants are also included.

MS-Fit, MOWSE, Prot-ID, Expasy tools, and PeptideSearch are some of the database-search programs that can be used with peptide-mass fingerprinting data. The molecular masses of peptides in the query are matched against the theoretical peptide-mass values created by *in silico* digestion of each protein entry in the database with the specific protease that was used in the experimental step (e.g., trypsin). In a typical search algorithm, the cleavage database is first selected, and the criteria of the data search are defined [42,45]. The criteria used can be the species, the mass-matching tolerance, an approximate upper value of the molecular mass of the protein, the pI value of the protein, a minimum number of matches, the number of cleavage sites that might have been missed by the trypsin digestion, and the type of cysteine modification. From a list of the experimentally

TABLE 8.2. Partial List of Available Databases

Database	Type	Internet Address	Institution[a]
GenBank	Nucleotide	www.ncbi.nlm.nih.gov/Genbank/index.html	NCBI
EMBL	Nucleotide	www.embl-heidelberg.de/services/index.html	EMBL
NCBI	Nucleotide	www.ncbi.nlm.nih.gov	NCBI
GenPept	Protein	www.ncbi.nlm.nih.gov/Genbank/index.html	NCBI
Swiss-Prot	Protein	www.ebi.ac.uk/swissprot/	EBI
TrEMBL	Protein	www.expassy.ch	
PIR	Protein	www.pir.georgetown.edu/pirwww/pirhome.html	NBRF
OWL	Protein	www.leeds.ac.uk/bmb/owl/owl.html	
IPI	Protein	www.ebi.ac.ukIPI/IPIhelp.html	EBI
dbEST	Expressed-sequence tag	www.ncbi.nlm.nih.gov/dbEST/index.html	NCBI

[a] NCBI, National Center for Biotechnology Information; EMBL, European Molecular Biotechnology Laboratory; EBI, European Bioinformatics Institute; and NBRF, National Biomedical Research Foundation.

derived $[M + H]^+$ values of the peptides, the output of the search gives a ranked list of most likely candidates. The database sequence that produces the best score has the highest probability of being the protein. Mismatches of the peptide candidates listed can reveal the possibility of mutation, PTMs, or coeluting/comigrating proteins. Four to six peptide masses are usually sufficient to provide a successful search [42,47]. The database-search selectivity can be enhanced by keeping the mass-matching tolerance low (<3 ppm) and by increasing the mass measurement accuracy (to within ±0.5 Da). The mass measurement data acquired with delayed-extraction TOF–MS and FT–ICR–MS meet this requirement easier than those acquired by quadrupole or ion-trap instruments.

To identify proteins on the basis of the MS/MS sequence ion data, search algorithms such as MASCOT, SEQUEST, ProFound, MS-Tag, PeptideSearch, X1Tandem, OMSSA, Parallax, DBDigger, GlobalLynx, and Spectrum Mill have been developed [64]. MASCOT and SEQUEST are the most widely used. In these search engines, the pattern of fragment ions observed is matched with the fragment ion patterns that are calculated theoretically from the database entries. For example, the SEQUEST algorithm first simplifies the acquired MS/MS spectrum. Next, it identifies amino acid sequences in the database with the measured mass of the peptide ion selected, and predicts the fragmentation pattern that is expected for each sequence [49,51]. The spectra are subjected to fast Fourier transform, and each virtual spectrum is matched with the experimentally observed MS/MS spectrum to produce a cross-correlation score. The highest scoring amino acid sequences are reported. This program can also identify PTMs by assuming that each putative modification site is modified and unmodified in one pass through the database. It should be pointed out that some programs tend to assign sequence ions to low-intensity signals. Therefore, one should always examine the spectral match to verify that the most abundant ions are identified on the basis of the sequence and that the sequence ions are of relatively high abundance.

In the peptide-tag technique, a noisy MS/MS spectrum that contains partial sequence ion information can be used to search databases [65]. In this approach, a short stretch of sequence (called a *tag*), together with the molecular mass of the preceding (m_1) and tailing region (m_2), is considered to be a unique signature of the peptide. This approach is used with the PeptideSerach algorithm. A search on a single sequence tag is often sufficient to identify a protein. The technique is also applicable to the ESI source-formed fragments and to peptides that carry a PTM.

The databases listed in Table 8.2 are maintained by independent research groups and can be accessed through the Internet. Some of these databases are nucleotide databases, and some protein databases that have been translated from the nucleotide databases. They are updated frequently and contain comprehensive lists of protein and nucleotide sequences. The choice of a particular database is dictated by the extensiveness of sequence entries, low redundancy rate, low error rate, and a high degree of annotation. A custom-made database can be assembled into a composite nonredundant database by including entries from all protein databases and translated-nucleotide databases (into protein sequences) and removing any duplicate entries.

8.7. QUANTITATIVE PROTEOMICS

The objective of quantitative proteomics is to identify differentially expressed proteins in a biological sample. Differential expression of proteins is caused by a disease state, stress due to external factors (drugs, toxins, etc.), or experimental manipulation. Quantitative proteomics can help identify biomarkers of a particular disease and aid in an early diagnostic intervention and prevention of a disease. Several strategies have been developed for quantitative proteomics; some are exclusively gel-based approaches, and others require mass spectrometry measurements [66].

Quantification by Two-Dimensional Gel Electrophoresis 2-DE is still considered the gold standard for quantitative proteomics [57,67]. This procedure relies on differential expression of proteins in control and test samples. Proteins from the two samples are separated by the 2-DE protocol, followed by staining with a suitable dye (e.g., CCB or SYPRO Ruby). Next, images of the stained spots are acquired, and spot densities are measured and compared by image analysis software to provide differential expression of proteins. Gel-to-gel variations can affect the precision of quantification by this procedure.

Quantification by Two-Dimensional Differential Imaging Gel Electrophoresis 2D-DIGE is a more precise version of the 2-DE approach. This method uses multicolored cyanine dyes to label proteins in the two experimental samples (control and test) and subsequent separation of the labeled proteins on a single gel [68]. The control sample is labeled with Cy-5 dye and the test sample with Cy-3 dye (GE Healthcare). The two samples are then mixed in equal concentrations and co-separated on the same 2-DE gel. The separated protein spots are visualized and co-matched by scanning the gel at two different wavelengths.

Quantification by Isotope-Coded Affinity Tags ICAT is a first-generation approach to quantify the relative levels of proteins in control and test samples by mass spectrometry [69]. In this technique, illustrated in Figure 8.8, the two samples are treated with an ICAT reagent, which consists of essentially three functional components: a thiol functional group, a linker group, and a biotin moiety. This treatment selectively alkylates cysteine residues with either a light or a heavy isotope (e.g., d_0 or d_8) label. The two samples are combined and tryptic-digested. Next, the labeled cysteine peptides are affinity purified on an avidin column and analyzed by LC/ESI–MS and MS/MS. The ratio of ion abundances of the heavy versus light isotope-labeled peptides in the LC/ESI–MS experiment provides the relative quantitative measurement of the two proteomes. An improvement over the original approach is to use ^{13}C-labeled reagents (instead of deuterium-labeled) and acid-cleavable linkers. The ^{13}C-labeled reagents minimize differential elution of peptides labeled with heavy and light isotopes to provide more accurate quantification. The acid-cleavable linkers improve the MS/MS performance. A major disadvantage of ICAT is that it is applicable to cysteine-containing proteins only;

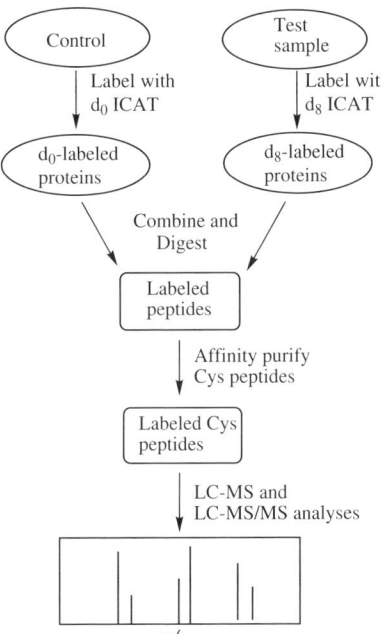

Figure 8.8. ICAT quantification protocol.

fortunately, nearly 90% of proteins contain cysteine. The cost of ICAT reagents is another discouraging factor in the use of this methodology.

Another ICAT reagent, known as *HisTag reagent*, used to tag Cys residues, has been reported recently [70]. It is a 10-mer derivatized peptide $H_2N-(His)_6-Ala-Arg-Ala-Cys(2-thiopyridyl\ disulfide)-CONH_2$ and has a trypsin cleavable site (−Arg−Ala−) that facilitates removal of the His_6-tag during the final digestion step. The thiopyridyl disulfide moiety acts as a thiol reactive group and d_4 label is incorporated in Ala-9 residue.

Quantification by iTRAQ Reagents These reagents are a set of four isobaric amine-specific labeling reagents (114, 115, 116, or 117) available from Applied Biosystems; their use allows a multiplexing of up to four different samples in a single LC−MS/MS experiment [71]. An iTRAQ reagent consists of a reporter group, a balance group, and a peptide reactive group (PRG). Unlike ICAT reagents, the iTRAQ labeling occurs at the peptide level. The PRG covalently connects the iTRAQ reagent with the primary amino group of lysines and the N-terminus of peptides. In the typical protocol shown in Figure 8.9, the control and treated protein samples are reduced, alkylated, and digested with trypsin. The digested samples are reacted with four different iTRAQ reagents. The two samples are combined and analyzed by LC/ESI−MS/MS. Because of the isobaric nature of the iTRAQ reagents, the labeled peptides cannot be distinguished in a normal ESI−MS experiment, but they produce distinct reporter ions at m/z 114, 115, 116, and 117 in the ESI−MS/MS experiment to allow analysis of up to four different samples.

Quantification by the Proteolytic ^{18}O-Water Labeling Approach This method, also known as enzymatic labeling, relies on trypsin-catalyzed ^{18}O labeling

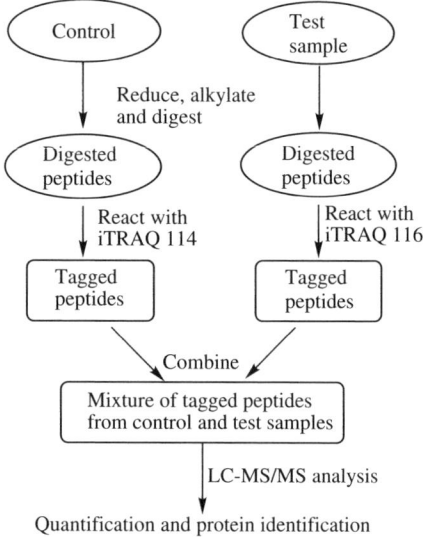

Figure 8.9. iTRAQ ICAT quantification protocol.

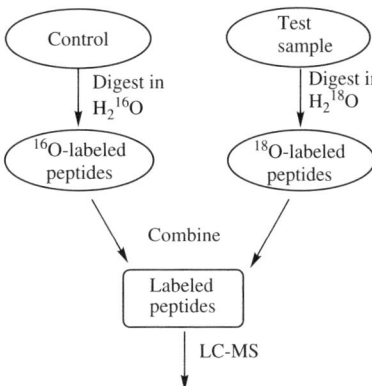

Figure 8.10. Protocol for quantification of proteins by ^{18}O-water labeling approach.

of C-terminal carbonyl oxygens [72]. As shown in Figure 8.10, the control sample is trypsin-digested in ordinary water, whereas the test sample is digested in ^{18}O-water to produce a 4-Da mass shift due to the incorporation of two ^{18}O atoms in the cleaved peptide. The peptides from the two digests are pooled and analyzed by LC/MS. Protein quantification is achieved by measuring the ratio of peak heights in an extracted-ion chromatogram of the coeluting labeled and unlabeled peptides. The labeling efficiency can be improved when the labeling and digestion steps are performed separately [73]. In this postdigestion labeling approach, the digested sample is dried completely and redissolved in water or ^{18}O-water in the presence of immobilized trypsin. One potential advantage of the ^{18}O labeling procedure is that each tryptic peptide of a protein, except its C-terminal peptide, can be labeled. In his way a protein can be quantified using multiple targets, a process that would enhance the confidence level in quantification.

Quantification with the Stable-Isotope Labeling by Amino Acids in Cell Culture (SILAC) Approach SILAC is a metabolic approach in which labeling occurs by supplementing the cell culture media with the stable isotope forms of essential amino acids [74,75]. In a typical procedure (Figure 8.11), the control and treated and/or diseased cell lines are grown in two different media; one is enriched with light isotope–containing amino acids and the other with heavy isotope–containing amino acids (e.g., Lys or d_4-Lys). The two media are mixed in a 1:1 ratio; proteins are extracted, purified, and digested. The peptides are analyzed by LC/MS or LC–MS/MS. Signal intensities of the labeled and unlabeled peptides provide quantitative information. Examples of the frequently used heavy isotope–labeled amino acids in the culture media are ^{15}N–Lys, ^{13}C^{15}N–Lys, ^{15}N–Arg, ^{13}C–Lys, ^{13}C–Arg, and ^{13}C–Leu.

Quantification by the Label-Free Approach In this method, the protein digests of the two samples are analyzed by LC/ESI–MS [76]. The extracted-ion chromatograms from the two samples are aligned, and those peptides ions with

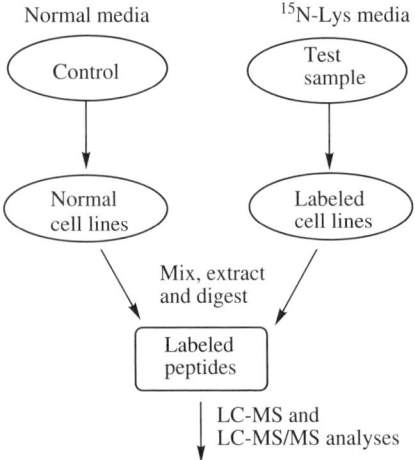

Figure 8.11. Protocol for quantification of proteins by the SILAC method.

nearly the same mass, charge, and retention time are quantified on the basis of differences in the signal intensities. For the success of this method, a high level of reproducibility is essential and necessitates the need for strict control in sample handling and of LC/MS analysis conditions.

Quantification by the AQUA Approach AQUA (*a*bsolute *qua*ntification) is a fundamentally different approach from those discussed above. As the name implies, it provides absolute quantification, but it is currently feasible for only one or a few target proteins [77,78]. Similar to the quantification of small molecules discussed in Section 14.3.3, quantification is based on the use of a stable isotope–labeled internal standard. The method requires *prior* identity of the target protein. A unique "signature peptide" is selected from the proteolytic digest of the target protein, and the stable isotope–labeled analog of that peptide is synthesized by incorporating heavy isotope–labeled amino acids (e.g., ^{13}C, ^{15}N). This labeled peptide, known as an *AQUA peptide*, acts as an internal standard. The sample is spiked with a known amount of the AQUA peptide and is analyzed by LC-selected reaction monitoring (SRM). The different masses of the endogenous signature peptide and the AQUA peptide are resolved by SRM tandem MS analysis, and the amount of the native peptide is calculated from the known amount of the AQUA peptide and the ratio of the area under the peaks. The AQUA approach can also be combined with MALDI–MS.

8.8. BIOMARKER DISCOVERY

Protein biomarker discovery is an important field that can aid in clinical diagnosis of a range of diseases. A defective gene can be translated into an abnormal protein that can manifest clinically and can serve as a biomarker of specific diseases. Ongoing efforts using proteomics are helping to discover novel biomarkers of a specific disease state. The search for a biomarker eventually would help in drug discovery process. In his section we describe a proteome-based approach to biomarker discovery using a prototype example of drug-induced idiosyncratic hepatoxicity [79].

A typical biomarker discovery involves three steps: (1) identification of biomarkers, (2) prioritization of markers, and (3) validation of prioritized markers. In the first step of this specific example, the cell culture was treated separately by five drugs, three that can induce idiosyncratic hepatoxicity and the remaining two nontoxic. The proteins were analyzed using the MUDPIT proteomics approach (see Figure 8.12a). A set of 700 potential biomarkers were identified. In the second

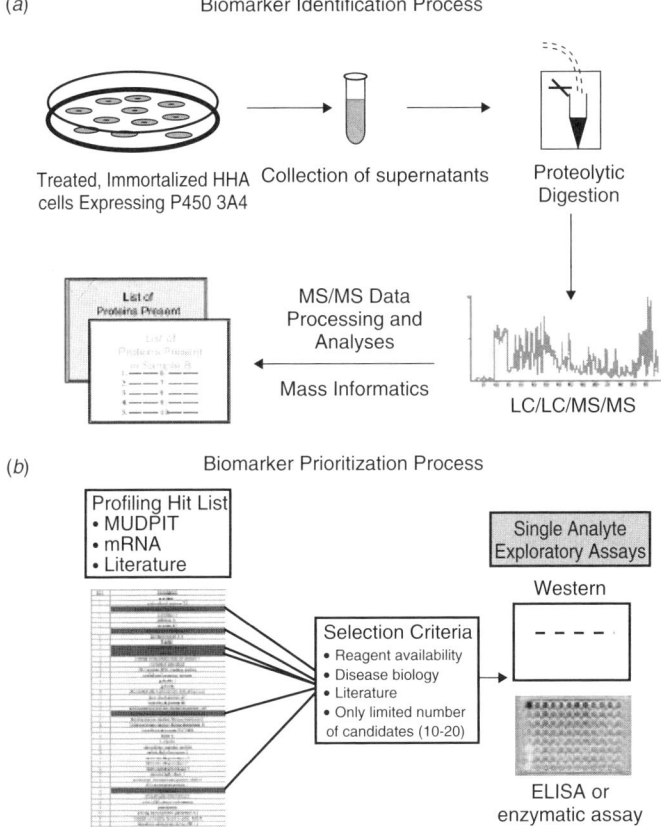

Figure 8.12. Typical biomarker discovery process: (*a*) Steps involved in a biomarker identification process; (*b*) a biomarker prioritization process selects a subset of biomarkers based on certain criteria (e.g., literature precedent, biological relevance, and reagent availability, etc.). (Reproduced from ref. 79 by permission of Elsevier Science, copyright © 2005.)

step, the list was narrowed to 15 candidates on the basis of several criteria (see Figure 8.12b). Primarily, those proteins were chosen that were present at highest concentration in the toxic drug–treated cell culture and were least expressed when treated with nontoxic drugs. Other criteria were the biological relevance and availability of a reagent set for assay development. Two lead hepatoxicity biomarkers, 14-3-3 ZETA and MIF, were selected for validation by developing immunoassays.

8.9. DE NOVO PROTEIN SEQUENCING

The computer database search is applicable to proteins of either known sequences or when their ESTs are available in the database. Several situations require identification of unknown proteins. In such cases, sufficient sequence information must be generated. If partial sequence information is available, it can be used to design oligonucleotide primers for the polymerase chain reaction (PCR) cloning of cDNA. From the sequence of the cDNA, the sequence of the target peptide can be confirmed. Such partial sequence information can be generated by the LC–MS/MS sequencing of peptides in the enzymatic digest of a protein. Tagging the peptide with ^{18}O can facilitate de novo sequence interpretation of MS/MS spectra [80]. Partial sequence information can also be obtained via a database search.

8.10. DETERMINATION OF THE AMINO ACID SEQUENCE OF PEPTIDES

Peptides are important molecules in biochemistry, medicine, and physiology because they play significant roles in a multitude of biochemical and neurological processes, including hormones, neurotransmitters, analgesics, cytokines, growth factors, and diagnostic biomarkers for various clinical disorders. Some peptides have shown clinical activity as inhibitors (of angiotensin-converting enzyme, renin, HIV protease, etc.) and hormones (e.g., oxytocin, leutinizing hormone-releasing hormone, vasopressin, calcitonin). Bioactive peptides are the endogenous products of the metabolic processing of a large precursor protein (Figure 8.3) or can be synthesized chemically in a laboratory setting.

The sequence determination of peptides is crucial to understand their biological functions, to profile metabolic changes in a particular peptide family that can occur as a result of pathological stress or therapeutical treatment, and to assess the purity of newly synthesized peptides. As discussed above, sequencing of peptide from a protein digest is a major step in identification of proteins.

In the past, sequencing of peptides was accomplished using the Edman degradation procedure (see Section 8.2) [3]. Currently, this objective is more often accomplished by interpretation of a peptide's product-ion spectrum. As discussed above, one easy way is to submit the mass spectral data to the database search. Often, this search fails to provide an unequivocal identification of a peptide or protein. Therefore, one needs to resort to a manual interpretation of the mass spectrum. This procedure, known as de novo peptide sequencing, requires a detailed, complete, and accurate understanding of the peptide fragmentation rules and the guidelines for interpretation of mass spectra. Several de novo peptide sequencing algorithms exist to automate spectral interpretation [81,82, and references therein]. Most instrument manufacturers offer de novo sequencing programs. The following sections provide guidelines in de novo sequencing of peptides.

8.10.1. Peptide Fragmentation Rules

In the positive-ion mode, peptides exist as protonated ions, with a preference for the positive charge initially at the N-terminus or at the side chain of a basic residue (Arg, Lys, or His). The proton is usually strongly attached to the basic residues, but the N-terminal proton has a tendency to migrate all along the peptide backbone owing to internal solvation. Thus, the charge can be randomly localized on any one of the backbone amide bonds to provide a heterogeneous population of peptide ions in which the charge is located at different amide linkages. The energy deposited during an ionization or collision–activation event leads to fragmentation of the target peptide. The fragmentation rules of protonated peptides are well documented [83–87]. The site of protonation plays a major role in directing fragmentation. Proton transfer by internal solvation causes cleavage of various bonds all along the peptide backbone [88]. There are three different types of peptide backbone bonds: (1) the alkyl carbonyl bond (CHR–CO), (2) the peptide amide bond (CO–NH), and (3) the amino alkyl bond (NH–CHR) (Figure 8.13). In

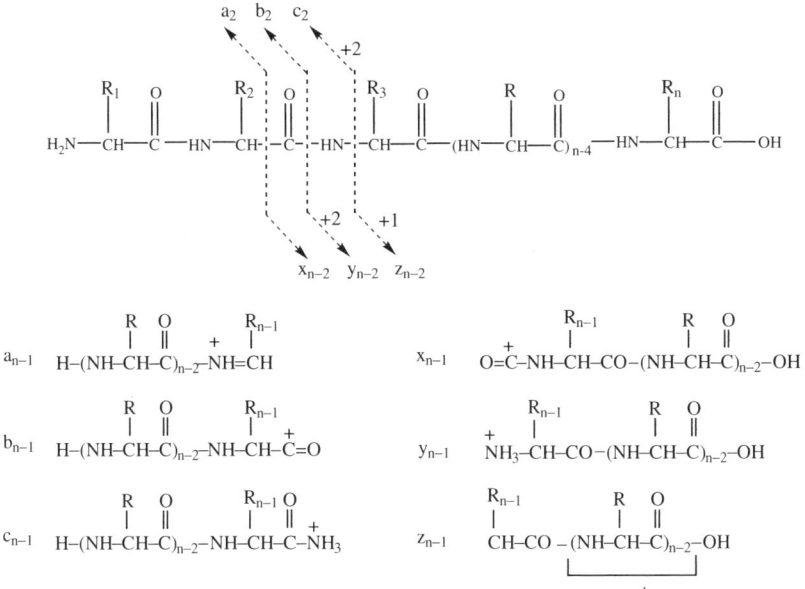

Figure 8.13. Nomenclature and structure of sequence-specific peptide ions. (Reproduced from C. Dass, *Principles and Practice of Biological Mass Spectrometry*, Wiley-Interscience, 2001.)

theory, each one of these three bonds is susceptible to cleavage. The charge might be retained on either N- or C-terminal fragments. Thus, in principle, six different sequence-specific ion series might be formed. The N-terminal charged fragments are designated by the symbols a_n, b_n, and c_n, and the corresponding C-terminal ions by x_n, y_n, and z_n, where n is the number of amino acid residues from the respective ends [89,90]. The structures of these six types of sequence-specific ions are shown in Figure 8.13. Of these six types of ions, b and y are formed more frequently, especially in the low-energy CID spectra of peptides. Often, a ions are also formed along with b ions (by the loss of CO from b). The mechanism of formation of these ions from a singly protonated peptide ion is depicted in Figure 8.14. The example shown is of the peptide Tyr–Leu–Met–Phe–Ser (MW = 659 Da). Although b-ions are known to exist in an oxazalone structure; for convenience, they are shown in acylium form [85]. In doubly charged ions (e.g., those formed by ESI of tryptic peptides) the first charge is firmly localized at the C-terminal basic residue (Arg or Lys) and the other charge can migrate all along the peptide backbone. Fragmentation of such a family of ions will produce singly charged b- and y-ions and doubly charged y-ions. Figure 8.15 rationalizes the formation of b- and y-ions from the doubly protonated precursor ion

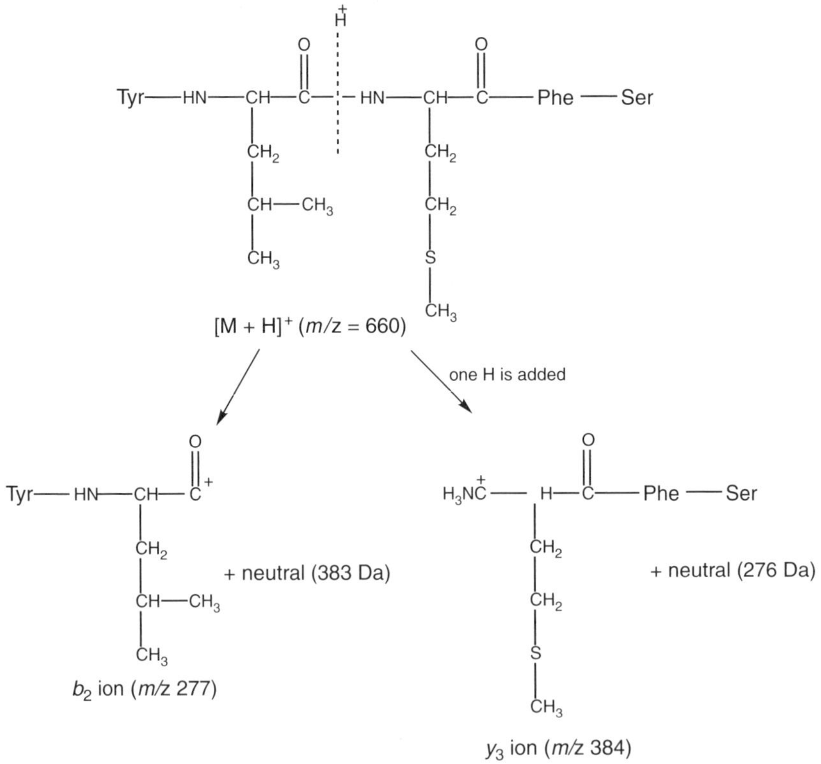

Figure 8.14. Formation of b- and y-ions from singly protonated peptide ions.

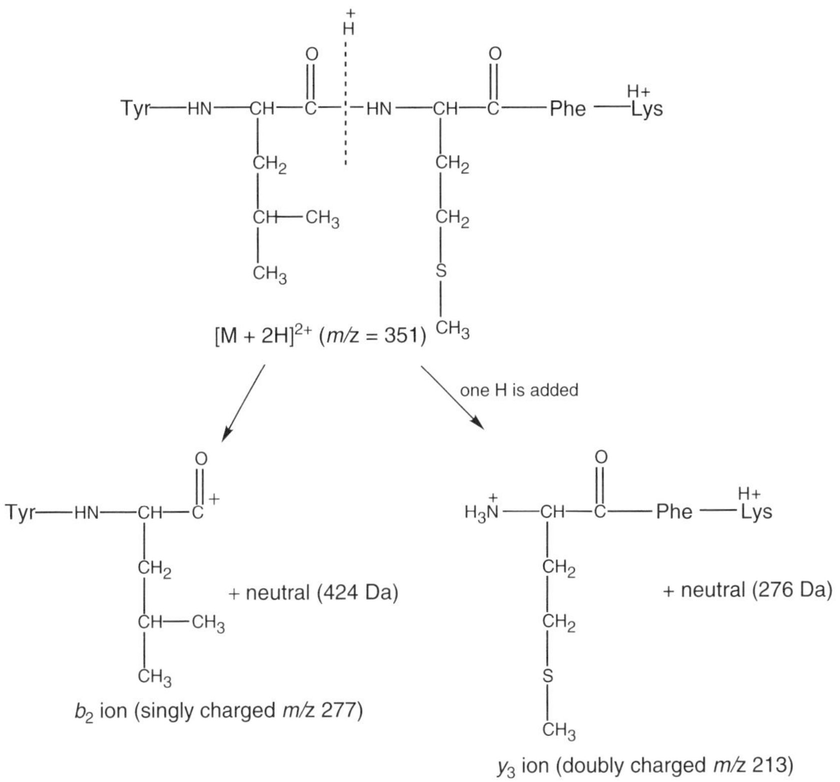

Figure 8.15. Formation of b- and y-ions from doubly protonated peptide ions.

$[M + 2H]^{2+}$ of Tyr–Leu–Met–Phe–Lys (MW = 700 Da). The b_2-ion is a singly charged ion of m/z 277, whereas the y_3-ion is doubly charged with $m/z = 213$.

Under high-energy CID, further cleavage of the nth side chain of the new C-terminal residue of these newly formed ions produces w_n, v_n, and d_n secondary ions (Figure 8.16). The w-ions are formed from z-ions via cleavage of the β,γ-bond, v-ions from y-ions via cleavage of the α,β-bond (i.e., cleavage of the entire side chain), and d ions from $(a + 1)$-ions via cleavage of the β,γ-bond [87,91]. The proclivity of w_n, v_n, and d_n ions is profound in the high-energy CID process [92]. Although not typical, they have also been observed in a conventional mass spectrum [93]. The w- and d-ions are of special significance because they can differentiate between Leu and Ile [87,92]. For example, the w-ions of Leu are always 14 Da lower in mass than those of Ile.

Two other types of sequence-specific ions are immonium ions and internal fragments. These ions are formed by cleavage of two bonds in the peptide backbone (see Figure 8.17). The immonium ions could undergo further fragmentation as shown in Figure 8.18 [94,95]. Internal fragments have either acylium- or immonium-type structure. The immonium ions and their fragments are seen

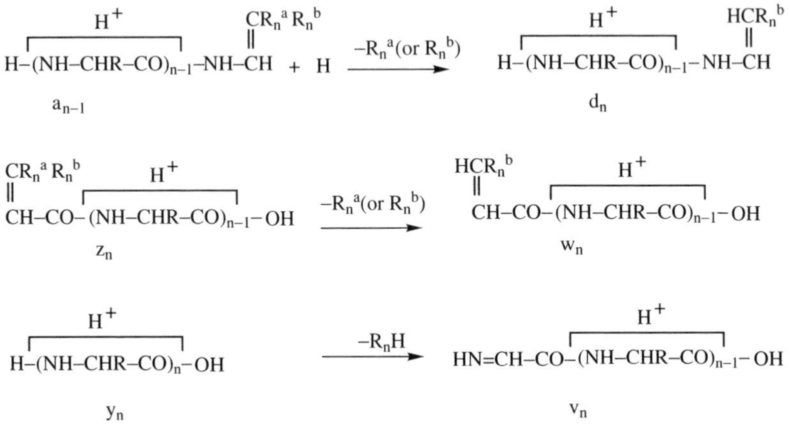

320 PROTEINS AND PEPTIDES: STRUCTURE DETERMINATION

$$\overbrace{H-(NH-CHR-CO)_{n-1}-NH-CH}^{H^+} + H\overset{CR_n^a R_n^b}{\underset{}{\|}} \xrightarrow{-R_n^a (\text{or } R_n^b)} \overbrace{H-(NH-CHR-CO)_{n-1}-NH-CH}^{H^+}\overset{HCR_n^b}{\underset{}{\|}}$$

$$a_{n-1} \hspace{5cm} d_n$$

$$\overset{CR_n^a R_n^b}{\underset{}{\|}}\overbrace{CH-CO-(NH-CHR-CO)_{n-1}-OH}^{H^+} \xrightarrow{-R_n^a (\text{or } R_n^b)} \overset{HCR_n^b}{\underset{}{\|}}\overbrace{CH-CO-(NH-CHR-CO)_{n-1}-OH}^{H^+}$$

$$z_n \hspace{5cm} w_n$$

$$\overbrace{H-(NH-CHR-CO)_n-OH}^{H^+} \xrightarrow{-R_n H} \overbrace{HN=CH-CO-(NH-CHR-CO)_{n-1}-OH}^{H^+}$$

$$y_n \hspace{5cm} v_n$$

Figure 8.16. Structure of sequence-specific side-chain fragments. (Reproduced from C. Dass, *Principles and Practice of Biological Mass Spectrometry*, Wiley-Interscience, 2001.)

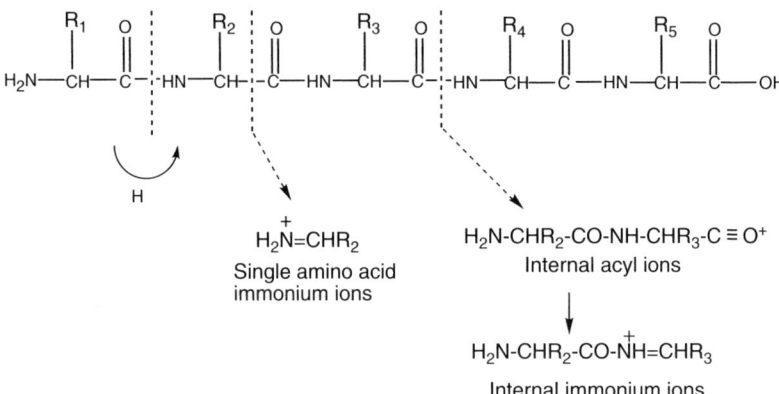

Figure 8.17. Formation of immonium ions and internal fragments.

in a mass spectrum at $m/z < 200$ and denoted by a single-letter code. Although these ions are diagnostic markers of the presence of certain amino acids, they do not provide the location of that amino acid in the sequence of the peptide. Similarly, although internal fragments (especially observed in proline-containing peptides) might, in some cases, provide needed sequence information, they are usually considered nuisances.

In addition, some neutral moieties that are specific to certain amino acid residues are lost from the six types of primary sequence ions. The most common losses are of ammonia and water from *b*- and *y*-ions. The loss of water is observed from the Ser-, Thr-, Asp-, and Glu-containing ions. Similarly, the loss

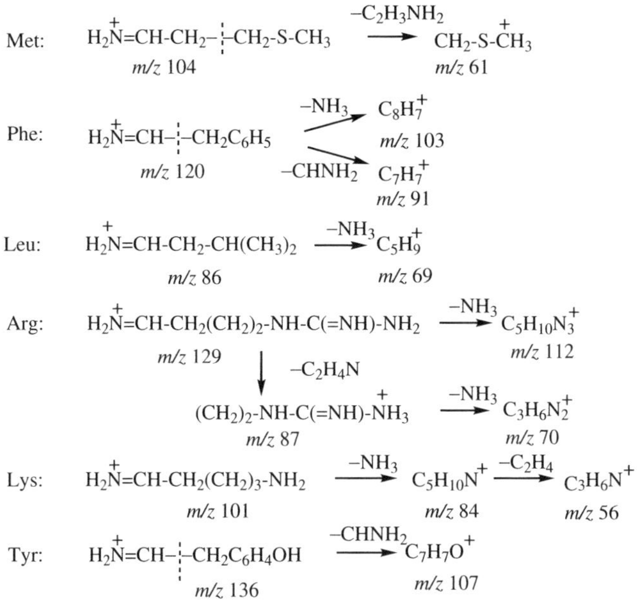

Met: $H_2\overset{+}{N}=CH-CH_2-\vdots-CH_2-S-CH_3$ $\xrightarrow{-C_2H_3NH_2}$ $\overset{+}{CH_2}-S-CH_3$
m/z 104   m/z 61

Phe: $H_2\overset{+}{N}=CH-\vdots-CH_2C_6H_5$
   $\xrightarrow{-NH_3}$ $C_8H_7^+$ m/z 103
   $\xrightarrow{-CHNH_2}$ $C_7H_7^+$ m/z 91
m/z 120

Leu: $H_2\overset{+}{N}=CH-CH_2-CH(CH_3)_2$ $\xrightarrow{-NH_3}$ $C_5H_9^+$
m/z 86   m/z 69

Arg: $H_2\overset{+}{N}=CH-CH_2(CH_2)_2-NH-C(=NH)-NH_2$ $\xrightarrow{-NH_3}$ $C_5H_{10}N_3^+$
m/z 129   m/z 112

$\downarrow -C_2H_4N$

$\overset{+}{(CH_2)_2-NH-C(=NH)-NH_3}$ $\xrightarrow{-NH_3}$ $C_3H_6N_2^+$
m/z 87   m/z 70

Lys: $H_2\overset{+}{N}=CH-CH_2(CH_2)_3-NH_2$ $\xrightarrow{-NH_3}$ $C_5H_{10}N^+$ $\xrightarrow{-C_2H_4}$ $C_3H_6N^+$
m/z 101   m/z 84   m/z 56

Tyr: $H_2\overset{+}{N}=CH-\vdots-CH_2C_6H_4OH$ $\xrightarrow{-CHNH_2}$ $C_7H_7O^+$
m/z 136   m/z 107

Figure 8.18. Fragmentation pathways of immonium ions.

of ammonia is diagnostic of the presence of Asn, Gln, Lys, and Arg residues in the peptide. The loss of 48 Da ($HSCH_3$) is exhibited by the Met-containing sequence ions, but if Met is in the oxidized form, the mass of the expelled neutral will be 64 Da. The Cys-containing ions show a loss of 34 Da (H_2S). If the protein was alkylated with iodoacetic acid prior to digestion, Cys will be in the carboxymethylated form and the neutral lost will be 92 Da ($HSCH_2COOH$). Also, loss of the C-terminal residue is observed to produce the (b_{n-1} + OH) ion.

Although, in principle, six types of primary sequence-specific ion series are likely to form, not all ion series are formed with an equal facility. Certain features in a peptide have an overwhelming influence on the yield of a particular type of sequence ion. For example, the presence of a residue with relatively high basicity in the gas phase, such as His, Trp, Arg, or Lys, near either terminus of a peptide favors charge retention by that terminal fragment. Also, cleavage in the backbone of a peptide at Pro and Phe residues tends to produce abundant y-ions [84].

Some helpful fragmentation rules are summarized below:

1. b- and y- Ions are formed more frequently, especially in the low-energy CID spectra of peptides.
2. High-energy CID will, in addition, produce w_n, v_n, and d_n ions.
3. The loss of water is observed from the Ser-, Thr-, Asp-, and Glu-containing ions.

4. The loss of ammonia is observed from the Asn, Gln, Lys, and Arg contained in the peptide.
5. When cleavage occurs before or after Arg, the $(y - 17)$ or $(b - 17)$ ion would be more abundant than the corresponding y- and b-ions.
6. The peptides that contain a basic residue at the C-terminus also show loss of that C-terminal residue to produce the $(b_{n-1} + 18)$ ion.
7. When the proton is bound to a basic residue, especially Arg, cleavage occurs selectively at Asp- or Glu-XXX.
8. Enhanced cleavage at XXX-Pro is often observed in multiply charged ions.
9. Fragmentation of multiply charged ions will produce a complementary b_n/y_m ion pair ($n + m$ = total residues in a peptide).
10. To determine whether the C-terminus residue is Arg or Lys, look for a y_1-ion.

▶ **Example 8.3** The peptide Thr–Gly–Met–Leu–Phe is sequenced by collision-induced dissociation. At what m/z values will b_3-, y_2-, and a_2-ions be seen in the positive-ion spectrum?

Solution The m/z of b-ions = 1 + residue masses. Therefore, the m/z of the b_3-ion = $1 + 101 + 57 + 131 = 290$. Similarly, the m/z of y-ions = 17 + residue masses + 2. Therefore, the m/z of the y_2-ion = $17 + 147 + 113 + 2 = 279$. The m/z of a-ions = 1 + residue masses − 28. Therefore, the m/z of the a_2-ion = $1 + 101 + 57 - 28 = 131$.

8.10.2. Mass Spectrometry Techniques for Sequence Determination of Peptides

The mass spectrometry techniques described below have been used at one time or other to sequence peptides.

Conventional FAB Mass Spectrometry During FAB ionization, sufficient energy is deposited into an ionized peptide to allow it to fragment into sequence-specific ions. This approach is applicable to pure peptides with fewer than 10 to 12 amino acid residues [84,93]. Fragmentation of peptides is enhanced when glycerol is used as a liquid matrix [96]. In addition, improved results in terms of sensitivity and sequence coverage are obtained with derivatization of peptides, such as by conversion to pyridinium salts, amides, iminothiolanes, ethyl triphenylphosphonium derivatives, quaternary ammonium salts, and other charged derivatives [97].

In-Source Collision-Induced Dissociation of ESI-Produced Ions Fragmentation of the intact molecular ions can be induced in the ion source by

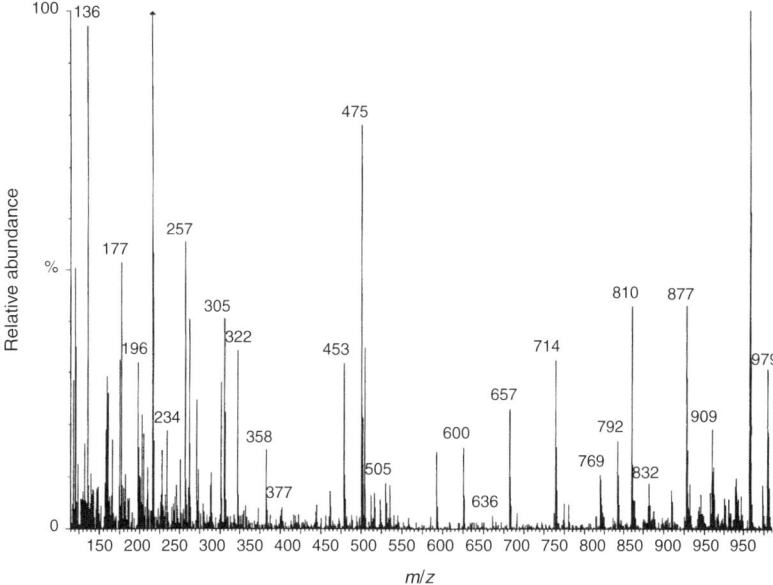

Figure 8.19. ESI in-source CID spectrum of Tyr(PO_4)–Gly–Gly–Phe–Met–Arg–Phe (MW = 956 Da). A complete y-ion series is seen at m/z 166, 322, 453, 600, 657, and 714.

adjusting the nozzle-skimmer cone voltage to impart an additional kinetic energy to the ions to induce their fragmentation in the transport region of the ion source. This method can be employed even with a single-quadrupole instrument. A high-purity sample is, however, a requirement for unambiguous assignment of the sequence by this procedure. The usefulness of this technique is demonstrated in Figure 8.19, which shows the ESI in-source CID spectrum of Tyr(PO_4)–Gly–Gly–Phe–Met–Arg–Phe (MW = 956 Da) at a nozzle-skimmer cone voltage of 80 V.

Ion-Source Decay and Post-Source Decay of MALDI-Produced Ions

Two methods are used to sequence MALDI-produced ions by TOFMS: One approach uses *ion-source decay*; sequence ions thus produced are characterized by operating the TOF instrument in the *delayed-extraction mode*. Fast dissociation in the source-accelerating region creates a wealth of sequence-specific fragment ions, which otherwise would be missed in the normal mode of operation of TOF (i.e., in which the laser pulse and ion-extraction pulse are both applied concurrently). By applying a delayed-extraction pulse, fragments can, however, be accumulated between pulses and mass-analyzed with an ordinary linear TOFMS instrument [98,99]. This process preferentially forms c_n-type N-terminus ions along with y_n and z_n C-terminus ions. Similar to FAB and ESI in-source CID, this approach requires a pure sample and is not applicable to a mixture of peptides.

For not-so-complex peptide mixtures, *post-source decay* (PSD) can be used, where metastable fragmentations in the first linear region of a reflectron–TOF mass spectrometer are monitored [100]. When an ion fragments in the flight tube, the resulting fragments retain the velocity of their precursor, but their kinetic energies are reduced in proportion to their mass. A linear TOF instrument will not make any distinction between the precursor and product ions because both reach the detector at the same time. The field-free region fragment ions can, however, be dispersed in time in terms of their kinetic energies and are mass analyzed with a reflectron–TOF–MS (see Figure 4.4).

Electron-Capture Dissociation ECD is applicable to ESI-produced multiply charged peptide ions [101,102]. It is most conveniently implemented in an FT–ICR instrument. Multiply protonated peptides $[M + nH]^{n+}$, capture a low-energy (<0.2 eV) electron to produce an odd-electron ion $[M + nH]^{(n-1)+\bullet}$, which dissociates rapidly via an energetic H$^\bullet$ transfer to the backbone carbonyl group to form c and $z^\bullet$ sequence-specific ions:

$$\begin{array}{c}
\text{O} \quad {}^\bullet\text{H}\text{-}\text{-}\text{-} \\
\parallel \\
\text{R}-\text{C}-\text{NH}-\text{CHR}'-
\end{array} \rightleftharpoons \begin{array}{c}
\text{OH} \\
| \\
\text{R}-\overset{\bullet}{\text{C}}-\text{NH}-\text{CHR}'-
\end{array} \longrightarrow \begin{array}{c}
\text{OH} \\
| \\
\text{R}-\text{C}=\text{NH} \\
c \\
+ \\
{}^\bullet\text{CHR}'- \\
z^\bullet
\end{array}$$

(8.4)

Minor amounts of $a^\bullet$- and y-ions are also produced when H$^\bullet$ attaches to the amide nitrogen:

$$\begin{array}{c}
\text{O} \quad {}^\bullet\text{H}\text{-}\text{-}\text{-} \\
\parallel \\
\text{R}-\text{C}-\text{NH}-\text{CHR}'-
\end{array} \longrightarrow \begin{array}{c}
\text{O} \\
\parallel \\
\text{R}-\text{C}-\text{NH}_2-\text{CHR}'- \\
a^\bullet
\end{array} \longrightarrow \begin{array}{c}
{}^\bullet\text{R} + \text{C}\equiv\text{O} \\
+ \\
\text{NH}_2\text{CHR}'- \\
y
\end{array}$$

(8.5)

ECD exhibits far more backbone cleavages than CID exhibits and thus can be used to sequence larger peptides.

Slimilar to ECD, electron-transfer dissociation (ETD) is another viable option for sequencing peptides (see Section 4.3.4 for detailed discussion).

Peptide Ladder Sequencing The N- or C-terminal residues can be cleaved sequentially when a peptide is reacted with an aminopeptidase (e.g., aminopeptidase M or leucine aminopeptidase) or a carboxypeptidase (e.g., carboxypeptidase Y), respectively [103]. These reactions generate a mixture of a nested set of fragments. Each contiguous fragment differs by a single amino acid residue. The molecular mass of the resulting truncated peptide is determined after each degradation cycle, and the sequence of the peptide is deduced from those molecular mass differences. The mass differences between successive peaks in the spectrum relate to specific amino acid residues lost during the proteolysis—the order of their occurrence determines the sequence of the peptide. For example, if methionine enkephalin (YGGFM) were to be sequenced by digestion with carboxypeptidase Y, the order of cleavage of amino acid residues would be M, F, G, and G, and the mass spectrum would contain ions at m/z 574.2 (YGGFM), 443.2 (YGGF), 296.1 (YGG), 239.1 (YG), and 182.1 (Y), as shown in Figure 8.20a. Similarly, the treatment of this peptide with aminopeptidase M would produce ions at m/z 574.2 (YGGFM), 411.1 (GGFM), 354.1 (GFM), 297.1 (FM), and 150.0 (M) (Figure 8.20b). One should, however, remember that often, some peptide bonds are not cleaved or sequential cleavage is so rapid that ions arising from the first cleavage are not seen.

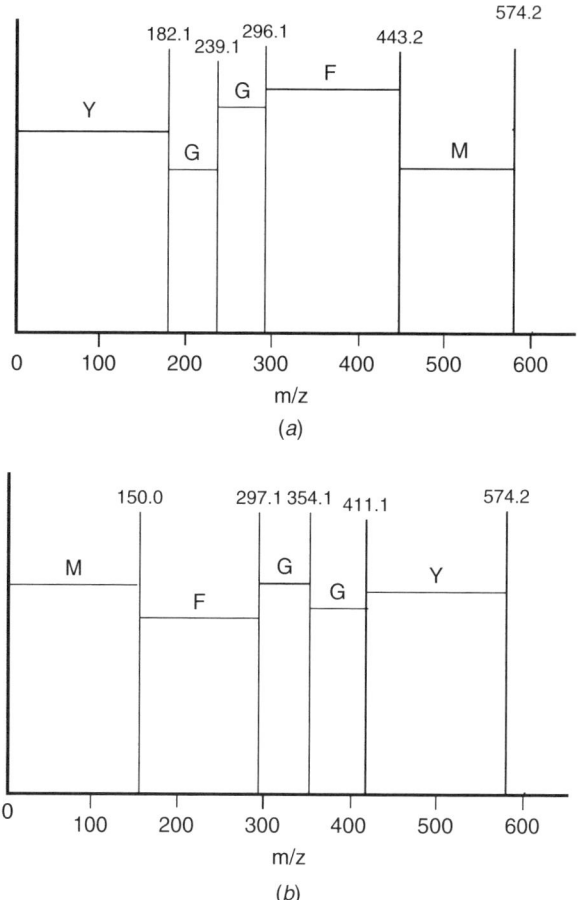

Figure 8.20. Theoretical mass spectra of YGGFM after digestion with (*a*) carboxypeptidase Y and (*b*) aminopeptidase M.

▶ **Example 8.4** The peptide of Example 8.3 (i.e., Thr–Gly–Met–Leu–Phe) is sequenced by the ladder sequencing technique. At what m/z values will ions be observed in the MALDI mass spectrum after digestion with carboxypeptidase Y?

Solution Carboxypeptidase Y is a C-terminal-acting enzyme. Therefore, ions produced are (Thr–Gly–Met–Leu–Phe + H)$^+$, (Thr–Gly–Met–Leu + H)$^+$, (Thr–Gly–Met + H)$^+$, (Thr–Gly + H)$^+$, and (Thr + H)$^+$. The corresponding m/z values are

$$(\text{Thr–Gly–Met–Leu–Phe} + \text{H})^+ = 1 + 101 + 57 + 131 + 113 + 147 + 17 + 1$$
$$= 568$$
$$(\text{Thr–Gly–Met–Leu} + \text{H})^+ = 568 - 147 = 421$$
$$(\text{Thr–Gly–Met} + \text{H})^+ = 421 - 113 = 308$$
$$(\text{Thr–Gly} + \text{H})^+ = 308 - 131 = 177$$
$$(\text{Thr} + \text{H})^+ = 177 - 57 = 120$$

Collision-Induced Dissociation–Tandem MS Tandem mass spectrometry, described in Chapter 4, is the most frequently used method for structure determination of peptides and other biomolecules [6,7]. A most common method of ion activation in tandem mass spectrometry involves collisions of fast-moving ions with a neutral gas to induce extensive fragmentation. Several other ion-activation methods were discussed in Section 4.3. Precursor ions are produced by FAB, MALDI, or ESI. With FAB it is important to reduce the ion-source fragmentations and increase the molecular ion signal. The use of dithiothreitol–dithioerythritol mixture as matrix and acidification of the matrix–sample mixture can help in this objective [96]. ESI has the advantage of producing multiply charged ions, which fragment more efficiently than singly charged ions [104]. CID of these ions will produce singly as well as multiply charged fragment ions. Recognition of multiply charged fragment ions is, however, not difficult. The isotopic peaks of an ion are separated from each other by the m/z value equal to $1/n$, where n is the charge state of the ion. For example, the isotopic peaks of singly, doubly, and triply charged ions will be separated by $m/z = 1$, 0.5, and 0.33, respectively.

High- and low-energy CID processes have both been advocated to obtain the amino acid sequence of peptides. The current preference is for low-energy CID due to the popularity of triple quadrupoles, quadrupole ion traps, LITs, LIT–orbitraps, and Q-TOFs as the tandem MS instrument. The MS/MS spectra of peptides in the two collision energy regimes are, however, not identical. In high-energy CID, the fragmentation pathways reflect the relatively high center-to-mass collision energies and the short time scale of the decomposition process. The formation of all six types of primary sequence-specific ions is possible in this energy regime. In addition, w_n, v_n, and d_n sequence-specific ions are also formed.

In contrast, the fragmentation patterns that are observed under low-energy CID are qualitatively similar to the conventional mass spectra and produce primarily b- and y- ions. The products formed in a low-energy process are due principally to charge-directed fragmentations. Also, w_n, v_n, and d_n sequence-specific ions are absent in the low-energy CID process.

Distinction between b- and y-type ions often becomes difficult in an MS/MS spectrum. ^{18}O labeling of the C-terminal carboxyl group of a peptide can simplify this task [81]. Labeling is performed during digestion of proteins in 1:1 $H_2^{16}O:H_2^{18}O$. Each y- ion that results from the fragmentation of the labeled peptides is split into a doublet of peaks of nearly the same abundance and separated by 2 Da; no such splitting occurs for b-ions. This signature facilitates "readout" of the amino acid sequence [81].

8.10.3. Guidelines for Obtaining the Amino Acid Sequence from a Mass Spectrum

The task of retrieving the amino acid sequence from a mass spectrum is tricky and time-consuming. It involves connecting certain ions into a sequence ion series. The following steps can be followed to find first either the y- or b-ion series:

1. The first step is to look for the presence of immonium ions at the low-mass end of the spectrum. This step will indicate the presence of certain amino acids. The masses of the amino acid residues along with the masses of their immonium ions are listed in Table 8.3. A hint of the presence of certain amino acids can also be gained from the loss of side-chain masses from the $[M + H]^+$ ion (also listed in Table 8.3) and also by the losses of neutral molecules such as ammonia and water.

2. Ignore the first few ions at the high-mass end of the mass spectrum because they are due to the uninformative (but energetically favorable) losses of neutral molecules, such as H_2O, NH_3, CO_2, and HCOOH, from the $[M + H]^+$ ion.

3. Next, identify pairs of ions that differ in mass by 28 Da. This information indicates putative members of b-and a-ion series.

4. Look for the b_2-ion in the low-mass region of the mass spectrum. This ion is invariably present in the low-energy CID spectra of peptides. Identity of the b_2-ion provides an idea of the first two amino acid residues, all possible combinations of which are listed in Table 8.4. The m/z values listed there are for singly charged ions and are equal to the sum of residue masses + 1. Caution should be exercised, however, as in some cases the combined mass of the two amino acid residues is equal to the residue mass of a single amino acid. These possibilities are listed as a footnote in Table 8.4. Knowledge of the b_2-ion also helps to recognize the y_{n-2} ion.

5. Next, it is important to recognize a specific sequence ion series. This objective can be achieved by identifying the $(n − 1)$th ion and then matching the mass difference between each of the fragment ion peaks with the

TABLE 8.3. Masses of Amino Acid Residues, Their Immonium Ions and Side Chains, and of Neutral Loss

Amino Acid	Symbol		Residue Mass (Da)	Immonium Ion Mass (Da)	Side-Chain Mass (Da)	Neutral Loss Da
Alanine	Ala	A	71.0371	44	15	—
Arginine	Arg	R	156.1011	129	100	17
Asparagine	Asn	N	114.0429	87	58	17
Aspartic acid	Asp	D	115.0269	88	59	18
Cysteine	Cys	C	103.0092	76	47	34[a]
Glutamic acid	Glu	E	129.0426	102	73	18
Glutamine	Gln	Q	128.0586	101	72	17
Glycine	Gly	G	57.0215	30	—	—
Histidine	His	H	137.0589	110	81	—
Isoleucine	Ile	I	113.0841	86	57	—
Leucine	Leu	L	113.0841	86	57	—
Lysine	Lys	K	128.0950	101	72	17
Methionine	Met	M	131.0405	104	75	48[b]
Phenylalanine	Phe	F	147.0684	120	91	—
Proline	Pro	P	97.0528	70	—	—
Serine	Ser	S	87.0320	60	31	18
Threonine	Thr	T	101.0477	74	45	18
Tryptophan	Trp	W	186.0793	159	130	—
Tyrosine	Tyr	Y	163.0633	136	107	—
Valine	Val	V	99.0684	72	43	—

[a] The S-Carboxymethyl derivative of Cys will lose 92 Da.
[b] Oxidized Met will lose 64 Da.

residue masses. Only the residue masses of 20 naturally occurring amino acids are considered (Table 8.3). This strategy is explained in Figure 8.21.

6. For example, look for the y_{n-1}-ion in the high-mass region of the spectrum. This task becomes easier once the b_2-ion is identified. Now, only two possibilities exist for the loss of the residue mass to form the y_{n-1}-ion from the $[M + H]^+$ ion. Further members of the y-ion series are recognized from the mass difference between the y_{n-1}-ion and the next ion in the spectrum, and matching it with the residue masses. Often, the last two residues in the sequence are difficult to recognize because of the absence of corresponding ions in the spectrum. In that case, listing of the mass values (1 + sum of two residue masses) in Table 8.4 can provide a clue. This mass value can be calculated from the m/z of the y_2-ion (see Example 8.5). The y_1-ion is easy to recognize in the spectrum of a tryptic peptide; this ion will appear at m/z 147 or 175 due to Lys or Arg, respectively.

7. Once the y-ions are identified, look for the corresponding b/y-ion pairs, keeping in mind that the mass of b + y ions = the peptide mass + 2.

DETERMINATION OF THE AMINO ACID SEQUENCE OF PEPTIDES

TABLE 8.4. Masses of b_2-Ions[a]

	G	A	S	P	V	T	C	I/L	N	D	K/Q	E	M	H	F	R	Y	W
G	115																	
A	129	143																
S	145	159	175															
P	155	169	185	195														
V	157	171	187	197	199													
T	159	173	189	199	201	203												
C	161	175	191	201	203	205	207											
I/L	171	185	201	211	213	215	217	227										
N	172	186	202	212	214	216	218	228	229									
D	173	187	203	213	215	217	219	229	230	231								
K/Q	186	200	216	226	228	230	232	242	243	244	257							
E	187	201	217	227	229	231	233	243	244	245	258	259						
M	189	203	219	229	231	233	235	245	246	247	260	261	263					
H	195	209	225	235	237	239	241	251	252	253	266	267	269	275				
F[b]	205	219	235	245	247	249	251	261	262	263	276	277	279	285	295			
R	214	228	244	254	256	258	260	270	271	272	285	286	288	294	304	313		
Y	221	235	251	261	263	265	267	277	278	279	292	293	295	301	311	320	327	
W	244	258	274	284	286	288	290	300	301	302	315	316	318	324	334	343	350	373

[a] GG = N = 114; GA = K/Q = 128; GV = R = 156; GE = AD = SV = W = 186.
[b] Oxidized methionine will also show this combination.

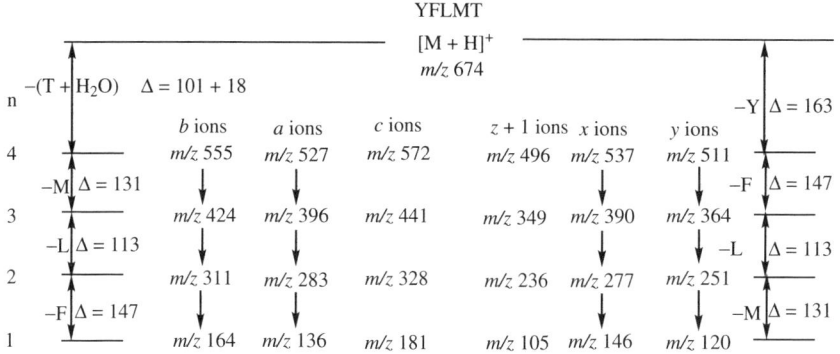

Figure 8.21. The m/z values of the six-sequence ion series in the spectrum of YFLMT.

8. Recognize all members of the b-ion series.
9. Tally the mass of the peptide from the amino acid sequence thus identified.
10. One can also proceed to find the b-ion series first if it is known that the peptide is derived by trypsinization of a protein. We would look for a y_1-ion at m/z 147 or 175 that will indicate the presence of Lys or Arg, respectively, at the C-terminus. Then we could identify b_{n-1}-ion

and would continue to recognize other b-ions in the series by finding the mass difference between the b_{n-1}-ion and the next ion in the spectrum and matching it with the residue masses. Once a b-ion is known, the corresponding complementary y-ion can easily be identified.

▶ **Example 8.5** Interpret the spectrum shown in Figure 8.22 that was acquired with a QIT. This MS/MS spectrum is of a singly protonated precursor ion at m/z 684 (i.e., the peptide mass is 683 Da).

Solution Because of the low-mass cutoff limit of the QIT, no ion below m/z 200 is seen. Therefore, immonium ion information is missing. The ion pairs that differ in mass by 28 Da are m/z 250/278, 397/425, and 510/538. These ions are probable members of the a- and b- ion series. Inspection of the low-mass region shows that m/z 221 could be the b_2-ion. From Table 8.4, the first two residues are identified as YG–(or GY–). The y_{n-1}-ion is at either m/z $(684 - 57 = 617)$ or $(684 - 163 = 521)$. The y_{n-2} ion will be seen at m/z $(683 - 221 + 2) = 464$ (calculated from the mass of the b_2-ion and using the formula "the mass of $b + y$ ions = the peptide mass + 2"). This m/z value confirms that m/z 521 is due to the y_{n-1}-ion and that the first two residues from the N-terminus are YG. Next, two members of the y ion series are found to be of m/z 407 and 260 because the mass difference between the y-ions matches the residue masses of 57 and 147 Da. The first four amino acids from the N-terminus are identified as Y–G–G–F–X–X. The sum of the residue masses of the last residues is = $(260 - 17 - 2) = 241$. Because this mass is greater than the mass of a single

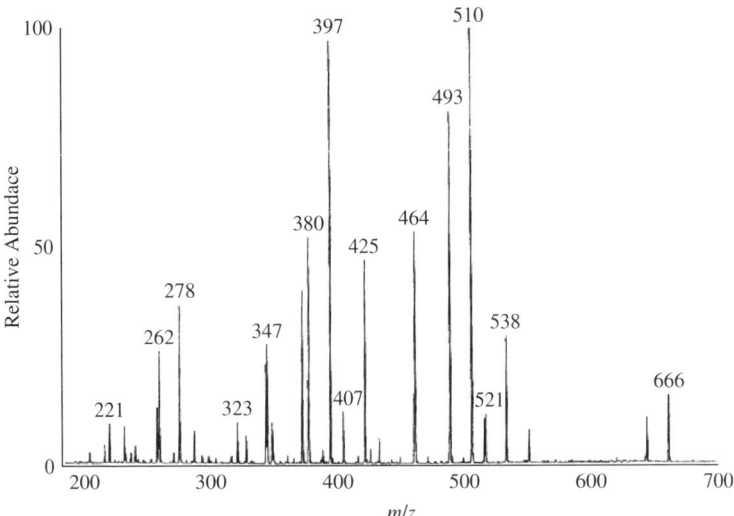

Figure 8.22. ESI–MS/MS spectrum of an unknown peptide.

residue, there is possibility of at least two residues. The entry $241 + 1 = 242$ in Table 8.4 can reveal the identity of the y_1/y_2 pair. The possible combinations are K/L(or I) and Q/L(or I). Now, look for probable members of the b-ion series. These are m/z 164, 221, 278, 425, and 538. The first four are identified from the corresponding b/y-ion pair combinations (e.g., 164/521, 221/464, 278/407, and 425/260) and the last one from the probable a/b-ion pair (e.g., m/z 510/538). Thus, the last two amino acids from the N-terminus are Leu (ILe) and Lys, and the complete amino acid sequence of the peptide is YGGFLK.

▶ **Example 8.6** Identify the peptide whose MS/MS spectrum is shown in Figure 8.23. This spectrum is of a doubly protonated precursor ion at m/z 574.3 of a tryptic peptide. The spectrum was acquired with a QIT.

Solution The peptide molecular mass is calculated to be $= 1146.6$ Da (M $= 2 \times 574.3 - 2 \times 1.0078$). Because the precursor ion is of m/z 574.3, the low-mass cut-off limit has also shifted to a lower m/z value (compared to the spectrum in Figure 8.22). Now the signal at m/z 175.1 is recognized as a y_1-ion and reveals the presence of Arg at the C-terminus. The corresponding b_{n-1}-ion is of m/z 973.4. The prominent ion pairs that differ in mass by 28 Da are of m/z 973.4/945.4 and 844.4/816.4, which suggest that ions of m/z 973.4 and 844.4 could be the members of the b-ion series. The mass difference of 129.0 between these two ions points to Glu as the next residue from the C-terminus. The corresponding y-ion is of m/z 304.3. The m/z 715.4 is identified as the next b-ion; the

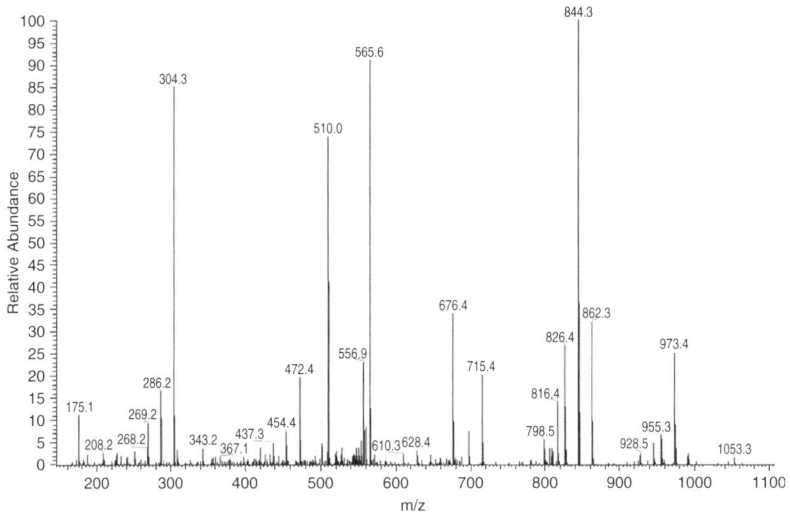

Figure 8.23. ESI–MS/MS spectrum of an unknown peptide.

mass difference between m/z 844.4 and 715.4 matches the residue mass of Glu. The next b-ion identified is of m/z 628.4; the mass difference between m/z 715.4 and 628.4 matches the residue mass of Ser. The next b-ion could be m/z 571.3 because it is 57.1 Da lower in mass. The corresponding y-ion is not seen. The next two b-ions identified are of m/z 472.4 and 343.2. The corresponding y-ions are present at 676.4 and 805.4, respectively. Inspection of the low-mass region shows that the m/z 286.2 is a probable b_2-ion. From Table 8.4 the first two residues from the N-terminus could be Trp–Val–, Val–Trp–, Arg–Glu, or Glu–Arg–. The presence of m/z 269.2 as a satellite peak (i.e., the loss of NH_3) suggests Glu–Arg–as the possible first two residues. The corresponding y_{n-2}-ion is seen at m/z 862.3. Thus, the b-ion series is observed at m/z 973.4, 844.3, 715.4, 628.4, 571.3, 472.4, 343.2 and 286.2, and the amino acid sequence identified in the peptide is Glu–Arg–Gly–Glu–Val–Gly–Ser–Glu–Glu–Arg. The members of y-ion series are present at m/z 862.3, 805.4, 676.4,–, 520.2, 433.2, 304.3, and 175.1. The amino acid sequence derived from this series is X–X–Gly–Glu–X–X–Ser–Glu–Glu–Arg. Further confirmation of the peptide sequence derived comes from the presence of ions at the m/z values that correspond to the $(b_n - 17)$ and $(b_n - 18)$ ions. For example, the ions of m/z 955.3, 826.4, 697.4, and 454.4 are all 18 Da lower than the m/z values of the respective b-ions and thus indicate the presence of Glu and Ser at those places in the sequence of the peptide.

OVERVIEW

The objective of this chapter was to describe the structure and functions of proteins and peptides and methods that are in common use to identify these biopolymers. Characterization of proteins in body tissues, fluids, and cellular system is critical from the viewpoint of human health. Currently, proteomics is playing a major role in analyzing the protein content of biological samples. The four basic steps of proteome analysis are sample preparation, protein separation, mass spectrometry analysis, and database search. Sample preparation involves homogenization of the tissue samples to rupture the cells and extract proteins, followed by enrichment of proteins. To reduce the complexity of this mixture, the extract may be divided into subcellular fractions before the separation step. Fractionation of proteins based on two-dimensional get electrophoresis is at the heart of proteomics. The technique first separates proteins in the IEF mode on the basis of differences in their pI values and in the second dimension on the basis of molecular size by SDS–PAGE. Alternative approaches that make use of orthogonal chromatographic separation techniques (e.g., RP–HPLC, SCX, SEC) are also becoming popular. For mass spectrometry analysis, the protein fractions are cleaved into smaller peptides. Two proteomics methods have gained wide acceptance. The first method creates peptide-mass fingerprints by measuring the molecular masses of the peptides in the protein digest with MALDI–TOF–MS. This approach has been integrated with 2-DE-based fractionation of proteins. The second method is based on sequence ion information. In a popular version of this approach, known as MUDPIT, the homogenized sample is digested directly without any prior fractionation of the complex mixture of proteins. The resulting complex peptide digest is analyzed directly with 2D–LC–ESI–MS/MS. In the final step of the proteome analysis, the mass information in the peptide-mass fingerprints or sequence ion information is correlated with the theoretical values created from each protein entry in the database.

Quantitative proteomics is performed to determine the differential expression of proteins in two samples. This analysis, as an example, can provide comparative information of the proteomes of diseased and healthy subjects as a guide to biomarker discovery and to identify potential drug targets. A majority of the methods discussed in this chapter rely on isotopic labeling of the two samples in question with light and heavy isotopes and comparing the MS response.

In the final section of the chapter we discuss procedures to obtain the amino acid sequence of peptides, which is a vital component of protein identification. The fragmentation rules of peptides and guidelines for interpretation of peptide-mass spectrum are also discussed. By following these systematic steps, de novo sequencing of peptides and hence of a protein can be achieved.

EXERCISES

8.1. The molecular mass of a protein that has three disulfide bonds and three cysteine residues is 10,525 Da. It is first reduced with 2-mercaptoethanol and then alkylated with iodoacetic acid. What will be the new molecular mass of the protein?

8.2. The cytochrome c (60 to 104) fragment (KEETLMEYLENPKKYIPGTK-MIFAGIKKKTEREDLIAYLKKATNE) was treated with cyanogen bromide. How many fragments are produced and at what m/z values will they be observed in the positive-ion MALDI spectrum?

8.3. A protein of the sequence Phe–Tyr–Ser–Met–Asp–Ile–Leu–Lys–Trp–Val–Ala–Met–Glu–Val–Asp–Tyr–Gly–Pro–Arg–Thr–Asn–Pro–Glu–Ala–Leu–Phe–Ile–Tyr–Arg–Thr–Ser–Met–Tyr–Trp–Gln–Leu–Met is treated first with endoproteinase Glu–C and then with trypsin. How many peptide fragments will be generated by this treatment, and what will be their molecular mass?

8.4. A protein of the sequence AGTKVVLPIRCTFIHGYAKWMpTGYARN-QVSAKAND (pT refers to phosphothreonine) is digested with trypsin. The T_4-tryptic fragment was analyzed for its amino acid sequence by CID–MS/MS technique. Calculate the m/z values of all of the sequence-specific y-, c-, and a-ions that can be formed from this tryptic peptide.

8.5. How do bottom-up proteomics differ from top-down proteomics?

8.6. Outline the basic steps that are used in the identification of proteins by proteomics.

8.7. How is proteome analysis performed with the peptide-mass fingerprinting approach?

8.8. How is proteome analysis performed with the sequence-ion analysis method?

8.9. Outline the salient points of the MUDPIT proteome analysis approach.

8.10. How can the differential expression of proteins in two samples be obtained by the ICAT method?

8.11. Describe how the differential expression of proteins is determined by the SILAC method.

8.12. How are proteins quantified by the AQUA method?

8.13. Mark the b_2-, x_2-, z_3-, y_4-, and a_3-ions in the following sequence.

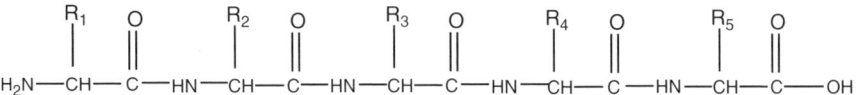

8.14. Write the structure of b_3, y_3, and a_3 ions that are derived from the sequence shown in Exercise 8.13.

8.15. A peptide of sequence Phe–Ala–Met–Leu–Tyr is ionized by MALDI and sequenced by collision-induced dissociation technique. At what m/z values will the b_3-, c_2-, y_3-, y_2-, and a_4 ions be seen in the positive-ion spectrum?

8.16. The ESI–MS/MS spectrum of the doubly charged precursor ion of the peptide Phe–Glu–Tyr–Met–Leu–Thr–Gly–Ile–Lys was acquired by a Q-TOF instrument. What will be the m/z values of y_4^{2+}, y_3^+, y_6^+, and $(b_6^+ - H_2O)$ ions?

8.17. At what m/z values will the single amino acid immonium ions for Phe, Tyr, Met, and Lys be observed in the spectrum of the peptide of Exercise 8.16?

8.18. What approach is used to distinguish b- and y-ions in the CID spectrum of a tryptic peptide?

8.19. A peptide of sequence TFIHGYA is treated separately with carboxypeptidase and aminopeptidase. Calculate the molecular masses of the peptide fragments that are generated in these two reactions.

8.20. The sequence of an unknown peptide was determined with the peptide ladder sequencing technique. After treatment with an aminopaptidase and

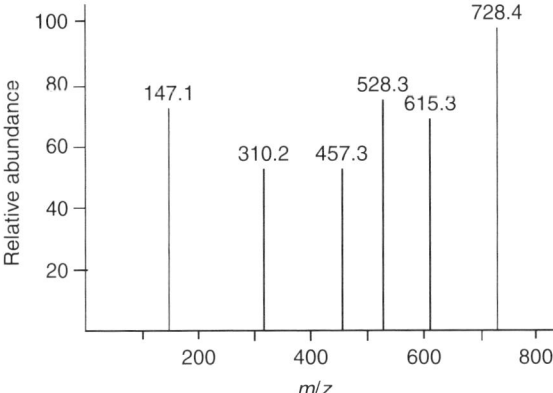

Figure 8.24. MALDI mass spectrum of an unknown peptide.

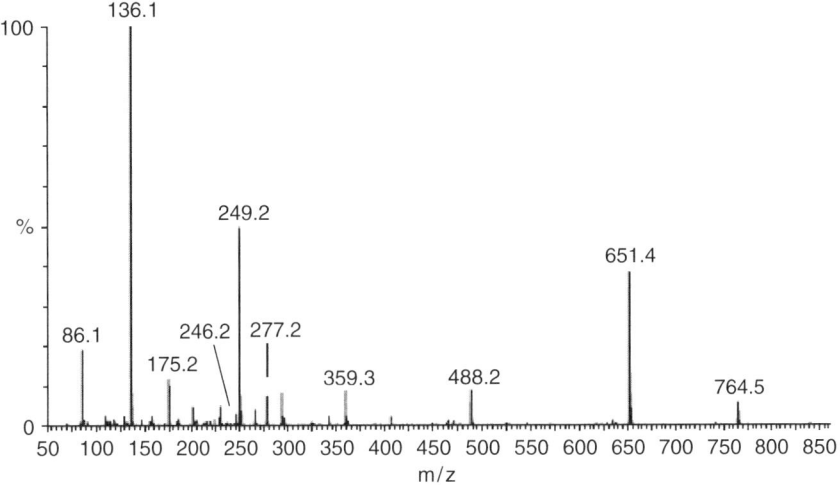

Figure 8.25. ESI-MS/MS spectrum of an unknown of Exercise 8.21. (Reproduced from ref. 105 by permission of Elsevier, copyright © 2005.)

MALDI–TOF–MS analysis of the ladders generated, the spectrum shown in Figure 8.24 was obtained. Determine the sequence of the protein.

8.21. Identify the sequence of a peptide whose ESI–MS/MS spectrum is shown in Figure 8.25.

8.22. The isotopic cluster of an ion is observed at m/z 574.3, 574.8, and 575.3. Determine the multiplicity (singly, doubly, etc.) of the ion.

ADDITIONAL READING

1. P. James, ed., *Proteome Research: Mass Spectrometry*, Springer-Verlag, New York, 2000.
2. M. Kinter, E. N. and Sherman, *Protein Sequencing and Identification Using Tandem Mass Spectrometry*, Wiley, New York, 2000.
3. I. M. Rosenberg, *Protein Analysis and Purification: Benchtop Techniques*, Birkhauser, Boston, MA, 1996.
4. W. S. Hancock, ed., *New Methods in Peptide Mapping for the Characterization of Proteins*, CRC Press, Boca Raton, FL, 1996.
5. C. Dass, *Principles and Practice of Biological Mass Spectrometry*, Wiley-Interscience, New York, 2000.
6. M. Hamdan and P. G. Righetti, *Proteomics Today: Protein Assessment and Biomarkers Using Mass Spectrometry, 2D-Electrophoresis, and Microarray Technology*, Wiley, Hoboken, NJ 2005.

REFERENCES

1. C. L. Brooks, M. Karplus, B. M. Pettitt, *Proteins: A Theoretical Perspective of Dynamics Structure and Thermodynamics*, Wiley, Hoboken, 1998.
2. P. Edman, Method for determination of the amino acid sequence in peptides, *Acta Chim. Scand.* **4**, 283–293 (1950).
3. P. Edman and G. Begg, A protein sequenator, *Eur. J. Biochem.* **1**, 80–91 (1967).
4. R. Aebersold and D. Goodlett, Mass spectrometry in proteomics, *Chem. Rev.* **101**, 269–295 (2001).
5. R. Aebersold and M. Mann, Mass spectrometry-based proteomics, *Nature* **422**, 198–206 (2003).
6. D. F. Hunt, J. R. Yates, et al., Protein sequencing by tandem mass spectrometry, *Proc. Natl. Acad. Sci. USA* **83**, 6233–6237 (1986).
7. K. Biemann and S. A. Martin, Mass spectrometric determination of the amino acid sequence of peptides and proteins, *Mass Spectrom. Rev.* **6**, 1–75 (1987).
8. B. Sundqvist and R. D. Macfarlane, Californium-252-plasma desorption mass spectrometry, *Mass Spectrom. Rev.* **2**, 421–460 (1985).
9. M. Dreger, Proteome analysis at the level of subcellular structures, *Eur. J. Biochem.* **270**, 589–599 (2003).
10. N. L. Anderson and N. G. Anderson, The human plasma proteome: History, character, and diagnostic prospects, *Mol. Cell Proteom.* **1**, 845–67 (2002).
11. K. Bjorhall, T. Miliotis, and P. Davidsson, Comparison of different depletion strategies for improved analysis in proteomic analysis of human serum samples, *Proteomics* **5**, 307–317 (2005).
12. H. Zischka, G. Weber, et al., Improved proteome analysis of *Saccharomyces cerevisiae* mitochondria by free-flow electrophoresis, *Proteomics* **3**, 906–16 (2003).
13. P. H. O'Farrell, High resolution two dimensional electrophoresis of proteins, *J. Biol. Chem.* **250**, 4007–4021 (1975).

14. J. Klose and U. Kobalz, Two-dimensional electrophoresis of proteins: an updated protocol and implications for a functional analysis of a genome, *Electrophoresis* **16**, 1034–1059 (1995).

15. H. Towbin, T. Staehelin, and J. Gordon, Electrophoretic transfer of proteins from polyacrylamide gels to nitrocellulose sheets: procedure and some applications *Proc. Natl. Acad. Sci. USA* **76**, 4350–4354 (1979); A. Klapper, B. MacKay, and M. D. Rash, Rapid high resolution Western blotting: from gel to image in a single day, *BioTechniques* **12**, 650–654 (1992).

16. F. Van, B. Subramanian, A. Nakeff, T. Barder, S. J. Parus, and D. M. Lubman, A comparison of drug-treated and untreated HCT-116 human colon adenocarcinoma cells using a 2-D LC mapping method based upon chromatofocusing pI fractionation, *Anal. Chem.* **75**, 2299–2308 (2003).

17. S. D. Patterson and R. Aebersold, Mass spectrometric approaches for the identification of gel-separated proteins, *Electrophoresis* **16**, 1791–1814 (1995).

18. M. Galvani, E. Bordini, C. Peubelli, and M. Hamdan, Effect of experimental conditions on the analysis of sodium dodecyl sulphate polyacrylamide gel electrophoresis separated proteins by matrix-assisted laser desorption/ionisation mass spectrometry, *Rapid Commun. Mass Spectrom.* **14**, 18–25 (2000).

19. R. R. Ogorzalek Loo, C. Mitchell, et al., Sensitivity and mass accuracy for proteins analyzed directly from polyacrylamide gels: implications for proteome mapping, *Electrophoresis* **18**, 382–390 (1997).

20. J. S. Cottrell, Protein identification by peptide mass fingerprinting, *Pept. Res.* **7**, 115–118 (1994).

21. J. Gao, S. Friedrichs, A. R. Dongre, and J. Opiteck, Guidelines for the routine application of the peptide hit technique, *J. Am. Soc. Mass Spectrom.* **16**, 1231–1238 (2005).

22. R. K. Blackburn and R. J. Anderegg, Characterization of femtomole levels of proteins in solution using rapid proteolysis and nanoelectrospray ionization mass spectrometry, *J. Am. Soc. Mass Spectrom.* **8**, 483–494 (1997).

23. R. W. Nelson, The use of bioreactive probes in protein characterization, *Mass Spectrom. Rev.* **16**, 353–376 (1997).

24. P. Matsudaira, Sequence from picomole quantities of proteins electroblotted onto polyvinylidene difluoride membranes, *J. Biol. Chem.* **262**, 10035–110038 (1987).

25. J. Rosenfeld, J. Capdevielle, J. C. Guillemot, and P. Ferrera, In-gel digestion of proteins for internal sequence analysis after one- or two-dimensional gel electrophoresis, *Anal. Biochem.* **203**, 173–179 (1992).

26. R. Aebersold, J. Leavitt, R. A. Saavedra, L. E. Hood, and S. B. H. Kent, Internal amino acid sequence analysis of proteins separated by one- or two-dimensional gel electrophoresis after in situ protease digestion on nitrocellulose, *Proc. Natl. Acad. Sci. USA* **84**, 6970–6974 (1987).

27. S. D. Patterson, D. Hess, T. Yungwirth, and R. Aebersold, High-yield recovery of electroblotted proteins and cleavage fragments from a cationic polyvinylidene fluoride-based membrane, *Anal. Biochem.* **202**, 193–202 (1992).

28. W. S. Hancock, C. A. Bishop, and M. T. W. Hearn, The use of high pressure liquid chromatography (hplc) for peptide mapping of proteins, IV, *Anal. Biochem.* **89**, 203–212 (1978).

29. B. A. Johnson, J. M. Shirokawa, W. S. Hancock, M. W. Spellman, L. J. Basa, and D. W. Aswad, Formation of isoaspartate at two distinct sites during in vitro aging of human growth hormone, *J. Biol. Chem.* **264**, 14262–14271 (1989).
30. P. Lecchi and R. M. Caprioli, Matrix-assisted laser desorption mass spectrometry for peptide mapping, in W. S. Hancock, ed., *New Methods in Peptide Mapping for the Characterization of Proteins*, CRC Press, Boca Raton, FL, 1996, pp. 219–240.
31. G. Li, M. Waltham, et al., Rapid mass spectrometric identification of proteins from two-dimensional polyacrylamide gels after in gel proteolytic digestion, *Electrophoresis* **18**, 391–402 (1997).
32. S. K. Chowdhury, V. Katta, and B. T. Chait, Electrospray ionization mass spectrometric peptide mapping: a rapid, sensitive technique for protein structure analysis, *Biochem. Biophys. Res. Comm.* **167**, 686–692 (1990).
33. R. Moore, L. Licklider, D. Schumann, and T. D. Lee, A Microscale electrospray interface incorporating a monolithic, poly(styrene-divinylbenzene) support for online liquid chromatography/tandem mass spectrometry analysis of peptides and proteins, *Anal. Chem.* **70**, 4879–4884 (1998).
34. J. Wahl, H. R. Udsethand, and R. D. Smith, Capillary electrophoresis–mass spectrometry in peptide mapping; in W. S. Hancock, ed., *New Methods in Peptide Mapping for the Characterization of Proteins*, CRC Press, Boca Raton, FL, 1996, pp. 143–179.
35. C. Dass, Recent developments and applications of high-performance liquid chromatography–electrospray ionization mass spectrometry, *Curr. Org. Chem.* **3**, 193–209 (1999).
36. M. T. Davis and T. D. Lee, Rapid protein identification using a microscale electrospray LC/MS system on an ion trap mass spectrometer, *J. Am. Soc. Mass Spectrom.* **9**, 194–201 (1998).
37. B. Küster and M. Mann, ^{18}O-labeling of N-glycosylation sites to improve the identification of gel-separated glycoproteins using peptide mass mapping and database searching, *Anal. Chem.* **71**, 1431–1440 (1999).
38. Q. Yand, A. J. Tomlinson, and S. Naylor, Membrane preconcentration CE, *Anal. Chem.* **71**, 183A-189A (1999).
39. M. R. Wilkins, C. Pasqualli, et al., From proteins to proteomes: large scale protein identification by two-dimensional electrophoresis and amino acid analysis, *Bio-Technology* **14**, 61–65 (1996).
40. J. R. Yates III, Mass spectrometry and the age of the proteome, *J. Mass Spectrom.* **33**, 1–19 (1998).
41. N. L. Kelleher, Top-down proteomics, *Anal. Chem.*, **76**, 197A-203A (2004).
42. W. J. Henzel, T. M. Billeci, J. T. Stults, S. C. Wong, C. Grimley, and C. Watanbe, Identifying proteins from two-dimensional gels by molecular mass searching of peptide fragments in protein sequence databases, *Proc. Natl. Acad. Sci. USA* **90**, 5011–5015 (1993).
43. D. J. C. Pappin, P. Hojrup, and A. J. Bleasby, Rapid identification of proteins by peptide-mass fingerprinting, *Curr. Biol.* **3**, 327–332 (1993).
44. B. Thiede, W. Höhenwarter, et al., Peptide mass fingerprinting, *Methods* **35**, 237–247 (2005).

45. J. R. Yates, S. Speicher, P. R. Griffin, and T. Hunkapillar, Peptide mass maps: a highly informative approach to protein identification, *Anal. Biochem.* **214**, 397–408 (1993).
46. M. Mann, P. Hojrup, and P. Roepstorff, Use of mass spectrometric molecular weight information to identify proteins in sequence databases, *Biol. Mass Spectrom.* **22**, 338–345 (1993).
47. P. James, M. Quadroni, E. Carafoli, and G. Gonnet, Protein identification by mass profile fingerprinting, *Biochem. Biophys. Res. Commun.* **195**, 58–64 (1993).
48. J. R. Yates, J. K. Eng, A. L. McCormack, and D. M. Schieltz, Method to correlate tandem mass spectra of modified peptides to amino acid sequences in the protein database, *Anal. Chem.* **67**, 1426–1436 (1995).
49. J. K. Eng, A. L. McCormack, and J. R. Yates III, An approach to correlate tandem mass-spectral data of peptides with amino acid-sequences in protein database, *J. Am. Soc. Mass Spectrom.* **5**, 976–989 (1994).
50. J. R. Yates III, and A. L. McCormack, Mining genomes with MS, *Anal. Chem.* **68**, 534A-540A (1996).
51. J. R. Yates III, S. F. Morgan, C. L. Gatlin, P. R. Griffin, and J. K. Eng, Method to compare collision-induced dissociation spectra of peptides: potential for library searching and subtractive analysis, *Anal. Chem.* **70**, 3557–3565 (1998).
52. T. P. Conrads, G. A. Anderson, T. D. Veenstra, L. Paša-Tolić, and R. D. Smith, Utility of accurate mass tags for proteome-wide protein identification, *Anal. Chem.* **72**, 3349–3354 (2000).
53. Y. Shen, N. Tolic, et al., Ultra-sensitive proteomics using high-efficiency on/line micro-SPE-nanoLC-nanoESI MS and MS/MS, *Anal. Chem.* **77**, 144–154 (2004).
54. A. D. Norbeck, M. E. Manroe, J. N. Adkins, K. K. Anderson, D. S. Daly, and R. D. Smith, The utility of accurate mass and LC elution time information in the analysis of complex proteome, *J. Am. Soc. Mass Spectrom.* **16**, 1239–1249 (2005).
55. B. J. Cargile and J. L. Stephenson, Jr., An alternative to tandem mass spectrometry: isoelectric point and accurate mass for the identification of peptides, *Anal. Chem.* **76**, 267–275 (2004).
56. S. P. Gygi, G. L. Corthals, Y. Zhang, Y. Rochon, and R. Aebersold, Evaluation of two-dimensional gel electrophoresis-based proteome analysis technology, *Proc. Natl. Acad. Sci. USA*, **97**, 9390–9395 (2000).
57. A. Gorg, W. Weiss, and M. J. Dunn, Current two-dimensional electrophoresis technology for proteomics, *Proteomics* **4**, 3665–3685 (2004).
58. A. Qualtieri, E. Urso, et al., Proteomics of bovine myelin sheath: characterization of truncated form of PO by MALDI–TOF/TOF mass spectrometry, *J. Am. Soc. Mass Spectrom.* **17**, 107–123 (2006).
59. D. A. Wolters, M. P. Washburn, and J. R. Yates III, An automated multidimensional protein identification technology for shotgun proteomics, *Anal. Chem.* **73**, 5683–5690 (2001).
60. T. Kislinger, A. O. Gramolini, D. H. MacLennan, and A. Emili, Multidimensional protein identification technology (MUDPIT): technical overview of a profiling method optimized for the comprehensive proteomic investigation of normal and diseased heart tissue, *J. Am. Soc. Mass Spectrom.* **16**, 1207–1220 (2005).
61. C. Delahunty and J. R. Yates III, Protein identification using 2D–LC–MS/MS, *Methods* **35**, 248–255 (2005).

62. B. J. Cargile, D. L. Talley, and J. L. Stephenson, Jr., Immobilized pH gradients as a first dimension in shotgun proteomics and analysis of the accuracy of pI predictability of peptides, *Electrophoresis* **25**, 936–945 (2004).

63. P. R. Jalili and C. Dass, Proteome analysis in the bovine adrenal medulla using liquid chromatography with tandem mass spectrometry, *Rapid Commun. Mass Spectrom.* **18**, 1877–1884 (2004).

64. K. Boutilier, M. Ross, et al., Comparison of different search engines using validated MS/MS test datasets, *Anal. Chim. Acta* **534**, 11–20 (2005).

65. M. Mann and M. Wilm, Error-tolerant identification of peptides in sequence databases by peptide sequence tags, *Anal. Chem.* **66**, 4390–4399 (1994).

66. M. J. McCass, Quantitative for proteomics, *Anal. Chem.* **77**, 294A-302A (2005).

67. E. Marengo, E. Robotti, F. Antonucci, D. Cecconi, N. Campostrini, and P. G. Righetti, Numerical approaches for quantitative analysis of two-dimensional maps: a review of commercial software and home-made systems, *Proteomics* **5**, 654–666 (2005).

68. M. Unlu, M. E. Morgan, and J. S. Minden, Difference gel electrophoresis: a single gel method for detecting changes in protein extracts, *Electrophoresis* **18**, 2071–2077 (1997).

69. S. P. Gygi, B. Rist, et al., Quantitative analysis of complex protein mixtures using isotope-coded affinity tags, *Nat. Biotechnol.* **17**, 994–999 (1999).

70. J. V. Olsen, J. R. Andersen, et al., Hys-Tag: a novel proteomic quantification tool applied to differential display analysis of membrane proteins from distinct areas of mouse brain, *Mol. Cell Proteom.* **3**, 82–92 (2004).

71. P. L. Ross et al., Multiplexed protein quantitation in *Saccharomyces cerevisiae* using amine-reactive isobaric tagging reagents, *Mol. Cell. Proteomics*, **3**, 1154–1169 (2004).

72. X. D. Yao, A. Freas, J. Remirez, P. A. Demirev, and C. Fenselau, Proteolytic O-18 labeling for comparative proteomics: model studies with two serotypes of adenovirus, *Anal. Chem.* **73**, 2836–2842 (2001).

73. X. D. Yao, C. Afonso, and C. Fenselau, Dissection of proteolytic ^{18}O labeling: endoprotease-catalyzed ^{16}O-to-^{18}O exchange of truncated peptide substrates, *J. Proteome Res.* **2**, 147–152 (2003).

74. S. E. Ong, B. Blagoev, I. Kratchmarova, D. B. Kristensen, H. Steen, A. Pandey, and M. Mann, Stable isotope labeling by amino acids in cell culture, *Mol. Cell Proteom.* **1**, 376–386 (2002).

75. N. Ibarrola, D. E. Klume, M. Gronborg, A. Iwahori, and A. Pandey, A proteomic approach for quantitation of phosphorylation using, stable isotope labeling by amino acids in cell culture, *Anal. Chem.* **75**, 6043–6049 (2003).

76. P. V. Bondarenko, D. Chelius, and T. A. Shaler, Identification and relative quantification of protein mixtures by enzymatic digestion followed by capillary reversed-phase liquid chromatography tandem mass spectrometry, *Anal. Chem.* **74**, 4741–4749 (2002).

77. S. A. Gerber, J. Rush, O. Stemman, M. W. Krishner, and S. P. Gygi, Absolute quantification of proteins and phopshoproteins from cell lysates by tandem MS, *Proc. Natl. Acad. Sci. USA* **100**, 6940–6945 (2003).

78. D. S. Kirkpatrick, S. A. Gerber, and S. P. Gygi, The absolute quantification strategy: a general procedure for the quantification of proteins and post-translational modifications, *Methods* **35**, 265–273 (2005).

79. J. Gao, L.-A. Garulacan, et al., Biomarker discovery in biological fluids, *Methods* **35**, 291–302 (2005).
80. A. Shevchenko, O. N. Jensen, et al., Linking genome and proteome by mass spectrometry: large scale identification of yeast proteins from two-dimensional gels, *Proc. Natl. Acad. Sci. USA* **93**, 14440–14445 (1996).
81. W. M. Hines, et al., Pattern-based algorithm for peptide sequencing from tandem high-energy CID mass spectra, *J. Am. Soc. Mass Spectrom.* **3**, 326–361 (1992).
82. M. T. Olson, J. A. Epstein, and A. L. Yergey, De novo peptide sequencing using exhaustive enumeration of peptide composition, *J. Am. Soc. Mass Spectrom.* **17**, 1041–1049 (2006).
83. I. A. Papayannopoulos, The interpretation of collision-induced dissociation tandem mass spectra of peptides, *Mass Spectrom. Rev.* **14**, 49–73 (1995).
84. C. Dass and D. M. Desiderio, Fast-atom-bombardment mass spectrometry analysis of opioid peptides, *Anal. Biochem.* **163**, 52–66 (1987).
85. T. Yalcin, C. Khouw, I. G. Csizmadia, M. R. Peterson, and A. G. Harrison, The structure and fragmentation of b_n ($n \geq 3$) ions in peptide spectra, *J. Am. Soc. Mass Spectrom.* **7**, 233–242 (1996).
86. X.-J. Tang and R. K. Boyd, An investigation of fragmentation mechanisms of doubly charged protonated tryptic peptides, *Rapid Commun. Mass Spectrom.* **6**, 651–657, (1992).
87. R. S. Johnson, S. A. Martin, and K. Biemann, Collision-induced fragmentation of $(M + H)^+$ ions of peptides: side chain specific sequence ions, *Int. J. Mass Spectrom. Ion Proc.* **86**, 137–154 (1988).
88. V. H. Wysocki, G. Tsaprailis, L. L. Smith, and L. A. Breci, Mobile and localized protons: a framework for understanding peptide dissociation, *J. Mass Spectrom.* **35**, 1399–1406 (2000).
89. P. Roepstorff and J. Fohlman, Proposal for a common nomenclature for sequence ions in mass spectra of peptides, *Biomed. Mass Spectrom.* **11**, 601 (1984).
90. K. Biemann, Contributions of mass spectrometry to peptide and protein structure, *Biomed. Environ. Mass Spectrom.* **16**, 99–111 (1988).
91. R. S. Johnson, S. Martin, K. Biemann, J. T. Stults, and J. T. Watson, Novel fragmentation process of peptides by collision-induced decomposition in a tandem mass spectrometer: differentiation of leucine and isoleucine, *Anal. Chem.* **59**, 2621–2625 (1987).
92. J. T. Stults and J. T. Watson, Identification of a new type of fragment ion in the collisional activation spectra of peptides allows leucine/isoleucine differentiation, *Biomed. Environ. Mass Spectrom.* **14**, 583–586 (1987).
93. C. Dass and P. Mahalakshmi, Amino acid sequence determination of phosphoenkephalins using liquid secondary ionization mass spectrometry, *Rapid Commun. Mass Spectrom.* **9**, 1148–1154 (1995).
94. W. Heerma and W. Kulik, Identification of amino acids in the fast atom bombardment mass spectra of peptides, *Biomed. Environ. Mass Spectrom.* **16**, 155–159 (1988).
95. C. Dass, *Rapid Commun. Mass Spectrom.* Characterization of phosphorylated amino acids by fast atom bombardment mass spectrometry, **3**, 264–266 (1989).
96. C. Dass, The role of a liquid matrix in controlling FAB-induced fragmentation, *J. Mass Spectrom.* **31**, 77–82 (1996).

97. K. D. Roth, K Z. H. Huang, N. Sadagopan, and J. T. Watson, Charge derivatization of peptides for analysis by mass spectrometry, *Mass Spectrom. Rev.* **17**, 255–274 (1998).
98. M. L. Vestal, P. Juhasz, and S. A. Martin, Delayed extraction matrix-assisted laser desorption time-of-flight mass spectrometry, *Rapid Commun. Mass Spectrom.* **9**, 1044–1050 (1995).
99. R. M. Whittal and L. Li, High-Resolution Matrix-Assisted Laser Desorption/Ionization in a Linear Time-of-Flight Mass Spectrometer, *Anal. Chem.* **67**, 1950–1954 (1995).
100. B. Spengler, Post-source decay analysis in matrix-assisted laser desorption/ionization mass spectrometry of biomolecules, *J. Mass Spectrom.* **32**, 1019–1036 (1997).
101. E. Mirgorodskya, P. Roepstroff, and R. A. Zubarev, Localization of O-glycosylation sites in peptides by electron capture dissociation in a Fourier Transform mass spectrometer, *Anal. Chem.* **71**, 4431–4436 (1999).
102. R. A. Zubarev, N. A. Kruger, et al., *J. Am. Chem. Soc.* **121**, 2857–2862 (1999).
103. R. J. Cotter, *Time-of-Flight Mass Spectrometry: Instrumentation and Applications in Biological Research*, American Chemical Society, Washington, DC, 1997.
104. J. A. Loo, J. Edmonds, and R. D. Smith, Tandem mass spectrometry of very large molecules: serum albumin sequence information from multiply charged ions formed by electrospray ionization, *Anal. Chem.* **63**, 2488–2499 (1991).
105. R. S. Johnson, M. T. Davis, J. A. Taylor, and S. D. Patterson, Informatics for protein identification by mass spectrometry, *Methods* **35**, 223–236 (2005).

CHAPTER 9

PROTEINS AND PEPTIDES: POSTTRANSLATIONAL MODIFICATIONS

A *posttranslational modification* (PTM) is defined as any modification of a protein structure that occurs after synthesis of the protein via translation of the messenger RNA. For example, nearly every eukaryotic protein is known to undergo proteolytic processing and modifications of functional groups of selected amino acids. The side chains of aliphatic amino acids (e.g., alanine, glycine, isoleucine, leucine, valine) and of phenylalanine do not participate in PTM, but active groups such as an amine, thiol, thioether, carboxylic acid, and hydroxyl, are susceptible to covalent modifications in a biological milieu. More than 300 PTMs are known. Some of the common ones are listed in Table 9.1 along with the residue mass and change in the residue mass upon modification relative to the unmodified amino acid. Often, those modifications have a significant influence on the activity and specificity of proteins. Also, they might play a role in stabilizing the protein structure and regulating the enzyme activity. Therefore, the characterization of PTMs has attracted much attention. Currently, mass spectrometry is playing an ever-increasing role in the analysis of posttranslationally modified proteins and peptides. Mass spectrometry-based techniques measure the precise change in the residue mass upon modification. For example, a phosphorylated protein–peptide is detected on the basis of an increase in the mass by 79.97 Da for each phospho unit from the corresponding

---

Fundamentals of Contemporary Mass Spectrometry, by Chhabil Dass
Copyright © 2007 John Wiley & Sons, Inc.

TABLE 9.1. Selected Posttranslational Modifications of Proteins

Modification	Modified Amino Acid	Mass of Modified Residue (Da) (Monoisotopic)	Mass Shift (Da) (Monoisotopic)
β-Elimination in:			
Serine	Dehydroalanine	69.02	−18.01
Threonine	Dehdroamino-2-butyric acid	83.04	−18.01
Acetylation of:			
Lysine	N-acetyllysine	170.11	+42.01
N-terminal			
Biotinylation of:			
Lysine			+226.08
N-terminal			
Bromination of tyrosine	Bromotyrosine	240.97	+77.91
Carbamyl	N-terminal		+43.01
Carboxylation of:			
Aspartic acid	β-Carboxyaspartic acid	159.03	+44.00
Glutamic acid	γ-Carboxyglutamic acid	173.04	+44.00
Chlorination of tyrosine	Chlorotyrosine	197.02	+33.96
Cysteinylation of cysteine	S-cysteinylcysteine	222.2	+119.00
Deamidation of:			
Asparagine			−0.984
Glutamine			−0.984
Disulfide bond formation			−2.02
Formylation	N-terminal		+27.99
Geranylgeranlylation of cysteine	S-geranylgeraniol cysteine	375.4	+272.4
Glutathionylation of cysteine	S-glutathionylcysteine	408.08	+305.07
Hydroxylation of:			
Proline	γ-Hydroxyproline	113.05	+16.0
Asparagine	β-Hydroxyasparagine	130.04	+16.0
Aspartic acid	β-Hydroxyaspartic acid	131.03	+16.0
Lysine	δ-Hydroxylysine	144.10	+16.0
Carbamyl lysine		171.11	+43
Methylation of:			
Aspartic acid	Aspartic acid methyl ester	129.05	+14.02
Glutamic acid	Glutamic acid methyl ester	143.06	+14.02

TABLE 9.1. (*continued*)

Modification	Modified Amino Acid	Mass of Modified Residue (Da) (Monoisotopic)	Mass Shift (Da) (Monoisotopic)
Nitration of tyrosine	Nitrotyrosine	208.1	+45.1
Oxidation of:			
Methionine	Methionine sulfoxide	147.04	+16.00
Cysteine	Cysteic acid	151.01	+48.00
Palmitoylation of cysteine	Palmitoylcysteine	341.3	+238.3
Phosphorylation of:			
Serine	Phosphoserine	167.00	+79.97
Threonine	Phosphothreonine	181.00	+79.97
Aspartic acid	Phosphoaspartic acid	195.00	+79.97
Lysine	Phospholysine	208.07	+79.97
Histidine	Phosphohistidine	217.03	+79.97
Tyrosine	Phosphotyrosine	243.03	+79.97
Sulfation of tyrosine	Sulfotyrosine	243.03	+79.97
Attachment of:			
Hexoses (Gal, Glc, Man)			+162.05
Hexosamines (GalN, GlcN)			+161.07
N-Acetylhexosamines (GalNAc, GlcNAc)			+203.08
Proteolysis			+18.01
Formation of:			
Homoserine	Homoserine	101.05	+14.02
Pyroglutamic acid	Pyroglutamic acid	111.03	−18.01

nonphosphorylated analog. Mass spectrometry–based methods have the advantages of sensitivity and speed of analysis. Characterization of some important PTMs, such as disulfide bonds, phosphorylation, and glycosylation, is the subject of this chapter.

DISULFIDE BONDS IN PROTEINS

One of the most common PTMs of proteins is the formation of disulfide bonds by oxidation of the cysteinyl thiols. This functionality is critical for biological activity of proteins because it plays an important role in the folding–unfolding processes and in stabilizing the correct three-dimensional structure of proteins [1]. The disulfide bonds are also essential for the expression of the activity of small neuropeptides (e.g., oxytocin and vasopressin). Identification of disulfide bonds can thus provide essential information on the structure and function of proteins.

9.1. TRADITIONAL APPROACHES TO IDENTIFY DISULFIDE BONDS

A well-established conventional protocol to locate disulfide bonds in proteins includes cleaving a protein between the half-cysteinyl residues in a manner that leaves the disulfide bonds intact, isolating the cysteine-containing peptides, and performing either amino acid analysis or sequence analysis [2]. Alternatively, the disulfide bonds are reduced sequentially to thiols and alkylated with radiolabeled iodoacetic acid [1,3]. The completely reduced protein is cleaved enzymatically, and the disulfide bonds are detected by identifying the radiolabeled peptides.

9.2. MASS SPECTROMETRY–BASED METHODS TO IDENTIFY DISULFIDE BONDS

Currently, mass spectrometry has assumed a greater role in characterizing the disulfide-bridge-containing proteins and peptides [4–15]. A general mass spectrometry–based protocol is illustrated in Figure 9.1. Important steps of this protocol are discussed below.

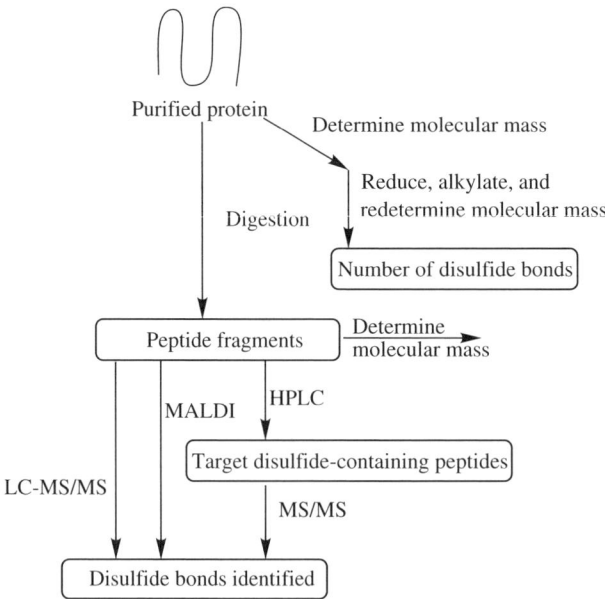

Figure 9.1. General scheme to analyze disulfide-containing proteins.

9.2.1. Determination of the Number of Disulfide Bonds

To determine the number of disulfide bonds, the protein is reduced and alkylated as described in Section 8.5.1 by reacting it with either dithiothreitol or 2-mercaptoethanol, followed by reaction with iodoacetic acid. The molecular masses of the native protein (M_{nat}) and reduced and alkylated protein (M_{r+a}) are determined with MALDI–MS or ESI–MS. S-Carboxymethylation increases the mass of each cysteine residue by 59 Da. From the change in the molecular mass, the number of cysteine residues (N_{cys}) and hence the number of disulfide bonds (N_{S-S}) can be estimated using

$$N_{cys} = \frac{M_{r+a} - M_{nat}}{59} \qquad N_{SH} = \frac{M_a - M_{nat}}{59 - 1} \qquad N_{S-S} = \frac{N_{cys} - N_{SH}}{2} \quad (9.1)$$

If free sulfhydryl groups are present, they must be taken into account for these calculations. The number (N_{SH}) of free sulfhydryl groups, if not known, is determined by following the same procedure but without performing the reduction step, and from the determined molecular mass of the alkylated protein (M_a).

▶ **Example 9.1** The molecular mass of a protein is 10,275 Da. Upon reduction with dithiothreitol and alkylation with iodoacetic acid, its mass increased to 10,865 Da. In a separate experiment, the protein was treated only with iodoacetic acid. The molecular mass of the protein was found to be 10,391 Da. Calculate the number of disulfide bonds in this protein.

Solution First, the number of free sulfhydryl groups must be calculated from the increase in mass that resulted from alkylation of the protein. Therefore, $N_{SH} = (10,391 - 10,275)/58 = 2$. The second step is to calculate the number of cysteine residues from the increase in mass upon reduction and alkylation reactions. $N_{cys} = (10,865 - 10,275)/59 = 10$. Finally, the number of disulfide bonds $= (10 - 2)/2 = 4$.

9.2.2. Generation of Disulfide-Containing Peptides

A disulfide-containing polypeptide chain needs to be cleaved into smaller fragments prior to mass spectrometry analysis. The cleavage of a protein generates two types of disulfide-containing peptides, intrachain (**a**; or intramolecular) and interchain (**b**; or intermolecular):

The cleavage of disulfide-containing proteins requires special attention. First, the generated peptides should contain only a single disulfide bond. Second, no disulfide exchange should occur during the cleavage reaction. Specific and nonspecific bond-cleaving reagents can both be used, but the digestion must be performed at pH 7.0 or less to avoid the exchange of disulfide bonds. Cleavage

TABLE 9.2. Reagents and Conditions for Cleaving Disulfide-Containing Proteins

Cleaving Agent	Digestion Conditions [Buffer; pH; Temperature (°C)]	Time (h)
Cyanogen bromide	88% formic acid	24
Trypsin	50 mM $CH_3CO_2NH_4$; 6.5; 37	24
Trypsin	0.1% TFA	5
Chymotrypsin	Water; 7.0; 35	6
Thermolysin	50 mM NH_4HCO_3; 8.0; 37	3
Pepsin	5% formic acid; 37	6
Endoproteinase Glu-C	Water; 35	14
	0.1 M $CH_3CO_2NH_4$; 4.0; 37	20

with specific reagents (e.g., CNBr, trypsin, and endoproteinase Glu–C) has the advantage that the C-terminal residues are known from the cleavage specificity of the reagent. If the specific reagent fails to cleave a protein between half-cysteinyl residues, partial acid hydrolysis or nonspecific proteases (e.g., pepsin) can be used. Mild acidic conditions employed in the reaction of nonspecific reagents also help to preserve disulfide bonds [5]. Disulfide bonds also remain intact during digestion with CNBr and endoproteinase Glu–C, but disulfide interchange might occur at the pH where trypsin has the highest activity. Table 9.2 lists various cleaving reagents and the conditions modified appropriately for disulfide bond–containing proteins.

$$\underset{(a)}{\overset{\text{}}{S-S}} \qquad \underset{(b)}{\text{—S—S—}}$$

9.2.3. Identification of Disulfide-Containing Peptides by FAB–MS

The following methods have been developed in conjunction with FAB–MS analysis [6–8].

Comparative Peptide Mapping This procedure is used when the protein is cleaved with specific cleaving agents and the sequence of the protein is known. A simple molecular mass measurement of the peptides in a protein digest will identify the position of a disulfide bond in the protein.

Reduction of Disulfide Bonds The disulfide-bridged peptides can also be identified by analyzing the protein digest before and after the dithiothreitol reduction of disulfide bonds. The peaks due to disulfide-containing peptides will disappear, and new peaks related to reduced peptides will appear. If the peptide contains an intramolecular disulfide bond, its molecular ion will shift by 2 Da after irradiation with the FAB beam:

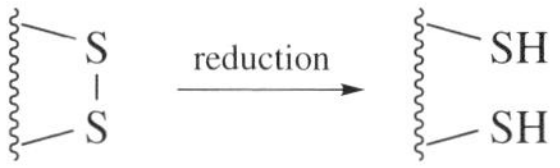

$$\begin{matrix} -S \\ | \\ -S \end{matrix} \xrightarrow{\text{reduction}} \begin{matrix} -SH \\ \\ -SH \end{matrix} \qquad (9.2)$$

For interchain-linked peptides, the molecular ions of both of the constituent chains will be observed:

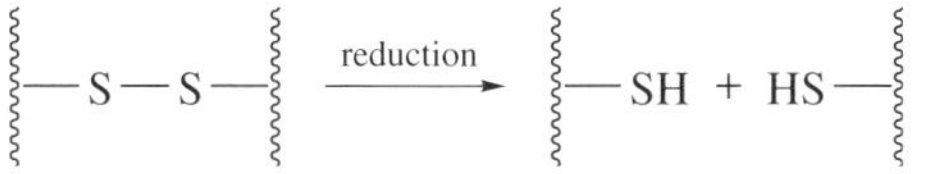

$$-S-S- \xrightarrow{\text{reduction}} -SH \;+\; HS- \qquad (9.3)$$

In a variation of this procedure, an S—S bond in a peptide can be reduced in situ during prolonged (>5 min) irradiation of the peptide–matrix (dithiothreitol) by the fast atom beam.

On-Probe Oxidation In this procedure, thiol- and disulfide-containing peptides are treated with performic acid on the FAB probe to convert each cysteine residue to cysteic acid; a concomitant increase of 98 Da in the mass of a peptide that contains an intramolecular disulfide bond results [9]:

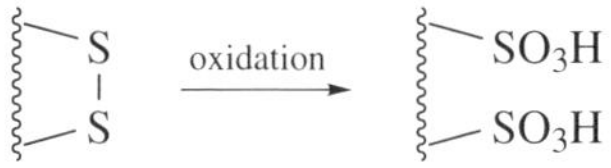

$$\begin{matrix} -S \\ | \\ -S \end{matrix} \xrightarrow{\text{oxidation}} \begin{matrix} -SO_3H \\ \\ -SO_3H \end{matrix} \qquad (9.4)$$

The signal due to the interchain disulfide bond-containing peptides disappears. Instead, two new peaks at 49 Da higher than the mass of the constituent peptides are observed:

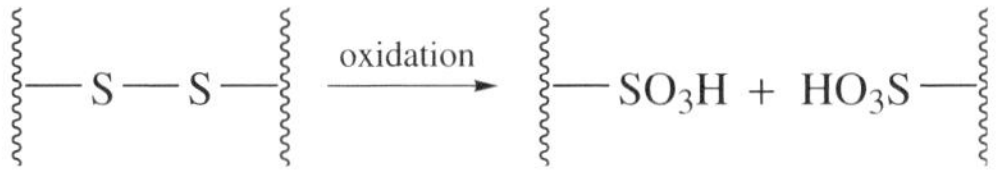

$$-S-S- \xrightarrow{\text{oxidation}} -SO_3H \;+\; HO_3S- \qquad (9.5)$$

One drawback of this procedure is that methionine and tryptophan residues are also prone to oxidation by performic acid.

9.2.4. Identification of Disulfide-Containing Peptides by MALDI–MS

As with the FAB analysis, disulfide-containing peptides are detected by comparative peptide mapping (i.e., by analyzing the protein digest before and after chemically reducing disulfide bonds). Also, if the sequence of the protein is known, the site of disulfide bonds can be ascertained from the peptide masses measured in a protein digest. Another MALDI-based approach relies on prompt fragmentation in the ion source. This *ion-source decay* (ISD) leads to cleavage of disulfide bonds and thus provides important information on their location

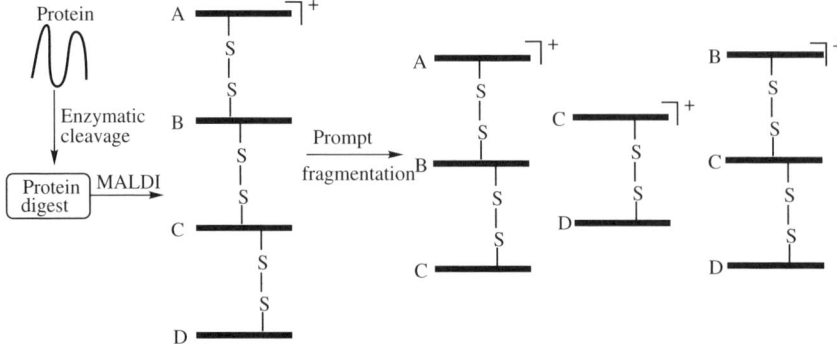

Figure 9.2. MALDI–ISD analysis of multichain disulfide-linked peptides.

in the polypeptide chain [10]. ISD is also an effective means to identify disulfide connectivity in multichain disulfide-linked peptides. As usual, the protein is first cleaved enzymatically or chemically under conditions that preserve disulfide bonds, and the protein digest is analyzed under in-source fragmentation conditions [11]. As shown in Figure 9.2, the multichain polypeptides fragment spontaneously at the disulfide linkages to yield fragments that contain various combinations of disulfide-linked peptide chains, from which it becomes feasible to establish the correct disulfide connectivity.

9.2.5. Identification of Disulfide-Containing Peptides by Electron-Capture Dissociation

It is also feasible to employ ECD to identify disulfide-containing peptides, but it requires FT–ICR–MS instrumentation. ECD of the ESI-produced multiply charged ions of proteins and peptides leads to cleavage of the disulfide bond to provide important sequence information [12]. The capture of an electron by a multiply charged protein ion releases an H atom, which is captured by the disulfide bond to cause its dissociation:

$$\text{RS-SR}' + {}^{\bullet}\text{H} \longrightarrow \text{R-S(H)}{\bullet}\text{S-R}' \longrightarrow \text{RSH} + {}^{\bullet}\text{SR}' \qquad (9.6)$$

9.2.6. Identification of Disulfide-Containing Peptides by Tandem MS

Amino acid sequence determination of the disulfide-containing peptides in a protein digest is another means to identify unambiguously the location of disulfide bonds. The MS/MS approach obviates the need to purify the protein digest into individual peptide fragments. With high-energy collision-induced dissociation (CID), cleavage occurs through the disulfide bridge and on either side of the bridge:

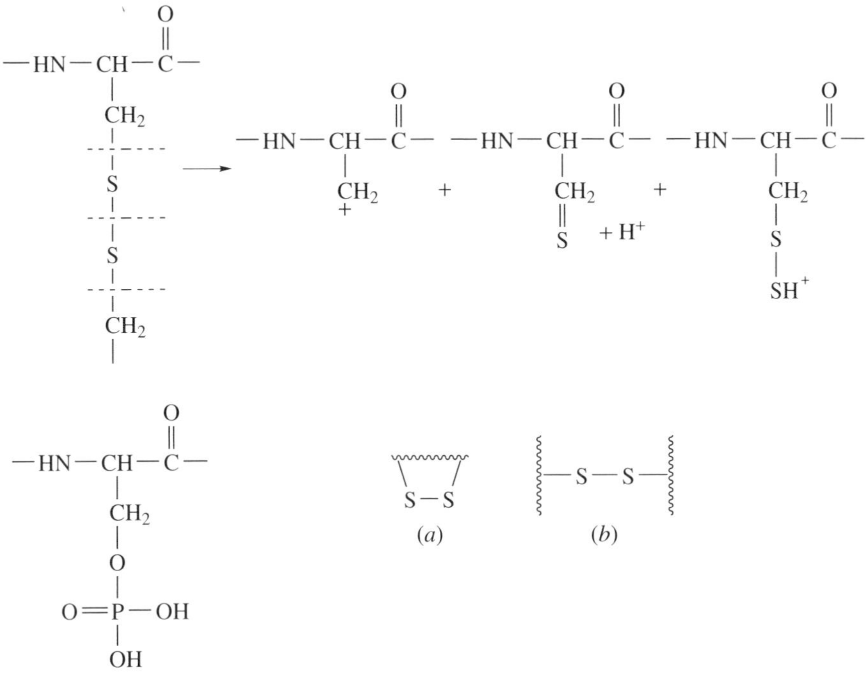

$$
\begin{array}{c}
\text{[Structural diagram of disulfide bond cleavage producing three peptide fragments with } +H^+ \text{]}
\end{array}
$$

(a) S—S (b) —S—S—

(9.7)

The spectrum contains a triplet of signals 32 and 34 Da apart to provide a mass spectrometry signature of the disulfide-containing peptide. This aspect is illustrated in Figure 9.3, which is the CID–MS/MS spectrum of a peptide that was generated by thermolysin digestion of the des–Val–Val–TGF–α protein [13]. Ions of m/z 968, 1001, and 1033 are related to cleavages in and around the disulfide bond. *Post-source decay* (PSD) can also provide information about the location of disulfide bonds [14]. PSD is a type of tandem mass spectrometry in which by monitoring the metastable decay of selected precursor ions in a reflectron–TOF instrument, sequence ion information is obtained (see Section 4.5.3). Similar to high-energy CID, symmetric and nonsymmetric cleavages of disulfide bonds both occur (i.e., at the S–S bond and both sides of the S–S bond).

In another elegant approach, MALDI–ISD is used in conjunction with tandem mass spectrometry (MS/MS) on a TOF/TOF instrument [15]. ISD first provides a screen for disulfide-containing peptides in a protein digest, and MS/MS subsequently confirms the sequence of the peptide [15]. ISD cleaves the disulfide bond to yield two protonated ionic species. The MALDI spectrum contains signals due to these protonated fragments as well as that due to the undissociated precursor. Software identifies the disulfide-linked precursor in the MALDI spectrum of the complex protein digest through information contained in Eq. (9.8), which states that the m/z of the precursor ion is given by the sum of the m/z

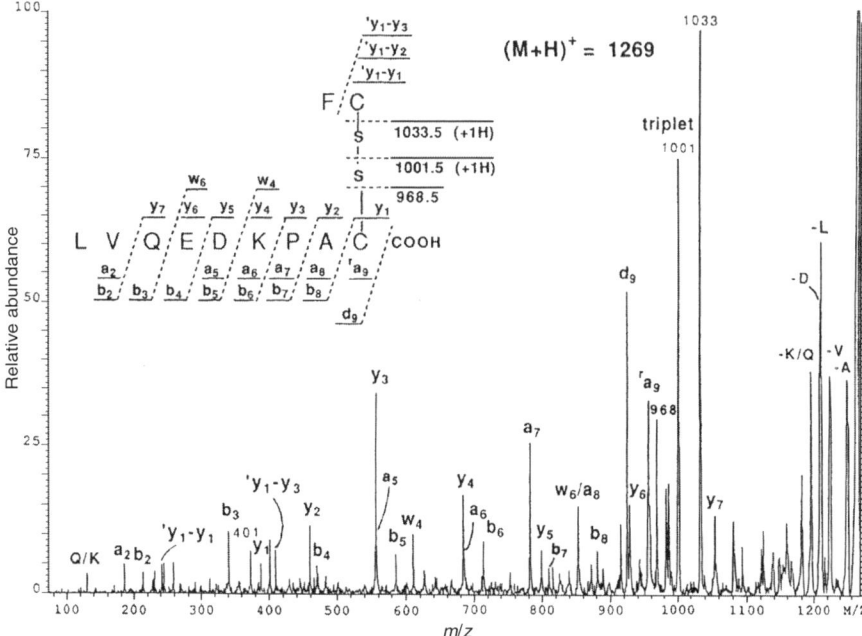

Figure 9.3. CID–MS/MS spectrum of a disulfide bridge–containing peptide that was generated by thermolysin digestion of the des–Val–Val–TGF–α protein. (Reproduced from ref. 13 by permission of Elsevier, copyright © 1992.)

of both ISD fragments minus the mass of two hydrogen atoms and of one proton:

$$m/z \text{ (fragment 1)} + m/z \text{ (fragment 2)} - (2H + H^+) = m/z \text{ of the precursor ion} \tag{9.8}$$

The precursor identified is automatically mass-selected and subjected to MS/MS analysis in a TOF/TOF tandem instrument to confirm the existence of a disulfide bond.

ANALYSIS OF PHOSPHOPROTEINS AND PHOSPHOPROTEOMICS

Protein phosphorylation is arguably one of the most important and ubiquitous PTMs known to occur in proteins; nearly 30% of all mammalian proteins are phosphorylated at a given time [16,17]. Often, phosphorylation acts as a molecular switch that controls many cellular processes, such as proliferation, differentiation, metabolism, signal transduction, and adaptation to environmental stress. Phosphorylation also plays an essential role in the function of many proteins, hormones, neurotransmitters, and enzymes. An abnormal regulation of phosphorylation often results in various diseases. Phosphorylation most commonly occurs on serine, threonine, and tyrosine residues (termed O-phosphorylation sites), but it is not limited to these residues. Phosphorylation of aspartate, glutamate (acylphosphates), arginine, lysine, histidine (N-phosphates), and cysteine (S-phosphates) is also known [18]. The discussion in this chapter is limited to the phosphorylation of serine, threonine, and tyrosine residues only. O-phosphorylation is mediated by certain substrate-specific phosphotransferase enzymes called *protein kinases*, which catalyze transfer of the terminal phosphate moiety on a nucleoside triphosphate (ATP or GTP) to the nucleophilic hydroxyl group of serine, threonine, and tyrosine residues of proteins. A large number of protein kinases have been identified, and each contains a consensus sequence for recognition and protein phosphorylation [16]. Some kinases exhibit activity for serine and threonine [16,17], whereas others are specific to tyrosine residues [16,19]. Phosphorylation is reversed by another group of enzymes collectively known as *protein phosphatases*.

9.3. ^{32}[P] LABELING FOR THE ANALYSIS OF PHOSPHOPROTEINS

Analysis of the entire complement of phosphoproteins is termed *phosphoproteomics*. Because of the unusual importance of phosphorylation, the field of phosphoproteomics has gained prominence in recent years. Knowledge of phosphorylation provides insight into the mechanism of regulation of kinases and phosphatases. The determination and location of phosphate groups in proteins present formidable challenge and require a highly sensitive and specific experimental approach.

A typical traditional approach to detect phosphorylation sites relies on the incorporation of a radiolabel into cells and tissues by incubating them with radioactive ^{32}[P]-phosphate. The labeled proteins are isolated by immunoprecipitation or acid extraction–precipitation and are separated using SDS–PAGE or chromatographic techniques [16]. The radiolabeled protein identified is digested into smaller fragments that are subsequently fractionated on RP–HPLC. The phosphorylation sites are identified via Edman sequencing of the fractionated peptides. The phosphoamino acid content of the protein thus identified is determined by subjecting it to complete hydrolysis, followed by quantitation of the phosphoamino acids released.

This protocol has several shortcomings. Besides being a potential radioactive hazard, the entire procedure is tedious, labor intensive, prone to sample losses, and often fails to identify the exact site of phosphorylation. In some cases, complete incorporation of ^{32}[P] fails to occur. Also, the multiplicity of the phosphorylated species within a cell creates problems in achieving the steady state. The identification of phosphorylated amino acids by the Edman procedure is also problematic. Under the harsh conditions employed in the Edman degradation, the phosphate ester bonds to serine and threonine are less stable. Both residues undergo β-elimination of H_3PO_4 to form dehydroalanine and β-methyldehydroalanine, respectively, and phenylthiohydantoin (PTH)–dithioerythreitol as by-products.

9.4. MASS SPECTROMETRY PROTOCOL FOR THE ANALYSIS OF PHOSPHOPROTEINS

Currently, mass spectrometry has become an increasingly indispensable technique to characterize phosphoproteins and phosphopeptides [18,20–22]. In the past, FAB–MS has enjoyed some degree of success [23–25]. Current applications rely primarily on ESI–MS, LC–MS, MALDI–MS, and MS/MS techniques. Mass spectrometric methods to characterize phosphorylated proteins/peptides have several decisive advantages over traditional methods. For example, they are inherently faster and highly sensitive, provide an unambiguous identity of a phosphopeptide and the location of the phosphate group in it, and avoid the need for hazardous radioactive isotope labeling. Despite these advantages, mass spectrometry characterization of phosphorylation is still a challenging proposition. First, the level of protein phosphorylation often is very low. Second, the mass spectrometry response due to phosphopeptides is considerably lower than that of nonphosphopeptides. Also, the MS/MS spectrum of phosphopeptides is difficult to interpret.

A general protocol for the analysis of phosphoproteins is illustrated in Figure 9.4. Various steps of this protocol are:

- Cleavage of purified phosphoproteins
- Fractionation of the peptide fragments in the digest

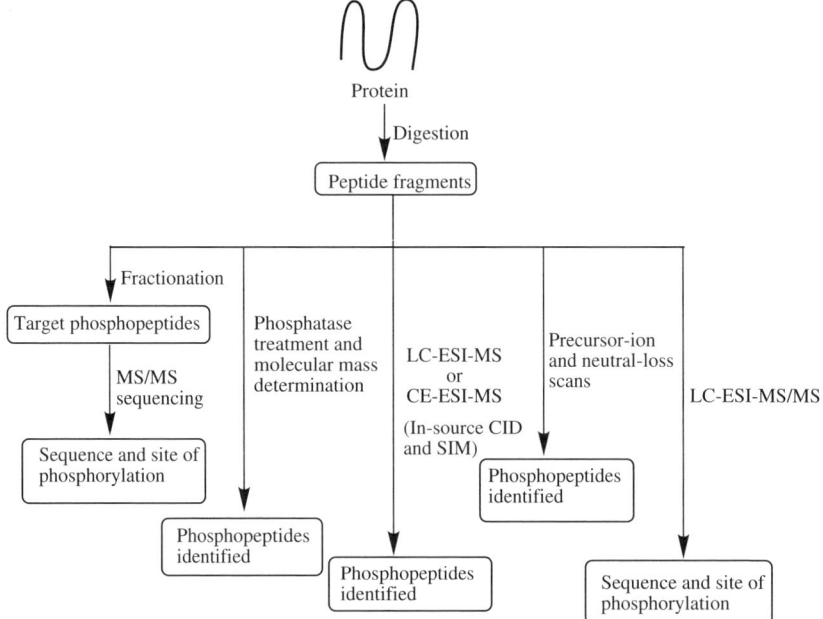

Figure 9.4. General protocol for the analysis of phosphoproteins.

- Reversed-phase HPLC
- Capillary electrophoresis
- Immobilized metal ion affinity chromatography (IMAC)
- Chemical tagging
- The use of antibodies
- Identification of phosphopeptides
  - Peptide mapping
  - Phosphatase treatment
  - In-source collision-induced dissociation
  - Precursor-ion scan
  - Neutral-loss scan
- Identification of phosphorylation sites
  - Collision-induced dissociation-tandem MS
  - Post-source decay-tandem MS
  - Electron-capture and electron-transfer dissociation
  - Infrared multiphoton dissociation

9.4.1. Cleavage of Purified Phosphoproteins

As is the normal practice in protein analysis, the determination and location of phosphate groups in proteins also requires cleavage of phosphoproteins into small manageable peptide fragments. To achieve this objective, phosphoproteins are digested with site-specific proteases and/or chemical reagents. The procedures described in Section 8.5.2 for unmodified proteins might be followed. A novel development in this field is to use a β-elimination reaction to convert phosphoserine and phosphothreonine to dehydroalanine and β-methyldehydroalanine, respectively [26]. Subsequent Michael addition reaction of these residues with cysteamine produces aminoethylcysteine and β-methylaminoethylcysteine, respectively (see Figures 9.6 and 9.7 for similar reactions). These products are isosteric with lysine and thus can be cleaved with proteases that are specific to lysine (e.g., Lys-C and trypsin). This protocol has the advantage that proteolysis would produce a peptide in which phosphorylation sites would always be present at the C-terminal residue that can easily be recognized by the presence of y_1-ion in the MS/MS spectra.

9.4.2. Fractionation of Peptide Fragments in the Digest

As mentioned above, main concerns in the mass spectrometry analysis of phosphopeptides in a complex protein digest are their low amounts and signal suppression in the presence of unphosphorylated peptides. To alleviate these concerns, the isolation and enrichment of phosphopeptides from the bulk of unphosphorylated peptides are highly desirable. A convenient means to fractionate phosphopeptides is RP-HPLC. Offline and online procedures with mass spectrometry have both been developed. Procedures that use capillary columns are given preference because of low flow rates and less sample requirements. Capillary electrophoresis (CE) is another separation technique that has enjoyed success in reducing the complexity of protein digests. This high-resolution separation technique also deals with low flow rates and requires lower sample volumes.

Immobilized Metal Ion Affinity Chromatography In recent years, IMAC has gained wide acceptance as a method that can selectively enrich phosphopeptides from complex mixtures [27,28]. This technique exploits the affinity of a negatively charged phosphate group for positively charged transition metal ions. The metal ions most often used are Fe^{3+} and Ga^{3+}, although Cu^{2+}, Zn^{2+}, Ni^{2+}, and Ti^{4+} ions have also been employed. The metal ions are first immobilized on a chromatographic support via tethered chelating agents. Typical examples of the chelating agents are iminodiacetate (IDA), nitrilotriacetate (NTA), tris(carboxymethyl)ethylenediamine (TED), and tris(2-aminoethyl)amine; the first two are more popular choices in the mass spectrometry analysis of phosphopeptides. Unlike the complexes of divalent metal ions with IDA, the IDA–Fe^{3+} complex has a net positive charge and involves ion exchange as the primary binding mechanism. The strong interaction of the IDA–Fe^{3+} complex (Figure 9.5) with the phosphate group is explained by the formation of two coordination bonds that involve a four-membered ring complex. An unwelcome reaction of the IDA–Fe^{3+} complex with the COOH group also occurs, but with lesser affinity because only one single bond is formed. Phosphopeptides bind selectively to the chelating support when an acidic solution (pH 3 to 5) is passed through the column. The elution of the bound phosphopeptides from the IMAC column is achieved with a basic pH buffer. A titanium oxide–based solid-phase packing has also been used as a precolumn to isolate phosphopeptides selectively [29]. In this column also, phosphopeptides are trapped under acidic conditions and subsequently are desorbed under alkaline conditions.

The nonspecific binding of acidic amino acid residues with the IMAC packing is a major concern of the IMAC approach. In addition, IMAC exhibits a preferential binding to multiply phosphorylated peptides and recovery of phosphopeptide is variable. To prevent nonspecific binding of the COOH group,

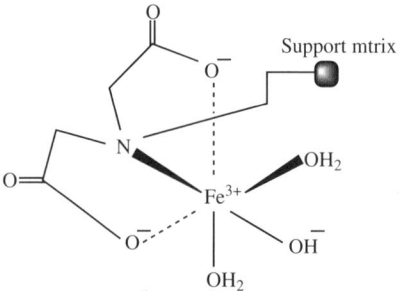

Figure 9.5. Octahedral Fe(III)–IDA–water complex.

methyl esterification of all COOH groups in peptides prior to IMAC separation has been recommended [30].

Chemical Tagging For selective isolation of phosphopeptides, two chemical-tagging approaches have been pursued [31–33]. In the first approach, which is applicable only to phosphoserine- and phosphothreonine-containing peptides, a chemical tag is created by removing the phosphate group as H_3PO_4 from phosphoserine (**1**) and phosphothreonine residues via an alkaline-induced β-elimination reaction and replacing it with 1,2-ethanedithiol (**2**; EDT) via a Michael-like addition reaction (Figure 9.6). This step is followed by the attachment of biotin on the thiol via the sulfhydryl-reactive group. The biotin-tagged peptides are isolated by avidin column chromatography. One concern of this approach is the poor recovery of biotin-tagged peptides, owing to a strong biotin–avidin interaction. In a variation of this procedure that overcomes the poor efficiency of the tagged peptides, the thiol tag introduced also acts as a ligand for affinity purification via disulfide exchange with an activated thiol resin (**3**; Figure 9.6) [34].

The second chemical-tagging approach, which in addition is also applicable to phosphotyrosine-containing peptides, uses a more complex chemistry to tag phosphopeptides [33]. It involves the attachment of 1-amine-2-thioethane to the phosphate group via a carbodiimide condensation reaction. The modified phosphopeptides are captured by an iodoacetyl resin and subsequently released with an acidic solution.

Use of Antibodies Antibodies are useful tools to isolate relevant proteins from a complex mixture by immunoprecipitation, provided that specific antibodies are available for each protein of interest. Although antibodies have been raised against all three O-phosphorylated residues, successful use of this technique has been demonstrated primarily for phosphotyrosine-containing peptides and proteins [35–37].

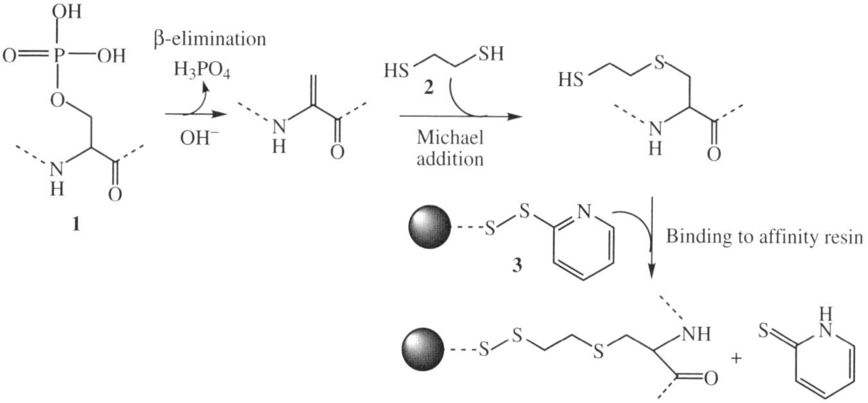

Figure 9.6. Protocol to purify phosphopeptides by a chemical-tag approach.

9.4.3. Determination of the Average Number of Phosphate Groups

A convenient way to determine the average number of phosphate groups in a protein is to measure the molecular mass of the intact protein. The protein is digested with a phosphatase and the molecular mass is remeasured. The difference in the two measured masses is divided by 80 to provide the average number of phosphate groups in the protein.

9.4.4. Identification of Phosphopeptides

Peptide Mapping If the sequence of the protein is known, molecular mass determination of the peptide fragments in the unfractionated protein digest can rapidly identify the presence of phosphopeptides. Protein digest is analyzed by MALDI–MS (or ESI–MS), and phosphopeptides are recognized from the mass shift of 80 Da (or a multiple of 80 Da).

Phosphatase Treatment The specificity of phosphatases for the removal of the phosphate group has also been exploited to identify phosphopeptides and proteins selectively [38]. The molecular mass of phosphopeptides decreases by 80 Da for each phospho unit after the phosphatase treatment. The reaction is monitored by MALDI–MS [39]. The peptide maps can be compared before and after phosphatase treatment to identify phosphopeptides in the digest [40]. Sensitivity is a major issue in such experiments. One way to improve sensitivity is to perform dephosphorylation on the MALDI target [40]. Another way is to use immobilized phosphatase packed in a small-diameter column in-line with an LC/ESI–MS or CE/ESI–MS system [39,41].

In-Source Collision-Induced Dissociation This method is implemented with an online LC/ESI–MS instrumentation and makes use of fragmentation of phosphopeptides in the transport region of the ESI source. At a higher skimmer potential, CID of phosphopeptides results in the formation of a phosphate marker ion of m/z 79 (i.e., the PO_3^- ion) [42,43]. In a typical analysis, the skimmer potential is first increased, and phosphopeptides in a digestion mixture are detected selectively during LC elution by selected-ion monitoring (SIM) of the phosphate group marker ion. Next, a conventional full-scan spectrum is acquired at low skimmer potential to determine the molecular mass of each eluting fragment. The advantage of this selective screening of phosphorylated peptides is that only those HPLC peaks that exhibit a positive signal of the marker ion are collected and analyzed further for their sequence determination.

Precursor-Ion Scan This procedure makes use of tandem mass spectrometry, and is implemented mainly on a triple quadrupole instrument. Phosphopeptide ions are generated by nano-ESI [44,45]. Phosphoserine-, phosphothreonine-, and phosphotyrosine-containing peptides readily form a structurally diagnostic PO_3^- ion of m/z 79 under CID conditions; precursor-ion scan of m/z 79 in the negative-ion mode provides a ready means to identify phosphopeptides selectively in a complex mixture. With the precursor-ion scan, the mass spectrometer provides a search of all phosphopeptides that fragment to generate m/z 79. Good sensitivity is achieved under basic pH conditions. This elegant approach has been applied to analysis of the phosphopeptide maps from two-dimensional gel-separated proteins. The detection of low-abundance phosphopeptides is still a problem and requires HPLC fractionation. However, HPLC fractionation of peptides, which occurs optimally at acidic pH, is not compatible with the negative-ion precursor-scanning mode. To overcome this incompatibility, a multidimensional ESI–MS-based approach has been developed [46]. A quadrupole-time-of-flight hybrid (Q–TOF) tandem instrument has also been used for precursor-ion scan phosphopeptide detection (see Section 4.3.5) [47,48].

In a variation of the negative-ion precursor-ion scan procedure, phosphopeptides are identified in the positive-ion mode by monitoring a diagnostic immonium ion [48,49]. This method is applicable only to phosphotyrosine-containing peptides and requires a high-resolution Q–TOF instrument for unambiguous monitoring of phosphotyrosine immonium ion of m/z 216.043. The procedure has been termed *phosphotyrosine-specific immonium-ion (PSI) scanning* [49].

Another variation of the negative-ion precursor-ion scan procedure is the derivatization of phosphorylated side chains to convert them to a form that, under CID, yields a phosphate group mass marker [50]. Phosphoserine and phosphothreonine are converted to S-pyridylethylcysteine and S-pyridylethyl-β-methylcysteine, respectively, upon Ba(OH)$_2$-catalyzed β-elimination of the phosphate group, followed by the reaction with 2-[4-pyridyl]ethanethiol (Figure 9.7). The thioether bond of these derivatives readily cleaves under CID to produce the pyridylethyl ion (C$_5$H$_4$N–CH$_2$CH$_2^+$) of m/z 106; the positive-ion precursor scanning of this ion selectively detects phosphoserine- and phosphothreonine-containing peptides.

A similar derivatization strategy also exists for the selective detection of phosphoserine- and phosphothreonine-containing peptides in the positive-ion

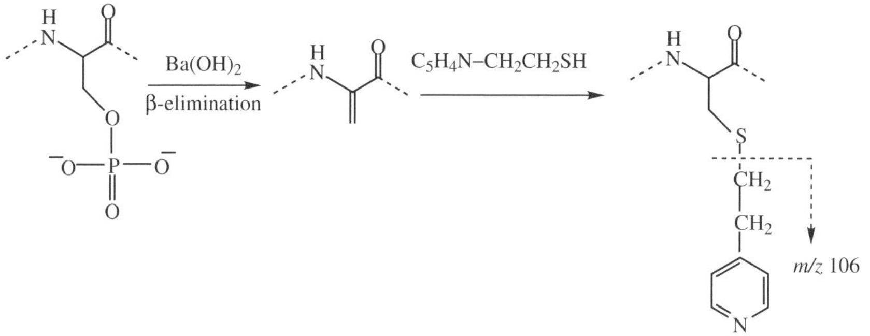

Figure 9.7. Modification of a phosphoserine-containing peptide using base-catalyzed β-elimination, followed by reaction with 2-(4-pyridyl)ethanethiol for the positive-ion precursor-ion scanning.

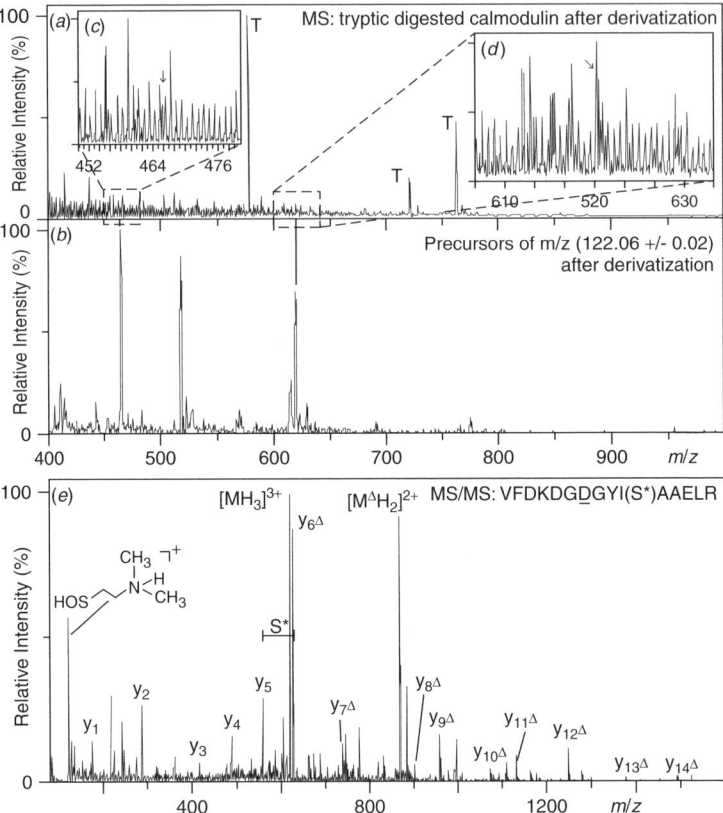

Figure 9.8. Mass spectrometric analysis of a tryptic digest of derivatized phosphorylated calmodulin; (*a*) nano-ESI mass spectrum, (*b*) precursor-ion scan of the marker ion m/z 122.06, and (*e*) product-ion spectrum of the triply charged precursor of m/z 620.31 provides the identity of the phosphorylation site at Ser-101 within the tryptic peptide T_{91-105} (VFDKDGNGYIS*AAELR). (Reproduced from ref. 51 by permission of Elsevier Science, copyright © 2002 American Society for Mass Spectrometry.)

precursor-scanning mode [51]. β-Elimination/Michael addition is performed with 2-dimethylaminoethanethiol [$(CH_3)_2N-CH_2CH_2-SH$], followed by oxidation of the thioether derivative to sulfoxide by treatment with H_2O_2. Under low-energy CID, the sulfoxide undergoes a facile β-elimination to produce a marker ion of m/z 122.06; the precursor-ion scan of this unique m/z ion provides selective detection of phosphoserine- and phosphothreonine-containing peptides. Figure 9.8 demonstrates the successful application of this concept. It shows mass spectrometric analysis of a tryptic digest of derivatized phosphorylated calmodulin.

Neutral-Loss Scan This procedure is similar to precursor-ion scanning except that it monitors a characteristic neutral loss using a triple-quadrupole tandem instrument. All three types of O-phosphorylated peptides exhibit the loss of 98

Da due to the expulsion of H_3PO_4 (and/or $HPO_3 + H_2O$) when activated collisionally [42]. This reaction can be monitored in the positive-ion mode via a neutral-loss scan to identify phosphopeptides unambiguously [52,53].

9.4.5. Identification of Phosphorylation Sites

Once the identity of phosphopeptides in a digestion mixture is revealed, it is imperative to recognize which amino acid residue (Ser, Thr, or Tyr) is phosphorylated. This task is accomplished by determining the amino acid sequence of peptide fragments in the protein digest after their fractionation into individual components by RP–HPLC, either off- or online with tandem mass spectrometry [34]. The tandem MS methods described below have proved to be practical for this purpose.

Collision-Induced Dissociation Tandem Mass Spectrometry A highly popular approach to identifying the location of phosphorylated residues is the CID of ESI-produced peptide ions on a triple quadrupole, ion trap, linear ion trap, and Q-TOF instrument [46,54,55]. As usual, *b*- and *y*-ion series are the common sequence-specific ions in CID spectra. Loss of the phosphate group from *b*- and *y*-ions is also observed. The residue masses of phosphoserine, phosphothreonine, and phosphotyrosine are 167, 181, and 243, respectively. These mass values are unique, as they do not match the residue masses of any other naturally occurring amino acids (see Table 8.3). Thus, on the basis of these unique mass differences, the exact site of phosphorylation can be ascertained quickly and accurately. Complications arise in those cases where the loss of the phosphate group (as HPO_3 or H_3PO_4) from the precursor ion dwarfs the *b*- and *y*-ion series. Specifically, phosphoserine- and phosphothreonine-containing peptides undergo the facile loss of H_3PO_4 via β-elimination [54,56].

One proposal to increase the visibility of the *b* and *y* sequence-specific ions is to derivatize phosphopeptides [57]. For example, the modification of the N-terminus of the peptide by tris[(2,4,6-trimethoxyphenyl)phosphonium]acetyl enhances the abundance of N-terminal ions and provides complete sequence information. Another derivatization strategy makes use of gas-phase reaction with either trimethoxyborane, diisopropoxymethylborane, or diethylmethoxyborane; the result is an enhancement in the yield of phosphate-containing ions in the CID spectrum [58]. Sequence analysis can also be improved by using β-elimination chemistry [51,59]. As described earlier (Section 9.4.4), phosphoserine and phosphothreonine can be converted to *S*-ethylcysteine and *S*-ethyl-β-methylcysteine, respectively, upon base-catalyzed β-elimination of the phosphate group and reaction with an ethanethiol (Figure 9.7). CID of the modified peptides results in more evenly distributed sequence-specific fragment ions. Upon CID, the modified serine and threonine residues undergo gas-phase β-elimination to produce dehydroalanine and 2-aminodehydobutyric acid, respectively, which are identified by their residue masses of 69 and 83 Da in the CID spectrum. For example, in Figure 9.8*e*, the formerly phosphorylated serine can be identified on the basis the 69-Da mass difference between y_6- and y_5-type product ions [51].

Another feasible proposal to increase the visibility of sequence-specific ions is a "pseudo-MS^n" procedure that involves collisional activation of the product ions that are formed by the neutral loss of H_3PO_4 [60]. In this way, the principal neutral-loss product ions are converted into a variety of second-generation sequence-specific ions. A complete MS/MS spectrum is the composite of the product ions that are derived from the original precursor as well as from the neutral-loss products. This pseudo-MS/MS spectrum increases the number and abundance of fragment ions.

▶ **Example 9.2** Calculate the m/z values of b_9^+, y_5^+, and y_8^{2+} sequence-specific ions that are formed upon CID of the phosphopeptide LFTGHPEpTLEK.

Solution m/z of b ions = 1.01 + residue masses. Therefore, the m/z value of b_9 ion = 1.01 + 113.08 + 147.07 + 101.05 + 57.02 + 137.06 + 97.05 + 129.04 + 181.01 + 113.08 = 1076.46. Similarly, the m/z value of y ions = 17.0 + residue masses + 2.02. Therefore, the m/z of y_5 ion = 17.0 + 128.09 + 129.04 + 113.08 + 181.01 + 129.04 + 2.02 = 699.28, and the m/z of y_8^{2+} ions = (17.0 + residue masses + 2.02 + 1.01)/2. Therefore, the m/z of y_8^{2+} ion = (17.0 + 128.09 + 129.04 + 113.08 + 181.01 + 129.04 + 97.05 + 137.06 + 57.02 + 2.02 + 1.01)/2 = 495.71.

Post-source Decay Tandem Mass Spectrometry The potential of PSD has also been exploited to identify phosphopeptides [61,62]. This technique is used for MALDI-generated ions in a reflectron TOF instrument. PSD can also distinguish any serine–threonine phosphorylation from tyrosine phosphorylation. Phosphotyrosine-containing peptides mostly yield the $[MH - HPO_3]^+$ (loss of 80 Da) ion, whereas phosphoserine- and threonine-containing peptides produce $[MH - H_3PO_4]^+$ (loss of 98 Da) and $[M - HPO_3]^+$ ions.

Electron-Capture Dissociation and Electron-Transfer Dissociation As mentioned earlier in the chapter, ECD involves excitation of the mass-selected protonated ion by the capture of subthermal energy (<0.2 eV) electrons and the subsequent fragmentation of the odd-electron ion $[M + nH]^{(n-1)+\bullet}$. ECD results in the homolytic cleavage of the $N-C_\alpha$ bond in peptides to form abundant c- and $z^\bullet$-type amino acid sequence-specific ions. Because there is no loss of the phosphate group, the phosphorylated residue can be recognized easily on the basis of an increase in the m/z value of the sequence-specific ion [63–65].

Electron-transfer dissociation (ETD) is a variation of ECD in which an ion–ion reaction is conducted in a quadrupole linear ion trap to transfer an electron to the multiply protonated peptide cation [66]. For example, anthracene anions that are generated in a CI source have been used as electron donors. Analogous to the ECD process, the transfer of an electron induces fragmentation in the peptide backbone to form c- and $z^\bullet$-type sequence-specific ions that usually retain the

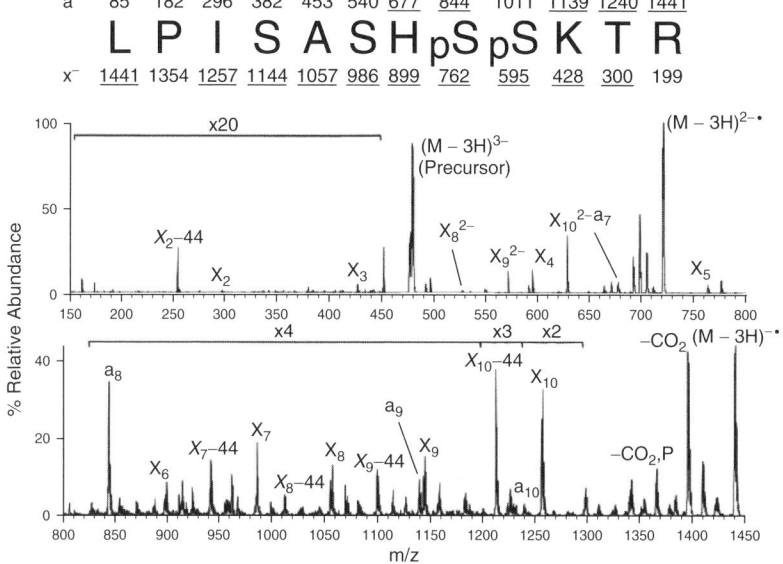

Figure 9.9. Tandem mass spectrum of the triply deprotonated phosphopeptide LPISAS-HpSpSKTR that was activated by ion/ion reactions with $Xe^{+\bullet}$. (Reproduced from ref. 67 by permission of Elsevier Science, copyright © 2005 American Society for Mass Spectrometry.)

phosphate group. ETD has also been observed with multiply deprotonated peptide anions [67]. In this case, an ion–ion reaction occurs between a mass-selected multiply deprotonated peptide anion and a xenon radical cation ($Xe^{+\bullet}$). Electron abstraction by $Xe^{+\bullet}$ generates an odd-electron charge-reduced peptide ion. Fragmentation of the peptide ion is driven by free-radical chemistry to produce a- and x-type sequence-specific ions. Figure 9.9 shows the tandem mass spectrum of a triply deprotonated phosphopeptide LPISASHpSpSKTR that was activated by ion–ion reactions with $Xe^{+\bullet}$. The two phosphoserine residues can be identified in the spectrum by the residue mass difference of 167 Da between the x_5-, x_4-, and x_3- ions. It is envisioned that with this approach, phosphopeptides can be characterized directly without resorting to an enrichment strategy.

Infrared Multiphoton Dissociation (IRMPD) IRMPD is also well suited to providing identification and sequence analysis of phosphopeptides in the positive [55] and negative polarity modes [68,69]. Because the phosphate group is a powerful chromophore for 10.6-μm photons (e.g., emitted from a CO_2 laser), ion activation with these photons can be used for selective detection of phosphopeptides in a mixture. The activation energy for the dissociation of phosphopeptides by IRMPD is lower than that of unmodified analogs [69]. As shown in Figure 9.10, which is the spectrum of tyrosine kinase receptor JAK2, IRMPD allows more sequence coverage than does the CID technique [55].

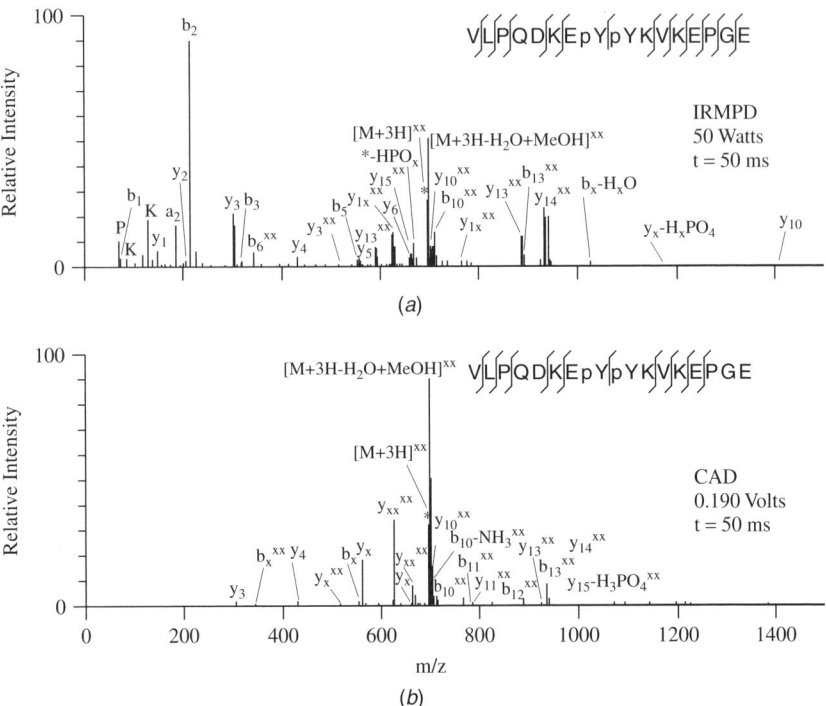

Figure 9.10. IRMPD (*a*) and CID (*b*) mass spectra of tyrosine kinase receptor JAK2. (Reproduced from ref. 55 by permission of Elsevier Science, copyright © 2004 American Society for Mass Spectrometry.)

Thus, a variety of mass spectrometry techniques are available that can selectively detect phosphoproteins and peptides in complex samples and can provide the site of phosphorylation in a protein.

ANALYSIS OF GLYCOPROTEINS

Glycosylation is another common posttranslational modification of proteins. In this modification, a carbohydrate chain (called a *glycan*) is attached to specific sites of a protein. It is estimated that over 50% of all mammalian proteins in eukaryotic systems are glycosylated at some point during their existence. The carbohydrate side chains of a glycoprotein play several critical roles in cell biology and have a profound influence in modulating the physiochemical (e.g., solubility and stability) and biological (e.g., immunological and proteolytic stability) properties of proteins [70]. Carbohydrate side chains serve as recognition markers for cell–cell and cell–molecule interactions and as receptors for viruses, bacteria, and parasites. In addition, the sugar units may orient glycoproteins in membranes and help to determine the destination of a glycoprotein. Carbohydrate side chains also shield significant areas of proteins from protease activity and thus enhance their stability. Glycosylation can also affect the secondary structure of peptides and proteins (e.g., rigidification of peptide chains from random coils to rodlike structures). O-Glycosylated regions of mucins and the extracellular regions of many membrane glycoproteins are examples of these extended conformations.

9.5. STRUCTURAL DIVERSITY OF GLYCOPROTEINS

The structural diversity of glycoproteins is enormous, especially with respect to glycans (i.e., the carbohydrate portion). Many more different oligosaccharides can be formed from four sugar units than oligopeptides from four amino acids. As a result, a single protein might exist in a range of glycoforms. The most commonly occurring monosaccharides in the carbohydrate side chain of mammalian proteins are D-mannose (Man or Hex), D-galactose (Gal or Hex), L-fucose (Fuc or dHex), N-acetylglucosamine (GlcNAc or HexNAc), N-acetylgalactosamine (GalNAc or HexNAc), and N-acetylneuraminic acid (sialic acid, NeuAc, or NANA). Glucose (Glc) and xylose are sugar units encountered less commonly.

Two major forms of glycoproteins are the N- and O-glycosylated proteins. In N-glycosylated proteins, sugar residues are attached to the nitrogen atom of the primary amide of an asparagine residue (Figure 9.11). This residue can accept a carbohydrate unit if it is a part of the motif Asn–X–Ser(Thr or Cys), where X is any amino acid except proline. O-glycosylated proteins, in which the carbohydrate portion is linked to the polypeptide chain through the oxygen atoms of serine and threonine side chains, are less common. Glycosylation of hydroxyproline, hydroxylysine, and tyrosine residues does occur, but less frequently. Although no consensus sequence exists, O-glycosylation occurs in sequence regions that have a high density of Ser and Thr residues.

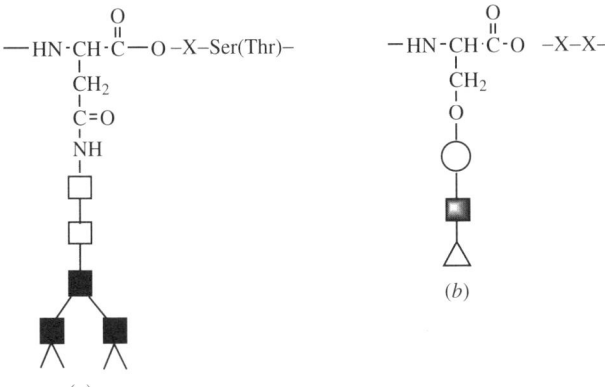

Figure 9.11. Structure of N- and O-linked glycoproteins (open square, GlcNAc; solid square, Man; shaded square, Gal; open circle, GalNAc; triangle, NeuAc).

9.6. ANALYSIS OF GLYCOPROTEINS

Because the carbohydrate moieties of glycoproteins are not encoded in DNA sequences, their structural information can only be gleaned by direct analytical measurements. A quick indication whether or not a protein is glycosylated is obtained during their separation on sodium dodecyl sulfate (SDS)–polyacrylamide gel electrophoresis (PAGE) gels. Normally, glycoproteins appear as broad fuzzy bands. On treatment with glycosidases, these bands, however, become sharper [71].

The detailed structural characterization of glycoproteins must include the determination of glycosylation sites and the type of glycosidic linkage, the position of the glycosidic linkage (i.e., 1 → 2, 1 → 4, 1 → 6, etc.), the anomeric configuration of each sugar, the sequence and branching of each monomer, and the identity of each sugar (Man, Gal, etc.).

A general strategy for a complete analysis of glycoproteins is illustrated in Figure 9.12. The following steps are involved in this strategy:

- Molecular mass determination of intact glycoproteins
- Identification of glycosylation
  - Comparative peptide mapping
  - Monitoring the carbohydrate-specific marker ions
  - Precursor-ion scan
- Site of glycosylation
  - Collision-induced dissociation tandem MS
  - Ladder sequencing
  - Electron-capture dissociation
  - Top-down sequencing of intact glycoprotein
- Structural characterization of the oligosaccharide side chains after their release from glycoproteins (discussed in Chapter 11)

9.6.1. Molecular Mass Determination of Glycoproteins

The first step in the characterization of a glycoprotein is to measure its molecular mass. Often, this information can provide a direct global assessment of glycosylation [72]. If the mass of the unglycosylated protein is known (e.g., from its cDNA sequence information), the glycan mass can be deduced by difference. ESI–MS and MALDI–MS are well suited to determining the molecular mass of these biopolymers. Individual glycoforms can, however, be resolved for small proteins (< 40 kDa) only. It is feasible to obtain molecular mass information from the ESI spectrum of a heterogeneous sample of glycoproteins. Each glycoform and nonglycosylated protein will produce its own series of multiply charged ions. With the increase in the sites of glycosylation, however, the number of probable glycoforms also increases. Such a heterogeneous mixture of glycoproteins will yield a complex ESI spectrum from which it may be difficult to

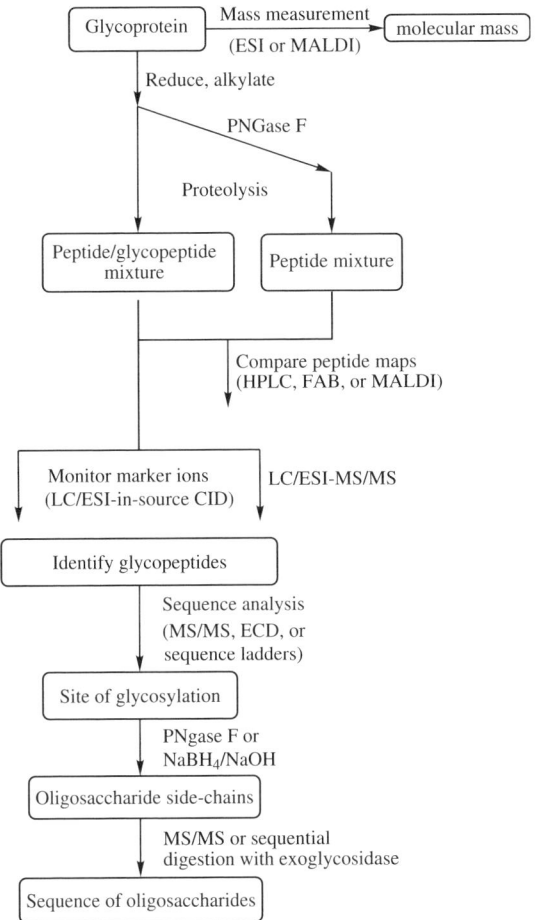

Figure 9.12. General protocol to characterize glycoproteins.

retrieve unambiguous molecular mass information. For this reason, MALDI–MS is preferred over ESI–MS. MALDI produces a simpler mass spectrum because each glycoform yields a singly charged molecular ion. Highly heterogeneous and high-mass glycoproteins are difficult to analyze even with MALDI because with the increase in the heterogeneity and molecular mass of a glycoprotein, the MALDI signal becomes broad and less intense.

A convenient way to directly analyze glycoproteins by MALDI–MS is to immobilize them on aminophenylboronic acid–derivatized magnetic beads [73]. With this approach, isolation of glycoproteins from a complex mixture becomes relatively simple. The protein-immobilized magnetic beads can be spotted directly onto a MALDI sample plate along with a matrix, thereby obviating the need for an additional separation step.

9.6.2. Identification of Glycosylation

To detect the presence of glycosylation, two approaches are in common use: comparative peptide mapping and selective detection of glycopeptides.

Comparative Peptide Mapping In this approach, the peptide maps are analyzed before and after the release of glycans [74]. The primary structure of the target glycoprotein must be known from its cDNA sequence. N-Linked glycans are released by treating the glycopeptides with an enzyme PNGase (see Section 11.3.1). The appearance of new peaks in the peptide map of the PNGase-treated fraction indicates that glycosylated peptides were present in the protein digest. From the m/z value of a new peptide, its location in the intact protein sequence can be ascertained. In addition, this knowledge, along with the consensus sequence Asn–X–Ser(Thr), can also predict the possible sites of N-linked glycosylation, provided that the peptide released has only a single glycosylation site. Peptide maps are generated as described in Chapter 8. In brief, the glycoprotein is first reduced and alkylated, and treated with trypsin. Other enzymes listed in Table 8.1 can also be used to prepare different-length peptide fragments. Mass spectrometry analysis is accomplished using ESI and MALDI, and less effectively, with FAB.

Selective Detection by Monitoring the Carbohydrate-Specific Marker Ions The selective detection of glycopeptides in a protein digest is performed by monitoring carbohydrate-specific marker ions [75–77]. As discussed in Section 9.4.4, for phosphopeptides, the marker ions are generated by in-source CID of the molecular ions of glycopeptides and include those oxonium ions that are formed at m/z 163 (Hex), 204 (HexNAc), 274 and 292 (NeuAc), and 366 (Hex–HexNAc). Most structural types of N- and O-linked carbohydrate side chains can be detected by monitoring the ion of m/z 204. In practice, during an HPLC/ESI–MS separation, the entire mass range is scanned at a low orifice potential to record the intact molecular ions of peptides and glycopeptides. The analysis is repeated at a high orifice potential, and the instrument is switched to the SIM mode to record only the suspected carbohydrate-specific oxonium ions. A *stepped-orifice voltage-scanning technique*, in which the full-scan and SIM recording are performed in a single HPLC injection also exists for selective detection of glycopeptides [75]. The ion current is monitored in the entire mass range (e.g., 150 to 2000 Da). The orifice potential is stepped up to generate carbohydrate-specific oxonium ions but only during the scan of the low-mass end (150 to 400 Da). The rest of the mass range is recorded at a normal low orifice potential to detect molecular ions of the peptide fragments of the protein digest. These steps are repeated back and forth during the entire HPLC separation.

Selective Detection by Precursor-Ion Scan The precursor-ion scan of specific marker ions (e.g., m/z 204) is another option for selective detection of glycopeptides [77]. For example, the MS-2 of a tandem instrument can be set to monitor m/z 204, and the MS-1 is scanned to detect those glycopeptides that fragment to yield the *m/z* 204 oxonium ion. This technique provides improved selectivity but is less sensitive than the stepped-orifice voltage-scanning procedure.

Once glycopeptides have been detected, the corresponding RP–HPLC fractions are collected and treated further with exoglycosidases to release specific sugar residues. The molecular mass is determined again, and the difference in the two masses provides the identity of the carbohydrate residue that was attached to the glycopeptide.

9.6.3. Site of Glycosylation

Tandem Mass Spectrometry Analysis A detailed analysis of the glycosylation sites can be performed by the amino acid sequencing of intact glycopeptides [78]. The direct MS/MS of glycopeptides, however, has certain limitations. First, the mass spectrometry response is poor, owing to the presence of carbohydrate side chains. Second, the fragment ion information is not very useful because glycopeptides usually fragment at the more labile peptide–carbohydrate linkages. Therefore, it is advantageous to free the peptide from the carbohydrate side chains prior to MS/MS analysis. In this strategy, glycopeptide fractions are isolated by RP–HPLC and are treated with PNGase or $NaBH_4$/NaOH to release N- or O-linked glycans, respectively; peptides that are free of carbohydrate chains are analyzed by MS/MS to obtain the amino acid sequence of the glycopeptides [79–82]. MALDI in conjunction with post-source decay can also provide information on the site of glycosylation [83].

Peptide Ladder Sequencing The treatment of a glycopeptide with a carboxypeptidase or aminopeptidase to generate peptide ladders, followed by MALDI–MS analysis of those ladders, is another feasible approach to identifying the site of glycosylation. This approach may not be effective with multiglycosylation sites because the enzymatic activity of these proteases is impaired at or near the site of glycosylation [84].

Electron-Capture Dissociation (ECD) ECD mass spectrum can also identify the glycosylated sites in peptides. In this "mild" activation process, the carbohydrate chains remain attached to the sequence-specific *c*- and *z*-type ions, thereby providing direct evidence of glycosylation in glycosylated peptides [85].

Top-Down Sequence Analysis of the Whole Protein via Ion–Ion Reactions In favorable cases, multiply charged ESI-produced ions of an intact glycoprotein can be subjected to CID in a quadrupole ion trap to identify the glycosylation sites [86]. The charge state of the protein ions can be manipulated by an ion–ion reaction with $[M - F^-]$ and $[M - CF_3^-]$ ions derived from perfluoro-1,3-dimethylcyclohexane. The charge reduction of the CID product ions by ion–ion reactions also simplifies the MS/MS spectrum and facilitates interpretation of the spectrum. Direct CID analysis of an intact glycoprotein has the advantages that it avoids extensive purification and proteolytic digestion steps and provides rapid identification of glycosylation sites.

OVERVIEW

In this chapter we discussed mass spectrometry–based approaches to identifying certain important PTMs, such as disulfide bond formation, phosphorylation, and glycosylation. The first step in PTM identification is to cleave the protein into smaller peptide segments which are analyzed by ESI–MS and MALDI–MS. One convenient way to identify a PTM is to measure the molecular mass of each peptide component of the protein digest. Disulfide bridge–containing peptides are identified by subjecting the putative peptides to reduction or oxidation reactions and measuring their molecular mass again. Other methods to identify disulfide bridge–containing peptides are based on gas-phase fragmentation in either the MALDI ion source or field-free regions of tandem MS instruments.

Of the several methods currently available to identify phosphorylation, SIM of the phosphate group marker ion (m/z 79), and precursor-ion scanning of m/z 79 (for all three O-phosphorylation sites) in the negative-ion mode and of m/z 216 (for phosphotyrosine-containing peptides only) in the positive-ion mode are the most effective approaches. CID, PSD, ECD, and IRMPD are some of the ion-activation and ion-fragmentation methods that can be employed to generate sequence-specific ions with the aim to provide information on the exact location of phosphorylation. Prior to mass spectrometry analysis, it is beneficial to enrich phosphopeptides with IMAC.

Glycoproteins are also characterized by using a similar strategy: protein digestion to cleave the protein into smaller peptide segments, identification of the cleaved N- and O-linked glycopeptides, determination of the site of glycosylation, and complete structure determination of the carbohydrate portion (this last aspect is discussed in Chapter 11). Monitoring of carbohydrate-specific marker ions, such as the ions of m/z 163 (Hex), 204 (HexNAc), 274 and 292 (NeuAc), and 366 (Hex–HexNAc) and the precursor-ion scan of m/z 204, are effective means to identify glycopeptides in a protein digest. The site of glycosylation is determined with tandem MS methodologies.

EXERCISES

9.1. Describe how to determine the number of disulfide bonds in a protein.

9.2. Ribonuclease A contains several disulfide bonds but no free sulfhydryl group. The molecular mass of the protein measured before and after reduction and alkylation (with iodoacetic acid) was found to be 13,682 and 14,155 Da, respectively. How many disulfide bonds does this protein contain?

9.3. It is beneficial to isolate phosphopeptides selectively from a complex mixture of peptides in a protein digest prior to mass spectrometry analysis. Describe how this step is performed.

9.4. Outline various methods that are used to identify phosphopeptides selectively in a complex protein digest.

9.5. What procedures can be employed to determine the site of phosphorylation in a peptide?

9.6. Which mass-marker ion is suitable for the selective identification of phosphoserine-, phosphothreonine-, and phosphotyrosine-containing peptides in the negative-ion mode?

9.7. Selective detection of phosphopeptides in a protein digest is usually performed via the negative-ion precursor-ion scanning technique. Two alternative positive-ion precursor-ion scanning methods were also discussed in this chapter. Discuss the concepts of these methods.

9.8. A phosphopeptide is sequenced by acquiring a CID–MS/MS spectrum of its $[M + H]^+$ ion of m/z 1296.6. A complete set of y^+ ions is observed at m/z 147.1, 260.2, 407.3, 478.3, 609.3, 790.4, 919.4, 1034.4, and 1133.5. Deduce the amino acid sequence of this peptide.

9.9. The measured molecular masses of a protein before and after the digestion with a phosphatase are 19,587 and 19,106 Da. Calculate the number of phosphate groups in the protein.

9.10. Glycopeptides can also be identified in a complex protein digest by monitoring the ion current due to certain mass-marker ions. Name those marker ions.

9.11. Explain the concept of the stepped-orifice voltage-scanning technique for selective identification of glycopeptides.

REFERENCES

1. T. E. Creighton, Disulfide bond formation in proteins, in F. Wold and K. Moldave, eds. *Methods in Enzymology*, Vol. 107, Academic Press, San Diego, CA, 1984, pp. 305–329.
2. P. E. Staswick, M. A. Hermodson, and N. C. Nielsen, Identification of the cystines which link the acidic and basic components of the glycinin subunits, *J. Biol. Chem.* **259**, 13431–13435 (1984).
3. W. R. Gray, F. A. Luque, et al., Conotoxin GI: disulfide bridges, synthesis, and preparation of iodinated derivatives, *Biochemistry* **23**, 2796–2802 (1984).
4. D. Chelius, M. E. Huff Wimer, and P. V. Bondarenko, Reversed-phase liquid chromatography in line with negative ionization electrospray mass spectrometry for the characterization of the disulfide-linkages of an immunoglobulin gamma antibody, *J. Am. Soc. Mass Spectrom.* **17**, 1590–1598 (2006).

5. Z. Zhou and D. L. Smith, Assignment of disulfide bonds in proteins by partial acid hydrolysis and mass spectrometry, *J. Protein Chem.* **9**, 523–532 (1990).
6. D. L. Smith and Y. Sun, Protein cross-linkages; in D. M. Desiderio, ed., *Mass Spectrometry of Peptides*, CRC Press, Boca Raton, FL, 1991, pp. 275–287.
7. D. L. Smith and Z. Zhou, Strategies for locating disulfide bonds in proteins; in J. A. McCloskey, ed., *Methods in Enzymology*, Vol. 193, Academic Press, San Diego, CA, 1990, pp. 374–389.
8. A. Tsarbopoulos, J. Varnerin, et al., Mass spectrometric mapping of disulfide bonds in recombinant human interleukin-13, *J. Mass Spectrom.* **35**, 446–453 (2000).
9. Y. Sun and D. L. Smith, Identification of disulfide-containing peptides by performic acid oxidation and mass spectrometry, *Anal. Biochem.* **172**, 130–138 (1988).
10. S. D. Patterson and V. Katta, Prompt fragmentation of disulfide-linked peptides during matrix-assisted laser desorption ionization mass spectrometry, *Anal. Chem.* **66**, 3727–3732 (1994).
11. J. Qin and B. T. Chait, Identification and characterization of posttranslational modifications of proteins by MALDI ion trap mass spectrometry, *Anal. Chem.* **69**, 4002–4009 (1997).
12. R. A. Zubarev, N. A. Kruger, et al., Electron capture dissociation of gaseous multiply charged proteins is favored at disulfide bonds and other sites of high hydrogen atom affinity, *J. Am. Chem. Soc.* **121**, 2857–2862 (1999).
13. M. F. Bean and S. A. Carr, Characterization of disulfide bond position in proteins and sequence analysis of cysteine-containing peptides by tandem mass spectrometry, *Anal. Biochem.* **201**, 216–226 (1992).
14. M. D. Jones, S. D. Patterson, and H. S. Lu, Determination of disulfide bonds in highly bridged disulfide-linked peptides by MALDI MS with post-source decay, *Anal. Chem.* **70**, 136–143 (1998).
15. V. Schnaible, S. Wefing, A. Resemann, D. Sukau, A. Bücker, S. Wolf-Kummeth, and D. Hoffmann, Screening of disulfide bonds in proteins by MALDI in-source decay and LIFT–TOF/TOF–MS, *Anal. Chem.* **74**, 4980–4988 (2002).
16. T. Hunter and B. M. Sefton, eds., *Methods in Enzymology*, Vol. 200, Academic Press, San Diego, CA, 1991.
17. A. M. Edelman, D. K. Blumenthal, and E. G. Krebs, Protein serine/threonine kinases, *Annu. Rev. Biochem.* **56**, 567–613 (1987).
18. A. Sickmann and H. E. Meyer, Phopsphoamino acid analysis, *Proteomics* **1**, 200–206 (2001).
19. T. Hunter and J. A. Cooper, Protein tyrosine kinases, *Annu. Rev. Biochem.* **54**, 897–930 (1985).
20. D. T. McLachlin and B. T. Chait, Analysis of proteins and peptides by mass spectrometry, *Curr. Opin. Chem. Biol.* **5**, 591–602 (2001).
21. D. E. Kalume, H. Molina, and A. Pandey, Tackling the phosphoproteome: tools and strategies, *Curr. Opin. Chem. Biol.* **7**, 64–69 (2003).
22. B. A. Garcia, J. Shabanowitz, and D. F. Hunt, Analysis of protein phosphorylation, *Methods* **35**, 256–264 (2005).
23. C. Fenselau, D. N. Heller, M. S. Miller, and H. B White III, Phosphorylation sites in riboflavin-binding protein characterized by fast atom bombardment mass spectrometry, *Anal. Biochem.* **150**, 309–314 (1985).

24. B. W. Gibson, A. M. Falick, et al., Liquid secondary ionization mass spectrometric characterization of two phosphoserine-containing peptides, *J. Am. Chem. Soc.* **109**, 5343–5348 (1987).
25. C. Dass and P. Mahalakshmi, Amino acid sequence determination of phosphoenkephalins using liquid secondary ionization mass spectrometry, *Rapid Commun. Mass Spectrom.* **9**, 1148–1154 (1995).
26. Z. A. Knight, B. Shilling, et al., Phosphospecific proteolysis for mapping sites of protein phosphorylation, *Nat. Biotechnol.* **21**, 1047–1054 (2003).
27. W. Zhou, B. A. Merrick, M. G. Khaledi, and K. B. Tomer, Detection and sequencing of phosphopeptides affinity bound to immobilized metal ion beads by matrix-assisted laser desorption mass spectrometry, *J. Am. Soc. Mass Spectrom.* **11**, 273–282 (2000).
28. S. Li and C. Dass, Iron(III)-immobilized metal ion affinity chromatography and mass spectrometry for the purification and characterization of synthetic phosphopeptides, *Anal. Biochem.* **270**, 9–14 (1999).
29. M. W. H. Pinske, M. P. Uitto, M. J. Hillhorst, B. Ooms, and J. R. Albert, Selective isolation at the femtomole level of phosphopeptides from proteolytic digests using 2D–nanoLC–ESI–MS/MS and titanium oxide precolumns, *Anal. Chem.* **76**, 3935–3943 (2004).
30. S. B. Ficarro, M. L. McCleland, P. T. Stukenberg, D. J. Burke, M. M. Ross, J. Shabanowitz, and D. F. Hunt, Phosphoproteome analysis, by mass spectrometry and its application to *Saccharmyces cerevisiae*, *Nat. Biotechnol.* **20**, 301–305 (2002).
31. Y. Oda, T. Nagasu, and B. T. Chait, Enrichment analysis of phosphorylated proteins as a tool for probing the phosphoproteome, *Nat. Biotechnol*, **19**, 379–385 (2001).
32. M.B. Goshe, T. P. Conrads, et al., Phosphoprotein isotope-coded affinity tag approach for isolating and quantitating phosphopeptides in proteome-wide analysis, *Anal Chem.* **73**, 2578–2586 (2001).
33. H. Zhou, J. D. Watts, and R. Aebersold, A systematic approach to the analysis of protein phosphorylation, *Nat. Biotechnol.* **19**, 375–378 (2001).
34. D. T. McLachlin and B. T. Chait, Improved β-elimination-based affinity purification strategy for enrichment of phosphopeptides, *Anal Chem.* **75**, 6826–6836 (2003).
35. M. S. Kalo and E. B. Pasquale, Multiple in vivo tyrosine phosphorylation sites in EphB receptors, *Biochemistry* **38**, 14396–14408 (1999).
36. A. Pandey, A. V. Podtelejnikov, B. Blagoev, X. R. Bustelo, and M. Mann, Analysis of receptor signaling pathways by mass spectrometry: identification of Vav-2 as a substrate of the epidermal and platelet-derived growth factor receptors, *Proc. Natl. Acad. Sci. USA* **97**, 179–184 (2000).
37. S. Kane, H. Sano, et al., A method to identify serine kinase substrates. Akt phophorylates a novel adipocyte protein with a Rab GTPase-activating protein (GAP) domain, *J. Biol. Chem.* **277**, 22115–22118 (2002).
38. T. T. Yip and W. Hutchens, Mapping and sequence-specific identification of phosphopeptides in unfractionated protein digest mixtures by matrix-assisted laser desorption/ionization time-of-flight mass spectrometry, *FEBS Lett.* **308**, 149–153 (1992).
39. L. N. Amankawa, K. Harder, F. Jirik, and R. Aebersold, High-sensitivity determination of tyrosine phosphorylated peptides by on-line enzyme reactor and electrospray ionization mass spectrometry, *Protein Sci.* **4**, 113–125 (1995).

40. M. R. Larsen, G. L. Sørensen, S. J. Fey, P. M. Larsen, and P. Roepstorff, Phosphoproteomics: evaluation of the use of enzymatic de-phosphorylation and differential mass spectrometric peptide mass mapping for site specific phosphorylation assignment in proteins separated by gel electrophoresis, *Electrophoresis* **22**, 223–238 (2001).
41. P. K. Jensen, L. Posa-Tolic, et al., Probing proteomes using capillary isoelectric focusing-electrospray ionization Fourier transform ion cyclotron resonance mass spectrometry, *Anal. Chem.* **71**, 2076–2084 (1999).
42. M. J. Huddleston, R. S. Annan, M. F. Bean, and S. A. Carr, Selective detection of phosphopeptides in complex mixtures by electrospray liquid chromatography/mass spectrometry, *J. Am. Soc. Mass Spectrom.* **4**, 710–717 (1993).
43. X. Zhu and C. Dass, Analysis of phosphoenkephalins by combined high performance liquid chromatography and electrospray ionization mass spectrometry, *J. Liq. Chromatogr. Rel. Technol.* **22**, 1635–1647 (1999).
44. M. Wilm, G. Neubauer, and M. Mann, Parent ion scans of unseparated peptide mixtures, *Anal. Chem.* **68**, 527–533 (1996).
45. S. A. Carr, M. J. Huddleston, and R. S. Annan, Selective detection and sequencing of phosphopeptides at the femtomole level by mass spectrometry, *Anal. Biochem.* **239**, 180–292 (1996).
46. R. S. Annan, M. J. Huddleston, R. Verma, R. J. Deshaies, and S. A. Carr, A multidimensional electrospray ionization mass spectrometry-based approach to phosphopeptide mapping, *Anal. Chem.* **73**, 393–404 (2001).
47. C. Borchers, C. E. Parker, L. J. Deterding, and K. B. Tomer, Preliminary comparison of precursor scans and liquid chromatography on a hybrid quadrupole time-of-flight tandem mass spectrometer, *J. Chromatogr. A* **854**, 119–130 (1999).
48. R. H. Bateman, R. Carruthers, et al., A novel precursor ion discovery method on a hybrid quadrupole orthogonal acceleration time-of-flight (Q-TOF) mass spectrometer for studying protein phosphorylation, *J. Am. Soc. Mass Spectrom.* **13**, 792–803 (2002).
49. H. Steen, B. Kuster, M. Fernandez, A. Pandey, and M. Mann, Detection of tyrosine phosphorylated peptides by precursor ion scanning quadrupole TOF mass spectrometry in positive ion mode, *Anal. Chem.* **73**, 1440–1448 (2001).
50. M. Quadroni, Specific detection and analysis of phosphorylated peptides by mass spectrometry, in P. James, ed., *Proteome Research: Mass Spectrometry*, Springer-Verlag, Berlin, 2000, pp. 188–206.
51. H. Steen and M. Mann, A new derivatization strategy for the analysis of phosphopeptides by precursor ion scanning in positive ion mode, *J. Am. Soc. Mass Spectrom.* **13**, 996–1003 (2002).
52. A Tholey, J. Reed, and W. D. Lehmann, Electrospray tandem mass spectrometry studies of phosphopeptides and phosphopeptide analogues, *J. Mass Spectrom.* **34**, 117–123 (1999).
53. A. Schlosser, R. Ripkorn, D. Bossemeyer, and W. D. Lehmann, Analysis of protein phosphorylation by a combination of elastase digestion and neutral loss tandem mass spectrometry, *Anal. Chem.* **73**, 170–176 (2001).
54. J. P. DeGnore and J. Quin, Fragmentation of phosphopeptides in an ion trap mass spectrometer, *J. Am. Soc. Mass Spectrom.* **9**, 1175–1188 (1998).
55. M. C. Crowe and J. S. Brodbelt, Infrared multi-photon dissociation (IRMPD) and collisionally activated dissociation of peptides in a quadrupole ion trap (QIT) with

selective IRMPD of phosphopeptides, *J. Am. Soc. Mass Spectrom.* **15**, 1581–1592 (2004).

56. S. C. Moyer, R. J. Cotter, and A. S. Woods, Fragmentation of phosphopeptides by atmospheric pressure MALDI and ESI/ion trap mass spectrometry, *J. Am. Soc. Mass Spectrom.* **13**, 274–283 (2002).

57. N. Sadagopan, M. Malone, and J. T. Watson, Effect of charge derivatization in the determination of phosphorylation sites in peptides by electrospray ionization collision-activated dissociation tandem mass spectrometry, *J. Mass Spectrom.* **34**, 1279–1282 (1999).

58. S. Gronert, K. H. Li, and M. Horiuchi, Manipulating the fragmentation patterns of phosphopeptides via gas-phase boron derivatization: determination of phosphorylation sites in peptides with multiple serines, *J. Am. Soc. Mass Spectrom.* **16**, 1905–1914 (2005).

59. H. Jaffe and H. C. Pant, Characterization of serine and threonine phosphorylationin beta-elimination/ethanethiol addition-modified proteins by electrospray ionization tandem mass spectrometry and database searching, *Biochemistry* **37**, 16211–16224 (1998).

60. M. J. Schroeder, J. Shabanowitz, J. C. Swartz, D. F. Hunt, and J. J. Coon, A neutral loss activation method for improved phosphopeptide sequence analysis by quadrupole ion trap mass spectrometry, *Anal. Chem.* **76**, 3590–3598 (2004).

61. R. S. Annan and S. A. Carr, Phosphopeptide analysis by matrix-assisted laser desorption time-of-flight mass spectrometry, *Anal. Chem.* **68**, 3413–3421 (1996).

62. S. Metzer and R. Hoffmann, Studies on dephosphorylation of phosphotyrosine-containing peptides during post-source decay in matrix-assisted laser desorption/ionization, *J. Mass Spectrom.* **35**, 1165–1177 (2000).

63. A. Stensballe, O. N. Jensen, J. V. Olsen, K. F. Haselmann, and R. A. Zubarev, Electron-capture dissociation of singly and multiply phosphorylated peptides, *Rapid Commun. Mass Spectrom.* **14**, 1793–1800 (2000).

64. S. D. H. Shi, M. E. Hemling, S. A. Carr, D. M. Horn, I. Lindh, and F. W. McLafferty, Phosphopeptide/phosphoprotein mapping by electron capture dissociation mass spectrometry, *Anal. Chem.* **73**, 19–22 (2001).

65. M. J. Chalmers, W. Kolch M. R. Emmet, A. G. Marshall, and H. Mischak, Identification and analysis of phosphopeptides, *J. Chromatogr. B* **803**, 111–120 (2004).

66. J. E. P. Syka, J. J. Coon, M. J. Schroeder, J. Shabanowitz, and D. F. Hunt, Peptide and protein sequence analysis by electron transfer dissociation mass spectrometry, *Proc. Natl. Acad. Sci. USA* **101**, 9528–9533 (2004).

67. J. J. Coon, J. Shabanowitz, D. F. Hunt, and J. E. P. Syka, Electron transfer dissociation of peptide anions, *J. Am. Soc. Mass Spectrom.* **16**, 880–882 (2005).

68. J. W. Flora and D. C. Muddiman, Selective, sensitive, and rapid phosphopeptide identification in enzymatic digests using ESI–FTICR–MS with infrared multi-photon dissociation, *Anal. Chem.* **73**, 3305–3311 (2001).

69. J. W. Flora and D. C. Muddiman, Determination of relative energies of activation for the dissociation of aromatic versus aliphatic phosphopeptides by ESI–FTICR–MS and IRMPD, *J. Am. Soc. Mass Spectrom.* **15**, 121–127 (2004).

70. A. Varki, Biological roles of oligosaccharides—all of the theories are correct, *Glycobiology* **3**, 97–130 (1993).

71. R. Kornfeld and S. Kornfeld, Assembly of asparagine-linked oligosaccharides, *Annu. Rev. Biochem.* **54**, 631–664 (1985).
72. R. S. Rush, P. L. Derby, D. M. Smith, C. Merry, G. Rogers, M. F. Rohde, and V. Katta, Microheterogeneity of erythropoietin carbohydrate structure, *Anal. Chem.* **67**, 1442–1452 (1995).
73. J. H. Lee, Y. Kim, M. Y. Ha, E. K. Lee, and J. Choo, Immobilization of aminophenylboronic acid on magnetic beads for the direct determination of glycoproteins by matrix-assisted laser desorption ionization mass spectrometry, *J. Am. Soc. Mass Spectrom.* **16**, 1456–1460 (2005).
74. K. F. Medzihradszky, D. A. Maltby, S. C. Hall, C. A. Settineri, and A. L. Burlingame, Characterization of protein N-glycosylation by reversed-phase microbore liquid chromatography/electrospray mass spectrometry, complementary mobile phases, and sequential exoglycosidase digestion, *J. Am. Soc. Mass Spectrom.* **5**, 350–358 (1994).
75. S. A. Carr, M. J. Huddleston, and M. F. Bean, Selective identification and differentiation of N- and O-linked oligosaccharides in glycoproteins by liquid chromatography–mass spectrometry, *Protein Sci.* **2**, 183–196 (1993).
76. M. J. Huddleston, M. F. Bean, and S. A. Carr, Collisional fragmentation of glycopeptides by electrospray ionization LC/MS and LC/MS/MS: methods for selective detection of glycopeptides in protein digests, *Anal. Chem.* **65**, 877–884 (1993).
77. P. A. Scindler, C. A. Settineri, X. Collet, C. J. Fielding, and A. L. Burlingame, Site-specific detection and structural characterization of the glycosylation of human plasma proteins lecithin:cholesterol acyltransferase and apolipoprotein D using HPLC/electrospray mass spectrometry and sequential glycosidase digestion, *Protein Sci.* **4**, 791–803 (1995).
78. S. H. J. Bauer, X. Y. Zhang, W. Van Donngen, and M. Claeys, Chromogranin A from bovine adrenal medulla: molecular characterization of glycosylations, phosphorylations, and sequence heterogeneities by mass spectrometry, *Anal. Biochem.* **274**, 69–80 (1999).
79. K. F. Medzihradszky, B. L. Gillece-Castro, R. R. Townsend, A. L. Burlingame, and M. R. Hardy, Structural elucidation of O-linked glycopeptides by high energy collision-induced dissociation, *J. Am. Soc. Mass Spectrom.* **7**, 319–328 (1996).
80. M. J. Kieliszewski, M. O'Neil, J. Leykam, and R. Orlando, Tandem mass spectrometry and structural elucidation of glycopeptides from a hydroxyproline-rich plant cell wall glycoprotein indicate that contiguous hydroxyproline residues are the major sites of hydroxyproline O-arabinosylation, *J. Biol. Chem.* **270**, 2541–2549 (1995).
81. F.-G. Hanisch, B. N. Green, R. H. Bateman, and J. Peter-Katalinic, Localization of O-glycosylation sites MUC1 tandem repeats by QTOF ESI mass spectrometry, *J. Mass Spectrom.* **33**, 358–362 (1998).
82. K. Hirayama, R. Yuji, N. Yamada, K. Kato, Y. Arato, and I. Shimada, Complete and rapid peptide and glycopeptide mapping of mouse monoclonol antibody by LC/MS/MS using ion trap mass spectrometry, *Anal. Chem.* **70**, 2718–2725 (1998).
83. S. Goletz, B. Thiede, F. G. Hanisch, M. Schultz, J. Peter-Katalinic, J. Muller, O. Seitz, and U. Karsten, A sequencing strategy for the localization of O-glycosylation sites of MUC1 tandem repeats by PSD–MALDI mass spectrometry, *Glycobiology* **7**, 881–896 (1997).

84. D. H. Patterson, G. E. Tarr, F. E. Ragnier, and S. A. Martin, C-terminal ladder sequencing via matrix-assisted laser desorption mass spectrometry coupled with carboxypeptidase Y time-dependent and concentration-dependent digestions, *Anal. Chem.* **67**, 3971–3978 (1995).
85. E. Mirgorodskya, P. Roepstroff, and R. A. Zubarev, Localization of O-glycosylation sites in peptides by electron capture dissociation in a Fourier transform mass spectrometer, *Anal. Chem.* **71**, 4431–4436 (1999).
86. G. E. Reid, J. L. Stepheson, Jr., and S. A. McLuckey, Tandem mass spectrometry of ribonuclease A and B: N-linked glycosylation site analysis of whole protein ions, *Anal. Chem.* **74**, 577–583 (2002).

CHAPTER 10

PROTEINS AND PEPTIDES: HIGHER-ORDER STRUCTURES

The crucial functions of proteins and peptides in living systems are based on their very specific three-dimensional (3-D) structures. The four structural levels—primary, secondary, tertiary, and quaternary structures—were defined in Chapter 8. The native state of a protein is relatively stable and is a highly ordered compact form that contains surface pockets and interior cavities that are sites for the specific binding and chemical modification of biological ligands. The three-dimensional structure of a folded protein is the result of a combination of several noncovalent interactions, such as short-range repulsive forces, electrostatic forces, van der Waals interactions, hydrophobic interactions, and hydrogen bonding. The native state of a protein has significantly lower energy than that of other allowed compact states. The changes in a localized as well as in a globular structure are frequently used in nature to regulate the function of enzymes and receptors [1]. To understand the function of a protein fully, it is essential to identify these changes in terms of atomic locations, folding–unfolding dynamics, and thermodynamics. The availability of high-resolution structures (i.e., identity of atoms in a protein within <2 Å) is also helpful in rational drug design.

To study folding–unfolding dynamics and to identify intermediate conformers, proteins are usually denatured and unfolded to a random coil structure by increasing the temperature and/or pH, or by subjecting them to high concentrations of detergents, organic solvents, or certain chaotropic compounds, such as urea and guanidinium chloride. Unfolded proteins can refold to their native state once the denaturant is removed. Disulfide bridge–containing proteins are denatured by the action of disulfide bond-breaking reagents (e.g., dithiothreitol).

Fundamentals of Contemporary Mass Spectrometry, by Chhabil Dass
Copyright © 2007 John Wiley & Sons, Inc.

The temperature at which thermal unfolding occurs varies from protein to protein and represents the strength of hydrogen-bonding and hydrophobic interactions. Denaturation by a change of pH is triggered by ionizing and charging moieties on the side chains. Coulombic repulsions among these charge centers destabilize the folded structure and lead to swelling and ultimately to unfolding of the protein molecule.

A variety of techniques are at the disposal of researchers to analyze a protein's higher-order structure, dynamics, and conformational changes [2]. These methods include X-ray crystallography [3], circular dichroism (CD) [4,5], nuclear magnetic resonance (NMR) [6,7], Fourier-transform (FT) infrared (IR) [8,9], Raman [10,11], ultraviolet (UV)–visible absorption [9,12,13], and fluorescence spectroscopies [9,14]. Of these, one- and two-dimensional NMR have found a greater success in the detection of changes in protein structures. Electron microscopy, neutron diffraction, ultracentrifugation, and liquid chromatography have also been used in this field. Besides instrumental methods, computational algorithms can also provide details of a three-dimensional protein structure [15].

With the development of relatively softer ionization techniques (ESI and MALDI), mass spectrometry has emerged as an option for the determination of conformational changes in proteins [2,16–23]. It is believed that the solution-phase structure of a protein is largely preserved during ionization by these two methods. Therefore, the ESI and MALDI mass spectra of a protein reflect the features of its aqueous solution chemistry. The multiple-charging feature of ESI is also a valuable asset because it allows the study of much larger proteins. In this chapter a broad outline of the commonly used mass spectrometry–based techniques is presented.

Mass spectrometry offers several distinct advantages, such as sensitivity, protein stability, and extended molecular mass, and also provides information complementary to NMR. The solubility and purity of a protein are of less concern. Furthermore, exchange rates of the most rapidly exchanging amide hydrogens can be determined. Therefore, mass spectrometry can reveal structural details on transient or folding intermediates. These species may not be accessible by other conventional techniques mentioned above (e.g., NMR, CD, fluorescence) because they provide information that represents an average of entire protein ensembles. The NMR approach, on the other hand, is superior with respect to resolution; also the precious protein sample is not destroyed.

10.1. CHARGE-STATE DISTRIBUTION

The application of mass spectrometry to the study of a protein's higher-order structures requires a protocol in which differences in the m/z ratio associated with each distinct conformation of the protein could be monitored. One common mass spectrometry methods for conformational analysis is charge-state distribution (CSD) [24]. This method can be implemented only with the ESI mode of analysis. CSD provides a qualitative, low-resolution picture of the conformational state of a protein. Positive-ion ESI generates a distribution of multiply charged ions of the type $[M + nH]^{n+}$. The shape of this distribution is dependent on the conformation of a protein under the experimental conditions of the study. In general, more basic sites are available in a protein for protonation in the unfolded flexible conformation, whereas in the tightly folded native state, those basic sites may be close together or buried in the hydrophobic core of the protein and thus have less opportunity for any proton attachment. As a consequence, fewer charge states will be formed. Once the protein unfolds, those buried basic sites become accessible to solvents to produce a greater degree of charging and a shift in the ESI mass spectrum from higher to lower m/z values. Thus, CSD can provide a means to monitor protein stability and folding–unfolding kinetics.

In the seminal work, it was demonstrated that changes in the ESI–CSD profile of cytochrome c can be correlated with fluctuations in its conformation induced by the pH of the solution [24]. Disruption in the native state of a protein with a change in the solution temperature, addition of denaturing agents, and cleavage

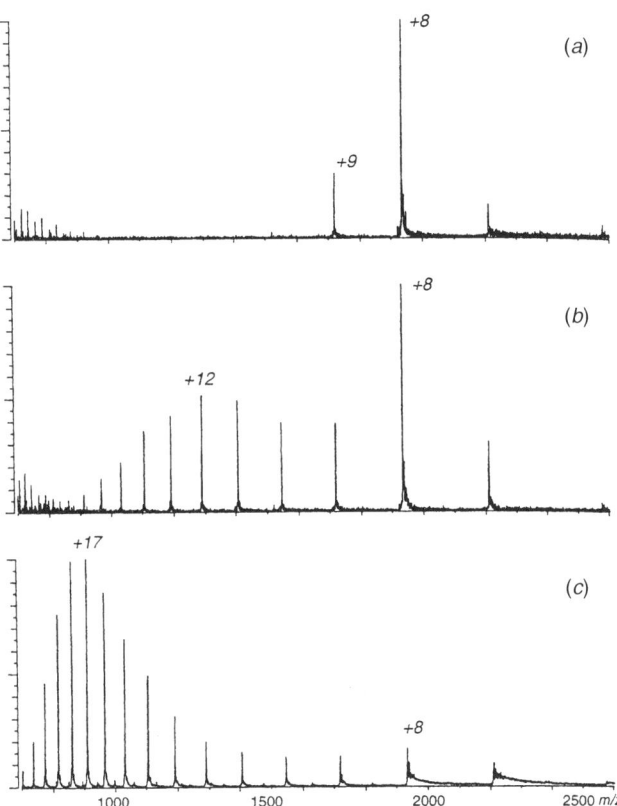

Figure 10.1. Positive-ion CSD of CRABP I: at pH (a) 7.4, (b) 3.5, and (c) 2.5. (Reproduced from ref. 25 by permission of the American Chemical Society, Washington, DC, copyright © 2001.)

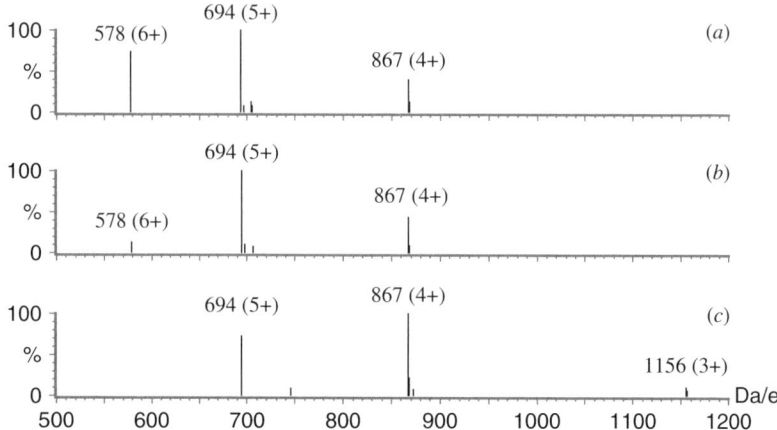

Figure 10.2. CSD in the ESI mass spectra of human β-endorphin in (a) water, (b) 20% and (c) 60% methanol. (26). (Reproduced from ref. 26 by permission of Wiley-Interscience, copyright © 2001.)

of the disulfide linkages is also reflected in the charge distribution in the ESI mass spectra. Figure 10.1 shows the acid-induced unfolding of cellular retinoic acid–binding protein I (CRABP I) [25]. At a pH above 5.0, the protein exists mainly in the native state, which is reflected in a single narrow charge-distribution profile that consists of only three charge states (+7, +8, and +9). Unfolding of the proteins begins at pH 4.5. At more acidic pH, the CSD shifts to lower m/z values (i.e., at higher charge states) to reflect the dominance of the open structure.

The power of the charge-state analysis to elucidate structures of peptides has also been demonstrated [26,27]. As shown in Figure 10.2 for β-endorphin, a 31-amino acid peptide, the CSD is shifted to lower charge states in the presence of a helicity-inducing organic solvent (Figure 10.2c), methanol, compared to that observed in water (Figure 10.2a), where the peptide exists predominantly in an open-structure form.

The CSD method is straightforward and simple to implement, but the results should be interpreted with caution because the reproducibility of the ESI ion profile might fluctuate due to changes in various experimental variables, such as pH, temperature, solution composition, and the ESI ion-source settings (e.g., gas flows and voltages).

▶ **Example 10.1** How many charge states will be observed from the polypeptide chain SYSMEHFRWGKPVGKKRRPVKVYPNGAEDESAEAFPLEF, and at what m/z values?

Solution Because there are eight basic residues (four K, three R, and one H) and one terminal amino group, a maximum of 9+ charge states might be observed in the ESI mass spectrum. The monoisotopic molecular mass of this polypeptide = 4539 Da. The m/z values of all of the 9+ charge states are:

$$\text{The } m/z \text{ of 1+ charge state} = [M+H]^+ = (4539+1) = 4540.$$

$$\text{The } m/z \text{ of 2+ charge state} = \frac{[M+2H]^{2+}}{2} = \frac{(4539+2)}{2} = 2270.5$$

$$\text{The } m/z \text{ of 3+ charge state} = \frac{[M+3H]^{3+}}{3} = \frac{(4539+3)}{3} = 1514.1$$

The m/z values of the remaining 4+, 5+, 6+, 7+, 8+, and 9+ charge states are calculated similarly; the values are 1135.8, 908.9, 757.6, 649.5, 568.4, and 505.4, respectively.

10.2. HYDROGEN–DEUTERIUM EXCHANGE TO STUDY CONFORMATIONAL STATES OF PROTEINS

Hydrogen–deuterium exchange (HDX) is the process of replacing labile hydrogen atoms in proteins with deuterium from the surrounding solvent. The HDX approach was introduced in the 1950s [28]. A protein contains three different types of hydrogens (see Figure 10.3): (1) nonexchanging alkyl hydrogens; (2) fast-exchanging labile hydrogens such as those at the N- and C-termini and those attached to side-chain functional groups (e.g., OH, SH, and NH), which are less suitable for conformational studies; and (3) slow-exchanging amide hydrogens, which are located all along the polypeptide backbone, except in proline. Amide hydrogens are of particular interest in HDX experiments because they exchange at measurable rates, whereas functional group hydrogens exchange too rapidly to be of any practical utility. The rates of exchange of labile hydrogens depend on whether or not they are involved in hydrogen bonding and are buried within the protein core or are accessible to solvent. Thus, amide hydrogens that are part of secondary structures (e.g., α-helix and β-sheet) and of the hydrophobic core of a protein exchange slowly, whereas the surface hydrogens exchange at

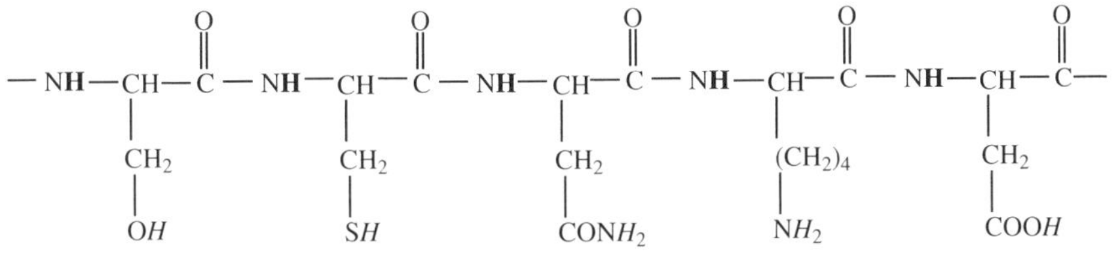

Figure 10.3. Three different types of hydrogens in a polypeptide chain: normal-font hydrogens are nonexchanging alkyl hydrogens, *italic* hydrogens are fast-exchanging labile hydrogens, and **bold** hydrogens are slow-exchanging amide hydrogens.

a faster rate (sometimes by a factor of 10^8). Thus, the distinct values of HDX rates can provide a sensitive method of determining the conformational states of a protein. A faster exchange is indicative of a more open structure, and a slower exchange is related to a tightly folded compact state. Also, more hydrogens are available for isotopic exchange in an unfolded conformer than in the native structure.

▶ **Example 10.2** How many exchangeable hydrogens are present in the peptide H_2N–Tyr–Ala–Gly–Phe–Leu–Pro–Arg–COOH?

Solution Exchangeable hydrogens include the amide hydrogens and labile hydrogens at the termini and in the side-chain functional groups. Therefore, the total exchangeable hydrogens are = 2 at the N-terminus + 1 in Tyr + 1 in Ala + 1 in Gly + 1 in Phe + 1 in Leu + 0 in Pro + 5 in Arg + 1 at the C-terminus = 13.

Traditionally, HDX has been studied with multidimensional NMR spectroscopy [6,7,29]. Katta and Chait were the first to combine H/D exchange with mass spectrometry to probe the conformational changes in proteins [30]. It was demonstrated that upon denaturation of bovine ubiquitin, not only did the ion profile shift to a higher charge state but the incorporation of deuterium was also higher in the unfolded structure [30]. Since then, HDX–MS has become a standard protocol to probe changes in protein structures [2,16–23]. Mass spectrometry provides an additional dimension in the HDX analysis in terms of the mass of proteins. The mass of a protein increases by 1.0063 Da (from 1.0078 Da to 2.0141 Da) for the replacement of each hydrogen atom with a deuterium; that change can be determined readily by mass spectrometry. ESI–MS [2,16–23,26,27,31,32] and MALDI–MS [33–35] can both be used for this purpose.

The mechanism of amide–hydrogen exchange in proteins is well understood. The exchange could be catalyzed by H^+ and OH^- ions both. The rate constant for hydrogen exchange (k_{ex}) is the sum of the rates for the acid- (k_H) and base-catalyzed (k_{OH}) reactions

$$k_{ex} = k_H[H^+] + k_{OH}[OH^-] + k_w \qquad (10.1)$$

The rate constant for water solvent, k_w, is very small and is normally neglected. The slowest exchange occurs at pH 2 to 3. At more basic pH values, the exchange rate increases rapidly as a function of the OH^- ion activity. The exchange rate varies by a factor of 10 for each unit change in pH. The exchange rate also increases as the pH goes below 2. The exchange rate is also affected by temperature; it is 10-fold lower at 0°C than at 25°C. Because of the slowest exchange rate, the solution at pH 2 to 3 and 0°C is considered at "quench conditions." In a normal experiment, the exchange can be performed at neutral pH and 25°C (faster exchange) and arrested by bringing the solution to quench conditions (the slowest exchange) to provide a measure of the exchange level.

10.2.1. Folding and Unfolding Dynamics of Proteins

In solution, native proteins fluctuate continuously around the average conformation even under physiological conditions. This dynamic process is the reason that even inaccessible hydrogens such as those involved in hydrogen bonding and buried deep in the hydrophobic core might undergo isotope exchange. The exact nature of solvent accessibility during protein-unfolding events is a matter of debate. In the *solvent-penetration model* it is believed that for HDX of the protein interior to occur, solvent enters the protein core through transiently formed channels and cavities [36]. In the *local unfolding model*, exchange occurs on the surface of the protein when small segments of the protein unfold and expose the exchangeable sites to the solvent [37,38]. Thus, amide–hydrogen exchange in proteins can be explained in terms of one of two rate-limiting processes, or both [38,39]; one involves fluctuations in the native folded (N) state [Eq. (10.2)], and the second corresponds to global unfolding [Eq. (10.3)]

$$N(H) \xrightarrow{k_N} U(D) \tag{10.2}$$

$$N(H) \underset{k_{-1}}{\overset{k_1}{\rightleftharpoons}} U(H) \xrightarrow{k_2} U(D) \underset{k_{-1}}{\overset{k_1}{\rightleftharpoons}} N(D) \tag{10.3}$$

where k_N is the exchange rate from the folded native state, k_1, and k_{-1} are the rates of the protein unfolding and refolding processes, and k_2 is the exchange rate of amide hydrogens for the completely unfolded (U) protein. In Eq. (10.2), hydrogen exchange takes place directly from the folded protein and involves only those amide hydrogens that are accessible to the deuterated aqueous medium. The experimentally observed exchange rate constant for a peptide amide linkage by this process is given by

$$k_N = \beta k_2 \tag{10.4}$$

where β is the probability that the amide hydrogens make contact with D_2O and OD^- catalyst. The value of β is 1 for those peptides in which all amide hydrogens are exposed to solvents, and also for completely unfolded proteins.

The global unfolding process is described by Eq. (10.3). It involves a reversible folding and unfolding of small regions of the protein as well as of the entire protein. Isotopic exchange takes place only after the protein has undergone complete unfolding. The rate constant (k_U) of the overall process is given by

$$k_U = \frac{k_2 k_1}{k_{-1} + k_2} \tag{10.5}$$

Two extreme situations are encountered. In one, the protein folding is much faster than the isotopic exchange rate (i.e., $k_{-1} \gg k_2$), as at a neutral pH and in the absence of denaturants. Under these conditions, opening and refolding of a protein will occur many times before any isotopic exchange takes place. This exchange is referred to as an *EX2 mechanism* [39], and the exchange rate is given by $k_{ex} = k_2 k_1 / k_{-1}$. In the second situation (e.g., under denaturing conditions), the protein-refolding rate is much slower than the isotopic-exchange rate (i.e., $k_{-1} \ll k_2$). This exchange mechanism is referred to as an *EX1 mechanism*, and $k_{ex} = k_l$. These two processes can be distinguished from the characteristic mass spectrometry data. When the EX2 mechanism is operative, a single peak will be observed in the mass spectrum. In contrast, two distinct peaks, one for the protonated form and the other for the deuterated form of the protein, would result after a short exchange period for EX1 mechanism. Thus, under these conditions, distinct protein conformers can be identified.

10.2.2. Experimental Measurements of Amide Hydrogen Isotopic Exchange

Protein folding–unfolding experiments are of two types:

1. *Equilibrium experiments*. The conformational changes are monitored as a function of pH, the concentration of denaturant, and temperature. Measurements are made only after equilibrium has been established.
2. *Kinetic experiments*. The solution conditions are altered rapidly and the conformational changes are followed as a function of time.

Amide–hydrogen isotopic exchange is performed in either the *exchange-in* or *exchange-out format*. In the former, the protiated form of the protein is exchanged isotopically with an excess of the deuterated solvent, whereas in the latter format, the fully deuterated protein is back-exchanged with the protiated solvent.

Hydrogen Exchange for Detection of Global Changes in Proteins To detect global changes, hydrogen exchange is performed on an intact protein. This experiment will provide the exchange rate averaged over all amide hydrogens in a protein. Unfolding (k_1) and folding (k_{-1}) exchange kinetics can both be studied. The experiments are conducted in either a continuous- or a pulse-labeling mode.

Continuous-Labeling Technique In a typical procedure for measurement of the rate constant for the folding process in the continuous-labeling mode, the protein is dissolved in the buffered H_2O of the pH desired. The protein is unfolded by heating, by lowering the pH to 2.0, or by adding a suitable denaturing agent. The protein is lyophilized and redissolved in H_2O. The exchange is initiated by diluting this solution 100-fold with D_2O that has been adjusted to the desired pD. The samples are withdrawn at several different times and each is analyzed by ESI–MS or MALDI–MS.

The number of exchanged hydrogens (H_x) in the protein can be calculated from the m/z ratio of the most abundant isotopic peak of the nth charge state with the following expression:

$$H_x = M_D - M = n[(m/z)_D - D^+] - M \qquad (10.6)$$

where M is the molecular mass of the nondeuterated species, M_D the measured molecular mass of the partially deuterated species, and D the mass of a deuteron (2.0141 Da). The deuterium content (in %) is defined as

$$[D] = \frac{H_x}{H_t} \times 100 = \frac{M_D - M}{M_T - M} \times 100 \qquad (10.7)$$

where H_t is the total number of exchangeable hydrogens and M_T is the molecular mass of the fully deuterated species.

▶ **Example 10.3** The conformational state of a protein of molecular mass 10,520 Da was studied using the HDX protocol. Upon deuteration, its 10+ charge state was observed at m/z 1060. Calculate the number of hydrogens exchanged.

Solution Substitution of these values into Eq. (10.6) gives

$$H_x = 10(1060 - 2.014) - 10{,}520 = 59.86$$

To study the kinetics of the unfolding process, the lyophilized folded protein is dissolved in D_2O at a pD value where folded and unfolded states coexist. The samples are withdrawn at different intervals and are analyzed by ESI–MS. A particular charge state of the folded protein is selected and the deuterium incorporation is estimated from these measurements by the increase in mass of this peak. The abundance of the deuterated peak is the measure of the amount of the unfolded state. The change in abundance of the native (F) and deuterated peaks (U) is monitored at different times. The plot of either a simple U term, or the expression F/(F + U), versus the time of reaction can provide an estimate of the rate constant (k_1) for the unfolding process.

Time-Resolved ESI–MS Alternatively, the exchange can be studied in continuous-flow apparatus online with an ESI source [22,26,27]. This approach has been termed *time-resolved ESI-MS* [22]. Equilibrium condition experiments [26,27] and kinetic experiments [22] can both be conducted with this apparatus. The exchange is initiated by injecting the protein or peptide solution in the flowing exchange solution [26,27] or by mixing the labeling and protein solutions in a mixer as shown in Figure 10.4. The exchange period (milliseconds to seconds) can be varied by changing the flow rate and the dimensions of the

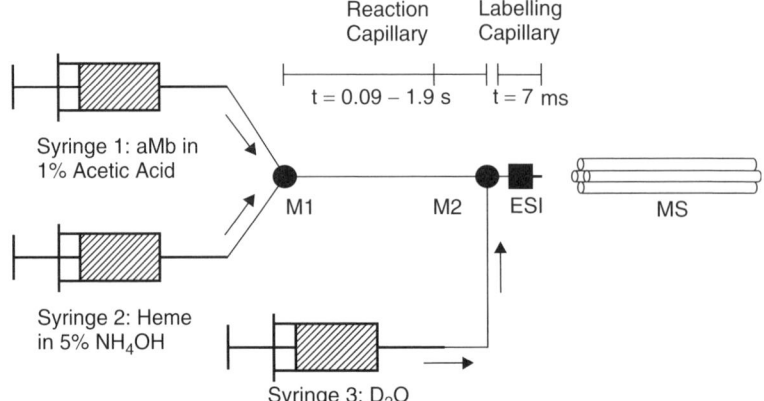

Figure 10.4. Continuous-flow apparatus. Mixing the labeling and protein solutions takes place at mixer M2. (Reproduced from ref. 40 by permission of the American Chemical Society, Washington, DC, copyright © 2002.)

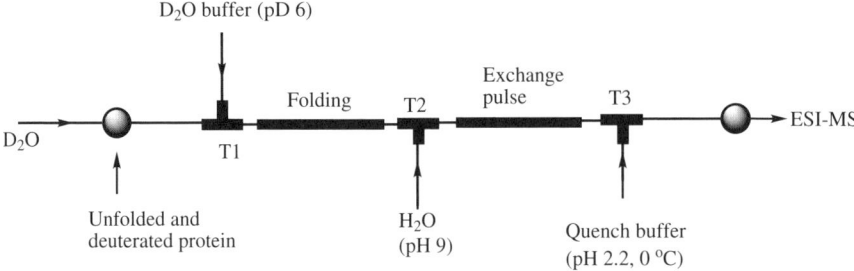

Figure 10.5. Quench-flow apparatus for pulse-labeling HDX. (Reproduced from C. Dass, *Principles and Practice of Biological Mass Spectrometry*, Wiley-Interscience, 2001.)

labeling capillary. A continuous-flow apparatus can also be used to monitor the CSD profil [22] and pulse-labeling HDX experiments [40].

Pulse-Labeling Technique In this technique the protein is exposed to rapid exchange conditions for a very short period while it folds [41]. This approach is well suited to determining the isotopic exchange rates of rapidly exchanging amide hydrogens. Pulse labeling is usually studied in a quench-flow apparatus [22,40,41]. Typically, the protein is dissolved in D_2O in the presence of a strong denaturant. The unfolded protein is injected into a quench-flow system similar to that shown in Figure 10.5, and the folding is initiated by mixing (at T1) the injected protein solution with an excess volume of buffered D_2O solution (pD 5 to 6). The length of the folding time, Δt_{fold} (milliseconds to seconds), can be controlled by varying the size of the folding tube. After this folding period, an exchange pulse is applied by mixing (at T2) the protein solution with a basic pH

(>7.0) H_2O solution. At this pH, the exposed hydrogens will exchange instantaneously. The exchange period, Δt_{ex}, can also be varied by changing the length of the labeling tube. Labeling is terminated when the quench-buffer (pH 2.5, 0°C) is introduced into a mixing tee marked T3. The exchange pattern can be monitored either by collecting the sample and subsequent off-line ESI–MS, or by connecting the quench-flow apparatus directly with the ESI source.

Information on Localized Regions of Proteins HDX on an intact protein allows global changes to be assessed readily but fails to provide any information on the dynamic behavior of individual structural segments within a protein. To identify the localized structural changes, greater spatial resolution is required. Two methods are available to obtain structural details in small regions of proteins: the fragmentation/mass spectrometry method and the gas-phase ion dissociation method.

Fragmentation/Mass Spectrometry Method In this method, pioneered by Smith and colleagues, the deuterated protein is cleaved into small segments by pepsin digestion under quench conditions, followed by mass spectrometry analysis of the protein digest [42–44]. A typical continuous-labeling procedure is illustrated in Figure 10.6. In this protocol, the exchange-in is performed for a defined time interval by incubating the folded protein in a D_2O solution at an appropriate pD value. At the end of the defined time interval, the exchange

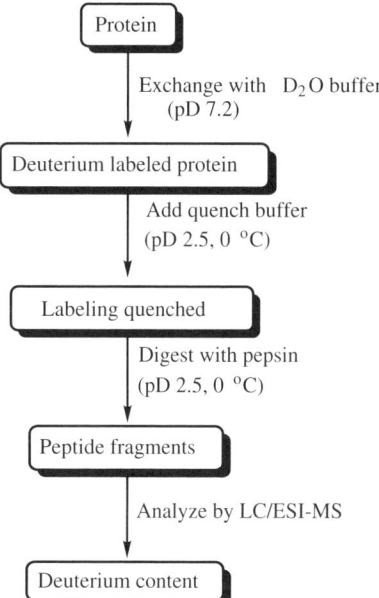

Figure 10.6. Outline of the fragmentation/mass spectrometry method to determine structural details in small regions of proteins [44]. (Reproduced from C. Dass, *Principles and Practice of Biological Mass Spectrometry*, Wiley-Interscience, 2001.)

process is quenched by reducing the pD value to 2.5 and the temperature to 0°C. The protein is digested with pepsin, which is chosen because it has the maximum proteolytic activity at the quenching pH and temperature. The deuterium level of each peptide fragment of the digest is measured with HPLC online with ESI–MS under conditions that minimize the hydrogen exchange at the peptide amide linkages [44]. The extent of the deuterium incorporation is determined by comparing the ESI mass spectra with a reference spectrum of the same peptide that contains no deuterium and another with 100% deuterium. The pulsed-labeling approach can also be used to study the isotope-exchange reaction in short segments of a protein [44]. The apparatus shown in Figure 10.4 is modified to add a pepsin column and an additional mixing tee for online peptide digestion.

Gas-Phase Ion Dissociation Method Further spatial details at the individual amino acid residue level can be obtained by the gas-phase ion dissociation of the intact polypeptide or of the small peptide segments that are generated via pepsin digestion of the isotopically exchanged protein. CID of the peptide segments produces mainly b- and y-ion series. If a specific amide hydrogen undergoes HDX, a shift in the m/z value will be observed for that sequence-specific ion in the CID–MS/MS spectrum. The number of exchanged hydrogens (H_x) in b- and y-ions can be calculated as follows:

$$b_j^{n+}: \quad H_x = n[(m/z)_D(b_j^{n+}) - (m/z)_H(b_j^{n+})] \quad (10.8)$$

$$y_j^{n+}: \quad H_x = n[(m/z)_D(y_j^{n+}) - (m/z)_H(y_j^{n+}) - 1] - 1 \quad (10.9)$$

where $(m/z)_H$ and $(m/z)_D$ refer to the m/z values of the ions from nondeuterated and partially deuterated species, respectively [45].

The usefulness of this approach is demonstrated in identifying the secondary structure of methionine enkephalin (YGGFM) [46]. A controversy has persisted in the literature as to whether this peptide forms a $5 \rightarrow 2$ β-turn or a $4 \rightarrow 1$ β-turn structure. By acquiring the CID–MS/MS spectrum of the deuterated peptide in a 50% trifluoroethanol-d_1/D_2O solvent mixture, it was revealed that methionine enkephalin forms a $5 \rightarrow 2$ β-turn structure.

▶ **Example 10.4** A CID–MS/MS spectrum of angiotensin II was acquired after performing HDX. The b_4^+ and y_7^{2+} sequence-specific ions were observed at m/z values of 544.4 and 474.1, respectively. The corresponding ions from the nondeuterated species are seen at m/z of 534.2 and 466.5, respectively. Calculate the number of exchanged hydrogens in these ions.

Solution Substitution of these m/z values into Eqs. (10.8) and (10.9) gives

$$H_x(b_4^+) = (1)(544.4 - 534.2) = 10.2$$
$$H_x(y_7^{2+}) = (2)[(474.1 - 466.5) - 1] - 1 = 12$$

10.3. CHEMICAL CROSS-LINKING AS A PROBE FOR THE THREE-DIMENSIONAL STRUCTURE OF PROTEINS

Mass spectrometry, in conjunction with chemical cross-linking, has evolved into a sensitive probe to obtain information on the tertiary structure of proteins [47–50]. In this procedure a protein is modified covalently with bifunctional cross-linking reagents of different spacer arm lengths, and subsequently, digested proteolytically. The cross-linked peptides are identified by comparing the peptide-mass map of the digest from the cross-linked protein with the digest from the unmodified protein. MS/MS analysis of the cross-linked peptides further corroborates their sequence and provides the identity of the modified residues. The maximum distance between the cross-linked residues is assigned on the basis of the cross-linker arm length. This information provides low-resolution distance constraints that can be used to construct the three-dimensional structure of a protein by computational methods. The name MS3D has been coined for the study of protein three-dimensional structure by using chemical cross-linking chemistry and mass spectrometry [47].

Experiments can be designed to target different amino acid residues (e.g., Cys, Tyr, or Lys) and to select chemical cross-linkers with distinct solubility characteristics and of varying spacer arm lengths. A typical example is the use of bis(sulfosuccinimidyl) suberate (BS^3), disuccinimidyl suberate (DSS), and disuccinimidyl glutarate (DSG) with spacer arm lengths of 11.4, 11.4 and 7.4 Å, respectively, to determine distant constraints in bovine serum albumin [49]. BS^3 is a water-soluble compound that can access lysines in hydrophilic or surface regions, whereas DSS and DSG are water-insoluble and thus are suitable to cross-link lysines in hydrophobic regions.

10.4. ION MOBILITY MEASUREMENTS TO STUDY PROTEIN CONFORMATIONAL CHANGES

Ion mobility spectrometry (IMS), in conjunction with mass spectrometry measurements, has been also employed to determine the conformation and folding–unfolding kinetics of proteins [51–53]. In an IMS instrument, ionic species are separated on the basis of differences in the cross-sectional area by allowing them to drift in a long tube under the influence of an electrical field gradient (see Section 3.10). Because of their distinct shapes, different conformations will drift at different rates in the IMS drift tube. The folded conformers have a smaller collision cross section and thus will have higher mobilities than those of larger open structures. The mobility of a protein also depends on its charge state. An IMS-based instrument for conformational studies consists of an ESI source, a mass-selecting quadrupole, an IMS drift tube, and a mass-measuring quadrupole. In a typical experimental protocol, a protein is ionized with ESI, and a short pulse of a particular charge state is mass-selected and injected into the drift tube. Ions that exit the drift tube are analyzed with a second quadrupole.

392 PROTEINS AND PEPTIDES: HIGHER-ORDER STRUCTURES

From the drift-time distributions obtained in this setup, information about the number of isomers can be obtained. Some examples of IMS–mass spectrometry in conformation and folding–unfolding studies are for cytochrome c [51], apomyoglobin [51], lysozyme [52], and ubiquitin [53].

OVERVIEW

Determination of higher-order structures of proteins and peptides is another important area that can benefit from the structure-specific and high-mass ability of mass spectrometry. Several mass spectrometry-based approaches are available that provide information complementary to that obtained by NMR. These techniques rely primarily on CSD in the ESI mass spectrum and hydrogen–deuterium isotopic exchange as monitored by ESI–MS or MALDI–MS. In contrast to NMR, which monitors average exchange levels to reveal an average population of protein structures, ESI–MS can differentiate coexisting protein structures. Distinct CSD profiles are observed from folded and unfolded open structures: A folded structure shows fewer charge states; the CSD profile of an unfolded protein contains more charged states and a shift to lower m/z values. HDX-based methods make use of slow-exchanging backbone amide hydrogens of a protein. These hydrogens exchange at a much slower rate if they are buried within the hydrophobic core of the protein or are involved in the secondary structure formation. The extent of isotope exchange can be ascertained conveniently by the increment of mass that results upon deuteration. ESI–HDX–MS can be used to detect global changes as well as study the behavior of individual structural segments. The latter aspect is revealed by pepsin digestion of the exchanged protein. CID–MS/MS of the peptide segments generated provides further spatial details at the individual amino acid residue level. In this respect, ESI–MS might compete with multidimensional NMR techniques.

EXERCISES

10.1. How many charge states will be observed in the ESI spectrum of the polypeptide chain PDKDFIVNPSDLVLDNKAALRDYLRQINEYFAII-GRPRF, and at what m/z values?

10.2. How many total exchangeable hydrogens and slow-exchanging backbone amide hydrogens are present in the polypeptide H_2N–SEEPPISLDLTFH-LLREVLEMARAEQLAQQAHSNRK–OH?

10.3. A protein of molecular mass 8562 Da contains 35 exchangeable hydrogens. Find the molecular mass of the fully deuterated protein.

10.4. The conformational state of the protein of Exercise 10.3 was studied using the HDX–ESI–MS protocol. Upon deuteration, the m/z value of its 7+

charge state was found to be 1228.43. Calculate the number of hydrogens exchanged.

10.5. Describe a protocol that can provide details of isotope exchange in a protein at the individual amino acid residue level.

10.6. Upon incubation with D_2O, the mass of a polypeptide was found to be 4805.0 Da. The molecular masses of nondeuterated and fully deuterated species are 4754.5 and 4830.9 Da, respectively. Calculate the percent deuterium content.

10.7. A protein, after incubation with D_2O, was subjected to pepsin digestion. In a CID–MS/MS spectrum of a peptide segment, b_9^{2+}, y_6^+, and y_8^{2+} sequence-specific ions were observed at m/z of 591.1, 772.5, and 519.3, respectively. The theoretical m/z values of the corresponding ions from nondeuterated species are m/z 583.3, 763.4, and 513.3, respectively. Calculate the number of hydrogens exchanged in these ions.

ADDITIONAL READING

1. I. A. Kaltashov and S. J. Eyles, *Mass Spectrometry in Biophysics: Conformation and Dynamics in Biomolecules*, Wiley, Hoboken, NJ, 2005, p. 458.

REFERENCES

1. T. E. Creighton, *Protein Structures and Molecular Properties*, W. H. Freeman, New York, 1984.
2. X. Cai and C. Dass, Conformational analysis of proteins and peptides, *Curr. Org. Chem.* **7**, 1841–1854 (2003).
3. T. E. Creighton, Protein folding, *Biochem. J.* **270**, 1–16 (1990).
4. G. D. Fasman, *Circular Dichroism and the Conformational Analysis of Biomolecules*, Plenum Press; New York, 1996.
5. N. Berova, K. Nakanishi, and R. W. Woody, *Circular Dichroism: Principles and Applications*, Wiley, New York, 2000.
6. Y. Bai, T. R. Sosmick, L. Mayne, and S. W. Englander, Protein folding intermediates: native-state hydrogen exchange, *Science* **269**, 192 (1995).
7. A. E. Ferentz and G. Wagner, NMR spectroscopy: a multifaceted approach to macromolecular structure, *Q. Rev. Biophys.* **33**, 29–65 (2000).
8. H. B. Osbome and E. Nabedryk-Viala, Infrared measurement of peptide hydrogen exchange in rhodopsin, *Methods Enzymol.* **88**, 676–680 (1982).
9. R. Vogel and S. Friedrich, Conformation and stability of alpha-helical membrane proteins, 2: Influence of pH and salts on stability and unfolding of rhodopsin, *Biochemistry* **41**, 3536–3545 (2002).
10. J. T. Pelton and L. R. McLean, Spectroscopic methods for analysis of protein secondary structure, *Anal. Biochem.* **277**, 167 (2000).

11. A. Ozdemir, L. K. Lednev, and S. A. Asher, Comparison between UV Raman and circular dichroism detection of short helices in bombolitin III, *Biochemistry* **41**, 1893–1896 (2002).
12. J. A. Thomson, B. A. Shirley, G. R. Grimsley, and C. N. Pace, Conformational stability and mechanism of folding of ribonuclease T1, *J. Biol. Chem.* **264**, 11614–11620 (1989).
13. J. Juneja and J. B. Udgaonkar, Characterization of the unfolding of ribonuclease A by a pulsed hydrogen exchange study: evidence for competing pathways for unfolding, *Biochemistry* **41**, 2641–2654 (2002).
14. M. R. Eftink, The use of fluorescence methods to monitor unfolding transitions in protein, *Biophys. J.* **66**, 482–501 (1994).
15. J. Skolnick and A. Kolinski, Computational studies of protein folding, *Comput. Sci. Eng.* **3**, 40–50 (2001).
16. D. L. Smith, Y. Deng, and Z. Zhang, Probing the non-covalent structure of proteins by amide hydrogen exchange and mass spectrometry, *J. Mass Spectrom.* **32**, 135–146 (1997).
17. D. L. Smith and Z. Zhang, Probing noncovalent structural features of proteins by mass spectrometry, *Mass Spectrom. Rev.* **13**, 411–429 (1994).
18. S. D. Maleknia and K. M. Downard, Radical approaches to probe protein structure, folding, and interactions by mass spectrometry, *Mass Spectrom. Rev.* **20**, 388–401 (2001).
19. A. Miranker, C. V. Robinson, S. E. Radford, and C. M. Dobson, Investigation of protein folding by mass spectrometry, *FASEB J.* **10**, 93–101 (1996).
20. J. R. Engin and D. L. Smith, Investigating protein structure and dynamics by hydrogen exchange MS, *Anal. Chem.* **73**, 256A–265A (2001).
21. I. Kaltashov and S. J. Eyles, Crossing the phase boundary to study protein dynamics and function: combination of amide hydrogen exchange in solution and ion fragmentation in gas phase, *J. Mass Spectrom.* **37**, 557–565 (2002).
22. L. Konermann and D. A. Simmons, Protein-folding kinetics and mechanisms studied by pulse-labeling and mass spectrometry, *Mass Spectrom. Rev.* **22**, 1–26 (2003).
23. X. Yan, J. Watson, P. S. Ho, and M. L. Deinzer, Mass spectrometric approaches, *Mol. Cell. Proteom.* **3(1)**, 10–23 (2004).
24. S. K. Chowdhury, V. Katta, and B. T. Chait, Probing conformational changes in proteins by mass spectrometry, *J. Am. Chem. Soc.* **112**, 9012–9013 (1990).
25. A. Dobo and I. A. Kaltashov, Detection of multiple protein conformational ensembles in solution via deconvolution of charge-state distributions in ESI MS, *Anal. Chem.* **73**, 4763–4773 (2001).
26. H. Lin and C. Dass, Conformational changes in β-endorphin as studied by electrospray ionization mass spectrometry, *Rapid Commun. Mass Spectrom.* **15**, 2341–2346 (2001).
27. H. Lin and C. Dass, A mass spectrometry investigation of the conformational changes in adrenocorticotropic hormones, *Eur. J. Mass Spectrom.* **8**, 381–387 (2002).
28. A. Hvidt and K. Linderstrom-Lang, Exchange of hydrogen atoms in insulin with deuterium atoms in aqueous solutions, *Biochem. Biophys. Acta* **14**, 574–575 (1954).
29. A. D. Robertson and R. L. Baldwin, Hydrogen exchange in thermally denatured ribonuclease, *Biochemistry* **30**, 9907–9914 (1991).

30. V. Katta and B. T. Chait, Conformational changes in proteins probed by hydrogen-exchange electrospray-ionization mass spectrometry, *Rapid Commun. Mass Spectrom.* **5**, 214–217 (1991); V. Katta and B. T. Chait, Hydrogen/deuterium exchange electrospray ionization mass spectrometry: a method for probing protein conformational changes in solution, *J. Am. Chem. Soc.* **115**, 6317–6321 (1993).
31. S. W. Englander, Hydrogen exchange and mass spectrometry: a historical perspective, *J. Am. Soc. Mass Spectrom.* **17**, 1481–1489 (2006).
32. D. A. Simmons, S. D. Dunn, and L. Konermann, Conformational dynamics of partially deuterated myoglobin studied by time-resolved electrospray mass spectrometry with online hydrogen-deuterium exchange, *Biochemistry* **42**, 5896–5905 (2003).
33. J. G. Mandell, A. M. Falick, and E. A. Komives, Measurement of amide hydrogen exchange by MALDI-TOF mass spectrometry, *Anal. Chem.* **70**, 3897–3995 (1998).
34. I. D. Figueroa and D. H. Russell, Matrix-assisted laser desorption/ionization hydrogen/deuterium exchange studies to probe peptide conformational changes, *J. Am. Soc. Mass Spectrom.* **10**, 719–731 (1999).
35. A. Nazabal, M. Laguerre, J.-M. Schmitter, J. Vaillier, S. Chaignepain, and J. Velours, Hydrogen/deuterium exchange on yeast ATPase supramolecular protein complex analyzed at high sensitivity by MALDI mass spectrometry, *J. Am. Soc. Mass Spectrom.* **14,** 471–481 (2003).
36. C. Woodward, I. Simon, and E. Tüchsen, Hydrogen exchange and the dynamic structure of proteins, *Mol. Cell Biochem.* **48**, 135–160 (1982).
37. S. W. Englander, L. Mayne, and J. N. Rumbley, Submolecular cooperativity produces multi-state protein unfolding and refolding, *Biophys. Chem.* **101–102**, 57–65 (2002).
38. S. W. Englander and N. R. Kallenbach, Hydrogen exchange and structural dynamics of proteins and nucleic acids, *Q. Rev. Biophys.* **16,** 521–655 (1984).
39. D. W. Miller and K. A. Dill, A statistical mechanical model for hydrogen exchange in globular proteins, *Protein Sci.* **4**, 1860–1873 (1995).
40. D. A. Simmons and L. Konermann, Characterization of transient protein folding intermediates during myoglobin reconstitution by time-resolved electrospray mass spectrometry with on-line isotope-pulse labeling, *Biochemistry* **42,** 1906–1914 (2002).
41. A. Miranker, C. V. Robinson, S. E. Radford, R. T. Aplin, and C. M. Dobson, Detection of protein transient populations by mass spectrometry, *Science* **262,** 896–900 (1993).
42. G. Thevenon-Emeric, J. Kozlowski, Z. Zhang, and D. L. Smith, Determination of amide hydrogen exchange rates in peptides by mass spectrometry, *Anal. Chem.* **64,** 2456–2458 (1992).
43. Z. Zhang and D. L. Smith, Determination of amide hydrogen exchange by mass spectrometry: a new tool for protein structure elucidation, *Protein Sci.* **2**, 522–531 (1993).
44. Y. Deng, Z. Zhang, and D. L. Smith, Comparison of continuous and pulsed labeling amide hydrogen exchange/mass spectrometry for studies of protein dynamics, *J. Am. Soc. Mass Spectrom.* **10**, 675–684 (1999).
45. M. Kraus, K. Janek, M. Bienert, and E. Krause, Characterization of intermolecular b-sheet peptides by mass spectrometry and hydrogen isotope exchange, *Rapid Commun. Mass Spectrom.* **14**, 1094–1104 (2000).
46. X. Cai and C. Dass, Structural characterization of methionine and leucine enkephalins by hydrogen/deuterium exchange and electrospray ionization tandem mass spectrometry, *Rapid Commun. Mass Spectrom.* **19**, 1–8 (2005).

47. B. Schilling, R. H. Row, B. W. Gibson, X. Guo, and M. M. Young, MS2Assign, automated assignment and nomenclature of tandem mass spectra of chemically cross-linked peptides, *J. Am. Soc. Mass Spectrom.* **14**, 834–850 (2003).
48. A. Sinz, Chemical cross-linking and mass spectrometry for mapping three-dimensional structures of proteins and protein complexes, *J. Mass Spectrom.* **38**, 1225–1237 (2003).
49. B. X. Huang, C. Dass, and H.-Y. Kim, Probing three-dimensional structure of bovine serum albumin by chemical cross-linking and mass spectrometry, *J. Am. Soc. Mass Spectrom.* **15**, 1237–1247 (2004).
50. B. X. Huang, C. Dass, and H.-Y. Kim, Probing conformational changes of bovine serum albumin due to unsaturated fatty acid binding by chemical cross-linking and mass spectrometry, *Biochem. J.* **387**, 695–702 (2005).
51. K. B. Shelimov and M. F. Jarrold, Conformations, unfolding and refolding of apomyoglobin in vacuum: an activation barrier for gas-phase protein folding, *J. Am. Chem. Soc.* **119**, 2987–2994 (1997).
52. S. J. Valentine, J. Anderson, A. E. Ellington, and D. E. Clemmer, Disulfide-intact and -reduced lysozyme in gas-phase: conformations and pathways for folding and unfolding, *J. Phys. Chem.* **101**, 3891–3900 (1997).
53. R. W. Purves, D. A. Barnnett, B. Ells, and R. Guevermont, Investigation of bovine ubiquitin conformers separated by high-field asymmetric waveform ion mobility spectrometry: cross section measurements using energy-loss experiments with a triple quadrupole mass spectrometer, *J. Am. Soc. Mass Spectrom.* **11**, 738–745 (2000).

CHAPTER 11

CHARACTERIZATION OF OLIGOSACCHARIDES

Besides proteins, several other classes of molecules exist that are critical to human health and emphasize the need for sensitive compound-specific techniques for their characterization. Three important classes of biomolecules—oligosaccharides, lipids, and oligonucleotides—are the subject of this and the next two chapters. Each represents a range of diverse molecules that play a variety of important roles in various body functions. For example, oligonucleotides are involved in storage, transmission, and processing of genetic information. Oligosaccharides participate in cell–cell recognition, and lipids are the storehouse of energy. With the development of MALDI and ESI, mass spectrometry is playing a pivotal role in the characterization of all three classes of compounds. In this chapter, mass spectrometry–based approaches to the analysis of oligosaccharides are highlighted.

Oligosaccharides (commonly known as carbohydrates) are complicated biomolecules. They exist in the body as independent units or are conjugated with other biomolecules, such as proteins and lipids. Structures of the most common form of sugar units that constitute carbohydrate side chains of mammalian proteins and lipids are shown in Figure 11.1. These sugar units include D-mannose (Man or Hex), D-galactose (Gal or Hex), L-fucose (Fuc or dHex), N-acetylglucosamine (GlcNAc or HexNAc), N-acetylgalactosamine (GalNAc or HexNAc), and N-acetylneuraminic acid (sialic acid, NeuAc, or NANA). Some less common sugar residues are glucose, xylose, and ribose.

In common with other *omic* words, the *glycome* is defined as the complete set of carbohydrates in cells, tissues, or organisms, and *glycomic* is the field of

Fundamentals of Contemporary Mass Spectrometry, by Chhabil Dass
Copyright © 2007 John Wiley & Sons, Inc.

study of glycomes. Because of the fact that carbohydrates bind to many disease markers in human blood, their characterization is a highly significant field of study. The type of carbohydrate on the surface of a cell can be related to disease. Carbohydrates exert their influence in the body via a selective interaction with proteins on cell surfaces.

11.1. STRUCTURAL DIVERSITY IN OLIGOSACCHARIDES

Before proceeding further, it is pertinent to define a few terms that are commonly used in carbohydrate chemistry:

- *Monosaccharide:* Monosaccharides are simple sugars with the general formula $C_nH_{2n}O_n$. They are the basic building blocks of carbohydrates. In oligosaccharides, monosaccharides exist in the ring-closed form.
- *Oligosaccharide:* An oligosaccharide is a carbohydrate polymer that is formed by combining several monosaccharides (sugar units) through glycosidic bonds.
- *Anomeric center:* The pyranose ring of sugar monomers contains a stereochemical center at carbon-1 (C1) called an *anomeric center*. In the β-anomer, the OH group on C1 is in the "up position" (i.e., *cis* to the CH$_2$OH on C5), and in the α-anomer, it is in the "down position" (see Figure 11.1).
- *Glycosidic bond:* A glycosidic bond is an acetal bond that links monosaccharides in a carbohydrate chain. This bond is formed by joining the anomeric carbon (C1) of a nonreducing-end sugar unit to any one of the four OH groups (i.e., 2, 3, 4, or 6) on the reducing-end sugar. Examples of various glycosidic linkages are shown in Figure 11.2.

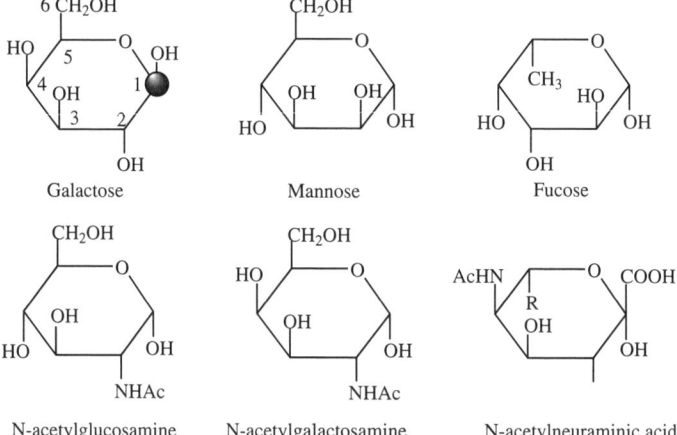

Figure 11.1. Structure of common sugar residues present in oligosaccharides. The carbon 1 (shown in galactose) is an anomeric carbon. Carbons 2, 3, 4, and 5 are stereochemical centers.

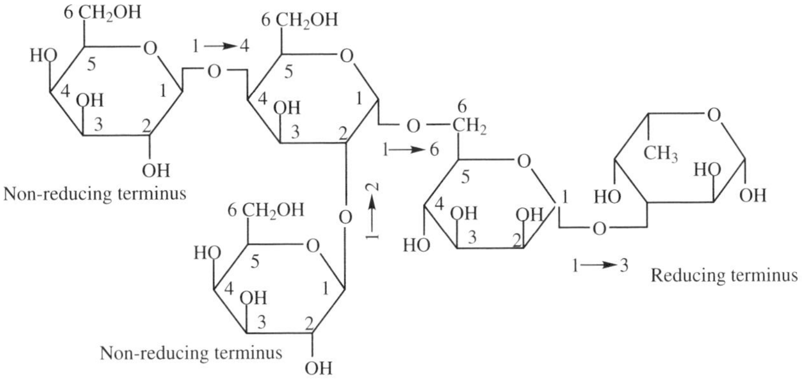

Figure 11.2. Diversity in glycosidic linkages is depicted in an oligosaccharide, Galβ 1–4(Galβ 1–2)Galα 1–6Manα 1–3Fuc. The β-glycosidic linkage is in the "up position" and the α-linkage in the "down position."

- *Reducing end:* The terminus that has a free anomeric center (a hemiacetal linkage) is known as the reducing end. This center can be reduced to an alcohol via the reaction of the sugar residue with a reducing agent. That end is used to form a covalent bond with proteins or lipids (see Figure 11.2). The other end of the glycan is called the nonreducing end.
- *Glycan:* This is the term widely used for carbohydrate chains connected to a protein or lipid.
- *Composition:* Composition provides the identity of constituent sugar monomers in an oligosaccharide.

Oligosaccharides have a potential for great structural diversity. Unlike amino acids, which combine in a linear unbranched fashion via amide bonds to form peptides, sugar units in oligosaccharides can be linked through a variety of hydroxyl groups to form a range of glycosidic linkages. In addition, a carbohydrate chain might contain numerous stereochemical centers, an anomeric center, and extensive branching. Carbons 2 to 5 are stereochemical centers. Furthermore, each glycosylation site has a heterogeneous population of oligosaccharides that are attached to the same structural class. Consequently, many more different oligosaccharides can be formed (e.g., >10,000) from four sugar units than from oligopeptides (e.g., 24) from four amino acids.

The International Union of Pure and Applied Chemistry (IUPAC) has developed nomenclature to represent sugar units in carbohydrates (shown in Figure 11.3*a*). For simplicity, researchers also use a set of geometric symbols to represent sugar units, but these symbols are not consistent. A set of consistent new geometric symbols and other nomenclature that has evolved recently as an effort of the Consortium for Functional Glycomics (CFG), a Scripps Research Institute-based organization, is also shown in Figure 11.3 with respect to an N-linked glycan. According to the CFG's nomenclature, NeuAc is designated

400 CHARACTERIZATION OF OLIGOSACCHARIDES

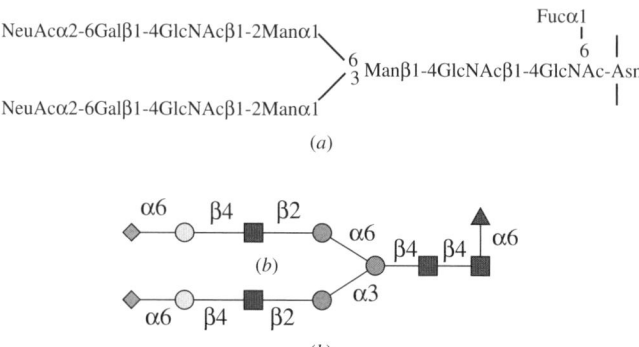

Figure 11.3. Representation of sugar units in a carbohydrate chain: (*a*) IUPAC nomenclature; (*b*) CFG nomenclature. (From *C & E News*, August 8, 2005, p. 46.)

as purple diamond; Gal as yellow circle; GlcNAc as dark blue square; Man as green circle; Glc as dark blue circle; GalNAc as yellow square; and Fuc as red triangle. Glycosidic linkages are designated α or β plus a number that indicates the carbon position in the second sugar that receives the glycosidic bond. The letters α and β stand for an axial or equatorial orientation, respectively, of the glycosidic carbon of the first sugar. Unlike in the IUPAC system, the position of the anomeric carbon of the first sugar is not indicated. The CFG symbols can still be recognized when printed in black and white, as is evident in Figures 11.3 and 11.4.

11.2. CLASSES OF GLYCANS

Three important classes of glycans exist in nature:

1. *Glycoprotein glycans.* A major class of oligosaccharides is that of glycoprotein glycans. As discussed in Section 9.5, two main types of glycoprotein glycans have been recognized, depending on whether the carbohydrate side chains are attached to proteins via an amide or acyl linkage: (1) the N-linked glycans are linked through the amide bond of asparagine residue if it is a part of the motif Asn–X–Ser(Thr), where X is any amino acid except proline, and (2) the O-linked glycans are linked through the oxygen atom of serine and threonine side chains. The carbohydrate units in N-linked oligosaccharides have a common inner-core structure that consists of two N-acetylglucosamines (GalNAc) and three mannose residues. The N-linked oligosaccharides are further classified into three common motifs: complex, high mannose, and hybrid; these structures are shown in Figure 11.4 [1]. The complex-type structures contain no mannose residues other than those present in the core pentasaccharide. The largest number of structural variations can be found in this subclass. Complex-type structures can have two to five antennas (or branches) that are formed by the addition of one or

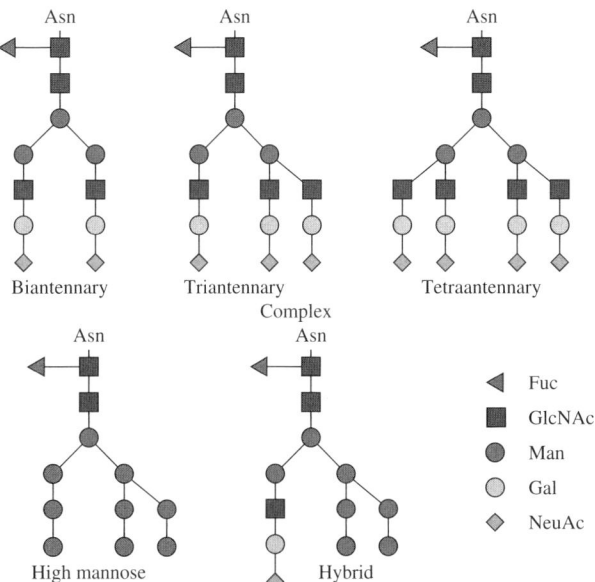

Figure 11.4. Three subclasses (complex, high mannose, and hybrid) of N-linked oligosaccharides; all contain a common core that is composed of two N-acetylglucosamine and three mannose residues. The geometrical designation reflects the CFG nomenclature.

more lactosoamines to the core structure. The antennas usually terminate in galactose or sialic acid. High-mannose structures contain several additional mannose residues in addition to those found in the inner core. Hybrid-type oligosaccharides are composed of complex- and high-mannose-type branches. The O-linked oligosaccharides are not well defined. Although a consensus sequence for their attachment is vague, O-glycosylation occurs in those sequence regions that have a high density of Ser and Thr residues. Also, unlike N-linked glycans, O-linked carbohydrate chains do not have a common inner core.

2. *Glycosphingolipid glycans.* Glycosphingolipids are complex lipids (see Section 12.4) in which an oligosaccharide head group is attached to a hydrophobic ceramide anchor. The oligosaccharide portion may contain one or several sugar units, such as glucose, galactose, N-acetylglucosamine, N-acetylgalactosamine, fucose, sialic acid, and glucuronic acids.

3. *Glycosaminoglycans.* These glycans are linear carbohydrate chains that are comprised of repeating disaccharide units.

11.3. MASS SPECTROMETRIC METHODS FOR COMPLETE STRUCTURE ELUCIDATION OF OLIGOSACCHARIDES

Currently, mass spectrometry has assumed a greater role in the characterization of oligosaccharides. This subject has been reviewed extensively [2–7]. A complete structure elucidation of carbohydrate side chains requires knowledge of the (1) monosaccharide composition, (2) sequence and branching of each monomer, (3) stereochemistry, (4) interglycosidic linkage positions, and (5) anomeric configuration of each glycosidic bond.

11.3.1. Release of Glycans

Prior to mass spectrometry analysis, the carbohydrate chain must be cleaved from glycopeptides and glycolipids.

Enzymatic Release To release N-linked carbohydrate chains, glycopeptides are treated with an endogycosidase. The ideal enzyme for this purpose is peptide N-glycosidase F [PNGase F; also known as peptide N^4-(N-acetyl-β-glucosaminyl) asparagine], an amidase that cleaves the intact glycan as the glucosylamine (i.e., the bond between Asn and GlcNAc is cleaved) and converts Asn to Asp. PNGase F has broad specificity and cleaves most glycans except fucose α1-3 linked to the reducing terminal of GlcNAc. In those situations, PNGase A can be used. Other endoglycosidases that can release glycans are endo-β-N-acetylglucosaminidase H and F (endo H and endo F), but they are less broadly active. Analogous enzymes for the general release of O-linked glycans are not known.

Chemical Release Chemical release is a nonselective process. Several chemical reagents are available to free N- and O-linked glycans, the most common being anhydrous hydrazine, which works effectively for O-linked glycans at 60°C and for N-linked glycans at 95°C. Trifluoromethanesulfonic acid has also been used for deglycosylation of glycopeptides. Reaction with this reagent leaves one or two pendant GalNAc residues with Asn. O-linked chains are released from the peptide backbone by base-catalyzed reductive β-elimination (i.e., by treatment with a $NaBH_4$–NaOH mixture).

Glycolipid glycans and glycosaminoglycans are also released chemically. Digestion with endoglycoceramidase will also release glycolipid glycans.

11.3.2. Derivatization of Carbohydrate Chains

Because oligosaccharides lack polar acidic or basic functional groups, they are relatively difficult to ionize. Therefore, a judicious purification step and/or derivatization strategies must be employed to enhance the mass spectrometry signal. Derivatization also leads to enhanced and well-defined fragmentation pattern. A common derivatization method involves permethylation or peracetylation [8]. This step also improves volatility and assists purification by solvent extraction. In addition, the mass increase upon pemethylation is not too high, a fact that enables larger oligosaccharides to be analyzed. Other derivatization procedures have aimed to introduce ionizable functional groups such as an amine: for example, by reductive amination with such aromatic amines as 2-aminoacridone (2-AMAC), 2-aminobenzamide (2-AB), 2-aminopyridine, 2-aminoquinoline, 4-aminobenzoic acid 2-(diethylamino)ethyl ester (ABDEAE), and many more [9–11, and references therein]. The reaction of these amines with reducing carbohydrates is fast and quantitative. Moreover, some of these derivatizing agents exhibit fluorescence that helps in their separation and detection by HPLC. A list of derivatizing agent, along with their structures and the mass increment that results on derivatization, is presented in Table 11.1. The mass increment is given by the molecular mass of the amine minus 16. Reductive amination produces an open-ring derivative of a carbohydrate, the structure of which is shown in Figure 11.5. The closed-ring derivative of these amines have also been studied [12].

11.3.3. Composition Analysis by GC/MS

Despite the development of ESI and MALDI modes of ionization, GC/MS is still a popular method for determination of the monosaccharide composition of carbohydrate chains [13]. This method involves breaking the oligosaccharide into constituent monosaccharides by hydrolysis or methanolysis. Following derivatization, the released monosaccharides are analyzed by GC/MS to identify sugar residues through their retention times and mass spectral profile. Common derivatization procedures are methylation, acetylation, and trimethylsilylation.

11.3.4. Linkage Analysis by GC/MS

A GC/MS method, commonly called *linkage analysis or methylation analysis*, has also proven to be a successful approach to determining interresidue linkage and branch points [14]. It involves complete alkylation of all free hydroxyl groups. The fully methylated oligosaccharides are hydrolyzed into monomers, which are further reduced and acetylated to provide partially methylated alditol acetates (PAMAs). Acetylation takes place at those positions that were newly exposed by hydrolysis and reduction. The GC retention times and fragmentation patterns can reveal the mode of attachment in the native sugar.

11.3.5. Rapid Identification by a Precursor-Ion Scan

A precursor-ion scan can provide rapid identification of oligosaccharides contained in a pool of glycans. Ionization is accomplished via ESI. Derivatization with 2-aminoacridone of the glycans released improves the analysis [15]. The precursors are identified by monitoring certain sugar-specific marker ions, such as m/z 204 and 306 (see Section 9.6.2). Several N-linked glycans have been identified with the precursor-ion scan of the 2-aminoacridone-labeled glycan pool of a glycoprotein from ovomucoid [15].

11.3.6. Composition Analysis by Direct Mass Measurement

The GC/MS approach, although of great utility in composition analysis, is time consuming and labor intensive and requires large sample amounts. Attention has now shifted to several desorption–ionization techniques, such as fast atom bombardment (FAB), MALDI, and ESI. In some cases it may be

TABLE 11.1. Derivatizing Agents for Reductive Amination of Carbohydrates

Amine	Mass Increment	Structure
2-Aminobenzamide	120	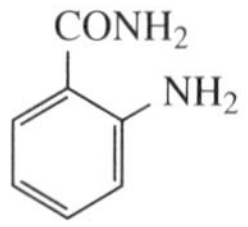
2-Aminobenzoic acid	121	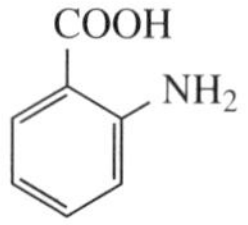
2-Aminopyridine	78	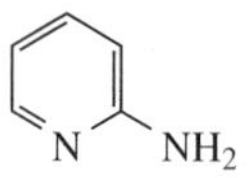
2-Aminoacridone	194	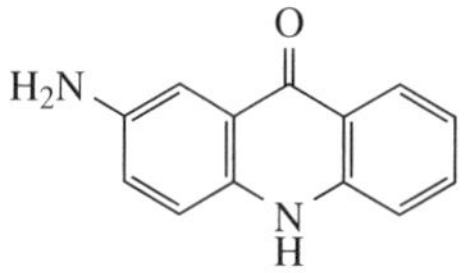
3-Aminoquinoline	128	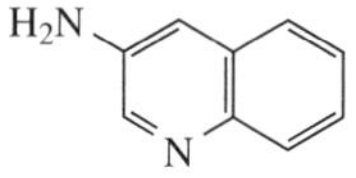
4-Aminobenzoic acid methyl ester	135	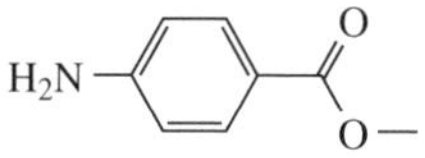
4-Aminobenzoic acid ethyl ester	149	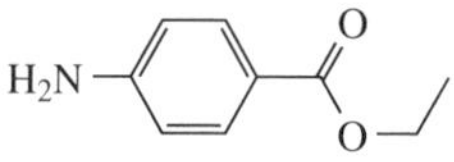
4-Aminobenzoic acid n-butyl ester	177	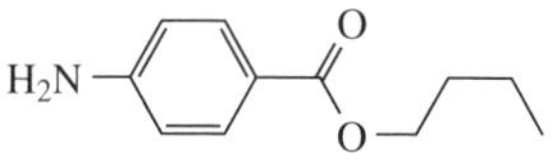
4-Amino-N-(2-diethylaminoethyl)benzamide	219	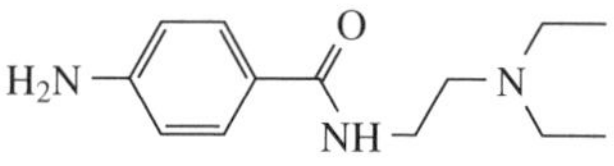
4-Amino-N-(2-diethylaminoethyl)benzoic acid	218	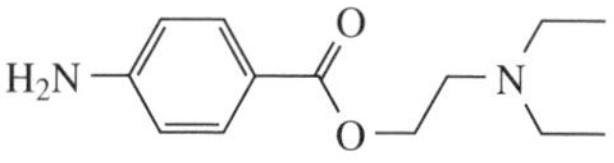
3-Acetylamino-6-aminoacridine	237	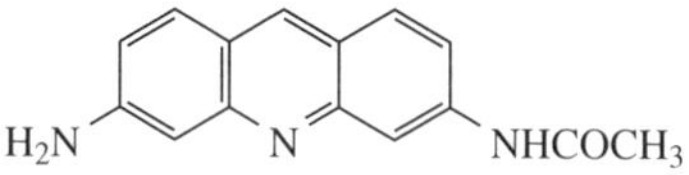

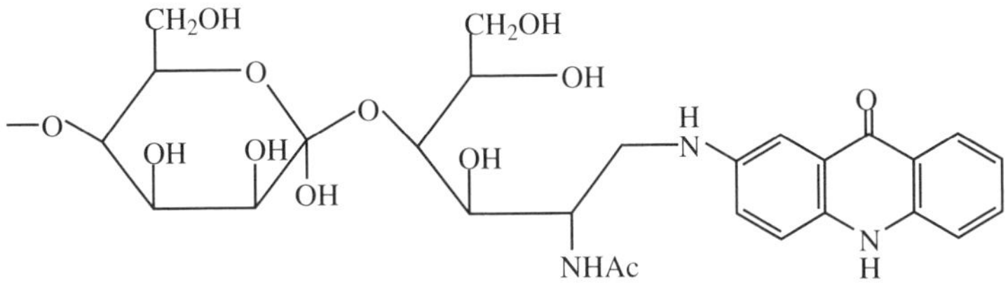

Figure 11.5. Structure of an open-ring derivative of 2-aminoacridone.

necessary to separate the mixture of oligosaccharides into individual fractions. High-performance ion-exchange chromatography (HPIEC) on pellicular quaternary amine–bonded resin is used for this purpose. In FAB, the molecular ions are formed as $[M + H]^+$ and $[M + Na]^+$ ions. FAB provides limited sensitivity and thus has largely been replaced by ESI and MALDI. 2,5-Dihydroxybenzoic acid is the most popular MALDI matrix for glycan analysis; 6-azo-2-thiamine is another preferred matrix. The use of some additives, such as 2-hydroxy-5-methoxy-benzoic acid, 1-aminoisoquinoline, and fucose, provides an improved ion signal. Neutral glycans are ionized primarily as $[M + Na]^+$ ions in MALDI. Fragmentation is observed in ion-source mode as well as in the post-source decay (PSD) mode. ESI also generates $[M + H]^+$ and $[M + Na]^+$ ions. Under negative-ion ESI, carbohydrates form $[M - H]^-$ and $[M + Na - 2H]^-$ anions. Negative ions are also stabilized by adduction with certain anions (e.g., nitrate ion).

The composition of a glycan can be derived from the molecular mass. This task is relatively easier for oligosaccharides because they are formed by the combination of a limited set of sugar residues (see Figure 11.1). Also, the majority of the glycans released have known structures. For example, the inner-core structure of the N-linked oligosaccharides is invariant and known. Matching the measured mass with the mass calculated by a combination of the known masses of common sugar residues (given in Table 11.2) can provide the composition of the glycan, but it will not pinpoint isobaric residues (e.g., galactose vs. mannose). The probable primary structures can, however, be narrowed down to a short list by searching the Complex Carbohydrate Structure Database (CCSD; http://www.ccrc.uga.edu) with the molecular mass measured [16].

▶ **Example 11.1** Calculate the m/z of the $[M + Na]^+$ adduct of the oligosaccharide shown in Figure 11.2.

Solution That oligosaccharide is composed of four hexose residues (three Gal and one Man). Therefore, the mass of the oligosaccharide $= (4)(162.053) + 18.011$ (the mass of H_2O) $= 666.223$ Da, and the m/z of the $[M + Na]^+$ adduct $= 666.2 + 23.0 = 689.2$.

TABLE 11.2. Common Monosaccharides and Their Accurate Masses

Monosaccharide (Abbreviation)	Examples (Abbreviation)	Formula	Mass		Residue Mass	
			Mono-isotopic	Average	Mono-isotopic	Average
Pentose	Ribose, xylose	$C_5H_{10}O_5$	150.053	150.14	132.042	132.116
Deoxyhexose (dHex)	Fucose (Fuc)	$C_6H_{12}O_5$	164.069	164.16	146.058	146.143
Hexose (Hex)	Glucose (Glc) Galactose (Gal) Mannose (Man)	$C_6H_{12}O_6$	180.063	180.16	162.053	162.142
Hexosamine	Glucosamine (GlcN)	$C_6H_{13}NO_5$	179.080	179.18	161.069	161.157
N-Acetylamino hexose (HexNAc)	N-Acetylglucosamine (GlcNAc) N-Acetyl-galactosamine (GalNAc)	$C_8H_{15}NO_6$	221.090	221.20	203.079	203.195
Hexuronic acid		$C_6H_{10}O_7$	194.043	194.15	176.032	176.126
N-Acetylneuraminic acid (NANA; NeuAc)	Sialic acid (SA)	$C_{11}H_{19}NO_9$	309.106	309.28	291.095	291.258
N-Glycoyl-neuraminic acid (GLcNeu)		$C_{11}H_{19}NO_{10}$	325.101	325.28	307.090	307.257

11.3.7. Structure Determination of Oligosaccharides by Sequential Digestion

Sequential Exoglycosidase Degradation Enzymatic degradation with exoglycosidases allows a very specific bond cleavage. Therefore, sequential exoglycosidase degradation can be a useful adjunct to determine the stereochemical sequence, linkage, and anomeric configuration of an oligosaccharide [17–19]. Conceptually, this procedure is similar to the peptide ladder sequencing approach (Section 8.10.2) and involves the sequential digestion of the oligosaccharide with an array of exoglycosidases to release specific terminal sugar units, followed by the measurement of the mass of the oligosaccharide chain with MALDI–MS before and after each digestion step. The shift in mass and knowledge of the specificity of the exoglycosidase provide information on the stereochemical sequence, linkages, and anomeric centers. Some commercially available exoglycosidases are listed with their specificity in Table 11.3.

A typical procedure is explained through an example of sequencing N-glycans that is released from a glycoprotein, human α1-acid glycoprotein (AGP) [19]. In this example, enzymatic release of the glycans and exoglycosidase sequencing are performed directly on a MALDI plate. AGP is deposited on three different spots, which are treated with different cocktails of endo- and exoglycosidases.

COMPLETE STRUCTURE ELUCIDATION OF OLIGOSACCHARIDES

TABLE 11.3. Commonly Used Exoglycosidases for Sequential Degradation of Oligosaccharides

Enzyme	Origin	Specificity
Neuraminidase	*Arthrobacter ureafaciens*	NeuAcα2 →(6 > 3, 8)
	Newcastle virus	NeuAcα2 → 3, 8
α-Galactosidase	Coffee bean	Galα1 → 3, 4, 6
β-Galactosidase	Jack bean	Galβ1 → (6 > 4 > 3)
	Diplococcus pneumonia	Galβ1 → 4
	Chicken liver	Galβ1 → 3, 4
N-acetyl-β-D-glucosaminidase	*Diplococcus pneumonia*	GlcNAcβ1 → 2
	Jack bean	GlcNAcβ1 → 2, 3, 4, 6
	Chicken liver	GlcNAcβ1 → 3, 4
α-Mannosidase	Jack bean	Manα1 →(2,6) > 3
	Aspergillus saitoi	Manα1 → 2
α-Fucosidase	Chicken liver	Fuc α1 → 2, 3, 4, 6
	Bovine epididymus	Fucα1 → (6 > 2, 3, 4)

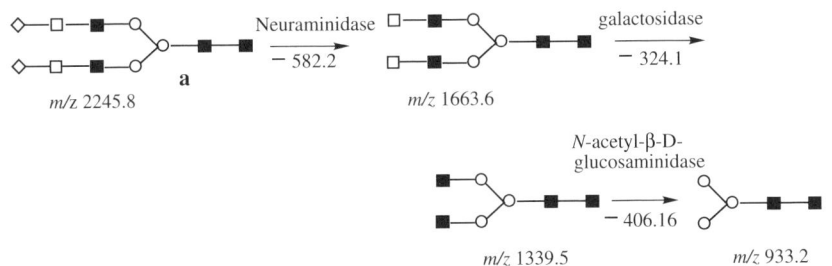

Figure 11.6. Sequencing of a glycan (**a**) by exoglycosidase degradation. Treatment of **a** with neuraminidase removes two sialic acid residues (diamond); *m/z* of the [**a** + Na]⁺ ion is reduced by 582.2 Da. Subsequent treatment with galactosidase and N-acetyl-β-D-glucosaminidase removes two galactose (open square) and two N-acetylglucosamine (solid square) residues to form ions at *m/z* 1339.5 and 933.2, respectively. The open circle represents mannose residues.

For example, the first spot is treated with PNGase F and neuraminidase, the second with PNGase F, neuraminidase, and β-galactosidase, and the third with PNGase F, neuraminidase, β-galactosidase, and N-acetyl-β-D-glucosaminidase. The resulting MALDI ions are shown in Figure 11.6 of Example 11.2.

▶ **Example 11.2** Explain the sequencing of the glycan of structure (**a**) in Figure 11.6 through the exoglycosidase degradation strategy.

Solution See Figure 11.6.

Chemical Degradation In a variation of the exoglycosidase degradation procedure, the oligosaccharide is treated with a strong base (e.g., NaOH) at an elevated temperature (ca. 60°C) [20]. This alkaline degradation, also known as the *peeling reaction*, cleaves a glycosidic bond at the reducing end through β-elimination to produce a new reducing end. The prevailing oligosaccharides are fragmented by MALDI, and if necessary, by CID. The linkage information is obtained from the cross-ring fragmentation from the new reducing end. The reaction is allowed to continue until every sugar residue in the oligosaccharide becomes a reducing ring so that it can be fragmented to provide complete linkage information. This methodology can also reveal branching points.

11.3.8. Tandem Mass Spectrometry for Structural Analysis of Carbohydrates

As with other classes of biomolecules, CID is an effective technique to use to obtain sequence information either directly from glycopeptides and glycolipids or from carbohydrate side chains after their liberation from these glycoconjugates [3, 8–12, 21–30]. CID also provides interresidue linkage and branching details. The sequencing of glycans from intact glycopeptides and glycolipids is also feasible because of the preferential cleavage of glycosidic linkages upon CID. The target analyte is methylated to improve fragmentation. The molecular ions are generated by FAB or ESI; the latter is a preferred mode of ionization because of its increased sensitivity, reduced matrix interference, and choice of several multiply charged precursors. A range of precursor ions, such as $[M + H]^+$, $[M - H]^-$, $[M + Li]^+$, $[M + Na]^+$, and $[M + 2Li - H]^+$, are formed that can be mass-selected for structural analysis. The $[M + H]^+$ ions provide less sensitivity, whereas the multiplicity (e.g., $[M + Li]^+$ or $[M + 2Li - H]^+$) of metal-ion adducts may complicate the spectra. Therefore, the attachment of a high-proton affinity site to the reducing terminal has been a successful strategy prior to the CID of glycans [9–12].

Nomenclature of Fragment Ions Two types of fragmentation occur in carbohydrates: (1) glycosidic cleavages that involve dissociation of bonds between the sugar rings, and (2) cross-ring cleavages that occur within the ring itself. In each case the charge will be retained on either the nonreducing or the reducing terminus. In addition, internal fragments are also formed which involve the cleavage of two glycosidic bonds, one from each end. Domon and Costello have proposed a nomenclature scheme to designate these fragment ions [31]. According to this scheme (Figure 11.7), ions that retain the charge on the nonreducing terminus (drawn on the left) are termed *A*, *B*, and *C*. The fragments that retain the charge on the reducing end (i.e., from the peptide or lipid end) are designated by the letters *X*, *Y*, and *Z*. Of these structures, *A* and *X* pertain to the cross-ring fragments and *B*, *C*, *Z*, and *Y* to glycosidic cleavages. The numbering is assigned from the nonreducing end as a subscript to *A*, *B*, and *C* ions, and from the reducing end for *X*, *Y*, and *Z* ions. The fragmentation within a

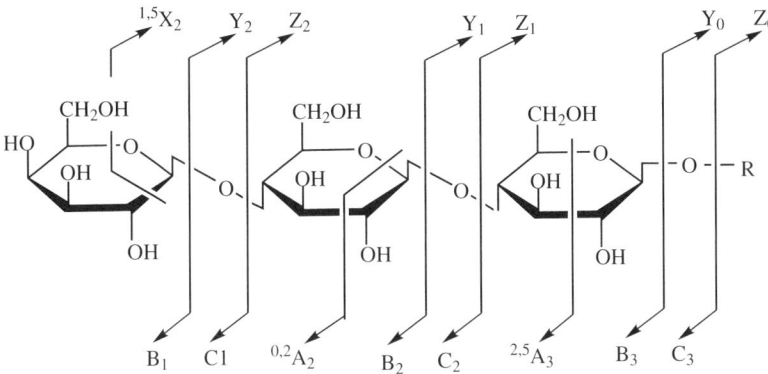

Figure 11.7. Scheme for nomenclature of carbohydrate ions proposed by Domon and Costello. The nonreducing terminus is on the left and reducing terminus on the right. (Reproduced from C. Dass, *Principles and Practice of Biological Mass Spectrometry*, Wiley-Interscience, 2001.)

carbohydrate ring is denoted by superscripts that precede the letter symbols; for example, $^{2,5}A_3$ indicates the cleavage of bonds 2 and 5 in ring 3. Branching is indicated by adding the subscripts α, β, and so on, to the letter symbols. Designation of internal fragments is not consistent. Some researchers have used these letter symbols separated by a slash to indicate the bonds that are cleaved: for example, $B_4/Y_{3\alpha}$, and $^{3,5}A_3/Y_4$ (see Figure 11.7). Other researchers have used a letter D for internal fragments, and the cleaved rings are denoted by numbering from the nonreducing terminus. For example, $D_{2\beta-3\beta}$ denotes a fragment formed by cleavage of the glycosidic bond between rings 2 and 3 and between 3 and 4 in the β-branch.

Thus, the *m/z* value of these sequence-specific ions is calculated as follows:
Positive-ion mode:

m/z of B ions = sum of residue masses + H (or Na)

m/z of C ions = sum of residue masses + 18 + H (or Na)

m/z of Y ions = sum of residue masses + 18 + H (or Na)

m/z of Z ions = sum of residue masses + H (or Na)

Negative-ion mode:

m/z of B ions = sum of residue masses − H

m/z of C ions = sum of residue masses + 18 − H

m/z of Y ions = sum of residue masses + 18 − H

m/z of Z ions = sum of residue masses − H

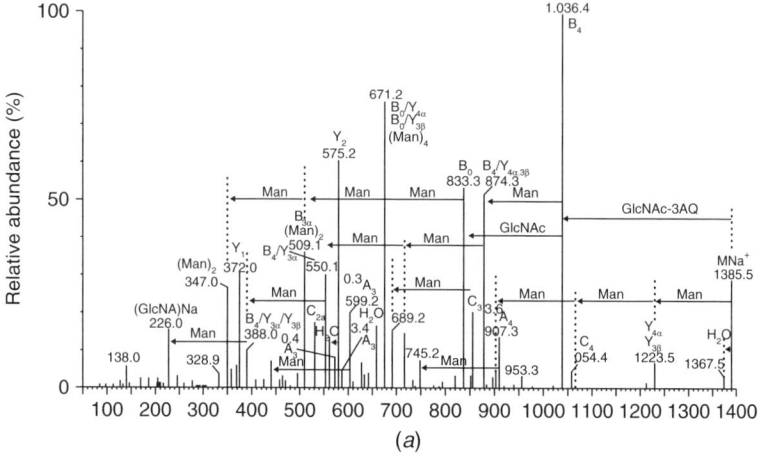

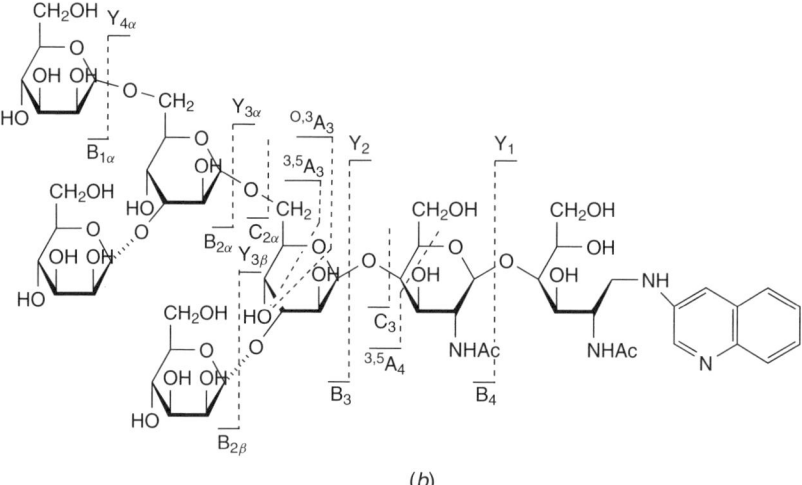

Figure 11.8. (*a*) CID mass spectrum of an ESI-produced [M + Na]$^+$ adduct of the 3-AQ derivative of (GlcNAc)$_2$(Man)$_5$, and (*b*) the fragmentation scheme. (Reproduced from ref. 9 by permission of Elsevier, copyright © 2000 American Society for Mass Spectrometry.)

Any mass increment due to derivatization should be taken into account in these calculations. For example, the *m/z* value of *Y* and *Z* ions will increase by 128 when the carbohydrate is derivatized by 3-aminoquinoline (see Figure 11.8).

▶ **Example 11.3** Designate B_2, C_3, Z_1, and Y_2 sequence-specific ions, cross-ring fragment $^{2,4}A_4$, and internal fragment Y_3/C_2 in the carbohydrate Galβ1–3GlcNACβ1–3Galβ1–4Glc.

Solution These ions are designated as follows:

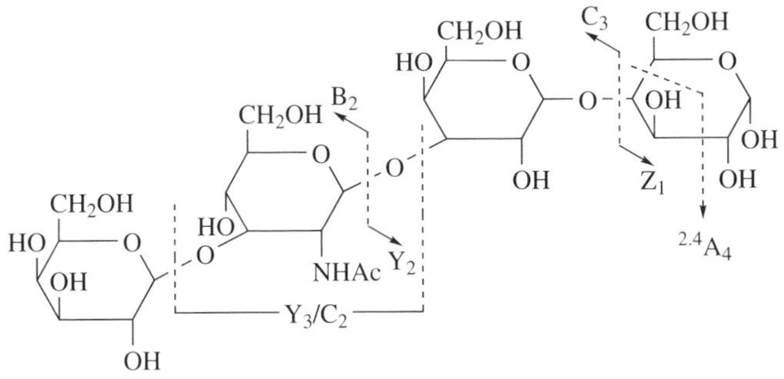

[Structural diagram of oligosaccharide with labeled fragmentation ions: B_2, C_3, Z_1, Y_2, Y_3/C_2, and $^{2,4}A_4$]

▶ **Example 11.4** Calculate the m/z values of the B_2, C_3, Z_1, and Y_2 ions that would be derived by CID of the $[M+H]^+$ ions of the carbohydrate Galβ 1–3GlcNACβ 1–3Galβ 1–4Glc.

Solution

m/z of B_2 ions = 162.05 + 203.08 + 1.01 = 366.14

m/z of C_3 ions = 162.05 + 203.08 + 162.05 + 18.01 + 1.01 = 546.20

m/z of Y_2 ions = 162.05 + 162.05 + 18.01 + 1.01 = 343.12

m/z of Z_1 ions = 162.05 + 1.01 = 163.06

Metal-Ion Adducts as Precursor Ions CID of the metal-ion adducts has been a useful tool to distinguish linkage positions in oligosaccharides. These applications have included cationization with alkali [9,20,21,32], alkaline earth [32], and transition metals [33].

Fragmentation of Derivatized Oligosaccharides To increase fragmentation it is often advantageous to derivatize the carbohydrate chain. A range of derivatizing agents (shown in Table 11.1) that can reductively aminate the reducing terminus reductively has been studied [9–12,34]. As an example, the CID spectrum of the $[M+Na]^+$ adduct of the 3-AQ derivative of $(GlcNAc)_2(Man)_5$ is shown in Figure 11.8. The spectrum is dominated by B- and Y-type sequence-specific ions that are formed by cleavage of glycosidic bonds. Linkage-revealing A-type cross-ring fragments are also reasonably abundant. For example, $^{3,5}A_4$ (m/z 907.3) is present 74 Da higher than the B_3 ion and thus provide an unequivocal indication of 1 → 4 linkage.

Post-source Decay Fragmentation of Oligosaccharides Post-source decay can also provide important structural information for glycans [28]. An example is presented in Figure 11.9, which is the MALDI–PSD spectrum of the $[M + Na]^+$ ion of a high-mannose glycan $(Man)_6(GlcNAc)_2$ generated by hydrazinolysis of ovalbumin. The fragmentation scheme is also depicted in this figure. Similar to the spectrum (Figure 11.8) of 3-AQ-derivatized $(GlcNAc)_2$ $(Man)_5$, this spectrum also contains B, Y, and cross-ring sequence-specific fragments. Internal fragments that arise from losses of one more mannose residues from the cross-ring fragments or B ions are also observed. From this spectrum, the sequence, branching, and linkage pattern can readily be deciphered.

Multistage MS/MS (MS^n) Experiments A complete understanding of sequence and linkage is obtained through multistage MS/MS (MS^n) experiments in a quadrupole ion trap (QIT) mass spectrometer [33,35–39]. In a single-stage MS/MS experiment, some structural features in glycans, such as N-acetyllactosamine antennas, neuraminic acids, and nonreducing terminal GlcNAc monosaccharides, usually suppress cross-ring and core saccharide cleavages [33]. Once these structural features are cleaved, the determination of branching patterns and intersaccharide linkages becomes an easier task. It has been demonstrated that linkage positions in four isomeric pentasaccharides, lacto-N-fucopentaoses, can be identified by acquiring the MS^2 and MS^3 spectra of cobalt-coordinated isomers [35]. The C-type ions participate in specific fragmentation reactions that allows differentiation among $1 \rightarrow 2$, $1 \rightarrow 3$, $1 \rightarrow 4$, and $1 \rightarrow 6$ linkage positions. In particular, this strategy is of great use for oligosaccharides that contain a large number of labile sugar units. This strategy has proven useful for analyzing the $GlcNAc_8Man_3$ carbohydrate from chicken ovalbumin [37]. In the MS^2 spectrum of the $[M + 2Na]^{2+}$ ion of this molecule, the fragmentation of the mannose core is suppressed in favor of the facile loss of HexNAc from the oligomer. The sequential loss of one GlcNAc residue in each stage of MS^7, however, produces an ionic species $GlcNAc_2Man_3(OH)_5$. Further MS/MS (i.e., overall MS^8 stages) of this species reveals the branching patterns of the GlcNAc residues. FT–ICR–MS has also been used to conduct multistage tandem mass spectrometry experiments to sequence permethylated oligosaccharides [39].

Fragmentation of Negative Ions CID of negative ions is also promising in terms of the structure elucidation of oligonucleotides [40–45]. Negative ions are generated in the form $[M - H]^-$, $[M + Cl^-]$, and $[M + NO_3^-]$ in the ESI source. Unlike the CID spectra of positive ions that fragment to produce B- and Y-type sequence-specific ions (Figure 11.8), these negative ions fragment to yield mainly A-type cross-ring cleavage products and C-type fragments [44]. This preference is shown in the CID spectra of underivatized and derivatized $[M + NO_3^-]$ adducts of an oligosaccharide $(Man)_5(GlcNAc)_2$ (Figure 11.10). The proclivity of the formation of A- and C-type ions is the result of the abstraction of a proton from various hydroxy groups during the formation of $[M - H]^-$ ions. The site where the proton is removed becomes susceptible to cleavage to produce the

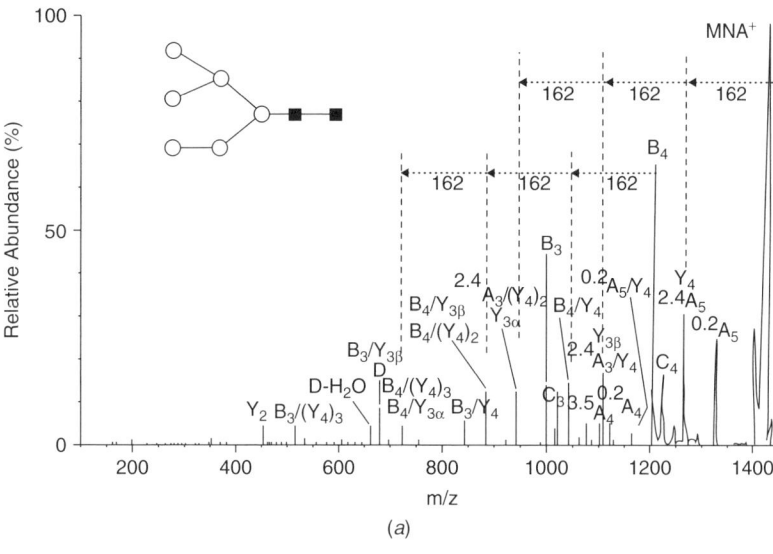
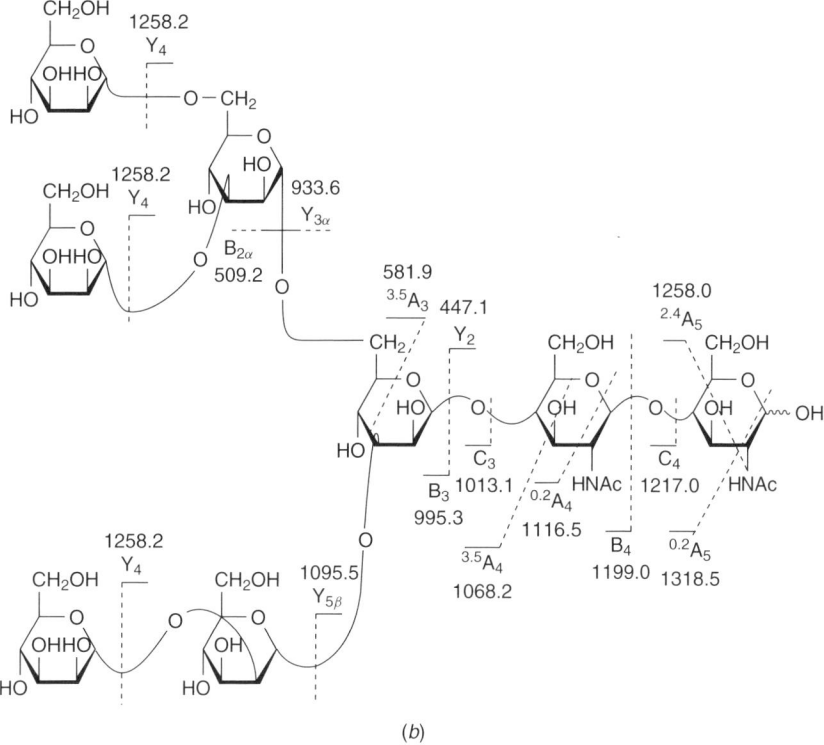

Figure 11.9. (*a*) PSD mass spectrum of the MALDI-produced [M + Na]$^+$ ions of a high-mannose glycan (Man)$_6$(GlcNAc)$_2$ that was released from ovalbumin, and (*b*) the fragmentation scheme. (Reproduced from ref. 28 by permission of Elsevier, copyright © 2000 American Society for Mass Spectrometry.)

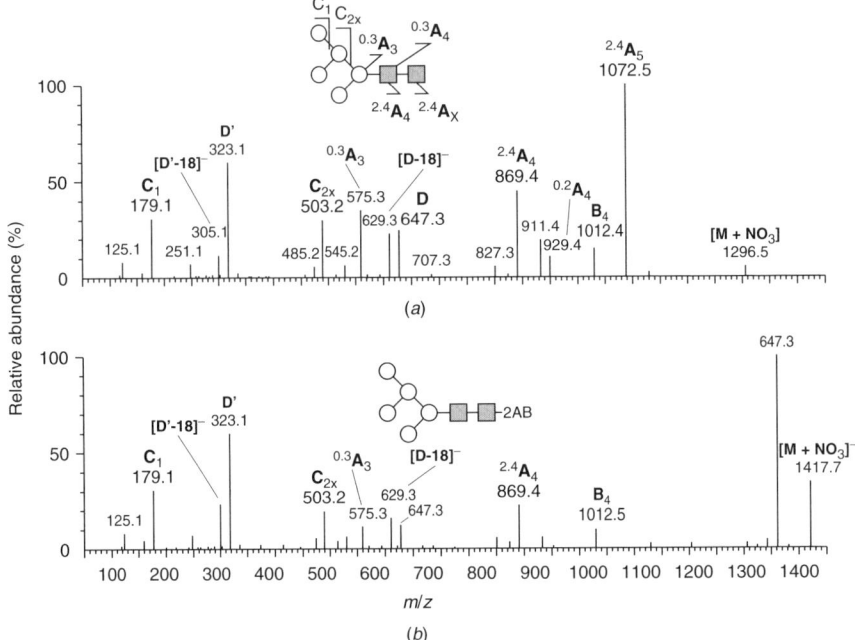

Figure 11.10. (a) CID mass spectrum of the ESI-produced [M + NO$_3^-$] ions of a high-mannose glycan (Man)$_5$(GlcNAc)$_2$; (b) is the corresponding spectrum of the 2-aminobenzamide-derivatized glycan. (Reproduced from ref. 44 by permission of Elsevier, copyright © 2005 American Society for Mass Spectrometry.)

signature mass spectra. Various types of structural information, such as sequence, composition of each antenna, and branching, can be derived from the negative-ion CID spectrum. Derivatization of oligosaccharides with acidic groups, such as that prepared from 2-aminobenzoic acid or esters of aminobenzoic acid, produces mainly C-type ions, due to localization of the negative charge [12].

ESI In-Source CID of Oligosaccharides In-source CID can also provide information on linkage and anomeric configurations [13,46]. Derivatization improves HPLC separation and fragmentation [46]. This aspect was demonstrated by the ESI in-source CID of p-aminobenzoic acid ethyl ester–labeled disaccharides and oligosaccharides [46].

Sequence and Branching Information from a CID–MS/MS Spectrum
To summarize, sequence and branching information is obtained through glycosidic bond cleavages, which are easily recognized in the CID spectrum by the presence of ions that are 18 Da apart (Figure 11.11). If the oligosaccharide is methylated prior to MS analysis, the identity of the reducing and nonreducing termini, as well as branching information, can be obtained from the appearance

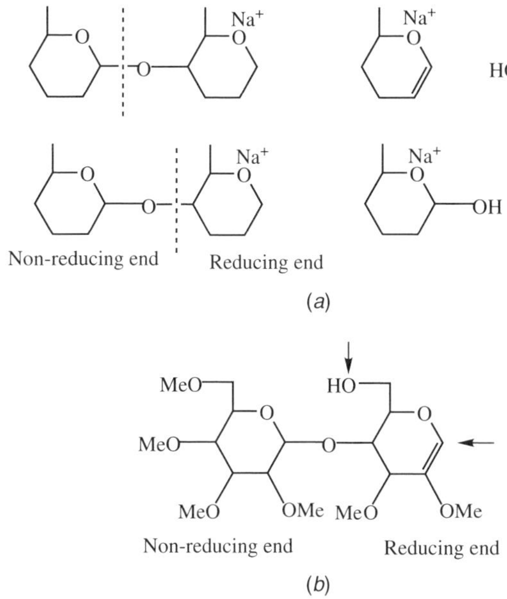

COMPLETE STRUCTURE ELUCIDATION OF OLIGOSACCHARIDES   415

Figure 11.11. (*a*) Glycosidic bond cleavages for sodiated oligomers. Cleavages occur on both sides of the glycosidic bond. The cleavage products appear at 18 Da apart in the spectrum. (*b*) A branch-point fragment: the bonds that break are indicated by arrows. (From ref. 5.)

of peaks at 14-Da lower intervals, due to the hydroxyl group that is exposed during fragmentation (indicated by an arrow in Figure 11.11) [3,5]. As an example, as shown below, the fully methylated linear structure (**b**) will show the Y_1 ion at *m/z* 259, whereas the branched carbohydrate (**c**) will yield an ion at *m/z* (259 − 14 = 245), and the ion at *m/z* 259 will be absent.

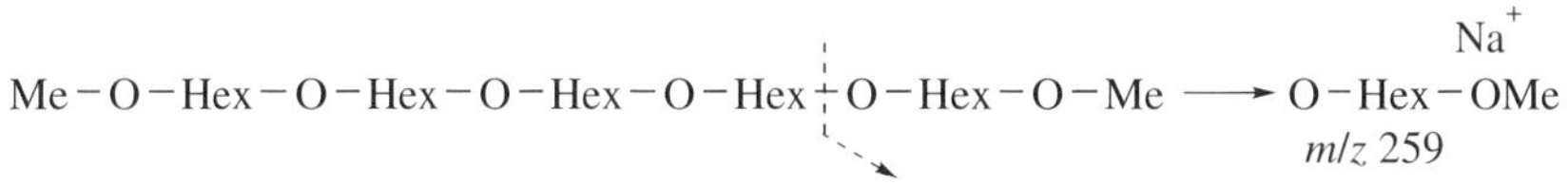

Me−O−Hex−O−Hex−O−Hex−O−Hex┼O−Hex−O−Me ⟶ O−Hex−OMe Na⁺
m/z 259

(**b**)

Me−O−Hex−O−Hex−O−Hex┼O−Hex−O−Me ⟶ O−Hex−OMe −14 Na⁺
            Hex−O   *m/z* 245

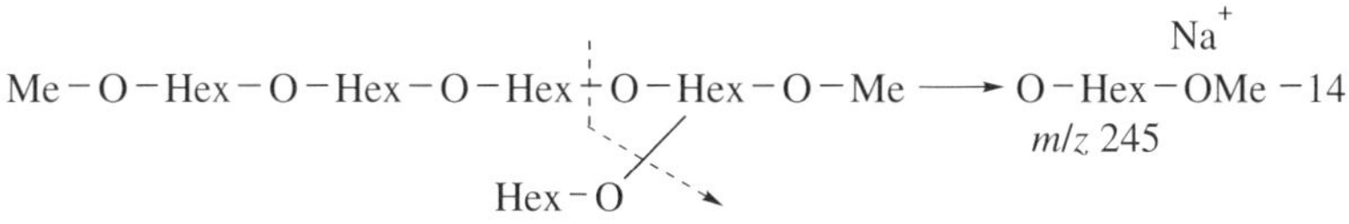

(**c**)

Linkage Analysis from a CID–MS/MS Spectrum   Linkage information can be obtained from the cross-ring cleavages around the linkage position. These fragments involve the rupture of two bonds in the pyranose ring. The ions thus

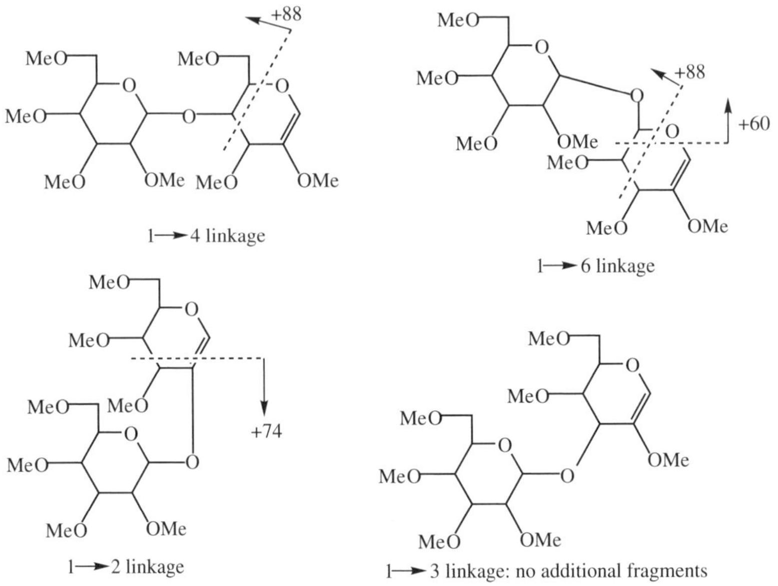

Figure 11.12. Identification of various linkages via cross-ring cleavages. (From ref. 5.)

formed appear as satellite peaks at specific mass increments above the primary glycosidic-bond cleavage products. Only ions that contain the nonreducing terminus are of use. For example, in the fully methylated carbohydrate, an increment of 88 Da indicates a 1 → 4 linkage, 74 Da a 1 → 2 linkage, and 88 and 60 Da a 1 → 6 linkage (see Figure 11.12). The 1 → 3 cross-ring cleavages usually are not observed. If the carbohydrate is not methylated, the corresponding ions will appear at an increment of 74 Da (1 → 4 linkage), 74 and 60 Da (1 → 6 linkage), and 60 Da (1 → 2 linkage).

OVERVIEW

Oligosaccharides are complicated molecules that are known to participate in cell–cell recognition. They exist in the body as independent units or are conjugated with proteins and lipids. The most common form of sugar units that constitute carbohydrate side chains of mammalian proteins and lipids include D-mannose, D-galactose, L-fucose, N-acetylglucosamine, N-acetylgalactosamine, and N-acetylneuraminic acid (sialic acid). Oligosaccharides exist in a wide range of structural diversity because sugar units can be linked through a variety of hydroxyl groups to form a range of glycosidic linkages and due to the presence of numerous stereochemical centers, an anomeric center, and extensive branching.

With the development of MALDI and ESI, mass spectrometry has assumed a pivotal role in the characterization of oligosaccharides. A complete structure elucidation of carbohydrate side chains includes monosaccharide composition, sequence, and branching of each monomer present, stereochemistry, interglycosidic linkage positions, and anomeric configuration of each glycosidic bond. Prior to mass spectrometry analysis, the carbohydrate chain is released by treating glycopeptides with PNGase F.

A popular method for composition analysis is to cleave an oligosaccharide into constituent monosaccharides by hydrolysis or methanolysis, and to analyze the monosaccharides released using GC/MS. The direct mass measurement of an oligosaccharide can also provide composition analysis. A common practice to sequence an oligosaccharide is to perform sequential digestion of the oligosaccharide with an array of exoglycosidases and mass measurement of the oligosaccharide chain before and after each digestion step. Another popular approach to sequence oligosaccharides is to perform CID–MS/MS of the ESI-produced ions. CID of positive ions produces primarily B- and Y-type sequence-specific ions, whereas negative ions yield mainly A-type cross-ring cleavage and C-type fragments. Often, multistage MS/MS analysis is required to obtain complete sequence and linkage information. CID–MS/MS can also provide branching information. Linkage information can be obtained from cross-ring cleavages around the linkage position (i.e., through A-type cross-ring fragments).

EXERCISES

11.1. Depict the structure of the following oligosaccharide in terms of geometric symbols that have been proposed by the CFG:

```
                          GlcNAcβ1
                             |
NeuAcα2–6Galβ1–4GlcNAcβ1–2Manα1              Fucα1
      GlcNAcβ1                                 |
              \6          4                    6
               \       6  Manβ1–4GlcNAcβ1–4GlcNAc–
              4      3
               Manα1
              2
             /
     GlcNAcβ1
```

11.2. Calculate the molecular weight of the oligosaccharide of Exercise 11.1.

11.3. Draw the structure of the following oligosaccharide:

```
   Manα1
        \6
         3 Manβ1–4GlcNAcβ1–4GlcNAc
        /
   Manα1
```

11.4. What is the common inner-core structure of N-linked oligosaccharides?

11.5. Distinguish among complex, high-mannose, and hybrid N-linked oligosaccharides.

11.6. Name suitable methods of cleaving the carbohydrate chain from glycopeptides.

11.7. Calculate the m/z of the $[M + H]^+$ ion of an oligosaccharide, $(Gal)_3(Man)_3(GlcNAc)_4(Fuc)$.

11.8. CID of the $[M + NO_3]^-$ ion is a promising proposition for sequencing oligosaccharides. Calculate the m/z of the $[M + NO_3]^-$ ion of the oligosaccharide $(NeuAc)(Gal)_2(Man)_3(GlcNAc)_4$.

11.9. Designate the $^{3,5}A_2$, $^{0,2}X_2$, B_3, Y_2, and Z_3/C_3 fragments in the structure of the following oligosaccharide:

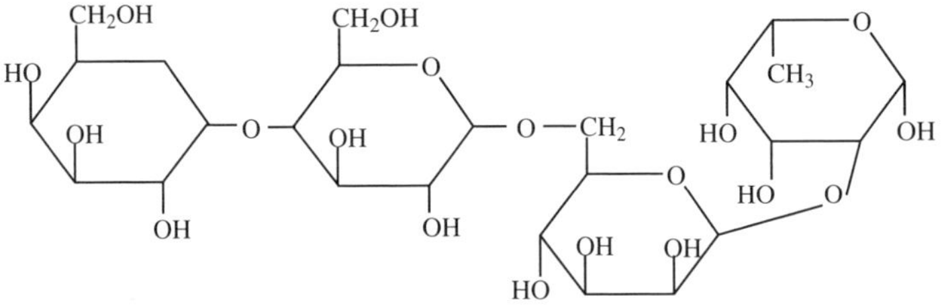

11.10. Calculate the m/z values of the $B_3, Y_2,$ and Z_3/C_3 ions that will be observed from fragmentation of the oligosaccharide of Exercise 11.9.

11.11. Calculate the m/z of the $[M + H]^+$ ions of the (**a**) 2-aminoacridone, (**b**) 2-aminobenzamide, (**c**) 2-aminoquinoline, and (**d**) 4-aminobenzoic acid 2-(diethylamino) ethyl ester derivatives of an oligosaccharide, $(Gal)(Man)_3(GlcNAc)_2$.

11.12. Plan a strategy to sequence the following oligosaccharide by a sequential exoglycosidase degradation procedure:

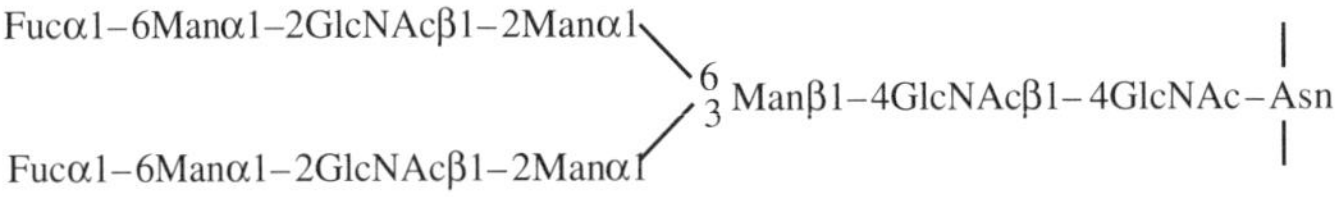

$$\text{Fuc}\alpha 1\text{-}6\text{Man}\alpha 1\text{-}2\text{GlcNAc}\beta 1\text{-}2\text{Man}\alpha 1\diagdown^6_3\text{Man}\beta 1\text{-}4\text{GlcNAc}\beta 1\text{-}4\text{GlcNAc-Asn}$$
$$\text{Fuc}\alpha 1\text{-}6\text{Man}\alpha 1\text{-}2\text{GlcNAc}\beta 1\text{-}2\text{Man}\alpha 1\diagup$$

At what m/z values MALDI signal would be obtained?

REFERENCES

1. P. H. Petra, P. R. Griffins, and J. R. Yates III, Complete enzymatic deglycosylation of native sex steroid-binding protein (SBP or SHBG) of human and rabbit plasma: effect on the steroid binding activity, *Protein Sci.* **1**, 902 (1992).

2. D. J. Harvey, Matrix-assisted laser desorption ionization mass spectrometry of carbohydrates, *Mass Spectrom. Rev.* **18**, 349–451 (1999).

3. V. N. Reinhold, B. B. Reinhold, and C. E. Costello, Carbohydrate molecular weight profiling, sequence, linkage and branching data: ES–MS and CID, *Anal. Chem.* **67**, 1772–1784 (1995).
4. J. Zaia, Mass spectrometry of oligosaccharides, *Mass Spectrom. Rev.* **23**, 161–227 (2004).
5. A. Hannemann and V. N. Reinhold, Oligosaccharide analysis by mass spectrometry, in *Encyclopedia of Biological Chemistry*, Elsevier, New York, 2004.
6. D. J. Harvey, Proteomic analysis of glycosylation: structural determination of N- and O-linked glycans by mass spectrometry, *Expert Rev. Proteom.* **2**, 87–101 (2005).
7. D. J. Harvey, Analysis of carbohydrates and glycoconjugates by matrix-assisted laser desorption/ionization mass spectrometry: an update covering the period 1999–2006, *Mass Spectrom. Rev.* **25**, 595–662 (2006).
8. N. Viseux, E. de Hoffmann, and B. Domon, Structural analysis of permethylated oligosaccharides by electrospray tandem mass spectrometry, *Anal. Chem.* **69**, 3193–3198 (1997).
9. D. J. Harvey, Electrospray mass spectrometry and fragmentation of N-linked carbohydrates derivatized at the reducing terminus, *J. Am. Soc. Mass Spectrom.* **11**, 900–915 (2000).
10. H. C. Birrell, J. Charlwood, I. Lynch, S. North, and P. Camilleri, A dual-detection strategy in the chromatographic analysis of 2-aminoacridone-derivatized oligosaccharides, *Anal. Chem.* **71**, 102–108 (1999).
11. S. Hanrahan, J. Charlwood, R. Tyldesley, J. Langridge, R. Bordoli, R. Bateman, and P. Camilleri, Facile sequencing of oligosaccharides by matrix-assisted laser desorption/ionisation on a hybrid quadrupole orthogonal acceleration time-of-flight mass spectrometer, *Rapid Commun. Mass Spectrom.* **15**, 1141–1151 (2001).
12. H. L. Cheng and G. R. Her, Determination of linkages of linear and oligosaccharides using cross-ring chromophore labeling and negative ion trap mass spectrometry, *J. Am. Soc. Mass Spectrom.* **13**, 1322–1330 (2002).
13. R. K. Merkle and I. Poppe, Carbohydrate composition analysis of glycoconjugates by Gc/MS, in J. K. Lennarz and G. W. Hart, eds., *Methods in Enzymology*, Vol. 230, Academic Press, San Diego, CA, 1994, pp. 1–15.
14. C. G. Hellerqvist, Linkage analysis by Lindberg method, in J. A. McCloskey, ed., *Methods in Enzymology*, Academic Press, Vol. 193, San Diego, CA, 1990, pp. 554–573.
15. J. Charlwood, J. Langridge, and P. Camilleri, Structural characterisation of N-linked glycan mixtures by precursor ion scanning and tandem mass spectrometric analysis, *Rapid Commun. Mass Spectrom.* **13**, 1522–1530 (1999).
16. R. Orlando and Y. Young, Analysis of glycoproteins, in B. S. Larsen and C. N. McEwen, eds., *Mass Spectrometry of Biological Materials*, Marcel Dekker, New York, 1998, pp. 215–245.
17. Y. Yang, and R. Orlando, Simplifying the exoglycosidase digestion/MALDI–MS procedures for sequencing N-linked carbohydrate side chains, *Anal. Chem.* **68**, 570–572 (1996).
18. D. J. Harvey, D. R. Wing, B. Küster, and I. B. H. Wilson, Composition of N-linked carbohydrate from ovalbumin and co-purified glycoproteins, *J. Am. Soc. Mass Spectrom.* **11**, 564–571 (2000).

19. Y. Mechref and M. V. Novotny, Mass spectrometric mapping and sequencing of N-linked oligosaccharides derived from submicrogram amounts of glycoproteins, *Anal. Chem.* **70**, 455–463 (1998).
20. M. T. Cancilla, S. G. Penn, and C. B. Lebrilla, Alkaline degradation of oligosaccharides coupled with matrix-assisted laser desorption/ionization Fourier transform mass spectrometry: a method for sequencing oligosaccharides, *Anal. Chem.* **70**, 663–672 (1998).
21. J. Lemoine, B. Fournet, D. Despeyrough, K. R. Jennings, R. Rosenberg, and E. de Hoffman, Collision-induced dissociation of alkali metal cationized and permethylated oligosaccharides: influence of the collision energy and of the collision gas for the assignment of linkage position, *J. Am. Soc. Mass Spectrom.* **4**, 197–203 (1993).
22. N. Viseux, E. de Hoffman, and B. Domon, Structural analysis of permethylated oligosaccharides by electrospray tandem mass spectrometry, *Anal. Chem.* **69**, 3193–3198 (1997).
23. P. H. Lipniunas, R. R. Reid Townsend, A. L. Burlingame, and O. Hindsgaul, High-energy collision-induced dissociation mass spectrometry of synthetic mannose-6-phosphate oligosaccharides, *J. Am. Soc. Mass Spectrom.* **7**,182–188 (1996).
24. D. J. Harvey, R. H. Bateman, and B. N. Green, High-energy collision-induced fragmentation of complex oligosaccharides ionized by matrix-assisted laser desorption/ionization mass spectrometry, *J. Mass Spectrom.* **32**, 168–187 (1997).
25. B. Küster, T. J. P. Naven, and D. J. Harvey, Effect of the reducing-terminal substituents on the high energy collision-induced dissociation matrix-assisted laser desorption/ionization mass spectra of oligosaccharides, *Rapid Commun. Mass Spectrom.* **10**, 1645–1651 (1996).
26. M. T. Cancilla, S. G. Penn, and C. B. Lebrilla, Collision-induced dissociation of branched oligosaccharide ions with analysis and calculation of relative dissociation thresholds, *Anal. Chem.* **68**, 2331–2339 (1996).
27. L. P. Brüll, V. Kovácik, J. E. Thoma-Oates, W. Heerma, and J. Haverkamp, Sodium-cationized oligosaccharides do not appear to undergo "internal residue loss" rearrangement processes on tandem mass spectrometry, *Rapid Commun. Mass Spectrom.* **12**, 1520–1532 (1998).
28. D. J. Harvey, Postsource decay fragmentation of N-linked carbohydrates from ovalbumin and related glycoproteins, *J. Am. Soc. Mass Spectrom.* **11**, 572–577 (2000).
29. D. J. Harvey, Collision-induced fragmentation of negative-ions from N-linked glycans derivatized with 2-aminobenzoic acid, *J. Mass Spectrom.* **40**, 642–655 (2005).
30. W. Chai, V. Piskarev, and A. M. Lawson, Branching pattern and sequence analysis of underivatized oligosaccharides by combined MS/MS of singly and doubly charged molecular ions in negative-ion electrospray mass spectrometry, *J. Am. Soc. Mass Spectrom.* **13**, 670–679 (2002).
31. B. Domon and C. E. Costello, A systematic nomenclature for carbohydrate fragmentation in FAB-MS/MS spectra of glycoconjugates, *Glycoconjugate J.* **5**, 397–409 (1988).
32. A. Fura and J. A. Leary, Differentiation of Ca^{2+}- and Mg^{2+}-coordinated branched trisaccharides isomers: an ESI and tandem mass spectrometry study, *Anal. Chem.* **65**, 2805–2811 (1996).
33. M. R. Asam and G. L. Glish, Tandem mass spectrometry of alkali cationized polysaccharides in a quadrupole ion trap, *J. Am. Soc. Mass Spectrom.* **8**, 987–995 (1997).

34. W. Mo, H. Sakamoto, A. Nishikawa, N. Kagi, J. I. Langridge, Y. Shimonishi, and T. Takao, Structural characterization of chemically derivatized oligosaccharides by nanoflow electrospray ionization mass spectrometry, *Anal. Chem.* **71**, 4100–4106 (1999).
35. S. König and J. A. Leary, Evidence for linkage position determination in cobalt coordinated pentasaccharides using ion trap mass spectrometry, *J. Am. Soc. Mass Spectrom.* **9**, 1125–1134 (1998).
36. U. Bahr, A. Pfenninger, and M. Karas, High-sensitivity analysis of neutral underivatized oligosaccharides by nanoelectrospray mass spectrometry, *Anal. Chem.* **69**, 4530–4535 (1997).
37. A. S. Weiskopf, P. Vouros, and D. J. Harvey, Electrospray ionization–ion trap mass spectrometry for structural analysis of complex N-linked glycoprotein oligosaccharides, *Anal. Chem.* **70**, 4441–4447 (1998).
38. D. M. Sheeley and V. N. Reinhold, Structural characterization of carbohydrate sequence, linkage, and branching in a quadrupole ion trap mass spectrometer: neutral oligosaccharides and N-linked glycans, *Anal. Chem.* **70**, 3053–3059 (1998).
39. T. Solouki, B. B. Reinhold, C. E. C. M. O'Malley, S. Guan, and A. G. Marshall, Electrospray ionization and matrix-assisted laser desorption/ionization fourier transform ion cyclotron resonance mass spectrometry of permethylated oligosaccharides, *Anal. Chem.* **70**, 857–864 (1998).
40. W. Chai, V. Piskarev, and A. M. Lawson, Branching pattern and sequence analysis of underivatized oligosaccharides by combined MS/MS of singly and doubly charged molecular ions in negative-ion electrospray ionization mass spectrometry, *J. Am. Soc. Mass Spectrom.* **13**, 670–679 (2002).
41. A. Pfenninger, M. Karas, B. Finke, and B. Stahl, Structural analysis of underivatized and neutral human milk oligosaccharides in the negative ion mode by nanoelectrospray MS^n, 1: Methodology, *J. Am. Soc. Mass Spectrom.* **13**, 1331–1340 (2002).
42. Y. Jiang and R. B. Cole, Oligosaccharide analysis using anion attachment in negative ion mode electrospray, *J. Am. Soc. Mass Spectrom.* **16**, 60–70 (2005).
43. D. J. Harvey, Fragmentation of negative ions from carbohydrates, 1: Use of nitrate and other anionic adducts for the production of negative ion electrospray spectra from N-linked carbohydrates, *J. Am. Soc. Mass Spectrom.* **16**, 622–630 (2005).
44. D. J. Harvey, Fragmentation of negative ions from carbohydrates; 2: Fragmentation of high-mannose N-linked glycans, *J. Am. Soc. Mass Spectrom.* **16**, 631–646 (2005).
45. D. J. Harvey, Fragmentation of negative ions from carbohydrates; 3: Fragmentation of hybrid and complex N-linked glycans, *J. Am. Soc. Mass Spectrom.* **16**, 631–646 (2005).
46. D. T. Li and G. R. Her, Structural analysis of chromophore-labeled disaccharides and oligosaccharides by electrospray ionization mass spectrometry and high-performance liquid chromatography/electrospray ionization mass spectrometry, *J. Mass Spectrom.* **33**, 644–652 (1998).

CHAPTER 12

CHARACTERIZATION OF LIPIDS

Lipids are a diverse group of biomolecules that are known to exist as thousands of distinct covalent entities, each with its unique structural and physical characteristics. Lipids are essential cellular constituents that have multiple distinct roles in cellular functions. They participate in the storage of energy, cell–cell communication, cell–cell recognition, and various human diseases. They provide a barrier that separates intracellular and extracellular compartments and a matrix for the interactions of membrane-bound proteins.

12.1. CLASSIFICATION AND STRUCTURES OF LIPIDS

Lipids are broadly classified into four major groups (Figure 12.1):

1. *Fatty acids and acylglycerols.* Fatty acids are carboxylic acids with a long hydrocarbon chain and a varying degree of unsaturation. They are designated by a special notation that includes the number of carbons, followed by a number that indicates the degree of unsaturation. For example, octadecenoic (oleic) acid (**1**) is denoted as 18:1 and hexadecanoic acid (**2**) as 16:0.

Fundamentals of Contemporary Mass Spectrometry, by Chhabil Dass
Copyright © 2007 John Wiley & Sons, Inc.

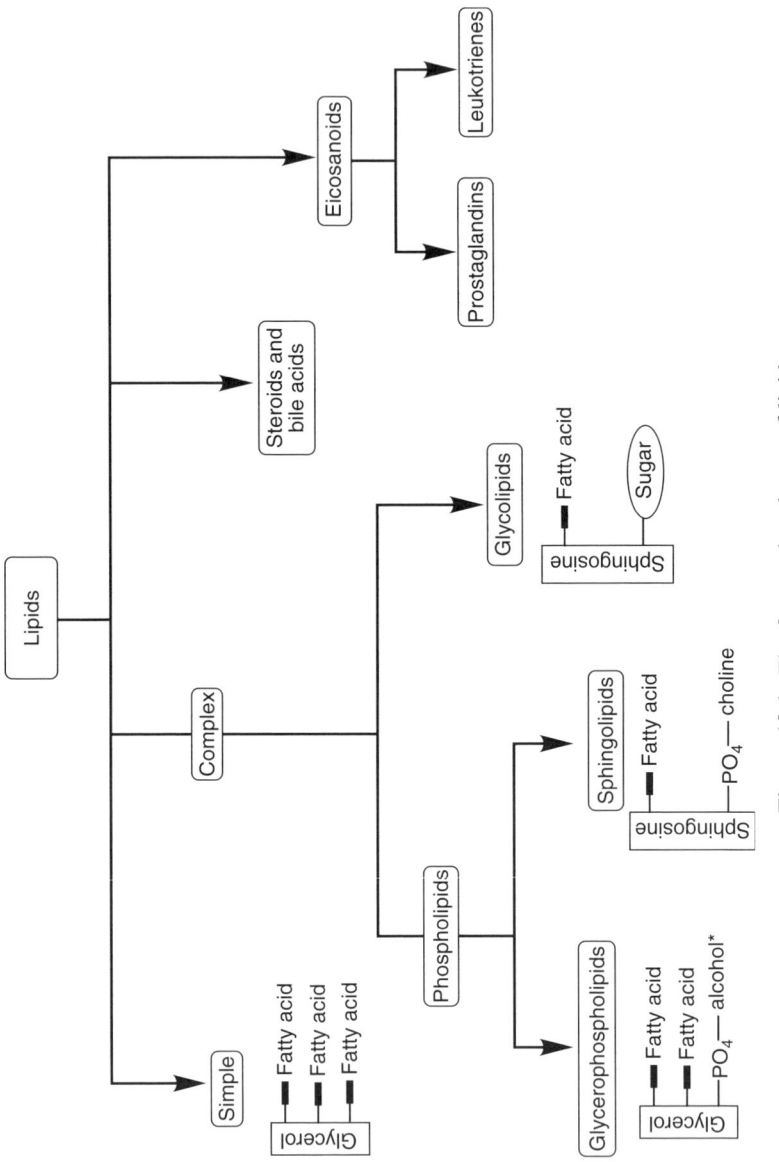

Figure 12.1. The four major classes of lipids.

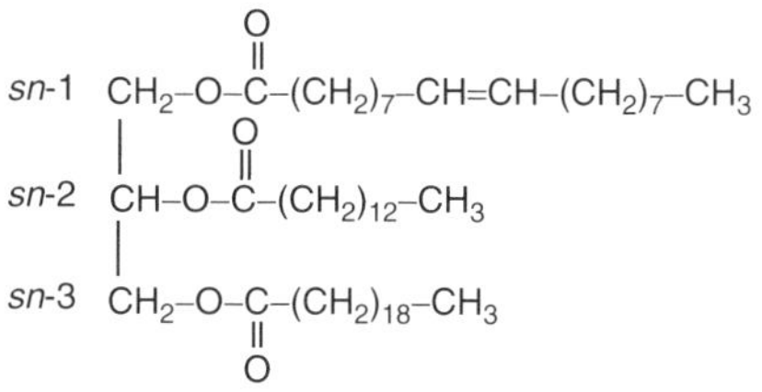

sn-1 $\text{CH}_2\text{-O-C(=O)-(CH}_2)_7\text{-CH=CH-(CH}_2)_7\text{-CH}_3$
sn-2 $\text{CH-O-C(=O)-(CH}_2)_{12}\text{-CH}_3$
sn-3 $\text{CH}_2\text{-O-C(=O)-(CH}_2)_{18}\text{-CH}_3$

Figure 12.2. Structure of a typical acylglycerol.

Acylglycerols are relatively simple lipid substances. Structurally, these nonpolar lipids are esters in which the alcohol component is glycerol and the acid components are long-chain fatty acids. In *triacylglycerols* (TAG; or triglycerides), all three OH groups on the glycerol backbone (i.e., at sn-1, sn-2, and sn-3 carbon atoms) are esterified (see Figure 12.2). They are the most abundant class of lipids and are significant components of fat storage cells. In mono- and diacylglycerols, one or two OH groups in glycerol are esterified by fatty acids, respectively.

2. *Complex lipids.* Complex lipids are the main constituents of all biological membranes. Two main groups of complex lipids are found in the body tissues: *phospholipids* and *glycolipids*. Phospholipids are also esters that are made of an alcohol, fatty acids, and a phosphate group. These polar lipids are further divided into two subgroups; glycerophospholipids (also called *phosphoglycerides*) and *sphingolipids*. These two subgroups differ structurally in that the alcohol in glycerophospholipids is glycerol, whereas in sphingolipids it is sphingosine (**3**).

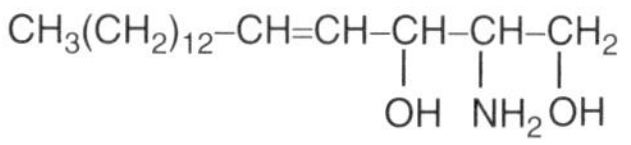

$\text{CH}_3(\text{CH}_2)_{12}\text{-CH=CH-CH-CH-CH}_2$
$\qquad\qquad\qquad\quad\ \ |\ \ \ \ |\ \ \ \ |$
$\qquad\qquad\qquad\ \text{OH}\ \ \text{NH}_2\ \text{OH}$

(**3**)

Structurally, glycerophospholipids are similar to triglycerides but have one of the fatty acids of the glycerol backbone (i.e., C3 or the sn-3 position) replaced by a phosphate group, which in turn is connected to another alcohol, called a *polar head group* (Figure 12.3). This alcohol defines the class of glycerophospholipids, the most abundant of which is glycerophosphocholine (GPC), in which the alcohol part is choline (**4**).

$\text{HO-CH}_2\text{CH}_2\text{-}\overset{+}{\text{N}}(\text{CH}_3)_3$

(**4**)

GPC is also known as *phosphatidylcholine* (PC). The general public knows it by the name of *lecithin*. Other less common classes of glycerophospholipids are glycerophosphoserine (GPS; also termed *phosphatidylserine*, PS), glycerophosphoethanolamine (GPE; also termed *phosphatidylethanolamine*, PE), glycerophosphoglycerol (GPG; also known as *phosphatidylglycerol*), and glycerophosphoinositol (GPI; also called *phosphatidylinositol*). The second alcohol

426 CHARACTERIZATION OF LIPIDS

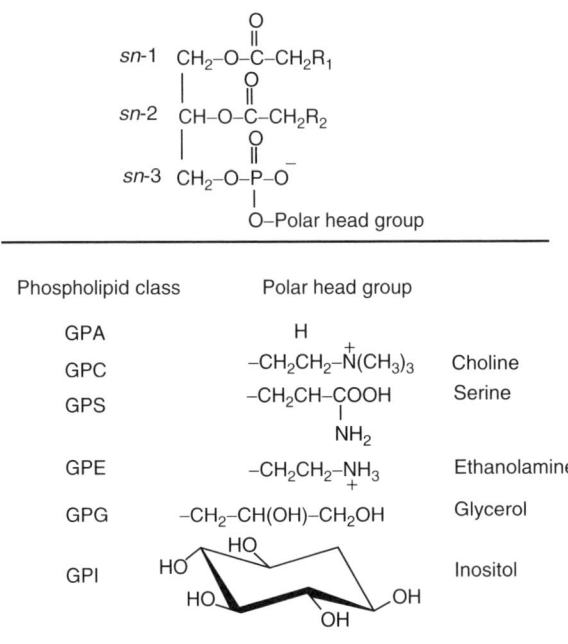

Figure 12.3. Structures of phospholipids. R_1 and R_2 are alkyl or alkenyl chains.

in these molecular species is replaced by serine, ethanolamine, glycerol, or inositol, respectively. Glycerophosphatidic acid (GPA) has no substitution on the phosphate group. The second glycerol in GPG can also be esterified to give an important compound called *cardiolipin*.

Sphingolipids, also called *sphingomyelins*, are also polar lipids that are composed of the *ceramide* backbone. Ceramide is a modified form of the sphingosine in which a long-chain fatty acid is connected to the amino group via an amide bond. Similar to GPC lipids, the polar head group in sphingomyelins is also phosphorylcholine, which is linked to the −OH group of ceramide (Figure 12.4).

Glycolipids are polar lipids that also contain the ceramide backbone but differ from sphingolipids in that the 1-hydroxyl group of the ceramide backbone is linked to one or more sugar units rather than to the phosphorylcholine entity (Figure 12.4). A subclass of glycolipids are the gangliosides, in which sialic acid constitutes the sugar portion.

3. *Steroids and bile acids*. The third major class of lipids is steroids. The structure of steroids is based on the cyclopentanoperhydrophenanthrene moiety (see Figure 12.5). Steroids exist in free form or as a conjugate of various molecules such as fatty acids, sulfuric acid, glucorinic acids, sugars, and amino acids. The most abundant steroid in the human body is cholesterol, which serves as a membrane component. It also acts as a precursor for the synthesis of other steroid hormones. Other known members of the steroid family are bile acids; they are also derivative of cholesterol.

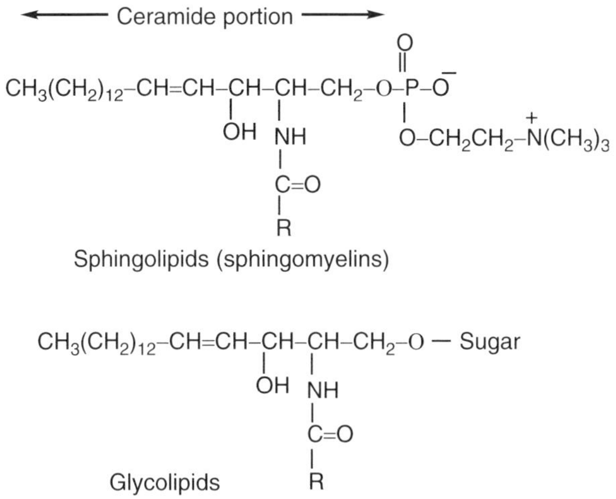

Sphingolipids (sphingomyelins)

CH₃(CH₂)₁₂–CH=CH–CH(OH)–CH(NH–C(=O)–R)–CH₂–O–P(=O)(O⁻)–O–CH₂CH₂–N⁺(CH₃)₃

← Ceramide portion →

Glycolipids

CH₃(CH₂)₁₂–CH=CH–CH(OH)–CH(NH–C(=O)–R)–CH₂–O—Sugar

Figure 12.4. Structures of sphingolipids and glycolipids.

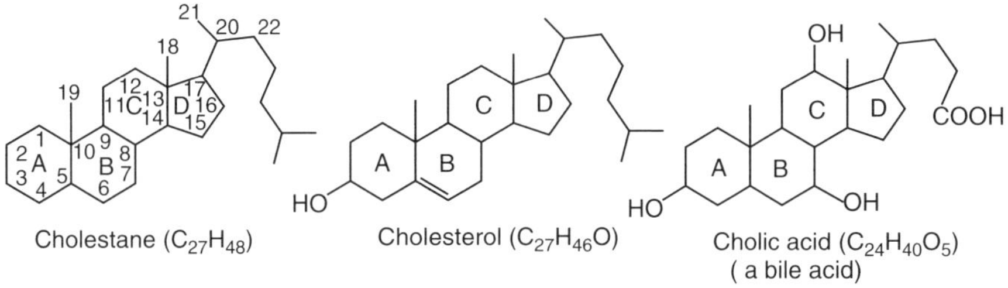

Figure 12.5. Structures of steroids and bile acids.
Cholestane ($C_{27}H_{48}$); Cholesterol ($C_{27}H_{46}O$); Cholic acid ($C_{24}H_{40}O_5$) (a bile acid)

4. *Eicosanoids.* Eicosanoids are a group of fatty acid lipids that are synthesized within cells from arachidonic acid through the action of several different oxygenases. Prostaglandins (PG) and leukotrienes (LT) are two common types of biologically active eicosanoids. PGs are derived from cyclooxygenases reaction and LTs by the 5-lipoxygenase enzyme. Arachidonic acid, in turn, is released in free form from phospholipids by the action of phospholipase A_2. Both types of molecules are actively involved in several physiological functions; they cause inflammation and fever and act as mediators of hormonal responses. Other known forms of eicosanoids are thromboxanes (TX), lipoxins (LX), oxo eicosanoids (oxo-ETE), hydroperoxy eicosanoids (HpETE), hydroxyeicosatetraenoic acids (HETE), and epoxyeicosatetraenoic acids (EET). Structures of eicosanoids are depicted in Figure 12.6. Lipoxins are arachidonic acid metabolites that contain a conjugated tetraene structure and three hydroxyl groups on the hydrocarbon chain.

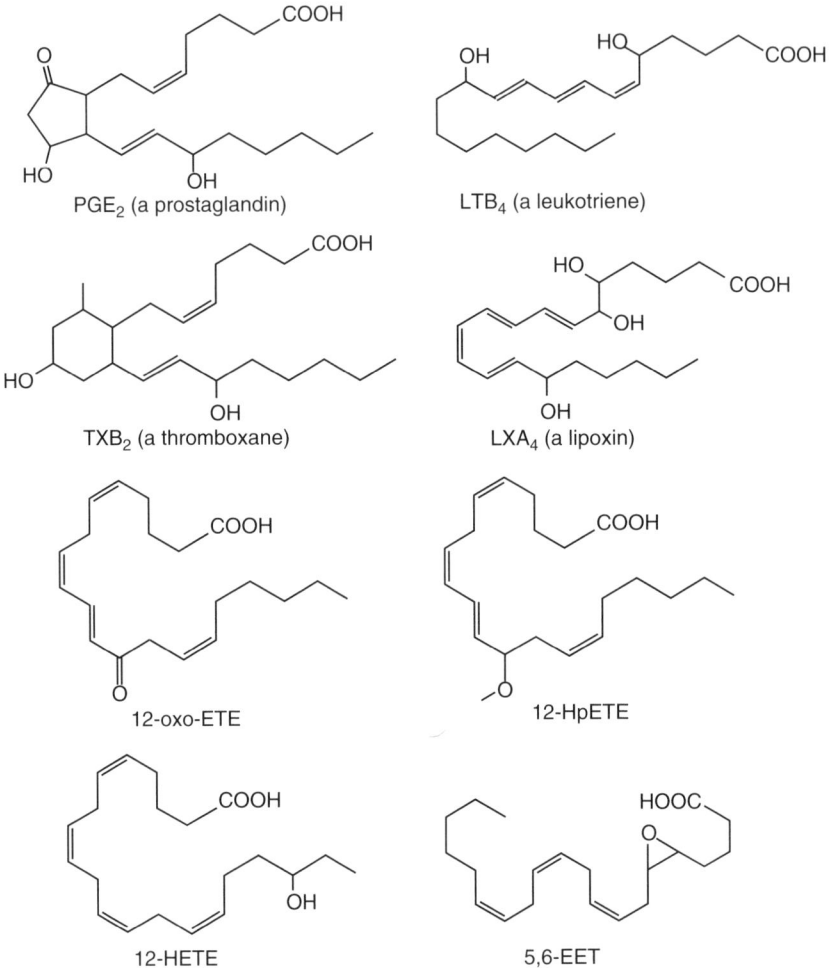

Figure 12.6. Structures of various types of eicosanoids.

12.2. MASS SPECTROMETRY OF FATTY ACIDS AND ACYLGLYCEROLS

12.2.1. Analysis of Fatty Acids

Fatty acids commonly occur in lipid extracts, where they exist in both free and esterified forms. Free fatty acids are isolated along with the neutral lipids. The esterified fatty acids can be released via hydrolysis of lipids. Negative-ion FAB and ESI are the most suitable modes of ionization for fatty acids. In combination with CID, a wealth of structural information has been gathered by using these two modes of ionization for fatty acids and other lipid species. Electron ionization (EI)–MS has also been one of the successful mass spectrometric approaches to the analysis of fatty acids. It is, however, imperative that these nonvolatile molecules be converted to methyl, trimethylsilyl, or *t*-butyldimethylsilyl (TBDMS) derivatives prior to EI–MS analysis. Methyl derivatives are of special interest because they can be identified readily on the basis of a characteristic McLafferty rearrangement ion at m/z 74 [1]. Although EI–MS is successful in the structural characterization of saturated fatty acids, it becomes highly improbable to locate the double-bond site from the EI mass spectrum of unsaturated fatty acids. The double bond tends to migrate upon ionization by EI and thus eludes its precise positional identity. Also, fragmentation is sparse in the high-mass region.

Charge-Remote Fragmentation *Charge-remote fragmentation* (CRF), developed by Gross and colleagues, has gained prominence as a means to characterize fatty acids [2–4]. This approach makes use of CID of FAB- and ESI-produced $[M + H]^+$, $[M + cation]^+$, $[M - H + 2cation]^+$, and $[M - H]^-$ species. Li^+ and Na^+ are the common cations, although alkaline earth metal ions have also been used [5]. With CRF analysis, the site of double bonds, alkyl substituents, and ring structures can be determined as demonstrated in Figure 12.7, which is the CID spectrum of $[M - H + Mg]^+$ ions of oleic acid. The major product ions in the high-energy CID spectra are of general formula $[M - C_nH_{2n+2}]^+$. For saturated fatty acids, cleavage of C—C bonds has been explained by 1,4-H_2 elimination with the formation of two unsaturated species (Scheme 12.1) [4]. A mechanism that includes homolytic C—C bond cleavage and formation of radicals may also be operative [6–8]. Subsequent elimination of an H-atom accounts for the formation of odd-mass terminally unsaturated product ions. A *charge-assisted process* has also been proposed as an additional mechanism to CRF to account for these even-electron ions [9]. This mechanism involves the transfer of hydrogen as H^- from the alkyl chain to the charge site at the carboxy terminus, leaving a charge at the site of hydrogen loss. This new charge site initiates a concerted cleavage of the alkyl chain to exihibit losses of either a H_2 and an alkene (Scheme 12.2, path *b*) or an alkane (Scheme 12.2, path *a*). In unsaturated fatty acids, the preferred site of fragmentation is the allylic C—C bond; thus, the site of the carbon–carbon double bond can readily be identified. Negative-ion

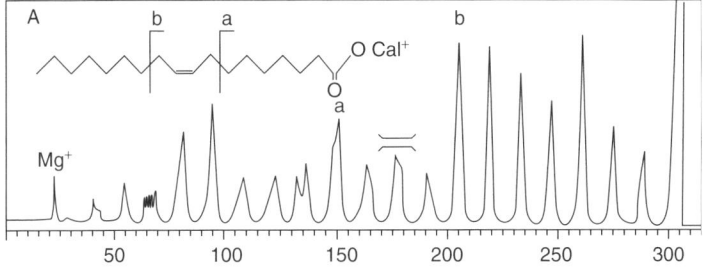

Figure 12.7. CID mass spectrum of FAB-produced $[M - H + Mg]^+$ ions of oleic acid. (Reproduced from ref. 5 by permission of Elsevier, copyright © 1990 American Society for Mass Spectrometry.)

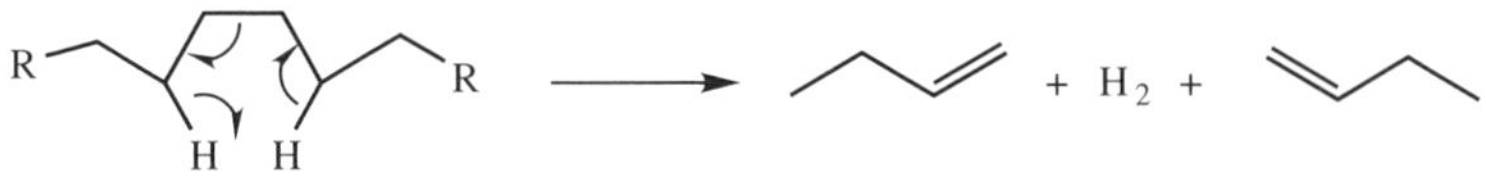

Scheme 12.1. Charge-remote fragmentation of fatty acids via a 1,4-hydrogen elimination.

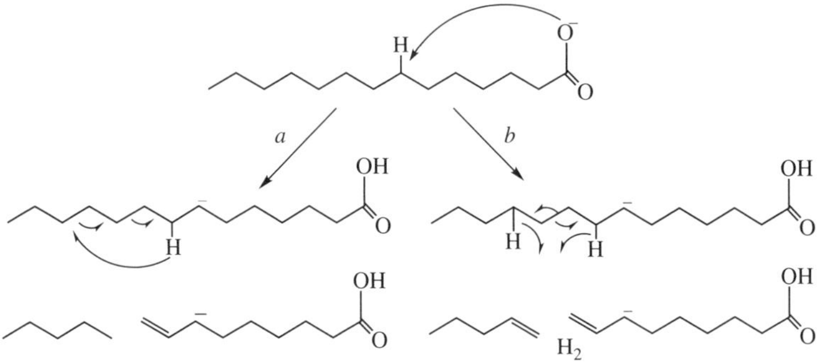

Scheme 12.2. Charge-assisted fragmentation of fatty acids. Path *a* shows the elimination of an alkane, and path *b*, the elimination of a H_2 molecule and an alkene. (From ref. 9.)

ESI, coupled with low-energy CID, although structurally less informative, can provide important structure-related data for polyunsaturated fatty acids [10] and polyhydroxy unsaturated fatty acids [11]. Fragmentation is directed by the position of the hydroxy group or a double bond. As a consequence, isomer-specific CID spectra are produced.

Derivatization of Fatty Acids Derivatized fatty acids have been analyzed to improve ionization and fragmentation characteristics. The FAB spectra of these derivatives are structurally more informative, and fragmentations predominantly occur via CRF processes. One such example is the application of aminoethyltriphenylphosphonium (AETPP) bromide derivatives to characterize fatty acids structurally by FAB–MS/MS [12]. Other examples include derivatives of aminonaphthalenesulfonic acid (ANSA), aminobenzenesulfonic acid (ABSA), picolinyl ester, N-methyl-2-alkylimidazoline (MIM), and dimethyl- (DMAE) and trimethylaminoethyl (TMAE) esters [8, and references therein]. ESI–MS/MS spectra of underivatized and ANSA derivatives of docosahexaenoic acid are compared in Figure 12.8.

12.2.2. Analysis of Acylglycerols

Acylglycerols are fatty acid derivatives of glycerol, of which TAGs are prominent. They are the most abundant class of lipids in nature. TAGs are highly concentrated stores of metabolic energy in animals and plants; fats and oils derived from these species are made exclusively of TAGs. There is growing evidence that the extraordinary rise in type II diabetes in the United States is

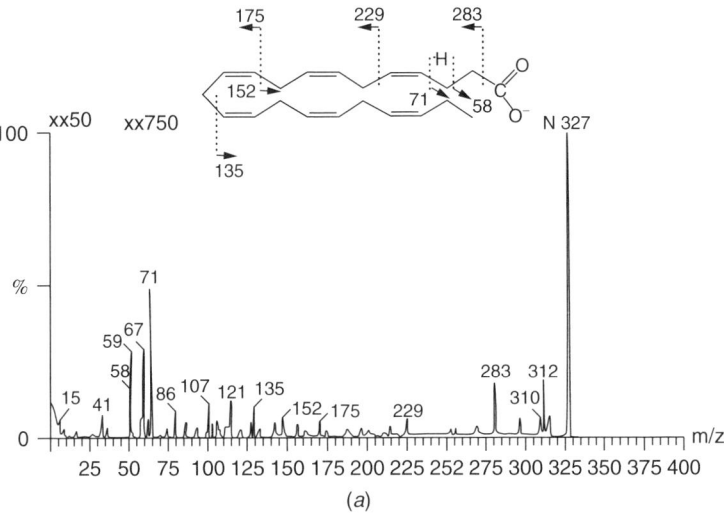

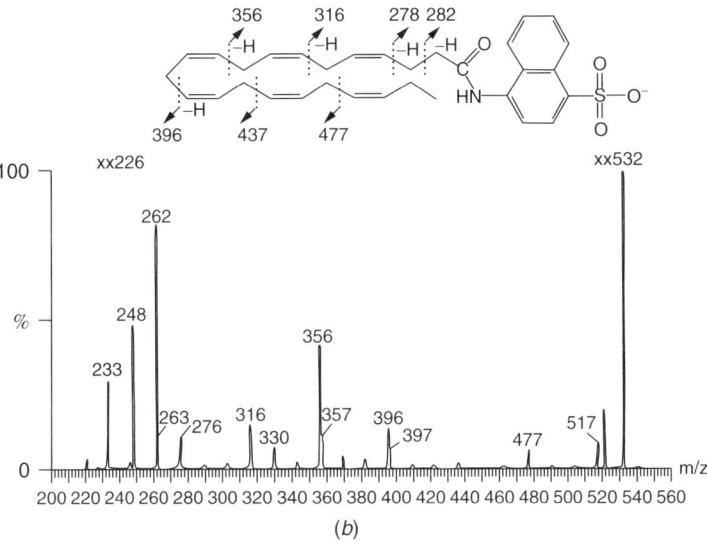

Figure 12.8. ESI–MS/MS spectra of [M − H]⁻ ions of the (*a*) nonderivatized and (*b*) ANSA-derivatized docosahexaenoic acid. (Reproduced from ref. 8 by permission of Wiley-Interscience, copyright © 2003.)

related to an elevation of TAGs in the diet. A complete structural investigation of TAGs involves identity of each acyl (i.e., the fatty acid substituent) and its location on the glycerol backbone. A range of mass spectrometry techniques has served to elucidate the structure of TAGs. For example, TAGs have been characterized by EI–MS on the basis of the presence of characteristic acylium ions

[M − RCOO]$^+$, [RCOO + 128]$^+$, [RCOO + 74]$^+$, and [RCO]$^+$ in the spectra. These ions allow calculation of the carbon number and degree of unsaturation on each acyl group. The location of double bonds and the identity of acyl groups on the glycerol backbone are, however, difficult to pinpoint by EI–MS [13]. To accomplish this task, high-energy CID of ESI- and FAB-produced [M + NH$_4$]$^+$ and [M + metal]$^+$ adducts has proven useful [14]. Charge-directed and charge-remote fragmentation products are both formed, from which it is feasible to deduce the carbon number and degree of unsaturation on each acyl group, the location of double bonds, and the position of the acyl groups on the glycerol backbone. Figure 12.9 compares the high-energy CID spectra of FAB- and ESI-produced [M + Na]$^+$ ions of 1-palmitoyl-2-oleoyl-3-stearoylglycerol. Low-energy CID of ESI-produced [M + Li]$^+$ ions also yields diacyl product ions that can reveal the identity and location of all three fatty acid substitutions in TAGs [15]. Similar strategies can be applied for the structural characterization of mono- and diacylglycerols.

A strategy that involves MS, MS/MS, and MS^3 experiments to characterize individual TAG molecular species from a biological extract has been developed [16]. In this approach, ESI mass spectra of [M + NH$_4$]$^+$ adducts permits determination of the molecular mass of each TAG species. CID of these adducts produces a diacyl product by the neutral loss of NH$_3$ and an acyl side chain

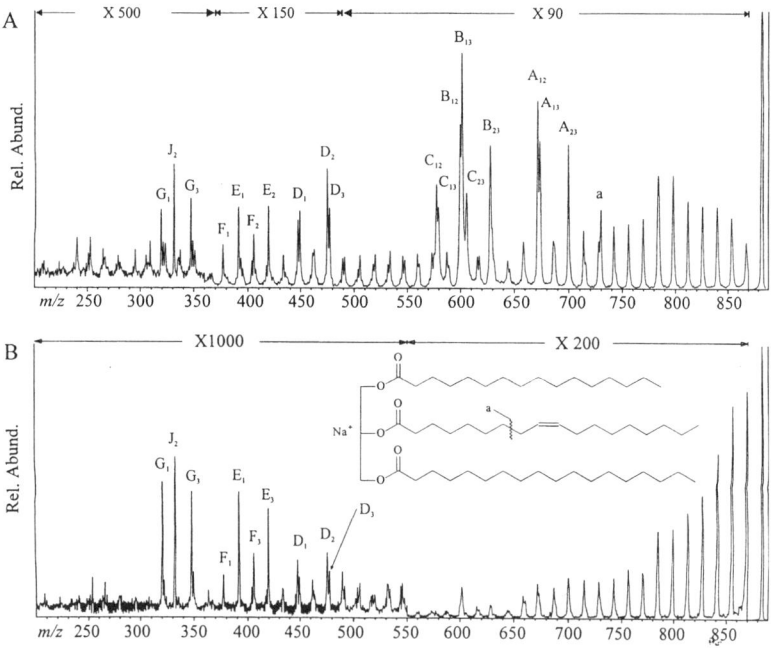

Figure 12.9. CID spectra of the (*a*) FAB- and (*b*) ESI-produced [M + Na]$^+$ ions of 1-palmitoyl-2-oleoyl-3-stearoylglycerol. (Reproduced from ref. 14 by permission of the American Chemical Society, Washington, DC, copyright © 1998.)

as a carboxylic acid. This information provides the identity of each acyl group. Subsequent MS^3 experiments on the diacyl product ions further confirmed the identity of acyl groups and molecular assignments of TAG species. Using this strategy, several TAG species were identified from the neutral fraction of lipid extract from RAW 264.7 cells [16].

12.3. MASS SPECTROMETRY OF PHOSPHOLIPIDS

Phospholipids are amphipathic molecules, meaning that they contain hydrophilic (polar head) and hydrophobic (tail) groups. Phospholipids can easily form the lipid bilayer that constitutes the plasma membrane of the living cells. The hydrophobic part of the bilayer forms an effective barrier to prevent the entry of ions and polar molecules into the cell. As shown in Figure 12.3, the structural diversity of phospholipids can be attributed to different polar head groups, the two fatty acid substituents, and their regioisomerism. The structural diversity and unique physicochemical properties of phospholipids make these compounds difficult to characterize. For complete characterization of phospholipids, identification of the fatty acyl groups as well as the nature of the polar head group are essential.

In traditional approaches, the fatty acyl profile is obtained after release of the sn-1 and sn-2 fatty acids from phospholipids via alkaline or phospholipase C hydrolysis, followed by analysis by GC or GC/MS of the fatty acid components released [17].

FAB, ESI, and MALDI have all proven to be highly successful in the analysis of phospholipids. FAB, especially, made it possible for the first time to analyze intact phospholipids directly by mass spectrometry [18–20]. In this mode of ionization, the hydrophobic tail facilitates their desorption from liquid matrices, and the polar head group allows facile ionization. All subclasses of glycerophospholipids except GPC lipids form negative ions more readily by FAB. The inherent positive charge on the choline moiety makes GPC lipids an easy target for positive-ion FAB analysis. The characteristic features of the positive-ion FAB spectra of glycerophospholipids are the presence of $[M + H]^+$ ions and the ions that are formed via cleavage of the phosphate ester bond. This cleavage produces the $[M + H - 98]^+$ ion (i.e., loss of H_3PO_4) in GPA, the phosphocholine ion (m/z 184) in GPC, and the $[M + H - 141]^+$ ion (loss of phosphoethanolamine) in GPE. GPC lipids are differentiated from other classes of phospholipids by the presence of the characteristic phosphocholine ion at m/z 184 (Figure 12.10a). In negative-ion FAB, the characteristic phosphocholine ion at m/z 184 is absent in the mass spectra [20]. Instead, the carboxylate anions (R_1COO^- and R_2COO^-) that are related to the sn-1 and sn-2 fatty acyl substituents are a characteristic signature of these spectra. In addition, the $[M - 15]^-$, $[M - 60]^-$, and $[M - 86]^-$ ions that represent the losses of CH_3, $[CH_3 + NH(CH_3)_2]$, and $[CH_3 + CH_2=CHN(CH_3)_2]$, respectively, from the choline moiety are also formed (Figure 12.10b) [19,21]. High-energy CID of these three types of ions yields carboxylate anions to provide further confirmation of fatty acid content of phospholipids.

High-energy CID of the FAB-produced $[M + H]^+$ ions of glycerophospholipids also yields structurally important fragment ions. For example, CID of the $[M + H]^+$ ions of GPC and GPE lipids form only a single ion that corresponds to m/z 184 (Figure 12.10c) and $[M + H - 141]^+$, respectively, whereas CID of the $[M - H]^-$ or $[M - 15]^-$ ions yields the abundant carboxylate anions shown in Figure 12.10d, which is the CID spectrum of the $[M - 15]^-$ ion of a diacylglycerophosphocholine (16:0a/18:1). In the CID spectra of $[M - 15]^-$, $[M - 60]^-$, or $[M - 86]^-$, the sn-2 carboxylate anion is always more abundant relative to the sn-1 carboxylate anion [19,20]. FAB–MS/MS has the advantage that it can provide information on the identity of these carboxylate anions [22].

In the positive-ion ESI mode, the protonated and sodiated molecular ion adducts are the primary phospholipids species [23]. In the negative-ion mode, except for GPC, all classes of phospholipids yield abundant $[M - H]^-$ ions. GPC lipids can be detected as $[M + Cl]^-$ ion adducts in this mode of ionization. Structural details are obtained by using ESI–MS/MS [24]. The ESI–MS/MS spectrum of the $[M - H]^-$ ion of a phosphatidylglycerol is shown in Figure 12.11. The characteristic fragments are seen at m/z 153, 171, 253, and 255.

ESI intrasource factors that influence selective ionization of lipid classes have been examined [26]. For example, GPC molecular species can be ionized selectively from other classes of lipids in the positive-ion mode after cationization with Li^+ ions. In contrast, GPE molecular species can be ionized selectively in the negative-ion mode after similar cationization with Li^+ ions.

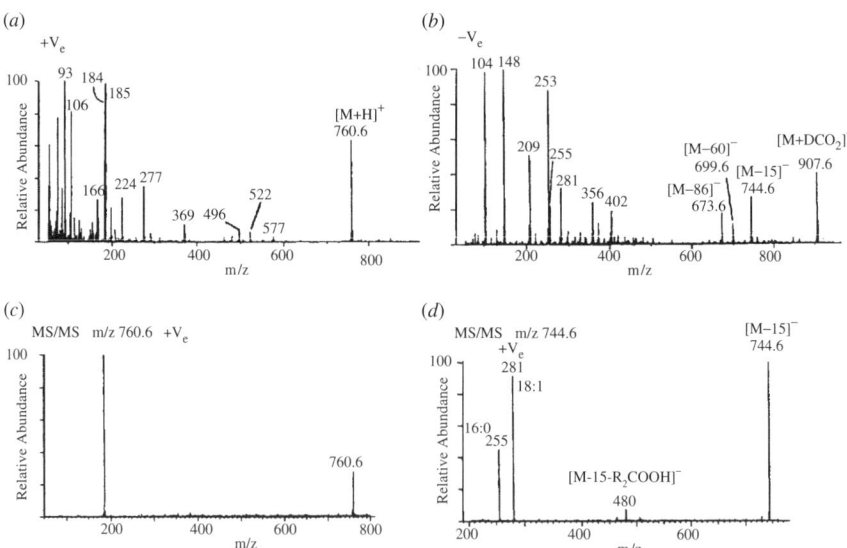

Figure 12.10. FAB-generated mass spectra of a diacylglycerophosphocholine: (a) positive ion, (b) negative ion, (c) CID of $[M + H]^+$ ion, and (d) CID of $[M - 15]^-$ ion. (Reproduced from ref. 20 by permission of Wiley-Interscience, copyright © 1994.)

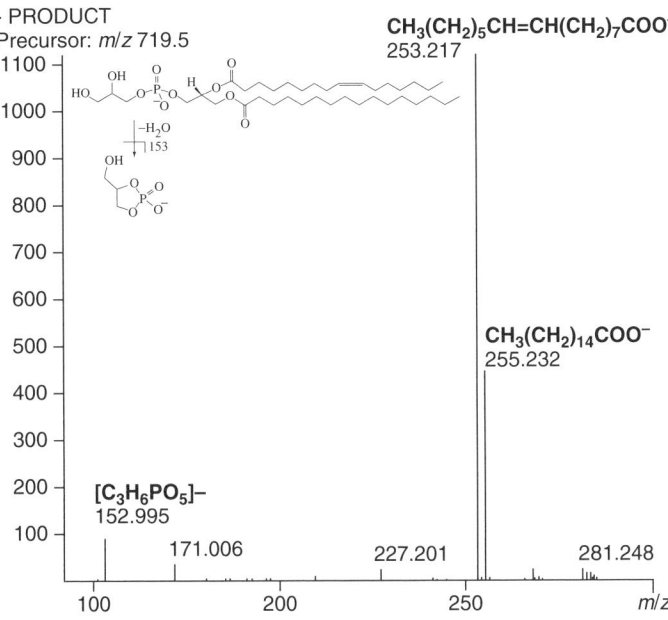

Figure 12.11. ESI–MS/MS spectrum of the [M − H]⁻ ion of a phosphatidylglycerol. (Reproduced from ref. 25 by permission of the American Chemical Society, Washington, DC, copyright © 2002.)

CID–MS/MS of lithiated adducts of GPC and GPE lipids exhibit losses of fatty acid substituents [27, 28]. The relative abundances of these ions can reveal the identity and position of the fatty acid substituents. Precursor-ion and neutral-loss scans can provide the identity of the polar head group and the fatty acid substituents. For example, using these scans, GPC species that contain stearate, arachidonate, linoleate, and palmitate have been identified from rat liver extracts [27].

MALDI has also been used to analyze several molecular forms of phospholipids (GPC, GPE, GPS, GPI, GPG, phosphatidic acid, and cardiolipin) [29]. In the positive-ion mode, all phospholipids yield signals due to the [M + H]⁺, [M + Na]⁺, and [M + K]⁺ ions. In-source fragmentation and PSD provide useful structure information [30]. With the exception of GPC, all phospholipids can also be analyzed in the negative-ion MALDI mode, where the molecular ion region is relatively noise-free and only the [M − H]⁻ signal is produced.

LC/ESI–MS is the preferred approach to identify phospholipids in complex mixtures of the lipid extracts from biological samples [23]. Phospholipids are resolved on a C18 column with water–methanol–hexane solvent that contains 0.5% NH_4OH, and the separated phospholipids are detected in the positive-ion mode as [M + H]⁺ and [M + Na]⁺ species. The relative response depends on the nature of the polar head group and not on the fatty acyl substituents; GPC lipids exhibit the most sensitive response, followed by GPE and GPS lipids.

Another approach to identifying phospholipids in complex lipid mixtures is to monitor neutral-loss and precursor-ion scans [25,31,32]. A specific polar head group exhibits a characteristic neutral loss and produces a characteristic fragment ion. For example, phosphatidylcholine, lysophosphatidylcholine, and sphingomyelin can be identified by observing the loss of $N(CH_3)_3$ (i.e., 59 Da loss) in the positive-ion ESI mode [32]. Similarly, the negative-ion ESI neutral-loss scan for 87 Da is a signature of phosphatidylserine [32]. Precursor-ion scans for m/z 164 ($[NaHO_3POC_2H_4NH_3]^+$ ion), 241 ($[PO_4C_6H_{10}O_4]^-$ ion), and 153 ($[HO_3POCH_2C(OH)CH_2]^-$ ion), provide the identity of phosphatidylethanolamine, phosphatidylinositol, and phosphatidyldiglycerol, respectively [32]. Another example of the usefulness of a precursor-ion scan is to identify the $[M+H]^+$ ions of phosphatidylcholine and sphingomyelin (via monitoring of m/z 184) and of the $[M+Li]^+$ ions of phosphatidylethanolamine and phosphatidylserine (via monitoring of m/z 148 and 192; $[LiHO_3POC_2H_4NH_3]^+$ and $[LiHO_3POCH_2CH(NH_3)COOH]^+$ ions, respectively) from an *E. coli* extract [25].

The behavior of sphingophospholipids under FAB, ESI, MALDI, and CID conditions is similar to that of GPC lipids [20,24,33]. This similarity is due to the presence of phosphocholine in both groups of phospholipids.

Location of Double Bonds in the Acyl Group of Phospholipids Beside FAB–MS [18], the position of the double bond of unsaturated fatty acid substituents can be determined by reacting phospholipids with OsO_4 to convert the double bond to a 1,2-dihydroxy derivative [34]. These derivatives are subsequently analyzed via negative ESI–MS/MS using low-energy CID. Ozonidation of the double bond can also provide useful information on its location [35].

Direct Tissue Analysis of Phospholipids To avoid the time-consuming sample preparation step, direct tissue analysis is an attractive proposition. MALDI–MS holds great promise for this type of analysis. In this approach the target tissues are cut into thin sections. Matrix solution is deposited directly on the tissue section and air-dried prior to MALDI–TOF analysis. Figure 12.12 shows the positive-ion MALDI mass spectrum of the cerebral cortex region [36]. Several phospholipid molecular species were detected in this spectrum. These include phosphatidylcholine 32:0, 34:0, 34:1, 36:1, and 40:6; phosphatidylethanolaminee 38:4 and 40:6; and sphingomyelin 18:0 and 24:0. The molecular ions were observed in the form of $[M+H]^+$, $[M+Na]^+$, and $[M+K]^+$ ions. MALDI–MS/MS of these species using a TOF/TOF instrument can verify the presence of a phosphocholine head group; acyl group–related fragments are not observed [37]. MALDI–MS/MS analysis of $[M+Li]^+$ ions, however, yields enough fragmentation to permit the identity and positional assignment of the acyl groups in phosphatidylcholines.

12.4. MASS SPECTROMETRY OF GLYCOLIPIDS

Glycolipids are ubiquitous constituents of cell membranes and are known to play a major role in cell–cell recognition. Much of the biological interest in

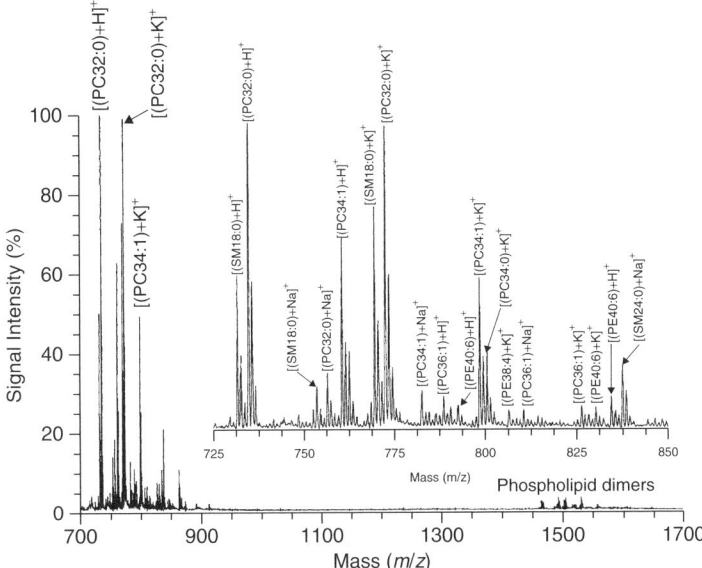

Figure 12.12. Lipid profile of the cerebral cortex region with MALDI–MS. (Reproduced from ref. 36 by permission of Elsevier, copyright © 2005 American Society for Mass Spectrometry.)

glycolipids centers on glycosphingolipids (GSL). Structurally, GSLs are made of a hydrophobic ceramide lipid tail to which an oligosaccharide head group is attached (Figure 12.4). The ceramide tail acts as an anchor to the cell membrane and the hydrophilic carbohydrate portion as a cell–cell recognition site. The oligosaccharide portion may contain one or several sugar units, such as glucose, galactose, N-acetylglucosamine, N-acetylgalactosamine, fucose, sialic acid, or glucuronic acids. Extensive heterogeneity in the structure of the ceramide portion also exists, due to the presence of a variety of fatty acids that differ in the alkyl chain length and in the presence or absence of a double bond and an OH group that are attached to the sphingosine base.

The structural diversity and labile nature of the bonds that link the various building blocks have posed a serious challenge to the characterization of glycolipids. Complete structural characterization of GSLs requires structure elucidation of the constituents of the ceramide and sugar portions. In a classical mass spectrometry method, glycolipids are cleaved into individual structural units and are analyzed by EI–MS or GC/MS after conversion to a suitable derivative, such as by the acetylation, permethylation, or permethylation–reduction reaction.

FAB and FAB–MS/MS have also played a major role in the analysis of GSLs [38]; the negative-ion mode provides better results than the positive-ion mode. Improved spectral content is obtained in the positive-ion FAB mode only when glycolipids are converted to permethylated, peracetylated derivatives. FAB produces a sufficient number of fragment ions from which the structural information can be retrieved. The negative-ion FAB of native GSLs yields the $[M - H]^-$, $[Cer]^-$, and sequence-specific carbohydrate ions. The positive-ion FAB of native GSLs yields mainly the $[M + H]^+$ or $[M + cation]^+$ and Cer^+ species. The positive-ion FAB of permethylated and peracetylated GSLs, in addition, produces sequence-specific carbohydrate ions.

CID provides an enhanced fragment ion yield [39,40]. Nomenclature similar to that used to designate peptide fragment ions has also been proposed [39]. The nomenclature for fragmentation in the carbohydrate portion was discussed in Chapter 11 (shown in Figure 11.7). Designation of the ceramide fragmentation is shown in Figure 12.13. Complementary information is obtained in the

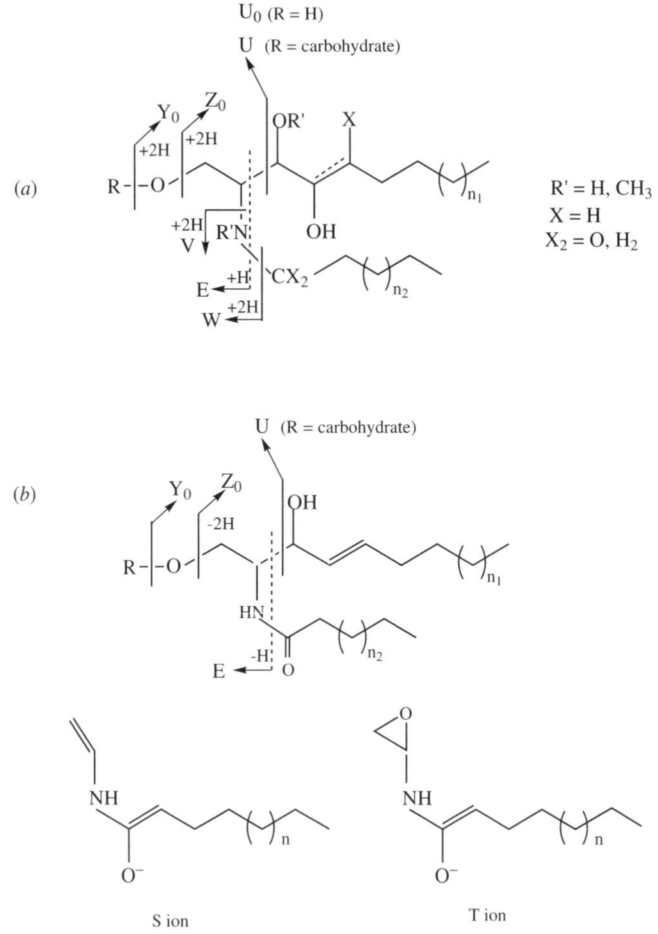

Figure 12.13. Nomenclature of ceramide fragment ions that are formed in the (*a*) positive-ion mode and (*b*) negative-ion mode. (Reproduced from C. Dass, *Principles and Practice of Biological Mass Spectrometry*, Wiley-Interscience, 2001.)

positive- and negative-ion analysis modes. For example, the CID mass spectra of the $[M + H]^+$ ion of the native GSLs contain mainly the ceramide-specific fragment, whereas the corresponding spectra of the $[M - H]^-$ ion are dominated by ions that are related to the structure of the carbohydrate moiety, with charge retention on the nonreducing end. Derivatization of glycolipids prior to CID–MS/MS analysis has been used to improve the yield of the $[M + H]^+$ and $[M - H]^-$ ions and to direct fragmentation into the useful structure-specific channels. Permethylation of amide and hydroxyl groups and borane reduction of amides are the common derivatization steps. Reduction of amides to amines creates an extra charge-retention site that can induce additional structure-specific fragment ions. The structure determination of GSLs via CID of the $[M + Li]^+$ ions has also been reported [41].

In MALDI–MS analysis of GSLs [33], the molecular ion signal is produced in the form of $[M + Na]^+$ species. Fragmentation is sparse and matrix dependent; α-CHCA produces more fragmentation. The loss of water, of an intact oligosaccharide moiety, and of an acylamide are the common modes of fragmentation reactions that are observed under MALDI conditions. MALDI can be performed directly on thin-layer chromatographic (TLC)–separated GSLs [42]. TLC plates are either exposed directly to a laser beam, or GSLs can be heat-transferred from TLC plates onto polymer membranes. The spectral quality is better for membrane-bound analytes. The use of ESI–MS/MS has also been explored to characterize glycolipids [43].

Mass Spectrometry of Gangliosides These molecules are a special class of complex glycosphingolipids in which the sugar portion contains at least one sialic acid. These amphiphilic compounds are localized primarily in the outer leaf of the cell membrane. A nomenclature has been developed to designate gangliosides [44]. According to this nomenclature, the gangliosides are grouped on the basis of the number of sialic acid residues (e.g., M for one, D for two, and T for three). Gangliosides with a tetrose chain (Galβ1 → 3GalNAcβ1 → 4Galβ1 → 4Glc) are designated as G1; the G2 series lacks the terminal galactose, and G3 is devoid of the terminal disaccharide, Galβ1 → 3GalNAc. Structures of GM1a and GM1b gangliosides that contain one sialic acid are shown in Figure 12.14.

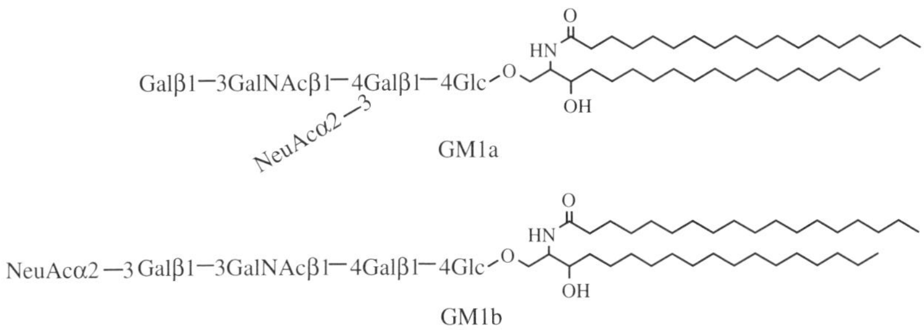

Figure 12.14. Structures of GM1a and GM1b gangliosides.

440 CHARACTERIZATION OF LIPIDS

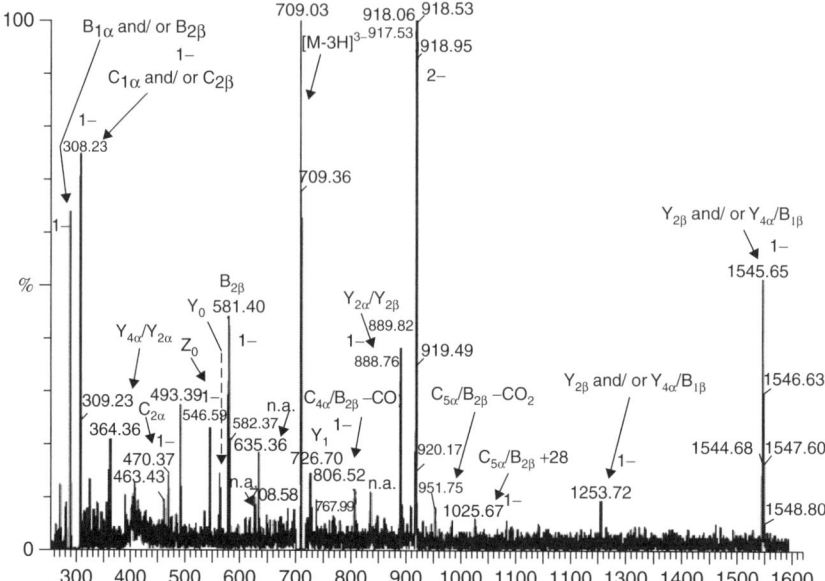

Figure 12.15. Automated nano-ESI-chip MS/MS spectrum of the GT1 species. (Reproduced from ref. 47 by permission of Elsevier, copyright © 2004 American Society for Mass Spectrometry.)

FAB, ESI, and MALDI all have proven to be effective approaches to analyze gangliosides [45]. ECD and IRMPD of the ESI-produced molecular ions of gangliosides can provide important structural information [46]. To analyze minute amounts of biological samples, chip-based nano-ESI–MS and nano-MS/MS have been used to characterize gangliosides from human cerebellum [47]. Figure 12.15, which is the MS/MS spectrum of the GT1 species, demonstrates the usefulness of this approach.

12.5. ANALYSIS OF BILE ACIDS AND STEROIDS

Steroids are hormones that are known to modulate DNA expression and protein expression. Adrenal hormones (e.g., aldosterone and cortisol) participate in the control of metabolism and of salt and water homeostasis. The other class of steroid hormones, gonadal hormones (e.g., estradiol, progesterone, and testosterone), influence mammalian sexual development and function. Several steroids also modulate neurotransmitter action, and some athletes have misused anabolic steroids to enhance their physical strength and performance. GC/MS has been used in conjunction with EI or CI as an ionization method for steroids to determine molecular mass, elemental composition, and length of the side chain of a steroid hydrocarbon. Although these techniques are highly sensitive, they require multistep derivatization of these thermally labile compounds. FAB–MS has proven to be a useful technique to analyze underivatized urinary steroids [48]. Currently, LC coupled to ESI–MS [49,50] or APCI–MS [49] and ESI–MS/MS [51] are the favored technique to analyze steroids. It has the advantages of sensitivity and less sample treatment. APCI–MS results in the formation of the $[M + H]^+$, $[M + H - H_2O]^+$, and $[M + H - 2H_2O]^+$-type ions; the extent of H_2O loss is compound-dependent [49]. MALDI has also shown a potential to analyze steroids, especially when derivatization is performed to enhance ionization and to increase molecular mass beyond the spectral region populated by the matrix-derived ions [52].

A simpler practice to analyze bile acids is to use negative-ion FAB–MS [53,54]. A Sep-Pak C_{18} cartridge can be used to isolate bile acids (e.g., from urine sample) prior to FAB–MS analysis. Currently, ESI–MS has largely supplemented FAB–MS to analyze polar bile acids. The coupling of ESI with MS/MS provides an additional level of specificity. For example, neutral-loss scans for 36 and 54 Da can detect di- and trihydroxy bile acids. LC/ESI–MS can be used for the sensitive detection of bile acids in biological extracts. Recently, cholic acid, chenodeoxycholic acid, and deoxycholic acid have been detected in the cytoplasmic fraction from a rat brain [55].

12.6. ANALYSIS OF EICOSANOIDS

Prostaglandins (PG) belong to the eicosanoid family. They are pharmacologically active molecules and are involved in cell–cell communication. Structurally, they are the oxygenated products of arachidonic acid. $PGF_{2\alpha}$, PGE_2, and PGD_2 are the major biologically active prostaglandins. Mass spectrometry has been pivotal in the structural analysis of these polyfunctional molecules [8]. Similar to other simple lipid molecules, after converting them to suitable derivatives, the structural characterization of PGs can be carried out with EI–MS, typically as the methyl ester trimethylsilyl ether [56]. EI–MS leads to extensive fragmentation that is centered around the derivative itself. The molecular ion is often not seen. A better strategy is to use FAB in conjunction with MS/MS. CID of the FAB-produced $[M - H]^-$, $[M - H + 2Li]^+$, and $[M - H + Ba]^+$ ions of prostaglandins generates a wealth of fragment ion information that is highly useful for their structure analysis [57]. Several subclasses of prostaglandins, such as those that belong to the PGA, PGB, PGD, PGE, and PGF series, have been studied with this approach. For example, CID of the $[M - H]^-$, $[M - H + 2Li]^+$, and $[M - H + Ba]^+$ ions of PGA_2 produces a range of compound-specific ions [57]. The loss of water is a very facile process from all three types of precursor ions. Compared to CID of the carboxylate anion, CID of the metalated PGA_2 produced a greater number of fragment ions. Negative-ion ESI–MS/MS can also be applied to characterize prostaglandins [58]. ESI produces an abundant $[M - H]^-$ ion of PGs; CID yields structure-specific ions. Figure 12.16 illustrates examples of the analysis of PGD_2 using this approach [59].

Leukotrienes (LTs) are lipid mediators derived enzymatically from arachidonic acid. Structurally, these eicosanoids contain a conjugated triene functionality. LTB_4 and LTC_4 are the two important leukotrienes. Mass spectrometry has played

442 CHARACTERIZATION OF LIPIDS

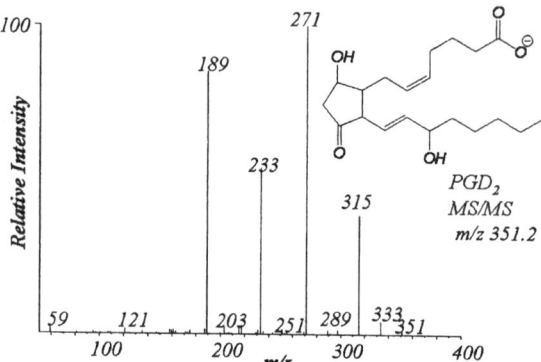

Figure 12.16. ESI–MS/MS mass spectrum of PGD$_2$. (Reproduced from ref. 59 by permission of Elsevier, copyright © 2005.)

a vital role in the structural characterization of these potent bioactive species and in the identification of their biosynthetic and metabolic pathways [7,59]. GC/MS and EI–MS of derivatized species [60], high- and low-energy FAB–MS/MS [61], and ESI–MS/MS [59,62] have all been used to analyze leukotrienes. Leukotriene B$_4$ (LTB$_4$), a potent neutrophil chemotactic agent, and its metabolites have been the subject of several mass spectrometry studies [62]. CID of ESI-produced [M − H]$^-$ ions, especially, has shown to yield unique spectra for leukotriene B$_4$ (LTB$_4$) and its metabolites, and these data can be used for the complete structural characterization of these molecular species [64]. Major fragmentation arises via cleavage of the C−C bond adjacent to the hydroxyl group.

LTC$_4$ and other sulfur-containing leukotrienes, such as LTD$_4$ and LTE$_4$, are adducts of LTA$_4$ and glutathione. These sulfidipeptide leukotrienes are difficult to characterize by EI–MS, but FAB and ESI–MS/MS can provide important structural information [59,62–64]. Figure 12.17 shows the positive-ion ESI–MS/MS spectra of three sulfur-containing leukotrienes, LTC$_4$, LTD$_4$, and LTE$_4$. The formation of compound-specific fragment ions is evident in these spectra [64]. The most abundant ion is produced by the cleavage of the glutathione C−S bond and charge retention by the peptide moiety.

Many other eicosanoids, such as thromboxanes, lipoxins, oxo eicosanoids, hydroperoxy eicosanoids, hydroxyeicosatetraenoic acids, epoxyeicosatetraenoic acids, and isoprostanes, are also biologically active molecules. ESI has emerged as an efficient ionization method for these classes of eicosanoids. An abundant carboxylate anion is produced in the negative-ion ESI mode [59]. Low-energy CID of the ESI-produced [M − H]$^-$ ions of these molecular species produces structurally distinct ions for these disparate molecular species.

12.7. LIPIDOMICS

The full complement of various chemically distinct covalent entities in cellular lipids is referred to as the *lipidome*. Lipidomics is the field of study that

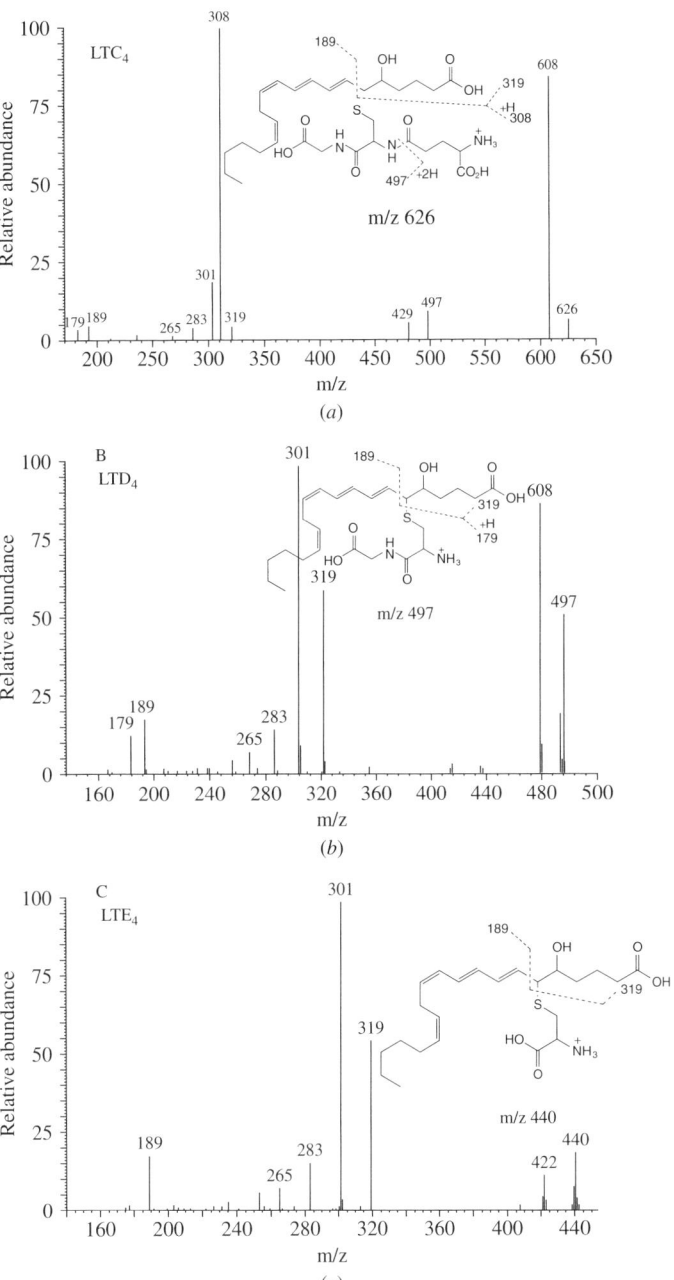

Figure 12.17. Positive ion ESI–MS/MS spectra of sulfur-containing leukotrienes: (a) LTC_4; (b) LTD_4; (c) LTE_4. (Reproduced from ref. 64 by permission of Elsevier, copyright © 2001 American Society for Mass Spectrometry.)

characterizes each distinct chemical species present in a cell's lipidomes, and it aims at understanding the implications of lipid diversity and how regulation of lipids affects cellular functions. Altered levels of cellular lipids can be indicative of a pathological state and environmental stress. Global lipidomics aims to provide a total lipid profile. In contrast, targeted lipidomics focuses on one class or subclass of lipids.

The field of lipidomics has benefited greatly from the advances made in mass spectrometry techniques discussed in this chapter for the analysis of different classes of lipids. ESI, especially, has made a major contribution to this field. A few lipidomics approaches that were developed recently are described here [65,66]. One, called *shotgun lipidomics*, involves the intrasource separation of lipids from a complex biological extract, multidimensional mass spectrometry, and computer-assisted array analysis. Lipids are separated into three main categories on the basis of their electrical properties without resorting to chromatography separation: anionic, weak anionic, and neutral. Anionic lipids contain an inherent negative charge under physiological conditions and include cardiolipin, GPG, GPS, GPI, GPA, cholesterol sulfate, and others. They yield an abundant signal in the negative-ion ESI mode at neutral pH. Weak anionic lipids also carry a net negative charge but only at alkaline pH, and include GPE, bile acids, eicosanoids, and others. These lipids produce an abundant signal in the negative-ion ESI mode in the presence of LiOH. The remaining lipids, such as GPC, sphingomyelins, acylglycerols, and cholesterol and its esters, are included in the neutral polar lipid category. All of these lipids can be analyzed in the positive-ion ESI mode after the addition of a small amount of LiOH. Thus, the selective ionization of lipids can be achieved by ESI. Figure 12.18 outlines the essential points of the intrasource separation–multidimensional MS strategy. Sample preparation involves homogenization of the biological sample, addition

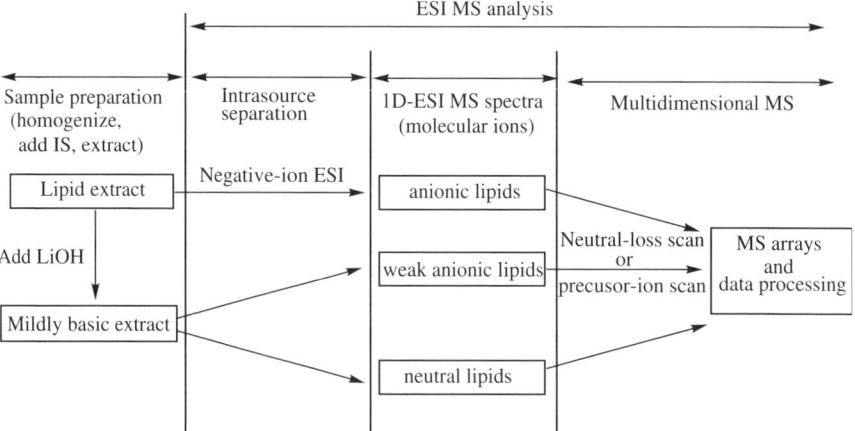

Figure 12.18. Shotgun lipidomics using ESI intrasource separation and multidimensional mass spectrometry of lipids from a complex extract. (From refs. 65 and 66.)

of an appropriate internal standard (IS), and solvent extraction. A portion of the extract is analyzed by negative-ion ESI–MS to profile anionic lipids. After the addition of a small amount of LiOH, the extract is analyzed via negative- and positive-ion ESI-MS to obtain lipid profiles of weak anionic and neutral lipids, respectively. These profiles contain molecular ion information of lipid classes. In the second-dimension MS analysis, neutral-loss and precursor-ion scanning is used to characterize individual building blocks (e.g., polar head groups and aliphatic chains) of lipids. Figure 12.19 shows a two-dimensional MS plot of weak anionic lipid species from a mouse myocardial extract. In this plot, several GPE species are identified in the ESI mass spectrum (*x*-axis). From the precursor-ion scanning of m/z 241.2, 255.2, 281.2, 283.2, 303.3, and 327.3, the acyl chains of ethanolamine-containing lipids (shown on the *y*-axis) can be identified. This approach can also be used to quantify lipid classes.

Another approach, *data-dependent acquisition* (DDA)-*driven lipid profiling* [67], analyzes an unfractionated sample using a Q–TOF instrument. The lipid extract is infused directly into a nanospray chip ion source of a Q–TOF

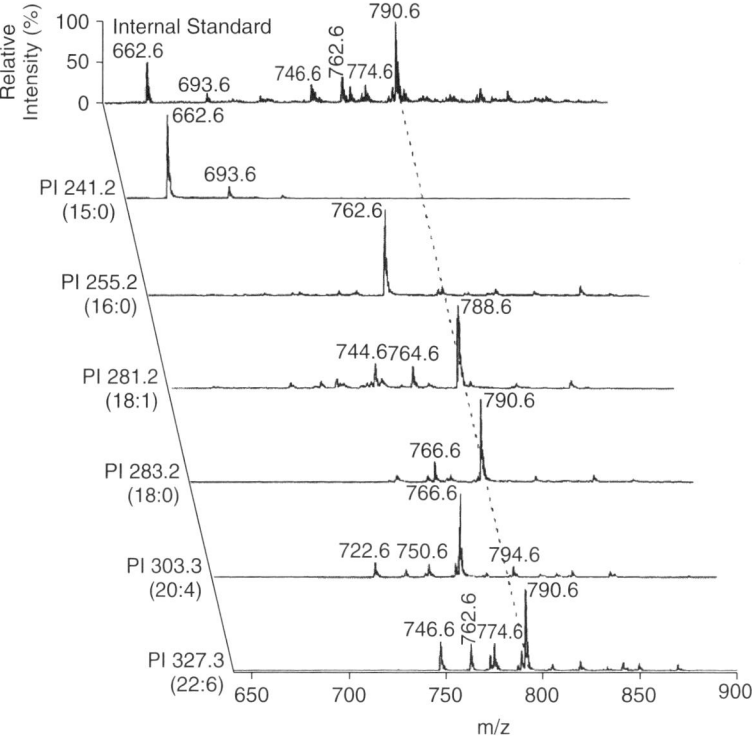

Figure 12.19. Two-dimensional MS analysis of a mouse myocardial lipid extract. GPE lipid species are identified in the ESI–MS spectrum (topmost panel). The lower five panels show precursor-ion (PI) scans that provide information on acyl moieties. (Reproduced from ref. 65 by permission of Wiley-Interscience, copyright © 2005.)

instrument. This procedure involves one MS survey scan by TOF followed by two successive MS/MS scans from the same precursor. The first MS/MS scan is acquired to collect low-m/z-range (100 to 500) fragments by setting the front-end quadrupole at unit mass resolution which minimizes the contribution of other neighboring lipid precursor ions. The second MS/MS spectrum is acquired by selecting the same precursor ion but at a low-resolution setting of the quadrupole, and products are collected in the high m/z range (500 to 700). The data are analyzed by dedicated LipidInspector software to provide information that emulates that of conventional precursor-ion and neutral-loss scanning data.

OVERVIEW

Lipids occur widely in nature. For examples, phospholipids are the main building blocks of biological membranes. Lipids exist in a variety of distinct forms, such as fatty acids, acylglycerols, phospholipids, sphingomyelins, glycosphingolipids, gangliosides, steroids, bile acids, prostaglandins, and leukotrienes. Some of lipids are structurally simple molecules, but others are complex molecules. Many are amphiphilic compounds.

Fatty acids and acylglycerols are relatively simple lipid substances. FAB and ESI are the most suitable modes of ionization for these lipid molecules. CID of FAB- and ESI-produced $[M + H]^+$, $[M + cation]^+$, $[M - H + 2cation]^+$, and $[M - H]^-$ have produced a wealth of structural information. Charge-directed and charge-remote fragmentation products are both formed for acylglycerols, which can help to deduce the carbon number and degree of unsaturation on each acyl group, location of double bonds, and position of the acyl groups on the glycerol backbone.

Phospholipids are structurally diverse amphiphilic molecules. A complete characterization of phospholipids requires the identity of fatty acyl groups and the polar head group. FAB, ESI, and MALDI, either alone or in conjunction with CID–MS/MS, have been highly successful in analyzing phospholipids. CID–MS/MS of the lithiated adducts of phospholipids, especially, can reveal the identity of the fatty acyl groups. The neutral-loss and precursor-ion scans of specific polar head group marker ions are useful in identifying phospholipids in complex lipid mixtures.

Structurally, glycosphingolipids and gangliosides are distinct from other classes of lipids; they are made of a hydrophobic ceramide lipid tail to which an oligosaccharide head group is attached. Characterization of glycosphingolipids and gangliosides is achieved by CID–MS/MS of the FAB-, ESI-, and MALDI-produced $[M + H]^+$ and $[M - H]^-$ ions. Structural information on ceramide and carbohydrate portions can be also derived.

FAB and ESI have also proven useful for characterizing steroids, bile acids, and all types of eicosanoids. Although positive- and negative-ion modes have both been used, the latter provide better detection sensitivity and information content, owing to the presence of carboxylate functionality in these molecules.

Through global lipidomics, each distinct lipid species present in a cell's lipidome can be identified. A shotgun lipidomics approach, which uses an ESI–intrasource separation of lipids from a complex extract, multidimensional mass spectrometry, and computer-assisted array analysis, is described.

EXERCISES

12.1. Describe the structural features of various phospholipid species.

12.2. How is an acyl-chain profile obtained for glycerophospholipids via GC/MS analysis?

12.3. What strategy is applied to characterize triacylglycerols?

12.4. In FAB–MS analysis, glycerophosphocholine lipid species can easily be distinguished from other glycerophospholipids. How?

12.5. What marker ions can be used in precursor-ion scans to identify a certain class of phospholipids in complex lipid mixtures?

12.6. What neutral loss is monitored to identify glycerophosphocholine and sphingomyelin species in the positive-ion mode?

12.7. Describe the structural features of gangliosides.

12.8. Negative-ion provides improved detection sensitivity for bile acids and eicosanoids. Give your reasoning.

12.9. What is the basis of an ESI–intrasource separation of lipids into distinct categories?

12.10. Outline the ESI intrasource separation-multidimensional MS lipidomics approach for profiling lipids in complex biological extracts.

REFERENCES

1. J. A. Zirrolli and R. C. Murphy, Low-energy tandem mass spectrometry of the molecular ion derived from fatty acid methyl esters: a novel method for analysis of branched-chain fatty acids, *J. Am. Soc. Mass Spectrom.* **4**, 223–229 (1993).
2. N. J. Jensen and M. L. Gross, Fast atom bombardment and tandem mass spectrometry for determining iso- and anteiso-fatty acids, *Lipids* **21**, 362–365 (1986).
3. N. J. Jensen and M. L. Gross, Mass spectrometry methods for structural determination and analysis of fatty acids, *Mass Spectrom. Rev.* **6**, 497–536 (1987).
4. N. J. Jensen, K. B. Tomer, and M. L. Gross, Gas-phase ion decompositions occurring remote to a charge site, *J. Am. Chem. Soc.* **107**, 1863–1868 (1985).
5. E. Davoli and M. L. Gross, Charge remote fragmentation of fatty acids cationized with alkaline earth metal ions, *J. Am. Soc. Mass Spectrom.* **1**, 320–324 (1990).

6. V. H. Wysoki and M. M. Ross, Charge-remote fragmentation of gas-phase ions: mechanistic and energetic considerations in the dissociation of long-chain functionalized alkanes and alkenes, *Int. J. Mass Spectrom. Ion Phys.* **104**, 197–211 (1991).
7. M. Claeys, L. Nizigiyimana, H. Van den Heuvel, I. Vedernikova, and A. Haemers, Charge-remote and charge-proximate fragmentation processes in alkali-cationized fatty acid esters upon high-energy collision activation, *J. Mass Spectrom.* **33**, 631–641 (1998).
8. W. J. Griffiths, Tandem mass spectrometry in the study of fatty acids, bile acids, and steroids, *Mass Spectrom. Rev.* **22**, 81–152 (2003).
9. D. J. Harvey, A new charge-associated mechanism to account for the production of fragment ions in the high-energy CID spectra of fatty acids, *J. Am. Soc. Mass Spectrom.* **16**, 280–290 (2005).
10. J. L. Kerwins, A. M. Weins, and L. H. Ericksson, Identification of fatty acids by electrospray ionization mass spectrometry and tandem mass spectrometry, *J. Mass Spectrom.* **31**, 184–192 (1996).
11. P. Wheelan, J. A. Zirrolli, and R. C. Murphy, Electrospray ionization and low energy tandem mass spectrometry of polyhydroxy unsaturated fatty acids, *J. Am. Soc. Mass Spectrom.* **7**, 140–149 (1996).
12. Y.-S. Chang and J. T. Watson, Charge-remote fragmentation during FAB–CAD–B/E linked-scan mass spectrometry of (aminoethyl)triphenylphosphonium derivatives of fatty acids, *J. Am. Soc. Mass Spectrom.* **3**, 769–775 (1992).
13. R. C. Murphy, Mass spectrometry of lipids, in F. Snyder, ed., *Handbook of Lipid Research*, Vol. 7, Plenum Press, New York, 1993, pp. 213–243.
14. C. Cheng, M. L. Gross, and E. Pittenauer, Complete structural elucidation of triacylglycerols by tandem sector mass spectrometry, *Anal. Chem.* **70**, 4417–4426 (1998).
15. F. F. Hsu and J. Turk, Structural characterization of triacylglycerols as lithiated adducts by ESIMS using low energy collisionally activated dissociation on a triple stage quadrupole instrument, *J. Am. Soc. Mass Spectrom.* **10**, 587–599 (1999).
16. A. M. McAnoy, C. C. Wu, and R. C. Murphy, Direct qualitative analysis of triacylglycerols by electrospray mass spectrometry using a linear ion trap, *J. Am. Soc. Mass Spectrom.* **16**, 1498–1509 (2005).
17. K. Eder, A. M. Reichlmayr-lais, and M. Kirchgessner, Studies on the methanolysis of small amounts of purified phospholipids for gas chromatographic analysis of fatty acid methyl esters, *J. Chromatogr.* **607**, 55–67 (1992).
18. N. J. Jensen and M. L. Gross, A comparison of mass spectrometry methods for structural determination and analysis of phospholipids, *Mass Spectrom. Rev.* **7**, 41–69 (1988).
19. N. J. Jensen, K. B. Tomer, and M. L. Gross, Fast atom bombardment and tandem mass spectrometry of phosphatidylserine and phosphatidylcholine, *Lipids* **21**, 580–588 (1986).
20. R. C. Murphy and K. A. Harrison, Fast-atom bombardment mass spectrometry of phospholipids, *Mass Spectrom. Rev.* **13**, 57–75 (1994).
21. J. A. Zirrolli, K. L. Clay, and R. C. Murphy, Tandem mass spectrometry of negative ions from choline phospholipid molecular species related to platelet activating factor, *Lipids* **26**, 1112–1116 (1991).

22. K. B. Tomer, N. J. Jensen, and M. L. Gross, Fast atom bombardment and tandem mass spectrometry for determining structural modification of fatty acids, *Anal. Chem.* **58**, 2429–2433 (1986).
23. H.-Y. Kim, T.-C. L. Wang, and Y.-C. Ma, Liquid chromatography/mass spectrometry of phospholipids using electrospray ionization, *Anal. Chem.* **66**, 3977–3982 (1994).
24. X. Han and R. W. Gross, Structural determination of picomole amounts of phospholipids via electrospray ionization tandem mass spectrometry, *J. Am. Soc. Mass Spectrom.* **6**, 1202–1210 (1995).
25. K. Ekroos, I. V. Chernushevich, K. Simons, and A. Shevchenko, Quantitative profiling of phospholipids by multiple precursor ion scanning on a hybrid quadrupole time-of-flight mass spectrometer, *Anal. Chem.* **74**, 941–949 (2002).
26. X. Han, K. Y. Jingyue, K. N. Fikes, H. Cheng, and R. W. Gross, Factors influencing the electrospray intrasource separation and selective ionization of glycerophospholipids, *J. Am. Soc. Mass Spectrom.* **17**, 264–274 (2006).
27. F.-F. Hsu, A. Bohrer, and J. Turk, Formation of lithiated adducts of glycerophosphocholine lipids facilitates their identification by electrospray ionization tandem mass spectrometry, *J. Am. Soc. Mass Spectrom.* **9**, 516–526 (1998).
28. F.-F. Hsu and J. Turk, Characterization of phosphatidylethanolamine as a lithiated adduct by triple quadrupole tandem mass spectrometry with electrospray ionization, *J. Mass Spectrom.* **35**, 596–606 (2000).
29. D. J. Harvey, Matrix-assisted laser desorption/ionization mass spectrometry of phospholipids, *J. Mass Spectrom.* **30**, 1333–1346 (1995).
30. K. A. Al-Saad, V. Zabrouskov, W. F. Siems, N. R. Knowls, R. M. Hannan, and H. H. Hill, Matrix-assisted laser desorption/ionization time-of-flight mass spectrometry of lipids: ionization and prompt fragmentation patterns, *Rapid Commun. Mass Spectrom.* **17**, 87–96 (2003).
31. M. J. Cole and C. G. Enke, Direct determination of phospholipids structure in microorganisms by fast atom bombardment triple quadrupole mass spectrometry, *Anal. Chem.* **63**, 1032–1038 (1991).
32. N. Pelizzi, S. Catinella, S. Barboso, and M. Zanoi, Different electrospray tandem mass spectrometric approaches for rapid characterization of phospholipids classes of Curosurf, a natural surfactant, *Rapid Commun. Mass Spectrom.* **16**, 2215–2220 (2002).
33. D. J. Harvey, Matrix-assisted laser desorption/ionization mass spectrometry of sphingo- and glycosphingo-lipids, *J. Mass Spectrom.* **30**, 1311–1324 (1995).
34. M. K. Moe, T. Anderssen, M. B. Strøm, and E. Jensen, Total structure characterization of unsaturated acidic phospholipids provided by di-hydroxylation of fatty acid double bonds and negative electrospray ionization mass spectrometry, *J. Am. Soc. Mass Spectrom.* **16**, 46–59 (2005).
35. K. A. Harrison and R. C. Murphy, Direct mass analysis of ozonides: application to unsaturated glycerophosphocholine lipids, *Anal. Chem.* **68**, 3224–3230 (1996).
36. S. N. Jackson, H.-Y. J. Wang, M. Ugarov, T. Egan, J. A. Schultz, and A. S. Woods, Direct tissue analysis of phospholipids in rat brain using MALDI–TOFMS and MALDI–ion mobility–TOFMS, *J. Am. Soc. Mass Spectrom.* **16**, 133–138 (2005).
37. S. N. Jackson, H.-Y. J. Wang, and A. S. Woods, In situstructural characterization of phosphatidylcholines in brain tissue using MALDI–MS/MS, *J. Am. Soc. Mass Spectrom.* **16**, 2052–2056 (2005).

38. J. Peter-Kataliniç, Analysis of glycoconjugates by fast atom bombardment mass spectrometry [MS] and related MS techniques, *Mass Spectrom. Rev.* **13**, 77–98 (1994).
39. C. E. Costello and J. E. Vath, Tandem mass spectrometry of glycolipids, in J. A. McCloskey, ed. *Methods in Enzymology*, Vol. 193, Academic Press, San Diego, CA, pp 738–768.
40. B. Dammon and C. E. Costello, Structure elucidation of glycosphingolipids and gangliosides by high-performance tandem mass spectrometry, *Biochemistry* **27**, 1534–1543 (1988).
41. Q. Ann and J. Adams, Structure determination of ceramides and neutral glycosphingolipids by collisional activation of $[M + Li]^+$ ions, *J. Am. Soc. Mass Spectrom.* **3**, 260–263 (1992).
42. J. Guittard, X. L. Hronowski, and C. E. Costello, Direct matrix-assisted laser desorption/ionization mass spectrometric analysis of glycosphingolipids on thin layer chromatographic plates and transfer membranes, *Rapid Commun. Mass Spectrom.* **13**, 1838–1849 (1999).
43. W. Wang, Z. Liu, L. Ma, C. Hao, S. Liu, V. G. Voinov, and N. I. Kalinovskaya, Electrospray ionization multiple-stage tandem mass spectrometric analysis of diglycosyldiacylglycerol glycolipids from the bacteria *Bacillus pumilus, Rapid Commun. Mass Spectrom.* **13**, 1189–1196 (1999).
44. L. Svennerholm, Designation and schematic structure of gangliosides and allied glycosphingolipids, in *Biological Functions of Gangliosides*, Vol. CI, Elsevier Science, Amsterdam, The Netherlands, 1994, pp. R11–R14.
45. P. Juhas and C. E. Costello, Matrix-assisted laser desorption/ionization time-of-flight mass spectrometry of underivatized and permethylated gangliosides, *J. Am. Soc. Mass Spectrom.* **3**, 785–796 (1992).
46. M. A. McFarland, A. G. Marshall, C. L. Hendrichson, P. Fredman, J.-E. Manson, and C. L. Nilsson, Structural characterization of GM1 ganglioside by IRMPD, ECD, and EDD electrospray ionization FT–ICR MS/MS, *J. Am. Soc. Mass Spectrom.* **16**, 752–762 (2005).
47. A. Zamfir, Z. Vukeliç, L. Bindila, J. Peter-Kataliniç, R. Almeida, A. Sterling, and M. Allen, Fully-automated chip-based nanoelectrospray tandem mass spectrometry of gangliosides from human cerebellum, *J. Am. Soc. Mass Spectrom.* **15**, 1649–1657 (2004).
48. K. D. Setchell et.al., Identification of new inborn error in bile acid synthesis: mutation of oxysterol 7αhydoxylase gene causes severe neonatal liver disease, *J. Clin. Invest.* **102**, 1690–1703 (1998).
49. Y.-C Ma and H.-Y Kim, Determination of steroids by liquid chromatography/mass spectrometry, *J. Am. Soc. Mass Spectrom.* **8**,1010–1020 (1997).
50. S. Liu, J. Sjövall, and W. J. Griffiths, Analysis of oxosteroids by nano-electrospray mass spectrometry of their oximes, *Rapid Commun. Mass Spectrom.* **14**, 390–400 (2000).
51. F. Guan, L. R. Soma, et al., Collision-induced dissociation pathways of anabolic steroids by electrospray ionization tandem mass spectrometry, *J. Am. Soc. Mass Spectrom.* **17**, 477–489 (2006).
52. W. J. Griffiths, S. Liu, G. Alvelius, and J. Sjövall, Derivatization for the characterization of neutral steroids by electrospray ionization and MALDI tandem mass

spectrometry: the Girard P derivative, *Rapid Commun. Mass Spectrom.* **17**, 924–935 (2003).

53. K. B. Tomer, N. J. Jensen, M. L. Gross, and J. Whitney, Fast atom bombardment combined with tandem mass spectrometry for determination of bile salts and their conjugates, *Biomed. Environ. Mass Spectrom.* **13**, 265–272 (1986).

54. Y. Yang, W. J. Griffiths, H. Nazer, and J. Sjövall, Analysis of bile acids and bile alcohols in urine by capillary liquid chromatography–mass spectrometry using fast atom bombardment or electrospray ionization and collision induced dissociation, *Biomed. Chromatogr.* **11**, 240–255 (1997).

55. N. Mano et al., Presence of protein bound unconjugated bile acids in cytoplasmic fraction of rat brain, *J. Lipid Res.* **45**, 295–300 (2004).

56. B. S. Middleditch and D. M. Desiderio, Mass spectra of prostaglandins, IV: Trimethylsilyl derivatives of prostaglandins of the F series, *Anal. Biochem.* **55**, 509–526 (1973).

57. J. A. Zirrolli, E. Davoli, L. Bettazzoli, M. L. Gross, and R. C. Murphy, Fast atom bombardment and collision-induced dissociation of prostaglandins and thromboxanes: some examples of charge remote fragmentation, *J. Am. Soc. Mass Spectrom.* **1**, 325–335 (1990).

58. J. Hankin, P. Wheelan, and R. C. Murphy, Identification of novel metabolites of prostaglandin E2 formed by isolated rat hepatocytes, *Arch. Biochem. Biophys.* **340**, 317–330 (1997).

59. R. C. Murphy et al., Electrospray ionization and tandem mass spectrometry of eicosanoids, *Anal. Biochem.* **346**, 1–42 (2005).

60. M. A. Shirley and R. C. Murphy, Novel 3-hydroxylated leukotriene B4 metabolites from ethanol-treated rat hepatocytes, *J. Am. Soc. Mass Spectrom.* **3**, 762–768 (1992).

61. L. J. Deterding, J. F. Curtis, and K. B. Tomer, Tandem mass spectrometric identification of eicosanoids: leukotrienes and hydroxyeicosatetraenoic acids, *Biol. Mass Spectrom.* **21**, 597–609 (1992).

62. P. Wheelan, J. A. Zirrolli, and R. C. Murphy, Negative ion electrospray tandem mass spectrometric structural characterization of leukotriene B4 (LTB4) and LTB4-derived metabolites, *J. Am. Soc. Mass Spectrom.* **7**, 129–139 (1996).

63. W. J. Griffiths, Y. Yang, J. Sjövall, and J. A. Lindgren, Electrospray tandem mass spectrometry of cysteinyl leukotrienes, *Rapid Commun. Mass Spectrom.* **10**, 1054–1070 (1996).

64. J. M. Hevko and R. C. Murphy, Electrospray ionization and tandem mass spectrometry of cysteinyl eicosanoids: leukotriene C_4 and FOG_7, *J. Am. Soc. Mass Spectrom.* **12**, 763–771 (2001).

65. X. Han and R. W. Gross, Shotgun lipidomics: electrospray ionization mass spectrometric analysis and quantitation of cellular lipidomes directly from crude extract of biological samples, *Mass Spectrom. Rev.* **24**, 367–412 (2005).

66. X. Han and R. W. Gross, Shotgun lipidomics: multidimensional MS analysis of cellular lipidomes, *Expert Rev. Proteom.* **2**, 253–264 (2005).

67. D. Schwudke, J. Oegema, et al., Lipid profiling by multiple precursor and neutral-loss scanning driven by data-dependent acquisition, *Anal. Chem.* **78**, 585–595 (2006).

CHAPTER 13

STRUCTURE DETERMINATION OF OLIGONUCLEOTIDES

Nucleic acids, deoxyribonucleic acid (DNA), and ribonucleic acid (RNA) are well-known molecules that have important functional ramifications in a variety of fundamental biological processes. DNA is a crucial component of all life forms and is the carrier of genes. It has been called a storehouse of genetic information that is passed on from one generation to another. An entire complement of organism's genes is called a *genome*, and the scientific study of a genome and the roles that genes play is referred to as *genomics*. RNA is the initial product of all genes. The flow of genetic information from DNA occurs via RNA. One form of RNA, *messenger RNA* (mRNA), acts as a template to convey the genetic information from DNA for the purpose of synthesizing a range of proteins (called *proteomes*) that are required to maintain essential functions of cells and organisms. Two other forms of RNA are *ribosomal RNA* (rRNA) and *transfer RNA* (tRNA). Analogous to *genomics*, the term *transcriptomics* seeks to unravel all of the cellular messenger RNA transcripts of an organism, whereas the term *ribonomics* has been used to describe the subset of mRNAs that bind with proteins. Research on the genome is an important field of study because it helps us to understand and to treat genetic diseases. DNA testing has already found widespread use in forensic identity, paternity testing, cell-line identification, and characterization of impaired genes.

13.1. STRUCTURES OF NUCLEOTIDES AND OLIGONUCLEOTIDES

The molecular structure of DNA was unraveled in 1953 by Watson and Crick [1]. DNA is a linear polymer of nucleotides (nt) arranged in a doubly stranded helical

---

Fundamentals of Contemporary Mass Spectrometry, by Chhabil Dass
Copyright © 2007 John Wiley & Sons, Inc.

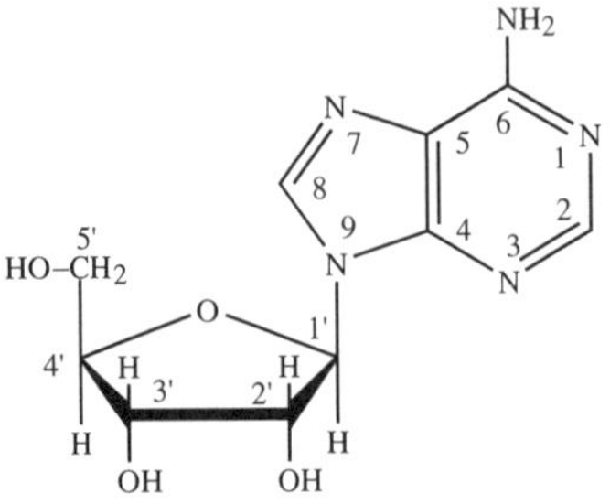

Nucleoside, adenosine

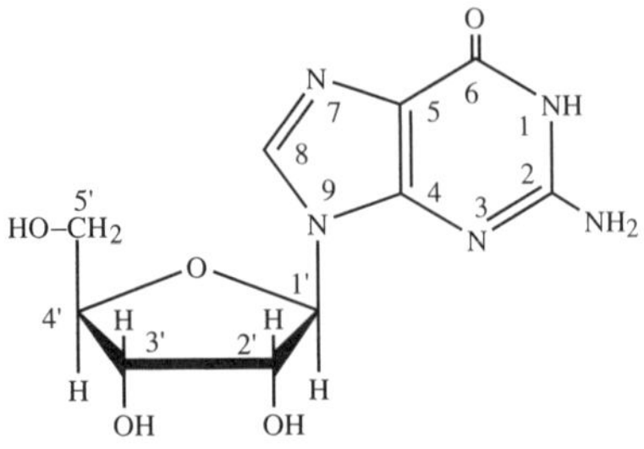

Nucleoside, guanosine

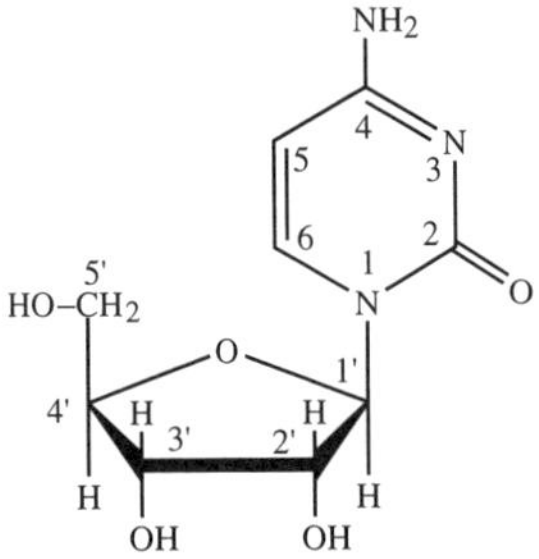

Nucleoside, cytidine

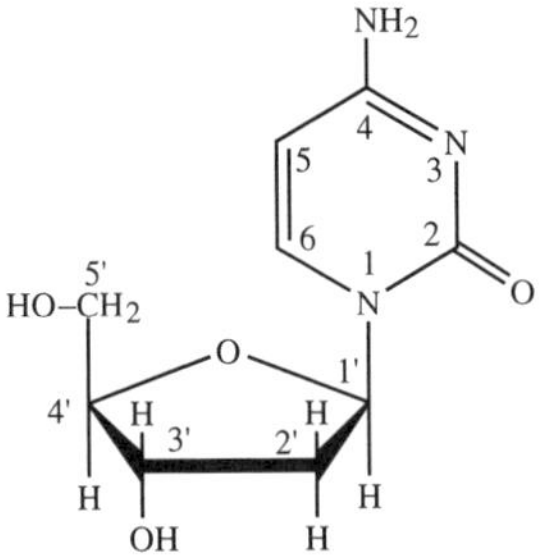

Nucleoside, deoxycytidine

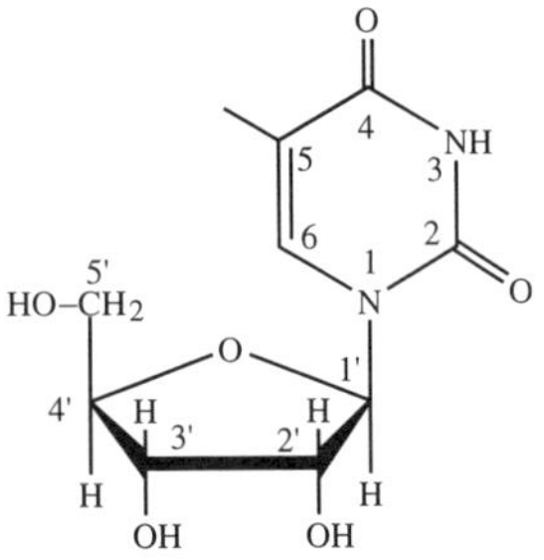

Nucleoside, thymidine

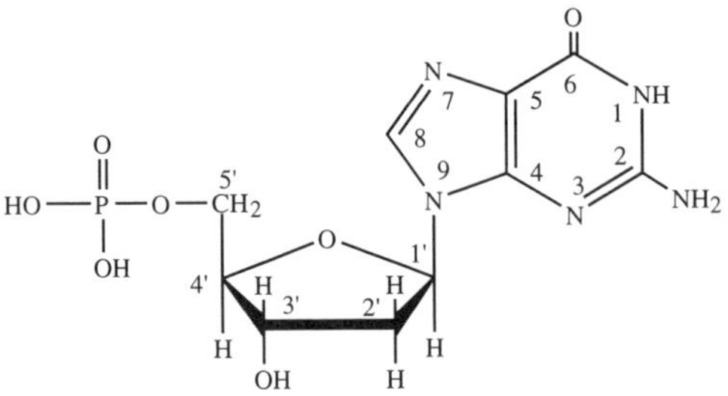

A nucleotide, deoxyguanosine monophosphate

Figure 13.1. Structures of nucleosides and nucleotides.

pattern (i.e., DNA is an oligonucleotide). Each nucleotide is composed of three simpler units: a heterocyclic base, a sugar, and a phosphate group (Figure 13.1). The DNA bases are identified as adenine (A), guanine (G), cytosine (C), and thymine (T); the first two (A and G) are derivatives of purine and the last two of pyrimidine (Figure 13.2). The sequence of these four bases in a DNA molecule is the determinant of genetic information that DNA molecule encodes. The sugar component in DNA is deoxy-D-ribose. A nucleotide is formed by joining the sugar to the N1 of the pyrimidine or the N9 of the purine base via the glycosidic C1′ carbon (Figure 13.1). The phosphate group is linked to the sugar via $-CH_2OH$ (i.e., at the 5′ position of the sugar residue). A nucleoside is the

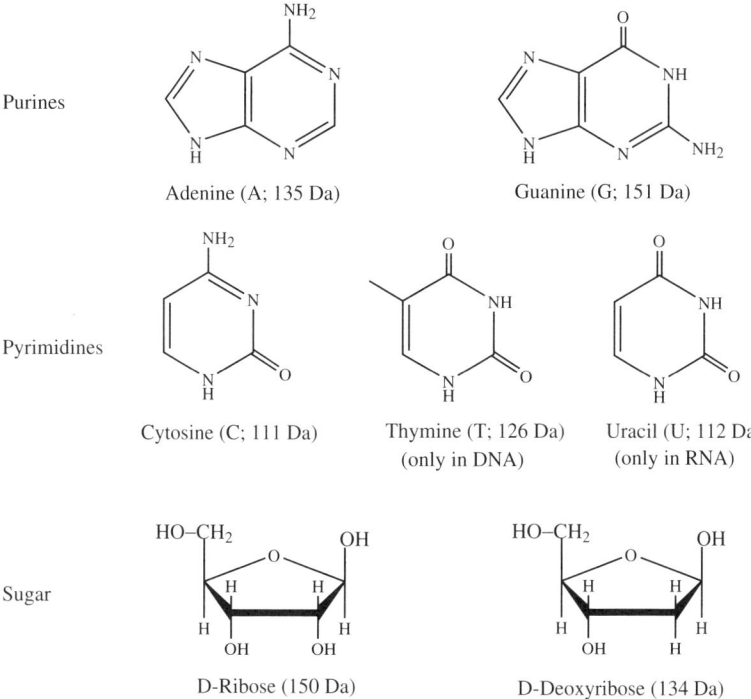

Figure 13.2. Structures of bases and sugars that are present in the DNA and RNA molecules.

product of the combination of a sugar and a base (Figure 13.1). RNA is also an oligonucleotide but differs from DNA in that it contains ribose as a sugar glycone (hence the name *ribonucleic acid*) and that thymine in DNA is replaced by uracil (U) in RNA (i.e., the RNA bases are A, G, C, and U). Table 13.1 provides a list of the common bases, nucleosides, and nucleotides that are present in DNA and RNA.

In an oligonucleotide chain, the sugar glycones act as the backbone and nucleobases (aglycones) as the side chain residues. The chain is elongated by connecting the 3'-carbon of one sugar to the 5'-of the next sugar via the phosphodiester bonds (see Figure 13.3). Thus, the backbone of a nucleic acid terminates in either a 3'- or a 5'-OH moiety. By convention, the primary sequence of oligonucleotides is read from the 5'-end. An oligonucleotide chain of a specific length is referred to as an *n-mer* (*n* is the number of nucleotide units). Shorthand notations have been coined to represent an oligonucleotide chain. In one of the formats (Figure 13.3), the sugar is denoted by a vertical line, the base by its letter abbreviation, and the phosphodiester bridge by the letter P either as such or within a circle. In another format, the sequence of an oligonucleotide is written in the letter format, starting with the 5'base. For example, the oligonucleotide of Figure 13.3 is written as 5'-*d*(pGTCA)- 3' or simply, as *d*(GTCA).

STRUCTURE DETERMINATION OF OLIGONUCLEOTIDES

TABLE 13.1. Common Bases and Nucleotides Present in DNA and RNA

Base		Nucleotide	
Name (Formula)	Monoisotopic; Mass (Average)	Name; (Formula); (Monoisotopic Mass)	Residue Mass Monoisotopic (Average)
DNA			
Adenine (A) ($C_5H_5N_5$)	135.054 (135.13)	Deoxyadenosine 5′-monophosphate (dAp) ($C_{10}H_{14}N_5O_6P$) (331.068)	313.058 (313.210)
Cytosine (C) ($C_4H_5N_3O$)	111.043 (111.10)	Deoxycytidine 5′-monophosphate (dCp) ($C_9H_{14}N_3O_7P$) (307.057)	289.046 (289.185)
Thymine (T) ($C_5H_6N_2O_2$)	126.043 (126.11)	Deoxythymidine 5′-monophosphate (dTp) ($C_{10}H_{15}N_2O_8P$) (322.057)	304.046 (304.197)
Guanine (G) ($C_5H_5N_5O$)	151.049 (151.13)	Deoxyguanosine 5′-monophosphate (dGp) ($C_{10}H_{14}N_5O_7P$) (347.063)	329.052 (329.210)
RNA			
Adenine (A) ($C_5H_5N_5$)	135.054 (135.13)	Adenosine 5′-monophosphate (Ap) ($C_{10}H_{14}N_5O_7P$) (347.063)	329.052 (329.210)
Cytosine (C) ($C_4H_5N_3O$)	111.043 (111.10)	Cytidine 5′-monophosphate (Cp) ($C_9H_{14}N_3O_8P$) (323.052)	305.041 (305.184)
Uracil (U) ($C_4H_4N_2O_2$)	112.027 (112.09)	Uridine 5′-monophosphate (Up) ($C_9H_{13}N_2O_9P$) (324.036)	306.025 (306.169)
Guanine (G) ($C_5H_5N_5O$)	151.049 (151.13)	Guanosine 5′-monophosphate (Gp) ($C_{10}H_{14}N_5O_8P$) (363.058)	345.047 (345.209)

In the double-helical structure of the DNA, the two oligonucleotide chains are complementary in their base sequences and are arranged in opposite directions, with the sugar–phosphate backbone outside and the bases directed inside. A base on one chain is connected via a hydrogen bond to another base on the other chain to form a complementary base pair (e.g., adenine–thymine and guanine–cytosine are complementary base pairs). The joining of complementary strands in DNA is termed *hybridization*. The stability of the double-stranded complexes increases with the length of the complementary sequence.

▶ **Example 13.1** Calculate the monoisotopic molecular mass of an oligonucleotide 5′-*d*(AGTCACG)-3′.

Solution The monoisotopic molecular mass of an oligonucleotide = 17.003 + (the sum of the monoisotopic residue masses) + 17.003 − 95.961. Therefore, the

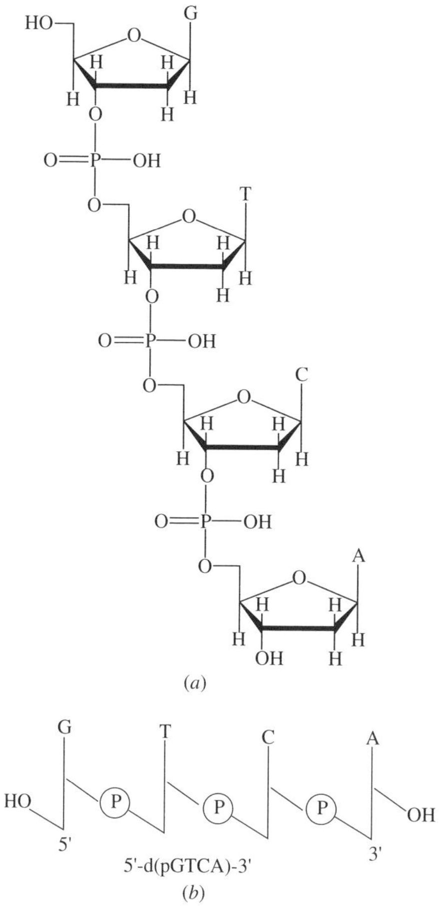

Figure 13.3. (*a*) Structure of a 4-mer DNA molecule, *d*(GTCA); (*b*) its shorthand notation.

monoisotopic molecular mass of 5′-*d*(AGTCACG)-3′ = 17.003 + (313.058 + 329.052 + 304.046 + 289.046 + 313.058 + 289.046 + 329.052) + 17.003 − 95.961 = 2104.403 Da.

13.2. MASS SPECTROMETRY ANALYSIS OF NUCLEOSIDES AND NUCLEOTIDES

Nucleotides are the biosynthetic precursors and also the metabolic products of oligonucleotides. The low thermal stability and highly polar nature of nucleosides and nucleotides have posed a formidable challenge to their analysis by

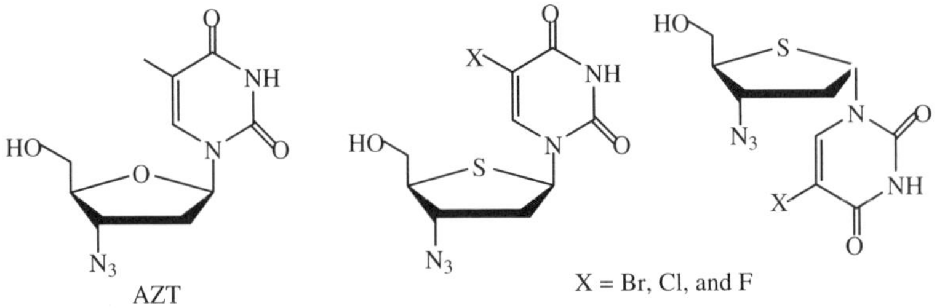

Figure 13.4. Structures of AZT and 3′-azido-2′,3′,4′-trideoxy-5-halogeno-4′-thio-β-D-uridine nucleosides.

mass spectrometric techniques. A common approach has been to derivatize these molecules by trimethylsilylation prior to their analysis by gas chromatography (GC)/MS with electron ionization (EI) or chemical ionization (CI) as a means of ionization [2,3]. Similar to other biopolymers, direct analysis of oligonucleotides is not feasible by GC/MS. Larger oligomers need to be degraded to constituent nucleosides by digestion with an enzyme nuclease P1, followed by alkaline phosphatase treatment [4]. Liquid chromatography (LC)/MS [5], fast atom bombardment (FAB)–MS [6], and electrospray ionization (ESI)–MS have largely replaced GC/MS because the derivatization step is not required. A representative example of the use of ESI, in conjunction with collision-induced dissociation (CID)–MS/MS, is the structural characterization of AZT and a series of 3′-azido-2′,3′,4′-trideoxy-5-halogeno-4′-thio-β-D-uridine nucleosides shown in Figure 13.4 [7]. These nucleosides are potent inhibitors of human immunodeficiency virus (HIV).

13.3. CLEAVAGE OF OLIGONUCLEOTIDES

For practical reasons, oligonucleotides must be cleaved to provide smaller components for mass spectrometric analysis [4]. Enzymatic hydrolysis and chemical digestion are effective means to degrade these biopolymers. Enzymatic hydrolysis uses base-specific endonucleases, nonspecific endonucleases, or exonucleases. An example of the base-specific endonucleases includes Rnase $T1$, which preferentially cleaves RNA at the 3′-side of all guanine residues to generate 3′-Gp-terminating oligonucleotides. Rnase U_2 is active at the 3′-side of all purine residues, but under specific conditions it can preferentially cleave at all adenine residues to generate oligonucleotides that terminate in 3′-Ap. Benzonase, nuclease $P1$, and phosphodiesterase I are nonspecific endonucleases. Exonucleases can cleave an oligonucleotide from either the 5′- or 3′-end. Calf spleen phosphodiesterase (CSP) is an exonuclease that cleaves the DNA in a stepwise manner from the 5′-end. Snake venom phosphodiesterase (SVP) is also an exonuclease, but it attacks the DNA at the 3′-end.

13.4. MOLECULAR MASS DETERMINATION OF OLIGONUCLEOTIDES

Molecular mass determination is an important component of oligonucleotide characterization. Knowledge of molecular mass can solve issues related to confirmation of the sequence of synthetic oligonucleotides, sizing of DNA fragments, detection of mutated genes, mapping of DNA sequencing reactions, mass measurement of polymerase chain reaction (PCR)–amplified products, screening of nucleoside oligomers, and determination of the base composition of oligonucleotides. ESI and matrix-assisted laser desorption/ionization (MALDI) are the two most appropriate modes of ionization frequently employed to accomplish these objectives.

13.4.1. Electrospray Ionization for Molecular Mass Determination

In the ESI mode, the molecular mass of oligonucleotides can be determined by observing the envelope of multiply charged ions. Although positive- and negative-ion ESI can both be used, negative-ion ESI-MS yields a better signal. In this mode, the spectrum contains a series of ions of the type $[M - nH]^{n-}$. The maximum observed charge state is equal to the number of phosphodiester bridges, or one less than the number of bases in an oligonucleotide. Because the phosphate group of the sugar unit and nucleobase aglycone contain metal ion–binding sites, clusters of the type $[M - (n+m)H + mNa/K]^{n-}$ are also observed. These metal ion clusters are detrimental in ESI–MS analysis because they reduce the analyte ion signal and adversely affect mass measurement accuracy.

Practical Considerations Careful control of instrument parameters and sample purity is essential for the success of ESI–MS analysis of oligonucleotides. The following approaches have been used successfully to purify oligonucleotide samples and suppress the alkali metal adduction:

1. Oligonucleotides are precipitated from an ammonium acetate–containing ethanol solution [8]. Upon precipitation, the alkali cations are replaced with ammonium ions to provide a significant improvement in signal intensity. A 48-mer oligonucleotide was analyzed by ESI–MS after this precipitation step, and the mass spectrum did not show any Na^+ ion clustering [8].

2. The addition of chelating agents such as *trans*-1,2-diaminocyclohexane-N,N,N',N'-tetraacetic acid (CDTA) and/or triethylamine (TEA), removes the bound transition-metal cations and Mg^{2+} ions from RNAs and enhances mass measurement accuracy [9]. Figure 13.5, the ESI spectrum of *E. coli* 5S rRNA (120 nt), demonstrates the utility of this approach. Before acquiring the ESI mass spectrum, the sample of rRNA was precipitated from 2.5 M ammonium acetate solution, and 500 pmol of CDTA and 10 µL of 0.1% TEA solution were added. Two major rRNA components were detected. In the absence of CDTA and TEA, no useful data could be obtained.

460 STRUCTURE DETERMINATION OF OLIGONUCLEOTIDES

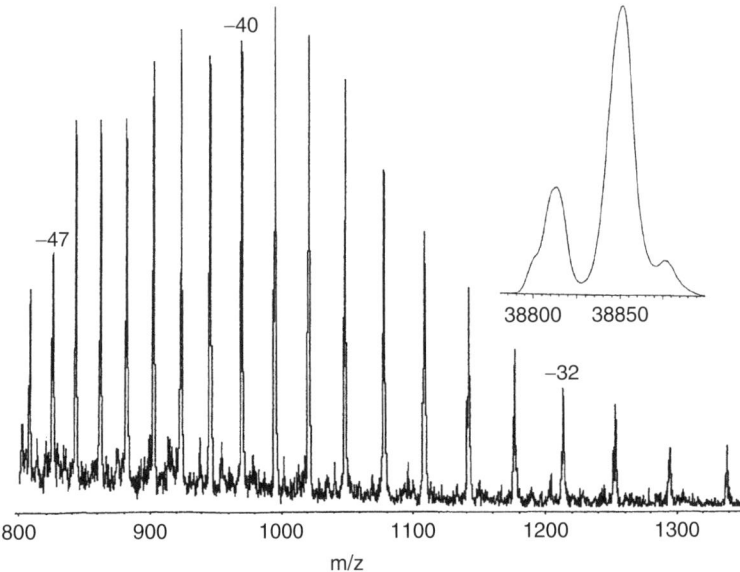

Figure 13.5. ESI mass spectrum of *E. coli* 5S rRNA (120 nt). Inset is the MaxEnt-derived molecular mass spectrum. (Reproduced from ref. 9 by permission of Elsevier Science, Copyright © 1995 American Society for Mass Spectrometry.)

3. The addition of strong organic bases such as trimethylamine (TMA), TEA, diisopropylamine, piperidine, or imidazole in place of ammonium acetate is another effective means to suppress the signal from alkali-adducted ions [10–12]. This strategy is particularly beneficial for phosphorothioate oligomers [10].

4. RP–HPLC either alone or in a mixed-mode format can effectively remove salts from oligonucleotide samples to provide an improvement in spectral quality [13,14]. When RP–HPLC is combined with FT–ICRMS, the mass of oligomers can be measured within ±0.001% accuracy.

5. Microdialysis is also a feasible option for purifying DNA fragments [15]. In this procedure the sample is injected into a regenerated hollow cellulose fiber and a countercurrent flow of the dialysis solvent is maintained through the annular space between a Teflon tube and the cellulose fiber.

▶ **Example 13.2** (a) At what m/z values do the 5−, 6−, 7−, and 8− charge states of an oligonucleotide 5′-d(ATGCCATGCGCAT)-3′ appear in a negative-ion ESI mass spectrum? (b) What maximum charge state is expected from this oligonucleotide?

Solution (a) The molecular mass of the oligonucleotide $= 17.00 + (3 \times A + 3 \times T + 3 \times G + 4 \times C) + 17.00 - 95.96 = 17.00 + (3 \times 313.06 + 3 \times 304.05$

$+ 3 \times 329.05 + 4 \times 289.05) + 17.00 - 95.96 = 3932.72$ Da.

$$m/z = \frac{(M - nH)^{n-}}{n}$$

Therefore,

$$m/z(5-) = \frac{(3932.72 - 5 \times 1.01)^{5-}}{5} = 785.53$$

$$m/z(6-) = \frac{(3932.72 - 6 \times 1.01)^{6-}}{6} = 654.44$$

$$m/z(7-) = \frac{(3932.72 - 7 \times 1.01)^{7-}}{7} = 560.81$$

$$m/z(8-) = \frac{(3932.72 - 8 \times 1.01)^{8-}}{8} = 490.58$$

(b) Because the oligonucleotide contains 12 phosphodiester bridges (or 13 bases), the maximum charge state observed is 12.

13.4.2. Matrix-Assisted Laser Desorption/Ionization for Molecular Mass Determination

Although MALDI has been highly successful in the analysis of proteins and peptides, its application to the field of nucleic acids has faced many challenges. The lack of an appropriate matrix, interference of sample impurities, and unexpected fragmentation of oligonucleotides upon bombardment with a laser beam are some of the hurdles that must be surmounted.

Choice of a Proper Matrix Sinapinic acid and 2,5-dihydroxybenzoic acid (DHB) are highly successful matrices in the analysis of proteins and peptides but are of little use for oligonucleotides. In the first application of MALDI–MS, oligonucleotides of only four to six bases could be detected [16]. Since then, several new matrix materials have emerged that have changed the scenario of MALDI analysis of oligonucleotides. Notable among these matrices is 3-hydroxypicolinic acid (HPA), which currently has become a standard matrix for the analysis of oligonucleotides [17]. Picolinic acid (PA) is also equally promising [18]. Other useful matrices are 3-aminopicolinic acid (APA) [19], 2′,4′,6′-trihydroxyacetophenone (THAP) [20], and 6-aza-2-thiothymine (ATT) [21a]. 3-Hydroxycoumarin is another matrix that has shown good results in the analysis of oligonucleotides [21b]. Large DNA oligomers (over one kilobase) have been analyzed effectively with 4-nitrophenol as the matrix, but it requires a cooled sample stage to prevent sublimation of the matrix [22]. Figure 13.6 shows the MALDI mass spectra of a single-stranded (ss)-oligonucleotide (468-mer with MM of 144,755 Da) and a double-stranded (ds)-oligonucleotide of over one kilobase pairs to demonstrate the usefulness of this approach. The ds-oligonucleotide is

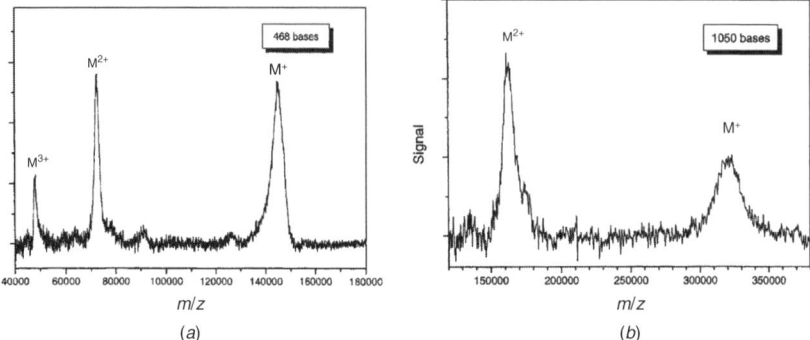

Figure 13.6. Positive-ion MALDI mass spectrum of (*a*) a 468-mer ss-oligonucleotide and (*b*) of a ds-oligonucleotide with 1050 base pairs. (Reproduced from ref. 22 by permission of Wiley-Interscience, copyright © 1999.)

seen as a denatured ss-species, with a molecular ion mass greater than 300 kDa. A 2:1 mixture of an anthranilic acid and nicotinic acid mixture has proven to be successful in the detection of oligonucleotides with excellent sensitivity [23]. Succinic acid and urea have been successful in IR–MALDI experiments [24].

Control of Fragmentation During MALDI Analysis Fragmentation of the oligomers during MALDI is a major reason for the limited mass range of MALDI analysis of DNA. Because fragmentation of the oligonucleotide chain is initiated by protonation of the base (followed by cleavage of the N-glycosidic bond, i.e., loss of the base and cleavage in the backbone at the 3'-CO bond) and because the proton affinity of the constituent bases differs, MALDI mass spectral response is a strong function of the base composition. For example, the proton affinity of thymine relative to the other three DNA bases is low. As a consequence, thymine is more stable with respect to the base loss and backbone fragmentation. RNA is more stable than DNA under MALDI conditions.

By manipulation of the matrix and laser wavelength, fragmentation of oligomers can be reduced significantly. PA and 3-HPA are both effective in this respect. These matrices have helped to analyze intact oligonucleotides with as many as 150 bases [18, 25]. The use of comatrices and addition of ammonium acetate [26] and organic bases [27] have also improved the extent of ion production and the stability of gas-phase oligonucleotide ions. High proton affinity of an organic base can serve as a proton sink to reduce fragmentation of oligonucleotides [28]. A mixture of 3-HPA with PA or diammonium citrate is currently the matrix of choice for analyzing larger oligonucleotides. The use of IR–MALDI in place of UV–MALDI also helps to control fragmentation and to increase the mass range of MALDI analysis. For example, with an optimized matrix–laser wavelength combination, the IR–MALDI mass spectra of a synthetic DNA, restriction enzyme fragments of plasmid DNA, and RNA transcripts up to 2180 nt have been reported [29].

Sample Preparation for MALDI Analysis   Sample impurity is also an issue in the MALDI analysis of oligonucleotides. Certain impurities and additives can reduce the signal intensity. Similar to ESI–MS, the adduction with cations (e.g., Na^+ or K^+) results in peak broadening and poor resolution. In addition, some impurities might adversely influence the matrix-crystallization process. Several purification schemes have been developed to eliminate sample impurities. Some approaches are similar to those discussed above for ESI–MS analysis. For example, alkali metal adduction is suppressed by the addition of chelating agents and ammonium salts such as ammonium acetate, diammonium citrate, or diammonium tartrate [20]. Use of an H^+ ion-exchange resin in situ permits the removal of alkali metal ions and results in substantial gain in sensitivity [30]. Ultrafiltration membranes can also be used to remove impurities [31].

Some approaches rely on extraction of oligomers from a sample solution onto a solid surface. In one such approach, a glass resin binds the DNA molecules in the presence of a chaotropic agent [31]. After washing the unbound buffer components, the bound DNA is eluted from the glass beads with deionized water. In another approach, NH_4^+-loaded cation exchange polymer beads are used to remove metal cations from the sample solution [24]. A solid surface with an attached DNA-binding polymer, such as polyethylene amine or poly(vinylpyrrolidone), can also be used effectively to extract oligonucleotides [32]. The extracted DNA is released from the surface by the addition of a MALDI matrix.

The removal of impurities from DNA samples can also be accomplished by using modified films of Nafian, nitrocellulose, or poly(diallyldimethylammonium chloride) as active substrates in place of normal metal MALDI sample supports [33–35]. These active surfaces are prepared by spreading the solution of a pure membrane on the stainless steel probe tip, followed by air drying. Impregnation of the active surface with ammonium hydroxide further improves the detection of large DNA molecules. With active substrates, the on-probe purification of nucleic acid samples is readily achieved [34]. The use of an active poly(diallyldimethylammonium chloride) substrate to analyze a mixture of oligonucleotides (TATTGCTTTAAAAACTCAAAA = QZ2 and TTTTGAGTT TTTAAAGCAATA = QZ1) is demonstrated in Figure 13.7.

13.4.3. Base Composition from an Accurate Mass Measurement

One of the important applications of the molecular mass measurement is to confirm the base composition of oligonucleotides. This approach is relatively less demanding for oligonucleotides because nucleic acids are composed of only four distinct types of base pairs. An extensive set of calculations has shown that compositions can be derived for up to 5-mer oligonucleotides, if the mass measurement value is known within ±0.01% [36]. The determination of the composition can be extended to larger oligonucleotides when additional constraints, such as knowledge of the number of residues, compositional limits, and limitation of compositional isomers, are applied and the mass measurement accuracy is improved further. As a specific example, if the chain length is known

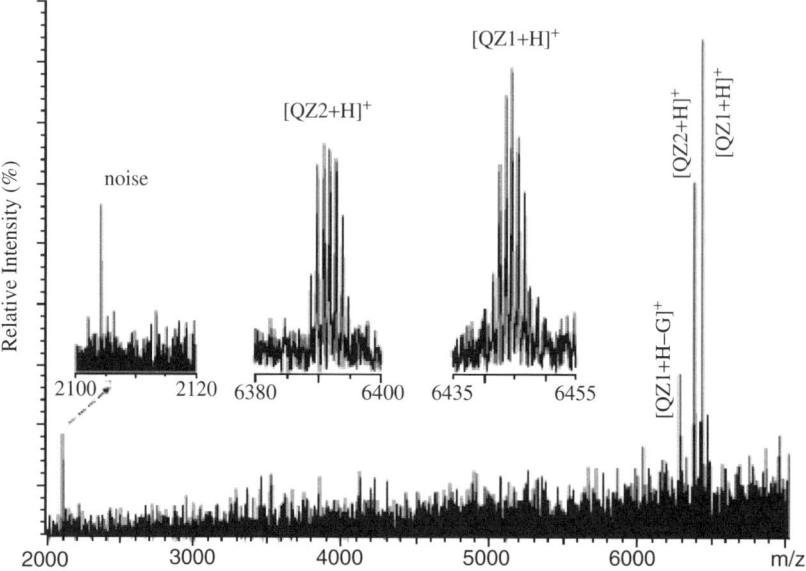

Figure 13.7. AP-MALDI mass spectrum of two 21-mer oligonucleotides, QZ1 and QZ2. (Reproduced from ref. 35 by permission of Elsevier Science, copyright © 2005 American Society for Mass Spectrometry.)

and the molecular mass is measured within an accuracy of better than 0.01%, compositional analysis is possible up to the 25-mer level. The increased mass measurement accuracy of FT–ICR–MS has been used effectively to determine the base composition of double-stranded oligonucleotides [37].

13.5. MASS SPECTROMETRY SEQUENCING OF OLIGONUCLEOTIDES

Mass spectrometry sequencing of oligonucleotides is not as common a practice as it is for proteins and peptides. However, the following mass spectrometry–based sequence determination strategies have gained acceptance:

- Gas-phase fragmentation for oligonucleotide sequencing
  - ESI in-source CID
  - ESI-infrared multiphoton dissociation (IRMPD)
  - MALDI in-source decay
  - MALDI post-source decay
  - CID–MS/MS
  - Ion–ion reactions
- Solution-phase techniques for oligonucleotide sequencing
  - Ladder sequencing

- Chemical cleavage coupled with mass spectrometry
- Sanger sequencing coupled with mass spectrometry

13.5.1. Gas-Phase Fragmentation for Oligonucleotide Sequencing

Several gas-phase sequencing techniques have attained success; all rely on the generation and mass analysis of sequence-specific fragment ions [38–40]. These methods have the advantage of simplicity and are potentially faster than solution-phase methods.

Gas-Phase Fragmentation Behavior and Fragmentation Nomenclature
The oligonucleotide backbone can cleave at four possible places in the phosphodiester bridge: two phosphoester bonds and two P—O bonds. In addition, the loss of a base is also observed. McLuckey and colleagues have proposed a systematic nomenclature (depicted in Figure 13.8), similar to that used for peptides, to designate these fragments [41]. The four possible fragments are named *a*, *b*, *c*, and *d* if the charge is retained by the 5′-terminus. The corresponding 3′-terminus fragments are called *w*-, *x*-, *y*-, and *z*-ions. A base is indicated as B_n, where n is the base position from the 5′-terminus. The loss of a specific base from a fragment ion is indicated in parentheses. For example, $[a_3 - B_3(T)]$ indicates an *a*-type ion at the 3-position from which thymine at the 3-position has been lost. Similarly, $[z_2 - B_3(T)]$ indicates a *z*-type ion at the 2-position, from which thymine at the 3-position has been lost.

Several ion activation techniques have been used to determine the gas-phase fragmentation behavior of oligonucleotides [41–49]. These studies have concluded that the loss of a base is a dominant fragmentation, and occurs via a

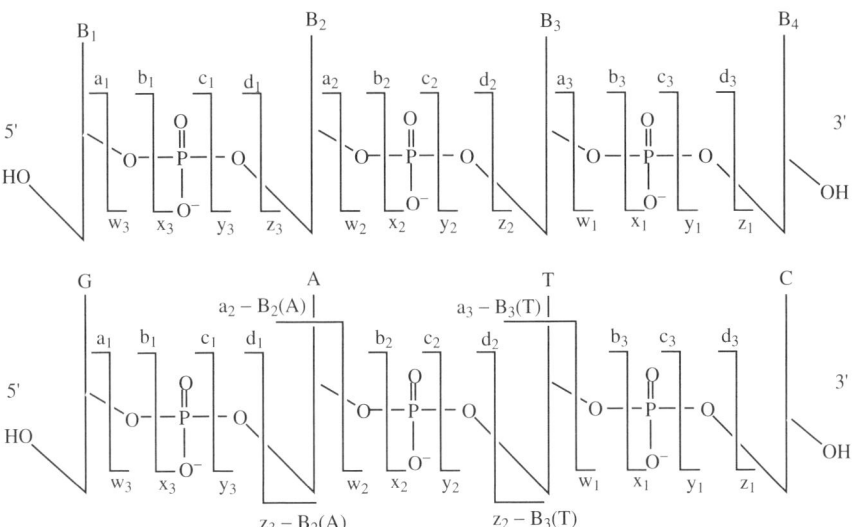

Figure 13.8. Nomenclature for the CID product ions. (From ref. 41.)

STRUCTURE DETERMINATION OF OLIGONUCLEOTIDES

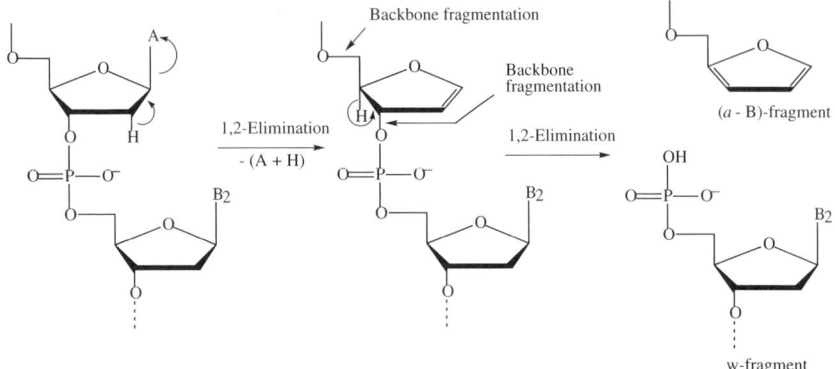

Figure 13.9. Mechanism for the loss of a base and subsequent cleavage of the 3' C—O bond. (Reproduced from C. Dass, *Principles and Practice of Biological Mass Spectrometry*, Wiley-Interscience, 2001.)

1,2-elimination reaction (Figure 13.9). If the charge resides on the base, the loss of that base occurs as an anion; otherwise, a neutral base is lost. Following the loss of the base, the 3'-C—O bond that once held the lost base is cleaved to yield complementary $(a - B_n)$- and w-type ions. These cleavages also involve the transfer of two hydrogen atoms, one to the base lost and the other to the complementary w-ion. In some cases, cleavage of the 5'-C—O bond might also occur to form complementary $(z - B_n)$- and d-type ions. The preference for the loss of a base as an anion follows the order A > T > G > C. The m/z values of some important sequence-specific anions are calculated by using the following simple rules:

$$m/z \text{ of } w^- \text{ ions} = 17 + \text{the residue mass}$$
$$m/z \text{ of } y^- \text{ ions} = 17 + \text{the residue mass} - 80$$
$$m/z \text{ of } a^- \text{ ions} = 17 + \text{the residue mass} - 96$$
$$m/z \text{ of } d^- \text{ ions} = 17 + \text{the residue mass}$$
$$m/z \text{ of } (a - B)^- \text{ ions} = 17 + \text{the residue mass} - 96 - B$$
$$(= 134, 110, 125, \text{ or } 150 \text{ for A},$$
$$\text{C, T, or G, respectively}) - 3$$

▶ **Example 13.3** At what m/z values will the w_3^-, w_5^{2-}, $(a_3 - B_3)^-$, and d_2^- ions be observed in the CID mass spectrum of 5'-(ATCTCGATC)-3'?

Solution

$$m/z \text{ of } w_3^- \text{ ion} = 17 + \text{sum of the residue masses of C,}$$
$$\text{T, and A} = 923.15$$

$$m/z \text{ of } w_5{}^{2-} \text{ ion} = \frac{17 + \text{sum of the residue masses of C, T, A, G and C} - 1}{2}$$

$$= 770.12$$

m/z of $(a_3 - B_3)^-$ ion = 17 + sum of the residue masses of A,

T, and C $- 96 - 110 - 3 = 714.13$

m/z of $d_2{}^-$ ion = 17 + sum of the residue masses of A and T = 634.11

In any gas-phase procedure, oligonucleotide sequencing is accomplished either by accounting w-type ion series alone or $(a - B_n)$- and w-type complementary ion series both. These complementary ions can provide bidirectional sequencing. The $(a - B_n)$-ions build a sequence from $5' \rightarrow 3'$ direction, whereas the w-ions provide the sequence from the $3' \rightarrow 5'$ direction. The following techniques have been developed to sequence oligonucleotides.

ESI In-Source CID By adjusting the nozzle-skimmer voltage, fragmentation can be induced in the ESI source interface region via collisions of the target ions with the buffer-gas molecules. This approach can be used to sequence oligonucleotides with a single-stage mass spectrometer but requires a pure sample and sufficient mass resolution and accuracy to identify unambiguously the resulting fragments. In this respect, implementation of the nozzle-skimmer dissociation concept on an FT–ICR mass spectrometer has the advantage of high-resolution accurate-mass data [13,48]. A mass accuracy of greater than 50 ppm and a resolving power of 10^5 for a 25-mer oligonucleotide have been demonstrated with this protocol [13]. The use of nozzle-skimmer dissociation to sequence an 8-mer oligonucleotide is demonstrated in Figure 13.10.

ESI–Infrared Multiphoton Dissociation IRMPD of ESI-generated ions can also provide sequence information. By combining the data from the nozzle-skimmer dissociation and IRMPD experiments, nearly complete sequence coverage for larger oligonucleotides can be obtained [48]. These two dissociation techniques yield complementary sequence ions. For example, IRMPD primarily produces w- and $(a - B_n)$-type sequence ions, whereas the nozzle-skimmer dissociation generates b-, c-, and d-type ions.

MALDI Ion-Source Decay As mentioned earlier, oligonucleotides are less stable under MALDI conditions. This feature has been exploited to obtain the sequence of an oligonucleotide. Fragmentation of MALDI-generated ions in a TOF–MS instrument has been classified into four different time scales: prompt, fast, fast metastable, and metastable [38]. The formation of these fragment ions depends largely on the laser power, the type of matrix used, and the base composition of the oligonucleotide. An increase in laser pulse energy above the threshold for molecular-ion production is required to detect a reasonable number

468 STRUCTURE DETERMINATION OF OLIGONUCLEOTIDES

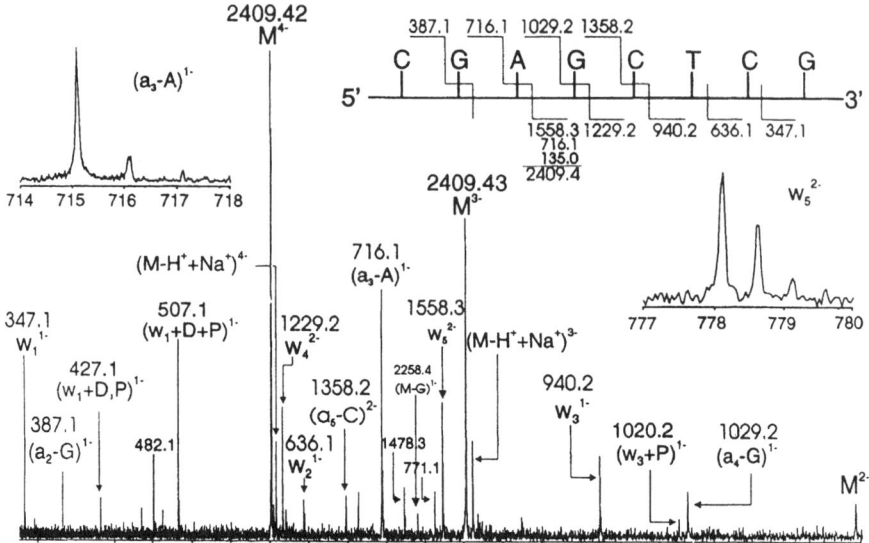

Figure 13.10. Nozzle-skimmer dissociation spectrum of an 8-mer oligonucleotide 5′-(CG AGCTCG)-3′. (Reproduced from ref. 48 by permission of the American Chemical Society, Washington, DC, copyright © 1994.)

of fast-dissociation ions. In principle, all four types of fragmentation can be used to provide sequence ion information for oligonucleotides. Prompt dissociation, which occurs within the time frame of a desorption event, can be monitored by a linear or reflectron TOF instrument to sequence oligonucleotides. Prompt dissociations of oligonucleotides have been observed with the IR laser/succinic acid and UV laser/2,5-DHBA combinations [50]. Fast dissociation occurs after the desorption event, but before an ion-acceleration step. The ions produced during fast-dissociation events can be mass-discriminated only with the delayed extraction (DE)–TOF–MS technique [51,52]. An 11-mer oligonucleotide 5′-d(CACACGCCAGT)- 3′ has been sequenced using this technique, with PA as the matrix and a 266-nm laser wavelength (see Figure 13.11). The spectrum contains mainly d-, w-, and y-type sequence ions. The MALDI-ion trap combination can also access these fast dissociations to sequence oligonucleotides [52,53]. The use of ionic liquid matrices, such as aniline–CHCA and N,N-diethylaniline–CHCA combinations, has proven advantageous for oligonucleotide sequencing [54]. These matrices provide an enhanced fragmentation of MALDI-formed ions of an oligonucleotide 5′-GGATTC-3′ in an FT–MS instrument as compared to that observed with HPA as the matrix.

MALDI Post-Source Decay The fast-metastable fragmentations have a decay-time constant on the order of the acceleration event and are considered a nuisance for the MALDI–TOF–MS analysis of large oligonucleotides. In contrast, the

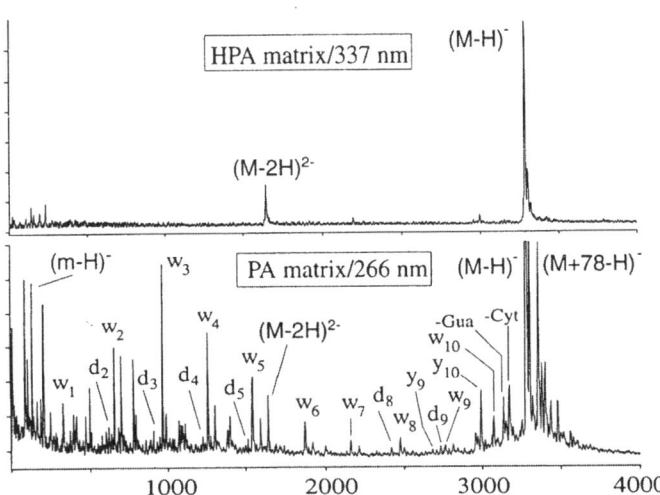

Figure 13.11. Sequencing of an 11-mer oligonucleotide, 5′-d(CACACGCCAGT)-3′ by DE-TOFMS. PA was used as a matrix. (Reproduced from ref. 52 by permission of the American Chemical Society, Washington, DC, copyright © 1996.)

field-free region metastable reactions are of great asset in sequencing oligonucleotides via the post-source decay (PSD) procedure [55]. The differentiation of isomeric photomodified oligonucleotides via PSD and ESI-MS/MS in a quadrupole ion trap has been discussed [56].

CID–MS/MS This approach has become a standard procedure in the structure elucidation of a wide variety of organic and biological compounds. In the past, FAB was used to produce molecular ions of nucleotides for subsequent MS/MS studies. High- and low-energy CID have both provided useful sequence information for small oligonucleotides [57]. The current applications rely heavily on CID of the ESI-generated multiply charged ions. These studies are performed ideally in a QIT instrument [41–44,58]. Sequencing of oligonucleotides with ESI–MS/MS is equally effective on triple-quadrupole [44,59,60], Q–TOF, and FT–ICR–MS [61] instruments. A typical example of oligonucleotide sequencing by FT–ICR–MS is shown in Figure 13.12, which is the MS/MS spectrum of an 8-mer oligonucleotide 5′-(CGAGCTCG)-3′ [48]. A triple quadrupole generally provides more complete sequence information than a QIT. The CID–MS/MS sequence strategy has been evaluated comparatively on Q–TOF, triple quadrupole, and QIT [62]. A Q–TOF provides more sequence-specific ions than the other two types of instrument. Also, assignment of the fragment-ion charge state is unambiguous on this instrument. Labeled oligonucleotides can also be sequenced with ESI–MS/MS, following their digestion with exonucleases [63].

Automatic Oligonucleotide Sequencing With an increase in the size of an oligonucleotide, the spectrum complexity also increases; therefore, the manual

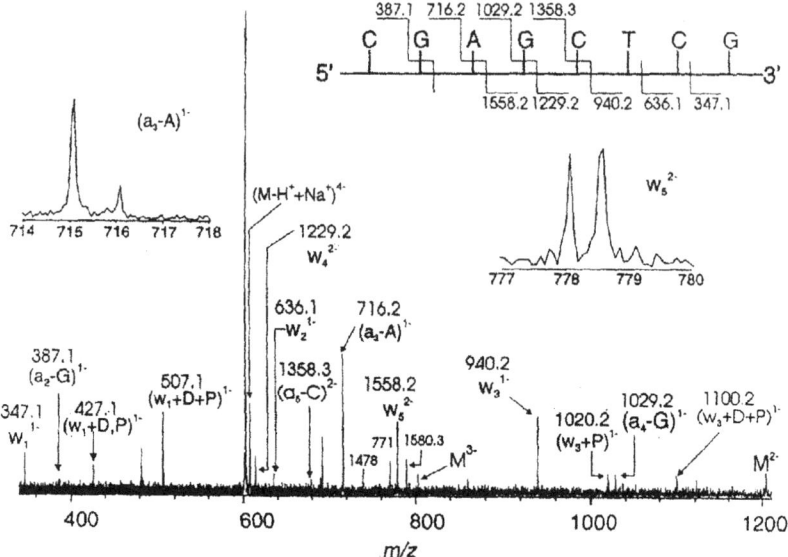

Figure 13.12. CID spectrum of an 8-mer oligonucleotide, 5'-(CGAGCTCG)-3'. (Reproduced from ref. 48 by permission of the American Chemical Society, Washington, DC, copyright © 1994.)

interpretation of the spectrum becomes painstaking and time consuming. Computer algorithms have been developed to interpret the MS/MS spectra automatically. One such approach is called a *simple oligonucleotide sequencing* (SOS) program [64]. In this program, the mass ladders are identified by sequentially adding each of the four possible nucleotide masses and searching the spectrum for the best match of ions expected. The sequence ion series, $(a - B_n)$- and w-ions, from both ends are considered. This algorithm can provide an automated determination of unknown sequences, but it is limited to the 20-mer level. Another program that has become available for computer-aided interpretation of CID spectra of ESI-produced multiply charged oligonucleotide ions is called *comparative sequencing* (COMPAS) [64–66]. This algorithm involves a comparison of the measured CID fragment ion spectrum to a set of m/z values that are predicted from a known reference sequence. Use of this algorithm to verify the sequence of an 80-mer oligonucleotide has been demonstrated.

Ion–Ion Reactions These reactions are conducted in ion trap–based instruments (e.g., in QIT or LIT) [67,68]. No external mode of ion activation is required in these exoergic reactions. The spectra obtained upon fragmentation of multiply charged precursors are usually too complex to interpret, but charge reduction by an ion–ion reaction simplifies the spectrum. Therefore, this approach can be applied to direct analysis of oligonucleotide mixtures [68]. For example, Figure 13.13 illustrates the analysis of a mixture that contains 14-, 17-, 18-, 19-, 20-, 22-, and 24-mer oligonucleotides. Charge reduction of the negatively

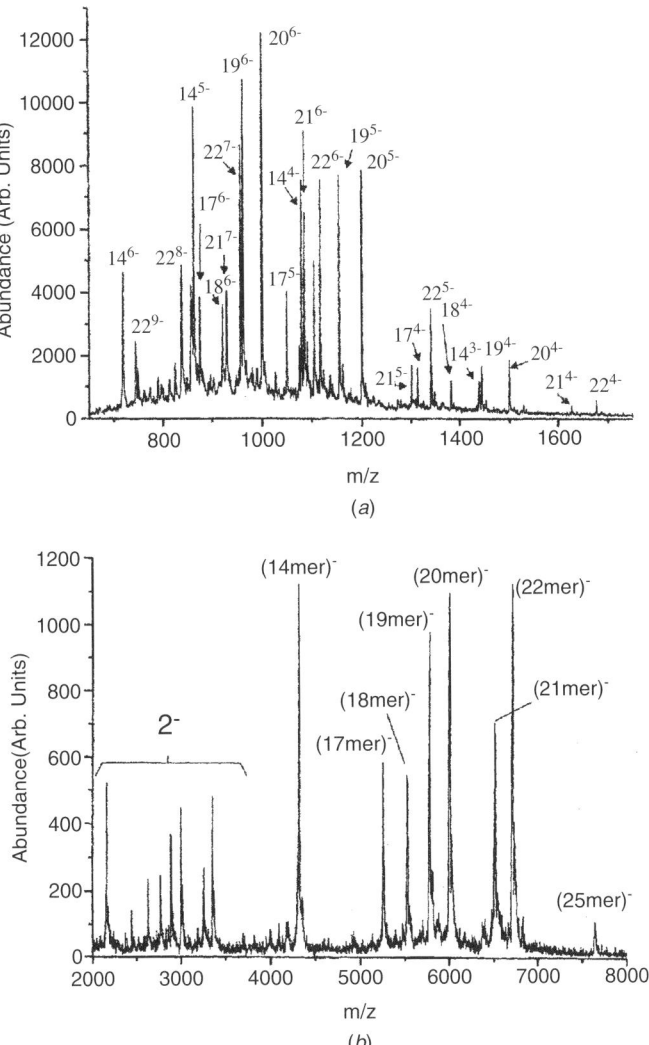

Figure 13.13. Negative-ion mass spectra of an eight-component oligonucleotide mixture: (*a*) nonreacted spectrum and (*b*) spectrum after ion/ion reaction with $O_2^{+\bullet}$. The ion/ion-reacted spectrum is much simpler then the nonreacted spectrum. (Reproduced from ref. 68 by permission of the American Chemical Society, Washington, DC, copyright © 2002.)

charged oligonucleotides by an ion–ion reaction with $O_2^{+\bullet}$ produced a much simpler spectrum than a nonreacted spectrum.

13.5.2. Solution-Phase Techniques for Oligonucleotide Sequencing

The cleavage of nucleic acids in solution with subsequent mass spectrometry detection of the digestion fragments is another possible option for sequencing oligonucleotides [39]. The following three techniques are commonly used for this purpose:

DNA Ladder Sequencing   The concept of ladder sequencing, which is well established in the field of protein and peptide sequencing, was first applied to oligonucleotide sequencing in 1993 [20]. An essential feature of this approach is the consecutive cleavage of an oligonucleotide to generate a ladder that contains a nested set of oligonucleotide fragments, followed by the mass measurement of each cleavage product. The sequence is deduced from the mass difference between successive peaks in the spectrum. MALDI–MS and ESI–MS are both highly useful to determine the mass of the ladders. Cleavage with an exonuclease such as snake venom phosphodiesterase (SVP or PDase I) and calf serum phosphodiesterase (CSP or PDase II) removes mononucleotides from an oligonucleotide sequentially by hydrolyzing the phosphodiester bond [69–73]. SVP attacks the 3′-terminus to generate 5′-ladders, whereas CSP cleaves the 5′-terminus to yield 3′-ladders. In oligodeoxyribonucleotides, mass differences of 289.046, 304.046, 313.058, or 329.052 Da that were observed indicates the presence of C, T, A, or G residue, respectively. With oligoribonucleotides, mass differences of 305.041, 306.025, 329.052, or 345.047 will be observed and will reveal the identity of C, U, A, or G, respectively. The concept of ladder sequencing is illustrated elegantly in Figure 13.14 with respect to the oligonucleotide 5′-ATCTCGATC-3′. Analysis of the sequence ladders using MALDI–DE-TOFMS has extended this approach to longer DNA segments because of the benefits of high resolution, increased mass accuracy, and increased detection sensitivity.

ESI–MS has also been used to measure the masses of the ladders [74]. However, the rapid buildup in concentration of nucleotides produces a strong nucleotide-cluster ion signal, such as dimers and trimers of C, G, T, and A. This process leads to suppression of the signal of the ladder ions. This experimental difficulty is especially critical with oligonucleotides that have more than 25 bases. This problem can be solved by performing the exonuclease digestion in the presence of an enzyme alkaline phosphatase, which converts nucleotides into noninterfering nucleosides [75,76]. An example of the use of this strategy is shown in Figure 13.15, which is the overlaid reconstructed molecular-weight spectrum of a 33-mer oligonucleotide, d(GCCAGGGTTTTCCCAGTCACGATGCAGAATTCA), from SVP and bovine alkaline phosphatase digestion [75].

Chemical Cleavage and Mass Spectrometry   The sequence ladders can also be generated by chemical digestion of oligonucleotides. Acid hydrolysis, base hydrolysis, and alkylation are commonly used chemical digestion reactions, but they are largely nonspecific [39]. In principle, all three types of bonds in the oligonucleotide backbone can be cleaved. For example, the glycosidic linkage is susceptible to acid hydrolysis. This reaction is more specific for DNA [77], whereas base hydrolysis is more specific for RNA and methylphosphonate-linked

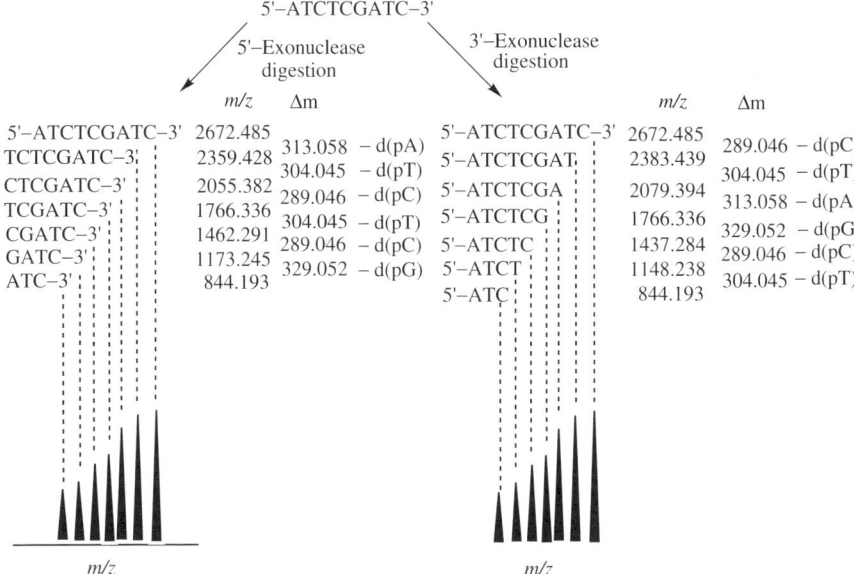

Figure 13.14. Basic concept of the ladder sequencing protocol. The oligonucleotide 5′-ATCTCGATC- 3′ is treated with SVP or CSP to generate 5′- or to generate 3′-ladders, respectively. (From ref. 73.)

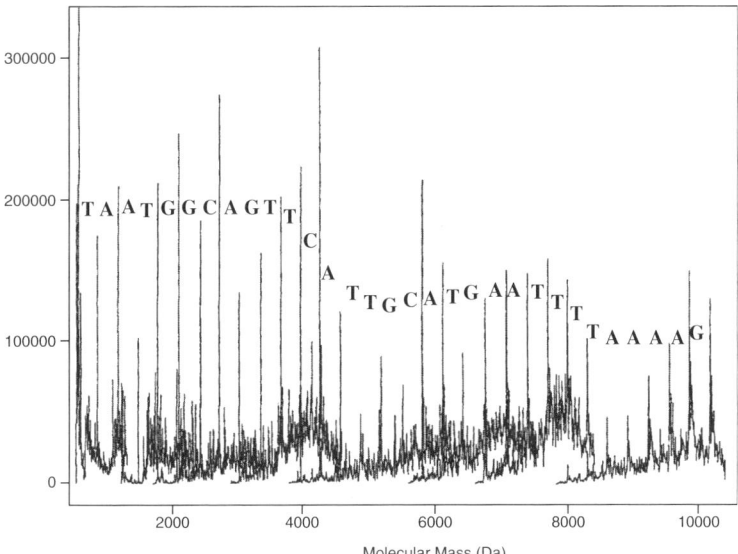

Figure 13.15. Example of the ladder sequencing method. Overlaid reconstructed molecular weight spectrum of a 33-mer oligonucleotide, d(GCCAGGGTTTTCCCAGTCACG ATGCAGAATTCA), from SVP and bovine alkaline phosphatase digestion. (Reproduced from ref. 75 by permission of Elsevier, copyright © 2001.)

oligonucleotides [69,78]. Alkylation is more specific for phosphorothioate-modified oligonucleotides; cleavage occurs at the site of modification. An example of the use of alkylation strategy is the sequencing of a 10-mer phosphorothioate oligonucleotide, d(GCATGACTCA) [79].

Maxam–Gilbert Protocol This method deploys the base-specific chemical cleavage of ^{32}P end-labeled DNA [80]. The reaction is performed in three steps: in the first step, the specific base is chemically modified; in the second step, the modified base is removed; in the third step, both phosphate groups are removed from that sugar to cleave the DNA. A complete procedure involves four different base-specific reactions (e.g., for G, G + A, C, and C + T residues) to create a nested set of four radiolabeled fragments, which are then separated by gel electrophoresis in four parallel lanes. The specificity of the base modification depends on the reagent, reaction time, and reaction temperature. For example, reaction with dimethyl sulfate methylates the N7 position of guanine. Hydrazine is used to create C- and (C + T)-specific fragments. Subsequent removal of the base and cleavage of the DNA is performed with piperidine. From the order of appearance of the fragments on the gel, the sequence of the oligonucleotide is deduced.

The Maxam–Gilbert protocol has been combined with mass spectrometry [81]. In this modified procedure, DNAs are labeled with biotin rather than with ^{32}P radiolabeling, and base modification and cleavage steps are performed as usual. The biotin-containing DNA fragments are separated from other fragments by their capture on streptovidin-coupled magnetic beads. The four sets of nested fragments are released from the beads by the hot ammonia treatment and are analyzed by MALDI–TOFMS. Figure 13.16 is an example of the use of this protocol to sequence a 30-mer DNA [81]. Panels A and B are the MALDI spectra of the individual four sequence-specific reactions, and panel C is the overlay of all four spectra to provide composite sequence information.

Sanger Sequencing and Mass Spectrometry The Sanger dideoxy method [82] has also been adapted for the mass spectrometry sequencing of oligonucleotides [31,39,83–86]. The Sanger method also creates four different sets of oligonucleotides ladders, one for each of the DNA bases (i.e., A, C, G, and T). The region of the DNA to be sequenced is cloned by the PCR [82]. A single-stranded fragment that is complementary to the strand to be synthesized is obtained from a restriction enzyme digest to serve as a primer for the synthesis. The synthesis is carried out by incubation with the four deoxyribose nucleosides and a $2',3'$-dideoxy analog of one of the four bases (e.g., ddA for the *A*-ladders). Addition of the dideoxy analogs in the synthesis scheme terminates the chain to create ladders of different length. In the conventional approach, gel electrophoresis separates the four complete sets of chain-terminated ladders, which are detected by either radiolabeling or fluorescent tagging. From the order of appearance, the sequence of the oligonucleotide is deduced.

When the Sanger sequencing is performed in conjunction with mass spectrometry, the gel separation and detection of the radiolabeled or photoaffinity-tagged

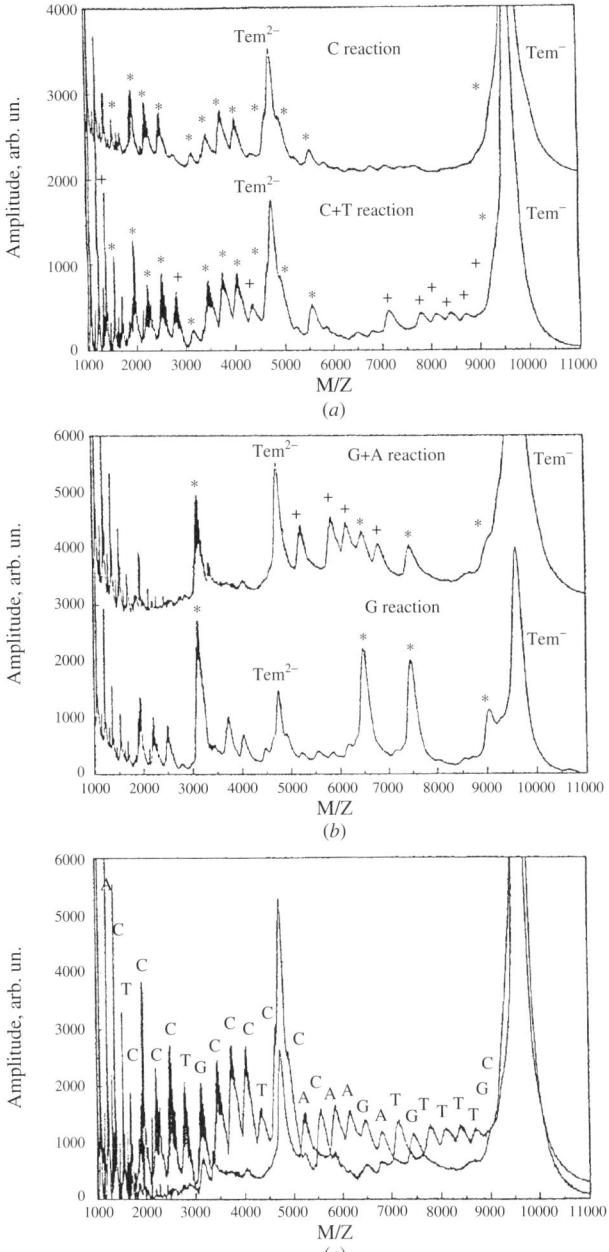

Figure 13.16. Example of sequencing of DNA using the Maxam–Gilbert method. MALDI spectra of the sequencing reactions are shown in (*a*) and (*b*), and the composite spectra of the 30-mer DNA is shown in (*c*). (Reproduced from ref. 81 by permission of the American Chemical Society, Washington, DC, copyright © 1999.)

products are eliminated, and instead, MALDI–TOF mass spectra are acquired of each set of terminated ladders [31,39,83–86]. The four spectra are overlaid to provide the sequence of the oligonucleotide. Although the mass spectrometry-based approach is rapid and much more specific than gel electrophoresis, its sensitivity is poor.

OVERVIEW

DNA and RNA are the two well-known nucleic acids. Structurally, they are oligomers of nucleotides; the difference is that the sugar component of DNA is deoxy-D-ribose, and in RNA it is ribose. The DNA bases are adenine, guanine, cytosine, and thymine; in RNA, they are adenine, guanine, cytosine, and uracil.

An important application of mass spectrometry is the molecular mass determination of oligonucleotides. ESI and MALDI have both served well in this aspect. A possible use of this information is to derive the base composition of oligonucleotides. In the ESI mode, a series of ions of the type $[M - nH]^{n-}$ and their alkali metal ion clusters are the molecular ion signals observed. The alkali metal ion clusters are a nuisance and must be suppressed by purifying the sample prior to mass spectrometry. The presence of alkali metal cations is also detrimental in the MALDI analysis of oligonucleotides. In addition, a proper choice of a matrix is critical to suppress the laser beam–induced fragmentation of oligonucleotides. 3-Hydroxypicolinic acid, picolinic acid, 3-aminopicolinic acid, 2′,4′,6′-trihydroxyacetophenone, and 6-aza-2-thiothymine have gained acceptance as standard matrices to analyze oligonucleotides.

Several gas-phase fragmentation techniques have found a niche as a possible means to sequence oligonucleotides. These include ESI in-source CID (nozzle-skimmer voltage to induce fragmentation), IRMPD in an FT–ICR MS instrument, MALDI in-source decay and post-source decay, and CID–MS/MS of the ESI-produced ions. CID generates complementary $(a - B_n)$- and w-type ions, which can provide bidirectional sequencing from the $5' \rightarrow 3'$ direction and $3' \rightarrow 5'$ direction, respectively.

Three solution-phase techniques to sequence oligonucleotides are available, including generation of DNA sequence ladders, the Sanger dideoxy method, and a chemical cleavage method (e.g., the Maxam–Gilbert method). In the sequence ladder approach, oligonucleotides are cleaved with an exonuclease, such as snake venom phosphodiesterase or calf serum phosphodiesterase, to remove mononucleotides from the oligonucleotide sequentially. Treatment with these enzymes generates the 5′- and 3′-ladders, respectively. The sequence is deduced from the mass difference between successive peaks in the MALDI or ESI spectrum. The sequence ladders can also be generated by chemical digestion (e.g., acid hydrolysis, base hydrolysis, or alkylation) of oligonucleotides. In the Sanger dideoxy method, four different sets of ladders of oligonucleotides, one for each DNA base (i.e., A, C, G, and T), are synthesized by the polymerase chain reaction, and MALDI mass spectra are acquired of each set of the terminated ladders. The four spectra are overlaid to provide the sequence of the oligonucleotide.

EXERCISES

13.1. Calculate the monoisotopic molecular mass of the oligonucleotide 5'-d(ACGGATCCG)-3'.

13.2. Write the structure of the oligonucleotide 5'-d(GATCG)-3' in shorthand notation form.

13.3. Calculate the m/z value of the $[M - 8H^+ + 2Na^+]^{6-}$ ion of the sodium salt of the 9-mer oligonucleotide 5'-d(ACGGATCCG)-3'.

13.4. At what m/z value will the $[M - 5H^+]^{5-}$ ion appear in the negative-ion ESI spectrum of an octamer 5'-d(TCGGATCG)-3'?

13.5. Which matrices will you use to acquire a MALDI mass spectrum of oligonucleotides?

13.6. Draw the structure of the oligonucleotide 5'-d(CATGC)-3'.

13.7. Mark the position of the w_3, a_3, d_2, $z_2 - B_4(G)$, and $a_2 - B_2(A)$ ions in the structure of the oligonucleotides of Exercise 13.6.

13.8. At what m/z values will the a_3^-, w_3^{2-}, $(a_4 - B_4)^-$, and d_2^- ions be observed in the CID mass spectrum of 5'-d(ACTGC)-3'?

13.9. The oligonucleotide 5'-d(TCACTGC)-3' is treated with snake venom phosphodiesterase. At what m/z values will the 5'-sequence ladders be seen in the negative-ion MALDI mass spectrum?

13.10. The oligonucleotide of Exercise 11.9 is digested with a 5'-acting exonuclease calf serum phosphodiesterase. What m/z values of the ions will be observed in the MALDI mass spectrum?

13.11. The purity of the sample is a critical factor in the ESI–MS analysis of oligonucleotides, especially that the sample must be freed from metallic cations. How are these cationic impurities removed?

13.12. An oligonucleotide was digested with a 5'-acting exonuclease calf serum phosphodiesterase. The MALDI mass spectrum contains ions of m/z values of 1775.37, 1471.30, 1182.20, 869.22, 540.12, and 227.10. What is the sequence of the oligonucleotide?

REFERENCES

1. J. D. Watson and F. H. Crick, Molecular structure of nucleic acids: a structure for deoxyribose nucleic acid, *Nature* **171**, 737–738 (1953).
2. K. H. Schram, Preparation of trimethylsilyl derivatives of nucleic acid components for analysis by mass spectrometry, *Methods Enzymol.* **193**, 791–796 (1990).
3. J. A. McCloskey, Electron ionization mass spectra of trimethylsilyl derivatives of nucleosides, *Methods Enzymol.* **193**, 825–842 (1990).

4. P. F. Crain, Preparation of enzymatic hydrolysis of DNA and RNA for mass spectrometry, *Methods Enzymol.* **193**, 781–792 (1990).
5. J. A. McCloskey, Analysis of RNA hydrolyzates by liquid chromatography–mass spectrometry, *Methods Enzymol.* **193**, 796–824 (1990).
6. L. Grotjahn, H. Bloecker, and R. Frank, Mass spectroscopic sequence analysis of oligonucleotides, *Biomed. Mass Spectrom.* **12**, 514–524 (1985).
7. J. Banoub, E. Gentil, et al., Electrospray tandem mass spectrometry of 3'-azido-2',3'-dideoxythymidine and of a novel series of 3'-azido-2',3',4'-trideoxy-4'-thio-5-halogenouridines and their respective α-anomers, *Spectroscopy* **12**, 69–83 (1994).
8. J. T. Stults and J. C. Marsters, Improved electrospray ionization of synthetic oligodeoxynucleotides, *Rapid Commun. Mass Spectrom.* **5**, 359–363 (1991).
9. P. A. Limbach, P. F. Crain, and J. A. McCloskey, Molecular mass measurement of intact ribonucleic acids via electrospray ionization quadrupole mass spectrometry, *J. Am. Soc. Mass Spectrom.* **6**, 27–39 (1995).
10. N. Potier, A. van Dorsselaer, Y. Cordier, O. Roch, and R. Bischoff, Negative electrospray ionization mass spectrometry of synthetic and chemically modified oligonucleotides, *Nucleic Acids Res.* **22**, 3895–3903 (1994).
11. M. Greig and R. H. Griffey, Utility of organic bases for improved electrospray mass spectrometry of oligonucleotides, *Rapid Commun. Mass Spectrom.* **9**, 97–102 (1995).
12. K. Deguchi, M. Ishikawa, T. Yokotura, I. Ogata, S. Ito, T. Mimura, and C. Ostrander, Enhanced mass detection of oligonucleotides using reverse-phase high-performance liquid chromatography/electrospray ionization ion-trap mass spectrometry, *Rapid Commun. Mass Spectrom.* **16**, 2133–2141 (2002).
13. D. P. Little, T. W. Thannhauser, and F. W. McLafferty, Verification of 50- to 100-mer DNA and RNA sequences with high-resolution mass spectrometry, *Proc. Natl. Acad. Sci. USA* **92**, 2318–2322 (1995).
14. R. B. Van Breen, Y. C. Tan, J. Lai, C. R. Huang, and X. Zhao, Immobilized thymine chromatography–mass spectrometry of oligonucleotides, *J. Chromatogr. A* **806**, 67–76 (1998).
15. D. C. Muddiman, D. S. Wunschel, C. Liu, L. Pasa-Tolic, K. F. Fox, A. Fox, G. A. Anderson, and R. D. Smith, Rapid and precise characterization of PCR products from bacilli using electrospray ionization FTICR mass spectrometry, *Anal. Chem.* **68**, 3705–3712 (1996).
16. B. Spengler, Y. Pan, R. J. Cotter, and L.-S. Kan, Molecular weight determination of underivatized oligodeoxyribonucleotides by positive-ion matrix-assisted ultraviolet laser-desorption mass spectrometry, *Rapid Commun. Mass Spectrom.* **4**, 99–102 (1990).
17. J. Wu, A. Steding, and C. H. Becker, Matrix-assisted laser desorption time-of-flight mass spectrometry of oligonucleotides using 3-hydroxypicolinic acid as an ultraviolet-sensitive matrix, *Rapid Commun. Mass Spectrom.* **7**, 142–147 (1993).
18. K. Tang, N. I. Tarrnenko, S. L. Allman, C. H. Chen, L. Y. Ch'ang, and K. B. Jacobson, Picolinic acid as a matrix for laser mass spectrometry of nucleic acids and proteins, *Rapid Commun. Mass Spectrom.* **8**, 673–677 (1994).
19. N. I. Tarrnenko, K. Tang, S. L. Allman, L. Y. Ch'ang, and C. H. Chen, 3-Aminopicolinic acid as a matrix for laser desorption mass spectrometry of biopolymers, *Rapid Commun. Mass Spectrom.* **8**, 1001–1006 (1994).

20. U. Pieles, W. Zürcher, M. Schär, and H. E. Moser, Matrix-assisted laser desorption ionization time-of-flight mass spectrometry: a powerful tool for the mass and sequence analysis of natural and modified oligonucleotides, *Nucleic Acids Res.* **21**, 3191–3196 (1993).
21a. P. Lecchi, H. M. T. Le, and L. K. Pannell, 6-Aza-2-thiothymine: a matrix for MALDI spectra of oligonucleotides, *Nucleic Acids Res.* **23**, 1276–1277 (1995).
21b. Z. Zhang, L. Zhou, et al., 3-Hydroxycoumarin as a new matrix for matrix-assisted laser desorption/ionization time-of-flight mass spectrometry of DNA, *J. Am. Soc. Mass Spectrom.* **17**, 1665–1668 (2006).
22. H. Lin, J. M. Hunter, and C. H. Becker, Laser desorption of DNA oligomers larger than one kilobase from cooled 4-nitrophenol, *Rapid Commun. Mass Spectrom.* **13**, 2335–2340 (1999).
23. L.-K. Zhang and M. L. Gross, Matrix-assisted laser desorption/ionization mass spectrometry methods for oligonucleotide: improvements in matrix, detection limits, quantification, and sequencing, *J. Am. Soc. Mass Spectrom.* **11**, 854–865 (2000).
24. E. Nordhoff, A. Ingendoh, R. Cramer, A. Overberg, B. Stahl, M. Karas, and F. Hillenkamp, Matrix-assisted laser desorption/ionization mass spectrometry of nucleic acids with wavelengths in the ultraviolet and infrared, *Rapid Commun. Mass Spectrom.* **6**, 771–776 (1992).
25. K. J. Wu, T. A. Shaler, and C. H. Becker, Time-of-flight mass spectrometry of underivatized single-stranded DNA oligomers by matrix-assisted laser desorption, *Anal. Chem.* **66**, 1637–1645 (1994).
26. G. J. Currie and J. R. Yates III, Analysis of oligodeoxynucleotides by negative-ion matrix-assisted laser desorption mass spectrometry, *J. Am. Soc. Mass Spectrom.* **4**, 955–963 (1993).
27. T. A. Simmons and P. A. Limbach, The use of a co-matrix for improved analysis of oligonucleotides by matrix-assisted laser desorption/ionization time-of-flight mass spectrometry, *Rapid Commun. Mass Spectrom.* **11**, 567–572 (1997).
28. P. A. Limbach and T. A. Simmons, Influence of co-matrix proton affinity on oligonucleotide ion stability in matrix-assisted laser desorption/ionization time-of-flight mass spectrometry, *J. Am. Soc. Mass Spectrom.* **9**, 668–675 (1998).
29. S. Berkenkamp, F. Kirpekar, and F. Hillenkamp, Infrared MALDI mass spectrometry of large nucleic acids, *Science* **281**, 260–262 (1998).
30. G. J. Langley, J. M. Herniman, N. L. Davies, and T. Brown, Simplified sample preparation for the analysis of oligonucleotides by matrix-assisted laser desorption/ionization time-of-flight mass spectrometry, *Rapid Commun. Mass Spectrom.* **13**, 1717–1723 (1999).
31. T. A. Shaler, Y. Tan, J. N. Wickham, K. J. Wu, and C. H. Becker, Analysis of enzymatic DNA sequencing reactions by matrix-assisted laser desorption/ionization time-of-flight mass spectrometry, *Rapid Commun. Mass Spectrom.* **9**, 942–947 (1995).
32. I. P. Smirov, L. R. Hall, P. I. Ross, and L. A. Haff, Application of DNA-binding polymers for preparation of DNA for analysis by matrix-assisted laser desorption/ionization mass spectrometry, *Rapid Commun. Mass Spectrom.* **15**, 1427–1431 (2001).
33. J. Bai, Y.-H. Liu, X. Liang, Y. Zhu, and D. M. Lubman, Procedures for detection of DNA by matrix-assisted laser desorption/ionization mass spectrometry using a modified nafion film substrate, *Rapid Commun. Mass Spectrom.* **9**, 1172–1176 (1995).

34. Y.-H. Liu, J. Bai, X. Liang, D. M. Lubman, and P. J. Venta, Use of a nitrocellulose film substrate in matrix-assisted laser desorption/ionization mass spectrometry for DNA mapping and screening, *Anal. Chem.* **67**, 3482–3490 (1995).
35. K. A. Kellersberger, E. T. Yu, S. I. Merenbloom, and D. Fabris, Atmospheric pressure MALDI–FTMS of normal and chemically modified RNA, *J. Am. Soc. Mass Spectrom.* **16**, 199–207 (2005).
36. S. C. Pomerantz, J. A. Kowalack, and J. A. McCloskey, Determination of oligonucleotide composition from mass spectrometrically measured molecular weight, *J. Am. Soc. Mass Spectrom.* **4**, 203–209 (1993).
37. D. A. Aaserud, N. L. Kellehur, D. P. Little, and F. W. McLafferty, Accurate base composition of double-strand DNA by mass spectrometry, *J. Am. Soc. Mass Spectrom.* **7**, 1266–1269 (1996).
38. E. Nordhoff, F. Kirpekar, and P. Roepstorff, Mass spectrometry of nucleic acids, *Mass Spectrom. Rev.* **15**, 67–138 (1996).
39. P. A. Limbach, Indirect mass spectrometric methods for characterizing and sequencing oligonucleotides, *Mass Spectrom. Rev.* **15**, 297–336 (1996).
40. K. K. Murray, DNA sequencing by mass spectrometry, *J. Mass Spectrom.* **31**, 1203–1215 (1996).
41. S. A. McLuckey, G. J. Van Berkel, and G. L. Glish, Tandem mass spectrometry of small, multiply charged oligonucleotides, *J. Am. Soc. Mass Spectrom.* **3**, 60–70 (1992).
42. S. Habibi-Goudarzi and S. A. McLuckey, Ion trap collisional activation of the deprotonated deoxymononucleoside and deoxydinucleoside monophosphates, *J. Am. Soc. Mass Spectrom.* **6**, 102–113 (1995).
43. S. A. McLuckey and S. Habibi-Goudarzi, Decomposition of multiply charged oligonucleotide anions, *J. Am. Chem. Soc.* **115**, 12085–12095 (1995).
44. S. A. McLuckey, G. Viadyanathan, and S. Habibi-Goudarzi, Charged vs. neutral nucleobase loss from multiply charged oligoribonucleotide anions, *J. Mass Spectrom.* **30**, 1222–1229 (1995).
45. F. Kirpekar, E. Nordhoff, K. Kristiansen, P. Roepstorff, S. Hahner, and F. Hillenkamp, 7-Deaza purine bases offer a higher ion stability in the analysis of DNA by matrix-assisted laser desorption/ionization mass spectrometry, *Rapid Commun. Mass Spectrom.* **9**, 525–531 (1995).
46. J. Gross, A. Leisner, F. Hillenkamp, S. Hahner, M. Karas, J. Schäfer, F. Lützenkirchen, and E. Nordhoff, Investigations of the metastable decay of DNA under ultraviolet matrix-assisted laser desorption/ionization conditions with post-source-decay analysis and hydrogen/deuterium exchange, *J. Am. Soc. Mass Spectrom.* **9**, 866–878 (1998).
47. J. Krause, M. Scalf, and L. M. Smith, High resolution characterization of DNA fragment ions produced by ultraviolet matrix-assisted laser desorption/ionization using linear and reflecting time-of-flight mass spectrometry, *J. Am. Soc. Mass Spectrom.* **10**, 423–429, (1999).
48. D. P. Little, R. A. Chorush, J. P. Spier, M. W. Senko, N. L. Kelleher, and F. D. McLafferty, Rapid sequencing of oligonucleotides by high-resolution mass spectrometry, *J. Am. Chem. Soc.* **116**, 4893–4897 (1994).

49. K. X. Wan, J. Gross, F. Hillenkamp, and M. L. Gross, Fragmentation mechanisms of oligonucleotides studied by H/D exchange and electrospray ionization mass spectrometry, *J. Am. Soc. Mass Spectrom.* **12**, 193–205 (2001).
50. A. Meyer, M. Spinelli, J.-L. Imbach, and J.-J. Vasseur, Analysis of solid-supported oligonucleotides by matrix-assisted laser desorption/ionization time-of-flight mass spectrometry, *Rapid Commun. Mass Spectrom.* **14**, 234–242 (2000).
51. M. L. Vestal, P. Juhasz, and S. A. Martin, Delayed extraction matrix-assisted laser desorption time-of-flight mass spectrometry, *Rapid Commun. Mass Spectrom.* **9**, 1044–1050 (1995).
52. P. Juhasz, M. T. Roskey, I. P. Smirnov, L. A. Haff, M. L. Vestal, and S. A. Martin, Applications of delayed extraction matrix-assisted laser desorption ionization time-of-flight mass spectrometry to oligonucleotide analysis, *Anal. Chem.* **68**, 941–946 (1996).
53. E. A. Stemmler, M. V. Buchanan, G. B. Hurst, and R. L. Hettich, Analysis of modified oligonucleotides by matrix-assisted laser desorption/ionization Fourier transform mass spectrometry, *Anal. Chem.* **67**, 2924–2930 (1995).
54. J. J. Jones, S. M. A. B. Batoy, R. Liyanage, J. O. Lay, Jr., and C. L. Wilkins, Ionic liquid matrix-induced metastable decay of peptides and oligonucleotides and stabilization of phospholipids in MALDI FTMS analysis, *J. Am. Soc. Mass Spectrom.* **16**, 2000–2008 (2005).
55. G. Talbo and M. Mann, Aspects of the sequencing of carbohydrates and oligonucleotides by matrix-assisted laser desorption/ionization post-source decay, *Rapid Commun. Mass Spectrom.* **10**, 100–103 (1996).
56. Y. Wang, J.-S. Taylor, and M. L. Gross, Differentiation of isomeric photomodified oligodeoxynucleotides by fragmentation of ions produced by matrix-assisted laser desorption ionization and electrospray ionization, *J. Am. Soc. Mass Spectrom.* **10**, 329–338 (1999).
57. R. L. Cerny, K. B. Tomer, M. L. Gross, and L. Grotjahn, Fast atom bombardment combined with tandem mass spectrometry for determining structures of small oligonucleotides, *Anal. Biochem.* **165**, 175–182 (1987).
58. R. H. Griffey, M. J. Greig, et al., Characterization of oligonucleotide metabolism in vivo via liquid chromatography/electrospray tandem mass spectrometry with a quadrupole ion trap mass spectrometer, *J. Mass Spectrom.* **32**, 305–313 (1997).
59. J. P. Barry, P. Vouros, A van Schepdael, and S.-J. Law, Mass and sequence verification of modified oligonucleotides using electrospray tandem mass spectrometry, *J. Mass Spectrom.* **30**, 993–1006 (1995).
60. J. Ni, M. A. A. Mathews, and J. A. McCloskey, Collision-induced dissociation of polyprotonated oligonucleotides produced by electrospray ionization, *Rapid Commun. Mass Spectrom.* **11**, 535–540 (1997).
61. D. P. Little, D. J. Aaserud, G. A. Valaskovic, and F. W. McLafferty, Sequence information from 42-108-mer DNAs (complete for a 50-mer) by tandem mass spectrometry, *J. Am. Chem. Soc.* **118**, 9352–9359 (1996).
62. J. Ni and K. Chan, Sequence verification of oligonucleotides by electrospray quadrupole time-of-flight mass spectrometry, *Rapid Commun. Mass Spectrom.* **15**, 1600–1608 (2001).

63. H. Wu, R. L. Morgan, and H. Aboleneen, Characterization of labeled oligonucleotides using enzymic digestion and tandem mass spectrometry, *J. Am. Soc. Mass Spectrom.* **9**, 660–667 (1998).
64. J. Rozenski and J. A. McCloskey, A SOS: a simple interactive program for ab initio oligonucleotide sequencing by mass spectrometry, *J. Am. Soc. Mass Spectrom.* **13**, 200–203 (2002).
65. H. Oberacher, W. Parson, P. J. Oefner, B. M. Mayer, and C. G. Huber, Applicability of tandem mass spectrometry to the automated sequencing of long-chain oligonucleotide, *J. Am. Soc. Mass Spectrom.* **15**, 510–522 (2004).
66. H. Oberacher, B. Wellenzohn, and C. G. Huber, Comparative sequencing of nucleic acids by liquid chromatography tandem mass spectrometry, *Anal. Chem.* **74**, 211–218 (2002).
67. S. Petri and S. A. McLuckey, Recent developments in ion/ion chemistry of high-mass multiply charged ions, *Mass Spectrom. Rev.* **24**, 931–958 (2005).
68. S. A. McLuckey, J. Wu, J. L. Bundy, J. L. Stephens, Jr., and G. B. Hurst, Oligonucleotide mixture analysis via electrospray and ion–ion reactions, in a quadrupole ion trap, *Anal. Chem.* **74**, 976–984 (2002).
69. T. Keough, J. D. Shaffer, M. P. Lacy, T. A. Riley, W. B. Marvin, M. A. Scurria, J. A. Hasselfield, and E. P. Hesselberth, Detailed characterization of antisense DNA oligonucleotides, *Anal. Chem.* **68**, 3405–3412 (1996).
70. C. M. Bentzley, M. V. Johnston, B. S. Larsen, and S. Gutteridge, Oligonucleotide sequence and composition determined by matrix-assisted laser desorption/ionization, *Anal. Chem.* **68**, 2141–2146 (1996).
71. C. M. Bentzley, M. V. Johnston, and B. S. Larsen, Base specificity of oligonucleotide digestion by calf spleen phosphodiesterase with matrix-assisted laser desorption ionization analysis, *Anal. Biochem.* **258**, 31–37 (1998).
72. W. P. Bartoloni, C. M. Bentzley, M. V. Johnston, and B. S. Larsen, Identification of single stranded regions of DNA by enzymatic digestion with matrix-assisted laser desorption/ionization analysis, *J. Am. Soc. Mass Spectrom.* **10**, 521–528 (1999).
73. I. P. Smirnov, M. T. Roskey, P. Juhasz, E. J. Takach, S. A. Martin, and L. A. Haff, Sequencing oligonucleotides by exonuclease digestion and delayed extraction matrix-assisted laser desorption ionization time-of-flight mass spectrometry, *Anal. Biochem.* **238**, 19–25 (1996).
74. R. P. Glover, G. M. A. Sweetman, P. B. Framer, and G. C. K. Roberts, Sequencing of oligonucleotides using high performance liquid chromatography and electrospray mass spectrometry, *Rapid Commun. Mass Spectrom.* **9**, 97–102 (1995).
75. H. Q. Wu and H. Aboleneen, Improved oligonucleotide sequencing by alkaline phosphatase and exonuclease digestions with mass spectrometry, *Anal. Biochem.* **290**, 347 (2001).
76. A. P. Null, L. M. Benson, and D. C. Muddiman, Enzymatic strategies for the characterization of nucleic acids by electrospray ionization mass spectrometry, *Rapid Commun. Mass Spectrom.* **17**, 2699–2706 (2003).
77. E. Nordhoff, R. Cramer, M. Karas, F. Hillenkamp, F. Kirpekar, K. Kristiansen, and P. Roepstorff, Ion stability of nucleic acids in infrared matrix-assisted laser desorption/ionization mass spectrometry, *Nucleic Acids Res.* **21**, 3347–3357 (1993).
78. T. Keough, T. R. Baker, R. L. M. Dobson, M. P. Lacy, T. A. Riley, M. A. Scurria, J. A. Hasselfield, and E. P. Hesselberth, Antisense DNA oligonucleotides, II: The

use of matrix-assisted laser desorption/ionization mass spectrometry for the sequence verification of methylphosphonate oligodeoxyribonucleotides, *Rapid Commun. Mass Spectrom.* **7**, 195–200 (1993).

79. L. M. Polo, T. D. McCarley, and P. A. Limbach, Chemical sequencing of phosphorothioate oligonucleotides using matrix-assisted laser desorption/ionization time-of-flight mass spectrometry, *Anal. Chem.* **69**, 1107–1112 (1997).

80. A. M. Maxam and W. Gilbert, Sequencing end-labeled DNA with base-specific chemical cleavages, *Methods Enzymol.* **65**, 499–560 (1980).

81. M. R. Isola, S. L. Allman, V. V. Golovlov, and C.-H. Chen, Chemical cleavage sequencing of DNA using matrix-assisted laser desorption/ionization time-of-flight mass spectrometry, *Anal. Chem.* **70**, 2266–2269 (1999).

82. F. Sanger, S. Nicklen, and A. R. Coulson, DNA sequencing with chain-terminating inhibitor, *Proc. Natl. Acad. Sci. USA* **74**, 5463–5467 (1977).

83. L. M. Smith, The future of DNA sequencing, *Science* **262**, 530–532 (1993).

84. M. C. Fitzgerald, L. Zhu, and L. M. Smith, The analysis of mock DNA sequencing reactions using matrix-assisted laser desorption/ionization mass spectrometry, *Rapid Commun. Mass Spectrom.* **7**, 895–897 (1993).

85. M. T. Roskey, P. Juhasz, I. P. Smirnov, E. J. Takach, S. A. Martin, and L. A. Haff, DNA sequencing by delayed extraction-matrix-assisted laser desorption/ionization time of flight mass spectrometry, *Proc. Natl. Acad. Sci. USA* **93**, 4724–4729 (1996).

86. S. Mouradian, D. R. Rank, and L. M. Smith, Analyzing sequencing reactions from bacteriophage M13 by matrix-assisted laser desorption/ionization mass spectrometry, *Rapid Commun. Mass Spectrom.* **10**, 1475–1478 (1996).

CHAPTER 14

QUANTITATIVE ANALYSIS

Apart from identification of a compound in a sample (qualitative analysis), often it becomes essential to measure its amount quantitatively. The aim of quantitative analysis is to provide accurate and reliable determination of the amount of a target analyte in a real-world sample. Quantitative analysis is the cornerstone of many health-related fields, such as clinical chemistry, pharmaceutical science, forensic science, and environmental science. Knowledge of the accurate amount of a biological compound in a cell culture or in a particular organ provides information that leads to an understanding of its functional role in various neurological and pathophysiological events. The development of a drug and its clinical trial heavily rely on accurate quantitative analysis. The management of various forms of illnesses requires quantitative analysis of endogenous biomolecules in extracts of body tissues and fluids. Quantification of drugs of abuse and their metabolites is a key factor in the management of an overdose patient. From a health point of view, it is important to know the accurate amounts of toxic chemicals in the air that we breath, the water we drink, and the food we consume.

Quantitative analysis is performed to provide absolute quantification or relative quantification. In absolute quantification, the aim is to determine the amount or concentration of the analyte in absolute terms (i.e., in terms of per unit mass or per unit volume of the sample). In contrast, the aim of relative quantification is to determine the amount or concentration of the analyte relative to another analyte or of the same analyte relative to another sample.

Fundamentals of Contemporary Mass Spectrometry, by Chhabil Dass
Copyright © 2007 John Wiley & Sons, Inc.

14.1. ADVANTAGES OF MASS SPECTROMETRY

Mass spectrometry has attained a unique position in analytical chemistry as a quantitative analysis tool, especially when coupled with high-resolution separation devices. Quantitative analysis of a variety of molecules is performed routinely with mass spectrometry–based methods at unprecedented high-sensitivity. It has several desirable features that make it the most sought-after analytical technique for quantitative analysis. These features include:

- Applicability to most types of compounds
- High level of sensitivity, including the capability to detect a single ion and detection limits in the attomole to femtomole range.
- High level of specificity because measurements are made on molecular and fragment ions and both are highly compound-specific parameters
- Coupling with high-resolution separation devices, providing an opportunity to analyze real-world samples
- Use of stable isotope-labeled internal standards of the same structure
- Wide dynamic range
- High speed of analysis
- Automation and high-throughput analysis

The field of quantitative analysis has been reviewed [1,2]. Quantitative analysis must be performed in compliance with good laboratory practices (GLPs) as defined by various regulatory agencies. This aspect has been discussed in a report in the *Journal of American Society for Mass Spectrometry* [3].

14.2. DATA ACQUISITION

Quantitative analysis by mass spectrometry involves establishing the accurate correlation of the compound-related ion signal with the amount of analyte. Although quantitative analysis has been performed by acquiring mass spectrometry data in the narrow-mass-range and full-scan modes [4], most applications have used either the selected-ion monitoring (SIM) or selected-reaction monitoring (SRM) modes of data acquisition. The sensitivity, specificity, type of information desired, and cost of instrumentation are the criteria that influence the choice of a particular technique. The detection sensitivity of a full-scan mode, in which the data are acquired in a wide mass range, is compromised significantly because large amounts of scan time and sample ion current are wasted. This scan mode, however, has the advantages that the identity of the analyte ion can be confirmed readily, and the presence and absence of any interfering signal can be ascertained. Scanning in the narrow mass range improves the detection sensitivity to some extent and still provides the isotopic pattern of the sample ion. This scan mode is a compromise between the full-scan and SIM modes.

14.2.1. Selected-Ion Monitoring

The SIM mode is a technique frequently used to record the sample ion current in quantitative analysis. In this procedure, the ion current from only one or a few selected ions, rather than a complete mass spectrum, is repetitively recorded and accumulated. The data system spends more time in recording the ion current at the m/z values selected to impart a significant improvement in detection sensitivity versus a full scan. A 1000-fold gain in detection sensitivity can be realized in switching from a full-scan mode to SIM, but at the expense of specificity because the identity of the ion selected may be questionable. As an example, an SIM of one ion with a dwell time of 0.1 s per ion per cycle would result in a gain of signal-to-noise (S/N) ratio of 31.6 over a full-scan recording in a 1000-u mass range in 1 s and sampling at every 0.1 u (i.e., $[0.1/(0.1/1000)]^{1/2} = 31.6$). Another benefit of the SIM mode is that a chromatographic peak shape can be defined more accurately because of the acquisition of more data points.

The type and number of ions that are selected for the SIM procedure depend on the nature of the compound. An important criterion in choosing the ion is its abundance. In most cases, the molecular ion, which usually is the most abundant ion, satisfies this criterion. In the absence of a prominent molecular ion, any other high-mass fragment ion next in abundance could serve the purpose. The ion chosen for SIM measurements must originate exclusively from the analyte; otherwise, the molecular specificity of the procedure would be severely compromised. This situation is a common occurrence with biological and environmental samples, which often are heterogeneous even after extensive chromatographic separation. Furthermore, the background signal due to matrices and solvents used in some ionization techniques might also produce an interfering signal. The selectivity of the SIM procedure can be improved when (1) m/z the value of the selected analyte ion is away from that of the interfering ion(s), (2) more than one analyte ion is monitored, (3) measurement is done at higher mass resolution to eliminate the contribution from any interfering ions, and finally, (4) SIM is coupled with a chromatography separation device. Monitoring of more than one ion and increasing the resolution, however, would lower the detection sensitivity. If these changes are of no avail, the SRM approach must be used in place of SIM.

14.2.2. Selected-Reaction Monitoring

Conceptually, SRM is similar to acquiring a product-ion scan spectrum in tandem mass spectrometry (MS/MS) (Section 4.2). Instead of recording the complete spectrum, however, the ion current due to a specific precursor–product pair is monitored. As with the product-ion scan, the molecular ion of the analyte is mass-selected by MS-1 and allowed to undergo collision-induced dissociation (CID) to a unique product ion; the ion current due to this exclusively derived ion is monitored by MS-2. The term *multiple reaction monitoring* (MRM) refers to the monitoring of more than one reaction, either from the same precursor or from more than one precursor. The following three criteria are applied in selecting a precursor–product reaction in SRM analysis: (1) a fragmentation reaction chosen for quantification must be unique to the analyte; (2) the precursor–product reaction should provide a strong ion current signal; and (3) an isotope-labeled internal standard of the same compound should be available in which the stable isotope is present at a site that yields the labeled fragment ion.

An obvious advantage of SRM over SIM is the increased level of detection specificity because of the exclusive structural link that is maintained during the analysis between the mass-selected precursor ion and its product ions. Also, in SRM, chemical noise is virtually eliminated. The latter aspect might improve the detection sensitivity in some cases. In general, the detection sensitivity of SRM is less than that of SIM because of the distribution of the precursor ion current into several of its CID products. The higher cost of the tandem instrumentation is another discouraging factor in favor of SRM. Increased specificity, however, outweighs these disadvantages.

Instrument operation in SRM measurements is exactly identical to that used in the product-ion analysis (Section 4.2) except that instead of acquiring a full-scan spectrum, only the products selected are monitored. A triple-quadrupole instrument is ideally suited for SRM experiments, although magnetic-sector and QIT instruments have also been used.

14.3. CALIBRATION

Calibration of the instrument is the first essential step of a quantitative procedure. Like any other analytical instrument, the response of a mass spectrometer is not absolute and might deviate with time. In addition, the sample matrix has a variable influence on the mass spectrometry response. Calibration involves determination of the correlation between a known concentration of the analyte and the resulting mass spectrometry signal. In ideal situations, the sample and the analyte standards are both analyzed under identical experimental conditions. Depending on the levels of accuracy and precision that are required, the calibration might be performed by one of the methods described below.

14.3.1. External Standard Method

External calibration is the simplest to perform. It involves preparation and analysis of several standards of accurately known concentrations of the analyte. The mass spectrometry response of these standards is measured from the signal intensity or the area under the chromatography peak; and those data are plotted against the concentration of each standard to obtain a calibration curve. Three to five replicate standards of each concentration must be analyzed to optimize precision and for sample-to-sample comparison. To avoid sample carryover, standard samples are analyzed from lowest to highest concentration. Ideally, the calibration graph should be linear and encompass the concentration range of the unknown samples. From the instrument response and the concentration of the standards, the calibration curve or regression line can be generated (see Figure 14.1). Usually, a linear regression

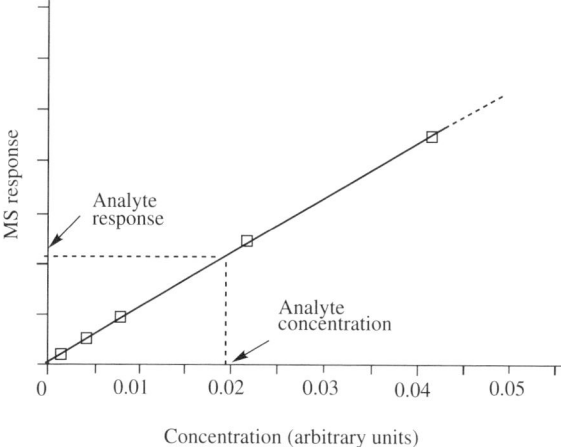

Figure 14.1. Typical external calibration curve.

model is used for regression analysis. Once the regression line is known, the concentration of the unknown can be calculated from the sample response.

The accuracy and precision of this method are limited because the standards and the samples are analyzed at two different times, and within that space of time, the mass spectrometry response might drift. Also, the matrix of the calibration standards is difficult to match with the sample matrix.

14.3.2. Standard Addition Method

The *standard addition method* must be adopted when the sample matrix is very complex and the instrumental fluctuations are difficult to control. In this method, the unknown sample is first analyzed. A known concentration of the standard solution of the analyte is added to this unknown sample, and the mass spectrometry response is measured again to provide the response factor (i.e.; the response per unit concentration). The concentration of the unknown is calculated by multiplying the signal intensity of the unknown with the response factor. This single-point calibration is, however, less precise. To enhance the precision of this method, several increments of the standard solution of the analyte are added to a fixed amount of the unknown sample. After each addition, the mass spectrometry response is also measured, and the calibration curve is obtained as in the external standard method (Figure 14.2). The x-axis intercept of the calibration curve provides the concentration of the analyte in the unknown sample.

14.3.3. Internal Standard Method

The internal standard method is the most widely used approach in mass spectrometry quantification. The method is especially useful when the amount of the sample introduced and the instrument response may vary from run to run. In this

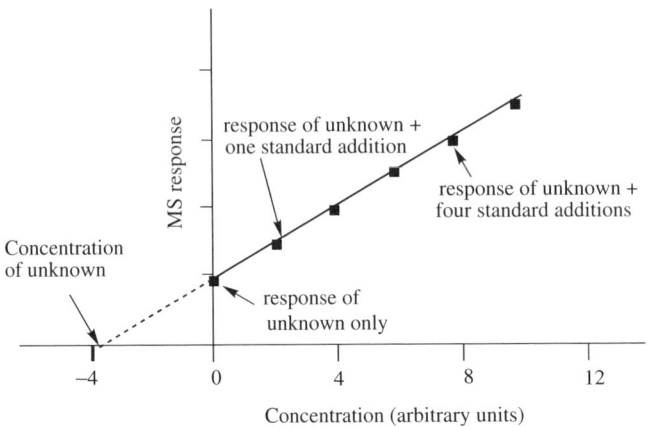

Figure 14.2. Calibration by standard addition method.

method, a known and constant amount of an internal standard (IS) is added to all calibration solutions of known concentrations of the analyte. The amount of internal standard should be above the limit of quantification but not high enough to suppress ionization of the target analyte. The mass spectrometry signal due to the internal standard and the analyte is measured, and the ratio of their signals is plotted versus the analyte concentration to provide a calibration curve similar to that shown in Figure 14.1. The internal standard is also added to the unknown sample at a very early stage of sample handling (e.g., during the extraction step). This method yields a high level of accuracy and precision in quantitative analysis because an internal standard acts as a self-correcting system; it can compensate for any fluctuations in the mass spectrometry response and the sample losses that might occur in various sample-handling and chromatographic steps. Neither the external standard method nor the standard addition method account for these errors.

The following criteria should be applied in the selection of an internal standard: (1) its physical and chemical properties should exactly match those of the analyte so that any losses are accounted for accurately (2) its chromatographic characteristics are similar to those of the analyte so that the two elute close to each other; (3) its molecular mass should be distinct from that of the analyte; (4) its mass spectrometry behavior (i.e., ionization efficiency, fragmentation behavior, detector response, etc.) should mirror that of the analyte; and (5) it should not be a constituent of the sample itself.

Types of Internal Standards Three types of internal standards have found a niche in quantitative analysis. The first includes those compounds that have chemical and physical properties similar to those of the analyte. The only requirements are that it should not be isobaric to the analyte and it must elute from a chromatographic column at a different time. This type of internal standard is easy to find, but it does not provide the level of precision and accuracy that is required for many applications. The second type of internal standard includes those compounds that are structurally homologous to the analyte. For example, [Ala]^2-leucine enkephalin was used as an internal standard in the quantification of leucine enkephalin from tissue samples [5]. A homologous compound usually has a greater match between its chromatographic and mass spectrometry characteristics with those of the analyte molecules, and it is also cost-effective. A great majority of assays for pharmaceutical drugs and their metabolites use this type of internal standard.

The most ideal internal standard is a stable isotope-labeled analog of the analyte because its chemical and physical properties are virtually identical to those of the analyte. Its cost can, however, be prohibitive. Internal standards labeled with ^{2}H, ^{13}C, ^{15}N, ^{18}O, ^{34}S, and ^{37}Cl have been used in quantitative assays. The ^{2}H-labeled internal standards, however, have the advantages of lower cost and ease of availability. Other factors that must be considered in the selection of an internal standard are the position of the isotope label and a shift in the mass of the internal standard. The isotope label must be in nonexchangeable positions, and the signal of the internal standard should be moved away from the isotopic pattern of the analyte. A shift of 4 to 5 u in the mass of the internal standard is reasonable for compounds of greater than 500 u.

The use of deuterated internal standards has been reported for the quantification of endogenous peptides, prostaglandins, pharmaceutical drugs, and the drugs of abuse. In the quantification of methionine enkephalin [1,5–7] and β-endorphin [1,5–7], the deuterium was incorporated in the phenyl ring of the phenylalaline and Ile^{22} residues, respectively. The phenyl ring–deuterated IS was also used in the quantification of μ-opioid receptor agonist Tyr–D–Arg–Phe–Lys–NH$_2$ from ovine plasma [8].

14.4. VALIDATION OF A QUANTITATIVE METHOD

The process of method validation (i.e., evaluation of the assay) affects the quality of the quantitative data directly [9; *A Guide to Analytical Method Validation*, Waters Corporation]. Through method validation, it is assured that the method developed is acceptable. Issues involved in the validation of a mass spectrometry method for quantitative analysis are similar to those in any other analytical technique. The validation involves undertaking a series of studies to demonstrate the limit of detection (LOD); limit of quantitation (LOQ); linear range; specificity; within-day precision and accuracy; and day-to-day precision and accuracy, specificity, and robustness of the method. All of these parameters must be determined with those commonly accepted good laboratory practices criteria that are applicable in the validation of analytical methods.

- *Limit of detection:* LOD is defined as the lowest analyte concentration that can be detected in the sample (Figure 14.3). It is the analyte concentration that yields an S/N ratio ≥ 3.

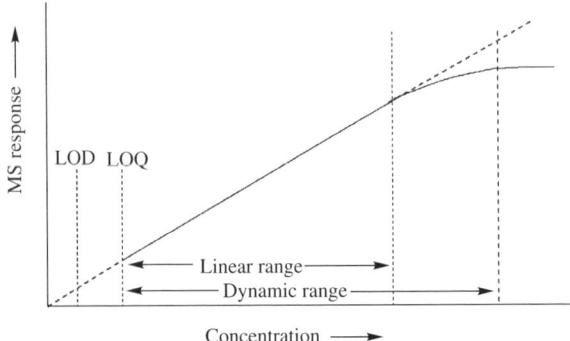

Figure 14.3. Graphical depiction of limit of detection (LOD), limit of quantification (LOQ), linear range, and dynamic range.

- *Limit of quantitation:* LOQ is the lowest analyte concentration that can be determined at an acceptable level of accuracy and precision. It is usually the analyte concentration that yields S/N = 10.
- *Linear range:* The linear range is defined as the concentration range where the analyte response is directly proportional to concentration. To determine this range, a calibration curve is prepared by analyzing five concentrations of standard solutions between 50 and 150% of the target analyte concentration. For an acceptable linearity range, the correlation coefficient (r^2) should be greater than 0.999 and the y-intercept less than 2% of the target concentration response.
- *Accuracy:* Accuracy is the measure of the closeness of the measured value to the true value. Accuracy is determined by comparing the test result with that obtained for a standard reference material or comparison to an existing well-characterized method.
- *Precision:* The precision of a method is the degree of mutual agreement among a series of measurements from multiple sampling of the sample. The method should be validated for short-term repeatability, day-to-day reproducibility, instrument-to-instrument reproducibility, and laboratory-to-laboratory reproducibility. The precision is expressed as variance, standard deviation, or coefficient of variance.
- *Specificity:* The specificity of a method is its ability to measure only the target analyte accurately in the presence of any other compounds in the sample matrix.
- *Robustness:* Robustness is the ability of the method to remain unaffected by small changes in method parameters (e.g., pH, temperature.).

14.5. SELECTED EXAMPLES

Most quantitative analyses are performed by coupling a chromatographic separation device with mass spectrometry. As enumerated in Chapter 5, the online combination of a chromatography system with mass spectrometry has several distinct advantages. Briefly, it provides enhanced sensitivity, selectivity, and speed of analysis. The opportunity for automation allows a large number of samples to be analyzed unattended. The feasibility of quantification of coeluting components or isobaric compounds is another useful feature. As a consequence, stable isotope–labeled analogs can be used as internal standards. Despite the fact that the analyte and its stable isotope–labeled analog coelute, they can be readily distinguished on the basis of their distinct m/z values. The literature contains an extensive list of examples of the use of mass spectrometry for the quantification of environmental compounds, biological compounds, pharmaceutical drugs, and their metabolites. A few representative examples are presented here to illustrate how a quantitative analysis is typically performed with gas chromatography (GC)/MS, liquid chromatography (LC)/MS, and matrix-assisted laser desorption/ionization (MALDI)–MS techniques.

14.5.1. Applications of Gas Chromatography/Mass Spectrometry

GC/MS has gained an excellent reputation in quantitative analysis. The technique is superior to LC/MS in terms of chromatographic resolution, and the background chemical noise is lower. It is, however, applicable to small, thermally stable, and relatively volatile compounds, and in many situations, an additional chemical derivatization step is required. Ionization is accomplished by electron ionization (EI) or chemical ionization (CI). A recent representative application is the simultaneous screening and quantification of amphetamine-type stimulant (ATS) drugs in body fluids (urine, blood, serum, and oral fluid) [10]. The drugs were extracted and derivatized with heptafluorobutyric anhydride (HFBA) in a single step by adding toluene + HFBA + IS. Derivatization improves chromatographic separation and shifts the m/z of the target ion away from a low-mass biological matrix background. The derivatized samples were subjected to GC/MS analysis. Several drugs (a total of 15: amphetamine, methamphetamine, ephedrine, norephedrine, etc.) were quantified by recording their individual ion current in the SIM mode. In most cases, Target ions monitored, were either molecular ions or the most abundant fragment ions. Calibration curves showed a linear response with $r^2 > 0.98$.

14.5.2. Applications of Liquid Chromatography/Mass Spectrometry

The development of a robust and reliable online LC/MS combination via atmospheric pressure ionization (API) has greatly advanced the science of quantitative analysis of polar and thermally labile compounds. The speed of analysis, simplicity, and enhanced selectivity are the highlights of this combination. HPLC is more conveniently coupled to mass spectrometry via atmospheric-pressure chemical ionization (APCI) and electrospray ionization (ESI) interfaces.

Quantification with LC/ESI–MS The use of LC/ESI–MS and SIM is demonstrated in the quantitative determination of a peptide drug, Ac–Arg–Pro–Asp–

Pro–Phe–NH$_2$, in human and rabbit plasma. In this procedure, after the addition of deuterated IS, the sample was desalted, proteins were precipitated, and the drug was extracted with solid-phase extraction (SPE). The HPLC separation was performed in the isocratic mode with a mobile phase that consists of 1% methanol and 20% acetonitrile in 10 mM ammonium formate (pH 5.2). A flow splitter was used to allow only 5% of the flow to enter the ESI source.

Quantification with LC/ESI–MS/MS The use of LC/ESI–MS/MS for quantification is exemplified by a procedure for the determination of amoxycillin and its metabolites ($5R, 6R$- and $5S, 6R$-amoxycilloic acid and amoxycillinpiperazine-2,5-dione) in animal tissues [11]. Amoxycillin is an antibiotic that is prescribed for the treatment of a respiratory, gastrointestinal, and urogenital infection. In the procedure reported, drugs were extracted from animal tissues by liquid extraction and subjected to SPE. The IS ampicillin was added to the tissue sample prior to the extraction step. The samples were analyzed with LC/ESI–MS/MS. MS signal was recorded in the SRM mode by monitoring the transitions m/z 365.8 → 207.9 (for amoxicillin), m/z 383.8 → 189.0 (for $5R, 6R$- and $5S, 6R$-amoxycilloic acid), m/z 365.8 → 160.0 (for amoxicillinpiperazine-2,5-dione), and m/z 349.9 → 106.0 (for ampicillin). Calibration graphs showed a linear response in the concentration range 25 to 500 ng g^{-1} ($r^2 \geq 0.9974$). The method was validated by measuring a set of required parameters, such as linearity, precision, accuracy, LOQ, LOD, and specificity.

Another example of quantification with the LC–MS/MS combination is the determination of mevinolinic acid, which is a major metabolite of lavastatin in human blood [12]. Clinical trials have shown that lavastatin lowers the serum level of cholesterol by acting as a potent inhibitor of 3-hydroxy-3-methylglutoryl-coenzyme A. Mevinolinic acid is the active form of lavastatin in this functional role. In this method, methylmevinolinic acid was used as an IS. CID of the [M–H]$^-$ ions of the analyte and methylmevinolinic acid produces a common fragment ion at m/z 319. The negative-ion MRM acquisition mode was used to monitor the precursor–product reactions of the analyte and internal standard. With this method, quantification of mevinolinic acid can be performed at the 50-pg/mL level.

Quantification with Microchip-Based ESI A microfabricated ESI source has been coupled to a triple-sector quadrupole for quantification of methylphenidate (Ritalin; m/z 234.2) in human urine samples [13]. Methylphenidate-d_3 (m/z 237.2) was used as an IS. The urine extracts were infused directly through the ESI source via the microfluidic channel. Quantification was performed by monitoring the SRM transitions m/z 237.2 → 84.1 and 234.2 → 84.1. The standard calibration curve was linear over the range 2 0.4 to 800 ng/mL of methylphenidate ($r^2 = 0.999$). The method also provided acceptable values of precision and accuracy.

14.5.3. Applications of MALDI–MS

Although not a common practice, quantification of a range of molecules, especially of peptides, has been performed with MALDI–MS [14–18]. A major problem with MALDI is its low shot-to-shot reproducibility. Many sample preparation variables that are difficult to control affect reproducibility. Incorporation of an IS in the quantification scheme could alleviate this problem to some extent. The IS chosen and analyte should have cocrystallization property similar to that of the MALDI matrix. Also, their desorption–ionization efficiency and the extent of metastable decomposition should be similar.

A typical example of the use of MALDI–MS is the simultaneous quantification of human cardiac α- and β-myosin heavy chain (MyHC) proteins in human atrial tissue [19]. These two isoforms of MyHC are expressed in the human heart. The method relies on the generation of two tryptic peptides, α-MyHC (726–741; ILNPVAIPEGQFIDSR) and β-MyHC (724–739; ILNPAAIPEGQFIDSR), one from each isoform, and the measurement of the ratio of their ion currents to provide a relative amount of each peptide. To determine absolute values, a synthetic analog (ILNPAAVPEGQFIDSR) of the two peptides was used as an IS. A known amount of the IS peptide (2 pmol) was added to varying amounts of α-MyHC (726–741) (0 to 6 pmol) and β-MyHC (726–741) (0 to 4 pmol) to prepare standard curves for the two peptide isoforms. α-Cyano-4-hydroxycinnamic acid (α-CHCA) served as the matrix. The ion current due to the two peptide isoforms and the IS peptide was monitored by acquiring the MALDI spectrum in a narrow m/z range of 1735 to 1780 (see Figure 14.4). The standard curves for the two peptide isoforms are shown in Figure 14.5. The accuracy of the method was determined by comparing the results with those obtained from an established method. To quantify α- and β-MyHC proteins in atrial samples, the tissue extract was separated by gel electrophoresis; the protein spots were digested and supplemented with IS prior to MALDI analysis.

MALDI has also been used to quantify small molecules. This aspect has been demonstrated by the analysis of lysergic acid diethyl amide (LSD) with atmospheric pressure (AP) MALDI [20]. LSD is a hallucinogen. Its quantification is highly important from forensic and clinical point of views. In this method, LSD was extracted and precleaned from urine samples by SPE and quantified by recording the ion current due to m/z 327 → 226 and 324 → 223 transitions. The standard curves were plotted by analyzing several LSD standard solutions, each of which was spiked with LSD-d_3 as an IS. A linear calibration curve was obtained over the concentration range 1 to 100 ng mL^{-1} with $r^2 = 0.9917$. The results were comparable with those of an existing HPLC/ESI–MS method.

OVERVIEW

In this chapter, quantitative applications of mass spectrometry were discussed. In this respect, mass spectrometry is distinctly superior over most other analytical techniques. Mass spectrometry–based methods are more specific and highly sensitive. In combination with high-resolution separation devices, the task of quantitation of real-world samples becomes much easier. A mass spectrometry signal is acquired in the SIM or SRM mode. In SIM, the ion current due to one or more compound-related ions is recorded, whereas in SRM, precursor–product

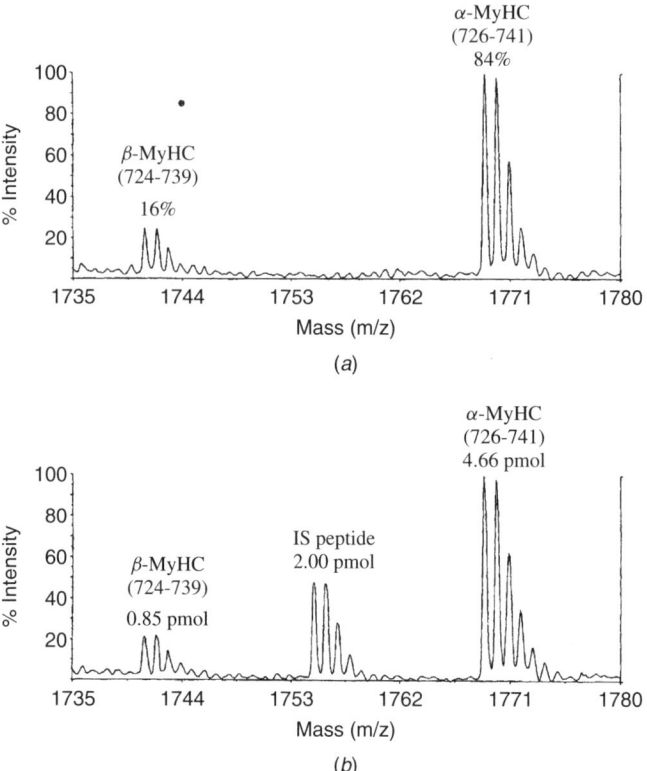

Figure 14.4. MALDI–TOF mass spectra: (*a*) relative quantification of a human atrial sample; (*b*) use of the peptide IS for absolute quantification. (Reproduced from ref. 19 by permission of the American Chemical Society, Washington, DC, copyright © 2004.)

transitions are monitored. SIM provides better detection sensitivity, whereas the SRM approach is more specific.

For precise and accurate quantification, it is essential to obtain a calibration curve to accurately define the relation between a known concentration of the analyte and the mass spectrometry signal. Calibration is performed with the external calibration, standard addition, or internal standard method. The last method is more accurate because an internal standard can account for deviation in the mass spectrometry response and the sample losses that might occur in various sample-handling and chromatographic steps. An internal standard is any compound that has chemical and physical properties similar to those of the analyte or homologous to the analyte or a stable isotope–labeled analog of the analyte. The last type of standard provides more accurate results because its chemical and physical properties are virtually identical to those of the analyte.

Any newly developed method must be validated according to the acceptable criteria. To validate a method, LOD, LOQ, linear range, specificity, precision and accuracy, specificity, and robustness of the method should be determined.

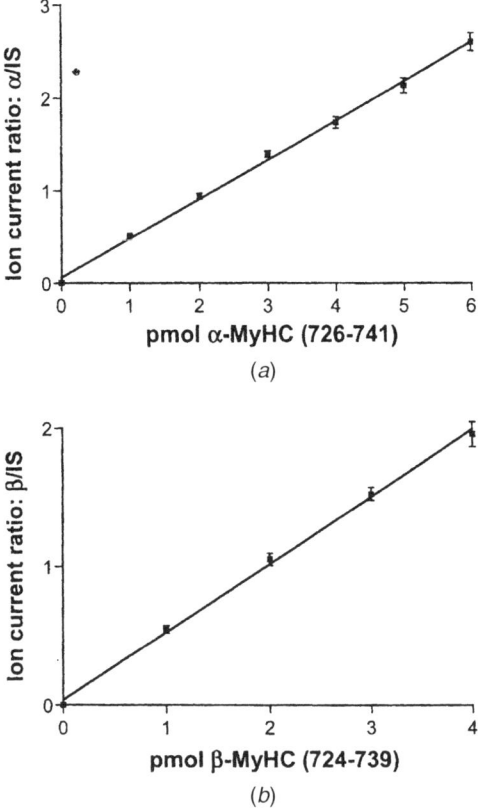

Figure 14.5. Standard curves for (*a*) α-MyHC (each point is an average of 10 measurements; $r^2 = 0.994$) and (*b*) β-MyHC ($r^2 = 0.998$). (Reproduced from ref. 19 by permission of the American Chemical Society, Washington, DC, copyright © 2004.)

Quantitative analysis is performed primarily with an online combination of a chromatographic separation device with mass spectrometry. For example, GC/MS is used for small, thermally stable, relatively volatile compounds, and LC/MS for nonvolatile compounds. MALDI–MS can also be used to quantify nonvolatile compounds.

EXERCISES

14.1. What are the advantages of using SIM over a full-scan mode of data acquisition in quantitative measurements?

14.2. Three different methods are used to perform a calibration in quantitative determination. Describe these procedures.

14.3. Why is an internal standard added to a calibration standard?

14.4. What type of internal standard is most appropriate in mass spectrometry quantification?

14.5. Why is the SRM procedure preferred over SIM?

14.6. Validation of the method developed is an important aspect of quantitative analysis. What parameters must be included in validation studies?

14.7. Enumerate your reasons for selecting an online LC/MS combination for quantitative analysis.

ADDITIONAL READING

1. M. W. Duncan, P. J. Gale, and A. L. Yergey, *The Principles of Quantitative Mass Spectrometry*, Rockpool Productions, Denver, CO, 2006.
2. B. J. Millard, *Quantitative Mass Spectrometry*, Wiley-Interscience, New York, 1977.
3. P. Traldi, F. Mogno, I. Lavagnini, and R. Seraglia, *Quantitative Applications of Mass Spectrometry*, Wiley, Hoboken, NJ, 2006.

REFERENCES

1. X. Zhu and D. M. Desiderio, Peptide quantification by tandem mass spectrometry, *Mass Spectrom. Rev.* **15**, 213–240 (1997).
2. D. M. Desiderio and X. Zhu, Quantitative analysis of methionine enkephalin and β-endorphin in the pituitary by liquid secondary ion mass spectrometry and tandem mass spectrometry, *J. Chromatogr. A* **794**, 85–96 (1998).
3. R. K. Boyd, J. D. Henion, M. Alexander, W. L. Budde, J. D. Gilbert, S. M. Musser, C. Palmer, and E. K. Zurek, Mass spectrometry and good laboratory practices, *J. Am. Soc. Mass Spectrom.* **7**, 211–218 (1996).
4. C. Dass, J. J. Kusmierz, and D. M. Desiderio, Mass spectrometric quantification of endogenous β-endorphin, *Biol. Mass Spectrom.* **20**, 130–138 (1991).
5. D. M. Desiderio, S. Yamada, F. S. Tanzer, J. Horton, and J. Trimble, High-performance liquid chromatographic and field desorption mass spectrometric measurement of picomole amounts of endogenous neuropeptides in biologic tissue, *J. Chromatogr.* **217**, 437–452 (1981).
6. J. J. Kusmierz, C. Dass, J. T. Robertson, and D. M. Desiderio, Mass spectrometric measurement of β-endorphin and methionine enkephalin in human pituitaries: tumors and post-mortem controls, *Int. J. Mass Spectrom. Ion Proc.* **111**, 247–262 (1991).
7. D. M. Desiderio, J. J. Kusmierz, X. Zhu, C. Dass, D. H. Hilton, J. T. Robertson, and H. S. Sacks, Mass spectrometric analysis of opioid and tachykinin neuropeptides in non-secreting and ACTH-secreting human pituitary adenomas, *Biol. Mass Spectrom.* **22**, 89–97 (1993).
8. O. O. Grigoriants, J.-L. Tseng, R. R. Becklin, and D. M. Desiderio, Mass spectrometric quantification of the μ opioid receptor agonist Tyr–D–Arg–Phe–Lys–NH2 (DALDA) in high-performance liquid chromatography-purified ovine plasma, *J. Chromatogr, B* **695**, 287–298 (1997).

9. J. M. Green, A practical guide to analytical method validation, *Anal. Chem.* **68**, 305A–309A (1996).

10. A. Kankaanpää, T. Gunnar, et al, Single-step procedure for gas chromatography–mass spectrometry screening and quantitative determination of amphetamine-type stimulants and related drugs in blood, serum, oral fluid, and urine, *J. Chromatogr. B* **810**, 57–68 (2004).

11. S. DeBaere, M. Cherlet, K. Baert, and P. De Backer, Quantitative analysis of amoxycillin and its major metabolites in animal tissues by liquid chromatography combined with electrospray ionization tandem mass spectrometry, *Anal. Chem.* **74**, 1393–1401 (2002).

12. R. E. Calaf, M. Carrascal, E. Gelpi, and J. Abian, Quantitative analysis of mevinolinic acid in human plasma by high-performance liquid chromatography coupled with negative-ion electrospray tandem mass spectrometry, *Rapid Commun. Mass Spectrom.* **11**, 75–80 (1997).

13. Y. Yang, L. Kameoka, et al, Quantitative mass spectrometric determination of methylphenidate concentration in urine using an ESI source integrated with a polymer microchip, *Anal. Chem.* **76**, 2568–2574 (2004).

14. H. Wei, et al., Identification and quantification of neuropeptides in brain tissues by capillary liquid chromatography coupled off-line to MALDI–TOF and MALDI–TOF/TOF-MS, *Anal. Chem.* **78**, 4342–4351 (2006).

15. M. Bucknell, K. Y. Fung, and M. W. Duncan, Practical quantitative biomedical applications of MALDI–TOF mass spectrometry, *J. Am. Soc. Mass Spectrom.* **13**, 1015–1027 (2002).

16. O. A. Migorodskaya, Y. P. Kozmin, et al, Quantitation of peptides and proteins by matrix-assisted laser desorption/ionization mass spectrometry using ^{18}O-labeled internal standards, *Rapid Commun. Mass Spectrom.* **14**, 1226–1232 (2000).

17. J. Gobom, K. O. Kraeuter, et al, Detection and quantification of neurotensin in human brain tissue by matrix-assisted laser desorption/ionization time-of-flight mass spectrometry, *Anal. Chem.* **72**, 3320–3326 (2000).

18. J. J. Corr, et al., Design considerations for high-speed quantitative mass spectrometry with MALDI ionization, *J. Am. Soc. Mass Spectrom.* **17**, 1129–1141 (2006).

19. S. M. Helmke, C.-Y. Yen, et al, Simultaneous quantification of human cardiac α- and β-myosin heavy chain proteins by MALDI-TOF mass spectrometry, *Anal. Chem.* **76**, 1683–1689 (2004).

20. M. Cui, M. A. McCooeye, C. Fraser, and Z. Mester, Quantitation of lysergic acid diethyl amide in urine using atmospheric pressure MALDI ion trap mass spectrometry, *Anal. Chem.* **76**, 7143–7148 (2004).

CHAPTER 15

MISCELLANEOUS TOPICS

In this chapter we review briefly a very broad and diverse range of topics not covered in previous chapters. These topics include enzyme kinetics, imaging mass spectrometry, identification of microorganisms, clinical mass spectrometry, forensic mass spectrometry, metabolomics, and screening of combinatorial libraries.

15.1. ENZYME KINETICS

Enzymes are proteins that catalyze specific reactions with a high degree of efficiency. The study of enzyme kinetics plays a critical role in understanding the chemical nature, biochemical activity, and biological significance of enzymes. Mass spectrometry is well suited to studying the mechanism of enzyme-catalyzed reactions, owing to its ability to determine unambiguously the structure of substrates, intermediates, and products in enzyme reactions [1,2]. Enzyme kinetic experiments can benefit from the high sensitivity, wide dynamic range, and well-established quantitation protocols of mass spectrometry.

15.1.1. Theory

According to the Michaelis–Menten theory, which is a widely accepted mechanism of an enzyme action, a substrate, S, is converted by an enzyme, E, to a product, P, through a reversible first step that leads to an intermediate enzyme–substrate complex, ES [3,4]. In other words, as shown in

$$E + S \underset{k_{-1}}{\overset{k_1}{\rightleftharpoons}} ES \overset{k_2}{\longrightarrow} E + P \qquad (15.1)$$

Fundamentals of Contemporary Mass Spectrometry, by Chhabil Dass
Copyright © 2007 John Wiley & Sons, Inc.

the enzyme first forms a complex with a low-energy barrier, which dissociates to form the product. In this equation, k_1 and k_{-1} are the rate constants of the first-step forward and reverse reactions, respectively, and k_2 is the rate constant of the second-step reaction. The enzyme reactions follow first-order kinetics, with the rate of the reaction, v, expressed as a function of either a decrease in the substrate concentration or an increase in the product concentration with time:

$$v = -\frac{d[S]}{dt} = \frac{d[P]}{dt} \tag{15.2}$$

The fundamental equation of enzyme kinetics is known as the Michaelis–Menten equation,

$$V_0 = \frac{V_{\max}[S]}{K_m + [S]} \tag{15.3}$$

in which the Michaelis constant, K_m, and the maximum velocity, $V_{\max}$, are the key kinetic parameters, and V_0 is the initial velocity. K_m is a measure of the enzyme activity and is equal to $(k_{-1} + k_2)/k_1$ in Eq. (15.3). Another key parameter is the catalytic constant, K_{cat}:

$$K_{\text{cat}} = \frac{V_{\max}}{[E_t]} \tag{15.4}$$

where $[E_t]$ is the total initial concentration of the enzyme. K_{cat}, also known as the *turnover number*, defines the number of catalytic cycles (i.e., the number of molecules of a substrate that the enzyme can complex per unit time and convert it into a product) that the enzyme can undergo in a unit time.

These parameters are measured experimentally. A typical enzyme reaction is conducted at a fixed amount of enzyme and by varying the concentration of the substrate. The plot of the initial rate versus the substrate concentration is a rectangular hyperbola through the origin (Figure 15.1). Initially, the rate increases rapidly and is proportional to [S] (i.e., the reaction is first order), but it levels off asymptotically at a higher [S] value (i.e., the reaction becomes pseudo zero order). At this point the enzyme becomes saturated with the substrate molecules, and the rate approaches the limiting value, $V_{\max}$.

A more convenient plot is a double-reciprocal plot known as the Lineweaver–Burk plot. [4,5] To draw this plot, the Michaelis–Menten equation [Eq. (15.3)] is transformed to

$$\frac{1}{V_0} = \frac{K_m}{V_{\max}[S]} + \frac{1}{V_{\max}} \tag{15.5}$$

and $1/V_0$ is plotted versus $1/[S]$ to yield a straight line (Figure 15.2), from which one can obtain more accurate values of K_m and $V_{\max}$. The slope of this line is equal to $(K_m/V_{\max})$, and the x- and y-intercepts are equal to $-1/K_m$ and $1/V_{\max}$, respectively. One disadvantage of this plot is that it gives a grossly misleading representation of the experimental error in V_0. An improvement over this plot is the Hanes–Woolf plot, which graphs $[S]/V_0$ versus $[S]$ (the straight-line equation

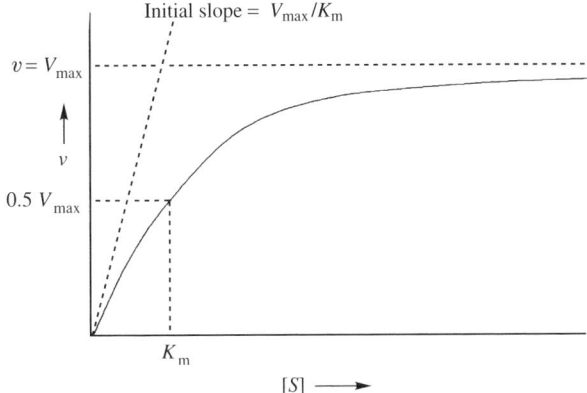

Figure 15.1. Michaelis–Menten plot showing the dependence of the rate v versus the substrate concentration $[S]$.

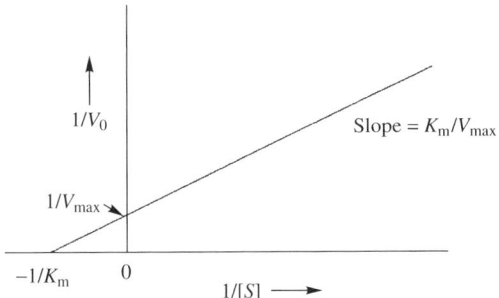

Figure 15.2. Lineweaver–Burk plot.

is obtained by multiplying both sides of Eq. (15.5) by [S]) [4]. This plot is a better reflection of V_0 errors.

Although enzyme reactions are highly specific, inhibition of the enzymes do occur. *Inhibitors*, substances that decrease the rate of an enzyme-catalyzed reaction, are classified as competitive, noncompetitive, uncompetitive, or mixed [4]. Each type can be characterized by deviation from the Lineweaver–Burk plot of the corresponding uninhibited reaction. Competitive inhibitors compete for the active sites with the substrate and slow down the enzyme reaction; they increase K_m but have no affect on V_{max}. Noncompetitive inhibitors bind reversibly to the enzyme at a site different from the active site, but one that is necessary for the enzyme action. These inhibitors decrease V_{max}, but K_m is unaffected. Uncompetitive inhibitors are known to bind reversibly to the enzyme–substrate complex to form an inactive enzyme–substrate-inhibitor complex. A decrease in K_m and V_{max} by the same factor is observed (i.e., the Lineweaver–Burk plot is parallel to the plot of the uninhibited reaction). In mixed-type inhibitors, more than one of the foregoing mechanisms operate, and K_m and V_{max} values are both altered.

504 MISCELLANEOUS TOPICS

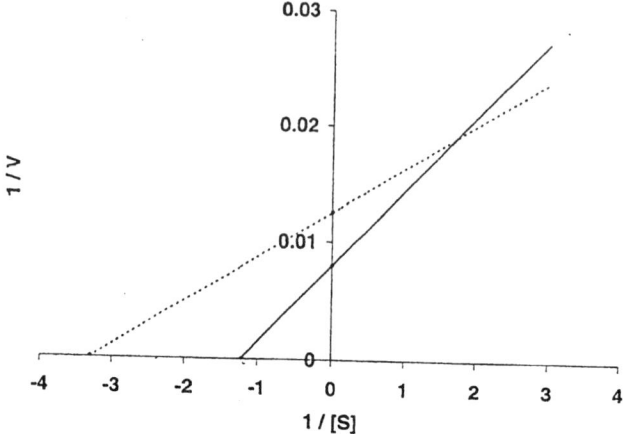

Figure 15.3. Lineweaver–Burk plots for the inhibition of aminopeptidase-catalyzed proteolysis of methionine enkephalin by acetylated methionine enkephalin: inhibited reaction (solid line) and the noninhibited reaction (dotted line). (Reproduced from ref. 6 by permission of Elsevier Science, copyright © 2000.)

Kinetic experiments for inhibition reactions are also conducted as mentioned above for the noninhibited reaction (i.e.; by varying the substrate concentration in the presence of a fixed concentration of the inhibitor) [6]. Figure 15.3, in which the data for the inhibition of the aminopeptidase-catalyzed proteolysis of methionine enkephalin by acetylated methionine enkephalin are plotted, shows an example of a mixed type of inhibition [6].

15.1.2. Reaction Monitoring

Depending on the rate of the enzyme reaction, the reaction mixture can be sampled for mass spectrometry measurements in one of the following three ways [1,2].

Off-Line Reaction Monitoring For slow enzyme reactions and long-lived intermediates, off-line reaction monitoring is more convenient. In these methods, the known amounts of an enzyme and a substrate are mixed together and incubated at physiological conditions. Aliquots are withdrawn at predetermined intervals, and the reaction is quenched immediately. The time course of the reaction can be monitored with mass spectrometric analysis immediately or later, at a more convenient time, by using either fast atom bombardment (FAB) [7], continuous-flow (CF)–FAB [7], matrix-assisted laser desorption/ionization (MALDI) [8], or electrospray ionization (ESI) [9–11]. Established quantitation procedures can be employed to monitor the concentration of the reactant or product (usually, the latter) (see Chapter 14). As an example, an appropriate internal standard that has no affinity for the enzyme can be added to the reaction mixture to improve

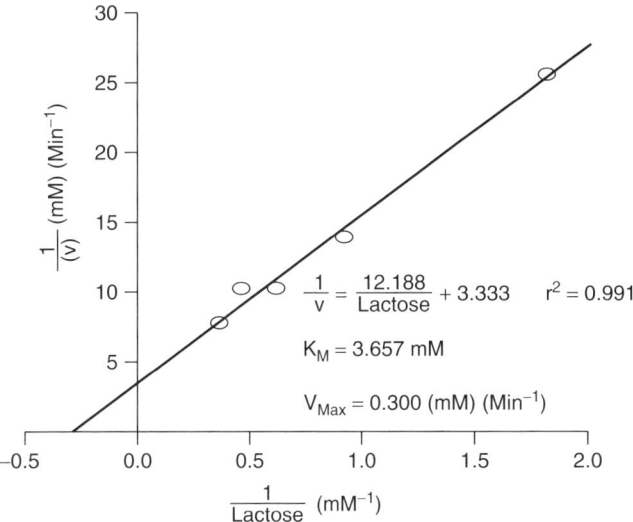

Figure 15.4. Lineweaver–Burk plot for the hydrolysis of lactose by galactosidase. (Reproduced from ref. 12 by permission of Elsevier, copyright © 1995.)

the accuracy of the measurement, and the signal is usually recorded in the selected-ion monitoring (SIM) mode. Often, sample cleanup is required prior to mass spectrometric analysis. In this respect, methods that employ the hyphenated techniques, such as liquid chromatography (LC)/MS and capillary electrophoresis (CE)/MS, have the advantage of simultaneous sample cleanup during the analysis. Figure 15.4 illustrates the use of LC/MS to study the hydrolysis of lactose by galactosidase [12]. Kinetic data for this double-reciprocal plot were obtained by withdrawing aliquots at 1-min intervals over 10 min. For each initial substrate concentration (in the range 0.505 to 2.912 mM), initial velocities were calculated from the plot of unreacted lactose versus the sampling time.

Continuous Online Monitoring   In this method the reaction mixture is introduced directly into the ion source of a mass spectrometer to provide continuous real-time monitoring of the enzyme reaction. The time resolution of the reaction is very short and is on the order of the data acquisition cycle of the mass spectrometry system. This procedure makes it possible to characterize the short-lived reaction intermediates. Because of the online sampling, additional reaction-quenching and sample-cleanup steps are dispensed. The solution composition, however, imposes a restriction on the choice of the ionization method unless a chromatographic separation device is coupled online to clean up the reaction mixture. The coupling of chromatographic would, however, adversely affect the accessible time resolution of the sampling rate.

At present, ESI is an ideal choice for continuous online monitoring of enzyme reactions [13–15]. The electrospray syringe itself can be converted into a reaction vessel. This arrangement has been employed to study the hydrolysis of O-nitrophenyl-β-D-galactopyranoside by lactase [13] and of an N-acetylglucosamine hexamer by hen egg white lysozyme [14]. In the past, CF–FAB has been used to study the time course of the proteolytic digestion of peptides [16].

Rapid-Mixing or Stop-Flow Monitoring To elucidate the mechanism of very fast enzyme reactions, it is critical to monitor the initial events before the system has a chance to attain a steady state. In many cases, the intermediate transient species exist for only few milliseconds. The fast enzyme reactions can be monitored readily by combining rapid mixing and quenching devices with mass spectrometry. The multiplex detection and fast scanning capabilities of time-of-flight (TOF) and Fourier transform (FT)–ion cyclotron resonance (ICR) mass spectrometers make them ideal for this combination.

Two devices are in common use to study the transient species in enzyme reactions: a rapid-mixing (or stop-flow apparatus) and a quench-flow apparatus. These devices are similar to the ones discussed in Section 10.2.2 to study the folding/unfolding of proteins via a hydrogen/deuterium-exchange process. For instance, a rapid-mixing apparatus consists of two syringes, one for the delivery of the enzyme solution and the other for the substrate solution (Figure 15.5). These

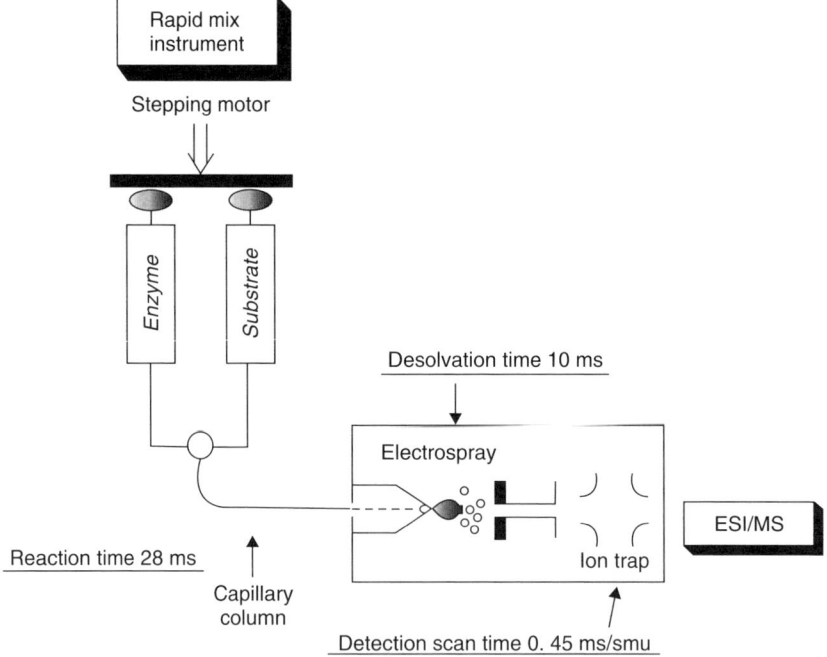

Figure 15.5. Schematic of a rapid-mixing apparatus coupled on-line with ESI–ion trap MS. (Reproduced from ref. 17 by permission of the American Chemical Society, Washington, DC, copyright © 1997.)

syringes are driven by computer-controlled stepper-motors. The two solutions are mixed in a mixing chamber to initiate the reaction, and the mixture is immediately pushed into the ESI ion source. Examples of enzyme reactions studied by this type of apparatus are the reactions of shikimate-3-phosphate and phosphoenol pyruvate with the enzyme ESPS synthase [17]. The quench-flow apparatus is also similar, except that the rapid mixing is followed by the addition of a quenching solution after a desired interval.

15.2. IMAGING MASS SPECTROMETRY

An important current application of mass spectrometry is the direct imaging of molecules embedded within a biological tissue with the objective to determine their chemical composition [18–20]. Desorption/ionization techniques of secondary-ion mass spectrometry (SIMS) and MALDI have been used successfully to accomplish this objective. The techniques are complementary. For example, SIMS can probe samples to a shallow depth (ca. 100 nm) with nanometer-level lateral resolution; MALDI can probe to a greater depth but has a lower spatial resolution. With these imaging approaches, two-dimensional images can be retrieved through a series of position-correlated mass spectra. Imaging allows a variety of analytes in their native and unmodified form to be probed in tissues. It is also possible to correlate ion images with histological features that had been observed with optical spectroscopy [21]. Imaging protocols provide an opportunity to collect information at the tissue level on healthy and diseased cells, with the possibility of diagnosis of various diseases, such as stroke, cancer, Alzheimer's, and others.

This field, known as *imaging mass spectrometry* (IMS), is a two-step process: mapping and imaging. *Mapping* is defined as the distribution of a target analyte in a given tissue, and *imaging* is presentation of the raw data in a picture form. An image displays ion abundance versus the spot position for any mass or set of masses. In general for mapping experiments, a beam of ions or photons is focused accurately at a predetermined location on the tissue, and the secondary ions emitted are detected with a TOF mass spectrometer. The point of focus is moved in a raster pattern to obtain a two-dimensional map of the chemical species. Signal intensities could be collected from a large numbers of points (e.g., 200×200 points) on the sample. By using computer software, this large set of data can be converted to a digital image.

Sample preparation is a major concern in acquiring representative data and must meet the following requirements: (1) target analytes should not change their in vivo position during sample preparation, (2) the tissue must be sectioned in plane slices for SIMS and MALDI analyses, (3) the tissue sections must be mounted on conducting surfaces (e.g., silicon wafers, metal slides, or conducting polymers), (4) no spreading or smearing of the sample analyte should occur during the addition of the MALDI matrix, and (5) the sample must be vacuum compatible.

15.2.1. Imaging with SIMS

In SIMS imaging, the tissue sample is bombarded with a primary-ion beam [18,22,23]. SIMS has the advantages of enhanced surface sensitivity and submicrometer imaging capability. In addition, the sample can be imaged "as received," without the addition of a matrix. SIMS provides analysis of metallic cations and low-mass molecular species. The instruments used are microprobes. A typical triple quadrupole-based instrument is shown in Figure 15.6. It consists of a primary-ion gun and a secondary-ion source to focus the secondary ions from the sample into the mass analyzer. TOF- and quadrupole-based instruments are also popular. For raster scanning, the beam is deflected by applying voltages to two pairs of plates. The primary beam is generated by a liquid metal ion (LMI) gun. For example, a Ga^+ ion beam can be focused on a spot of less than 50 nm. Other monoatomic ions that have served as a primary-ion beam are Xe^+, Cs^+, and In^+. Alternatively, molecular ion beams from SF_6 and Cs_xI_y have also been used. Current interest has focused on the use of cluster ion beams of Au_3^+, Bi_3^+, and C_{60}^+ projectiles [22]. Cluster-ion bombardment enhances secondary-ion yields nearly 300-fold and improves the quality of images. Because of very little subsurface damage with C_{60}^+ ion bombardment, sequential removal of the sample layers occurs to allow improved depth profiling and three-dimensional molecular imaging.

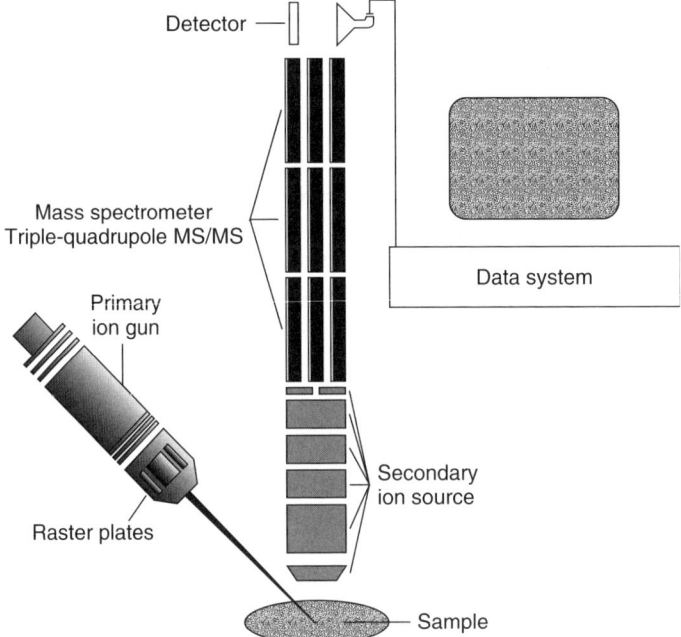

Figure 15.6. Schematic of an imaging SIMS microprobe. (Reproduced from ref. 18 by permission of the American Chemical Society, Washington, DC, copyright © 1997.)

Two variations of conventional SIMS have evolved. In one, known as *matrix-enhanced SIMS*, the MALDI sample preparation protocol is used to coat the sample surface with an organic acid (e.g., 2,5-dihydroxybenzoic acid) or other suitable matrix material to improve ionization efficiency [24,25]. This approach has been used to image brain tissue samples from freshwater snails (*Lymnaea stagnalis*). Another variation employs a coating of a thin layer of gold or other metals to enhance analytical signals [26]. This method, known as *metal-assisted SIMS*, provides images with improved spatial and chemical resolution.

15.2.2. Imaging with MALDI–MS

MALDI, in combination with TOF-MS, has emerged as a valuable technique to provide images of tissue samples. MALDI is effective for compounds in the m/z range 500 to over 100,000 [19,20]. It has made a tremendous contribution especially, in the imaging of peptides and proteins in biological samples. A typical experimental protocol for MALDI profiling and imaging of tissue samples is illustrated in Figure 15.7. Profiling experiments measure target compounds from several discrete sample spots; typically, up to 10 spots on a given tissue are compared. In an imaging experiment, the image of either the entire sample or a selected area is obtained. In both experiments, frozen tissue sections of 10 to 20 μm thickness are cut at $-15°C$ and placed on a MALDI target plate. For profiling analysis, the matrix (e.g., sinapinic acid for protein analysis) is applied to tissue sections as discrete spots. Each spot is irradiated with a laser beam (e.g., a 335-nm nitrogen laser). For the imaging experiment, the entire sample is coated with a matrix. Several discrete spots are selected for laser-beam irradiation. Resolution of the image is a function of the size and spacing of the *laser spots* (also called *pixels*). A high-resolution image might contain many thousand pixels. Typically, each spot is irradiated with 30 to 200 laser shots. The signal intensity for each m/z value at each spot is integrated and used to construct a two-dimensional ion density map. Thus, several hundred images can be obtained from a single acquisition.

In addition to frozen tissue sections, tissue blotting and laser-capture microdisection methodologies can also be used. In the former method, the proteins are transferred by blotting freshly dissected tissues onto an active C_{18}-coated surface. Another surface of choice for blotting is a carbon-filled polyethylene membrane. In the laser-capture microdisection method, a specific population of cells from a stained tissue is transferred onto an ethylene–vinyl acetate transfer film. A matrix is applied to the isolated cells through a narrow capillary.

Application of a matrix to tissue sections is a critical step in MALDI imaging. As mentioned above, the target analytes should not change their in vivo position and should not spread or smear during use of the MALDI matrix. The following procedures have been developed to achieve a uniform coating of a matrix:

1. Airspray coating, in which the matrix solution is sprayed over the sample with a pneumatic airbrush. Several coatings are applied and the sample is dried between applications.

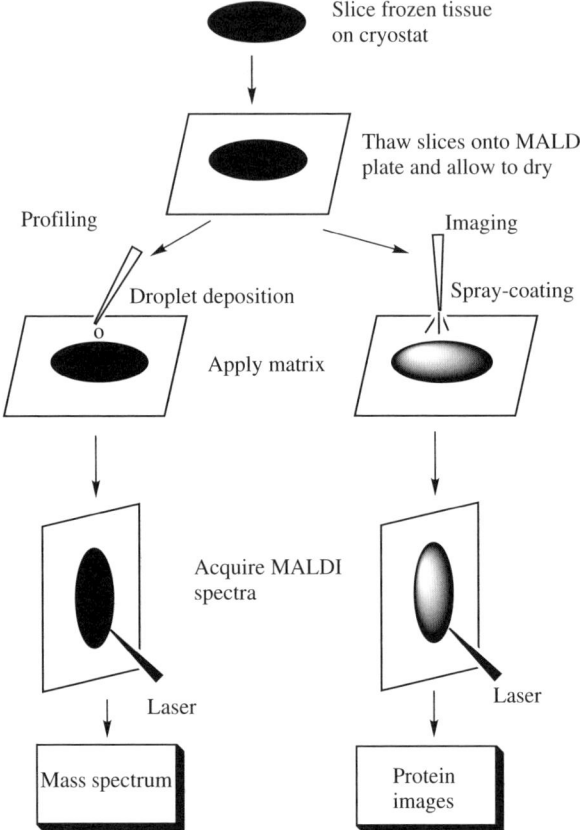

Figure 15.7. Experimental protocol for MALDI profiling and imaging of mammalian tissue samples. (From ref. 19.)

2. In electrospray deposition, the matrix solution is electrosprayed through a needle that is attached to a two-dimensional plotter. The needle is moved across the sample in a regular pattern. This procedure provides better control of the amount of the matrix deposited and yields a homogeneous coating.
3. In the sliding-drop method, a large volume of matrix is deposited along a side of the tissue sample and is distributed over the tissue using a thin spatula. In this procedure there is a possibility of migration of analytes from their in vivo position.

Several studies have demonstrated the ability of MALDI to profile and image phospholipids, peptides, and proteins in tissue samples [19].

15.3. ANALYSIS OF MICROORGANISMS

The identification of microorganisms is an important step in the protection of public health and crops and in civil-, criminal-, and terror-related forensic analyses. Illness due to microbial infection is an ever-increasing human health hazard. Thus, the accurate identification of microorganisms can be a powerful tool to diagnose diseases, monitor potential contamination in foods, and recognize biological and environmental hazards. *Chemotaxonomy*, a field of microbial identification, is based on the detection of a subset of cellular components (i.e., biomarkers). This field can not only distinguish among related organisms, but can also identify species and strains. The basis of identification is the presence of unique biomarkers. Any chemical component in cells, such as lipids, oligosaccharides, proteins, and DNA/RNA, can serve as biomarkers to classify microorganisms. Protein biomarkers have the advantage of less interference from other low-mass species that are inherently more abundant in the bacterial cells. In this endeavor, mass spectrometry, especially MALDI–MS, can provide rapid and accurate information to bacterial taxonomy. The advantages of MALDI–MS are the direct analysis of bacterial cells, the ability to analyze large molecules such as proteins, and low amounts of sample. In the past, FAB, laser desorption, plasma desorption, and pyrolysis GC/MS have been used to analyze bacteria [27].

15.3.1. Bacterial Identification

One way to identify bacterial species is to analyze the extracts of bacteria [28,29]. Several pathogenic bacterial strains, including *Bacillus anthracis*, *Yersinia pestis*, and *Brucella melitensis*, can be distinguished from nonpathogenic species on the basis of characteristic protein ions in the mass range 2400 to 10,000 Da [28]. Various *Bacillus* species (e.g., *anthracis*, *subtilis*, *cerus*, and *thuringiensis*) also exhibit distinct mass spectrometry data. The strain-specific biomarkers for six different strains of *Helicobacter pylori* were obtained with MALDI–MS [29].

Another approach is to analyze intact bacteria rather than the extract, using MALDI–MS [30]. This technique, usually termed *whole-cell MALDI–TOF–MS*, affords rapid identification of bacterial strains. **Whole-cell** signifies that the cells have not been processed for isolation of any chemical species (e.g., proteins) but that the intact cell was subjected directly to MALDI–MS analysis. In a typical procedure the cells are suspended in a MALDI matrix solution and the mixture is deposited on the MALDI target. The bacterial spectra depend on a number of experimental factors (e.g., sample preparation method, solvents, matrix composition) and microbial variables (culture medium and cell growth time). Despite these variations, a number of peaks are conserved that can serve as potential biomarkers [31]. MALDI spectra contain characteristic ions in the m/z range 3000 to 10,000. Marker ions in this fingerprint region usually are adequate to differentiate among bacterial strains. As an example, two different species

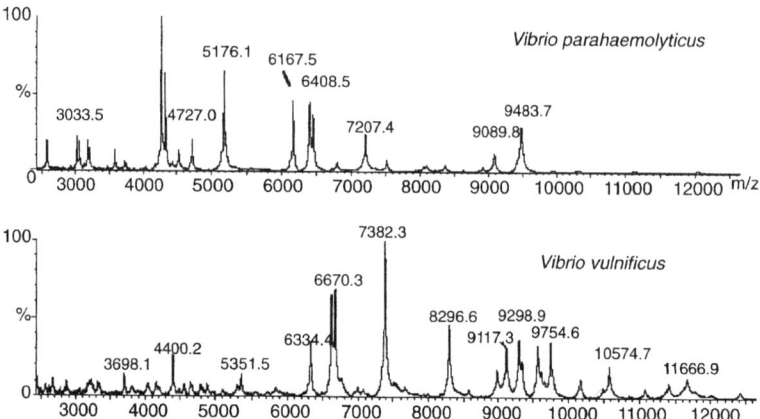

Figure 15.8. Whole-cell MALDI–TOF spectra of two different *Vibrio* species. (Reproduced from ref. 32 by permission of Wiley-Interscience, copyright © 2006.)

(*V. parahaemolyticus* and *V. vulnificus*) of *Vibrio* bacteria can be distinguished on the basis of their distinct MALDI spectra (Figure 15.8).

Improved sample stability can extend the mass range of biomarker detection beyond 10,000 Da [33,34]. For example, a biomarker for *H. pylori* has been detected at m/z 58,268 and for *H. mustelae* at m/z 49,608 and 57,231 [34]. Similarly, characteristic biomarkers have been observed at m/z 10,074 and 25,478 for *Campylobacter coli*, at m/z 10,285 and 12,901 for *C. jejuni*, and at m/z 10,726 and 11,289 for *C. fetus* [34]. Thus, *Campylobacter* species could be readily distinguished from *Helicobacter* species.

Proteomics has also been used to determine differences among closely related bacterial species [35]. That method involves protein extraction, two-dimensional gel electrophoresis of the extracted proteins, proteolytic digestion of proteins, mass spectrometric sequence determination of the peptides in a digest, and identification of proteins via database search (see Section 8.6). Protein expression is compared to provide a distinction among various closely related bacterial strains. Another method is based on a comparison of intact protein profiles of individual bacteria strains [36]. The protein extract is analyzed by LC/ESI–MS and the data are deconvoluted automatically into a single peak to provide protein profiles that contain the nominal masses of intact proteins. A rapid identification method is also available in which a whole bacterial cell suspension is infused directly into an ESI source. This technique, termed *direct infusion ESI–MS*, has been used to discriminate among 36 strains of aerobic endospore-forming bacteria consisting of six *Bacillus* species and one *Brevibacillus* species [37].

It is also feasible to use profiles of small molecules, such as fatty acids, phospholipids, glycolipids, and peptides, to distinguish among various bacterial species. For example, *Francisella tularensis*, *Brucella melitensis*, *Yersinia pestis*, *Bacillus anthracis*, and *Bacillus cerus* species have been discriminated on the basis of their distinct fatty acid profiles [38]. The samples can be ionized by EI, positive-ion CI, or negative-ion CI. The positive-ion CI mode produces the greatest amount of differentiation among the four genera of bacteria when combined with principle component analysis (PCA).

15.3.2. Analysis of Viruses

Viruses are made up of molecules of nucleic acids that are encased in an envelope of proteins called *capsid proteins*. When they are part of the outermost shell, envelope proteins are referred to as *coat proteins*. These proteins usually exist as a noncovalent association of protein subunits and are responsible for an array of functions, including cell attachment, cell entry, and RNA release. The study of viruses can shed light on their molecular biology and pathology and on their possible interactions with antibodies. Similar to mass spectrometry identification of bacteria, viruses are characterized on the basis of unique marker ions [39–45]. Because capsid proteins are unique to each virus, they are good candidate as biomarkers.

One mass spectrometric approach has used ESI–MS either alone or online with LC. Viral proteins from intact viruses can be characterized without preliminary extraction or a disruption step [43]. Using this methodology, cricket paralysis virus (CrPV) has been identified on the basis of four unique capsid proteins: VP1, VP2, VP3, and VP4. It is also possible to detect viral peptides after gradual enzymatic digestion of the viral surface [42].

MALDI–MS is another approach that can be used successfully to characterize viruses. This technique has been used to identify tobacco mosaic virus U2, MS2 bacteriophage, and Venezuelan equine encephalitis (VEE) virus [40]. The spectrum contains molecular ions of the coat proteins that are highly characteristic of species and strains. MALDI–TOF–MS has also been used to analyze a chemically modified recombinant hepatitis B surface antigen (HBsAg) [41]. HbsAg is a very large molecule composed of multiple copies of a protein subunit called *S antigen*, the glycosylated form of this subunit, and a lipid membrane. MALDI–MS analysis is able to distinguish between protein subunits (S antigen) and biotin-labeled S antigen.

Another example of the use of MALDI–MS is to characterize Sindbis virus [44]. A glycoprotein was identified as a biomarker unique to this virus. The glycoprotein was released by treating the suspension of the Sindbis virus with a solvent mixture of 0.5% *n*-octyl glucoside (OG) and 0.5% trifluoroacetic acid (TFA). TFA disrupted the lipid bilayer and solubilized the capsid protein and OG helped to solubilize integral membrane glycoproteins. Figure 15.9 shows the MALDI–MS spectrum of the virus. Peaks due to a capsid protein and glycoproteins are clearly visible.

15.4. CLINICAL MASS SPECTROMETRY

Clinical testing provides patient-derived chemically relevant data to diagnose a disease. Mass spectrometry has made a valuable contribution to the routine diagnosis of a variety of clinical disorders [46]. A list of such applications would

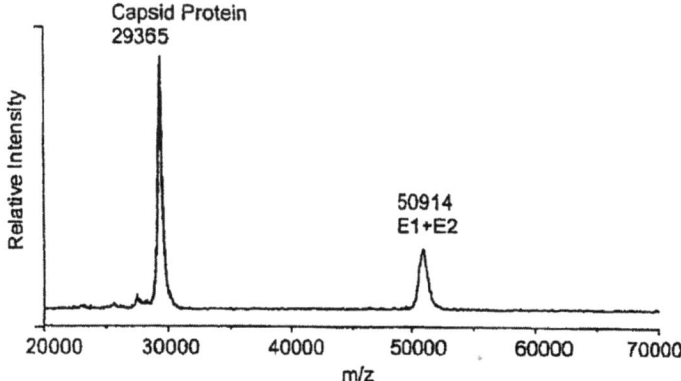

Figure 15.9. MALDI–MS spectrum of the Sindbis virus. The peak at m/z 50,914 is due to E1 and E2 glycoproteins. (Reproduced from ref. 44 by permission of the American Chemical Society, Washington, DC, copyright © 2001.)

be too large, however, and a discussion of all those applications is beyond the scope of this chapter. Nevertheless, a few notable examples are presented here.

15.4.1. Low-Molecular-Mass Compounds as Biomarkers of Disease

Metabolites are compounds that are the end products of cellular regulatory and metabolic processes. They are mostly low-molecular-mass compounds (LMC; <1000 Da). Defective genes might produce abnormal proteins with altered enzymatic activity, which may have adverse metabolic consequences. Impaired metabolic activity could lead to cellular damage and to a disease. Thus, the concentration levels of abnormal metabolites can be taken as evidence of a specific disease. Mass spectrometry has provided an important platform to identify and quantify several classes of LMC, with the aim of diagnosing metabolic disorders. As an example, GC/MS is used to obtain metabolic profiles of a diverse range of compounds, such as organic acids, free fatty acids, homocysteine, caritines, bile acids, cholesterol, steroids, phenolic compounds, and biogenic amines. Urine and plasma are common body fluids that are used for LMC analysis. Often, derivatization is required. LC/MS is also used for some of these polar compounds. Lipids are characterized with LC/MS.

A typical example of the use of GC/MS is in the detection of acidemia (defined as an accumulation of organic acids in cells or body fluids), which can manifest in infants. Propionic acidemia can induce lethargy, vomiting, acidosis, hypoglycemia, and possibly death in infants. To diagnose these disorders, organic acids are extracted from body fluids, derivatized by trimethylsilylation, and profiled by GC/MS [46].

15.4.2. Analysis of DNA to Diagnose Genetic Disorders

Cystic fibrosis (CF) is a common form of an autosomal recessive genetic disease within the Caucasian population. The mutant gene lacks three base pairs (bp) in exon 10, resulting in the loss of Phe at codon 508 (ΔF508). MALDI–MS can be used in the molecular diagnosis of mutant alleles. In this procedure, the DNA from exon 10 of the gene from carrier and normal patients is amplified by a polymerase chain reaction (PCR) and cleaved with MseI enzyme to produce a 59-bp fragment from the control CF gene and a 56-bp fragment from the mutant gene [47]. Negative-ion MALDI analysis showed that the 56-bp fragment was present in carrier and CF patients. Thus, detection of this fragment forms the basis of a diagnostic test of CF.

15.4.3. Proteins as Biomarkers of Disease

A defective gene can be translated into an abnormal protein, which may have clinical implications if the altered amino acid composition affects the protein activity. Thus, proteins can serve as biomarkers of specific diseases. A specific example of the use of mass spectrometry is the diagnosis of familial transthyretin amyloidosis (ATTR), a hereditary degenerative disease that is caused by the mutation of a single amino acid in transthyretin [48]. The detection and identification of transthyretin variants can thus provide a definitive diagnosis of ATTR. HPLC/ESI–MS was used to identify the transthyretin mutants, and MALDI–MS of the digests of the isolated HPLC fractions unequivocally detected the site of mutation. The analysis of serum from ATTR patients successfully detected the variants Val30 → Met, Val22 → Ile, and Ser23 → Asn [48].

A mass spectrometry-based procedure exists to detect rare diseases that are caused by a deficiency in specific enzymes [49]. This strategy involves an assay of the enzyme with its synthetic substrate conjugate and analysis by ESI–MS. For example, to diagnose glycogen-storage disease, which is caused by a β-galactosidase deficiency, the substrate conjugates were incubated with cultured fibroblasts from control and affected subjects, labeled internal standards (deuterated conjugates that lacked the sugar moiety) were added, and biotinylated components were separated from the mixture with biotin-binding streptovidin. After multiple washings, the biotinylated components were released and analyzed by ESI–MS. Very little enzyme product was detected in cells from patients with β-galactosidase deficiency compared to cells from controls.

Characterization of Hemoglobin Variants as Disease Markers Normal adult hemoglobin is a tetrameric protein that is composed of two α chains and two β chains. Structural variations in the chains of hemoglobin are responsible for hemoglobin disorders, which can manifest themselves in the form of a number of disease states, such as sickle cell anemia. For nearly 20 years, mass spectrometry–based methods have provided an accurate and rapid means to detect and characterize the structural changes that occur in the hemoglobin chains. Most of these methods rely on a mapping of tryptic digests to detect mutations. For example, CE has been combined with ESI to identify mutants of hemoglobin [50]. Comparing the peptide maps of the control and mutant proteins allows detection of the site mutation.

In a MALDI procedure, peptide maps from the hemoglobin samples were generated using trypsin-activated bioreactive MALDI probes [51]. Samples were obtained directly from whole human blood without further purification. The digestion was carried out by depositing a few microliters of diluted (1000×) blood on the active probe and heating the probe (37°C, 15 min). This technique is rapid and can detect some elusive tryptic peptides, such as αT12, αT13, and βT12.

MALDI–TOF–MS has the potential to detect intact hemoglobin variants [52]. In this procedure, 1 μL of a blood sample was collected on a nonporous polyurethane membrane, dried, and washed twice with water. The matrix solution was added, and after drying of the matrix, the sample was analyzed with DE–MALDI–TOF–MS. Figure 15.10 compares the DE-MALDI mass spectra of normal and Hb Shepherds Bush hemoglobin variant b74(E18)Gly → Asp. This variant, known to enhance oxygen affinity, expresses itself in patients as

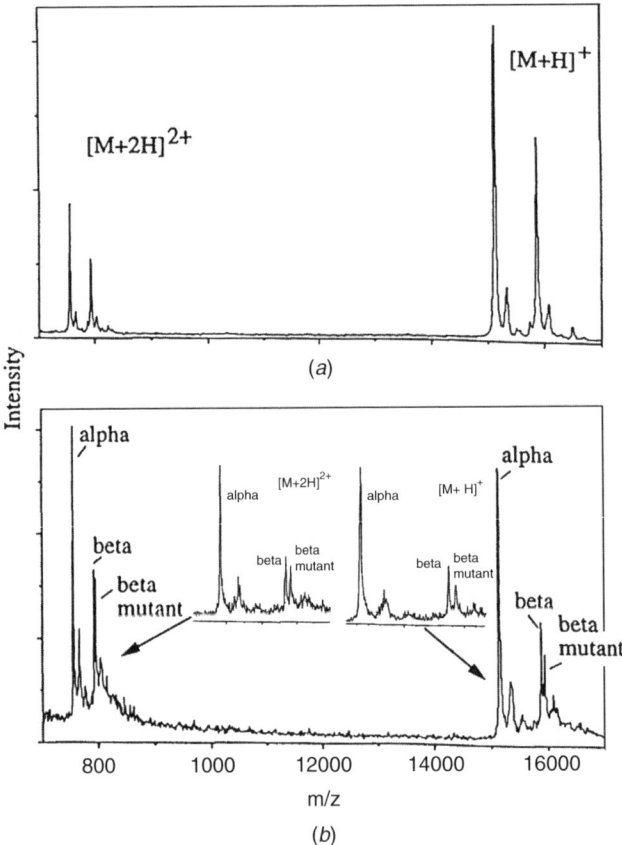

Figure 15.10. DE–MALDI–MS spectrum of (a) control and (b) Hb Shepherds Bush hemoglobin variant b74(E18)Gly → Asp. (Reproduced from ref. 52 by permission of the American Chemical Society, Washington, DC, copyright © 1998.)

mild hemolytic anemia. The variant b74(E18)Gly → Asp is indicated by a mass shift of 58 Da in the β-chain of hemoglobin.

Intact hemoglobin variants can also be detected with ESI-MS. The ESI mass spectrum contains a series of multiply charged ions that usually complicate the detection of a protein mixture. The data can, however, be software-transformed to provide a single peak that is unique to a specific protein. This approach has been used to identify hemoglobin from a blood sample from a newborn sickle-cell carrier [46].

15.5. METABOLOMICS

Metabolomics is the field of analysis of low-molecular-mass endogenous and exogenous metabolites. As mentioned above, metabolites are low-molecular-mass compounds that are produced by cellular and metabolic processes. The entire set of low-molecular-mass compounds is termed the *metabolome*. The concentration of each LMC can be correlated to the response of a biological system to environmental stress and pathology. Therefore, metabolomics has the potential to identify novel biomarkers for specific diseases. Two examples of the use of mass spectrometry for global metabolomic studies are presented here. One study employed GC/MS to profile LMCs in human blood plasma [53]. Blood plasma is a better sample choice because it is a source of a wide variety of chemically diverse LMCs. Most LMCs exist in plasma as noncovalent complexes with protein. In this study, deproteinization and extraction of LMCs were effected by adding methanol to the sample. The extract was derivatized with the usual GC derivatization reagents (e.g., N-methyl-N-trimethylsilyltrifluoroacetamide and trimethylchlorosilane) and analyzed by GC/TOF–MS. The chromatogram contained more than 500 peaks that belonged to different classes of LMCs (e.g., amines, amino acids, fatty acids, sterols, disaccharides). By adding 10 stable isotope–labeled internal standards that were representative of these diverse classes of LMCs, quantitative data on several endogenous compounds could be collected.

The second example represents a large-scale human metabolomics study that was performed with LC/MS [54]. The aim of this study was to identify potential biomarkers from lipid profiles of some 600 human plasma samples. Lipids were extracted from plasma samples and subjected to LC/ESI–MS analysis. Several different classes of lipids, such as phosphatidylcholines, lysophosphatidylcholines, triglycerides, diglycerides, sphingomyelins, and cholesterol esters were the target of this study. To detect small differences in metabolic profiles, statistical methods were used to process this large set of data. Partial least-squares discriminant analysis of the data could locate potential biomarkers.

15.6. FORENSIC MASS SPECTROMETRY

Forensic science deals with solving a crime. Analytical chemistry, especially mass spectrometry, has been at the forefront in gathering evidence of the crime. In addition, the early detection of explosives and biohazards can prevent catastrophic events. The profession of a forensic investigator has been glamorized by the recent popularity of certain TV shows. DNA fingerprinting is a commonly used method that can link or exonerate a suspect from the crime [55]. A typical forensic science laboratory might participate in the analysis of illicit drug samples or specimens, such as urine, blood, hair, plastic containers, tapes, and banknotes, that contain these drugs [56]. Other common materials analyzed to solve a crime include explosives, fire debris, paints, glass, bloodstain, firearms, gunpowder, clothing, blood alcohol, and forged documents. The field is vast and difficult to cover in this small section, but a few representative applications are highlighted.

15.6.1. Analysis of Banned Substances of Abuse

Because drug testing for controlled substances bears heavily on a person's reputation, the method to detect controlled substance must be accurate and must follow international guidelines, which advocate a method that must include chromatographic separation to isolate the analyte from other matrix components and mass spectrometric detection with either SIM of three compound-specific ions on a single mass analyzer or with selected-reaction monitoring (SRM) of two specified precursor–product reactions. GC/MS and LC/MS methods that follow these guidelines are used routinely to detect and quantify cannabinoids, narcotics, cocaine, amphetamines, and other substances of abuse [57].

Similarly, routine screening of banned anabolic androgenic steroids in urine samples is performed regularly using GC/MS and LC–MS/MS [58–60], in conjunction with SIM of a selected number of target ions or SRM transitions from a list of known steroids. New designer steroids can, however, escape detection by these routine screening procedures. One possible solution is to use precursor-ion scanning to detect unknown anabolic androgenic steroids [61]. The strategy has wide applicability because MS/MS of steroids produces common product ions. Another practical approach is based on LC with parallel bioactivity detection and mass spectrometry identification [62]. A typical procedure involves enzymatic deconjugation of steroids in the urine sample, solid-phase extraction (SPE), and LC separation of steroids. The effluent is split toward two identical 96-well fraction collectors. A bioassay is performed on one well plate, and identical fractions on the other well plate that yield bioactivity are analyzed for their accurate mass measurement. Elemental compositions calculated from accurate masses are used to search the electronic substance databases to identify the steroids.

Screening for alcohol abuse is also performed using the LC–MS/MS–SRM approach. Two metabolites of alcohol, ethyl glucuronide (EtG) and ethyl sulfate (EtS), are prime candidates for urine testing of alcohol drinkers [63,64]. These markers can be detected up to 36 hours following alcohol consumption. For the LC–ESI–MS/MS detection of EtS, the $[M - H]^-$ ion of m/z 125 was mass-selected as the precursor, and its product ions of m/z 97 ($[HSO_4]^-$) and 80 (SO_3^-) were detected via SRM [63]. D_5–Ets was used as the internal standard. Simultaneous detection of two markers can provide an effective workplace-monitoring program for alcohol consumption.

15.6.2. Analysis of Explosives

Civil aviation, mass transport, and urban centers have become increasingly vulnerable to terrorist activities. Explosives are a common form of weapon used by terrorists, with devastating effects. The detection of explosives in luggage, mail, and vehicles and on suspects is a one form of prevention. Highly sensitive, specific, and rapid analytical techniques and devices are required to detect explosives. Mass spectrometry is a major player in explosive detection because of its sensitivity and wide range of species recognition. For example, MS-based techniques are used to detect ignitable liquid residues (ILRs) in fire debris to provide evidence of arson. Gasoline is commonly used to set a fire. GC-MS/MS has been used successfully to analyze gasoline components in fire debris to provide ILR evidence [65]. The detection limits can be improved by incorporating solid-phase microextraction (SPME).

Ion mobility spectrometry (IMS)–based methods have been developed to detect explosives (see Section 3.10 for the basic principle of IMS). Explosive detectors based on this technology have been deployed at airports worldwide [66]. In these instruments, ions are allowed to drift in a field gradient against the flow of an inert gas. The drift time of the analyte-specific ion identifies the target analyte. For the detection of nitro-based explosives, the sample can be collected on cellulose or Teflon filters, which are pressed against the hot anvil of a detector to release the desorbed vapors into the IMS analyzer. Ionization of the vapors is accomplished by Cl^- ion attachment. Bench-scale analyzers have been designed to detect up to 40 explosives in about 8 s. Nitroglycerine, trinitrotoluene (TNT), cyclotrimethylene–trinitramine (RDX), pentaerythrite tetranitrate (PETN), and cyclotetramethylene–tetranitramine (HMX) are some of the nitro-based explosives regularly detected by these analyzers.

15.6.3. Analysis of Glass and Paints

Automotive accidents are investigated by analyzing glass and paints. One way to match the glass from a broken windshield found at a hit-and-run accident scene is to compare elemental profiles. Inductively coupled plasma (ICP) mass spectrometry is an ideal method for this purpose [67]. Acid digestion is used for sample preparation. The use of laser ablation, coupled with ICP–MS, obviates the need for the laborious and time-consuming acid-digestion step [68].

15.6.4. Authenticity of Questioned Documents

Another forensic use of mass spectrometry is to verify the authenticity of questioned documents. The age of an ink can answer this dilemma. This task can be accomplished by laser desorption/ionization (LDI) mass spectrometry [69]. Methyl violet is a typical dye in ballpoint inks. It is relatively unstable under light and undergoes demethylation upon exposure to light. In a case study, an LDI spectrum showed an abundant amount of $M^{+\bullet}$ ion of m/z 372.2 and less $[M - CH_3]^+$ ion of m/z 358.2. The light-exposure-aged specimen showed a decrease in abundance of the

M$^{+\bullet}$ ion and an increase in the abundance of a series of demethylated products. A plot of the relative peak area versus time could be correlated with the age of the document.

15.6.5. Mass Spectrometry in Bioterror Defense

Mass spectrometry is also playing a pivotal role on the bioterror front [70]. Currently, a great need exists for the rapid and positive identification of biological threat agents such as bacteria, toxins, and viruses. The MS-based approaches discussed in Section 15.3 can be applied to identify deadly microorganisms. Recently, the potential of shotgun proteomics has been demonstrated to identify biological threat agents in complex environmental matrices [71,72]. This methodology could discriminate a target species from others in a large database of background species.

Another example of the use of mass spectrometry to detect toxins is to identify ricin, a highly toxic protein that inhibits cell protein synthesis. Ricin is produced from the seeds of *Ricinus communis* plants (known commonly as castor beans) [73]. Structurally, ricin is made of A- and B-chains linked by a disulfide bridge. The toxicity of ricin is due primarily to the A-chain, which acts as an RNA *N*-glycosidase, which leads to ribosome incapacitation and ultimately to cell death. Ricin was identified from crude castor bean extracts using LC–MS/MS. The extract was denatured, reduced, and alkylated prior to trypsin digestion. Ricin identification was based on the detection of marker peptides in the digest. These markers include T5, T7, T11, T12, and T13 from the A-chain and T3, T5, T14, T19, and T20 from the B-chain. MS/MS can provide the amount and sequence of each marker for irrefutable evidence. For quick screening of ricin in crude extracts, MALDI–MS can be used to provide the molecular mass profile of the marker peptides.

15.7. SCREENING COMBINATORIAL LIBRARIES

Combinatorial chemistry is a fast-growing technology for synthesis of compounds on a grand scale and to test them rapidly for desirable properties. The technology has become especially attractive for rapid drug development and synthesis of biologically active molecules. It allows the synthesis of hundreds of thousands of compounds within a time period that would allow preparation of only a few compounds by conventional one-by-one synthesis and one-by-one testing protocol. The pharmaceutical industry and material science have greatly benefited from this technology.

In a typical combinatorial chemistry drug-discovery protocol, a large array of compounds, called *combinatorial libraries*, are produced and tested for high-affinity binding to enzymes, receptors, or antibodies. The compounds that provide a "hit" with the target are selected and further developed into "leads." Further optimization of leads narrows the potential drug candidates to one or two compounds that show a significantly high pharmacological activity and can be carried on to the next stages of human trial and approval. In contrast, the traditional approach to drug discovery has involved the synthesis of individual compounds, their isolation, and characterization through a series of chemical and biochemical assays. This one-at-a-time approach is expensive and time consuming. With the combinatorial chemistry protocol, libraries of peptides [74], oligonucleotides [75], carbohydrates, oligocarbamate [76], and other small organic compounds have been prepared [77].

15.7.1. Combinatorial Synthetic Procedures

Combinatorial libraries are prepared by the (1) parallel synthesis of arrays, (2) split-pool method, (3) biological method, or (4) spatially addressable parallel synthesis [74,78–80]. Parallel synthesis is carried out by the simultaneous synthesis of an array of different compounds. Several methods are available. In the *multipin method*, the peptide synthesis is carried out on polyethylene rods that have attached protected amino acids [81]. The amino acid sequence of a synthesized peptide on a particular pin depends on the order in which the amino acids are added. The number of products synthesized is the same as the number of pins. Another version of parallel synthesis, known as the *teabag method*, uses resin-filled bags in place of pins [74]. By pooling the resin portions from the appropriate bags, followed by redistribution and further coupling with a specific amino acid, a peptide library can be synthesized. The *SPOT method* uses a cellulose paper membrane as a solid support, which acts as an open reactor. Respective reagent solutions are pipetted onto several spots to synthesize as many peptides as the spots chosen [74,82].

In the *split-pool procedure*, also known as the *portioning-mixing method*, the solid support is first divided into as many equal portions as the number of amino acids in the peptide's sequence [74,75,83]. Each portion is coupled individually to only one amino acid. All portions of the resin are mixed, and the entire process of splitting and combining is repeated until all amino acids have been combined. An exorbitantly large number of derivatives can be synthesized at a time.

The *biological method* is based on cloning synthetic DNA that encodes random peptide sequences [74,84]. The concept is similar to the split-pool method except that the synthesis is carried out in a host bacterium. The DNA replication is achieved by infecting the host bacterium with phages that had been prepared by inserting selected oligonucleotides into their DNA. Each phage clone carries a different coat protein, the sequence of which depends on the oligonucleotide inserted.

In the *spatially addressable parallel synthesis method*, the libraries are synthesized on the surface of a glass slide that has been functionalized photolithographically with a photosensitive group [74,85]. Masking of some areas of the plate and irradiating the rest of the glass slide allow synthesis in the selected unmasked areas only. This selective photodeprotection permits the synthesis of different peptides on a single glass plate.

15.7.2. Screening Methods

Combinatorial chemistry generates an astronomical number of compounds in one synthesis. The screening of all the members of a library, with the intent to identify active components, is a major enterprise of combinatorial chemistry. The analytical procedure must be fast, have a high sample throughput, and be able to handle complex samples with a high data-processing speed. That is one area where mass spectrometry can play an important role because of its low-sample requirements and its ability to distinguish closely related compounds on the basis of their molecular mass and fragmentation pattern. Mass spectrometry has the further advantage that it can be combined with high-resolution separation devices to allow automated analysis of complex library components.

The most popular methods used to screen all the members of a library are based on a specific reaction of the library members (called *ligands*) with enzymes, receptors, or antibodies. These affinity-based methods use either immobilized receptors, immobilized ligands, or binding activity in solution [77,86]. A variety of solid supports, such as beads, microtiter wells, or chromatography supports, are convenient for the immobilization of receptors. The solid-support libraries are screened directly for the binding affinity of ligands tethered to the labeled receptor. The identity of the ligand is revealed by reading the encoding tag for its specified location or by the structural analysis procedure. The solution-phase libraries are tested one at a time with classical bioassays for the biological activity desired. The libraries that are synthesized on solid supports can also be tested in solution after their release from the solid support.

Testing of each component of the library, one at a time, is the desired goal of any screening method but has constraints of time and resources. One way to increase the sample throughput is to perform several parallel analyses. An alternative approach to speed up the screening process is to reduce the number of samples for analysis by selecting only a random number of library members [87]. An improvement over this protocol is *iterative deconvolution* [88]. In one such approach, called the *mimotope approach* [89], a pool of compounds of soluble libraries is tested first; based on this outcome, a smaller pool of compounds is synthesized and sublibraries are retested. The process is repeated until a compound with the highest activity is identified.

Mass spectrometry–based techniques to identify combinatorial libraries make use of ESI–MS [90], HPLC/ESI–MS [91], MALDI–MS [92], and a combination of ESI with IMS [93,94] and FT–ICR–MS [95,96]. The last one has the advantage that the molecular mass of the analyte can be measured at ultrahigh resolution to allow the separation of nominal isobaric ions. A better fingerprint of the library can be obtained at high mass resolution.

Analysis of Support-Bound Libraries Irrespective of the synthetic method used for constructing a library, its screening relies on two strategies: in one, testing for biological activity is performed when the library members are still attached to the solid support that was used for synthesis. In the second approach, the library compounds are released from the solid support and subjected to

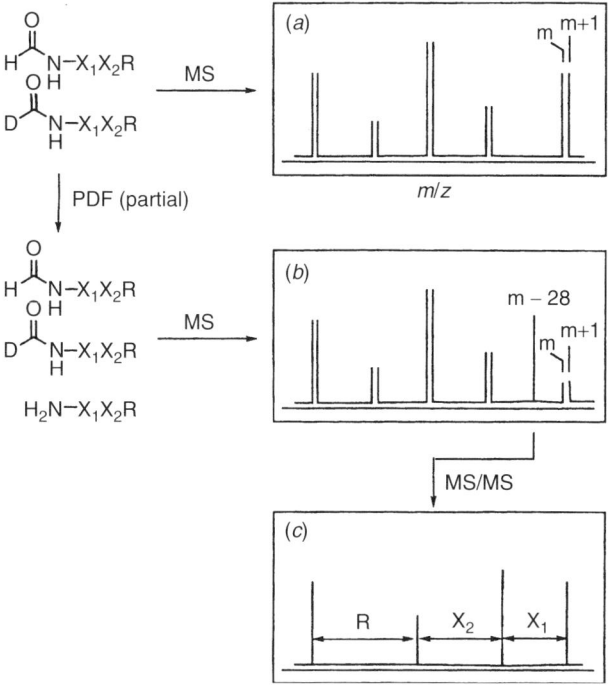

Figure 15.11. Enzyme substrate screening protocol. (Reproduced from ref. 95 by permission of Wiley-Interscience, copyright © 2001.)

biological testing in solution. An example of this second approach is identification of enzyme substrates [95]. In this method, a 361-member N-terminally tripeptide library was synthesized on a solid-phase support by the split-pool synthesis method. To identify the library members, the N-terminal formyl group was added by treating the resin with an equimolar mixture of HCOOH and DCOOD in the presence of dicyclohexylcarbodiimide. The peptides were released from the resin and screened with ESI–FT–ICR–MS analysis. The screening protocol is depicted in Figure 15.11. In the mass spectrum, each member of the tripeptide library produced a doublet (one for normal isotope and one for deuterated peptide) separated by 1.0063 Da. The peptide library was treated with the enzyme *E. coli* peptide deformylase (PDF) to deformylate those peptides that were a substrate of the enzyme. These peptides were identified in the spectrum by the loss of a formyl group (i.e., a shift in the ion signal by 28 and 29 Da). These members were selected and analyzed further by MS/MS to determine their sequence.

One Bead–One Compound Analysis In this strategy the products that are bound to the solid support are tested for their binding affinity, and the beads that exhibit activity are selected for further analysis by MALDI–MS, ESI–MS, or SIMS–TOF imaging after their release from these beads. In one simple approach, the bead-bound peptides interacted with the target molecules that had been tagged with a chromophore or a fluorescence probe and the active peptides were recognized by a distinct color change. The peptides anchored via methionine were released from the SPPS resin beads with a cyanogen bromide (CNBr) treatment and were analyzed with MALDI–TOF–MS [97]. Alternatively, the peptides are anchored to the resin via acid-labile or photolabile linkers, and released with TFA vapors or ultraviolet irradiation, respectively [90].

In Situ Monitoring of Support-Bound Libraries This protocol is applicable to encoded libraries, in which chemical tags are incorporated to reflect the chemical history of any member of a library. The explicit information carried by these tags is encoded to identify each library component [74]. This strategy is useful for peptidomimetic and small organic molecule libraries. The building blocks of the encoding tags were attached to the beads in parallel with the building blocks of organic libraries. Specific sequences of peptides or oligonucleotides can serve as the identifying tags. An alternative approach is binary coding, in which halobenzenes that are attached to a varying length of hydrocarbon chain are used as tags [98]. The hydrocarbon chains were linked to the beads via a cleavable spacer. The presence of the coding tag after cleavage was determined by GC-electron-capture detector analysis.

A photolabile handle is also a helpful for in situ monitoring of support-bound products by MALDI–TOF–MS. A single-laser shot can cleave the product from the SPPS resin bead and simultaneously allow it to desorb/ionize into the gas phase [99]. SIMS also affords in situ monitoring of support-bound peptide intermediates, and has the advantage that the functionalization of resin with a photolabile handle, so necessary for MALDI–MS analysis, is not required [100].

Library Affinity Selection-Mass Spectrometry (LAS–MS) In this approach the library components are reacted with agarose gel–bound receptor in solution [101]. After washing the beads, peptides of increasing receptor affinity are released by adding to the beads elution buffers of different pH values; the peptides released are identified with ESI–MS or ESI–MS/MS.

Affinity Capillary Electrophoresis (ACE)-Mass Spectrometry In this ingenious approach, the separation and characterization steps are combined by adding the receptor to the electrophoresis buffer [102]. When a solution of ligands was passed through the ACE column, the ligands that had affinity for the receptor were retained and were subsequently analyzed with ESI–MS.

Immunoaffinity Extraction (IAE) This method combines immunoaffinity extraction with two HPLC columns online with either mass spectrometry or tandem mass spectrometry. These systems have been used to characterize benzodiazepine libraries [103]. To isolate them from the library, benzodiazepines were trapped in an IAE column that was packed with antibodies specific to these molecules.

A pH change in the mobile phase eluted these components onto a reversed-phase C_{18} column, whereby benzodiazepines were separated from the antibody. The trapped benzodiazepines were back-flushed onto a C_8 column for separation and were identified with ESI–MS or ESI–MS/MS.

Pulsed Ultrafiltration-Mass Spectrometry In this approach, the receptor was placed in an ultrafiltration membrane, and a pulse of library compounds was applied to this cell [104]. The active components were retained in the ultrafiltration cell, and the unretained components and the binding buffer were washed away to the waste. The bound components were liberated and eluted from the membrane with a flow of methanol or an acidic solvent. ESI–MS or ESI–MS/MS provided the identity of the components liberated.

Bioaffinity Characterization Mass Spectrometry (BACMS) This approach also relies on the formation of noncovalent complexes between the ligands and a target but requires no isolation step prior to their mass spectrometry analysis [105]. Instead, the ligand-target complexes formed in the solution were directly introduced into an ESI-FTMS instrument. In the FTMS cell, collisional excitation first dissociates the electrosprayed complex, and the free ligands are trapped and characterized.

ESI–MS/MS using an FT–ICR mass spectrometer also enables direct screening of dynamic combinatorial library (DCL) [106]. The protein-ligand complexes are selectively trapped and dissociated to facilitate selective identification of the DCL ligands.

High-Throughput Screening Protocols A variety of high-throughput screening protocols have made the evaluation of combinatorial libraries practical as well as efficient [107]. A fully automated parallel LC/MS system was designed to allow simultaneous analysis and purification of combinatorial libraries [108]. It comprised of four parallel LC columns, two for analytical and two for preparative use, and a dual ESI interface. This system allowed more than one sample to be analyzed at a time.

A MALDI–FT–MS-based automated system for rapid screening of a large array of compounds is also available [109]. In this approach, the library compounds were mixed with a suitable matrix and were deposited on an auto-indexed multiple-sample disk. The matrix–sample mixture was ionized by irradiation with UV nitrogen laser (337 nm), and mass calibrant ions were generated by electron ionization of perfluorotributyl amine.

The high-throughput screening of combinatorial libraries can also be accomplished with flow-injection analysis (FIA)–MS [110]. All operation steps of analysis, including input of the sample information, setup of the mass spectrometry analysis parameters, compound checking, and results reporting to the customer are automated. Samples are placed in a 96-well sample unit and analyzed by ESI–MS or APCI–MS.

The coupling HPLC with IMS–TOF–MS also makes possible the high-throughput screening of combinatorial libraries [111]. With this combination, several members of a peptide library that have identical retention times can be distinguished on the basis of their gas-phase mobility in the IMS.

ADDITIONAL READING

1. C. L. Wilkins and J. O. Lay, Jr., eds., *Identification of Microorganisms by Mass Spectrometry*, Wiley-Interscience, New York, 2006, pp. 1–352.

REFERENCES

1. D. Fabris, Enzyme kinetics, in M. L. Gross and R. M. Caprioli, eds., *The Encyclopedia of Mass Spectrometry*, Vol. 2, Elsevier Science, Amsterdam, The Netherlands, 2005, pp. 327–338.
2. D. Fabris, Mass spectrometric approaches for the investigation of dynamic processes in condensed phase, *Mass Spectrom. Rev.* **24**, 30–54 (2005).
3. M. Michaelis and M. L. Menten, Kinetics of invertase action, *Biochem. Z.* **49**, 333–369 (1913).
4. A. Cornish-Bowden, *Fundamentals of Enzyme Kinetics*, Portland Press, London, 1995.
5. H. Lineweaver and J. Burk, The determination of enzyme dissociation constants, *J. Am. Chem. Soc.* **56**, 658–666 (1934).
6. D. S. Jayawardene and C. Dass, The effect of N-terminal acetylation and the inhibition activity of acetylated enkephalins on the aminopeptidase M-catalyzed hydrolysis of enkephalins, *Peptides* **20**, 963–970 (1999).
7. R. M. Caprioli, Analysis of biochemical reactions with molecular specificity using fast atom bombardment mass spectrometry, *Biochemistry* **27**, 513–521 (1988).
8. C. M. Bentzley and B. S. Larsen, Base specificity of oligonucleotide digestion by calf spleen phosphodiesterase with matrix-assisted laser desorption ionization analysis, *Anal. Biochem.* **259**, 31–37 (1998).
9. B. Bothner et al., Monitoring enzyme catalysis with mass spectrometry, *J. Biol. Chem.* **275**, 13455–13459 (2000).
10. N. Pi, J. I. Armstrong, C. R. Bertozzi, and J. A. Leary, Kinetic analysis of NodST sulfotransferase using electrospray ionization mass spectrometry, *Biochemistry* **41**, 13283–13288 (2002).
11. N. Pi and J. A. Leary, Determination of enzyme/substrate specificity constant using multiple substrate ESI-MS assay, *J. Am. Soc. Mass Spectrom.* **15**, 233–243 (2004).
12. F. Y. L. Hsieh et al., Kinetic monitoring of enzymic reactions in real time by quantitative high-performance liquid chromatography–mass spectrometry, *Anal. Biochem.* **229**, 20–25 (1995).
13. E. D. Lee, W. Muck, J. D. Henion, and T. Covey, Real-time reaction monitoring by continuous-introduction ion-spray tandem mass spectrometry, *J. Am. Chem. Soc.* **111**, 4600–4604 (1989).
14. B. Ganem, Y. T. Li, and J. D. Henion, Observation of noncovalent enzyme–substrate and enzyme–product complexes by ion-spray mass spectrometry, *J. Am. Chem. Soc.* **113**, 7818–7819 (1991).

15. J. Zaia, D. Fabris, et al., Monitoring metal ion flux in reactions of metallothionein and drug-modified metallothionein by electrospray mass spectrometry, *Protein Sci.* **7**, 2398–2404 (1998).
16. R. M. Caprioli, On-line fast atom bombardment analysis of dynamic biological systems, *Biomed. Environ. Mass Spectrom.* **16**, 35–39 (1988).
17. A. A. Paiva, R. F. Tilton, et al., Detection and identification of transient enzyme intermediates using rapid mixing, pulsed-flow electrospray mass spectrometry, *Biochemistry* **36**, 15472–15476 (1997).
18. P. J. Todd, J. M. McMahon, R. T. Short, and C. A. McCandish, Organic SIMS of biologic tissues, *Anal. Chem.* **69**, 528A–535A (1997).
19. P. Chaurand, S. A. Schwartz, and R. M. Caprioli, Profiling and imaging proteins in tissue sections by MS, *Anal. Chem.* **76**, 86A–93A (2004).
20. P. J. Todd, T. G. Schaaff, P. Chaurand, and R. M. Caprioli, Organic ion imaging of biological tissues with secondary ion mass spectrometry mass spectrometry and matrix assisted laser desorption/ionization, *J. Mass Spectrom.* **36**, 355–369 (2001).
21. P. Chaurand, S. A. Schwartz, et al., Integrating histology and imaging mass spectrometry, *Anal. Chem.* **76**, 1145–1155 (2004).
22. N. Winogard, The magic of cluster SIMS, *Anal. Chem.* **77**, 142A–149A (2005).
23. S. Perry and N. Winogard, High-resolution TOF-SIMS imaging of eukaryotic cells preserved in a trehalose matrix, *Anal. Chem.* **77**, 7950–7957 (2005).
24. A. F. M. Altelarr, J. V. Minnen, et al., Direct molecular imaging of *Lymnaea stagnalis* nervous tissue at subcellular spatial resolution by mass spectrometry, *Anal. Chem.* **77**, 735–741 (2005).
25. L. A. McDonnell, et al., Subcellular imaging mass spectrometry of brain tissues, *J. Mass Spectrom.* **40**, 160–168 (2005).
26. L. A. McDonnell, I. W. Fletcher, R. P. J. de Lange and R. M. A. Heeren, Higher sensitivity SIMS of biological molecules for high resolution chemical specific imaging, *J. Am. Soc. Mass Spectrom.* **17**, 1195–1202 (2006).
27. C. Fenselau, ed., *Mass Spectrometry for the Characterization of Microorganisms*, ACS Symposium Series, Vol. 541, American Chemical Society, Washington, DC, 1994.
28. T. Krishnamurthy, P. L. Ross, and U. Rajamani, Detection of pathogenic and non-pathogenic bacteria by matrix-assisted laser desorption/ionization time-of-flight mass spectrometry, *Rapid Commun. Mass Spectrom.* **10**, 883–887 (1996).
29. C. L. Nilsson, Fingerprinting of *Helicobacter pylori* strains by matrix-assisted laser desorption/ionization mass spectrometric analysis, *Rapid Commun. Mass Spectrom.* **13**, 1067–1071 (1999).
30. R. D. Holland, J. G. Wilkes, F. Rafii, J. B. Sutherland, C. E. Voorhees, and J. O. Lay, Jr., Rapid identification of intact whole bacteria based on spectral patterns using matrix-assisted laser desorption/ionization with time-of-flight mass spectrometry, *Rapid Commun. Mass Spectrom.* **10**, 1227–1232 (1996).
31. Z. Wang, L. Russon, L. Li, D. C. Roser, and S. R. Long, Investigation of spectral reproducibility in direct analysis of bacteria proteins by matrix-assisted laser desorption/ionization time-of-flight mass spectrometry, *Rapid Commun. Mass Spectrom.* **12**, 456–464 (1998).

32. J. O. Lay, Jr., and R. Liyanage, MALDI-TOF mass spectrometry of intact bacteria, in C. L. Wilkins and J. O. Lay, Jr., eds., *Identification of Microorganisms by Mass Spectrometry*, Wiley-Interscience, New York, 2006, pp. 125–152.

33. K. J. Welham, M. A. Domin, D. E. Scannell, E. Cohen, and D. S. Ashton, The characterization of micro-organisms by matrix-assisted laser desorption/ionization time-of-flight mass spectrometry, *Rapid Commun. Mass Spectrom.* **12**, 176–180 (1998).

34. M. A. Winkler, J. Uher, and S. Cepa, Direct analysis and identification of *Helicobacter* and *Campylobacter* species by MALDI–TOF mass spectrometry, *Anal. Chem.* **71**, 3416–3419 (1999).

35. S. J. Cordwell, A. S. Nouwens, and B. J. Walsh, Comparative proteomics of bacterial pathogens, *Proteomics* **1**, 461–472 (2001).

36. T. L. Williams, P. Leopold, and S. Musser, Automated postprocessing of electrospray LC/MS data for profiling protein expression in bacteria, *Anal. Chem.* **74**, 5807–5813 (2002).

37. S. Vaidyanathan, J. J. Rowland, D. B. Kell, and R. Goodcare, Discrimination of aerobic endospore-forming bacteria via electrospray ionization mass spectrometry, *Anal. Chem.* **73**, 4134–4144 (2001).

38. M. B. Beverly, F. Basile, K. J. Voorhees, and T. D. Hadfield, The effects of electron and chemical ionization modes on the MS profiling of whole bacteria, *J. Am. Soc. Mass Spectrom.* **10**, 747–758 (1999).

39. G. Siuzdak, Probing viruses with mass spectrometry, *J. Mass Spectrom.* **33**, 203–211 (1998).

40. J. J. Thomas, B. Falk, C. Fenselau, J. Jackman, and J. Ezzell, Viral characterization by direct analysis of capsid proteins, *Anal. Chem*, **70**, 3863–3867 (1998).

41. M. A. Winkler, N. Xu, H. Wu, and H. Aboleneen, MALDI–TOFMS of chemically modified recombinant hepatitis B surface antigen, *Anal. Chem.* **71**, 3416–3419 (1999).

42. J. K. Lewis, B. Bothner, J. J. Smith, and G. J. Siuzdak, Antiviral agents block breathing of the common cold virus, *Proc. Natl. Acad. Sc. USA*, **95**, 6774–6778 (1998).

43. D. Despeyroux, R. Phillpotts, and P. Watts, Electrospray mass spectrometry for detection and characterization of purified cricket paralysis virus (CrPV), *Rapid Commun. Mass Spectrom.* **10**, 937–941 (1996).

44. Y. J. Kim, A. Freas, and C. Fenselau, Analysis of viral glycoproteins by MALDI-TOF mass spectrometry, *Anal. Chem* **73**, 1544–1548 (2001).

45. Y. M. She, S. Haber, et al., Determination of the complete amino acid sequence for the coat protein of brome mosaic virus by time-of-flight mass spectrometry: evidence for mutations associated with change of propagation host, *J. Biol. Chem.* **246**, 20039–20047 (2001).

46. D. H. Chase, Mass spectrometry in clinical laboratory, *Chem. Rev.* **101**, 445–477 (2001).

47. L. Y. Ch'ang, K. Tang, et al., Detection of F508 mutation of the cystic fibrosis gene by matrix-assisted laser desorption/ionization mass spectrometry, *Rapid Commun. Mass Spectrom.* **9**, 772–774 (1995).

48. R. Theberge, L. Conners, M. Skinner, J. S. Kare, and C. E. Costello, Characterization of transthyretin mutants from serum using immunoprecipitation, HPLC/electrospray

ionization and matrix-assisted laser desorption/ionization mass spectrometry, *Anal. Chem.* **71**, 452–459 (1999).

49. S. A. Gerber, C. R. Scott, F. Tureček, and M. H. Gelb, Analysis of rates of multiple enzymes in cell lysates by electrospray ionization mass spectrometry, *J. Am. Chem. Soc.* **121**, 1102–1103 (1999).

50. R. M. Whittal, B. O. Kelly, and L. Li, Nanoliter chemistry combined with mass spectrometry for peptide mapping of proteins from single mammalian cell lysates, *Anal. Chem.* **70**, 5344–5347 (1998).

51. C. T. Houston and J. P. Reilly, Toward a simple, expedient, and complete analysis of human hemoglobin by MALDI-TOFMS, *Anal. Chem.* **71**, 3397–3404 (1999).

52. M. E. McComb, R. D. Oleschuk, A. Chow, W. Ens, K. G. Standing, H. Perreault, and M. Smith, Characterization of hemoglobin variants by MALDI–TOF MS using a polyurethane membrane as the sample support, *Anal. Chem.* **70**, 5142–5149 (1998).

53. A. Jiye, J. Trygg, et al., Extraction and GC/MS analysis of the human blood plasma metabolome, *Anal. Chem.* **77**, 8086–8094 (2005).

54. S. Bijlsma, I. Bobeldijk, et al., Large-scale human metabolomics studies: a strategy for data (pre-) processing and validation, *Anal. Chem.* **78**, 567–574 (2006).

55. A. J. Jeffreys, V. Wilson, and S. L. Thein, Individual-specific "fingerprints" of human DNA, *Nature* **316**, 76–79 (1985).

56. J. R. Almirall, Forensic chemistry education, *Anal. Chem.* **77**, 69A–72A (2005).

57. T. A. Brettell, J. M. Butler, and R. Saferstein, Forensic science, *Anal. Chem.* **77**, 3839–3860 (2005).

58. K. Deventer and F. T. Delbeke, Validation of a screening method for corticosteroids in doping analysis by liquid chromatography/tandem mass spectrometry, *Rapid Commun. Mass Spectrom.* **17**, 2107–2114 (2003).

59. C. Van Poucke and C. Van Peteghem, Development and validation of a multi-analyte method for the detection of anabolic steroids in bovine urine with liquid chromatography–tandem mass spectrometry, *J. Chromatogr. B* **772**, 211–217 (2002).

60. S. A. Hewitt, et al., Screening and confirmatory strategies for the surveillance of anabolic steroid abuse within Northern Ireland, *Anal. Chima. Acta* **473**, 99–109 (2002).

61. M. Thevis, H. Geyer, et al., Screening of unknown synthetic steroids in human urine by liquid chromatography/tandem mass spectrometry, *J. Mass Spectrom.* **40**, 955–962 (2005).

62. M. W. F. Nielen, T. F. H. Bovee, et al., Urine testing for designer steroids by liquid chromatography with androgen bioassay detection and electrospray quadrupole time-of-flight mass spectrometry identification, *Anal. Chem.* **78**, 424–431 (2006).

63. W. Weimmann, P. Schaefer, et al., Confirmatory analysis of ethyl glucuronide in urine by liquid chromatography/electrospray–tandem mass spectrometry, *J. Am. Soc. Mass Spectrom.* **15**, 188–193 (2004).

64. S. Dresen, W. Weimmann, and F. M. Wurst, Forensic confirmatory analysis of ethyl sulfate:a new marker for alcohol consumption by liquid chromatography/electrospray ionization–tandem mass spectrometry, *J. Am. Soc. Mass Spectrom.* **15**, 1644–1648 (2004).

65. P. M. L. Sandercock and E. Du Pasquier, Chemical fingerprinting of unevaporated automotive gasoline samples, *Forensic Sci. Int.* **134**, 1–10 (2003).

66. G. A. Eiceman and J. A. Stone, Ion mobility spectrometers in national defense, *Anal. Chem.* **78**, 424–431 (2006).
67. S. Montero, A. L. Hobbs, T. A. French, and J. R. Almirall, Elemental analysis of glass fragments by ICP-MS as evidence of association: analysis of a case, *Forensic Sci.* **48**, 1101–1107 (2003).
68. T. Trejos and. J. R. Almirall, Effect of fractionation on the forensic elemental analysis of glass using laser ablation inductively coupled plasma mass spectrometry, *Anal. Chem.* **76**, 1236–1242 (2004).
69. C. Weyermann, D. Kirsch, et al., Photofading of ballpoint dyes on paper studied by LDI and MALDI MS, *J. Am. Soc. Mass Spectrom.* **17**, 297–306 (2005).
70. K. Cottingham, MS on the bioterror front, *Anal. Chem.* **78**, 18–23 (2006).
71. J. P. Dworzanski, A. P. Snyder, et al., Identification of bacteria using tandem mass spectrometry combined with a proteome database and statistical scoring, *Anal. Chem.* **76**, 2355–2366 (2004).
72. N. C. VerBerkmoes, W. J. Harvey, et al., Evaluation of shotgun proteomics for identification of biological threat agents in complex environmental matrixes: experimental simulations, *Anal. Chem.* **77**, 923–932 (2005).
73. S. A. Fredriksson, A. G. Hulst, et al., Forensic identification of neat ricin from crude castor bean extracts by mass spectrometry, *Anal. Chem.* **77**, 1545–1555 (2005).
74. M. A. Gallop, R. W. Barret, W. J. Dower, S. P. A. Fodor, and E. M. Gorden, Applications of combinatorial technologies to drug discovery; 1: Background and peptide combinatorial libraries, *J. Med. Chem.* **37**, 1233–1251 (1994).
75. L. Gold, B. Polisky, O. Uhlenbeck, and M. Yarus, Diversity of oligonucleotide functions, *Annu., Rev. Biochem.* **64**, 763–797 (1995).
76. C. Y. Cho, E. J. Moran, S. R. Cherry, J. C. Stephaus, S. P. A. Fodor, C. L. Adams, A. Sundaram, J. W. Jacobs, and P. G. Schultz, An unnatural biopolymer, *Science* **261**, 1303 (1993).
77. J. A. Ellman and L. A. Thomson, Synthesis and applications of small molecule libraries, *Chem. Rev.* **96**, 555–600 (1996).
78. E. M. Gorden, R. W. Barret, W. J. Dower, S. P. A. Fodor, and M. A. Gallop, Applications of combinatorial technologies to drug discovery, 1: Background and peptide combinatorial libraries, *J. Med. Chem.* **37**, 1385–1401 (1994).
79. E. M. Gorden, M. A. Gallop, and D. V. Patel, Strategy and tactics in combinatorial organic synthesis: applications to drug discovery, *Acc. Chem. Res.* **29**, 144–154 (1996).
80. R. W. Armsrong, A. P. Combs, P. A. Tempest, S. D. Brown, and T. A. Keating, Multiple-component condensation strategies for combinatorial library synthesis, *Acc. Chem. Res.* **29**, 123–131 (1996).
81. H. M. Geysen, R. H. Meloen, and S. J. Barteling, Use of peptide synthesis to probe viral antigens for epitopes to a resolution of a single amino acid, *Proc. Natl. Acad. Sci. USA* **81**, 3998–4002 (1984).
82. R. Frank and R. Doering, Simultaneous multiple peptide synthesis under continuous flow conditions on cellulose paper discs on segmental solid supports, *Tetrahedron* **44**, 6031–6040 (1988).

83. A. Furka, F. Sebestyen, M. Asgedom, and G. Dibo, General method for rapid synthesis of multicomponent peptide mixtures, *Int. J. Peptide Protein Res.* **37**, 487–493 (1991).

84. G. P. Smith, Filamentous fusion phage: novel expression vectors that display cloned antigens on the virion surface, *Science* **228**, 1315–1317 (1985).

85. S. P. A. Fodor, J. L. Read, M. C. Pirrung, L. Stryer, A. T. Lu, and D. Solas, Light-directed, spatially addressable parallel chemical synthesis, *Science* **251**, 767 (1991).

86. M. A. Kelly, H. Liang, I.-I. Sytwu, I. Vlattas, N. L. Lyons, B. R. Brown, and L. P. Wennogle, Characterization of SH2-ligand interactions via library affinity selection with mass spectrometric detection, *Biochemistry* **35**, 11747–11755 (1996).

87. D. J. Ecker, T. A. Vickers, R. Hanecak, V. Driver, and K. A. Anderson, Rational screening of oligonucleotide combinatorial libraries for drug discovery, *Nucleic Acids Res.* **21**, 1853–1856 (1993).

88. J. Blake and L. Litzi-Davis, Evaluation of peptide libraries: an iterative strategy to analyze the reactivity of peptide mixtures with antibodies, *Bioconj. Chem.* **3**, 510–513 (1992).

89. H. M. Geysen, S. J. Rodda, and T. J. Mason, The antibody response to myoglobin, I: Systematic synthesis of myoglobin peptides reveals location and substructure of species-dependent continuous antigenic determinants, *Mol. Immunol.* **23**, 709–715 (1986).

90. B. B. Brown, D. S. Wagner, H. M. Geysen, A single-bead decode strategy using electrospray ionization mass spectrometry and a new photolabile linker: 3-amino-3-(2-nitrophenyl)propionic acid, *Mol. Diversity.* **1**, 4 (1995).

91. Y. M. Dunayevskiy, J.-J. Lai, C. Quinn, F. Talley, and P. Vouros, Mass spectrometric identification of ligands selected from combinatorial libraries using gel filtration, *Rapid Commun. Mass Spectrom.* **11**, 1178–1184 (1997).

92. C. L. Brummel, I. N. W. Lee, Y. Zhou, M. E. Hemling, and N. Winograd, A mass spectrometric solution to the address problem of combinatorial libraries, *Science* **264**, 399–401 (1994).

93. C. A. Srebalus, J. Li, W. S. Marshal, and D. E. Clemmer, Determining synthetic failures in combinatorial libraries by hybrid gas-phase separation methods, *J. Am. Soc. Mass Spectrom.* **11**, 352–355 (2000).

94. C. A. Srebalus Barnes and D. E. Clemmer, Assessment of purity and screening of peptide libraries by nested ion mobility-TOFMS: identification of RNase S-protein binders, *Anal. Chem.* **73**, 424–433 (2001).

95. P. Wang, D. Snavley, M. A. Freitas, and D. Pei, Screening combinatorial libraries for optimal enzyme substrates by mass spectrometry, *Rapid Commun. Mass Spectrom.* **15**, 1166–1171 (2001).

96. S.-A. Poulson, P. J. Gates, G. R. Cousius, and J. K. M. Sanders, Electrospray ionisation Fourier-transform ion cyclotron resonance mass spectrometry of dynamic combinatorial libraries, *Rapid Commun. Mass Spectrom.* **14**, 44–48 (2000).

97. R. S. Youngquist, G. R. Fuentes, M. P. Lacey, and T. Keough, Generation and screening of combinatorial peptide libraries designed for rapid sequencing by mass spectrometry, *J. Am. Soc. Mass Spectrom.* **117**, 3900–3906 (1995).

98. M. H. J. Ohlmeyer, R. N. Swanson, et al., Complex synthetic chemical libraries indexed with molecular tags, *Proc. Natl. Acad. Sci. USA* **90**, 10922–10926 (1993).

99. M. R. Carrasco, M. C. Fitzgerald, Y. Oda, and S. B. H. Kent, Direct monitoring of organic reactions on polymeric supports, *Tetrahedron Lett.* **38**, 6331–6334, (1997).

100. J. L. Aubaganac, C. Enjalbal, et al., Application of time-of-flight secondary ion mass spectrometry to in situ monitoring of solid-phase peptide synthesis on the multipin system, *J. Mass Spectrom.* **33**, 1094–1103 (1998).

101. M. A. Kelly, H. Liang, et al., Characterization of SH2-ligand interactions via library affinity selection with mass spectrometric detection, *Biochemistry* **35**, 11747–11755 (1996).

102. Y. H. Chu, D. P. Kirby, and B. L. Karger, Free solution identification of candidate peptides from combinatorial libraries by affinity capillary electrophoresis/mass spectrometry, *J. Am. Chem. Soc.* **117**, 5419–5420 (1995).

103. M. L. Nedved, S. Habibi-Goudarzi, B. Ganem, and J. D. Henion, Characterization of benzodiazepine "combinatorial" chemical libraries by on-line immunoaffinity extraction, coupled column HPLC-ion spray mass spectrometry–tandem mass spectrometry, *Anal. Chem.* **68**, 4228–4236 (1996).

104. R. B. van Breemen, C. R. Huang, D. Nikolic, C. P. Woodbury, Y. Z. Zhao, and D. L. Venton, Pulsed ultrafiltration mass spectrometry: a new method for screening combinatorial libraries, *Anal. Chem.* **69**, 2159–2164 (1997).

105. J. E. Bruce, G. A. Anderson, R. Cheng, X. Cheng, D. C. Gale, S. A. Hofstadler, B. L. Schwartz, and R. D. Smith, Bio-affinity characterization mass spectrometry, *Rapid Commun. Mass Spectrom.* **9**, 644–650 (1995).

106. S.-A. Poulsen, Direct screening of a dynamic combinatorial library using mass spectrometry, *J. Am. Soc. Mass Spectrom.* **17**, 1074–1080 (2006).

107. D. B. Kessel, Combinatorial chemistry and mass spectrometry in the 21st century drug discovery laboratory, *Chem. Rev.* **101**, 255–267 (2001).

108. L. Zeng and D. B. Kessel, Developments of a fully automated parallel HPLC/mass spectrometry system for the analytical characterization and preparative purification of combinatorial libraries, *Anal. Chem.* **70**, 4380–4388 (1998).

109. D. C. Tutko, K. D. Henry, B. E. Winger, H. Stout, and M. E. Hemling, Sequential mass spectrometry and MS^n analyses of combinatorial libraries by using automated MALDI–FT–MS, *Rapid Commun. Mass Spectrom.* **12**, 335–338 (1998).

110. G. Eckehard, R. Ramsey, and I. Lewis, High-throughput flow-injection analysis mass spectrometry with networked delivery of color-rendered results. 2. Three dimensional spectral mapping of 96-well combinatorial chemistry rack, *Anal. Chem.* **70**, 3227–3234 (1998).

111. C. A. S. Barnes and D. E. Clemmer, Ion mobility/time-of-flight analysis of combinatorial mixtures, *High-Throughput Analysis*, 187–216 (2003).

APPENDIX A

ABBREVIATIONS

2-AB	2-aminobenzamide
ABSA	aminobenzenesulfonic acid
ACE	affinity capillary electrophoresis
ADH	alcohol dehydrogenase
AE	appearance energy
AGP	α 1-acid glycoprotein
2-AMAC	2-aminoacridone
amol	attomole
AMS	accelerator mass spectrometry
ANSA	aminonaphthalenesulfonic acid
APA	aminopicolinic acid
APCI	atmospheric pressure chemical ionization
API	atmospheric pressure ionization
AP–MALDI	atmospheric pressure–MALDI
APPI	atmospheric pressure photoionization
AQUA	absolute quantification
AREX	axial resonant excitation
ATT	6-aza-2-thiothymine
ATTR	transthyretin amyloidosis
ATX	amphetamine-type stimulant
bp	base pair
BET	best emitter temperature
BIRD	blackbody-induced radiative dissociation

Fundamentals of Contemporary Mass Spectrometry, by Chhabil Dass
Copyright © 2007 John Wiley & Sons, Inc.

BS³	bis(sulfosuccinimidyl)suberate
BSA	bovine serum albumin
CCB	colloidal Coomassie Blue
CCD	charge-coupled device
CCSD	Complex Carbohydrate Structure Database
CD	charge detection
	circular dichroism
CD–EM	continuous-dynode electron multiplier
CE	capillary electrophoresis
	charge exchange
CEC	capillary electrochromatography
CE–CI	charge-exchange-chemical ionization
CEM	channel electron multiplier
CF	continuous-flow
	chromatofocusing
	cystic fibrosis
CFG	Consortium for Funtional Genomics
CGE	capillary gel electrophoresis
α-CHCA	α-cyano-4-hydroxycinnamic acid
CI	chemical ionization
CID	collision-induced dissociation
CIEF	capillary isoelectric focusing
CIT	capillary isotachophoresis
CNBr	cyanogen bromide
CoA	coenzyme A
CRF	charge-remote fragmentation
CRM	charge-residue model
CrPV	cricket paralysis virus
CSD	charge-state distribution
CSF	cerebrospinal fluid
CSP	calf spleen phosphodiesterase
CTL	cytotoxic T lymphocytes
CZE	capillary-zone electrophoresis
Da	dalton
DA–APPI	dopant-assisted atmospheric-pressure photoionization
DART	direct analysis in real time
dc	direct current
DCL	dynamic combinatorial library
DDA	data-dependent analysis
DE	delayed extraction
DESI	desorption electrospray ionization
DHB	2,5-dihydroxybenzoic acid
DIGE	differential imaging gel electrophoresis
DIOS	desorption–ionization on silicon
DNA	deoxyribonucleic acid

DRC	dynamic reaction cell
DSG	disuccinimidyl glutarate
DSS	disuccinimidyl suberate
DTE	dithioerythritol
DTT	dithiothreitol
EA	electron affinity
EC–CI	electron-capture-chemical ionization
ECD	electron-capture dissociation
ECF	extracellular fluid
EDT	ethanedithiol
EE	even electron
EI	electron ionization
EM	electron multiplier
endo H	endo-β-N-acetylglucosaminidase H endo H
ES	enzyme–substrate complex
ESA	electrostatic analyzer
ESI	electrospray ionization
EST	expressed sequence tags
ETD	electron-transfer dissociation
eV	electron volt
FAB	fast atom bombardment
FD	field desorption
FFE	field-flow electrophoresis
FFR	field-free region
FIA	flow-injection analysis
FIK	field ionization kinetics
FIS	field ion spectrometry
fmol	femtomole
FPD	focal plane detector
FSOT	fused-silica open tubular
FT	Fourier transform
FWHM	full width at half maximum
GalNAc	N-acetylgalactosamine
GlcNAc	N-acetylglucosamine
GC	gas chromatography
GD	glow discharge
GLP	good laboratory practices
GPA	glycerophosphatidic acid
GPC	glycerophosphocholine
GPE	glycerophosphoethanolamine
GPG	glycerophosphoglycerol
GPI	glycerophosphoinositol
GPS	glycerophosphoserine
GSL	glycosphingolipid
HBV	hepatitis B virus

HCD	heated capillary dissociation
HDX	hydrogen–deuterium exchange
H-ECD	hot electron-capture dissociation
HEL	hen egg lysozyme
ΔH_f	heat-of-formation
HFBA	heptafluorobutyric anhydride
HGP	human genome project
HOMO	highest occupied molecular orbital
HPA	3-hydroxypicolinic acid
HPIEC	high-performance ion-exchange chromatography
HPLC	high-performance liquid chromatography
IAE	immunoaffinity extraction
ICAT	isotope-coded affinity tags
ICP	inductively coupled plasma
ICR	ion cyclotron resonance
i.d.	internal diameter
IDA	iminodiacetate; iminodiacetic acid
IDM	ion-desorption model
IDMS	isotope dilution mass spectrometry
IE	ionization energy
IEF	isoelectric focusing
IKE	ion kinetic energy
ILR	ignitable liquid residues
IMAC	immobilized metal ion affinity chromatography
IMS	ion mobility spectrometry
	imaging mass spectrometry
IPD	ion-photon detector
IPG	immobilized pH gradient
IR	infrared
IRMPD	infrared multiphoton dissociation
IRMS	isotope ratio mass spectrometry
IS	internal standard
ISD	ion-source decay
IUPAC	International Union of Pure & Applied Chemistry
KE	kinetic energy
KER	kinetic energy release
LA	laser ablation
LAP	leucyl aminopeptidase
LAS	library affinity selection
LD	laser desorption
LDI	laser desorption/ionization
liquid SIMS	liquid-secondary-ion mass spectrometry
LIT	linear ion trap
LMC	low-molecular-mass compounds
LMI	liquid metal ion gun

LOD	limit of detection
LOQ	limit of quantitation
LSD	lysergic acid diethylamide
LT	leukotriene
LTQ	quadrupole linear ion trap
MAB	metastable atom bombardment
MALDI	matrix-assisted laser desorption/ionization
MCA	multichannel acquisition
MCP	microchannel plate
MEKC	miceller electrokinetic chromatography
MELDI	material-enhanced laser desorption/ionization
MeV	million electron volt
MIKES	mass-analyzed ion kinetic energy spectroscopy
MM	molecular mass
mmu	millimass unit
MPI	multiphoton ionization
MRM	selected-ion monitoring
MS	mass spectrometry
MSE	matrix suppression effect
MS/MS	tandem mass spectrometry
MUDPIT	multidimensional protein identification technology
MyHC	β-myosin heavy chain
m/z	mass-to-charge ratio
nano-ES	nanoelectrospray
NBA	3-nitrobenzyl alcohol
NBS	National Bureau of Standards
NC	nitrocellulose
Nd:YAG	neodynium/yttrium–aluminum–garnet
NMR	nuclear magnetic resonance
NOG	n-octyl pyranoglucoside
NRMS	neutralization–reionization mass spectrometry
NS	nozzle skimmer
nt	nucleotide
NTA	nitrilotriacetate
oa	orthogonal acceleration
OE	odd electron
OES	optical emission spectroscopy
OG	n-octyl glucoside
OPLC	overpressured-layer chromatography
OPO	optical parametric oscillator
PA	proton affinity
	picolinic acid
PAD	post-acceleration detector
PAGE	polyacrylamide gel electrophoresis
PAH	polycylic aromatic hydrocarbon

PAMA	partially methylated alditol acetates
PC	phosphatidylcholine
PCA	principle component analysis
PCD	plasma-coupled device
PCR	polymerase chain reaction
PD	plasma desorption
	photodissociation
PDB	Pee Dee Belemnite
PE	phosphatidylethanolamine
PEEK	polyether ether ketone
PEG	polyethylene glycol
PFK	perfluorokerosene
PFTBA	perfluorotributylamine
PG	prostaglandin
pI	isoelectric point
PIF	proteolysis-inducing factor
PIS	precursor-ion scan
PITC	phenylisothiocyanate
pmol	picomole
PNGase F	N-glycosidase F
PPG	polypropylene glycol
PRG	peptide reactive group
PS	phosphatidylserine
PSD	post-source decay
PSI	pulsed sample introduction
	phosphotyrosine-specific immonium ion
PTC	phenylthiocarbamyl
PTH	phenylthiohydantoin
PTM	posttranslational modification
PVDF	poly(vinylidene difluoride)
QET	quasi-equilibrium theory
QIT	quadrupole ion trap
RA	relative abundance
R+DB	rings plus double bonds
RDX	cyclotrimethylene–trinitramine
RE	recombination energy
REC	resonance electron capture
REMPI	resonance-enhanced MPI
rf	radio frequency
RI	resonance ionization
RIT	rectilinear ion trap
RNA	ribonucleic acid
RNase	ribonuclease
RP	reversed phase
	resolving power

R2PI	resonant two-photon ionization
RPMCs	rat peritoneal mast cells
RTOF	reflectron time of flight
SALDI	surface-assisted laser desorption/ionization
SCX	strong cation exchange
SDS	sodium dodecyl sulfate
SELDI	surface-enhanced laser desorption/ionization
SEM	secondary electron multiplier
SFC	supercritical-fluid chromatography
SH2	src homology 2
SID	surface-induced dissociation
SILAC	stable-isotope labeling by amino acids in cell culture
SIM	selected-ion monitoring
SIMS	secondary-ion mass spectrometry
SIS	superconductor–insulator–superconductor
SMOW	Standard Mean Ocean Water
S/N	signal-to-noise ratio
SORI	sustained off-resonance irradiation
SPE	solid-phase extraction
SPPS	solid-phase peptide synthesis
SRM	selected-reaction monitoring
SVP	snake venom phosphodiesterase
SWIFT	stored waveform inverse Fourier transform
TAG	triacylglycerol
TBDMS	t-butyldimethylsilyl
TEA	triethylamine
	transversely excited atmospheric
TED	tris(carboxymethyl)ethylenediamine
TFA	trifluoroacetic acid
Th	Thomson
TIC	total ion current
TIMS	thermal ionization mass spectrometry
TIS	timed-ion selector
TLC	thin-layer chromatography
TOF	time of flight
TSQ	triple-sector quadrupole
TX	thromboxane
UV	ultraviolet
VEE	Venezuelan equine encephalitis
VUV	vacuum ultraviolet
zmol	zeptomole
UPLC	ultra-performance liquid chromatography

APPENDIX B

PHYSICAL CONSTANTS, UNITS, AND CONVERSION FACTORS

Fundamental Physical Constants

Quantity	Symbol	Value
Speed of light (in vacuum)	c	2.99792458×10^8 m s^{-1}
Planck's constant	h	$6.6260755 \times 10^{-34}$ J · s
Avogadro's number	N	6.0221367×10^{23} particles mol^{-1}
Boltzmann's constant	k	$1.60217733 \times 10^{-19}$ J
Elementary charge	e	$1.60217733 \times 10^{-19}$ C
Unified mass (dalton)	u, Da	$1.6605402 \times 10^{-27}$ kg
Electron mass	m_e	$9.1093897 \times 10^{-31}$ kg
		5.4857989×10^{-4} u
Proton mass	m_p	$1.6726231 \times 10^{-27}$ kg
		1.007276 u
Neutron mass	m_n	$1.6749286 \times 10^{-27}$ kg
		1.008665 u
Molar gas constant	R	8.314510 J mol^{-1} K^{-1}
		0.08206 L · atm K^{-1} mol^{-1}

Fundamentals of Contemporary Mass Spectrometry, by Chhabil Dass
Copyright © 2007 John Wiley & Sons, Inc.

Common Scientific Units

Symbol	Unit	Also, expressed as:	Symbol	Unit	Also, expressed as
A	ampere		L	liter	
Å	ångstrom		M	mol L^{-1}	
atm	atmosphere		m	meter	
C	coulomb	s · A	s	second	
°C	degree Celsius		min	minute	
K	kelvin		h	hour	
cal	calorie		Hz	hertz	s^{-1}
eV	electron volt		J	joule	N · m
F	faraday	C V^{-1}	N	newton	m · kg s^{-2}
G	gauss		Pa	pascal	N m^{-2}
T	tesla	kg s^{-2} A^{-1}			

SI Prefixes for Common Multiplication Factors

Prefix	Symbol	Factor	Prefix	Symbol	Factor
exa	E	10^{18}	deci	d	10^{-1}
peta	P	10^{15}	centi	c	10^{-2}
tera	T	10^{12}	milli	m	10^{-3}
giga	G	10^{9}	micro	μ	10^{-6}
mega	M	10^{6}	nano	n	10^{-9}
kilo	k	10^{3}	pico	p	10^{-12}
hector	h	10^{2}	femto	f	10^{-15}
deca	da	10^{1}	atto	a	10^{-18}
			zepto	z	10^{-21}

Common Conversion Factors

Energy	1 erg = 10^{-7} J	Length	1 km = 0.6214 mile
	1 cal = 4.1840 J		1 m = 100 cm
	1 eV = 1.6022×10^{-19} J		1 inch = 2.54 cm
	1 eV = 3.8293×10^{-20} cal		1 Å = 1.00×10^{-10} m
	1 C · V = 1 J		
	1 kg · m^2 s^{-2} = 1 J	Pressure	1 atm = 101.325 kPa
			1 atm = 760 torr
Mass	1 kg = 1000 g		1 atm = 14.70 psi
	1 g = 1000 mg		1 atm = 101,325 N m^{-2}
	1 lb = 453.59 g = 16 oz		1 bar = 0.987 atm
			1 bar = 106 dyn cm^{-2}
Volume	1000 L = 1 m^3		1 torr = 1 mmHg
	1 L = 1000 cm^3		
	1 mL = 1000 μL		
	1 gal = 4 qt = 3.785 L		

APPENDIX C

ISOTOPES OF NATURALLY OCCURRING ELEMENTS AND THEIR ABUNDANCES

Atomic Number	Element	Symbol	Nominal Mass	%	Relative %	Isotopic Mass	Average Mass
1	Hydrogen	H	1	99.985	100	1.007825	1.00794
	Deuterium	D	2	0.015	0.015	2.014102	
2	Helium	He	3	0.00014	0.00014	3.016030	4.00260
			4	ca. 100	100	4.002603	
3	Lithium	Li	6	7.5	8.0108	6.015121	6.9410
			7	92.5	100	7.016003	
4	Beryllium	Be	9	100	100	9.012182	9.0122
5	Boron	B	10	19.9	24.84	10.012937	10.8110
			11	80.1	100	11.009305	
6	Carbon	C	12	98.90	100	12.000000	12.0110
			13	1.10	1.112	13.003355	
7	Nitrogen	N	14	99.63	100	14.003074	14.0067
			15	0.37	0.37	15.000109	
8	Oxygen	O	16	99.76	100	15.994915	15.9994
			17	0.04	0.04	16.999131	
			18	0.20	0.20	17.999160	

(*continued*)

Fundamentals of Contemporary Mass Spectrometry, by Chhabil Dass
Copyright © 2007 John Wiley & Sons, Inc.

Atomic Number	Element	Symbol	Nominal Mass	%	Relative %	Isotopic Mass	Average Mass
9	Fluorine	F	19	100	100	18.998403	18.9984
10	Neon	Ne	20	90.48	100	19.992439	20.1797
			21	0.27	0.298	20.993843	
			22	9.25	10.22	21.991384	
11	Sodium	Na	23	100	100	22.989768	22.9898
12	Magnesium	Mg	24	78.99	100	23.985042	24.3050
			25	10.00	12.66	24.985839	
			26	11.01	13.94	25.982595	
13	Aluminum	Al	27	100	100	26.981539	26.9815
14	Silicon	Si	28	92.21	100	27.976927	28.0855
			29	4.67	5.065	28.976495	
			30	3.10	3.336	29.973770	
15	Phosphorus	P	31	100	100	30.973762	30.9738
16	Sulfur	S	32	95.03	100	31.972072	32.0660
			33	0.75	0.789	32.971459	
			34	4.22	4.44	33.967868	
			36	0.02	0.021	35.967080	
17	Chlorine	Cl	35	75.77	100	34.968853	35.4530
			37	24.33	31.98	36.965903	
18	Argon	Ar	36	0.337	0.338	35.967546	39.9480
			38	0.063	0.0633	37.962732	
			40	99.600	100	39.962383	
19	Potassium	K	39	93.2581	100	38.963707	39.0983
			40	0.0117	0.0125	39.963999	
			41	6.7302	7.22	40.961825	
20	Calcium	Ca	40	96.941	100	39.962591	40.0780
			42	0.647	0.6674	41.958618	
			43	0.135	0.139	42.958766	
			44	2.086	2.152	43.955480	
			46	0.004	0.004	45.953689	
			48	0.187	0.193	47.952533	
21	Scandium	Sc	45	100	100	44.955911	44.9559
22	Titanium	Ti	46	8.0	10.84	45.952629	47.8800
			47	7.3	9.892	46.951764	
			48	73.8	100	47.947947	
			49	5.51	7.466	48.947871	
			50	5.4	7.317	49.944792	
23	Vanadium	V	50	0.25	0.251	49.947161	50.9415
			51	99.75	100	50.943962	
24	Chromium	Cr	50	4.345	5.185	49.946046	51.9961
			52	83.79	100	51.940509	
			53	9.50	11.34	52.940651	
			54	2.365	2.82	53.938882	
25	Manganese	Mn	55	100	100	54.938046	54.9380

ISOTOPES OF NATURALLY OCCURRING ELEMENTS AND THEIR ABUNDANCES

Atomic Number	Element	Symbol	Nominal Mass	%	Relative %	Isotopic Mass	Average Mass
26	Iron	Fe	54	5.9	6.43	53.939612	55.8470
			56	91.72	100	55.934939	
			57	2.1	2.29	56.935396	
			58	0.28	0.305	57.933277	
27	Cobalt	Co	59	100	100	58.933198	58.9332
28	Nickel	Ni	58	68.27	100	57.935346	58.6934
			60	26.10	38.23	59.930788	
			61	1.13	1.66	60.931058	
			62	3.59	5.26	61.928346	
			64	0.91	1.33	63.927968	
29	Copper	Cu	63	69.17	100	62.929958	63.5460
			65	30.83	44.57	64.927765	
30	Zinc	Zn	64	48.6	100	63.92914	65.3900
			66	27.9	57.41	65.926034	
			67	4.1	8.44	66.927129	
			68	18.8	36.68	67.924846	
			70	0.6	1.23	69.925335	
31	Gallium	Ga	69	60.11	100	68.925580	69.7230
			71	39.89	66.37	70.924700	
32	Germanium	Ge	70	20.5	56.16	69.924250	72.6100
			72	27.4	75.07	71.922079	
			73	7.8	21.37	72.923463	
			74	36.5	100	73.921177	
			76	7.8	21.37	75.921401	
33	Arsenic	As	75	100	100	74.921596	72.9216
34	Selenium	Se	74	0.9	1.8	73.922475	78.9600
			76	9.1	18.24	75.919212	
			77	7.6	15.23	76.919912	
			78	23.6	47.29	77.917309	
			80	49.9	100	79.916520	
			82	8.9	17.84	81.916698	
35	Bromine	Br	79	50.69	100	78.918336	79.9040
			81	49.31	97.28	80.916289	
36	Krypton	Kr	78	0.35	0.614	77.920401	83.8000
			80	2.25	3.947	79.916380	
			82	11.6	20.35	81.913482	
			83	11.5	20.18	82.914135	
			84	57.0	100	83.911507	
			86	17.3	30.35	85.910610	
37	Rubidium	Rb	85	72.17	100	84.911794	85.4678
			87	27.83	38.56	86.909187	
38	Strontium	Sr	84	0.56	0.68	83.913431	87.6200
			86	9.86	11.94	85.909267	
			87	7.00	8.5	86.908884	
			88	82.58	100	87.905619	

(continued)

Atomic Number	Element	Symbol	Nominal Mass	%	Relative %	Isotopic Mass	Average Mass
39	Yttrium	Y	89	100	100	88.905849	88.9059
40	Zirconium	Zr	90	51.45	100	89.904703	91.2240
			91	11.22	21.73	90.905643	
			92	17.15	33.33	91.905039	
			94	17.38	33.78	93.906314	
			96	2.80	5.44	95.908275	
41	Niobium	Nb	93	100	100	92.906377	92.9064
42	Molybdenum	Mo	92	14.84	61.50	91.906808	95.9400
			94	9.25	38.33	93.905085	
			95	15.92	65.98	94.905840	
			96	16.68	69.13	95.904678	
			97	9.55	39.58	96.906020	
			98	24.13	100	97.905406	
			100	9.63	39.91	99.907477	
43	Technetium	Tc	No natural abundance				
44	Ruthenium	Ru	96	5.54	17.53	95.907599	101.0700
			98	1.86	5.89	97.905267	
			99	12.7	40.19	98.909939	
			100	12.6	38.87	99.904219	
			101	17.1	54.11	100.905582	
			102	31.6	100	101.904348	
			104	18.6	58.86	103.905424	
45	Rhodium	Rh	103	100	100	102.905500	102.9055
46	Palladium	Pd	102	1.02	3.73	101.905634	106.4200
			104	11.14	40.76	103.904029	
			105	22.33	81.71	104.905079	
			106	27.33	100	105.903478	
			108	26.46	96.82	107.903895	
			110	11.72	42.88	109.905167	
47	Silver	Ag	107	51.84	100	106.905092	107.8682
			109	48.16	94.90	108.904757	
48	Cadmium	Cd	106	1.25	4.35	105.906461	112.4110
			108	0.89	3.10	107.904176	
			110	12.49	43.47	109.903005	
			111	12.80	44.55	110.904182	
			112	24.13	83.99	111.902758	
			113	12.22	42.53	112.904400	
			114	28.73	100	113.903357	
			116	7.49	26.07	115.904754	
49	Indium	In	113	4.3	4.49	112.904061	114.8200
			115	95.7	100	114.903880	

ISOTOPES OF NATURALLY OCCURRING ELEMENTS AND THEIR ABUNDANCES

Atomic Number	Element	Symbol	Nominal Mass	%	Relative %	Isotopic Mass	Average Mass
50	Tin	Sn	112	0.97	2.98	111.904826	118.7100
			114	0.65	1.99	113.902784	
			115	0.36	1.10	114.903348	
			116	14.53	43.58	115.901747	
			117	7.68	23.57	116.902956	
			118	24.22	73.32	117.901609	
			119	8.58	26.33	118.903310	
			120	32.59	100	119.902200	
			122	4.63	14.21	121.903440	
			124	5.79	17.77	123.905274	
51	Antimony	Sb	121	57.4	100	120.903821	121.7520
			123	42.6	74.22	122.904216	
52	Tellurium	Te	120	0.095	0.28	119.904048	127.6000
			122	2.59	7.65	121.903054	
			123	0.905	2.67	122.904271	
			124	4.79	14.14	123.902823	
			125	7.12	21.02	124.904433	
			126	18.93	55.89	125.903314	
			128	31.70	93.59	127.904463	
			130	33.87	100	129.906229	
53	Iodine	I	127	100	100	126.904476	126.9045
54	Xenon	Xe	124	0.10	0.37	123.905894	131.2900
			126	0.09	0.33	125.904281	
			128	1.91	7.10	127.903531	
			129	26.4	98.14	128.904780	
			130	4.1	15.24	129.903509	
			131	21.2	78.81	130.905072	
			132	26.9	100	131.904144	
			134	10.4	38.87	133.905395	
			136	8.9	33.09	135.907214	
55	Cesium	Cs	133	100	100	132.905439	132.9054
56	Barium	Ba	130	1.101	1.536	129.906284	137.3270
			132	0.097	0.135	131.905045	
			134	2.42	3.77	133.904493	
			135	6.59	9.2	134.905671	
			136	7.81	10.9	135.904559	
			137	11.32	15.8	136.905815	
			138	71.66	100	137.905235	
57	Lanthanum	La	138	0.09	0.09	137.907110	138.9055
			139	91.91	100	138.906347	
58	Cerium	Ce	136	0.19	0.214	135.907100	140.1200
			138	0.25	0.283	137.905980	
			140	88.48	100	139.905433	
			142	11.08	12.5	141.909241	

(*continued*)

APPENDIX C

Atomic Number	Element	Symbol	Nominal Mass	%	Relative %	Isotopic Mass	Average Mass
59	Praseodymium	Pr	141	100	100	140.907650	140.9077
60	Neodymium	Nd	142	27.13	100	141.907720	144.2400
			143	12.18	44.9	142.909810	
			144	23.80	87.7	143.910084	
			145	8.30	30.6	144.912570	
			146	17.19	63.4	145.913114	
			148	5.76	21.2	147.916890	
			150	5.64	20.8	149.920888	
61	Promethium	Pm	No natural abundance				
62	Samarium	Sm	144	3.1	11.6	143.912000	150.3600
			147	15.0	56.2	146.914895	
			148	11.3	42.3	147.914820	
			149	13.8	51.7	148.917181	
			150	7.4	27.7	149.917273	
			152	26.7	100	151.919729	
			154	22.7	85.0	153.922206	
63	Europium	Eu	151	47.8	91.6	150.919847	151.9600
			153	52.2	100	152.921226	
64	Gadolinium	64	152	0.20	0.81	151.919787	157.2500
			154	2.18	8.78	153.920862	
			155	14.80	59.6	154.922618	
			156	20.47	82.4	155.922119	
			157	15.65	63.0	156.923956	
			158.	24.84	100	157.924100	
			160	21.86	88.0	159.923051	
65	Terbium	Tb	159	100	100	158.925341	158.9254
66	Dysprosium	Ds	156	0.06	0.213	155.924280	162.5000
			158	0.10	0.355	157.924400	
			160	2.34	8.3	159.925193	
			161	18.90	67.0	160.926929	
			162	25.50	90.4	161.926795	
			163	24.90	88.3	162.928726	
			164	28.20	100	163.929172	
67	Holmium	Ho	165	100	100	164.930319	164.9303
68	Erbium	Er	162	0.14	0.416	161.928780	167.2600
			164	1.61	4.8	163.929199	
			166	33.60	100	165.930292	
			167	22.95	68.3	166.932047	
			168	26.80	79.8	167.932369	
			170	14.90	44.3	169.935461	

ISOTOPES OF NATURALLY OCCURRING ELEMENTS AND THEIR ABUNDANCES 549

Atomic Number	Element	Symbol	Nominal Mass	%	Relative %	Isotopic Mass	Average Mass
69	Thulium	Tm	169	100	100	168.934212	168.9342
70	Ytterbium	Yb	168	0.13	0.409	167.933890	173.0400
			170	3.05	9.6	169.934760	
			171	14.30	45.0	170.936324	
			172	21.90	68.9	171.936379	
			173	16.12	50.7	172.938208	
			174	31.80	100	173.938860	
			176	12.70	39.9	175.942563	
71	Lutetium	Lu	175	97.4	100	174.940771	174.9670
			176	2.6	2.67	175.942680	
72	Hafnium	Hf	174	0.162	0.46	173.940044	178.4900
			176	5.206	14.83	175.941406	
			177	18.606	53.01	176.943217	
			178	27.297	77.77	177.943696	
			179	13.269	38.83	178.945812	
			180	35.100	100	179.946545	
73	Tantalum	Ta	180	0.012	0.012	179.947462	180.9480
			181	99.988	100	180.947992	
74	Tungsten	W	180	0.12	0.39	179.946701	183.850
			182	26.3	85.67	181.948202	
			183	14.3	46.51	182.950220	
			184	30.7	100	183.950928	
			186	28.6	93.16	185.954357	
75	Rhenium	Re	185	37.4	59.74	184.952951	186.2070
			187	62.6	100	186.955744	
76	Osmium	Os	184	0.02	0.05	183.952488	190.2000
			186	1.58	3.85	185.953830	
			187	1.6	3.90	186.955741	
			188	13.3	32.44	187.955860	
			189	16.1	39.27	188.958137	
			190	26.4	64.39	189.958436	
			192.	41.0	100	191.961467	
77	Iridium	Ir	191	37.3	59.49	190.960584	192.2200
			193	62.7	100	192.962917	
78	Platinum	Pt	190	0.01	0.03	189.959917	195.0800
			192	0.79	2.34	191.961019	
			194	32.9	97.34	193.962655	
			195	33.8	100	194.964766	
			196	25.3	74.85	195.964926	
			198	7.2	21.30	197.967869	

Atomic Number	Element	Symbol	Nominal Mass	%	Relative %	Isotopic Mass	Average Mass
79	Gold	Au	197	100	100	196.966543	196.9665
80	Mercury	Hg	196	0.15	0.5	195.965807	200.5900
			198	10.0	33.56	197.966743	
			199	16.9	56.71	198.968254	
			200	23.1	77.52	199.968300	
			201	13.2	44.30	200.970617	
			202	29.8	100	201.970617	
			204	6.85	22.99	203.973467	
81	Thallium	Tl	203	29.524	41.89	202.97320	204.3830
			205	70.476	100	204.974401	
82	Lead	Pb	204	1.4	2.67	203.973020	207.2000
			206	24.1	45.99	205.974440	
			207	22.1	42.18	206.975872	
			208	52.4	100	207.976627	
83[a]	Bismuth	Bi	209	100	100	208.980374	208/9804
90	Thorium	Th	232	100	100	232.038054	232.0381
92	Uranium	U	234	0.0055	0.0055	234.040946	238.0300
			235	0.720	0.725	235.043924	
			238	99.2745	100	238.050784	

[a] From atomic number 84 onward, only Th and U have natural abundances.

APPENDIX D

REFERENCE IONS AND THEIR EXACT MASSES

m/z	m/z	m/z	m/z	m/z	m/z
			PFK		
51.00462	168.98882	330.97924	480.96966	630.96008	780.95005
68.99521	180.98882	342.97924	492.96966	642.96008	792.95005
80.99521	192.98882	354.97924	504.96966	654.96008	804.95005
92.99521	204.98882	366.97924	516.96966	666.96008	816.95005
99.99361	218.98563	380.97605	530.96647	680.95689	830.94731
111.99361	230.98563	392.97605	542.96647	692.95689	842.94731
118.99202	254.98563	404.97605	554.96647	704.95689	854.94731
130.99202	268.98244	416.97605	566.96647	716.95689	866.94731
142.99202	280.98244	430.97285	580.96327	730.95369	880.94411
149.99042	292.98244	442.97285	592.96327	742.95369	892.94411
154.99202	304.98244	454.97285	604.96327	754.95369	904.94411
	318.97924	466.97285	616.96327	766.95369	
		Perfluorotributylamine			
30.99840	118.99202	168.98882	225.99030	375.98072	537.97114
49.99379	130.99202	175.99249	230.98563	413.97753	575.96795
68.99521	149.99042	180.98882	263.98711	425.97753	613.96475

(*continued*)

Fundamentals of Contemporary Mass Spectrometry, by Chhabil Dass
Copyright © 2007 John Wiley & Sons, Inc.

552 APPENDIX D

m/z	m/z	m/z	m/z	m/z	m/z
92.99521	161.99042	213.99030	313.98391	463.97433	
113.99669	163.99349	218.98563	325.98391	501.97114	

Glycerol (Positive Ions)

57.03404	277.14986	645.33923	1013.52861	1381.71799	1749.90737
75.04460	369.19720	737.38658	1105.57596	1473.76533	
93.05517	461.24455	829.43392	1197.62330	1565.81268	
185.10251	553.29189	921.48127	1289.67064	1657.86002	

Cesium Iodide (Positive Ions)

132.90543	2471.1946	4809.4838	7147.7730	9486.0622	11824.351
392.71534	2731.0045	5069.2937	7407.5829	9745.8721	12084.161
652.52525	2990.8144	5329.1036	7667.3928	10005.682	12343.971
912.33516	3250.6244	5588.9135	7927.2027	10265.492	12603.781
1172.1451	3510.4344	5848.7235	8187.0126	10525.302	12863.591
1431.9550	3770.2442	6108.5334	8446.8226	10785.112	13123.401
1691.7649	4030.0541	6368.3433	8706.6325	11044.922	13383.211
1951.5748	4289.8640	6628.1532	8966.4424	11304.732	13643.021
2211.3847	4549.6739	6887.8631	9226.2523	11564.541	13902.831

Cesium Iodide (Negative Ions)

126.90448	2205.3838	4283.8630	6362.3423	8440.8216	10519.301
386.71439	2465.1937	4543.6729	6622.1522	8700.6315	10779.111
646.52430	2725.0036	4803.4829	6881.9621	8960.4414	11038.921
906.33421	2984.8135	5063.2928	7141.7720	9220.2513	11298.731
1166.1441	3244.6234	5323.1027	7401.5820	9480.0612	11558.541
1425.9540	3504.4333	5582.9126	7661.3919	9739.8711	11818.350
1685.7639	3764.2432	5842.7225	7921.2018	9999.6811	12078.160
1945.5738	4024.0531	6102.5324	8181.0117	10259.491	12337.970

Protonated Ions of Poly(ethylene glycol)

63.04460	415.25432	767.46404	1119.67376	1471.88348	1824.09320
107.07082	459.28054	811.49026	1163.69997	1515.90969	1868.11941
151.09703	503.30675	855.51647	1207.72619	1559.93591	1912.14563
195.12325	547.33297	899.54269	1251.75240	1603.96212	1956.17184
239.14946	591.35918	943.56890	1295.77862	1647.98834	2000.19805
283.17568	635.38540	987.59511	1339.80483	1692.01455	2044.22426
327.20189	679.41161	1031.62133	1383.83105	1736.04077	
371.22811	723.43783	1075.64754	1427.85726	1780.06698	

m/z	m/z	m/z	m/z	m/z	m/z
\multicolumn{6}{c}{Protonated Ions of Poly(propylene glycol)}					
56.02621	447.29340	795.54459	1143.79578	1492.04697	1840.29816
251.18585	505.33526	853.58645	1201.83764	1550.08883	1898.34002
309.22771	563.37713	911.62832	1259.87951	1608.13070	1956.38189
331.20967	621.41899	969.67018	1317.92137	1666.17256	2014.42375
389.25153	679.46086	1027.71205	1375.96324	1724.21443	
425.31140	737.50272	1085.75391	1434.00510	1782.25629	

Mass Calibration Standards for MALDI

Compound	Mass
α-CHCA dimer	379.0930
Porcine trypsin (54–57)	515.33
Porcine trypsin (108–115)	842.51
Angiotensin I	1295.6775
Angiotensin II	1045.5345
bradikinin	1,060.5692
Gramicidin-S	1,141.5
ACTH (18–39)	2,465.7
Bovine insulin	5,733.5
Horse heart cytochrome c	12360
Chicken lysozyme	1,4307
Horse heart myoglobin	1,6951
Bovine trypsinogen	2,3957
Bovine serum albumin	66,431

APPENDIX E

INTERNET RESOURCES

General Mass Spectrometry Tools
1. Isotopic distributor and mass spec plotter:
   http://www2.sisweb.com/mstools/isotope.htm
2. Mass spec generator:
   http://www2.sisweb.com/mstools/spectrum.htm
3. Exact molecular wt calculator:
   http://www.sisweb.com/reference/tools/exactmass.htm
4. Isotope simulators for Windows:
   http://members.aol.com/msmssoft/ (Iso Pro)

Mass Spectrometry Tools for Proteomics
5. Peptide mass fingerprinting (PMF) database search engines:
   http://ca.expasy.org/tools/#proteome
   http://www.matrixscience.com/cgi/search_form.pl?FORMVER=2& SEARCH=PMF
   http://phenyx.vital-it.ch/
6. Proteomics tools for mining sequence databases in conjunction with mass spectrometry experiments:
   http://prospector.ucsf.edu/
7. Various protein identification and characterization tools:
   http://ca.expasy.org/tools/#proteome

Fundamentals of Contemporary Mass Spectrometry, by Chhabil Dass
Copyright © 2007 John Wiley & Sons, Inc.

8. Calculation of a possible molecular weight from a pair of adjacent electrospray charge state masses:
   http://sx102a.niddk.nih.gov/software/electrospray.html
9. Calculation of molecular formula from molecular weight:
   http://sx102a.niddk.nih.gov/software/massToFormula.html
10. Database of posttranslational modifications DELTA MASS:
    http://www.abrf.org/index.cfm/dm.home
11. Some important links of protein mass spectrometry:
    http://proteome.nih.gov/links.html
12. Protein phosphorylation mass spectrometry links:
    http://www.phosphosite.org/Login.jsp?test=true& (phosphoprotein database)
    http://vigen.biochem.vt.edu/xpd/xpd.htm (searchable database for protein phosphorylation)
    http://phospho.elm.eu.org/ (phosphobase, a database of phosphorylation sites)
13. Comprehensive list of posttranslational modifications:
    http://www.aber.ac.uk/~mpgwww/Proteome/MS_Tut.html
14. Exclusive Web site dedicated to hydrogen deuterium exchange (HDX) mass spectrometry:
    http://www.hxms.com/hxms.htm

Online Tutorials for Mass Spectrometry

15. "What Is Mass Spectrometry" (prepared by the American Society for Mass Spectrometry):
    http://www.asms.org/whatisms/index.html
16. Introduction to mass spectrometry by Scripps Research Institute:
    http://masspec.scripps.edu/information/intro/index.html
17. Links to other tutorials:
    http://web.mit.edu/toxms/www/tutorials.htm
18. Tutorial on protein identification by peptide mass fingerprinting (PMF):
    http://www.aber.ac.uk/~mpgwww/Proteome/MS_Tut.html

Mass Spectrometry Societies

http://www.asms.org/ (United States)
http://www.latrobe.edu.au/ANZSMS/ANZSMS.html (Australia and New Zealand)
http://www.bmss.org.uk (UK)
http://www.csms.inter.ab.ca/ (Canada)
http://msig.ncifcrf.gov/msig-links.html (others)

Comprehensive Web Sites Dedicated to Mass Spectrometry

http://www.i-mass.com/
http://www.ionsource.com/
http://www.sisweb.com/portal/ms.htm

Addresses for Mass spectral Compilation Sites Are on Page 216

APPENDIX F

SOLUTIONS TO EXERCISES

CHAPTER 1

1.1. Mass spectrometry analysis involves three steps: ionization, separation and m/z analysis of ions, and detection of the ion current.

1.2. See page 6.

1.3. See page 6.

1.4. Nominal mass = $16 \times 12 + 13 \times 1 + 2 \times 14 + 1 \times 16 + 1 \times 35 = 284$ u.
Monoisotopic mass = $16 \times 12.000000 + 13 \times 1.007825 + 2 \times 14.003074 + 1 \times 15.994915 + 1 \times 34.968853 = 284.0716$ u.
Average mass = $16 \times 12.0110 + 13 \times 1.00794 + 2 \times 14.0067 + 1 \times 15.9994 + 1 \times 35.4527 = 284.7447$ u.

1.5. $m/z = (2051 + 3)/3 = 684.67$.

1.6. Mass in kg = 284 u $(1.66 \times 10^{-27}$ kg/1 u$) = 4.71 \times 10^{-25}$ kg.

CHAPTER 2

2.1. See Figure 2.1 and read page 17.

2.2. The ion source block at 6000 V, filament at 5950 V, electron trap at 6100 V, ion repeller at 6020 V, and exit slit at ground potential.

Fundamentals of Contemporary Mass Spectrometry, by Chhabil Dass
Copyright © 2007 John Wiley & Sons, Inc.

2.3. Electron capture, CI with specific reagent anions, FAB, MALDI, APPI, APCI, and ESI.

2.4. Read page 21.

2.5. (a) At 70-eV, the variation in the electron energy has little effect on the efficiency of ionization and subsequent fragmentation, and thus on the nature of the EI spectrum.

(b) CI is a softer ionization technique used to analyze highly labile molecules.

(c) Multiple-charging in ESI reduces the m/z ratio, so that low-mass range instrument can be used to analyze high-mass compounds.

(d) To serve as a solvent for the analyte and to absorb energy at the wavelength of laser beam for subsequent transfer to the analyte molecules.

(e) The matrix can absorb a large amount of energy from laser radiation and transfer it to the analyte molecules in a control manner. This process helps to desorb large molecules intact into the gas phase without causing thermal degradation.

(f) Owing to the reduced matrix-to-analyte ratio the matrix background is reduced drastically to provide better detection sensitivity.

(g) For accurate timing of the arrivals of ions at the detector, all ions must enter the flight tube of the TOF mass analyzer in the pulse mode. Because the ion formation in PD and MALDI also occurs in the pulse mode, they are ideal ion sources for TOFMS.

2.6. (a) CH_5^+ and $C_2H_5^+$ ions, (b) thermal electrons, (c) $N_2^{+\bullet}$ ions, (d) $Ar^{+\bullet}$ for charge-exchange CI and $C_4H_9^+$ for positive-ion CI, (e) OH^- ions.

2.7. The energy transferred during methane-, isobutene-, hydrogen-, water-, and methanol-CI is 68, 4, 99, 33, and 18 kcal/mol, respectively. Because fragmentation is proportional to the energy transferred, the [225]/[243] ratio increases in the order isobutene < methanol < water < methane < hydrogen.

2.8. $\lambda = \dfrac{hc}{E} = \dfrac{(6.626 \times 10^{-34} \ Js) \times (3.00 \times 10^8 \ m/s)}{9.02 \ eV} \times \dfrac{1 \ eV}{(1.602 \times 10^{-19} \ J)}$

$\times \dfrac{1 \ nm}{10^{-9} \ m} = <138 \ nm$

2.9. The energy of 650-nm photon is

$$E = h\upsilon = \dfrac{hc}{\lambda} = \dfrac{(6.626 \times 10^{-34} \ J \cdot s) \times (3.00 \times 10^8 \ m/s)}{650 \ nm \times (10^{-9} \ m/nm)}$$

$$\times \dfrac{1 \ eV}{(1.602 \times 10^{-19} \ J)} = 1.9 \ eV$$

Therefore, at least (9.02/1.9) 5 photons are required to ionize cyclopentene.

2.10. Only $10 - 9.02 = 0.98$ eV energy/photon is available for fragmentation, therefore energy per mole photon $= 0.98 \times 1.602 \times 10^{-19}$ CV $\times 6.02 \times 10^{23}$ photons/mole $\times (1$ J/1 CV$) \times ($kJ/1000 J$) = 94.51$ kJ/mole.

Because, energy required to cause fragmentation in organic compounds lies in the 200 to 600 kJ/mol range, the 10-eV photon beam is insufficient to cause fragmentation in cyclopentene.

2.11. In this example, a maximum of 15.98 eV excess energy can be transferred to the $M^{+\bullet}$ ion. Therefore, energy per mole electrons is given by $E = 15.98 \times 1.602 \times 10^{-19}$ CV $\times 6.02 \times 10^{23}$ electrons/mole $\times (1$ J/1 CV$) \times ($kJ/1000 J$) = 1541$ kJ/mole, which is sufficient to cause fragmentation in cyclopentene.

2.12. $n = (1618.1 - 1.0078)/(1658.5 - 1618.1) = 40.02 = 40$ charge state of m/z 1658.5. $M = 40(1658 - 1.0078) = 66,299.7$ Da.

2.13. In LDI, energy absorbed by the analyte molecules is difficult to control, leading to thermal degradation, whereas in MALDI, the matrix is designed to absorb energy from the laser and transfer it to the analyte molecules in a control manner to allow desorption of large molecules intact into the gas phase without causing thermal degradation.

2.14. Read page 38.

2.15. Read page 42.

2.16. Read page 41.

2.17. (**a**) EI, PI, MAB, FI, and APPI, (**b**) FD, PD, FAB, ESI, DESI, and MALDI, (**c**) PD, ESI, DESI, and MALDI, and (**d**) PD, ESI, DESI, and MALDI.

2.18. In ESI, ionization occurs due to electrospray process, whereas in DESI, it is the impact of a beam of ions, neutrals, etc. that causes ionization. Also, DESI is performed in ambient air, whereas ESI in the mass spectrometry ion source.

2.19. Energy of the 337-nm photon $= 3.68$ eV. Energy of the second photon should be $>8.83 - 3.68 = 5.15$ eV. Therefore, λ of this photon is <241 nm.

CHAPTER 3

3.1. (**a**) The mass-resolution of a single-focusing magnetic sector instrument is poor owing to kinetic energy dispersion of the ions. Incorporation of an ESA improves the resolution by providing an ion beam homogenous in kinetic energy.

(**b**) The Mattauch-Herzog geometry instrument detects all mass-resolved ions simultaneously all the time; no ion current is lost, whereas the Nier-Johnson instrument detects only one ion at a time.

(c) The mass-resolution of a quadrupole mass analyzer is proportional to the number of oscillations that ions make within the quadrupole field. Therefore, the ions are accelerated to very low velocities.

(d) For accurate timing of the arrivals of ions at the detector, all ions must enter the flight tube of the TOF mass analyzer precisely at the same time. Therefore, for simultaneously entry of all ions, the ion beam needs to be pulsed.

(e) A reflectron can correct for the energy dispersion of the ions, and thus it improves the mass-resolution of a of the TOF mass analyzer.

(f) Because of the multiplex detection of all ions by the focal-plane detectors, no ion current is lost, whereas the focal-plane detectors detect only one ion at a time.

3.2. (a) M(C_3H_6S) = 74.0190 Da; M($C_5H_{12}D$) = 74.1080 Da; RP = 832.2.

(b) M($C_{12}H_{24}$) = 168.1878; M($C_6H_4N_2O_4$) = 168.0171 Da; RP = 984.8.

(c) M($C_{18}H_{18}O$) = 250.1358 Da; M($C_{18}H_{20}N$) = 250.1596 Da; RP = 10,506.

3.3. Increase B, increase r, and decrease V.

3.4. 40,000 Da [*Hint:* Use Eq. 3.4, and multiply 2500 by $(2^2 \times 2^2)$].

3.5. $V = \dfrac{zeB^2r^2}{2m} = \dfrac{1 \times (1.602 \times 10^{-19}\ C) \times (1.5T)^2 \times (0.5\ m)^2}{2 \times 5000\ Da \times (1.66 \times 10^{-27}\ kg)} = 5428\ V$

3.6. $v = \sqrt{\dfrac{2zeV}{m}} = \sqrt{\dfrac{2 \times 1 \times 1.602 \times 10^{-19}C \times 6000\ V}{2000 \times 1.66 \times 10^{-27}\ kg}} = 2.41 \times 10^4\ m/s$

$t = \dfrac{L}{v} = \dfrac{1.0\ m}{2.41 \times 10^4\ m/s} = 4.15 \times 10^{-5}\ s = 41.5\ \mu s$

$v(2500) = 2.15 \times 10^4\ m/s$ and $t(2500) = 46.5\ \mu s$

$\Delta t = 5.0\ \mu s$

3.7. $r = \sqrt{\dfrac{2Vm}{qB^2}} = \sqrt{\dfrac{2 \times 6000\ V \times 78 \times 1.66 \times 10^{-27}\ kg}{2 \times 1.602 \times 10^{-19}\ C \times (1.25T)^2}} = 5.57 \times 10^{-2}\ m$
$= 5.57\ cm$

3.8. $v = \sqrt{\dfrac{2zeV}{m}} = \sqrt{\dfrac{2 \times 2 \times 1.602 \times 10^{-19}C \times 8000\ V}{78 \times 1.66 \times 10^{-27}\ kg}} = 1.99 \times 10^5\ m/s$

3.9. $\dfrac{m}{z} = \dfrac{8\ eV}{a_u(2\pi f)^2 r_0^2} = \dfrac{8 \times 1.602 \times 10^{-19}\ C \times 500\ V}{0.256 \times (2 \times 3.14 \times 0.5 \times 10^6\ s^{-1})^2 (0.01\ m)^2}$
$\left(\dfrac{1u}{1.66 \times 10^{-27}\ kg}\right) = 1529\ u$

3.10. Double-focusing magnetic sector, reflectron-TOF, and FT-ICRMS.

3.11. Read page 82.

3.12. Longer flight tube, higher acceleration potential, delayed-extraction, and reflectron-TOF.

3.13. Read page 83.

3.14. Read page 95.

3.15. $\omega_c = \dfrac{Bze}{2\pi m} = \dfrac{1.0T \times 1 \times 1.602 \times 10^{-19}\ C}{2 \times 3.14 \times 122 \times 1.66 \times 10^{-27}\ kg} = 1.26 \times 10^5\ Hz$
$= 0.126\ MHz$

3.16. $\omega_c = \dfrac{Bze}{2\pi m} = \dfrac{1.25T \times 2 \times 1.602 \times 10^{-19}\ C}{2 \times 3.14 \times 122 \times 1.66 \times 10^{-27}\ kg} = 3.1 \times 10^5\ Hz$
$= 0.31\ MHz$

3.17. Read page 91.

3.18. Use Eq. 3.17, in which $q_{max} = 0.908$, the lower mass limit is given by

$$\dfrac{m}{z} = \dfrac{4\,eV}{q_{max}(2\pi f)^2 r_0^2} = \dfrac{4 \times 1.602 \times 10^{-19}\ C \times 100\ V}{0.908 \times (2 \times 3.14 \times 0.5 \times 10^6\ s^{-1})^2 (0.01\ m)^2}$$
$$\left(\dfrac{1u}{1.66 \times 10^{-27}\ kg}\right) = 43\ u,$$

and the upper mass limit is $1294\ u$.

3.19. The yield of secondary electrons falls off exponentially as the velocity of the striking ions decreases; because the velocity of an ion is an inverse function of its mass, the detection efficiency decreases with an increase in the mass.

3.20. By using post-acceleration of ions.

CHAPTER 4

4.1. Read page 119.

4.2. In the tandem in-time instruments, all three steps of MS/MS operation are performed in the same region, but are separated in time, whereas in the tandem in-space instruments, these steps are carried out in different regions.

4.3. There is an incontrovertible link between the mass-selected precursor and its product ions.

4.4. Read page 121.

4.5. $E_{com} = [40/(1000 + 40)] \times 200 = 7.69$ eV.

4.6. Read page 123.

4.7. Read page 128.

4.8. From Eq. 3.4, $m_p/q = B_p^2 r^2/2V$ and $m_1/q = B_1^2 r^2/2V$, or $m_p/m_1 = B_p^2/B_1^2$
Similarly, from Eq. 3.5, $m_p v^2 = E_p rq$ and $m_1 v^2 = E_1 rq$, or $m_p/m_1 = E_p/E_1$
By equating these two terms, we get $m_p/m_1 = B_p^2/B_1^2 = E_p/E_1$ or $B^2/E =$ constant.

4.9. Read page 132.

4.10. Read page 133.

4.11. Because the flight-tube fragments continue to travel with the velocity of their precursor, the velocity of the b_4-ion is calculated using the $m/z = 556$. Therefore,

$$v = \sqrt{\frac{2zeV}{m}} = \sqrt{\frac{2 \times 1 \times 1.602 \times 10^{-19} \, C \times 20000 \, V}{556 \times 1.66 \times 10^{-27} \, kg}} = 8.33 \times 10^4 \, m/s$$

4.12. Read page 137.

4.13. The precursor is mass-selected by adjusting the field of the mass-resolving quadrupole, collisionally activated in the rf-only quadrupole, and the CID products are mass-analyzed with the TOF mass analyzer. For precursor-ion spectrum, read page 140.

4.14. Magnetic sector, triple quadrupole, Q-TOF, and magnetic sector-Q hybrid instrument.

4.15. (a) Product-ion spectrum by using B/E linked-field scan and MIKE scan and (b) Precursor-ion spectrum by using B^2/E and B^2E linked-field scans.

4.16. $m^* = m_1^2/m_p = (90)^2/108 = 75$.

4.17. The first mass-resolving quadrupole (Q_1) is adjusted to transmit the chosen precursor, which is collisionally activated in the rf-only quadrupole, and the second mass-resolving quadrupole (Q_3) is set to monitor the desired product ion.

4.18. By using ion guides made of rf-only multipoles.

4.19. Read page 124.

4.20. $E_1 = (m_1/m_p)E_p = (90/108) \times 400 = 333.3$ V/m.

4.21. Triple quadrupole, Q-TOF, QIT, LIT, FTICR, Q-FTICR, LTQ-orbitrap, and LTQ-FTICR.

CHAPTER 5

5.1. Because in SIM chromatogram the data system collects ion current due to only a few ions all the time.

5.2. Read page 154.

5.3. Read page 161.

5.4. Read page 163.

5.5. Read Section 5.8.1 on page 175.

5.6. In MEKC mode of CE, neutral molecules partition in the micelle to different extents according to their hydrophobicity, whereas no such fractionation of neutrals occur in CZE.

5.7. In CZE, the order of elution is cations (first), neutrals, and anions (last). In MEKC, the order of elution is anions, neutrals, and cations.

5.8. Read page 177.

CHAPTER 6

6.1. The accurate mass of benzoic acid is 122.0368 u. Therefore, mass measurement errors in ppm and mmu are:

$$ppm = \frac{122.0368 - 122.0343}{122.0368} \times 10^6 = 20.5 \; ppm$$

$$mmu = (122.0368 - 122.0343) \times 10^3 = 2.5 \; mmu$$

6.2. The accurate mass of naphthalene is 128.0626 u. Therefore, mass measurement errors in ppm would be:

$$ppm = \frac{mmu}{accurate\;mass} \times 10^3 = \frac{2}{128.0626} \times 10^3 = 15.6 \; ppm$$

6.3. The accurate masses of CF_3^+, $C_4H_{10}O$, $C_4H_6D_2O$, C_5H_{14}, or $C_3H_{10}N_2$ are 68.995209, 74.07317, 74.07007, 74.10955, and 74.08440 u. The calculated mass of the unknown is:

$$m_2 = m_1 \frac{V_1}{V_2} = (68.995209)\frac{100}{93.099} = 74.109506 \; u$$

Therefore, the unknown is C_5H_{14}.

6.4. (a) $C_{13}H_9OCl$: $[M+1] = 13 \times 1.11 + 9 \times 0.015 + 0.04 = 14.61$
$[M+2] = 1.01 + 0 + 0.20 + 32.0 = 33.21$
$[M+3] = 0.146 \times 33.21 = 4.85$

(b) $(CH_3)_4Si$: $[M+1] = 4 \times 1.11 + 12 \times 0.015 + 5.1 = 9.72$
$[M+2] = 0.1 + 0 + 0.20 + 3.4 = 3.5$

(c) C_5H_9NOS: $[M+1] = 5 \times 1.11 + 9 \times 0.015 + 0.37 + 0.04 + 0.79$
$$= 6.10$$
$$[M+2] = 0.15 + 0 + 0 + 0 + 0.20 + 4.4 = 4.75$$

6.5. $M = {}^{12}C_{13}{}^{1}H_9{}^{16}O{}^{79}Br$

$M + 1 = {}^{12}C_{12}{}^{13}C^1H_9{}^{16}O{}^{79}Br + {}^{12}C_{13}{}^{1}H_8{}^{2}H{}^{16}O{}^{79}Br$
$\quad + {}^{12}C_{13}{}^{1}H_9{}^{17}O{}^{79}Br$

$M + 2 = {}^{12}C_{11}{}^{13}C_2{}^{1}H_9{}^{16}O{}^{79}Br + {}^{12}C_{13}{}^{1}H_9{}^{18}O{}^{79}Br$
$\quad + {}^{12}C_{13}{}^{1}H_9{}^{16}O{}^{81}Br + {}^{12}C_{12}{}^{13}C^1H_8{}^{2}H{}^{16}O{}^{79}Br$
$\quad + {}^{12}C_{12}{}^{13}C^1H_9{}^{17}O{}^{79}Br + {}^{12}C_{13}{}^{1}H_8{}^{2}H{}^{17}O{}^{79}Br$

$M + 3 = {}^{12}C_{12}{}^{13}C^1H_9{}^{18}O{}^{79}Br + {}^{12}C_{12}{}^{13}C^1H_9{}^{16}O{}^{81}Br$
$\quad + {}^{12}C_{12}{}^{13}C^1H_8{}^{2}H{}^{17}O{}^{79}Br + {}^{12}C_{13}{}^{1}H_8{}^{2}H{}^{18}O{}^{79}Br$
$\quad + {}^{12}C_{13}{}^{1}H_8{}^{2}H{}^{16}O{}^{81}Br + {}^{12}C_{13}{}^{1}H_9{}^{17}O{}^{81}Br$

$M + 4 = {}^{12}C_{12}{}^{13}C^1H_8{}^{2}H{}^{18}O{}^{79}Br + {}^{12}C_{12}{}^{13}C^1H_8{}^{2}H{}^{16}O{}^{81}Br$
$\quad + {}^{12}C_{12}{}^{13}C^1H_9{}^{17}O{}^{81}Br + {}^{12}C_{13}{}^{1}H_8{}^{2}H{}^{17}O{}^{81}Br$
$\quad + {}^{12}C_{13}{}^{1}H_9{}^{18}O{}^{81}Br + {}^{12}C_{11}{}^{13}C_2{}^{1}H_9{}^{18}O{}^{79}Br$
$\quad + {}^{12}C_{11}{}^{13}C_2{}^{1}H_9{}^{16}O{}^{81}Br$

$M + 5 = {}^{12}C_{12}{}^{13}C^1H_9{}^{18}O{}^{81}Br + {}^{12}C_{13}{}^{1}H_8{}^{2}H{}^{18}O{}^{81}Br$
$\quad + {}^{12}C_{12}{}^{13}C^1H_8{}^{2}H{}^{17}O{}^{81}Br + {}^{12}C_{11}{}^{13}C_2{}^{1}H_8{}^{2}H{}^{18}O{}^{79}Br$
$\quad + {}^{12}C_{11}{}^{13}C_2{}^{1}H_8{}^{2}H{}^{16}O{}^{81}Br + {}^{12}C_{11}{}^{13}C_2{}^{1}H_9{}^{17}O{}^{81}Br$

6.6. Here, $a \sim 3$, $b \sim 1$ and $n = 2$, and $c \sim 1$, $d \sim 1$ and $m = 1$, and the binomial expansion is $(a+b)^2 \times (c+d) = a^2c + 2abc + b^2c + a^2d + 2abd + b^2d$.

By collecting all M ($^{35}Cl^{35}Cl^{79}Br$; a^2c), $M + 2$ ($^{35}Cl^{37}Cl^{79}Br = 2abc$; $^{35}Cl^{35}Cl^{79}Br = a^2d$), $M + 4(^{37}Cl^{37}Cl^{79}Br$; b^2c; $^{35}Cl^{37}Cl^{81}Br = 2abd$), and $M + 6$ ($^{37}Cl^{37}Cl^{81}Br = b^2d$) terms together gives, isotope abundance ratios $= 9 : (6+9) : (1+6) : 1 = 9 : 15 : 7 : 1$.

6.7. Because the molecular ion is of odd-mass, there is a likelihood of at least one nitrogen. Subtract 0.37 (the contribution of one nitrogen) from the $[M + 1]$ value, and calculate the number of carbons; that value is 6. From the $[M + 2]$ $(0.42 - 0.22 = 0.20)$, it is also clear that the compound also contains one oxygen. The contribution of 6 carbons, 1 oxygen and 1 nitrogen to the m/z value is 102. The number of hydrogen atoms is $115 - 102 = 13$. Therefore, the elemental composition is $C_6H_{13}NO$.

6.8. Because the molecular ion is of even-mass, therefore it may or may not contain any nitrogen. The calculated number of carbons is $(12.3/1.11)$ 11.

After subtraction the contribution of $^{13}C_2$ due to 11 carbons from the RA of M + 2 (1.31 − 0.73 = 0.58), one arrives at three oxygens in this compound. The contribution of 11 carbons and 3 oxygens to the m/z value is 180. The number of hydrogen atoms is 192 − 180 = 12. Therefore, the elemental composition is $C_{11}H_{12}O_3$. The possibility of two nitrogens is ruled out (the calculated carbons will be 10, nitrogens 2, and oxygens 3; that will exceed the m/z value of the molecular ion).

6.9. (a) C_3H_8S: would produce an odd-electron ion because its mass is even (76 u) and it contains no nitrogen.

(b) C_7H_7NO: would produce an odd-electron ion because its mass is odd (121 u) and it contains one nitrogen.

(c) C_5H_{12}: would produce an odd-electron ion because its mass is even (72 u) and it contains no nitrogen.

6.10. It contains no nitrogen because its mass is even; two nitrogens will take the mass beyond 134 u.

6.11. (a) $C_{13}H_9OCl$: "R + DB" = 13 − 10/2 + 0 + 1 = 9; yes
(b) C_3H_7COOH: "R + DB" = 4 − 8/2 + 0 + 1 = 1; yes
(c) $C_3H_7CONH_3$: "R + DB" = 4 − 10/2 + 1/2 + 1 = 0.5; no

6.12. At the C_2–C_3 bond

6.13. (a) $CH_2=OH^+$; (b) $CH_2=NH_2^+$

6.14.

6.15. Because loss of the largest group is favored, the order of relative abundances of the α-cleavage products is

$$C_2H_5C(CH_3)=O^+ > C_3H_7C(CH_3)=O^+ > C_2H_5C(C_3H_7)=O^+$$

6.16.

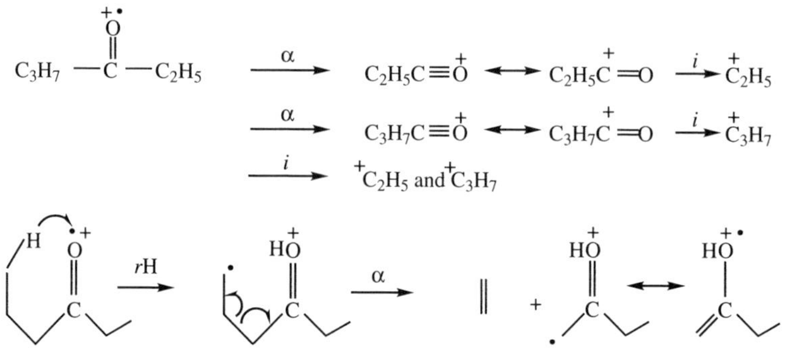

6.17.

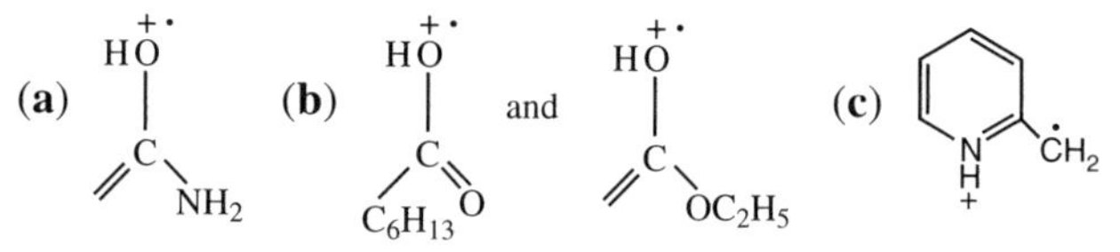

6.18.

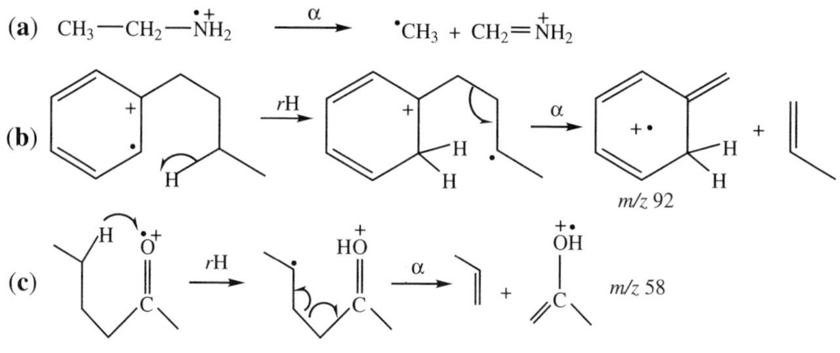

6.19.

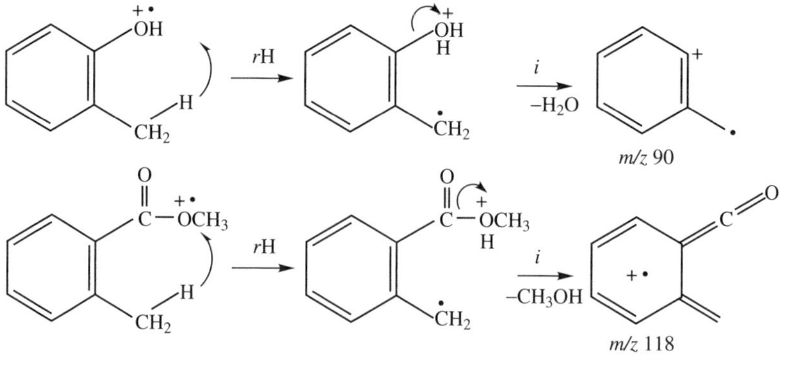

6.20.

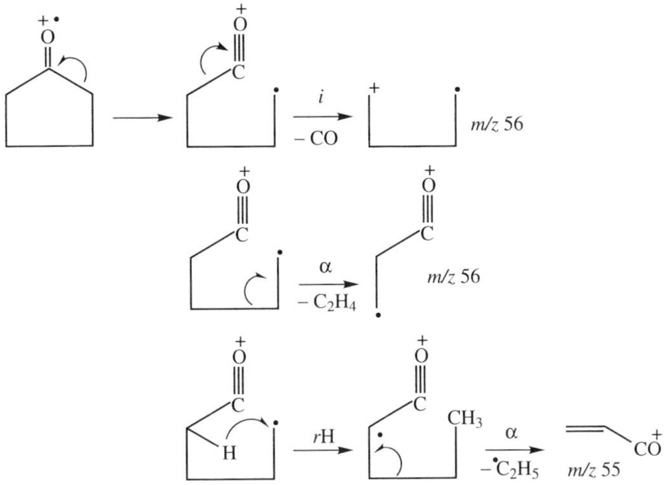

6.21. As shown below, RDA fragmentation can distinguish 4-phenyl cyclohaxene and 3-phenyl cyclohaxene:

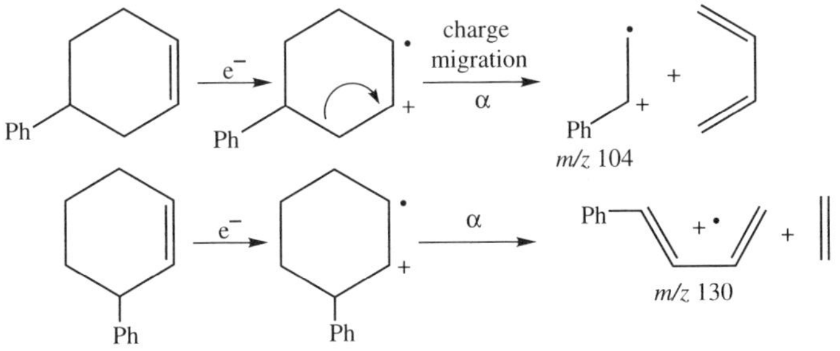

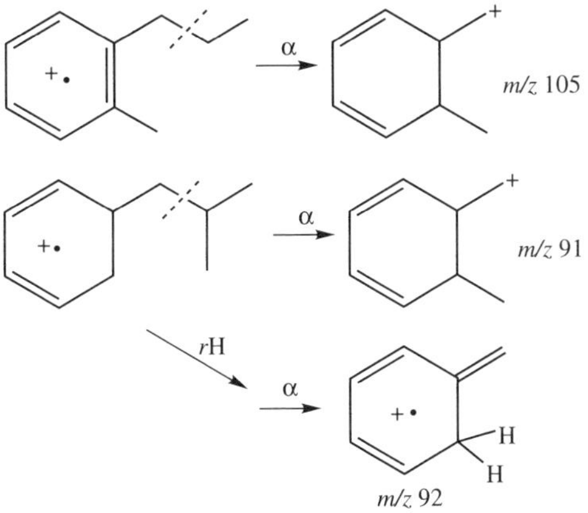

6.22.

$C_2H_5-\overset{\overset{+\bullet}{O}}{\underset{}{C}}-O-C_3H_7 \xrightarrow{\alpha} C_2H_5C\equiv\overset{+}{O} \quad m/z\ 57$

$\underset{H_5C_2}{\overset{\bullet+}{O}}\overset{}{\underset{}{C}}\overset{H}{\underset{O}{\diagup}}\overset{rH}{\longrightarrow} \underset{H_5C_2}{\overset{\overset{+}{OH}}{\underset{}{C}}}\overset{}{\underset{H}{\diagdown}}\overset{}{\underset{CH_2}{O}} \xrightarrow[\alpha]{rH} \underset{H_5C_2}{\overset{H\overset{+}{O}}{\underset{}{C}}}\overset{}{\underset{OH}{}} \quad m/z\ 75$

6.23.

[reaction scheme showing cyclohexylamine radical cation → ring opening → rH → α, −•C₃H₇ → CH=NH₂⁺ m/z 56]

6.24. 1-Chloropropane would undergo hydrogen rearrangement, followed by the loss of HCl to yield abundant m/z 42. In contrast, 2-chloropropane prefers the loss of •Cl radical to preferentially form m/z 43.

6.25. The odd m/z shows the presence of at least one nitrogen. After subtracting the contribution of one nitrogen on the abundance of M + 1 ion, the number of carbon atoms calculated are five. The shape of the molecular ion points to the presence of two chlorine atoms. This analysis suggests the elemental composition to be $C_5H_3NCl_2$. The "R + DB" is equal to 4. The successive losses of 35 and 36 u indicates that the aromatic ring is pyridine. The unknown is 2,6-dihydropyridine.

6.26. The $M^{+\bullet}$ ion is at m/z 134. From the normalized values of [M], [M + 1], and [M + 2], the number of carbons is calculated to be nine and oxygen as one; thus, the elemental composition is $C_9H_{10}O$, which is further confirmed by similar calculations for the cluster of ions at m/z 105 (its elemental composition is C_7H_5O) and the loss of 29 u from the $M^{+\bullet}$ ion. The "R + DB" value for $C_9H_{10}O$ and C_7H_5O is 5. From Table 6.7, m/z 105 corresponds to $C_6H_5C=O^+$, and m/z 77 to $C_6H_5^+$. Thus, the unknown is ethylphenylketone.

6.27. The spectrum has the characteristics of a straight-chain hydrocarbon; e.g., it shows a smooth distribution of clusters at the low-mass end, there is no break in this distribution, and the $C_nH_{2n+1}^+$ ion-series is clearly visible. The high-mass ion at m/z 156 suggests that the alkane is $C_{11}H_{24}$ (undecane).

6.28. 2-Pentanone will produce m/z 58 via McLafferty rearrangement, whereas no such product will be formed in the EI spectrum of 3-methyl-2-butanone.

6.29. Similar to the mechanism shown in Eq. 6.36, m/z 61 $[C(OH)_2CH_3]^+$ would be produced via the double-hydrogen rearrangement in ionized octyl acetate.

6.30. The m/z is formed by the α-cleavage, 122 via the McLafferty rearrangement, and 123 via the double-hydrogen rearrangement.

6.31. Read page 223.

CHAPTER 7

7.1. Because the ion production is higher at higher values of the filament work function and heating temperature, the rhenium filament heated to 2200°C will produce more positive ion current than tungsten filament heated to 2100°C.

7.2. Read Section 7.5.4 on page 271.

7.3. Read Sections 7.4 (page 267) and 7.5.1 (page 268), respectively.

7.4. Because photons of a well-defined wavelength are used for ionization.

7.5. $\delta = \left(\dfrac{R_{sample}}{R_{standard}} - 1 \right) \times 1000 = \left(\dfrac{2025.00 \times 10^{-6}}{2005.20 \times 10^{-6}} - 1 \right) \times 1000 = 9.874$

7.6. $R = \left(\dfrac{\delta}{1000} + 1 \right) \times R_{SMOW} = \left(\dfrac{-30.02}{1000} + 1 \right) \times (2005.20 \times 10^{-6})$
$= 1945.00 \times 10^{-6}$

CHAPTER 8

8.1. New molecular mass is $[10{,}525 + 58(3 \times 2 + 3)] = (10{,}525 + 522) = 11{,}047$ Da.

8.2. Cleavage of the protein with CNBr will produce three fragments: KEETLM, EYLENPKKYIPGTKM, and IFAGIKKKTEREDLIAYLKKATNE. The m/z values of their $[M + H]^+$ ions are 750.4, 1810.9, and 2779.6, respectively.

8.3. Six fragments: Phe-Tyr-Ser-Met-Asp-Ile-Leu-Lys = 1016.5; Trp-Val-Ala-Met-Glu = 635.3; Val-Asp-Tyr-Gly-Pro-Arg = 706.4; Thr-Asn-Pro-Glu = 460.2; Ala-Leu-Phe-Ile-Tyr-Arg = 782.5; Thr-Ser-Met-Tyr-Trp-Gln-Leu-Met = 1059.5 Da.

8.4. The sequence of T_4-tryptic peptide is WMpTGYAR. CID-MS/MS will produce the following sequence-specific y-, c-, and a-type ions:
y-ions: $175.1(y_1)$, $246.2(y_2)$, $409.2(y_3)$, $466.2(y_4)$, $647.2(y_5)$, and $778.3(y_6)$.
a-ions: $159.1(a_1)$, $290.1(a_2)$, $471.1(a_3)$, $528.2(a_4)$, $691.2(a_5)$, and $762.3(a_6)$.
c-ions: $204.1(c_1)$, $335.1(c_2)$, $516.2(c_3)$, $573.2(c_4)$, $736.2(c_5)$, and $807.3(c_6)$.

8.5. Read Section 8.6.1 on page 304.

8.6. Four basis steps used are sample preparation, separation and purification, mass spectrometry analysis, and database search (page 305).

8.7. Read page 305.

8.8. Read page 306.

8.9. Read page 307.

8.10. Read page 311.

8.11. Read page 313.

8.12. Read page 314.

8.13.

8.14.

$$\text{H}_2\text{N}-\underset{\underset{R_1}{|}}{\text{CH}}-\overset{\overset{O}{\|}}{\text{C}}-\text{HN}-\underset{\underset{R_2}{|}}{\text{CH}}-\overset{\overset{O}{\|}}{\text{C}}-\text{HN}-\underset{\underset{R_3}{|}}{\text{CH}}-\overset{\overset{O}{\|}}{\text{C}}\quad \text{H}_3\overset{+}{\text{N}}-\underset{\underset{R_3}{|}}{\text{CH}}-\overset{\overset{O}{\|}}{\text{C}}-\text{HN}-\underset{\underset{R_4}{|}}{\text{CH}}-\overset{\overset{O}{\|}}{\text{C}}-\text{HN}-\underset{\underset{R_5}{|}}{\text{CH}}-\overset{\overset{O}{\|}}{\text{C}}-\text{OH}$$

b_3 (left), y_3 (right)

$$\text{H}_2\text{N}-\underset{\underset{R_1}{|}}{\text{CH}}-\overset{\overset{O}{\|}}{\text{C}}-\text{HN}-\underset{\underset{R_2}{|}}{\text{CH}}-\overset{\overset{O}{\|}}{\text{C}}-\overset{+}{\text{HN}}=\underset{\underset{R_3}{|}}{\text{CH}}$$

a_3

8.15. $b_3 = 350.2$, $c_2 = 236.2$, $y_3 = 426.1$, $y_2 = 295.2$, and $a_4 = 435.2$.

8.16. The m/z values of y_4^{2+}, y_3^+, y_6^+, and $(b_6^+ - H_2O)$ ions are 209.64, 317.22, 662.39, and 767.34, respectively.

8.17. The m/z values the single amino acid immonium ions for Phe, Tyr, Met, and Lys will be 120.08, 136.08, 104.05, and 101.11, respectively.

8.18. Read page 327.

8.19. Ions from carboxypeptidase-treated peptide: $m/z = 808$, 737, 574, 517, 380, 267, and 120; ions from aminopeptidase-treated peptide: $m/z = 808$, 707, 560, 447, 310, 253, and 90.

8.20. LSAFYK.

8.21. YLYEIAR.

8.22. Doubly charged.

CHAPTER 9

9.1. Read Section 9.2.1 on page 347.

9.2. $N_{Cys} = (14155 - 13682)/59 = 8$; and $N_{S-S} = 8/2 = 4$.

9.3. IMAC and chemical tagging are used for selective isolation of phosphopeptides. For details read Section 9.4.2 on page 355.

9.4. Peptide mapping, in-source collision-induced dissociation, precursor-ion scan, and neutral-loss scan are employed to selectively identify phosphopeptides

9.5. To determine the sites phosphorylation peptides are sequenced using CID-MS/MS, PSD, ECD, and IRMPD techniques.

9.6. The m/z 79 (PO_3^- ion) in the negative-ion mode.

9.7. Alternatives to precursor-ion monitoring of m/z 79 in the negative-ion mode are (**a**) PIS of m/z 216.043 in the positive-ion mode for phosphotyrosine-containing peptides and (**b**) PIS of m/z 106 and 122 in the positive-ion mode for phosphoserine- and phosphothreonine-containing peptides. These ions are generated by CID of the derivatives of the phosphopeptides formed via β-elimination/Michael-type addition with 2-[4-pyridyl]ethanethiol and 2-dimethylaminoethanethiol, respectively.

9.8. The sequence of the peptide is Tyr-Val-Asp-Glu-Thr(-PO$_4$)-Met-Ala-Phe-Leu(Ile)-Lys-OH.

9.9. The number of phosphate groups = (19587 − 19106)/80 = 6.

9.10. The m/z 163 (Hex), 204 (HexNAc), 274 and 292 (NeuAc), and 366 (Hex-HexNAc) are the marker ions that are used for selective detection of glycopeptides.

9.11. See Section 9.6.2 on page 368.

CHAPTER 10

10.1. Average mass of this protein is 4594.17 Da. Because there are seven basic sites in the protein, a maximum of seven charge states will be observed. The corresponding m/z values are: 4595.18 (for 1+), 2298.09 (2+), 1532.40 (3+), 1149.55 (4+), 919.84 (5+), 766.7 (6+), and 657.32(7+).

10.2. Total exchangeable hydrogens = 71 and slow-exchanging backbone amide hydrogens = 33.

10.3. For each hydrogen exchanged with deuterium the mass increases by 1.006 Da. Therefore, the mass of the fully deuterated protein is (8562 + 35 × 1.006) = 8597.21 Da.

10.4. The number of exchanged hydrogens = 7(1228.43 − 2.014) − 8562 = 22.9.

10.5. Read page 389.

10.6. $[D] = \left(\dfrac{4805.0 - 4754.5}{4830.9 - 4754.5}\right) 100 = 66.1\%$.

10.7. Because it has maximum proteolytic activity at the HDX quenching conditions.

10.8. $H_x(b_9^{2+}) = 2(591.1 - 583.3) = 15.6$.
$H_x(y_6^+) = [(772.5 - 763.4) - 1] - 1 = 7.1$.
$H_x(y_8^{2+}) = 2[(519.3 - 513.3) - 1] - 1 = 9$.

CHAPTER 11

11.1.

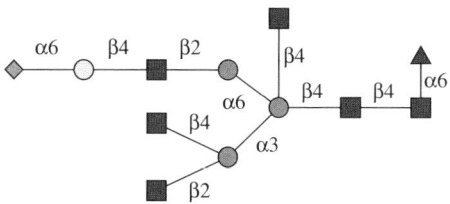

11.2. $291.095 + 146.058 + 4(162.053) + 6(203.079) + 18.011 = 2321.85$ Da.

11.3.

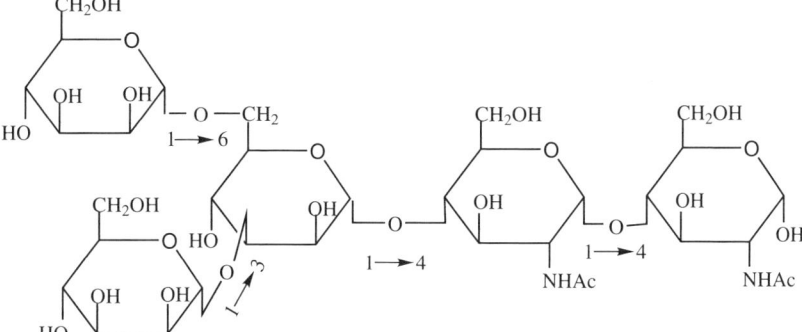

11.4.

$$\begin{array}{c} \text{Man}\alpha 1 \diagdown \\ {}^{6}_{3} \text{Man}\beta 1\text{-}4\text{GlcNAc}\beta 1\text{-}4\text{GlcNAc} \\ \text{Man}\alpha 1 \diagup \end{array}$$

11.5. Read Section 11.2 on page 400.

11.6. Read Section 11.3.7 on page 406.

11.7. $6(162.053) + 4(203.079) + 146.058 + 18.011 + 1.0078 = 1949.71$.

11.8. The $m/z = 291.095 + 5(162.053) + 4(203.079) + 18.011 + 61.988 = 1993.67$.

11.9.

[Structure diagram showing a tetrasaccharide with labeled fragmentation ions: $^{3,5}A_2$, Y_2, B_3, Z_3C_3, $^{0,2}X_2$. Features CH$_2$OH, OH, HO, CH$_3$, O—CH$_2$ groups on sugar rings.]

11.10. $B_3 = 162.053 + 162.053 + 162.053 + 1.008 = 487.167$.
$Y_2 = 162.053 + 162.053 + 18.011 + 1.008 = 343.125$.
$Z_3 = 162.053 + 162.053 + 162.053 + 1.008 = 487.167$.

11.11. (a) $= 162 + 3(162) + 2(203) + 194 + 1 = 1249$.
(b) $= 162 + 3(162) + 2(203) + 120 + 1 = 1175$.
(c) $= 162 + 3(162) + 2(203) + 128 + 1 = 1183$.
(d) $= 162 + 3(162) + 2(203) + 218 + 1 = 1273$.

11.12. The first MALDI spot is treated with PNGase F and α-fucosidase to release the glycan and to cleave two terminal fucose residues; the second spot is treated with PNGase F, α-fucosidase, and α-manosidase; finally, the third spot is treated with PNGase F, α-fucosidase, α-manosidase, and N-acetyl-β-D-glucosaminidase. The MALDI signal will appear at: m/z 1933.71 (the [M + H]$^+$ ion), $1933.71 - 292.12 = 1641.59$, $1641.59 - 324.10 = 1317.49$, and $1317.49 - 406.16 = 911.33$.

CHAPTER 12

12.1. Read Section 12.1 on page 425.

12.2. To obtain the fatty acyl profile the sn-1 and sn-2 fatty acids are first released from phospholipids via alkaline or phospolipase C hydrolysis and the fatty acid released are analyzed by GC or GC/MS.

12.3. Read Section 12.2.2 on page 432.

12.4. The characteristic feature of the positive-ion spectra of glycerophosphocholines is the presence of the phosphocholine ion at m/z 184.

12.5. Precursor-ion scans for m/z 164, 241, and 153 are used to identity phosphatidylethanolamine, phosphatidylinositol, and phosphatidyldiglycerol, respectively. Precursor-ion scan of m/z 184 can be used to identify phosphatidylcholine, and sphingomyelin, and of m/z 148 and 192 to identify the [M + Li]⁺ ions of phosphatidylethanolamine and phosphatidylserine.

12.6. Phosphatidylcholine, lisophosphatidylcholine, and sphingomyelin are identified by observing the loss of N(CH$_3$)$_3$ neutral (i.e., 59 Da loss) in the positive-ion ESI mode.

12.7. Read Section 12.4 on page 439.

12.8. Because of the presence of carboxylate functionality.

12.9. Read page 444.

12.10. Read page 444.

CHAPTER 13

13.1. MM = 17.003 + 2 × 313.058 + 3 × 289.046 + 3 × 329.052 + 304.046 + 17.003 − 95.961 = 2722.501 Da.

13.2.

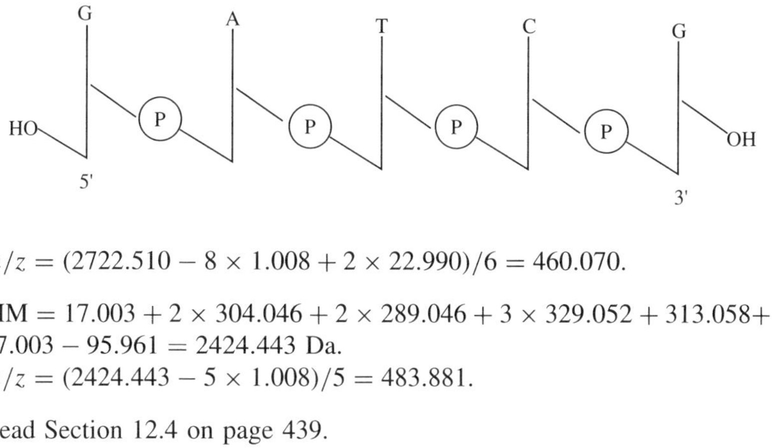

13.3. $m/z = (2722.510 − 8 × 1.008 + 2 × 22.990)/6 = 460.070$.

13.4. MM = 17.003 + 2 × 304.046 + 2 × 289.046 + 3 × 329.052 + 313.058 + 17.003 − 95.961 = 2424.443 Da.
$m/z = (2424.443 − 5 × 1.008)/5 = 483.881$.

13.5. Read Section 12.4 on page 439.

13.6.

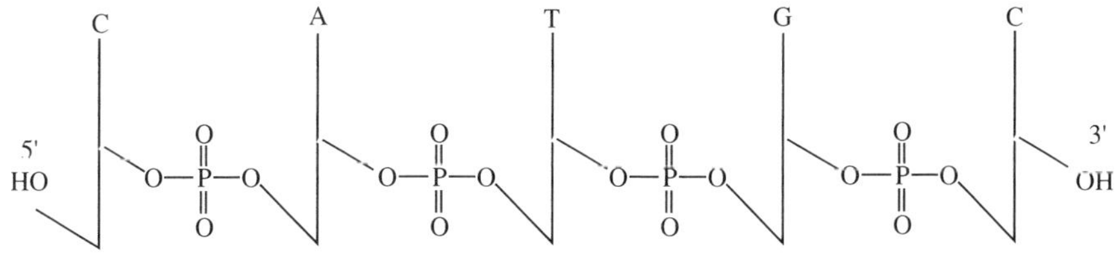

13.7.

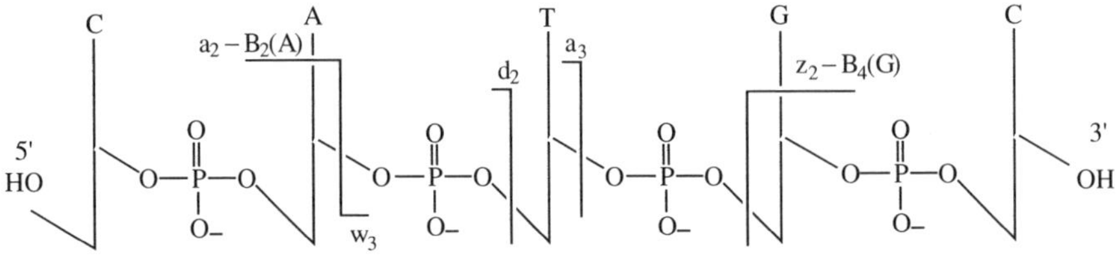

[Figure: oligonucleotide fragmentation diagram showing 5'-HO-C-A-T-G-C-OH-3' with cleavage labels $a_2 - B_2(A)$, a_3, d_2, w_3, $z_2 - B_4(G)$ marked on the phosphodiester backbone]

13.8. m/z of a_3^- ion = 17 + sum of the residue masses of C, A, and T − 96 = 827.1.

m/z of w_3^{2-} ion = (17 + sum of the residue masses of C, G, and T − 1)/2 = 464.1.

m/z of $(a_4 - B_4)^-$ ion = 17 + sum of the residue masses of C, A, T, and G − 96 − 150 − 3 = 993.1.

m/z of d_2^- ion = 17 + sum of the residue masses of C and A = 619.1.

13.9. 2055.38, 1756.34, 1437.28, 1133.24, 844.19, 531.13, and 242.09.

13.10. 2055.38, 1751.34, 1462.29, 1149.23, 860.19, 556.14, and 227.09.

13.11. Read Section 13.4.1 on page 459.

13.12. The oligonucleotide is 5'-d(TCAGAT)-3'.

CHAPTER 14

14.1. Improved detection sensitivity and well-defined chromatographic peak.

14.2. Read Section 14.3 on page 488.

14.3. Internal standard can compensate for any fluctuations in the mass spectrometry response and the sample losses that might occur in various sample-handling and chromatographic steps, thus improving precision and accuracy of the measurement.

14.4. The most suitable internal standard is a stable isotope-labeled analog of the analyte because its chemical and physical properties match well with those of the analyte.

14.5. Because of the increased level of detection specificity and lower chemical noise.

14.6. Read Section 14.4 on page 491.

14.7. The combination provides enhanced sensitivity, selectivity, speed of analysis, an opportunity for automation, feasibility of quantification of coeluting components or isobaric compounds, and use of the stable-isotope labeled analogs as internal standards.

INDEX

Accelerator mass spectrometry, 278
Acetebutol, 8
Acetophenone, 7, 9
N-Acetylgalactosamine, 365, 397, 437
N-Acetylglucosamine, 365, 397, 400, 437
N-Acetylneuraminic acid, 365, 397
Acylglycerols, 423
 analysis of, 430
Adenine, 454
Adenosine, 454
Affinity chromatography, 161
 coupling of with MS, 181
Aminopeptidase M
Atmospheric pressure chemical ionization, 46
 for coupling LC with MS, 171
Atmospheric pressure-MALDI, 43
Atmospheric pressure photoionization, 47
 for coupling LC with MS, 171
Atomic mass, 9
Average mass, 9

Bacterial identification, *see* Microorganisms
Benzonase, 458
Bile acids, 426
 analysis of, 440
Biomarkers of disease, 314, 514, 515
Blackbody-induced radiative dissociation, 126

Calf spleen phosphodiesterase, 458, 472
Calibration, *see* Quantitative analysis
Calibration ions, 201, 551–554
Calorimetric detector, 106
Capillary electrophoresis, 175
 coupling of with MS, 174, 177
 principles of, 175
Capillary electrophoresis/mass spectrometry, 174–181
Capillary electrochromatography, 177
Capillary gel electrophoresis, 176
Capillary isoelectric focusing, 173, 176
Capillary isotachophoresis, 177
Carbohydrates, *see* Oligosaccharides
Carboxypeptidase Y, 324
Cardiolipin, 426, 435
Ceramide, 426
CFG nomenclature for oligosaccharides, 399
Charge-exchange ionization, 128
Charge inversion reaction, 128
Charge-permutation reactions, 128, 253
Charge-remote fragmentation, 223, 429
Charge site-initiated fragmentation, 222
Charge-state distribution, 380
Chemical cross-linking, 391
Chemical ionization, 20–26
 atmospheric pressure-CI, 46

Fundamentals of Contemporary Mass Spectrometry, by Chhabil Dass
Copyright © 2007 John Wiley & Sons, Inc.

Chemical ionization (*Continued*)
  charge exchange, 24
  negative-ion, 25
  reagent ions for, 23
Chemotaxonomy, *see* Microorganisms
Choline, 425
Chromatofocusing, 297
Chromatography properties, 155
  capacity factor, 156
  plate height and plate number, 157
  selectivity factor, 156
α-Cleavage, 220, 245
Clinical mass spectrometry, 513–517
  biomarkers of disease, 514, 515
  genetic disorders, 514
  hemoglobin variants, 515
Collision-induced dissociation, 50, 124, 253
  for analysis of oligosaccharides, 412, 414, 415
  for disulfide-containing proteins, 350
  for sequencing oligonucleotides, 469
  for identification of phosphopeptides, 361
  for sequencing peptides, 326
Collision ionization, 128
Combinatorial libraries, 520–525
Complementary base-pairs, 456
Complex Carbohydrate Structure Database, 405
Coupling of CE with MS, 177
  via CF-FAB interface, 180
  via ESI interface, 177
  via MALDI, 180
Coupling of GC with MS, 159
Coupling of ICP with MS, 269
Coupling of LC with MS, 163
  via APCI interface, 171
  via APPI interface, 171
  via CF-FAB interface, 165
  via EI, 165
  via ESI interface, 168
  via MALDI, 172
  via moving-belt interface, 164
  via particle-beam interface, 167
  via thermospray interface, 166
Cycloalkanes, 228, 240
Cyclobutane, 228, 229
Cyclobutylamine, 228
Cyclodecane, 229
Cyclohexanone, 229, 231
  EI spectrum of, 232
Cyclohexene, 228
  EI spectrum of, 229
Cycloreversion reaction, 228

Cytosine, 454
Cytidine, 454

Database search and proteomics, 308
  databases, list of (table), 309
Data-dependent acquisition, 154, 302, 445
Data system, 7
Desorption-ESI (DESI), 55
Desorption/ionization on silicon (DIOS), 43
Detectors, 7, 103–108
  calorimetric, 106
  electron multiplier, 104
  Faraday cup, 103
  focal-plane, 107
  multichannel array, 108
  multichannel plate, 107
  photomultiplier, 105
  postacceleration, 105
2D-gel electrophoresis, 296, 310
Diels-Alder reaction, 228
Differentiation of isomeric structures, 232
Diketene, 228
Disulfide bonds in proteins, 298, 345–352
  identification of, 348
    using CID, 350
    using ECD, 350
    using FAB, 348
DNA, 290, 453
Double-hydrogen rearrangement, 226
Direct analysis in real time (DART), 56
Direct insertion probe, 19

Eicosanoids, 427
  analysis of, 441
Electron-capture dissociation, 127
  for sequencing peptides, 324
  for disulfide-containing proteins, 350
  for identification of phosphopeptides, 362
  for analysis of glycoproteins, 369
Electron ionization, 17–20
  advantages, 20
  for coupling GC with MS, 160
  for coupling LC with MS, 165
  for inorganic compounds, 263
  ionization efficiency, 18
  ion source (figure) 18
  limitations, 20
  sample introduction, 19
Electron-transfer dissociation, 128, 324
  for identification of phosphopeptides, 362
Electron-transfer reaction, 128
Electrospray ionization, 10, 48–54
  desorption-ESI (DESI), 55
  for acylglycerols, 432

INDEX **579**

for bacterial identification, 511
for combinatorial libraries, 522, 524
for coupling LC with MS, 168
for conformational analysis, 380
for fatty acids, 429
for glycolipids, 437
for HDX, 386
for lipidomics, 442
for mass measurement of macromolecules, 50
for oligonucleotides, 459, 467, 469
for phospholipids, 433
for quantitative measurements, 493
mechanism of, 52
nanoelectrospray, 54
Endo-β-N-acetyl-glucosaminidase F, for deglycosylation of proteins, 402
Endo-β-N-acetyl-glucosaminidase H, for deglycosylation of proteins, 298, 402
Endonuclaeases, 458
Energy-resolved mass spectrometry, 132
Enzyme kinetics, 501
  monitoring of, 504–507
Enzyme-substrate complex, 501
Ethyl butanoate, EI spectrum of, 244
Exonucleases, 458
Extracted ion chromatogram, 154

Faraday cup detector, 103
Fast atom bombardment, 32–35
  for acylglycerols, 432
  for disulfide-containing proteins, 348
  for fatty acids, 429
  for phospholipids, 433
  for glycolipids, 437
  for sequencing peptides, 322
  matrices for, 33
Fast-flow LC, 162
Fatty acids, 423
  analysis of, 428
  charge-remote fragmentation in, 429
  derivatization of, 430
Field desorption, 29
  of crude oil distillate, 30
Field-ion spectrometry, 102
Field ionization, 28
Field ionization kinetics, 254
Focal-plane detectors, 107
Forensic mass spectrometry, 517–520
  analysis of controlled substances, 518
  analysis of explosives, 519
  analysis of glass and paints, 519
  in bioterror defense, 520
Fourier-transform ion-cyclotron resonance mass spectrometers, 94–98

performance characteristics, 98
principle of, 95
tandem mass spectrometry with, 138
Fragment ions, list of, 217
Fragmentation processes, 216
  of alcohols, 240
  of aldehydes and ketones, 242
  of carboxylic acids, 242
  of cyclic compounds, 227
  of cycloalkanes, 229
  of esters, 243
  of ethers, 241
  of halogen-containing compounds, 246
  of hydrocarbons, 238
  of nitrogen-containing compounds, 244
  of sulfur-containing compounds, 246
Free-flow electrophoresis, 295

D-Galactose, 365, 397
Gamma-hydrogen rearrangement, 223. *See also* McLafferty rearrangement
Gangliosides, 426
  analysis of, 439
Gas chromatography, 158
  coupling of with MS, 159
  principles of, 158
Gas chromatography/mass spectrometry, 158–161
Genomics, 292, 453
Glow discharge ionization, 16, 267
Glycerophosphocholine, 425, 433, 517
Glycerophosphoethanolamine, 425, 433
Glycerophosphoglycerol, 425
Glycerophosphoinositol, 425
Glycerophosphoserine, 425
Glycerophospholipids, 425
Glycolipids, 425
  analysis of, 436
Glycomes, 397
Glycomic, 397
Glycoproteins, 364–370
  analysis of, 366
  analysis of via ECD, 369
  analysis of via ladder sequencing, 369
  analysis of via marker ions, 368
  analysis of via peptide mapping, 368
  analysis of via precursor-ion scan, 368
  analysis of via tandem MS, 369
  release of glycans, 402
N-Glycosidase F, *see* PNGase F
Glycosphingolipids, 401, 437, 439
N-Glycosylation, 365, 400
O-Glycosylation, 365, 400

580 INDEX

Guanine, 454
Guanosine, 454

Hanes-Woolf plot, 502
Heat-of-formation, 251
Hemoglobin variants, 515
Hydrogen/deuterium exchange, 383–390
Hyphenated separation techniques, 151
   benefits of, 152
   GC/MS, 158
   LC/MS, 161
Hybrid tandem mass spectrometers, 138–143
   LIT-oa-TOF, 140
   magnetic sector-oa-TOF, 140
   quadrupole-oa-TOF, 139
   quadrupole-FT-ICR, 140
   LTQ-FT-ICR, 141
   LTQ-orbitrap, 142

Imaging mass spectrometry, 507–510
   with MALDI, 507, 509
   with SIMS, 507, 508
Immobilized metal ion affinity chromatography, 355
Immonium ions, 319, 328
Immunodepletion columns, 295
Inductive (i)-cleavage, 222
Inductively coupled plasma mass spectrometry, 16, 268
   laser ablation-ICPMS, 273
   spectral interferences, 271
Inhibitors, of enzymes, 503
Infrared multiphoton dissociation, 126
   for identification of phosphopeptides, 363
   for sequencing oligonucleotides, 467
Inlet system, 6
In-source CID, 50
   for analysis of oligosaccharides, 414
   for sequencing oligonucleotides, 467
   for sequencing peptides, 322
   for identification of phosphopeptides, 358
Ion activation and dissociation, 123
   blackbody-induced radiative dissociation, 126
   collision-induced dissociation, 124
   electron-capture dissociation, 127
   infrared multiphoton dissociation, 126
   photodissociation, 126
   surface-induced dissociation, 125
   sustained off-resonance irradiation (SORI), 138
Ion-extraction chromatogram, 154
Ion mobility mass spectrometers, 101–103
   for protein conformational analysis, 391
Ion source, 6, 16

Ion-source decay MALDI
   for sequencing oligonucleotides, 467
   for sequencing peptides, 323
   for disulfide-containing proteins, 349, 351
Ionization
   electrospray ionization, 48–54
   fast atom bombardment, 32
   field ionization, 28
   field desorption, 29
   for gas-phase molecules, 17
   for condensed-phase molecules, 29, 45
   glow discharge ionization, 267
   laser desorption/ionization, 35
   matrix-assisted laser desorption/ionization, 35–44
   metastable atom bombardment, 28
   modes of, 15
   plasma desorption, 30
   spark-source ionization, 265
   thermal ionization, 264
   thermospray, 45
   why required, 15
Isoelectric focusing, 296
Isotope dilution mass spectrometry, 280
Isotopic abundances, 204, 543
Isotopic contributions of carbon atoms, 208
Isotopic masses, 9, 203, 543
Isotopic peaks, 203
Isotope ratio mass spectrometry, 275
IUPAC nomenclature for oligosaccharides, 399

[Ketene$^{+\bullet}$ + ketene] cycloadduct, 122
Kinetic-energy release, 252

Ladder sequencing
   for sequencing peptides, 325
   for analysis of glycoproteins, 369
Laser ablation-ICPMS, 273
Laser desorption/ionization, 35, 263
Leukotrienes, 427, 441
Lineweaver-Burk plot, 502
Linked-field scans, 130
Lipidomes, 442
Lipidomes, shotgun, 444
Lipids, 423–447
Liquid chromatography, 161
   adsorption chromatography, 161
   affinity chromatography, 161
   coupling of with MS, 163
   fast-flow LC, 162
   ion-exchange chromatography, 161
   partition chromatography, 161
   principles of, 161

size-exclusion chromatography, 161
ultra-performance LC, 163
Liquid chromatography/mass spectrometry, 161–174
coupling of LC with MS, 163
multi-dimensional LC/MS, 173
Liquid secondary-ion mass spectrometry, see FAB

Magnetic sector mass spectrometers, 70–75
double-focusing, 73, 129
electrostatic analyzer, 73
linked scans in, 131
Mattauch-Herzog geometry, 73
Nier-Johnson instrument, 74
performance characteristics, 74
principle of, 70
tandem mass spectrometry with, 129
Manhattan Project, 4
D-Mannose, 365, 397
MASCOT, 309
Mass-analyzed ion kinetic energy spectroscopy, 131
Mass analyzers, 6, 70–103
desirable features, 67
FT-ICR, 94
ion mobility, 101
magnetic sector, 70
orbitrap, 99
quadrupole, 75
quadrupole ion trap, 86
quadrupole linear ion trap, 92
rectilinear ion trap, 94
time-of-flight, 80
Mass calibration standards, 201, 511
Mass chromatogram, 154
Mass spectrometry
history, 3
desirable features, 5
basic components, 6
basic principles, 5
general applications, 11
Mass-to-charge ratio, 10, 17
Material-enhanced MALDI, 45
Matrix-assisted laser desorption/ionization, 35–44, 509
atmospheric pressure-MALDI, 43
for bacterial identification, 511
for combinatorial libraries, 522, 524
for conformational analysis, 380
for coupling LC with MS, 172
for disulfide-containing proteins, 349, 351
for imaging mass spectrometry, 509
for oligonucleotides, 461, 467, 474

for quantitative measurements, 493
for virus analysis, 513
material-enhanced MALDI, 45
matrices for, 38, 39 (table)
mechanism of ion formation, 37
of low-mass compounds, 42
sample preparation for, 41
Mattauch-Herzog geometry instrument, 73, 266
Maxam-Gilbert DNA sequencing, 474
McLafferty rearrangement, 223, 233, 235, 236, 242, 243, 245
"McLafferty + 1" rearrangement, 226, 243. See also Double-hydrogen rearrangement
Metabolomics, 292, 517
Metastable atom bombardment, 28
Metastable ions, 8, 249, 250, 252
3-Methyl-2-butanone, 232
EI spectrum of, 233
3-methyl-2-pentanone, 233
EI spectrum of, 234
4-methyl-2-pentanone, 233
EI spectrum of, 234
5-Methyltridecane, EI spectrum of 239
Miceller electrokinetic chromatography, 176, 181
Michaelis-Menten theory, 501
Microorganisms, 511–513
analysis of viruses, 513
bacterial identification, 511
chemotaxonomy, 511
Molecular formula
from accurate mass, 201
from isotopic peaks, 203, 208
Molecular ion, 7, 17
Molecular mass, 9, 17
measurement, 198
measurement of glycoproteins, 366
measurement of oligonucleotides, 459
measurement of proteins, 297
measurement via ESI, 51, 200
measurement via MALDI, 200
measurement via peak-matching, 200
Multichannel array detector, 108
Multichannel plate detector, 107
Multi-dimensional LC/MS, 173
separation of proteins, 297
Multiphoton ionization, 27
resonance-enhanced, 27
Multiple-reaction monitoring, 123, 487
Multiply charged ions, 10, 50
Multistage MS/MS, 90, 120, 137, 412

$NaBH_4$-NaOH mixture for release of glycans, 402

Nanoelectrospray, 54
Negative ions, 17, 25, 48, 57
Neutralization-reionization mass spectrometry, 128
Neutral losses, list of, 212
Neutral-loss scan, 123
  for identification of phosphopeptides, 360
Nitrogen rule, 211, 215
Nominal mass, 9
Normal phase HPLC, 162
Nozzle-skimmer (NS) dissociation, 50, 467
Nuclease $P1$, 458
Nucleosides, 454, 457
Nucleotides, 453, 457

Oligonucleotides, 453–476
  base composition of, 463
  cleavage of, 458
  fragmentation behavior of, 465
  molecular mass determination, 459
    via ESI-MS, 459
    via MALDI-MS, 461
  nomenclature of CID ions, 465
Oligonucleotides, sequencing of, 464–476
  gas-phase techniques, 465
    via CID, 469
    via in-source CID, 467
    via ion-ion reactions, 470
    via ion-source decay, 467
    via IRMPD, 467
    via post-source decay, 468
  solution-phase techniques, 471
    via chemical cleavage, 472
    via ladder sequencing, 472
    via Maxam-Gilbert protocol, 474
    via Sanger method, 474
Oligosaccharides, 366, 397–417
  anomeric center, 398
  CFG nomenclature, 399
  Complex Carbohydrate Structure Database, 405
  composition analysis by GC/MS, 403
  composition analysis by mass measurement, 403
  derivatization of, 402
  derivatizing agents (table), 404
  exoglycosidase degradation, 407
  fragment ion nomenclature, 408
  identification by precursor-ion scan, 403
  IUPAC nomenclature, 399
  linkage analysis by GC/MS, 403
  linkage analysis by CID, 415
Oligosaccharides, structure determination by CID-MS, 414
  by in-source CID, 414
  by post-source decay 412
  by sequential digestion, 406
  by tandem MS, 408
  via fragmentation of negative ions, 412
Orbitrap mass spectrometers, 99–101
Ortho-effect, 226, 235
Ortho rearrangement, 226
Overpressured-layer chromatography, 184

Peak matching mode, 199
Peak parking, 155
Pee Dee Belemnite standard, 276
3-Pentanone, 232
  EI spectrum of, 233
Peptide mass fingerprinting, see Peptide mass mapping
Peptide mass mapping, 298, 305, 358, 368
Peptides, 316–330
  fragmentation rules, 317
  fragmentation nomenclature, 317
  sequence determination, 316, 322
    guidelines for, 327
    via CID, 326
    via, ECD, 324
    via FAB, 322
    via in-source CID, 322
    via ion-source dissociation MALDI, 323
    via ladder sequencing, 324
    via post-source dissociation MALDI, 323
4-Phenylcyclohexene, 228
Phosphatidylcholine, see Glycerophosphocholine
Phosphatidylethanolamine, see Glycerophosphoethanolamine
Phosphatidylglycerol, see Glycerophosphoglycerol
Phosphatidylinositol, see Glycerophosphoinositol
Phosphatidylserine, see Glycerophosphoserine
Phosphoglycerides, see Glycerophospholipids
Phosphodiesterase I, 458
Phospholipids, 425
  analysis of, 433
  direct tissue analysis of, 436
Phosphopeptides
  fractionation of, 355
    via IMAC, 182, 356
    via chemical-tagging, 357
    via antibodies, 357
  identification of, 358
    via peptide mass mapping, 358
    via in-source CID, 358
    via precursor-ion scan, 358

via neutral-loss scan, 360
via CID, 361
via post-source decay, 362
via ECD and ETD, 362
via IRMPD, 363
phosphatase treatment, 358
Phosphoproteins, 352–364
analysis of, 353
via ^{32}[P] labeling, 353
mass spectrometry protocol, 354
cleavage of, 355
Phosphoproteomics, see Phosphoproteins
Photodissociation, 126
Photoionization, 26–28
atmospheric pressure-PI, 47
Photomultiplier, 105
Piperidine, 230, 245
EI spectrum of, 230
Planar chromatography/mass spectrometry, 183
Plasma desorption, 30
Plasmaspray ionization, 45
PNGase F, for deglycosylation proteins, 298, 368, 369, 402
Polymerase chain reaction, 316, 459, 474, 515
Postacceleration detector, 105
Post-source decay, 133
for analysis of oligosaccharides, 412
for phosphopeptides, 362
for sequencing oligonucleotides, 468
for sequencing peptides, 323
Posttranslational modifications of proteins, 290, 343–370
selected examples (table), 344
Precursor-ion scan, 122
for identification of phosphopeptides, 123, 358
for analysis of glycoproteins, 368
for identification of oligosaccharides, 403
Product-ion scan, 122
Prostaglandins, 427, 441
Proteins, 289–316
chromatofocusing of, 297
deglycosylation of, 298,
2D-gel electrophoresis of, 296
de novo sequencing, 316
folding-unfolding dynamics of, 385
free-flow electrophoresis of, 295
higher-order structures of, 290, 379–392
mass measurement of, 297
peptide mass mapping of, 298
purification of, 295
reduction–alkylation of, 299
sequence determination, 292
general protocol, 294

via cDNA sequencing, 293
via Edman sequencing, 293
subcellular fractionation of, 295
Proteins, biomarker discovery, 314
Proteins, cleavage of
of gel-separated proteins, 301
reagents for (table), 300
with N-Chloro- and N-bromosuccinimide, 300
with cyanogen bromide, 300
with proteolytic agents, 301
Proteins, conformational analysis of, 379–392
charge-state distribution for, 380
chemical cross-linking, for, 391
folding-unfolding dynamics, 385
hydrogen/deuterium exchange for, 383
via CID, 390
via continuous-labeling technique, 387
via fragmentation MS, 389
via pulse-labeling technique, 388
via time-resolved ESI-MS, 387
Proteome analysis, see Proteomics
Proteomics, 292, 303–314
bottom-up, 304
database search and 308
quantitative, 310
shotgun, 307
top-down, 304
via accurate mass tags, 306
via MudPIT approach, 307, 315
via peptide mass fingerprinting, 305, 307
via peptide sequence tags, 306, 307
Pulse-labeling technique, 388
Pyrimidine, 454
Purine, 454

Quadrupole mass spectrometers, 75–80
performance characteristics, 79
principle of, 76
rf-only, 80
tandem mass spectrometry with, 132
Quadrupole ion trap (QIT) mass spectrometers, 86–92
external ion injection, 91
mass-selective instability mode, 89
mass-selective stability mode, 90
resonance ion ejection, 90
performance characteristics, 79
principle of, 87
tandem mass spectrometry with, 136
Quadrupole linear ion trap (LIT) mass spectrometers, 86–92
Quantitative analysis 485–497
calibration methods, 488

Quantitative analysis (*Continued*)
  external standard method, 488
  internal standard method, 488
  standard addition method, 489
  validation procedures, 491
  via GC/MS, 493
  via LC/MS, 493
  via LC/MS/MS, 494
  via MALDI-MS, 494
  via microchip-based ESI, 494
Quantitative proteomics, 310–314
  via AQUA approach, 314
  via 2D gel electrophoresis, 310
  via 2D-DIGE, 310
  via ICAT, 311
  via ^{18}O-water labeling approach, 312
  via SILAC approach, 313
  via label-free approach, 313
  via iTRAQ reagents, 312
Quasi-equilibrium theory, 124, 247

Radical cation, 15, 17, 24, 27, 33, 48
Radical site-initiated fragmentation, 220
  in alkenes, 221
  in aromatic compounds, 221
  in ethers, 221
  in ketones, 221
  in alcohols, 221
  in amines, 221
RDX, 519
Rearrangement reactions, 223, 248
  double-hydrogen rearrangement, 226
  hydrogen rearrangement, 225
  McLafferty rearrangement, 223
  ortho rearrangement, 226
Reconstructed ion chromatogram, 154
Rectilinear ion trap, 94
Resolution, 67, 68
Resolving power, 68
Resonance electron capture, 25
Resonance ionization mass spectrometry, 16, 274
Retro-Diels–Alder reaction, 228, 235
  in cyclohexene, 228
  in 4-phenylcyclohexene, 228
  in 1- and 2-tetralols, 235
Reverse phase HPLC, 162
Ribonomics, 453
Rings plus double bonds, 214, 215
RNA, 453
Rnase T1, 458
Rnase U2, 458
RRKM, 124, 247

Saha-Langmuir equation, 265
Sample introduction, 19
Sanger DNA sequencing, 474
Satellite ions, 8
SDS-PAGE, 294, 296, 298, 366
Secondary-ion mass spectrometry, 31, 263
  for imaging mass spectrometry, 508
Selected-ion monitoring, 486, 487, 505, 518
  chromatogram, 154
Selected-reaction monitoring, 123, 314, 358, 486, 487, 518
SEQUEST, 309
Shotgun lipidomics, 444
Shotgun proteomics, 307
Sigma-bond (σ) cleavage, 219
Simple bond-cleavage reactions, 219
Snake venom phosphodiesterase, 458, 472
SMOW standard, 276
Spark-source ionization, 16, 265
Sphingomyelins, 426, 517
Sphingosine, 425
Steroids, 426
  analysis of, 440
Stevenson's rule, 219, 220
Stored waveform inverse Fourier transform (SWIFT), 97
Structurally diagnostic fragment ions, 216, 235, 218 (table)
Supercritical-fluid chromatography/mass spectrometry, 183
Surface-enhanced laser desorption/ionization, 44,
Surface-induced dissociation, 125
Sustained off-resonance irradiation (SORI), 138

Tandem mass spectrometry, 119
  charge-permutation reactions, 128
  instruments, 129
  ion activation, 123
  linked-field scans, 130
  multiple-reaction monitoring, 123
  multistage MS, 120
  neutral-loss scan, 123
  precursor-ion scan, 122
  principles of, 119
  product-ion scan, 122
  scan modes, 121
  selected-reaction monitoring, 123
Tetradecane, EI spectrum of 239
Tetrahydopyran, 230
Thermal ionization, 16, 264
Thermospray, 45–46
  for coupling LC with MS, 166

Thymidine, 454
Thymine, 454
Thin-layer chromatography/MS, 183
Thomson, 10
Thromboxanes, 427
Time-of-flight (TOF) mass spectrometers, 80–86
  coupling LC with, 171
  delayed extraction, 83
  orthogonal acceleration TOF, 85
  performance characteristics, 86
  post-source decay, 133
  principle of, 81
  reflectron TOF, 84
  tandem mass spectrometry with, 133
  time-lag focusing, 83
Total ion current (TIC) chromatogram, 154
Transcriptomics, 292, 453
Translational-energy spectroscopy, 130
Triacylglycerols, 425, 517

Unified atomic mass unit, 9
Uracil, 455

Vacuum system, 7
Viruses, *see* Microorganisms

Western blotting, 297

WILEY-INTERSCIENCE SERIES IN MASS SPECTROMETRY

Series Editors

Dominic M. Desiderio
*Departments of Neurology and Biochemistry
University of Tennessee Health Science Center*

Nico M. M. Nibbering
Vrije Universiteit Amsterdam, The Netherlands

John R. de Laeter • *Applications of Inorganic Mass Spectrometry*
Michael Kinter and Nicholas E. Sherman • *Protein Sequencing and Identification Using Tandem Mass Spectrometry*
Chhabil Dass • *Principles and Practice of Biological Mass Spectrometry*
Mike S. Lee • *LC/MS Applications in Drug Development*
Jerzy Silberring and Rolf Eckman • *Mass Spectrometry and Hyphenated Techniques in Neuropeptide Research*
J. Wayne Rabalais • *Principles and Applications of Ion Scattering Spectrometry: Surface Chemical and Structural Analysis*
Mahmoud Hamdan and Pier Giorgio Righetti • *Proteomics Today: Protein Assessment and Biomarkers Using Mass Spectrometry, 2D Electrophoresis, and Microarray Technology*
Igor A. Kaltashov and Stephen J. Eyles • *Mass Spectrometry in Biophysics: Confirmation and Dynamics of Biomolecules*
Isabella Dalle-Donne, Andrea Scaloni, and D. Allan Butterfield • *Redox Proteomics: From Protein Modifications to Cellular Dysfunction and Diseases*
Silas G. Villas-Boas, Jens Nielsen, Jorn Smedsgaard, Michael E. Hansen, and Ute Roessner-Tunali • *Metabolome Analysis: An Introduction*
Mahmoud H. Hamdan • *Cancer Biomarkers: Analytical Techniques for Discovery*
Chabbil Dass • *Fundamentals of Contemporary Mass Spectrometry*